THE
BIOCHEMISTRY OF GENE EXPRESSION
IN
HIGHER ORGANISMS

THE BIOCHEMISTRY OF GENE EXPRESSION IN HIGHER ORGANISMS

THE PROCEEDINGS OF A SYMPOSIUM SPONSORED BY THE INTERNATIONAL UNION OF BIOCHEMISTRY, THE AUSTRALIAN ACADEMY OF SCIENCE AND THE AUSTRALIAN BIOCHEMICAL SOCIETY

Edited by

J. K. POLLAK
Department of Histology and Embryology,
University of Sydney
Sydney, Australia

and

J. WILSON LEE
Wheat Research Unit, CSIRO,
North Ryde, N.S.W. Australia

D. REIDEL PUBLISHING COMPANY

DORDRECHT-HOLLAND/BOSTON-U.S.A.

Library of Congress Catalog Card Number 72-97960

ISBN 90 277 0289 6

Published by Australia & New Zealand Book Co. Pty. Ltd.,
P.O. Box 200, Artarmon, N.S.W., Australia

Distributors for Australia and New Zealand
Australia & New Zealand Book Co. Pty. Ltd.

Distributors for the U.S.A., Canada and Mexico
D. REIDEL PUBLISHING COMPANY, INC
306 Dartmouth Street, Boston, Mass. 02116, U.S.A.

Distributors for all other countries
D. REIDEL PUBLISHING COMPANY
P.O. Box 17, Dordrecht, Holland

Printed in Australia

Editorial Preface

The papers assembled in this volume are based on the symposium on "*The Biochemistry of Gene Expression in Higher Organisms*" which was held at the University of Sydney from May 14-19, 1972.

Many symposia have been held on the control of gene expression in prokaryotes but to date considerably less attention has been paid to eukaryotic organisms. It has been appreciated only recently that some of the information gained from the study of prokaryotes is directly applicable to eukaryotes; however, it is now realized that the principles of the control mechanisms of gene expression in these two classes of organism, differ considerably.

This symposium was organized in an effort to bring together workers from widely different fields concerned with gene expression, with the aim of circumscribing the current concepts and speculating on future developments in studies on the mechanisms which control and modulate gene expression, in the widest sense, in eukaryotes. This volume contains all the 36 papers presented at the symposium. In a few instances the sequence of contributions has been changed to provide the reader with a more logical presentation. In addition, three papers which were not actually presented at the symposium, have been included in this volume. These three papers were not read because last-minute hitches prevented speakers from attending.

The Editors wish to express their thanks to all the contributors who co-operated by supplying their manuscripts promptly. Unfortunately it is extremely difficult to recreate for the reader of this volume, the same spirit of enthusiasm, bonhomie and stimulation which permeated the meeting and the discussions which lasted well into the nights.

The symposium was cosponsored by the International Union of Biochemistry, the Australian Academy of Science and the Australian Biochemical Society to whom the organizers wish to express their gratitude. Particular thanks are also due to Associate Professor J. F. Williams and his committee who so ably handled all of the physical arrangements for the meeting.

J. K. POLLAK

J. WILSON LEE

Contents

Editorial Preface

CHROMOSOME STRUCTURE AND THE MANIPULATION AND ANALYSIS OF GENES

Chromosome structure and units of function in higher organisms 3
W. J. Peacock

Transgenosis of bacterial genes from *Escherichia coli* to cultures of haploid *Lycopersicon esculentum* and haploid *Arabidopsis thaliana* plant-cells 21
Colin H. Doy, Peter M. Gresshoff and Barry Rolfe

Sequences in genetic nucleic acids 38
F. Sanger

Mutations at the end of the iso-l-cytochrome *c* gene of yeast 56
Fred Sherman and John W. Stewart

TRANSCRIPTIONAL AND TRANSLATIONAL CONTROL MECHANISMS

RNA polymerases and transcriptive specificity in eukaryotic organisms 89
W. J. Rutter, P. W. Morris, M. Goldberg, M. Paule and R. W. Morris

The control of gene expression by membrane organization in *Saccharoinyces Cerevisiae* 105
Kevin A. Ward, S. Marzuki and J. M. Haslam

Transcription and translation in mammalian cells 117
S. Penman, R. Price, S. Perlman and R. Singer

Isolated chromatin in the study of gene expression 142
Sarah C. R. Elgin and J. Bonner

Structural modifications of histones in cultured mammalian cells 164
George R. Shepherd, Billie J. Noland, Julia M. Hardin and Paul Byvoet

The role of histones in avian erythropoiesis 177
V. L. Seligy, G. M. H. Adams and J. M. Neelin

Chromosomal components in relation to differentiation of avian red blood cells 191
R. Appels, R. Harlow, P. Tolstoshev and J. R. E. Wells

Mechanism of glucocorticoid hormone action and of regulation of gene expression in cultured mammalian cells 206
John D. Baxter, Guy G. Rousseau, Stephen J. Higgins and Gordon M. Tomkins

Regulation of transcription by glucocorticosteroids 225
C. E. Sekeris and W. Schmid
A control mechanism in gene expression of higher cells operating at the termination step in protein synthesis 242
I. T. Oliver
The mechanism and control of the biosynthesis of α-lactalbumin by the mammary gland 260
P. N. Campbell
Changes in protein synthesis and degradation involved in enzyme accumulation in differentiating liver 274
F. J. Ballard, M. F. Hopgood, Lea Reshef and R. W. Hanson

GENE EXPRESSION AND DEVELOPMENT

Obligatory requirement for DNA synthesis during myogenesis erythrogenesis and chondrogenesis 287
H. Holtzer, R. Mayne, H. Weintraub and G. Campbell
The dependence of gene expression on membrane assembly 305
C. G. Duck-Chong and J. K. Pollak
Patterns of gene activity in larval tissues of the blowfly *Calliphora* 320
J. A. Thomson
Hormonal and environmental modulation of gene expression in plant development 333
B. Kessler
DNA and RNA synthesis during growth by cell expansion in *Vicia faba* cotyledons 357
Adele Millerd

GENE EXPRESSION IN DIFFERENTIATED CELLS

Induction of δ-aminolevulinic acid synthetase in perfused rat liver by drugs, steroids, lead and adenosine-3′, 5′ -monophosphate 369
A. M. Edwards and W. H. Elliott
The anaemia-induced reversible switch from haemoglobin A to haemoglobin C in goats and sheep: the two haemoglobins are present in the same cell during the changeover 379
M. D. Garrick, R. F. Manning, M. Reichlin and M. Mattioli
Gene expression in liver endoplasmic reticulum 393
L. Ernster
The use of neurological mutants as experimental models 410
P. Mandel, J. L. Nussbaum, N. Neskovic, L. Sarlieve, E. Farkas and O. Robain

GENE EXPRESSION IN MITOCHONDRIA AND CHLOROPLASTS

The phenomenology of cytoplasmic genetics in yeast: a proposal for an autonomy of mitochondrial membranes and the determinism of nucleo-cytoplasmic genetic interactions 425
A. W. Linnane, C. L. Bunn, Neil Howell, P. L. Molloy and H. B. Lukins

Location of DNAs coding for various kinds of chloroplast proteins 443
S. G. WILDMAN, N. KAWASHIMA, D. P. BOURQUE, FLOSSIE WONG, SHALINI SINGH, P. H. CHAN, S. Y. KWOK, K. SAKANO, S. D. KUNG and J. P. THORNBER

Nuclear genes controlling chloroplast development in barley 457
KNUD W. HENNINGSEN, J. E. BOYNTON, D. VON WETTSTEIN and N. K. BOARDMAN

Gene expression in chloroplasts and regulation of chloroplast differentiation 479
ROBERT M. SMILLIE, N. STEELE SCOTT and D. G. BISHOP

Products of chloroplast DNA-directed transcription and translation 504
P. R. WHITFIELD, D. SPENCER and W. BOTTOMLEY

GENE EXPRESSION AND THE IMMUNE RESPONSE

The relevance of immunology to the biochemistry of gene expression 525
G. J. V. NOSSAL

The reaction of antigen with lymphocytes 532
G. L. ADA, M. G. COOPER and R. LANGMAN

Synthesis transport and secretion of immunoglobulin in lymphoid cells 542
FRITZ MELCHERS

Molecular and cellular mechanisms of clonal selection 555
GERALD M. EDELMAN

Antibody diversification: the somatic mutation model revisited 574
MELVIN COHN

An alternate mechanism for immune recognition 593
K. J. LAFFERTY

The clonal development of antibody forming cells 606
A. J. CUNNINGHAM

Immunoglobulin gene expression in murine lymphoid cells 612
NOEL L. WARNER and ALAN W. HARRIS

Structure and function of lymphocyte surface immunoglobulin 629
JOHN J. MARCHALONIS, ROBERT E. CONE, JOHN L. ATWELL and RONALD T. ROLLEY

INDEX 649

LIST OF ABBREVIATIONS

Chromosome Structure and the Manipulation and Analysis of Genes

Chromosome Structure and Units of Function in Higher Organisms

W. J. PEACOCK
Division of Plant Industry, Commonwealth Scientific and Industrial Research Organization, Canberra, A.C.T., Australia.

It is not my intention in this paper to present an exhaustive review of what is known of the structure of the eukaryotic chromosome; rather I thought in this conference I would emphasize the role of cytological and cytogenetic approaches to the problem, and demonstrate that they are a necessary adjunct to biochemical and molecular analyses. In the first part of the paper, in which I will consider the possibility that the eukaryotic chromosome is similar to the bacterial chromosome in having one continuous DNA molecule, I will largely discuss a cytological experiment I carried out in association with Dr. Sheldon Wolff (University of California, San Francisco) and Dr. Dan Lindsley (University of California, San Diego). In the second part of the paper I will largely confine myself to a discussion of the chromosomes of *Drosophila melanogaster,* a species in which recent genetic work, when considered with molecular data, has posed a problem which is perhaps the fundamental issue in chromosome research—what is the unit of functional organization in the chromosome?

IS THE CHROMOSOME A CONTINUOUS DNA MOLECULE?

The possibility that chromosomes of higher organisms may contain one continuous DNA molecule was first indicated by the autoradiographic analysis of chromosome duplication by Taylor *et al.* (1957). In this simple experiment root tip cells of *Vicia faba,* the broad bean, were provided with tritiated thymidine ($[^3H]dT$) during the DNA replication period of one cell cycle. Some of the cells were then examined by autoradiography at the next metaphase; others were permitted to proceed through a further DNA replication period in which the isotope was omitted, and examined at the succeeding metaphase—the second metaphase after incorporation of $[^3H]dT$. In the first metaphase both sister chromatids of each chromosome were labelled whereas in the second division, in each chromosome, only one of the two sister chromatids was labelled. An example of the second metaphase labelling pattern is shown in Fig. 1 which shows a cell from a Chinese Hamster tissue culture. The patterns found in the two successive metaphases imply that chromosomal DNA followed a semi-conservative mode of duplication with the unreplicated chromosome (chromatid) containing two subunits extending along its length.

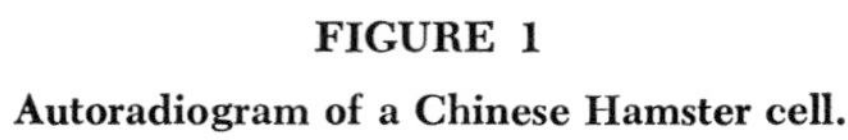

FIGURE 1

Autoradiogram of a Chinese Hamster cell.

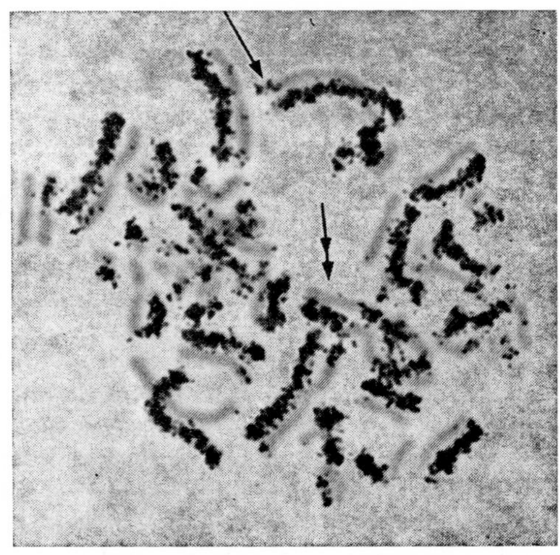

This cell is at the second metaphase after incorporation of [^{3}H]dT and shows segregation of label over sister chromatids. Most chromosomes are labelled along their full length. The arrow indicates a chromosome with one terminal sister chromatid exchange; the double arrow indicates a chromosome with three exchanges.

An exception to the labelling pattern described above is the finding of some chromosomes in the second metaphase with labelling over both sister chromatids. Peacock (1963) described such *isolabelling* in *Vicia faba* and pointed out that it may indicate that more than two strands existed along a chromosome, although in general the strands were parcelled together into two subunits. This possibility was in concert with other classes of cytological evidence which suggested a multineme structure to chromosomes (Wolff, 1969). Isolabelling has been found in a number of organisms (Deaven and Stubblefield, 1969; Darlington and Haque, 1969) and although it is certainly not a technical artifact as Thomas (1971) has asserted, it does not provide critical evidence for the existence of more than two subunits in the unreplicated chromosome. Peacock (1963) recognized that autoradiographic resolution could not exclude the possibility that isolabelling resulted from a number of breakage and reunion events between sister chromatids, and both Dupraw (1968) and Comings (1971) have suggested other ways in which isolabelling is compatible with only two chromatid subunits, given exchanges between sister chromatids. Sister chromatid exchanges are certainly visible in autoradiographic experiments (see Fig. 1) and provide a second exception to the all and none label segregation over second metaphase sister chromatids. However these switches of label are entirely consistent with the assumption of two subunits along a chromatid and in fact, Taylor (1958) used these exchanges in an attempt to evaluate the interpretation of the two subunits as the two complementary polynucleotide chains of a single DNA double helix. His rationale was that certain restrictions to rejoining of broken ends of the subunits would apply if the chromatid did in fact contain only one DNA molecule.

Polarity of Subunits

An affirmative answer to Taylor's question about the existence of "polarity" in the duplication of subunits came from a simple cytological experiment. Brewen and Peacock (1969) took Chinese Hamster cells through one DNA replication in the presence of [^{3}H]dT and then irradiated the culture to produce isochromatid breaks which in some cases resulted in sister chromatid union to form a dicentric chromatid (Fig. 2). Anaphase was suppressed with colcemid so that following a

FIGURE 2

Isochromatid breakage and reversed reunion in the two chromatids of a chromosome which had replicated in the presence of [^{3}H]dT.

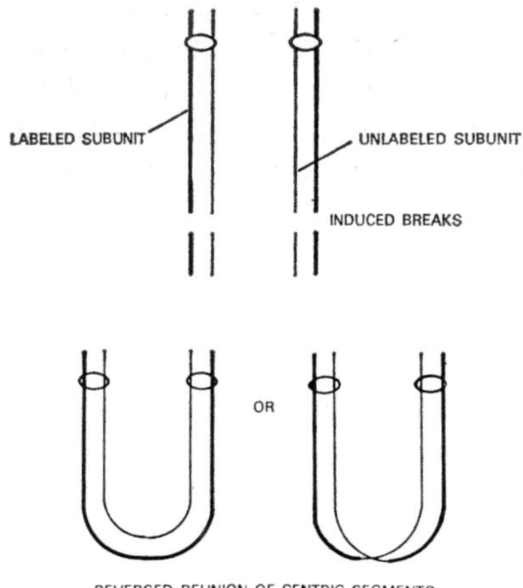

The left hand pattern of reunion of labelled-labelled and unlabelled-unlabelled subunits would be enforced if the two subunits of a chromatid differed in polarity. Equal frequencies of the two patterns would result if there were random reunion of subunits.

further replication, the dicentric chromatid would be represented as a dicentric chromosome. If the chromatid subunits did differ in structural polarity then the only possible reunion would be the labelled subunit of one chromatid joining to the labelled subunit of the other chromatid, and similarly, the two unlabelled subunits would be joined. This would result in the intercentromeric segment of the dicentric chromosome of the second metaphase being labelled in only one chromatid. If there were no restrictions on rejoining of subunits the expectation would be for dicentrics with a label switch in the middle of the intercentromeric segment to be equally as frequent as those with a restriction of label to one chromatid. Brewen and Peacock found that 105 of 137 dicentrics showed no switch of label between the two centromeres. The remaining dicentrics, including six having a medial label switch, were explicable in terms of the frequency of sister chromatid exchanges occurring in these cells. This result demonstrated that the chromatid subunits were not identical; the restriction to rejoining of subunits was precisely that expected for the polynucleotide chains of a DNA double helix.

Mention was made in the last section that Taylor (1958) had used sister chromatid exchanges in looking into the question of polarity. He reasoned that in tetraploid second metaphases, if polarity restrictions were enforced then an exchange which had occurred in the first cell cycle after label incorporation would be visible in two sister homologues as a twin exchange; exchanges occurring in the next cell cycle would appear in only one chromosome as single exchanges. Since the chromosome number doubled in this second cycle a ratio of one twin: two single exchanges would be expected. A lack of subunit polarity would have led to ten singles to each twin exchange. His data offered only tenuous support to a polarity restriction, but later work by Herreros and Gianelli (1967) in human cells, and by Geard and Peacock (1969) in *Vicia faba* provided extensive data in support of the polarity-consistent one twin:two single exchange ratio.

The dicentric experiment and the sister chromatid exchange analyses showed that reunion of chromatid subunits was restricted, both for reversed reunion and for non-reversed reunion, to subunits of a *like* polarity. It is important to note that these two analyses showed that this restriction was not enforced by the presence of isotope in one subunit; in the dicentric experiment the *like* subunits were both labelled, whereas in sister chomatid exchanges the *like* subunit restriction ensured rejoining of a labelled subunit to an unlabelled subunit.

Continuity of Subunit Polarity

A directional polarity difference in the two duplication subunits of a chromatid is entirely consistent with the notion that a chromatid is composed of one long DNA molecule. This would mean lengths of some centimetres in most eukaryote chromosomes and extending, for example in the Liliaceae, up to one metre or more. Long DNA molecules (ca. 0.1 mm) from higher organism chromosomes have been recorded by electron microscopy (Solari, 1965) and by audioradiography (Cairns, 1966; Huberman and Riggs, 1968), and there has been one report of a molecule of 2.2 cm which could account for the entire DNA of one of the smaller mammalian chromosomes (Sasaki and Norman, 1966). Recently velocity sedimentation data have been presented in support of a single DNA molecule per chromosome in yeast (Petes and Fangman, 1972).

If the higher organism chromosome does consist of one long molecule then the polarity of each subunit should remain unidirectional along the full length of the chromosome. Both the dicentric experiment and the sister chromatid exchange data demonstrated a polarity difference between the subunits at any given point along the chromosome but did not offer any evidence as to whether there was a continuity of polarity along the subunit. We have tested this latter point by examining the consequences of rejoining the subunits after breaks were induced in two different places along a chromosome.

Induction of Ring Chromosomes

If two breaks are induced, by irradiation, one in each half of an unreplicated chromosome, then the broken ends can rejoin to form a centric ring chromosome (Fig. 3); the terminal pieces may also rejoin to form a linear acentric fragment. Breaks which are both positioned in the same arm would produce an acentric ring chromosome. In either case, if there is continuity of subunit polarity along a chromosome then rejoining would be restricted in such a way as to form, upon

FIGURE 3

Induction of a ring chromosome.

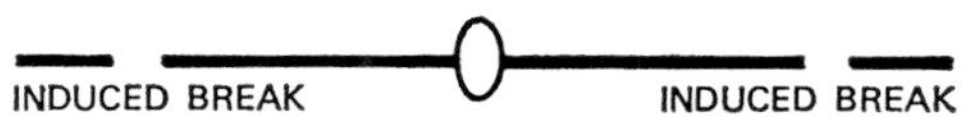

If one break is induced in each of the two arms of a chromosome then the ends may rejoin to form a ring chromosome and a linear fragment (not shown).

FIGURE 4

Polarity restriction in ring formation.

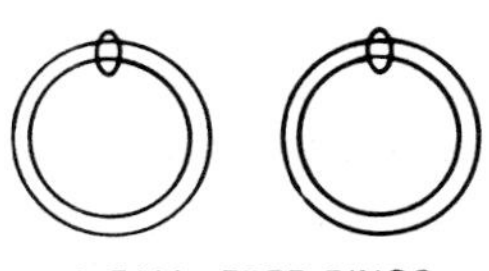

If there is continuity of subunit polarity along the length of a chromosome then rings induced prior to chromosome replication should produce two freely separating rings at the next anaphase.

DNA replication, two ring chromatids which would separate freely at the next anaphase. Each of these fall-free chromatids would be identical unit rings —*monomers* (Fig. 4).

We grew large numbers of Chinese Hamster cells and obtained highly synchronized populations of metaphase cells by the shake-off method (Terasima and Tolmach, 1961). The metaphase index of the population was better than 95% in all experiments. These cells were then irradiated prior to the initiation of DNA replication and, after a colcemid block of the next anaphase, were finally harvested as tetraploid cells in the second metaphase after ring chromosome induction. Complete details of the method are given elsewhere (Peacock *et al.*, 1972). Tetraploid second metaphase cells were used in lieu of anaphase of the preceding division because of the cytological difficulty in unequivocal analysis of anaphase separations; the freely separating ring chromatids of the anaphase would be scored as two identical ring chromosomes in the tetraploid cell.

FIGURE 5

Tetraploid Chinese Hamster cells with induced ring chromosomes.

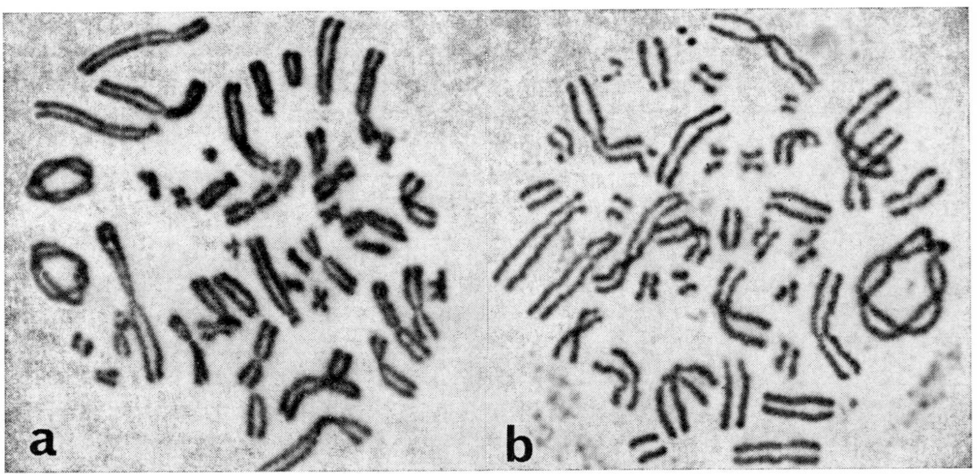

(a) This cell contains two identical ring chromosomes derived from the original induction of a single ring chromosome.
(b) This cell contains only a single ring chromosome.

Examination of the tetraploid cells showed that there were cells with two identical rings (Fig. 5a) as would be expected on the basis of continuous polarity of subunits, but there were also cells which contain only a single ring. Data from one experiment involving two irradiation treatments are given in Table 1. The occurrence of single rings represents a striking departure from the result expected for a chromosome composed of a single DNA molecule. However examination of other induced aberrations in the cells showed another unexpected category which suggested a mode of origin of the single rings which would be consistent with continuous polarity.

TABLE 1

Aberrations in tetraploid Chinese Hamster cells following irradiation of a synchronous G_1 population

Irradiation	Number of cells examined	Number of 2 identical rings	Number of single rings	Dicentric Chromosomes			
				Paired Asym-metric	Unpaired Asym-metric	Paired Sym-metric	Unpaired Sym-metric
300 rd	150	25	19	102	3	15	91
400 rd	200	39	48	207	10	33	225

Sister Union Origin of Single Rings

Some of the tetraploid cells contained paired linear dicentric chromosomes. An example is given in Fig. 6a. Such pairs of dicentrics are expected as a result of chromosome duplication following reunion of two individual breaks where each break is in a different chromosome. The unexpected finding was that of linear dicentric chromosomes which were not repeated twice in the tetraploid cell (Fig. 6a). These unpaired dicentrics were almost invariably symmetrical and are directly comparable to the mirror-image dicentrics induced by Brewen and Peacock (1969) in irradiation of post replication chromosomes. However labelling studies confirmed that in the present experiments the cells were irradiated well

FIGURE 6

Induced dicentric chromosomes in tetraploid Chinese Hamster cells.

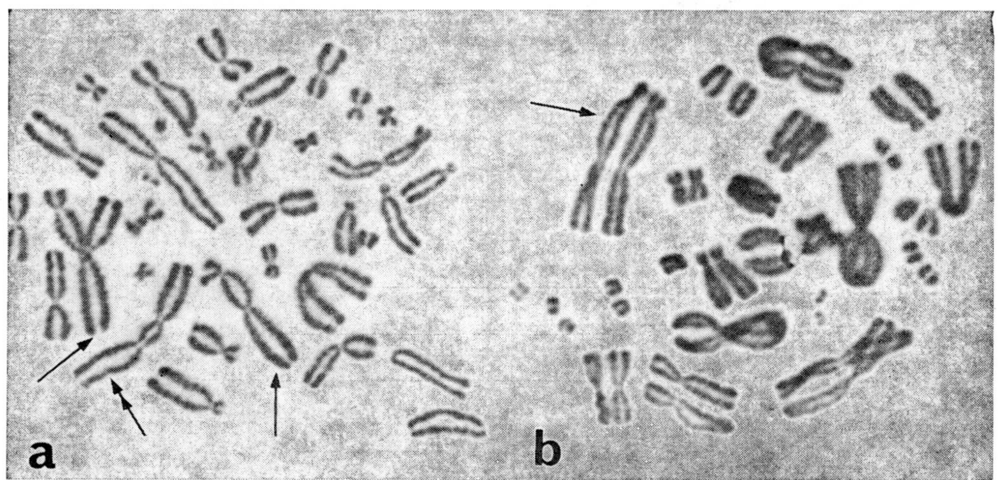

(a) Tetraploid cell showing a pair of asymmetrical dicentrics (single arrows) and an unpaired symmetrical dicentric (double arrow).

(b) Endotetraploid cell with paired sister homologues showing paired asymmetric dicentrics (single arrow) and four examples of reversed sister union resulting in the formation of unpaired symmetrical dicentrics.

before chromosome duplication; also no other post-replication chromosome aberrations were evident. The unpaired dicentrics must represent reversed reunion of the two subunits at a point of breakage; this is clearly shown in Fig. 6b, an endotetraploid cell in which sister homologues are maintained in a paired state.

The significance of the unpaired dicentrics in our present discussion is that they demonstrate a possible origin of the single rings. If two reunion events of this type occurred in a chromosome then a single chromatid ring would be present in the cell after chromosome duplication. If the events occurred in each arm of a chromosome a dicentric ring would be formed, the centromeres frequently being asymmetrically placed in the ring (Fig. 7a). In genetic terms these rings are *reversed dimers.* The frequency of ring formation by this method can be accurately estimated from the probability of sister subunit reunion as represented by the unpaired symmetric dicentrics, and a correction can be applied to the ring data. An example is given in Table 2 which shows the correction for the data in Table 1. Obviously this mode of ring formation contributes only a minor component to the single ring frequency and we are still left with approximately equal frequencies of the two different ring categories.

FIGURE 7

Induced dicentric rings in tetraploid Chinese Hamster cells.

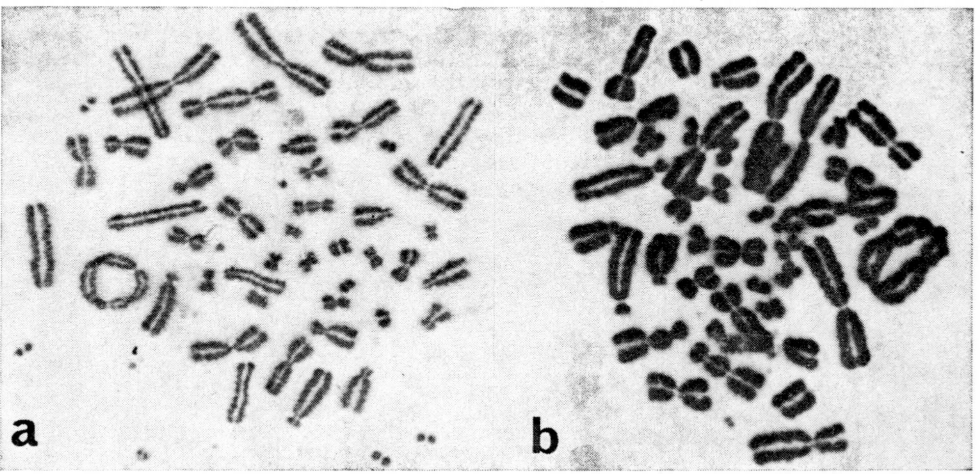

(a) Asymmetrically placed centromeres of a dicentric ring (reversed dimer) formed by two sister subunit reunion events.

(b) Symmetrically placed centromeres of a dicentric ring (probable tandem dimer).

Alternative Explanations for the Induced Rings

In the majority of single rings where centromeres can be clearly distinguished, they are symmetrically placed in the ring (Fig. 7b); it seems likely that these dicentrics are *tandem dimers* in contrast to the reversed dimers described above. They could arise from a ring formed according to the rules of polarity if sub-

TABLE 2

Calculation of number of single rings formed by sister subunit reunion
(data from Table 1)

	300 rd	400 rd
Frequency of unpaired symmetric dicentrics/cell	$\frac{91}{150} = 0.607$	$\frac{225}{200} = 1.125$
Probability of sister subunit reunion/chromosome (p) ..	0.036	0.066
Probability of two events/chromosome (p^2)	0.0013	0.044
Estimated rings formed in this manner	3.25	14.89
Corrected data—Two rings: Single rings	25:16	39:33
X^2 test against 1:1 expectation	$X^2 = 1.98$	$X^2 = 0.50$

sequently a sister chromatid exchange occurred correcting the two potentially fall-free chromatid rings into a single tandem dimer ring (Fig. 8). Unfortunately other origins can be postulated. For example if the subunits of the unreplicated chromosome were free to rejoin at random, equal frequencies of the single ring and the

FIGURE 8

Origin of a single dicentric ring from a ring formed according to the rules imposed by polarity.

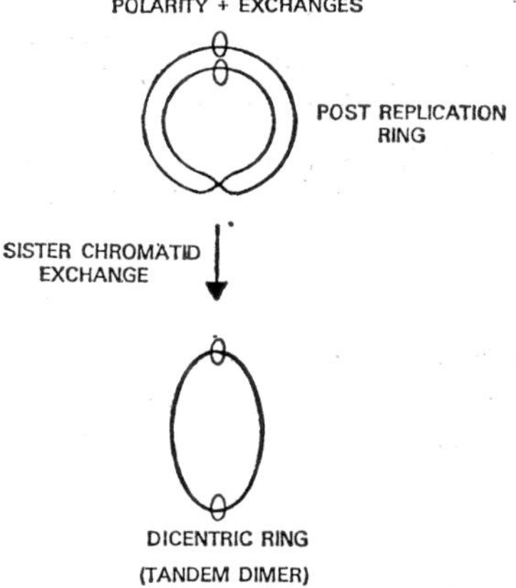

A sister chromatid exchange (or odd number of exchanges) converts two fall-free monomer rings into a single tandem dimer ring.

two ring categories of tetraploid cells would be expected. A third possibility is that the chromosome actually consists of a number of DNA molecules linked together, where the linkage is such that it results in mixed polarity along a subunit. In this case polarity differences in subunits would apply at any given point but there would not be a continuity of polarity along the length of the subunit.

A means of distinguishing between these three alternatives was found when the distributions of the two ring categories were plotted as a function of ring length. For example from an experiment in which 300 rd induced a corrected frequency of 136 cases of fall-free monomer rings and 92 tandem rings, the rings were either photographed or drawn by camera lucida and their circumference determined with the aid of a computerised device (details in Peacock *et al.*, 1972). The relative frequencies of the ring categories are shown in Fig. 9 in which the two smallest ring classes differ significantly from a one:one distribution. The extent of the excess of paired monomers is dependent on the size of the ring. This type of distribution is expected only on the basis of the first alternative presented above; both of the other models predict no change in relative frequencies with differing sizes of rings. The probability of exchange in the rings can be calculated to be of the order of 0.12 per micron of metaphase chromosome length.

FIGURE 9

Distribution of relative frequencies of ring types according to the size of the ring.

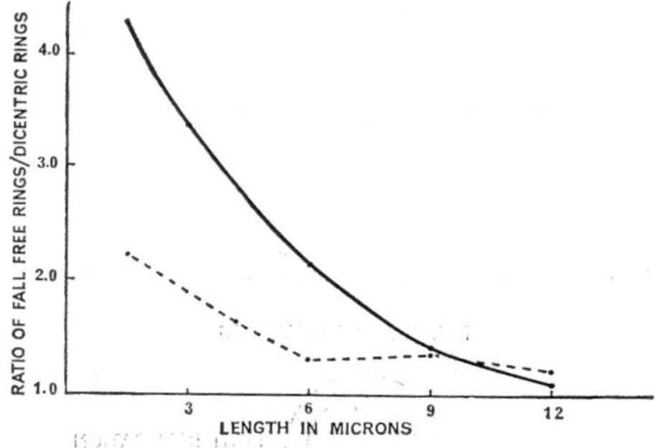

The broken line shows the actual data points. The continuous line demonstrates the general shape of the family of curves expected on the assumption of a length dependent conversion of monomer to dimer rings.

Conclusion

The induction of ring chromosomes from unreplicated chromosomes of Chinese Hamster cells has permitted a test of the assumption of continuous polarity of chromatid subunits. This assumption is obligate if the chromatid consists of a continuous DNA molecule. The results are consistent with continuity of polarity and in addition demonstrate that sister chromatid exchange is of frequent occurrence in these newly formed ring chromosomes. It should be emphasized that although the experiment is consistent with a continuous DNA molecule chromatid structure, it cannot distinguish against the possibility that the chromatid consists of many DNA molecules tandemly linked such that polarity of each subunit is unidirectional along the length of the chromatid. It does exclude models in which there is no polarity of subunits of the unreplicated chromosome or in which DNA molecules are linked together in mixed polarity.

GENETIC DEFINITION OF UNITS OF FUNCTION

I would now like to outline a problem which has recently been defined as a result of genetic fine structure analyses of *Drosophila* chromosomes. The problem arises when the number of detectable genetic units in a certain region of a chromosome is considered in relation to the amount of DNA in that region of the chromosome. General thinking about the nature of chromosomal DNA has changed radically since the demonstration that a sizable proportion of the genome of most higher organisms is composed of sequences which are repeated many times (Britten and Kohne, 1968). This finding contradicted the previously generally held view that the genome consisted solely of a collection of unique sequences, apart from the few cases of genetic duplication that had been identified. I think it is clear that the existence of genetic information from *Drosophila* has really sharpened the questions about the organization of unique and repeated sequences in the chromosome.

Genetic Fine Structure

Three areas of the *Drosophila melanogaster* genome have been subjected to fine structure analysis. Lifschytz and Falk (1968, 1969) have examined the region at the heterochromatin-euchromatin junction of the *X* chromosome, Hochman (1971) has analysed chromosome 4, and Judd *et al.* (1972) have concentrated on a small distal region of the *X* chromosome. I will briefly discuss this last study as an example of the technique. Judd *et al.* (1972) defined the region of analysis by using an *X* chromosome containing a particular deletion [$Df(1)w^{rJ1}$] in the region of the eye colour mutants *zeste* (*z*) and *white* (*w*). (Details of all the chromosomes and mutants mentioned in this discussion can be obtained from Lindsley and Grell (1968).) By treating *X* chromosomes with a mutagen and then making the treated chromosomes heterozygous to the deficient *X* chromosome, recessive mutants induced in the region delimited by the deficiency were isolated. The general method is illustrated in Fig. 10. After producing 121 single site mutations Judd *et al.* (1972) screened them by pairwise testing and found them to fall into 14 complementation units. They then mapped these complementation units, and the two genes *z* and *w*, by both deletion and recombination mapping. The procedures produced an unique linear sequence of 16 loci. Light microscope analysis shows that the deficiency deletes 15 bands from the *X* chromosome and an electron microscope study (Berendes, 1970) has provided evidence of an additional band making a correspondence of the number of bands and the number of complementation units. This correspondence becomes more significant when a comparison is made of the cytological extent of the deletions used in the mapping studies and the number of complementation loci exposed by each deletion. The data clearly show a one band:one complementation unit relation.

This same relation, first suggested by the work of Muller and Prokofyeva (1935), is almost certainly the case in other regions which have been examined. Lifschytz and Falk (1968, 1969) mapped 105 lethal mutations into 34 complementation groups in that part of the *X* chromosome delimited by a *ma-l* duplication on a *Y* chromosome. Cytological mapping in this region of the *X* chromosome

FIGURE 10

Isolation of mutants in the *X* chromosomes in a region defined by a deficiency.

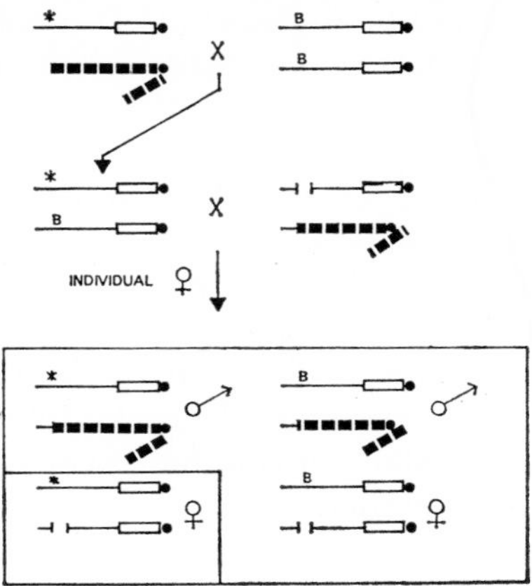

Males (*X/Y*, the *Y* chromosome being denoted by the heavy broken line) are irradiated and mated to females having both *X* chromosomes marked with a dominant gene (B). The resulting F_1 females are mated individually to males carrying an *X* chromosome with a deficiency (vertical bars) and a *Y* chromosome carrying a duplication to cover the *X* deficiency. Lethals induced on the irradiated *X* chromosome (*) in the region delimited are detected by an absence of the class of females in the left lower corner of the matrix.

is difficult but Schalet *et al.*, (1970) showed that seven of the complementation units (10-17) corresponded to eight bands of the salivary gland chromosome, and the other sections cannot deviate significantly from this same relationship. Similarly Hochman (1971) has identified approximately 150 recessive lethal mutations with 33 complementation units on chromosome 4, and in addition has found seven loci in which only morphological effects are known. This total of 40 functional units corresponds with approximately 50 bands in the right arm of chromosome 4. Two deficiencies on chromosome 4 delimit three regions along the chromosome and in these regions the correspondence of functional units to bands is 10 units:11-15 bands, eight units:8-9 bands and 22 units:26-31 bands respectively.

Since the existence of only one detectable functional unit per band holds for a distal and a proximal region of the *X* chromosome and for an autosome, it seems probable that it is a general rule for the genome. Lefevre and Green (1972) have recently noted that an apparent exception to this rule is the *w-spl* region of the chromosome, encompassing only four bands but from previous reports containing 14 loci. However they were able to show that most of the gene assignments were in error and that the only way this region violates the 1:1 relation of genes and bands is that it probably contains a duplication of a band.

DNA Content of a Band and the Paradox

Rasch *et al.* (1971) have estimated the amount of DNA in the unreplicated haploid genome of *Drosophila* to be about 0.18×10^{-12} g of DNA which corresponds to a molecular weight of approximately 1.1×10^{11} daltons. Since there are about 5,000 bands in the genome the average band should contain about 2.2×10^7 daltons in its unit strand, a molecular weight which corresponds to somewhere around 35,000 nucleotide pairs. This may be a slight over-estimate because centromeric heterochromatin has not been allowed for, but it is in good agreement with Rudkin's (1965) more direct estimate of about 30,000 nucleotide pairs in the average band. In the *z-w* region within the area analysed by Judd *et al.*, Rudkin's measurements show that there is an average of 25,000 nucleotide pairs per band. Thus the average band has enough DNA in its unit chromatid to code for 25-30 proteins each of approximately 300 amino acids, yet the genetic studies clearly indicate a single complementation unit per band. This is the paradox which genetic analysis has been instrumental in defining. What is the nature of the DNA in a band? This question is likely to have meaning for most other higher organisms since banded polytene chromosomes have been described in such diverse organisms as the Protozoa and flowering plants. Also it seems likely that the chromomeres exhibited in the oocytes of amphibians, and also in other animals, are comparable to the bands of polytene chromosomes.

Explanation of the Paradox

Many explanations of the gene-band paradox are currently possible. Some of the alternatives are preferable to others on present evidence but a unique explanation seems not yet to be evident. An obvious explanation would be that the mode of genetic analysis has detected only one class of mutation. If a majority of loci could not mutate to a condition which would yield an obvious morphological mutant or would interrupt normal viability and development, then they would not have been detected by these studies. If this is the case then one might wonder why in all bands there is only one locus which can mutate so as to be detectable as a lethal or morphological mutant. This latter point is true irrespective of the size (DNA content) of the band.

It seems certain in all of these studies, and particularly in that of Judd *et al.* (1972), that the detection of complementation units has reached a near saturation level. The probability that increased sample size would increase the number of complementation units is very low. Also in all three studies chemical mutagens were used as well as X-rays; in this case a majority of the isolated lethals mapped as point mutants, so that it is unlikely that all the mutants isolated were deletions which would give a false impression as to the number of functional units. Judd *et al.* (1972) have pointed out the desirability of acquiring information as to the revertibility of the lethals.

We are left with the obligate assumption that if there are several different functions coded within a band then their control is *cis*-dominant i.e. a mutation which impairs any one of the functions causes a block to all other functions. This effectively means that a cistron in higher organisms will on the average, contain enough genetic information for 20-30 separate functions. On the other hand the necessity to adopt a cis-dominant explanation of the enigma embraces explanations which do not assume 20-30 structural genes per cistron. Both Judd *et al.* (1972) and Lifschytz (1971) favour the interpretation that each chromomere contains a single structural gene with the remaining DNA being regulatory in function. Lifschytz has termed this concept of the band a "service unit" model. Many postulates can be made as to the components of such a "service unit"—sequences may be needed to provide control of the structural gene at transcriptional and translational levels as well as at other unspecified chromosomal levels. Nevertheless the fact remains that at present this concept is merely an "explanation" for the paradox and is not based on any hard data.

Judd *et al.* (1972) have also cautioned that the genetic data are consistent with an assumption of identity of the complementation units to the interbands. Since the ratio of DNA in bands and inter-bands is on the average 30:1 the inter-band does provide sufficient DNA for a coding sequence for a protein. Just this possibility has been proposed by Crick (1971) who has suggested that the coding DNA sequences are confined to the inter-band and that all of the DNA in the bands is concerned in the matter of control.

Another entirely different explanation to the paradox is that the 20-30 genes-worth of DNA in the average band represents 20-30 repeats of a single coding unit. Thus a chromosome would consist of a number of families of repeated DNA sequences. This view is favoured by Callan (1967) who derived the viewpoint largely from cytological studies of the lampbrush chromosomes in the oocytes of newts (Callan and Lloyd, 1960). In proposing that each gene was repeated many times Callan recognized that mutational and recombinational data introduced difficulties; he answered these difficulties with a Master-Slave concept where only the master gene participates in meaningful mutation and recombination. Thomas (1970) has also championed this suggestion largely on the basis of evidence which he believes demonstrates the existence of substantial frequencies of tandemly repeated DNA sequences. This evidence has been provided by cyclization of eukaryotic DNA fragments (Thomas *et al.*, 1970). In these experiments Thomas *et al.* found that 1-3 micron fragments of DNA from higher organisms could form circles if partially digested with exonuclease III or λ-exonuclease and then annealed (Fig. 11). *Necturus* (an amphibian) DNA yielded up to 35% circular structures in the population of DNA fragments, and trout and salmon sperm showed about 20% circular structures. Similar sized fragments of bacteriophage or bacterial DNA did not show any propensity to form circles. On the assumption that cyclization depends upon the existence of complementary sequences, Thomas *et al.* concluded that the eukaroytic DNA fragments were terminally repetitious. They argued that the efficiency of the cyclization process was such that the potential population of cyclizable fragments was greater than the observed frequency by at least a factor of two. Thus with potential frequencies of at least 50-75% they concluded the results could be plausibly interpreted to mean that tandem repeat sequences form a major component of the genome of higher organisms.

Thomas (personal communication) has found exonuclease treatment of *Drosophila* fragments to yield about 18% circular structures; this applies to *D. melanogaster*, *D. virilis* and *D. hydei*. He found almost the same frequency in each of the species when using DNA isolated from the polytene nuclei of the salivary glands. This finding is significant because of two further facts. One is that

FIGURE 11

Cyclization of a DNA fragment.

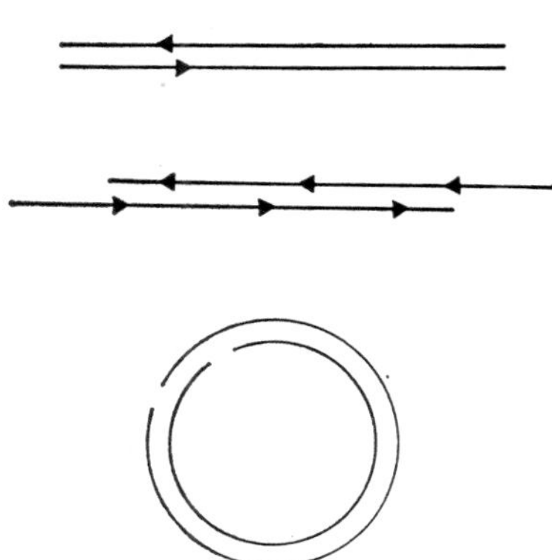

This diagram shows the manner in which a terminally repetitious DNA fragment can cyclize following limited digestion with exonuclease.

in the polytene chromosomes of *Drosophila* the centromeric heterochromatin is under-represented relative to the euchromatic portion of the chromosome; this has been demonstrated by Rudkin (1969) for *D. melanogaster* and by Berendes and Keyl (1967) for *D. hydei.* The second point is that the centromeric heterochromatin is the site of localization of much of the repetitive DNA sequences of the *Drosophila* genome.

Dickson *et al.* (1971) found in *D. hydei* that DNA extracted from embryos contained about 20% fast renaturing sequences, whereas DNA isolated from polytene chromosomes contained only 5% fast renaturing DNA. Coupling this finding with the demonstration by Berendes and Keyl (1967) mentioned above, that centric heterochromatin is grossly underrepresented in the polytene chromosomes, Dickson *et al.* concluded that most of the repeated DNA sequences in *D. hydei* are localized near the centromere, with the possibility of a small proportion of the remaining 5% fast renaturing DNA being distributed along the chromosome arms. This latter possibility is supported by the report of Hennig *et al.* (1970) that some satellite DNA sequences are distributed over various regions of the chromosomes in *D. hydei.*

A similar conclusion, that satellite sequences are largely located in the centromeric heterochromatin, has been reached by Gall *et al.* (1971) for *D. melanogaster.* This conclusion is based on *in situ* hybridization studies (also by Rae, 1970; Jones and Robertson, 1970; and Botchan *et al.* 1971).

Another Paradox

Thomas' finding that the polytene chromosome DNA, which is depleted of the centromeric fast-renaturing sequences, can cyclize to about the same extent as total embryonic DNA in *Drosophila* suggests that tandem repeat sequences are a widely distributed and common component of all chromosomes. On the other hand renaturation kinetic data suggest that 95% of the sequences in polytene chromosome DNA are represented only once in the genome. Pyeritz *et al.* (1971) have shown that the highly repetitious satellite DNA in the mouse shows up to 60% cyclized fragments after exonuclease treatment. In our laboratory we have found that the repetitious satellites of *D. melanogaster* also show a high frequency of circular structures after exonuclease treatment. This applies to the light satellites studied by Gall *et al.* (1971), Botchan *et al.* (1971) and Rae (1970) and to the very light "dAT" satellite (Blumenfeld and Forrest, 1971). A high frequency of circular structures is also formed by another highly repetitious fraction ($\rho = 1.705$) we have isolated, this accounting for approximately 3% of total DNA; this satellite, like the light satellites, hybridizes largely to the chromocentral region of the polytene nucleus. If all of these highly repetitious fractions are underrepresented in the polytene chromosome DNA, and this has not been rigorously shown as yet, then Thomas' results are difficult to explain since, *in toto,* these highly cyclizable fractions can account for a major portion of the 15% fast-renaturing embryonic DNA.

This paradox must be resolved if resolution between the alternative explanations as to the nature of the functional unit of the chromosome, the band, is to be achieved. If the kinetic data were not sufficiently sensitive to distinguish between a condition where 20 copies of a sequence exist and a condition where only one copy of each sequence is present, then the Master-Slave postulate could be tenable; so also could any model which assumed repeating noncoding sequences in the band. However, Laird (1971) has presented data which strongly argue for a single representation of the sequences of the slowly renaturing "unique" DNA in the unit chromatid of *Drosophila,* and he claims the accuracy of the technique is good within a factor of 2.

Wu *et al.* (1972) have claimed that short segments of "middle-repetitive" DNA are distributed throughout the *Drosophila* genome forming a fairly regular alternation with unique sequences. Possibly intercalated repetitive segments of this nature are responsible for the cyclized products found by Thomas. The cyclized products might also be largely produced by only certain parts of the genome which are well represented in the polytene chromosomes as well as in the nonpolytene chromosomes. Examples of known discrete repeated fractions are the DNA sequences coding for ribosomal RNA (including 5 S RNA) and for the t-RNA's; also Rae (1970) and Gall *et al.* (1971) noted a region of one or two bands on one of the autosome arms which hybridizes extensively with c-RNA prepared from the fast-renaturing DNA. However, these various fractions can only account for an extremely small portion of the genome and unless many others are found it seems improbable that they will account for Thomas' observations.

A direct approach to the problem would be to isolate the cyclized products away from the linear DNA fragments and subject them to *in situ* hybridization and to renaturation kinetic studies. Information as to the possibility that the circular structures are largely due to intercalated repetitive sequences could also be obtained if the circular structures were isolated, sheared and reannealed. It is to be expected that analyses along these lines, together with experiments designed to measure the extent to which "unique" DNA is transcribed will provide important restrictions on the possible models for the unit of function of the higher organism chromosome.

REFERENCES

Berendes, H. D. (1970) *Chromosoma 29,* 118

Berendes, H. D. and Keyl, H. G. (1967) *Genetics 57,* 1

Blumenfeld, M. and Forrest, H. S. (1971) *Proc. Nat. Acad. Sci. U.S. 68,* 3145

Botchan, M., Kram, R., Schmid, C. W. and Hearst, J. E. (1971) *Proc. Nat. Acad. Sci. U.S. 68,* 1125

Brewen, J. G. and Peacock, W. J. (1969) *Proc. Nat. Acad. Sci. U.S. 62,* 389

Britten, R. J. and Kohne, D. E. (1968) *Science 161,* 529

Cairns, J. (1966) *J. Mol. Biol. 15,* 372

Callan, H. G. and Lloyd, L. (1960) *Phil. Trans. Roy. Soc. 243,* 135

Callan, H. G. (1967) *J. Cell Sci. 2,* 1

Comings, D. E. (1971) *Nature new Biol. 229,* 24

Crick, F. (1971) *Nature (London) 234,* 25

Darlington, C. D. and Haque, A. (1969) *Heredity 24,* 273

Deaven, L. L. and Stubblefield, E. (1969) *Exptl Cell Res. 55,* 132

Dickson, E., Boyd, J. B. and Laird, C. D. (1971) *J. Mol. Biol. 61,* 615

Dupraw, E. J. (1968) *Cell and Molecular Biology* p. 571, Academic Press, New York and London

Gall, J. G., Cohen, E. H. and Polan, M. L. (1971) *Chromosoma 33,* 319

Geard, C. G. and Peacock, W. J. (1969) *Mutation Res. 7,* 215

Hennig, W., Hennig, I. and Stein, H. (1970) *Chromosoma 32,* 31

Herreros, B. and Gianelli, F. (1967) *Nature (London) 216,* 286

Hochman, B. (1971) *Genetics 67,* 235

Huberman, J. A. and Riggs, A. D. (1968) *J. Mol. Biol. 32,* 327

Jones, K. W. and Robertson, F. W. (1970) *Chromosoma 31,* 331

Judd, B. H., Shen, M. W. and Kaufman, T. C. (1972) *Genetics 71,* 139

Laird, C. D. (1971) *Chromosoma 32,* 378

Lefevre, G. and Green, M. M. (1972) *Chromosoma 36,* 391

Lifschytz, E. (1971) *Mutation Res. 13,* 35

Lifschytz, E. and Falk, R. (1968) *Mutation Res. 6,* 235

Lifschytz, E. and Falk, R. (1969) *Mutation Res. 8,* 147

Lindsley, D. L. and Grell, E. H. (1968) *Genetic Variations of Drosophila melanogaster,* Carnegie Inst. Wash. Publ. *627*

Muller, H. J. and Prokofyeva, A. A. (1935) *Proc. Nat. Acad. Sci. U.S. 21,* 16

Peacock, W. J. (1963) *Proc. Nat. Acad. Sci. U.S. 49,* 793

Peacock, W. J., Wolff, S. and Lindsley, D. L. (1972) *Chromosomes Today 4, Heredity (Suppl.)* in press

Petes, T. D. and Fangman, W. L. (1972) *Proc. Nat. Acad. Sci. U.S. 69,* 1188

Pyeritz, R. E., Lee, C. S. and Thomas, C. A. (1971) *Chromosoma 33,* 284

Rae, P. (1970) *Proc. Nat. Acad. Sci. U.S. 67,* 1018

Rasch, E. M., Barr, H. J. and Rasch, R. W. (1971) *Chromosoma 33,* 1

Rudkin, G. T. (1965) *Genetics 52,* 665

Rudkin, G. T. (1969) *Genetics (Suppl.) 61,* 227

Sasaki, M. S. and Norman, A. (1966) *Exptl Cell Res. 44,* 642

Schalet, A., Lefevre, G. and Singer, K. (1970) *Dros. Inf. Serv. 45,* 165

Solari, A. J. (1965) *Proc. Nat. Acad. Sci. U.S. 53,* 503

Taylor, J. H. (1958) *Genetics 43,* 515

Taylor, J. H., Woods, P. S. and Hughes, W. L. (1957) *Proc. Nat. Acad. Sci. U.S. 43,* 122

Terasima, T. and Tolmach, L. J. (1961) *Nature (London) 190,* 1210

Thomas, C. A. (1970) In *The Neurosciences: a second study program SSP.* The Rockefeller Press.

Thomas, C. A. (1971) *Annu. Rev. Gen. 5,* 237

Thomas, C. A., Hamakalo, B. A., Misra, D. M. and Lee, C. S. (1970) *J. Mol. Biol. 51,* 621

Wolff, S. (1969) *Int. Rev. Cytol. 25,* 279

Wu, J. R., Hurn, J. and Bonner, J. (1972) *J. Mol. Biol. 64,* 211

Transgenosis of Bacterial Genes from *Escherichia Coli* to cultures of Haploid *Lycopersicon Esculentum* and Haploid *Arabidopsis Thaliana* Plant Cells

COLIN H. DOY[a], PETER M. GRESSHOFF[a] and BARRY ROLFE[a]

Department of Genetics, Research School of Biological Sciences, Australian National University, Canberra, A.C.T. Australia

HAPLOID PLANT CELLS USED AS RECIPIENTS FOR BACTERIAL GENES

What may be a general method for the production of haploid callus cultures of higher plants (angiosperms) on fully defined and minimal media has recently been developed in this laboratory (Gresshoff and Doy, 1972a, 1972b). Cell lines from pollen mother cells of the diploid plants *Lycopersicon esculentum* (tomato) and *Arabidopsis thaliana* have remained haploid throughout a year of sub-culture. Thus the genotypes of these plants are now carried by haploid cultures analogous to those of vegetatively dividing eukaryotic micro-organisms.

A developmental sequence from a tomato anther *via* haploid callus to haploid tomato plantlet is illustrated in Figs. 1-3. The cell line *Lycopersicon esculentum* ANU-H27-1, see Gresshoff and Doy (1972b) derived from the callus illustrated in Fig. 2 is the basis for most of the work to be discussed. With continuous culture on a high auxin low kinetin medium [Gresshoff and Doy (1972a, 1972b) and see legend to Table 1] the typical callus consists of large, loosely attached cells (about 10^3 cells per mg wet weight) off-white in colour. The original mass doubling time at 27°C, using the optimal carbon source glucose, was 10-15 days, but is now about 4-5 days. Suspension cultures with a doubling time of about 18 hours have recently been established (Gresshoff and Doy, unpublished results).

Callus cultures were used as the recipient cells for 'infections' of specialized transducing phage carrying genes of the bacterium *Escherichia coli* (Fig. 12). We are not claiming that haploidy is essential for the present experiments but we do

[a]Our authorship policy is to reflect alphabetical order of surnames.

think that the physical nature of the cultures is important (see, Resistance to Transgenosis) and that haploidy may be important for some future examples of transgenosis. Unless otherwise stated, the experiments were done with the tomato cell line. Reasons for using specialized transducing phage are discussed later (see Discussion).

BACKGROUND TO EXPERIMENTS

The present state of knowledge of genotypic and phenotypic mechanisms of expression owes much to the development and application of genetical and biochemical methods (molecular biology) to cultures of pro- and eukaryotic microorganisms having an extended haploid phase. The life cycles may also include a natural diploid phase and/or give the possibility of creating 'artificial' diploid situations. These situations create the opportunity for genetic recombination and tests of dominance. The introduction of similar methods to higher plants would greatly facilitate the study of their biology at a more fundamental level than is generally possible at present and create opportunities for the directed modification of plant genotypes.

A potential difficulty in extending the methods of microbiology to higher plants is that there is little knowledge of gene redundancy in haploid cell lines. Indeed some haploid cultures have a natural gene redundancy in that the term haploid refers to the gametic chromosome number which may represent more than one copy of individual chromosomes. Such polyploids have frequently been used for the creation of haploid plants. This difficulty can be avoided by the use of true diploid plants as the source of haploid cells. This condition is met by our haploid cultures of tomato and arabidopsis. These diploid "parents" also have the advantage of a considerable genetical background. Our haploid cultures therefore avoid one possibility for gene redundancy but leave the possibility of redundancy due to multiple copies of the same gene on the same or different chromosomes. Whether or not this is a serious problem remains to be determined and will depend on whether or not gene duplication is a developmental process.

A recent study with tobacco (Carlson, 1970) suggests that it may be possible to select auxotrophic mutants from haploid cultures of higher plant cells. Auxotrophy was not complete. Phenotypic leakiness and the poor yield of mutants may reflect the known polyploid nature of the "diploid" parent. Until genetical tests are done it remains possible that the "mutants" represent epigenetic changes rather than a change within DNA, that is, they are not mutants at all.

Provided that stable genetical change can be introduced into haploid cell lines then various molecular and breeding-in experiments become possible. Methods of chromosome doubling already exist and hence means for the production of homozygous diploid cells. Similarly, the ability to control development and differentiation leads to the possibility of creating homozygous seed-bearing plants. These may have new characteristics either directly because of selected gene mutations or because of the elimination of lethal recessiveness.

The traditional ways of altering genomes are by mutation, or the introduction of genetic variation from closely related strains. We are attempting to develop these approaches with the haploid cultures (Gresshoff and Doy, unpublished). It seemed also that the many parallels between microbial and plant metabolism indicate a reservoir of microbial genes of potential use in the study and modification of plant genotypes and phenotypes. In terms of evolution, abrupt changes might follow if genes could be transferred from an unrelated species and be expressed in a new cellular environment.

The experiments we shall describe test the feasibility of using specialized transducing phage as a vector for the transfer of well characterized genes, coding for products of known function, from the bacterium *Escherichia coli* to haploid plant cells. Following transfer, the phenotypic expression of these genes would require transcription, translation and function or, in the example of transfer-RNA species, transcription and function, using apparatus provided by the plant cells. In order to mimic the evolutionary precept of survival of the fittest, we decided to judge transfer and expression on the overall criterion of growth in an environment previously lethal, or death in an environment previously optimal.

We have not regarded the measurement of enzymic activity in plant cells as being particularly informative unless evidence could be presented that the protein concerned was characteristic of *E. coli.* Thus in our original experiments we relied entirely on the biological criterion of life or death. Since the meeting on which this communication is based, we have been able to do specific biochemical and immunological tests for the presence of *E. coli* β-galactosidase. The positive outcome of these experiments (see later) is regarded as definitive molecular evidence in support of the biological evidence.

TRANSGENOSIS—DEFINITION AND REASONS FOR INTRODUCING A NEW TERM

The term *transgenosis* is defined as: *the transfer of genetic information from one cell to another, followed by phenotypic expression.* The term was introduced because we found difficulty in applying the terminology for gene transfer and expression derived from studies of bacteria and their viruses. There is already ambiguity in the usage of terms such as transformation by those working with other systems. We are reporting phenomena without experimental knowledge of the mechanisms that lead to them. In the absence of an exact determination of mechanisms, the use of established terms may be confusing, even misleading. Transgenosis therefore contains no implications of mechanisms and is not restrictive. For example, if a phenomenon was later found to be properly described as transduction, this term could be used, since transduction is a form of transgenosis. Transgenosis is a term that we regard as especially appropriate if the species of the donor and recipient cells are separated widely by evolution.

The present experiments have tested for gene transfer and functional phenotypic expression, not specifically for inheritance. Consequently, gene inheritance is not included in the above definition. If later studies should prove that gene replication and inheritance occurs then we would redefine transgenosis to include inheritance.

FIGURES 1-11

Plants as Haploid Micro-organisms

All Figures are reproduced from original Kodachrome transparancies.

Figures 1-3

From *Lycopersicon esculentum* anther *via* detailed haploid callus to haploid plantlet, see also Gresshoff and Doy (1972b) for methods and other details.

Figure 1

Haploid callus emerging from dying diploid anther tissue.

Figure 2

Haploid callus shortly after detachment from the anther and transfer for culture.

Figure 3

Haploid plantlets developing from callus after transfer to differentiation medium (growth medium 5 of Table 1, Gresshoff and Doy (1972b)).

Figure 4

An example of haploid *L. esculentum* callus transgenosised for gal^{+} and therefore surviving and growing slowly on galactose (2%) non-differentiation medium ([growth medium 3 of Table 1, Gresshoff and Doy (1972b)] and legend to Table 1). In this experiment all negative controls (see Table 1) had stopped growing and were pronounced dead (blackening, lack of protoplasmic streaming) some 15 weeks before this picture was taken. Note the difference in appearance from the callus of Fig. 2. Callus transgenosised for gal^{+} has a rather waxy appearance and never looks particularly healthy.

Figures 5 and 6

Examples of lac^{+} transgenosis of haploid *L. esculentum* callus compared with "negative" and "neutral" controls. Photographs taken at 13 weeks.

Figure 5

A comparison between transgenosised calluses and a "negative" control "inoculated" originally with phage suspension medium without phage (no ϕ) all on lactose (10%) non-differentiation medium ([growth medium 3 of Table 1, Gresshoff and Doy (1972b)] and legend to Tables 1 and 2). $Lac^{+}2$ is a callus inoculated with $\phi 80plac^{+}$. $\phi 80\lambda$ +/+ is a callus inoculated with a mixture of $\phi 80plac^{+}$ and $\lambda pgal^{+}$.

Figure 6

Similar to Fig. 5 except that a "neutral" $\phi 80$ control is included. In Figs. 5 and 6 the negative and neutral controls represent the original size of the calluses in all experimental variations. The increases in callus in the examples of transgenosis are at least 20-fold or at least 4 to 5 doublings. Callus is of better appearance than for gal^{+} transgenosis (Fig. 4) but not as good as callus grown on glucose medium. It should be appreciated that there is a delay of about 5 weeks before lac^{+} transgenosis is evident as growth and therefore there is considerable stress on the plant cells during this lag. Growth is never as fast as controls on glucose medium but may approximate to the original growth rate of newly isolated haploid callus. Some examples of calluses transgenosised for lac^{+} from the original experiments on lactose (4%) medium have continued to grow for many months after sub-culture on lactose medium. On this basis it seems possible, even probable, that there has been maintenance and replication of at least the *E. coli lac* Z^{+} gene within the plant cells. A callus similar to the +/+ example of Fig. 5 was used to obtain the β-galactosidase data incorporated in Table 2.

FIGURES 1-11

Plants as Haploid Micro-organisms

All Figures are reproduced from original Kodachrome transparencies.

1 2

3 4

5 6

FIGURES 1-11

Plants as Haploid Micro-organisms

All Figures are reproduced from original Kodachrome transparencies.

7

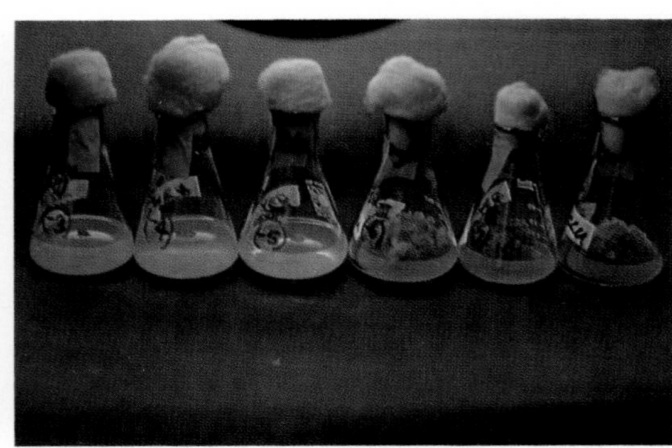

8

9

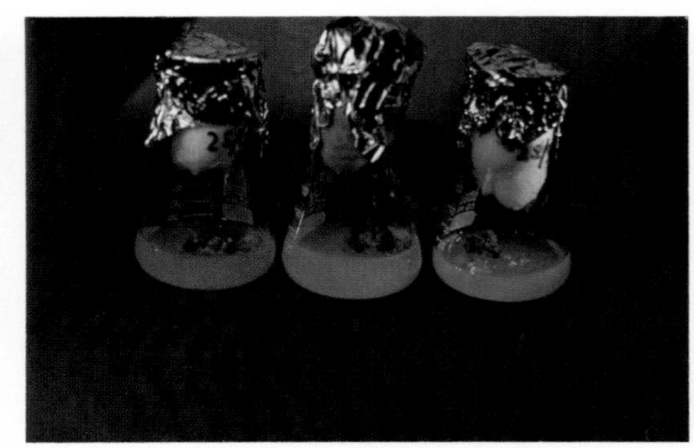

10

11

	Left to Right
Fig. 7	ϕ80
	ϕ80 sup F$^+$
	no ϕ
Fig. 10	ϕ80
	ϕ80 sup F$^+$
	no ϕ

Figures 7, 8 and 9

Examples of transgenosis of *supF*$^+$. All callus are on glucose (2%) non-differentiation medium [growth medium 4 of Table 1, Gresshoff and Doy (1972a) and growth medium 3 of Table 1, Gresshoff and Doy (1972b) and legend to Table 1].

Figures 7 and 8

Haploid *L. esculentum* at about 6 weeks.

Figure 7

A positive control without phage (no ϕ), a neutral control inoculated with phage ϕ80 (ϕ80) and an example of death as the result of inoculation with phage ϕ80p*supF*$^+$ (about 10^8 phage per callus inoculum). See also Fig. 13.

Figure 8

Titration of the ratio of phage ϕ80p*supF*$^+$ to haploid tomato cells required for death. The flask on the extreme right is a control without phage. The numbers on the flasks refer to dilutions (powers of 10) of an original suspension of phage ϕ80p*supF*$^+$. Hence, (-5) corresponds to 100 phage (P.F.U.) per haploid tomato cell in the original callus inoculum calculated on the basis that 1 mg wet wt. of callus contains 1000 cells and that all cells are exposed to phage. Thus, in the examples shown, there is no effect at a ratio of 10 : 1 (or less) (see also Fig. 14). There is variation in the result obtained at multiplicities between 10 and 100 : 1. At ratios greater than 100 the result is always death. It will be noted that blackening of the callus inoculum is greatest in the flask to the extreme left. See also Fig. 14.

Figure 9

Similar to Fig. 7. Haploid *Arabidopsis thaliana* callus on non-differentition medium [growth medium 4 of Table 1, Gresshoff and Doy (1972a)] at about 8 weeks. *SuIII* is an alternative nomenclature for *SupF*$^+$.

Figure 10

Resistence of differentiating haploid *A. thaliana* to ϕ80p*supF*$^+$. Except for the fact that the calluses are on differentiation media [growth medium 3 of Table 1, Gresshoff and Doy (1972a)] containing sucrose (2%) and therefore are greening as a preliminary to haploid plant formation (Gresshoff and Doy, 1972a) the conditions of inoculation are similar to those of Figs. 7 and 9. Differentiating calluses are completely resistant to ϕ80p*supF*$^+$.

Figure 11

Selection of a callus line of haploid *L. esculentum* which is resistant to ϕ80p*supF*$^+$. The callus is on glucose (2%) non-differentiation medium [growth medium 3 of Table 1, Gresshoff and Doy (1972b)] and was inoculated with ϕ80p*supF*$^+$ exactly as in Fig. 7. The bulk of the callus died but it gradually became apparent that a clone of resistant cells was developing. This clone grows rapidly and consists of a dense callus fairly difficult to cut for sub-culture. Although on non-differentiation medium (Gresshoff and Doy, 1972b) the callus resembles an early stage of normal haploid *L. esculentum* on differentiation medium [growth medium 5 of Table 1, Gresshoff and Doy (1972b)]. The callus has been sub-cultured many times under the same conditions and retains its characteristics of morphology and resistance to ϕ80p*supF*$^+$. Greening occurs in parts of the callus but plantlets have not developed on the non-differentiation medium (see Discussion).

OUTLINE OF EXPERIMENTAL PROCEDURE

Our experiments are, in principle, very simple. Calluses of the tomato cell line are unable to use the sugars galactose (2%) or lactose (2-10%) as general sources of carbon for growth. Further, galactose (2%) is toxic to growth on glucose (2%) medium. Thus, galactose or lactose media are lethal environments (gradual death of callus over about 3 weeks at 27°C) which we predicted might be corrected by an input of *E. coli* genes known to convert these sugars to metabolites anticipated to be usable by the plant cells. These genes are the well characterised galactose (*gal*) and lactose (*lac*) operons (Fig. 12). High titre preparations of specialized transducing phages (λ, φ80) carrying the required bacterial genes were applied to small pieces of haploid tomato callus (20-40 mg wet wt, about 10^8 phage (p.f.u.) per callus) on the appropriate medium, then incubated at 27°C. The test for successful transgenosis (in these examples, gene transfer, transcription, translation, enzyme function and appropriate metabolism) was the biological one of survival and growth where normally there was death. Controls of various kinds will be discussed with the experimental results below. Thus we were mimicking the evolutionary test of survival of the fittest—a fitness conferred by infection with phage carrying the appropriate genes.

Precautions against molecular or microbiological contamination

Precautions were taken to ensure that the phage preparations were not contaminated with enzymes, messengers or micro-organisms. Phage were grown on *E. coli* deleted for the genes in question and under conditions of repression (glucose medium) thus minimising the potential for carry-over of enzymes or messengers. Moreover, we were able to demonstrate the absence of detectable enzyme even in our highest titre preparations.

The phage preparations and calluses were routinely sampled and tested for contaminant micro-organisms in rich liquid medium (LB) (Bertani, 1951) incubated both aerobically and anaerobically at 27°C. This would detect most kinds of contaminants. Phage preparations were stored over chloroform and millipore-filtered before use. Aseptic methods were performed in sterile laminar flow hoods.

A range of micro-organisms (bacteria, yeasts, fungi) can grow on the media used for the culture of callus, but only when the normal culture medium was obviously contaminated (deliberately or accidentally) have we found contaminants in callus samples. Contaminated cultures were rejected. It should be emphasised that whenever contaminants (including deliberately added *E. coli*) occur, the result is to kill callus, even on glucose medium, not to assist callus growth. Moreover, for phage to infect a bacterium requires a specific process and infection with DNA requires spheroplasts or competent cells. These prerequisites are unlikely to be found except for the natural host *E. coli* and tests would have detected this micro-organism.

We envisage that the infection of plant cells whether by intact phage or DNA is by non-specific processes.

TRANSGENOSIS FOR SURVIVAL

Our first attempt at transgenosis was with the *E. coli gal* operon carried by λ phage. The overall result of the functions specified by this operon is to convert galactose to glucose 1-phosphate. We considered this a less than ideal system since the plant cells were required to convert glucose-1-phosphate to a more general carbon source such as glucose or glucose-6-phosphate. One would expect

FIGURE 12

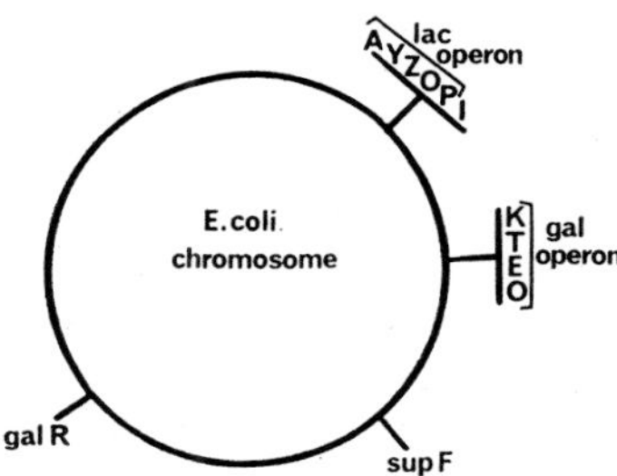

The chromosome of *Escherichia coli* showing the locus of genes related to the transgenosis experiments. The lactose (*lac*) operon: Genes concerned with the structures of three proteins, the Z gene (β-galactosidase), the Y gene (permease) and the A gene (thiogalactosidase transacetylase). Transcription into a single polygenic messenger RNA is initiated at the promoter (P) and the repressor protein binds to the operator (O) DNA. The repressor protein is encoded by the I gene. The structural genes carry information sufficient for the conversion of lactose to glucose plus galactose; they carry also a permease protein. Gene I and the sites P and O are involved in control. The ϕ80p*lac*$^+$ strain used for transgenosis carries all these genes except that a deletion extends partly into gene A (Iida *et al.*, 1970). The galactose (*gal*) operon: Genes concerned with the structures of three enzymic proteins, the gene E (UDP: glucose 4-epimerase), the gene T (UTP:D galactose-1-phosphate uridyl transferase), the gene K (ATP:D-galactose-phosphotransferase). Repressor protein binds to the operator (O) DNA. The repressor protein is encoded by the R gene which is remote from the *gal* operon and is therefore not carried by the specialized transducing phage strains used for transgenosis. The structural genes carry information sufficient for the conversion of galactose to glucose 1-phosphate. The λp*gal*$^+$ strain used for transgenosis carries all the operon, the λp*gal*$^-$ strain carries a nonsense (amber) mutation in the gene T (see text). Because of polarity it is probable that the phosphotransferase (kinase) is not made in addition to the defect in the transferase.

SupF (also known as ***SuIII***): A mutant gene coding for a transfer-RNA able to correct a nonsense (amber) mutation carried as a nonsense triplet in messenger-RNA.

This results in the synthesis of a protein which is otherwise missing or defective.

enzymes for these functions to be present in plant cells but these might be rate limiting for transgenosis. A better system was the *lac* operon since the products of β-galactosidase activity are glucose (directly available for general metabolism) and galactose (probably not available for general metabolism). However, we were not then able to obtain phage carrying *lac*. Phages λp*gal*$^+$ and λp*gal*$^-$ were kindly provided by Dr. M. Gottesman. These phages were, by coincidence of source, identical to those used by Merril *et al.* (1971) for the transgenosis of the gene (T) coding for UTP, D-galactose-1-phosphate uridyl transferase (transferase) E.C.2.7.7.10, from *E. coli* to defective human cells in culture.

The λp*gal*$^-$ strain was known to contain a mutation in the T gene such that the transferase function was absent. We characterised the T mutation further and

TABLE 1

Transgenosis of the *E. coli* galactose operon to haploid tomato callus ANU-H27-1 and the inability of the plant cells to effectively suppress a nonsense mutation.

General Carbon Source	Phage	Anticipated Alternative Results	Interpretation of Alternative Results	Actual Result
Galactose (2%)	None	Non-growth, gradual death over 3-5 weeks	Expected negative control	Non-growth, gradual death over 3-5 weeks
	$\lambda pgal^{+}$	Survival and growth	Effective transgenosis of *gal* genes	Survival and growth for at least 15 weeks longer than all other variables except those on glucose medium
		Non-growth, gradual death over 3-5 weeks	Negative experiment	
	$\lambda pgal^{-}$ (non-sense mutation in T gene)	Survival and growth	Efficient correction of nonsense mutation by plant cells, hence equivalence to $\lambda pgal^{+}$	
		Non-growth, gradual death over 3-5 weeks	Nonsense mutation not efficiently corrected	Non-growth, gradual death over 3-5 weeks
	$\phi 80$ or $\phi 80plac^{+}$	Non-growth, gradual death over 3-5 weeks	Neutral result, phage not carrying relevant genes	Non-growth, gradual death over 3-5 weeks
Glucose (2%)	None	Excellent growth	Expected positive control	Excellent growth
	Any of above phage	Excellent growth	Neutral result, phage do not prevent the primary growth response of callus	Excellent growth
		Inhibited growth	Phage interferes with primary growth ofcallus	

The derivation of haploid cell lines of *Lycopersicon esculentum* (tomato) and *Arabidopsis thaliana*, the nomenclature used to describe these cultures and the methods of controlling development and differentiation are described elsewhere (Gresshoff and Doy, 1972a, 1972b). The following is a description of the methods employed in the production and characterization of phage used in the transgenosis of the *E. coli gal* operon to haploid tomato callus ANU-H27-1. However, the methods are general and with changes of bacterial and phage strains and, for experiments with haploid *Arabidopsis thaliana*, the same growth medium is used also in the other transgenosis experiments. Originally (Gresshoff and Doy, 1972a) calluses were maintained on medium containing sucrose but later the carbon source was changed to glucose (2%) (Gresshoff and Doy, 1972b) except for *A. thaliana* on differentiation medium. Other modifications were also made to promote growth. In transgenosis experiments the carbon source was changed as appropriate (e.g. galactose (2%), lactose (10%) etc.). Haploid tomato callus ANU-H27-1 was maintained on DBM3 (Gresshoff and Doy, 1972b) but with naphthylene acetic acid (NAA) 8 mg/ml, kinetin 0.01 mg/l, penicillin G 1000 i.u./ml and the appropriate carbon source. Phages $\lambda pgal_8cI857$ ($K^+T^+E^+$) [$\lambda pgal^+$] and $\lambda pgal_8cI857$ ($K^+T^-E^+$) [$\lambda pgal^-$] were supplied by Dr. Max Gottesman (see also Merril *et al.*, (1971)) and $\phi 80lac^+$ by Dr. Y. Ohshima (Iida *et al.*, 1970). The λ phages were grown on *E. coli* KB5 (*gal*Δ *att*λΔ *sup*D) (Onodera *et al.*, 1970) at 39°C and $\phi 80$ phages were grown on *E. coli* RVSM (*lac*Δ) (supplied by Dr. J. Langridge) at 30°C. At the onset of lysis, chloroform was added to complete the lysis of phage infected cells (Reader and Siminovitch, 1971). The suspension was centrifuged at 12,000 × g for 10 min, the supernatant millipore filtered (0.45μ) and centrifuged in a Spinco 30 rotor, 27,000 r.p.m., 4 h. Phage pellets were resuspended overnight at 4°C in 3 ml of phage adsorption buffer (tris, 0.1M pH 7.4 containing $MgSO_4$, 0.02M) or MS1 (Gresshoff and Doy, 1972a, 1972b), twice millipore filtered (0.45μ) then assayed for phage (PFU) and for contaminants (see text). This procedure gave phage titres of about 10^{12} PFU/ml. Phage stock suspensions were diluted in MS1 (Gresshoff and Doy, 1972a, 1972b) as required for callus inoculation. The further characterisation of the mutation in the T gene of $\lambda pgal^-$ employed the *E. coli* strains S1652 (*gal*Δ *supD*$^+$), its parent S165 (*gal*Δ *his*$^-$ *supD*) (Saedler and Starlinger, 1967) and KB5-1 (*gal*Δ *supD*) (Rolfe and Onodera, 1972). Plaque and transduction assays with these bacteria were at 30°C on both EMB and minimal medium plates supplemented with galactose. On S1652 all $\lambda pgal^-$ phage plaques contained gal^+ transductants. Examples of these transductants were re-isolated and shown to retain the original $\lambda pgal^-$ phage indicating that the T^- mutation had not reverted during phage replication. In contrast, phage $\lambda pgal^-$ gave no gal^+ transductants on KB5 and KB5-1. Strain S1652 is known to suppress amber mutations in the *gal* operon and in phages T_4 and λ (Saedler and Starlinger, 1967). We conclude that the lesion in the T gene of *gal* is an amber (UAG) nonsense mutation.

showed that it is a nonsense (amber) codon. This was done by standard phage-*E. coli* methods (see legend to Table 1). Because of the polarity of the nonsense mutation within the *gal* operon, the product of gene K (ATP: D-galactose-phosphototransferase (kinase) E.C.2.7.1.6.) is also likely to be absent following translation. In the experiments of Merril *et al.* (1971) the $\lambda pgal^-$ strain was ineffective and thus we were able to conclude from the data of these authors that the human cell cultures were operationally suppressor negative.

The $\lambda pgal^-$ strain was regarded as not only a powerful control for transgenosis but also a test for nonsense suppression. At the level of DNA, the $\lambda pgal^+$ and $\lambda pgal^-$ strains differ in only a nucleotide pair and after transcription the m-RNA differs in only a single nucleotide. After translation, these minimal differences must be expressed as a severe operational defect of one, probably two, of the enzymes of galactose metabolism. Further, the nonsense mutation led to the prediction that if the plant cells were able to correct nonsense codons efficiently then $\lambda pgal^-$ would become equivalent to $\lambda pgal^+$. The *gal* experiments included many controls, classified as potentially positive, neutral or negative (Table 1). The results were strongly in favour of transgenosis of *gal* genes (Table 1). Phages not carrying specific *E. coli* genes were neutral and did not alter the result obtained in the absence of phage.

Fig. 4 shows an example of haploid tomato cells transgenosised for the *E. coli gal* operon. Such calluses (6 replicates) survived for at least 15 weeks longer than all negative controls but were of relatively poor appearance and grew at only 5-20% of the rate of callus on the optimal glucose medium. Increase of mass and of cell number was between 3 and 5-fold. Calluses sub-cultured to galactose, glucose or sucrose media continued to grow at the slow rate for a time but never recovered to the rates characteristic of normal callus. We think that the transgenosised calluses were only marginally assisted to live and that the slow growth and possibly the accumulation of toxins eventually led to a decline and death.

The second transgenosis for survival and growth of plant cells was that of the Z gene of the *lac*$^+$ operon from *E. coli.* A ϕ80p*lac*$^+$ phage strain was provided by Dr. Yasumi Ohshima. Our bacterial-phage tests (see legend to Table 2) confirmed that the ϕ80p*lac*$^+$ strain carried the *E. coli* gene (Z) for β-galactosidase. Genes of the *lac* operon, that is, the *E. coli* gene cluster, P, O, Z, Y, plus the regulator gene I are carried by this phage strain (Iida *et al.*, 1970). Our tests were restricted to proof of the presence of the gene for β-galactosidase. For the *lac* (Z gene) transgenosis experiments the control phages were ϕ80, λp*gal*$^+$ and λp*gal*$^-$. An additional variation for transgenosis was a mixture of ϕ80p*lac*$^+$ plus λp*gal*$^+$ (each 10^8 per callus inoculum). This mixture was used in the hope that the galactose as well as the glucose product of β-galactosidase would become available for growth. At least ten replicates were used per variation.

The results followed the same pattern as for the *gal* experiments. That is (with an exception to be discussed later) the only calluses to survive and grow were those inoculated with either ϕ80p*lac*$^+$ alone or in mixture with λp*gal*$^+$. It took about 5 weeks for the overall situation to clarify. Different experiments were done on 2%, 4% and 10% lactose medium and the growth rate of transgenosised cells is probably proportional to the lactose concentration. This suggests that active transport and accessibility systems were not produced. These operationally more complex functions would be expected to be more difficult to establish than enzymic functions. About 80% of calluses inoculated with ϕ80p*lac*$^+$ survived and grew. A comparison at 13 weeks between calluses non-inoculated with phage, or inoculated with ϕ80, or ϕ80p*lac*$^+$, or ϕ80p*lac*$^+$ plus λp*gal*$^+$, is given in Figs. 5 and 6. Transgenosised calluses have continued to grow although some deteriorate on sub-culture.

Transgenosis of *lac*$^+$, specifically of gene Z coding for β-galactosidase, was confirmed by measuring β-galactosidase activity and by the use of antiserum specific for *E. coli* β-galactosidase. This latter is known to protect the specific enzyme against heat denaturation (Melchers and Messer, 1970). A callus transgenosised for *E. coli lac*$^+$ (using ϕ80p*lac*$^+$ plus λp*gal*$^+$) and chosen at random, contained a high level of β-galactosidase when compared to that of other calluses sampled from either glucose or lactose medium (Table 2). The extent of protection of β-galactosidase in a crude extract of *E. coli* was identical to that of enzymes from the transgenosised plant cells but was less than that reported by the authors of the test (Melchers and Messer, 1970). On first examination it looked as if the β-galactosidase from transgenosised plant cells was more protected in the pre-

TABLE 2

β-Galactosidase from crude extracts of haploid tomato callus ANU-H27-1, specific activity and protection by specific and non-specific antisera

Source and nature of callus	Specific activity (units)	Activity (%) remaining after 60°C for 20 min in presence of rabbit antiserum	
		Specific for *E. coli* enzyme	non-specific
Normal callus, fast growing on glucose (2%)	158	<5	<5
Non-phage infected, slow growing control (rare) on lactose (10%)	144	<5	<5
Non-phage infected, non-growing control on lactose (10%)	264	<5	<5
Transgenosised for *lac*⁺ using *φ80lac*⁺ plus *gal*⁺, growing on lactose (10%)	6,850[a]	50-60	<0.01
Normal callus, fast growing on glucose (2%) plus *E. coli* extract[b] containing β-galactosidase	[b] —	50-60	<5

Before transgenosis, tests for the presence of the β-galactosidase gene carried by the phage φ80p*lac*⁺ were done by examining the plating characteristics of the phage on both RVSM and X64 mutants using lactose tetrazolium agar plates. Strain RVSM has a deletion of the lactose operon and mutant X64 contains an amber mutation in the gene coding for the β-galactosidase enzyme. Plaques of phage φ80p*lac*⁺ on these strains contain white *lac*⁺ transductants while the bacterial lawn remains red due to the inability to use lactose. Lysogens of RVSM (φ80p*lac*⁺) and X64 (φ80p*lac*⁺) were prepared by streaking of the colony growth from the centre of individual plaques and testing the isolated lysogens for their ability to use lactose as a carbon source. Strain X64 was also used in transduction experiments where phage infected cells were plated onto minimal medium plates supplemented with lactose and selection made for *lac*⁺ transductants. These tests confirmed the characteristics of the φ80p*lac*⁺ strain supplied by Dr. Ohshima (Iida *et al.*, 1970) especially the presence of gene Z⁺ (see Fig. 12). High titre phage preparations were prepared by the general procedures indicated in the text and legend to Table 1.

After 13 weeks, calluses sampled at random from the *lac* transgenosis experiment were examined for β-galactosidase activity and for interaction with antiserum specific for *E. coli* β-galactosidase. Specific activity and molecules of β-galactosidase were calculated using the procedure and constants suggested by Rotman (1970). The method of testing the ability of antisera to protect β-galactosidase activity at 60°C was that of Melchers and Messer (1970) with the exception that extracts were made and heat treatment carried out in sodium phosphate buffer 0.1M, pH 7.0 containing $MgSO_4$ 1mM, $MnSO_4$ 0.2 mM and mercaptoethanol 0.1M. Specific rabbit antiserum to pure *E. coli* β-galactosidase was obtained from Dr. J. Langridge, the enzyme was prepared by Dr. A. Millerd and the antiserum by Dr. W. Dudman. In our procedure, the amount of extract was selected to give a rate not limited by substrate and antiserum added in equivalent volume. The experiments were done both at relatively high and at relatively low levels of activity. Non-specific rabbit antiserum was provided by Dr. A. Cunningham. Effects of plant cell extracts on preparations of known *E. coli* enzymes were taken into account.

---

[a]Corresponds to 1.78×10^{13} molecules of β-galactosidase per mg protein or 2×10^{7} molecules per cell. [b]The amount of *E. coli* extract was adjusted to give rates comparable to the activity from transgenosised cells. Specific activity of the *E. coli* preparation, obtained from Dr. G. Miklos, was not determined.

sence of specific *E. coli* antiserum and lost more activity in the absence of specific antiserum than did the known *E. coli* enzyme. That is, enzyme from transgenosised cells behaved in the manner previously attributed to the *E. coli* enzyme (Melchers and Messer, 1970) but the known *E. coli* preparation did not. When an extract of a normal plant callus from glucose medium was mixed with an extract of *E. coli* the effect was to lower the total *E. coli* β-galactosidase activity by 35-40% and to enhance the loss of activity at 60°C. This apparent lack of conformity by the *E. coli* enzyme is presumably a reflection of our different buffer, extract and antisera preparations. There was no variation between the behaviour of known *E. coli* enzyme and that from transgenosised plant cells under similar experimental conditions.

One of the calluses with low and non-protected β-galactosidase activity (Table 2) is a non-phage infected control that grew slowly on lactose (10%) medium; about 5% of such controls do this. A possible explanation is that an available form of carbon was stored in the portion of callus used as the inoculum. Whatever the reason for the growth of these abnormal controls, they do not contain *E. coli* specific β-galactosidase and the β-galactosidase content was low.

A similar experiment has been done for a transgenosised callus from lactose (4%) medium maintained, after sub-culture, for 4 weeks in continuous light, a condition similar to that used for differentiation but on non-differentiation medium (Gresshoff and Doy, 1972a, 1972b). This callus had about 5-times the β-galactosidase activity of a normal callus from glucose (2%) medium (see later comment) but only the former was protected 50% by the specific antiserum.

Protection by specific antiserum is regarded as a definitive test for the presence of the β-galactosidase protein specified by the *E. coli* gene Z and made after transcription and translation in the plant cells. β-Galactosidase protected by specific antiserum is found only in calluses inoculated with ϕ80p*lac*$^+$.

It should be recalled that the phage preparations used for inoculation contained no detectable β-galactosidase. Only a low level of β-galactosidase was present in callus soon after inoculation with ϕ80p*lac*$^+$.

TRANSGENOSIS FOR DEATH

An independent and reciprocal test of transgenosis was the conversion of the optimal glucose (2%) medium into a lethal environment as the result of infection of callus with a ϕ80 phage carrying the *E. coli* nonsense suppressor mutant gene *SupF*$^+$ (known also as *SuIII*$^+$). These experiments derived from the observation that the nonsense mutation in the T gene of *gal* was not suppressed by the plant cells (Table 1). Watson (1970) has considered that since suppressor mutations lead to more mistakes in reading the genetic code they were unlikely to accumulate in higher organisms. We concluded that the failure of transgenosis with λp*gal*$^-$ is consistent with this idea and postulated that nonsense codons might be vital for the life of the plant cells. This was tested by transgenosis of *SupF*$^+$. Several independent preparations of ϕ80*supF*$^+$ and experiments with different batches of both haploid tomato and haploid *Arabidopsis thaliana* calluses gave results showing that phenotypic death of calluses occurred when the ratio of

FIGURE 13

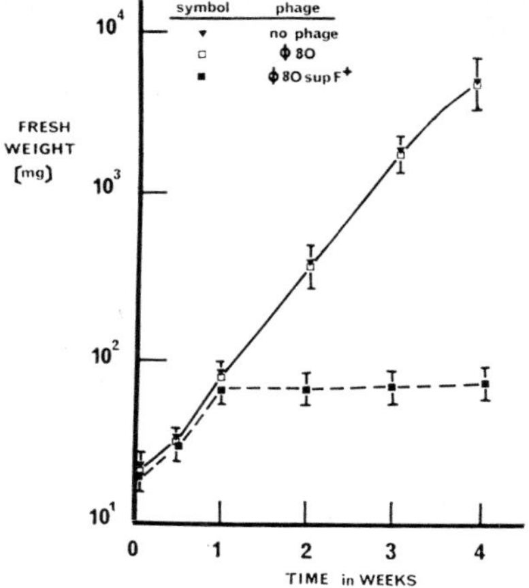

Inhibition (death) of haploid *L. esculentum* callus by phage ϕ80p*supF*+. Note that ϕ80 has no effect on callus growth, however, on continuous subculture, effects of such neutral phages begin to appear (unpublished results). Phages ϕ80p*supF*+ and ϕ80 were grown on *E. coli* Z 32 (try (amber) supplied by Dr. J. Langridge). See also Figs. 7 and 8.

FIGURE 14

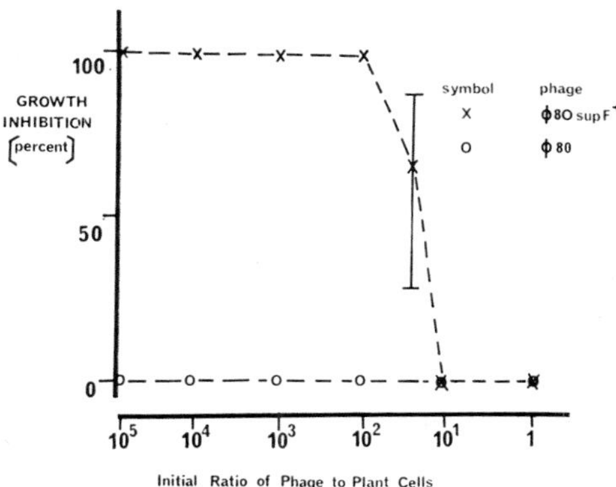

Titration of the ratio between ϕ80p*supF*+ phage and haploid *L. esculentum* cells required for inhibition (death). About 100 phage are required per callus cell to ensure callus "death". This assumes all callus cells are exposed to phage which may not be the case. See also Fig. 8.

ϕ80*supF*+ to that of callus cells was about 100 or greater (Figs. 7, 8, 9, 13, 14). As for the previous experiments, infection with a neutral phage (ϕ80) did not interfere with the initial growth of callus (Figs. 7 and 9).

RESISTANCE TO TRANSGENOSIS

Not all forms of callus are susceptible to killing by ϕ80*supF*+. The calluses so far discussed were grown on medium favouring non-differentiation (Gresshoff and Doy, 1972a, 1972b). Haploid *Arabidopsis thaliana* callus greening on differentiation medium (Gresshoff and Doy, 1972a) was unaffected by ϕ80*supF*$^+$ (Fig. 10). In one of 40-50 replicas of haploid tomato callus on non-differentiation medium, the bulk of the callus died following inoculation with ϕ80*supF*$^+$ but a rapidly growing callus grew from one small region (Fig. 11). On reinoculation with ϕ80*supF*$^+$, sub-cultures of this callus have remained resistant.

These cultures are phenotypically abnormal in that they commence greening on the non-differentiation medium. It is not known whether the resistant clone was developed as a consequence of the original infection with ϕ80*supF*$^+$ or whether this phage acted as a selective environment for a pre-existing clone.

The simplest and most probable general explanation for resistance to ϕ80*supF*$^+$ is that the physical nature of the callus cells prevents infection. Alternative explanations would involve a change in the cellular metabolism so that transgenosis is not established or else that nonsense codons are no longer vital to genetic expression. The involvement of nonsense codons is such a fundamental process that the latter seems an unlikely alternative.

DISCUSSION

The transgenosis of bacterial genes to plant cells has been achieved for all three gene systems tested. We are impressed by the way one experiment has logically led to further experiments and to the predicted results. The reliability of the biological test of life versus death has been reinforced by the biochemical and immunological test for the presence in transgenosised callus of a β-galactosidase characteristic of *E. coli* rather than of normal plant cells. The details of transgenosis remain to be determined, but for *gal*$^+$ and *lac*$^+$ there must have occurred infection by whole phage, or phage DNA, followed by gene maintenance, transcription, translation and effective function within and using the normal molecular and structural factors of the plant cells. Whether or not the transferred bacterial components of mechanisms of gene expression (that is, the promoter and operator loci, and for *lac* and the regulator gene I specifying the normal repressor protein) participate in expression within the plant cells is unknown. The relatively low level of *E. coli* β-galactosidase in the transgenosised callus held in the light, with therefore a possibility of glucose synthesis from CO_2, may hint at a glucose effect. In the example of *supF*$^+$, infection must have been followed by transcription, charging of t-RNA and then read out so that nonsense codons vital to life were converted into sense that was no sense at all for the plant cells.

It should be noted that at this stage we are not claiming gene inheritance. The use of specialized transducing phage, however, was partly on the basis that it may facilitate gene maintenance and inheritance within the cytoplasm, including organelles, as well as "normal" nuclear inheritance. The continued growth of some calluses transgenosised for *lac* does encourage the view that some form of inheritance has occurred.

Integration into the chromosome and inheritance through meiosis is not a remote possibility, Ledoux, Huart and Jacobs (1971) have reported that bacterial DNA may be taken up by seeds of *Arabidopsis thaliana* (L) Heyn and can be detected as part of the DNA of F_1 progeny. If these workers had used a known genetic marker, it is possible that they would have detected transgenosis in whole plants[a]. It would, of course, be very convenient to achieve transgenosis directly into whole plants. For the present experiments we have preferred to use cell cultures rather than whole plants. This preference is associated with greater biological simplicity and also the specific systems of gene transgenosis in which sugars are required to be sole sources of carbon.

On the basis of present knowledge, it now seems but a matter of time before bacterial genes are introduced, expressed and maintained in whole plants. This does not necessarily depend on phage-mediated transgenosis nor on the use of haploid cell lines.

We recognise that there are many potential methods for the infection of plant cells with bacterial genes.

Specialised transducing phages were used in these initial experiments because we considered that they provide a number of advantages favourable for transgenosis.

The integration of *E. coli* genes into the phage genome depends on the ability of the phage DNA to circularise and become attached to the bacterial chromosome at an attachment site (Rolfe, 1970). This attachment site can be moved to the vicinity of selected *E. coli* genes (wild or mutant) (Shimada *et al.,* 1972; Bernstein, Rolfe and Onodera, unpublished). The two genomes then combine and for a time exist as a single bacterial chromosome. When the phage genome subsequently again becomes autonomous the process of excision can result in a phage genome that now carries a few of the bacterial genes which were adjacent to the original attachment site. The altered phage genome is now built into complete phage and these can be obtained in high titre. Thus, in effect, one obtains a high titre of selected bacterial genes rather than a random assortment. These bacterial genes can be characterised by established *E. coli*-phage tests (for examples see legends to Tables 1 and 2).

An attempt at transgenosis can then be made either with the complete phage or a preparation of phage DNA. Our use of complete phage was mainly because it was the simplest procedure. It may be that the complete phage acts as a protection to the phage DNA during the initial stages of transgenosis. We have noted, however, that the total titre of complete phage (plaque forming units, P.F.U.) falls rapidly after the callus is inoculated and it may be that we are infecting with phage DNA released outside the plant cells. In collaboration with Dr. Clark-Walker, we are presently attempting the suppression experiments using phage DNA preparations.

The ability of the phage genome to circularize provides the potential for plasmid formation within the plant cells (including the organelles) with therefore an opportunity for cytoplasmic inheritance as well as nuclear inheritance.

---

[a]We have now seen a report where these authors claim that similar treatment with bacterial DNA of *A. thaliana* strains requiring thiamine or tryptophan for growth can result in the inheritance in progeny of the ability to grow without these nutrients (Ledoux, L., Huart, R and Jacobs, M. (1971) in *Informative Molecules in Biological Systems* (Ledoux, ed.) p. 159, North Holland Publishing Company, Amsterdam-London).

Presumably phage genes may be transgenosised as well as bacterial genes (see Sander (1964, 1967) for a claim that phage replication, modification and assembly can occur in a whole plant). In the long term, this is likely to be a disadvantage, although we think it may be possible to control the transgenosis of the specific phage genes.

The major advantage of transgenosis into haploid cells lies in the possibility of making and then breeding from homozygous individuals. In our particular experiments, there is no obvious advantage in haploidy. We have recently duplicated the *supF*$^{+}$ experiment with a diploid culture of tomato callus. It is possible, however, that our methods of culture produce cells physically well suited for infection.

We are frequently asked whether transgenosis (in this context including inheritance) has any practical application. This question is sometimes associated with expressions of concern about genetic engineering. The latter is frequently discussed in terms of animal, including human, cells and in this association is a very emotive question. As we see it in relation to plants there is less emotion and more immediate potential. From time to time evolution may have depended on transgenosis to achieve an advantageous result quickly. Information gained gradually in one cell might be acquired abruptly by an unrelated cell.

Since survival of the fittest is the supreme test of practicability, transgenosis may already have had practical consequences. A limit on natural transgenosis across evolution is that the necessary space-event coincidences would be very infrequent.

Transgenosis can be anticipated to lead to further important fundamental and, with a careful selection of bacterial genes, practical gains. Two practical objectives of importance that come to mind are (1) the transgenosis of bacterial nitrogen fixation and (2) the modification of the amino acid content of plant cells, either free or combined as protein, to improve the food value of plant material. In theoretical consideration of (1), we had designed a procedure for the transgenosis of the necessary *nif* genes from Klebsiella to *E. coli* as a first stage in transfer to plants. Unfortunately, these strains have not been made freely available from other workers. A group having suitable strains has now reported *nif* gene transgenosis from Klebsiella to *E. coli* (Dixon and Postgate, 1972). Clearly the transgenosis of nitrogen fixation to plants will be achieved in a laboratory having the right combination of bacterial and plant cell cultures. Possibly the greatest barrier to the establishment of nitrogen fixation in plant cells is the anaerobic requirement of the enzymic reaction for the conversion of molecular nitrogen to ammonia.

We regard our experiments as a beginning designed to test whether plant cells can make use of the information contained in bacterial genes. It appears that they can, and transgenosis now needs to be developed in detail at all levels.

One is particularly interested to know the fate and heritability of the bacterial genes. The work has already raised certain social questions concerned with the ethics of genetic engineering and also the free exchange of ideas and cultures. In this laboratory we shall try to cooperate with groups who have similar interests in achieving a common scientific goal with considered ethics.

ACKNOWLEDGEMENTS

Those who provided us with bacterial and phage strains or materials are thanked in the text. Mr. Neal Gowen and Mr. Peter Fokker are thanked for technical assistance and Mr. Barry Parr for help with photography.

REFERENCES

Bertani, G. (1951) *J. Bacteriol. 62,* 293

Carlson, P. (1970) *Science 168,* 487

Dixon, R. A. and Postgate, J. R. (1972) *Nature (London) 237,* 102

Gresshoff, P. M. and Doy, C. H. (1972a) *Australian J. Biol. Sci. 25,* 259

Gresshoff, P. M. and Doy, C. H. (1972b) *Planta. 107,* 161

Iida, Y., Kameyama, T., Ohshima, Y. and Horiuchi, T. (1970) *Mol. Gen. Genetics. 106,* 296

Ledoux, L., Huart, R. and Jacobs, M. (1971) *Eur. J. Biochem. 23,* 96

Melchers, F. and Messer, W. (1970) *Biochem. Biophys. Res. Commun. 40,* 570

Merril, C. R., Geier, M. R. and Petricciani, J. C. (1971) *Nature (London) 233,* 398

Onodera, K., Rolfe, B. and Bernstein, A. (1970) *Biochem. Biophys. Res. Commun. 39,* 969

Reader, R. W. and Siminovitch, L. (1971) *Virology 43,* 623

Rolfe, B. (1970) *Virology 42,* 643

Rolfe, B. and Onodera, K. (1972) *J. Membrane Biol 9,* 195

Rotman, M. B. (1970) in *The Lactose Operon,* (Beckwith and Zipser, eds.) p. 279. Cold Spring Harbor Laboratory

Saedler, H. and Starlinger, P. (1967) *Mol. Gen. Genetics 100,* 178

Sander, E. M. (1964) *Virology 24,* 545

Sander, E. M. (1967) *Virology 33,* 121

Shimada, K., Weisberg, R. A. and Gottesman, M. E. (1972) *J. Mol. Biol. 63,* 483

Watson, J. D. (1970) in: *Molecular Biology of the Gene,* 2nd Edition W. A. Benjamin, New York

Sequences in Genetic Nucleic Acids

F. SANGER
Medical Research Council Laboratory of Molecular Biology, Cambridge, England

The basic chemical material of the gene is DNA and all the information for the developing organism is contained in the form of a specific nucleotide sequence in this DNA. In order to understand the gene more fully and the way in which the information is encoded it is necessary to study these nucleotide sequences. This is however a science that is very much in its infancy and as yet we know almost nothing about sequences in the DNA of higher organisms. However methods are being developed and have been applied to the study of smaller nucleic acids—in particular to the RNA of some bacteriophages. These RNA bacteriophages (R17, MS2, Qβ etc.), which are probably the simplest organisms that exist, contain no DNA, and all the genetic information they contain is encoded in an RNA molecule of 3000-4000 residues long, which acts as a template for its own replication and for the production of three proteins (see Fig. 1). Thus in this

FIGURE 1

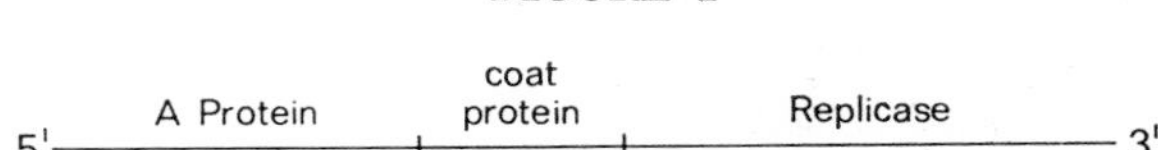

Diagram showing the position of the three protein cistrons on the R17 RNA.

lecture I shall not be dealing with the nucleic acids of higher organisms but will discuss various methods available and their application to the lowest organisms in the hope that this work is a preliminary to the study of detailed sequences in mammalian genes. I shall concentrate largely on work from our own laboratory.

FRACTIONATION TECHNIQUES

In the determination of sequences both in proteins and in nucleic acids the main technical problem is usually the fractionation of the complex mixtures of closely related oligomers that are produced on partial digestion, and our initial studies were directed to the development of efficient techniques for fractionating oligonucleotides. As we were interested in isolating sequences from relatively large RNAs, rapid and simple methods were necessary and two-dimensional techniques using paper or thin-layers were to be preferred. Because only small amounts can usually be applied to such systems, it was necessary to have a very sensitive method for the identification and estimation of the nucleotides. To achieve this we have used RNA that has been labelled *in vivo* with ^{32}P. This can readily be done with bacteriophages to obtain RNA of high specific activity. Every mononucleotide contains a P atom so that the position of oligonucleotides of a fractionation system can be determined by radioautography and they can be estimated by counting techniques.

The most useful method for degrading RNA is with ribonuclease T_1 which splits specifically at Gp residues. Two methods for fractionating ribonuclease T_1 digests are illustrated in Figs. 2 and 3. In both cases the first dimension is by high voltage electrophoresis on cellulose acetate at pH 3·5. This is a very efficient

FIGURE 2

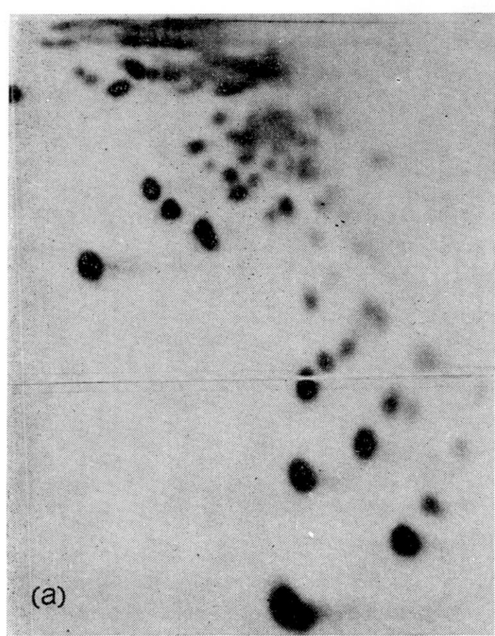

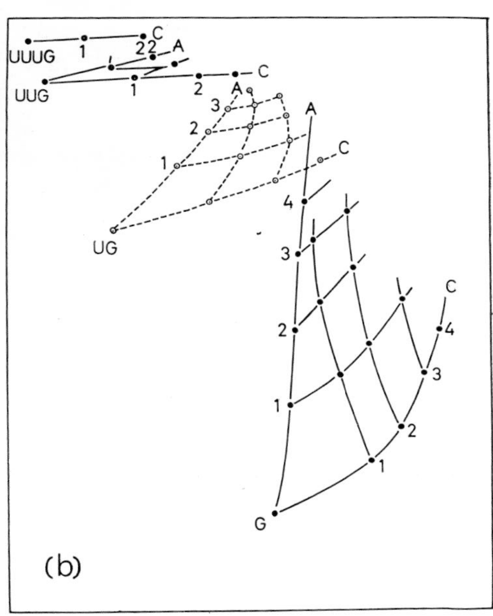

(a) Radioautograph of two-dimensional fractionation of a ribonuclease T_1 digest of ^{32}P-labelled ribosomal RNA (Sanger, *et al.*, 1965). First dimension high-voltage ionophoresis at pH 3.5 on cellulose acetate. Second dimension ionophoresis on DEAE paper in 7% formic acid.

(b) Diagram showing the relationship between the composition of a nucleotide and its position on the fingerprint illustrated in (a). The composition of a spot is determined from its position on the diagram as follows: Each oligonucleotide has one G residue as it was obtained by ribonuclease T_1 digestion. The spots marked G, UG, U_2G, U_3G at the origins of the graticules show the positions of these oligonucleotides. All spots within a graticule contain the same number of U residues (e.g. all spots within the graticule whose origin is occupied by UG contain one U residue). The numbers of A and C residues in a spot are then read off within the graticules by reference to the A and C axis respectively.

method of fractionation and oligonucleotides of almost any size move on this system as clearly defined spots. In the "fingerprint" shown in Fig. 2 separation in the second dimension is by ionophoresis on DEAE paper in 7% formic acid. This method is suitable for the separation of the small T_1 products which do not contain more than 3 Up residues. The position of the oligonucleotides on the fingerprint depends on their composition and Fig. 2b is a diagram showing this relationship. It can be used to determine the composition of a spot from its position. This method has formed the basis of most of our work on RNA.

The method illustrated in Fig. 3 employs a type of displacement chromatography on thin-layers of DEAE cellulose as the second dimension (Brownlee and

FIGURE 3

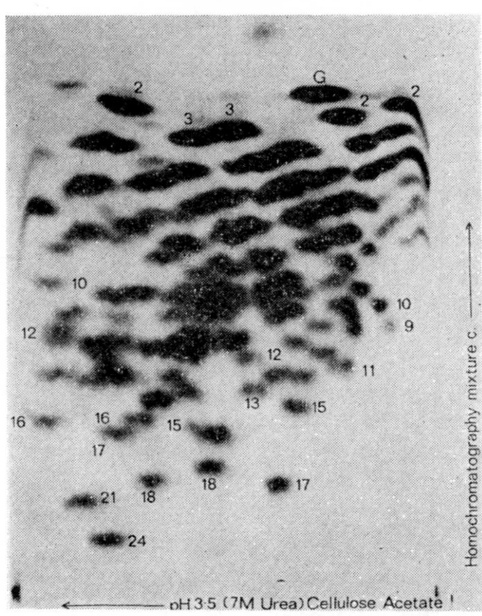

Radioautograph of a two-dimensional fractionation of a ribonuclease T_1 digest of ^{32}P-labelled R17 RNA. First dimension as in Fig. 2(a): second dimension thin-layer "homochromatography" on DEAE cellulose (Jeppesen, 1971). The numbers indicate the size (mono-nucleotide units) of the oligonucleotides.

Sanger, 1969). In this method, referred to as "homochromatography", a relatively concentrated and complex mixture of non-radioactive oligonucleotides is used to develop the chromatogram. These oligonucleotides form a series of fronts on which the corresponding radioactive nucleotides move up the chromatogram and separate, largely according to their size. Although the fractionation of smaller nucleotides is less efficient by this method it is useful for separating the larger ones.

R17 RNA

Fig. 3 shows a fingerprint of a ribonuclease T_1 digest of bacteriophage R17 (Adams, *et al.*, 1969; Jeppesen, 1971). Since there are about 800 Gp residues that are broken by the enzyme, one can expect a very complex mixture, and most of the spots at the top of the chromatogram, which contain the smaller oligonucleotides, are mixtures. However at the bottom of the chromatogram there is a number of clearly defined spots which represent the larger oligonucleotides in a pure form. These spots form a characteristic pattern that can be used as a specific characterization of the different phages (Nichols and Robertson, 1971; Robertson and Jeppesen, 1972).

Various micromethods have been devised for the determination of the sequences of the radioactive oligonucleotides, mostly depending on further degradation with other nucleases (Sanger, *et al.*, 1965; Barrell, 1971; Jeppesen, 1971). Using these, Jeppesen (1971) has determined the sequences in the large T_1 products isolated as in Fig. 3. One of these (spot 21) was of particular interest as its sequence corresponded, according to the genetic code, to a heptapeptide sequence in the coat protein (Fig. 5), showing that it was derived from the coat protein cistron (Adams *et al.*, 1969). This was the first time that an informational nucleotide sequence had been directly sequenced in this way and was a prelude to more extensive studies on the R17 RNA.

The two fingerprinting methods outlined above were suitable for the fractionation of oligonucleotides up to about 30 residues long but in order to study an RNA the size of R17 (3300 residues) it was necessary to isolate larger fragments. This was accomplished by carrying out very limited partial digestion with ribonuclease T_1 so that only about 5% of the Gp bonds were split, and fractionating the products by ionophoresis on acrylamide gel (Adams *et al.*, 1969). Such a fractionation is shown in Fig. 4. The presence of a limited number of well defined

FIGURE 4

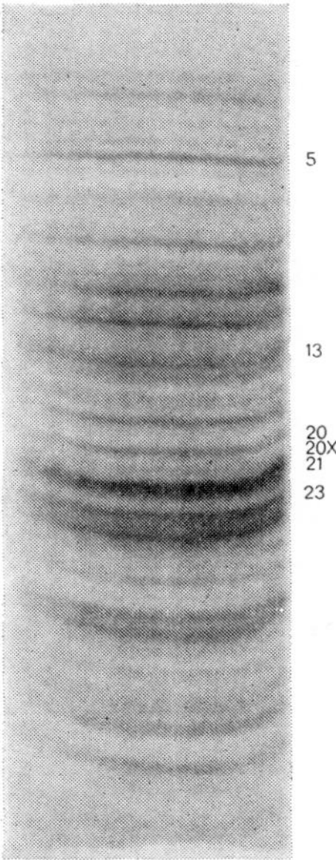

Radioautograph of a polyacrylamide gel electrophoresis of a partial ribonuclease T_1 digest of ^{32}P-labelled bacteriophage R17 RNA. (Adams *et al.*, 1969). (Figure reproduced with permission from *Biochem. J. 124*, 833 (1971))

bands was somewhat unexpected and showed that the partial digestion with ribonuclease T_1 was a relatively specific process. This specificity probably depends largely on the secondary structure of the RNA. Our studies indicated that much of the RNA chain is in the form of loops which are held together by base pairing. The double-stranded parts formed in this way are relatively resistant to ribonuclease T_1, which splits preferentially at the G bonds in single-stranded sections.

The acrylamide gel fractionation made it possible to isolate larger fragments (50-100 residues long) whose sequences could be determined. The results of these studies have been described elsewhere (Adams *et al.*, 1969; Steitz, 1969; Jeppesen, *et al.*, 1970; Nichols, 1970; Sanger, 1971; Cory, *et al.*, 1972; Jeppesen, *et al.*, 1972) and will not be discussed in detail here. They are summarized in Fig. 5, together with similar studies on the related bacteriophage MS2 by Fiers

FIGURE 5

```
(a)     p p p G–G–G–U–G–G–G–A–C–C–C–C–U–U–U–C–G–G–G–G–U–C–C–U–G–C–U–C–A–A–
(b)           ____________________________________________________________
(c)
(d)

(a)     C–U–U–C–C–U–G–U–C–G–A–G–C–U–A–A–U–G–C–C–A–U–U–U–U–U–A–A–U–G–U–C–U–
(b)     __________________________________________________________________
(c)
(d)

(a)     U–U–A–G–C–G–A–G–A–C–G–C–U–A–C–C–A–U–G–C–U–A–U–C–G–C–U–G–U–A–G–G–U–
(b)     __________________________________________________________________
(c)
(d)

(a)     A–G–C–C–G–G–A–A–U–U–C–C–U–A–G–G–A–G–G–U–U–U–G–A–C–C–U–A–U–G–C–G–A–
(b)     ______________________________________________
(c)(1)                                                        f– Met  – Arg  –
(d)                                                                     1

(a)     G–C–U–U–U–U–A–G–U–G– – – – – – (approximately 1000 residues) – – – – – –
(b)
(c)(1)  Ala  – Phe  – Ser  – – – –
(d)

(a)     – – –    A–G–G–C–A–A–C–G–G–C–U–C–U–C–U–A–A–A–U–A–G–A–G–C–C–C–U–C–A–
(b)
(c)
(d)

(a)     A–C–C–G–G–G–G–U–U–U–G–A–A–G–C–A–U–G–G–C–U–U–C–U–A–A–C–U–U–U–A–C–U–
(b)                                     ______________________________
(c)(2)                                  f–Met  – Ala  – Ser  – Asn  – Phe  – Thr  –
(d)                                              1                          5

(a)     C–A–G ____ ____ ____ ____ ____ ____ ____ ____ ____ ____
(b)     _____
(c)(2)  Gln  – Phe  – Val  – Leu  – Val  – Asn  – Asp  – Gly  – Gly  – Thr  – Gly  –
(d)                                 10                                 15

(a)     ____ ____ ____ ____ ____ ____ ____ ____ ____ ____ ____
(b)
(c)(2)  Asn  – Val  – Thr  – Val  – Ala  – Pro  – Ser  – Asn  – Phe  – Ala  – Asn  –
(d)                          20                                 25

(a)     ____ ____ ____ ____      G–A–U–C–A–G–C–U–C–U–A–A–C–U–C–G–C–G–C–
(b)                              _____________________________________U–
(c)(2)  Gly  – Val  – Ala  – Glu  – Trp  – Ile  – Ser  – Ser  – Asn  – Ser  – Arg  –
(d)                   30                                 35
```

FIGURE 5—continued

(a) U-C-A-C-A-G-G-C-U-U-A-C-A-A-A-G-U-A-A-C-C-U-G-U-A-G-C-G-U-U-C-G-U-
(b) ————————————————————————————
(c)(2) Ser - Gln - Ala - Tyr - Lys - Val - Thr - Cys - Ser - Val - Arg -
(d) 40 45

(a) C-A-G-A-G-C-U-C-U-G —— —— —— G-C-A-A-A-U-A-C-A-C-C-A-U-U-
(b) ——————— C-G-C-A-G-A-A-U-C ——————— C
(c)(2) Gln - Ser - Ser - Ala - Glu - Asn - Arg - Lys - Tyr - Thr - Ile -
(d) 50 55 60

(a) A-A-A-G-U-C-G-A-G-G-U-G-C-C-U-A-A-G-G-U-G-G-C-A-A-C-U-C-A-G-A-C-U-
(b) ——————— A ——————— C ———
(c)(2) Lys - Val - Glu - Val - Pro - Lys - Val - Ala - Thr - Gln - Thr -
(d) 65 70

(a) G-U-U-G-G-U-G-G-U-G-U-A-G —— —— —— —— —— G-C-A-U-G-G-
(b) ——————— A-G-C-U-U-C-C-U-G-U-A-G-C-C ———
(c)(2) Val - Gly - Gly - Val - Glu - Leu - Pro - Val - Ala - Ala - Trp -
(d) 75 80

(a) C-G-U-U-C-G-U-A-C-U-U-A-A-A-U-A-U-G-G-A-A-U-U-A-A-C-U-A-U-U-C-C-A-
(b) ——————— C ——— C ———
(c)(2) Arg - Ser - Tyr - Leu - Asn - Met - Glu - Leu - Thr - Ile - Pro -
(d) 85 90

(a) A-U-U-U-U-C-G-C-U-A-C-G-A-A-C-U-C-C-G- — —— —— —— ——
(b) ——————— U ——— A-C-U-G-C-G-A-G-C-U-U-A-U-U-
(c)(2) Ile - Phe - Ala - Thr - Asn - Ser - Asp - Cys - Glu - Leu - Ile -
(d) 95 100

(a) —— —— —— —— —— —— —— —— —— —— ——
(b) G-U-U-A-A-G-G-C-A-A-U-G G-A-U-G-G-A-
(c)(2) Val - Lys - Ala - Met - Gln - Gly - Leu - Leu - Lys - Asp - Gly -
(d) 105 110 115 -

(a) —— —— —— —— —— —— —— —— G-C-A-A-A-C-U-C-C-
(b) A-A-C-C-C-G-A-U-U-C-C-C-U-C-A-G-C-A-A-U-C-G-C-A ———
(c)(2) Asn - Pro - Ile - Pro - Ser - Ala - Ile - Ala - Ala - Asn - Ser -
(d) 120 125

(a) G-G-U-A-U-C-U-A-C-U-A-A-U-A-G-A-U-G-C-C-G-G-C-C-A-U-U-C-A-A-A-C-A-
(b) —— C ——————— C ———————
(c)(2) Gly - Ile - Tyr
(d) 129

(a) U-G-A-G-G-A-U-U-A-C-C-C-A-U-G-U-C-G-A-A-G-A-C-A-A-C-A-A-A-G — — —
(b) ——
(c)(3) f- Met - Ser - Lys - Thr - Thr - Lys — — —
(d) 1 5

(a) — — — — — — — — — — — (approximately 1500 residues) — — — — — — — — — — — — —
(b)
(c)
(d)

(a) — — — — — — — — — — — — — A-C-C-C-G-G-G-A-U-U-C-U-C-C-C-G-A-U-U-U-G-G-
(b) A-A-A-G-A-G-A-G-G ———————————
(c)
(d)

(a) U-A-A-C-U-A-G-C-U-G-C-U-U-G-G-C-U-A-G-U-U-A-C-C-A-C-C-C-A OH
(b) ———————————————————————
(c)
(d)

Nucleotide sequence in bacteriophage RNAs. (a) Sequence of bacteriophage R17 RNA. A broken line indicates that the sequence has not been determined. The sequence of spot 21 (Fig. 3) is indicated by double underlining, that of band 20 from the acrylamide gel fractionation (Fig. 4) by single underlining. (b) Sequence of bacteriophage MS2 RNA. An unbroken line indicates that the sequence is the same as in R17 RNA. (c) Amino acid sequence in the proteins from bacteriophage R17: (1) A protein; (2) Coat protein; (3) Replicase. (d) Position of amino acid residues in the respective protein sequences.

et al. (1971). They provided a firm proof by chemical methods of the genetic code, which had previously been determined by entirely different and less direct techniques. Within the limits of the data, it appeared that the code as used in these phages behaves in a redundant way. Most of the possible codons for each amino acid are found to be used. There are indications that a few may be rarer than others though there are not sufficient results at present to draw definite conclusions.

The work on the bacteriophages also showed the presence of secondary structure in the RNA. Most of the sequences found were such that they could be written in the form of base-paired loops (see Fig. 6), and it seems probable that

FIGURE 6

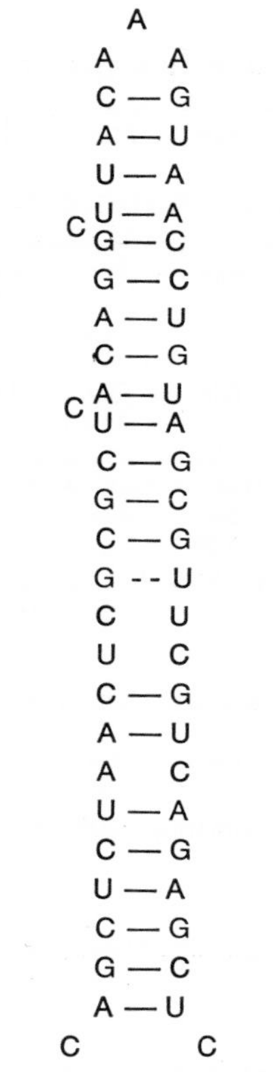

```
      A
    A   A
    C — G
    A — U
    U — A
  C U — A
    G — C
    G — C
    A — U
    C — G
  C A — U
    U — A
    C — G
    G — C
    C — G
    G -- U
    C   U
    U   C
    C — G
    A — U
    A   C
    U — A
    C — G
    U — A
    C — G
    G — C
    A — U
  C       C
AU          UG
```

The fragment from band 20 (single underlining in Fig. 5) written in the form of a loop to show base-pairing.

such loops exist in the RNA and may be involved in the biological properties of the messenger. Thus, for instance, the availability of sites on the RNA to the translation machinery probably depends on their secondary structure, and this in its turn may vary as translation proceeds. Such a dynamic situation may play an important role in control of protein synthesis and gene expression. Signals on the RNA that code for the starting and stopping of protein chains have also been detected. The work of Steitz (1969) on the ribosomal binding sites showed that all three cistrons started with AUG, which codes for the initiating formyl methionine transfer RNA (see Fig. 5). The three binding sites appeared to have no other feature in common, so it is still not clear what it is that is recognised by the protein synthesizing machinery to identify the three AUGs as initiation sites. There are of course many other AUGs in the RNA which never act as initiators. It may be that the three sites are recognised by different factors, or it may be that the secondary structure of the RNA is involved (see Lodish, 1970; 1971).

The sequence corresponding to the C-terminus of the coat protein was identified in one of the bands from the acrylamide gel ionophoresis of the partial ribonuclease T_1 digest (Nichols, 1970) and was shown to be followed by the two terminator codons U-A-A-U-A-G. Presumably the presence of the two may act as a defence mechanism against non-termination. If a single codon did not act quantitatively it could lead to the production of a large protein, which might have disastrous biological effects. From an overlap between the above sequence and that for the replicase binding site, it was possible to deduce the complete sequence between the two cistrons (see Fig. 5). This contains about 30 residues. The exact function of these is unknown but presumably they play a role in the control of protein synthesis and determine how it is that the replicase is made in lower amounts than the coat protein and is only made at the earlier stages of infection. This inhibition at later stages appears to be brought about by binding of the coat protein to the RNA, and presumably the intercistronic sequence is involved. Other untranslated regions are found at the 5′ and 3′ ends of the RNA (Adams and Cory, 1970; Cory, *et al.,* 1972). It is probably significant that the sequence of at least 117 residues at the 5′ end is the same in the three bacteriophages R17, MS2 and F2 (Nichols and Robertson, 1971; Fiers *et al.,* 1971) although in other parts of the chains they differ to the extent of about one change every 25 residues. This might suggest a particularly important function for this 5′ sequence, presumably in connection with gene expression. The nature of the signals involved is not clear at present but it is hoped that the further study of other untranslated regions in messenger RNAs and in DNA may throw some light on the language in which theses control signals is written.

DNA SEQUENCES

The study of DNA sequences has progressed considerably more slowly than that on RNA. One reason for this is that there are no small DNAs available on which methods can be developed. Early studies on RNA sequences were done on the t-RNAs, which contain 75-85 residues, and the 5S RNA, of 120 residues. The smallest pure DNAs, however, are the DNA bacteriophages (fd, F1, ϕX, etc.), which contain about 5000 residues. Another difficulty with DNA is the lack of an enzyme with a specificity for one base. Most of the work with RNA depends on the use of ribonuclease T_1, which is specific for G residues. Of the DNases that have been studied, some—such as the restriction enzymes—are extremely specific, splitting only a few bonds in a bacteriophage DNA, whereas the majority—such as pancreatic DNase—show very little specificity.

PYRIMIDINE SEQUENCES

The best defined and most quantitative method for the degradation of DNA is the depurination technique of Burton and Petersen (1960). In this reaction all the purine residues are split out of the chain, leaving runs of pyrimidines with a phosphate group on both ends. Ling (1972a) has shown that pyrimidine tracts may be fractionated on the two-dimensional system using homochromatography on thin-layers of DEAE cellulose in the second dimension. Fig. 7a shows such a

FIGURE 7

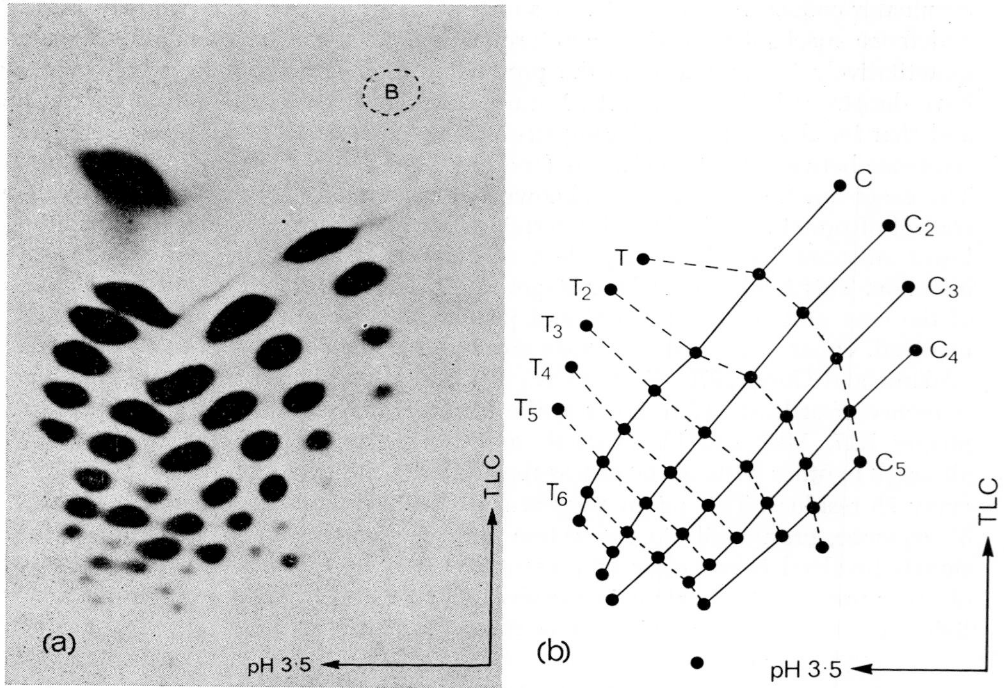

(a) Radioautograph of fractionation of depurination products from bacteriophage fd DNA. Separation method as in Fig. 3. (reproduced with permission from Ling, 1972a)

(b) Diagram showing the relationship between the composition of a nucleotide and its position on the fingerprint.

fingerprint of bacteriophage fd. There is no separation of isomers but all oligonucleotides having different compositions are separated. As is to be expected for a DNA of this size, most of the shorter pyrimidine sequences occur many times in the molecule, and the faster moving spots on the homochromatography dimension are mixtures of isomers. However there is a number of well separated slow moving spots which represent the longer pyrimidine tracts in a pure form. These spots form a characteristic pattern, which varies with different bacteriophages, and may thus be used as a fingerprinting method for the characterization of a DNA and for the detection of small differences between related phages. Fig. 8 shows a

FIGURE 8

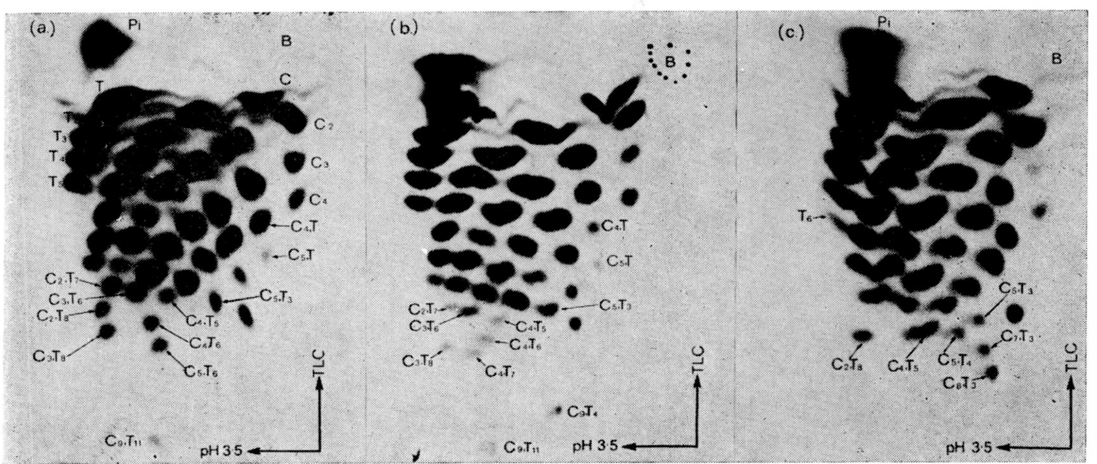

Fingerprints of depurination products of the DNA of three bacteriophages. (a) fd. (b) fl. (c) ϕX. (reproduced with permission from Ling, 1972b).

comparison of the fingerprints for three different phages (Ling, 1972b). Phages fd and fl are very similar in their biological properties and in the amino acid sequence of the single small coat protein. Their fingerprints following depurination are, however different, showing the sensitivity of the method. Bacteriophage ϕX is very different biologically from the other two and does not have a similar small coat protein. Detailed sequence determination of the pyrimidine tracts (Table 1) demonstrates the marked homology between the first two phages (e.g. both contain the same large oligonucleotide, C_9T_{11}), but also suggests a sequence homology between them and ϕX.

The rate of migration in the first dimension depends largely on the relative proportion of C and T residues, whereas migration in the second depends on size. Thus small oligonucleotides are at the top of the fingerprint, those rich in C to the right, and those rich in T to the left. This makes it possible to construct a graticule (Fig. 7b) showing the relationship between the composition of an oligonucleotide and its position on the fingerprint. With the exception of the largest oligonucleotide (C_9T_{11}), the compositions of all the spots could be determined in this way and they were confirmed by direct analysis. In order to determine the sequence of the larger pyrimidine tracts, Ling (1972a) subjected their dephosphorylated products to partial digestion with venom or spleen phosphodiesterase and separated the products on the same two-dimensional system. Venom phosphodiesterase degrades oligonucleotides sequentially from the 3′ end, liber-

TABLE 1

Composition	fd DNA	f1 DNA	ϕX 174 DNA
C_5, T_3	Present in more than 1 molar yield	Present in more than 1 molar yield	C C T T T C C C
C_5, T_4	NP	NP	C T T C C T C C T
C_4, T_5	C T T C C T C T T	Same as fd	Present in 3 molar yield: not sequenced
C_3, T_6	T T C C T T T C T; T T T C C T T C T	2 sequences: same as fd	NP
C_2, T_7	T C T T C T T T T; T T T C T T T C T	1 sequence: T C T T C T T T T	NP
C_7, T_3	NP	NP	T C C T C T C C C C
C_4, T_6	T C C T T C T C T T	Same as fd	NP
C_2, T_8	T T T T T C C T T T	NP	C T T T T T T T C T
C_8, T_3	NP	NP	C T C C C T C T C C C
C_5, T_6	T T T T T C C T C C C	NP	NP
C_4, T_7	NP	T T T T T C T T C C C	NP
C_3, T_8	C C T T T T T T T T C	Same as fd	NP
C_9, T_4	NP	C C T T C C C T C C C T C	NP
C_9, T_{11}	C T T T C T T C C C T T C C T T T C T C	Same as fd	NP

Sequences of large pyrimidine oligonucleotides from fd, fl and ϕX174 DNA. NP means that the oligonucleotide was not present in the phage DNA. Underlined sequences indicate homology between the pyrimidine oligonucleotides within the DNA of a single phage. Double-headed arrows indicate homology between the pyrimidine sequences in the DNA of different phages. (reproduced with permission from Ling, 1972b).

ating 5′ mononucleotides. Under suitable conditions of partial degradation one is left with a mixture of products of different length—starting from the 5′ end. Thus the oligonucleotide C-C-T-C-T should give C-C-T-C, C-C-T and C-C, as well as unchanged material and mononucleotides. Fractionation of such a mixture of products on the two-dimensional system then gives rise to a series of spots from which the sequence can be deduced as follows. The relative position of two oligonucleotides that differ by only one residue is determined by the nature of that one residue: if it is a pT, the smaller one will be above and considerably to the right of the larger one, whereas if it is a pC, it will be above and usually slightly to the left (see Fig. 7b). Fig. 9a shows the fingerprint obtained from

FIGURE 9

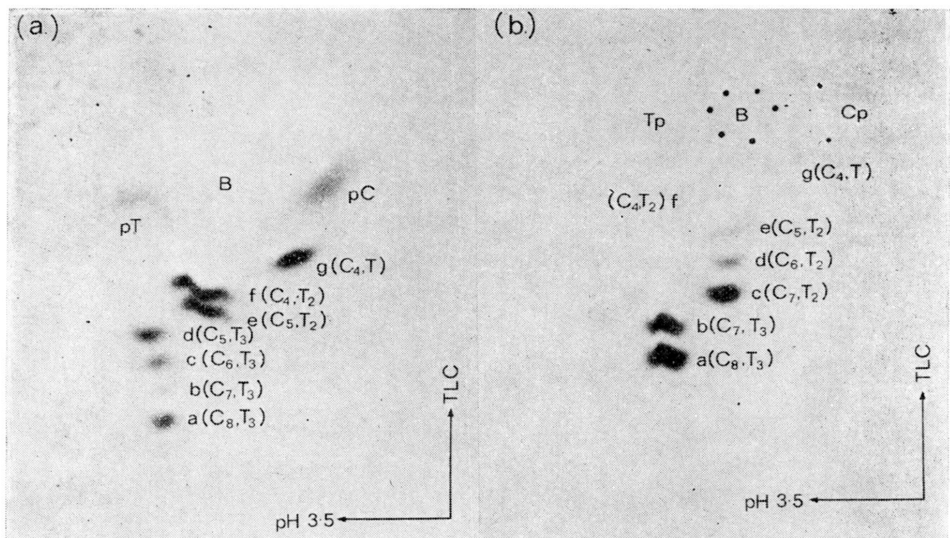

Two-dimensional separation of partial products of the digestion of nucleotide (C_8T_3) from φX DNA: (a) with snake venom phosphodiesterase, and (b) with spleen phosphodiesterase. Fractionation was as in fig. 3 (reproduced with permission from Ling, 1972b).

digestion of the oligonucleotide C_8T_3 from φX174 DNA with venom phosphodiesterase. The slowest moving spot 'a' is the unchanged oligonucleotide which is always mixed with the digest before running. The next one above it 'b' is slightly to the left and therefore represents the loss of a pC residue, which must be at the 3′ end. Similarly the next two splits are pC residues. However spot 'e' is definitely to the right of spot 'd' and represents a pT split off. In this way the sequence from the 3′ end is deduced from the fingerprint as . . . T-C-T-C-C-C-3′. Spleen phosphodiesterase degrades sequentially from the 5′ end and thus can be used in a similar way. Fig. 9b shows the degradation of the same oligonucleotide with this enzyme, giving the sequence 5′-C-T-C-C-C-T . . . . Combining the two results the whole sequence of C_8T_3 can be constructed as C-T-C-C-C-T-C-T-C-C-C. These methods were applied to all the unique large pyrimidine tracts from the three bacteriophages (Table 1), including the longest one (C_9T_{11}) from fd. The determination of its 20-nucleotide sequence is shown in Fig. 10.

FIGURE 10

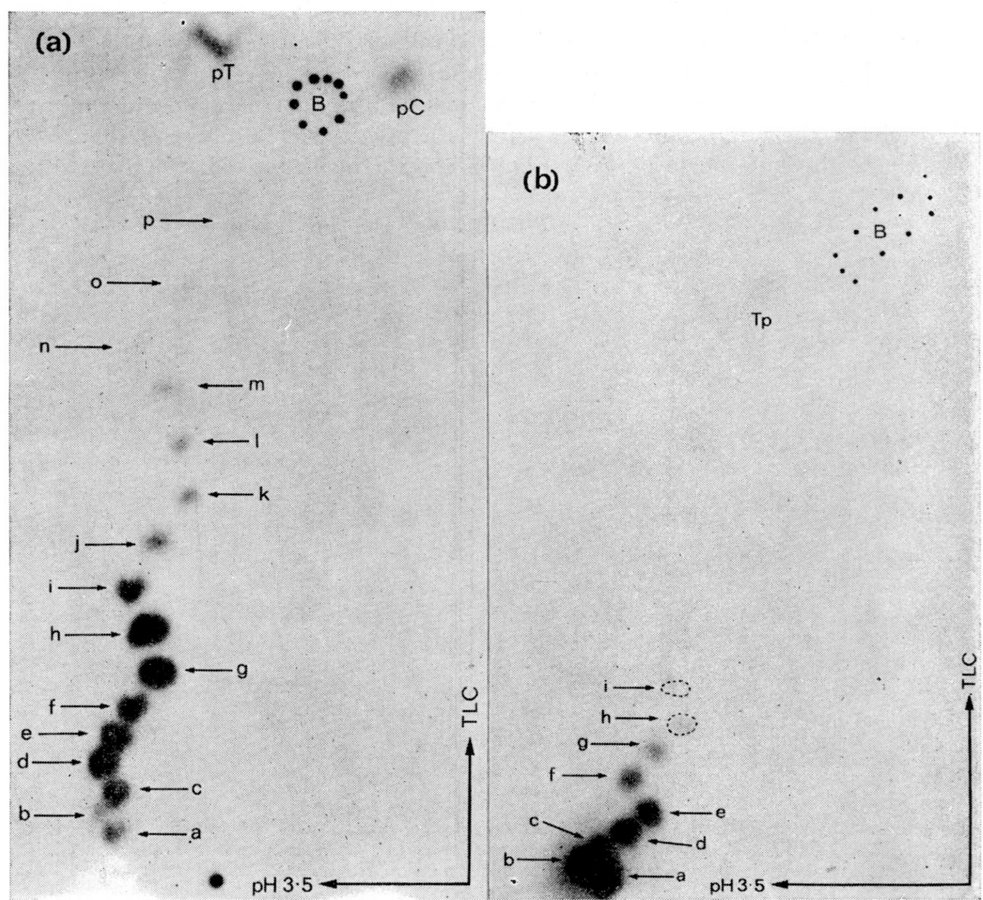

Two-dimensional separation of partial products of the digestion of nucleotide (C_9T_{11}) from fd DNA: (a) with snake venom phosphodiesterase, and (b) with spleen phosphodiesterase (reproduced with permission from Ling, 1972a).

DEOXYRIBONUCLEASES

In the sequence analysis of a large nucleic acid, degradation must take place in various stages. Thus initial fragmentation of the R17 RNA by limited partial digestion with ribonuclease T_1 gave fragments of about 50-100 residues long (see Fig. 4); further partial digestion gave fragments of 10-20 average length, and these were completely digested with ribonuclease T_1 or ribonuclease A to products of average size four and two residues respectively. The sequences of these oligonucleotides, particularly larger ribonuclease T_1 products, were then determined by further degradation by other enzymes and by alkali. Thus there were essentially four steps in the degradation, and it seems that another step at the beginning to give very specific large fragments, would be necessary if the complete sequence of an RNA bacteriophage is to be determined. In the above degradation series ribonuclease T_1 has played a central rôle and has been used effectively both for complete digestion and also for partial digestion to give specific degradation, dependent largely on the secondary structure of the RNA. Although many deoxyribonucleases have been isolated, none of them appears to have a unique specificity for one base or to be as generally useful as ribonuclease T_1. There are however other deoxyribonucleases that are likely to be useful at some stages of degradation. For the initial very specific digestion, the most hopeful enzymes are the restriction enzymes, which show a very limited specificity on double-stranded DNA. The most extensively studied is that from *Haemophilus influenzae* (Kelly and Smith, 1970). It appears to recognize a hexanucleotide sequence and splits the SV40 DNA neatly into 11 fragments (Danna and Nathans, 1971).

Another enzyme that promises to be useful for the initial partial degradation of DNA is the endonuclease IV of Sadowsky and Hurwitz (1969). This enzyme is isolated from T4 infected *E. coli* and may be concerned in the degradation of host DNA. Ling (1971) has studied its action on the small DNA bacteriophages (fd, ϕX). Fig. 11 shows an acrylamide gel fractionation of ^{32}P-labelled fd DNA

FIGURE 11

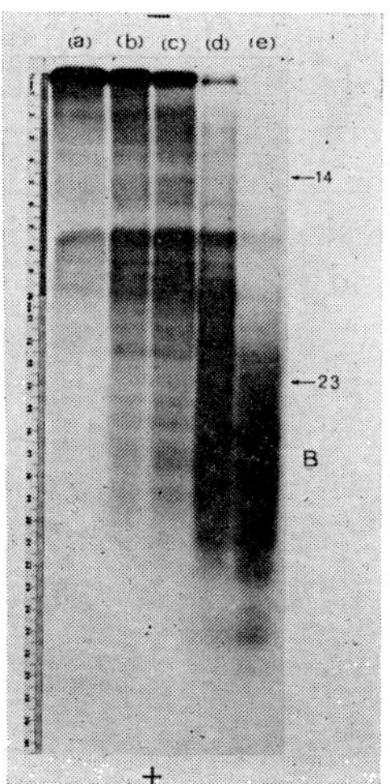

Radioautograph of polyacrylamide gel fractionation of ^{32}P-labelled fd DNA digested with various concentrations of Endonuclease IV (reproduced with permission from Ling, 1971).

digested under various conditions with endonuclease IV. Reasonably well resolved bands are formed showing that the digestion is fairly specific. The sizes of the products vary considerably but some of the largest may be several hundred nucleotides long. Sadowski has shown that the 5′ ends of the products are mostly pC residues, suggesting the enzyme is specific for the XpC bond, but it would appear that the specificity is probably somewhat more exacting as very few small nucleotides are produced.

Although most of the bands obtained from an endonuclease IV digest of the phage DNA are probably mixtures, they are considerably simpler than the original DNA and are suitable substrates for the development of further methods of degradation and sequence analysis. Ling has studied the depurination products of some bands and shown them to give characteristic and relatively simple fingerprints.

Most other deoxyribonucleases are relatively non-specific. Murray (1970) has investigated the products obtained by digestion with pancreatic DNase in detail, using fingerprinting techniques, and has shown that almost all possible small oligonucleotides are present. Other enzymes such as micrococcal DNase, neurospora DNase, etc. may have slightly greater specificity, but so far none of them has a specificity similar to that of ribonuclease T_1. No doubt these enzymes will be very useful for the study of sequences in pure oligodeoxynucleotides of 5-20 residues long if and when these can be obtained.

OTHER APPROACHES TO DNA SEQUENCING

Another approach to DNA sequencing is by copying DNA enzymically either with DNA or RNA polymerase. This approach was used very successfully for RNA by Weissmann and his colleagues (Billeter *et al.*, 1969; Goodman, *et al.*, 1970) in their studies on RNA of bacteriophage Qβ. Using the specific Qβ replicase to copy the minus strand RNA in pulse-labelling experiments, sequences of 175 residues from the 5′ end have been determined. The replicase starts copying from the 5′ end, so that if short times of incubation with labelled triphosphates are used only short lengths of RNA are synthesized. Such short lengths could be sequenced by the fingerprinting techniques, and there were two other useful advantages of this approach. One was that it was possible to order fragments—for instance T_1 digestion products—by their specific activity because those near the 5′ end would be labelled first and in short pulses would have the highest specific activity. The other advantage was that it was possible to use "nearest-neighbour" techniques. This helped considerably in the detailed sequence determination, though it required the carrying out of four separate parallel labelling experiments with the four triphosphates.

RNA polymerase can be used in two ways to copy DNA. In the presence of σ factor it will copy double stranded RNA starting at specific sites (promoters) and it is possible to determine sequences from these sites by pulse-labelling. This approach has been used very successfully by Blattner, *et al.*, (1971) to determine sequences in bacteriophage λ.

RNA polymerase can also be used to copy single-stranded DNA, but in this case initiation appears to occur much more randomly and σ factor is probably not involved. In preliminary studies on single stranded DNA phages we have digested transcription products with ribonuclease T_1 and obtained characteristic fingerprints. The advantage of this approach is of course that the established techniques for RNA sequencing can then be applied, though one must be certain that the polymerase is copying the template exactly.

DNA can also be copied using DNA polymerase and this can, in certain cases, be used as the basis for a pulse-labelling approach. The enzyme initiates copying at the junction of a double stranded DNA with the 3′ end of a single stranded DNA (see Fig. 12). It adds residues to this 3′ end using the longer fragment as a template and the shorter as a primer. The most successful use of this approach has been the determination of the sequences at the "sticky-ends" of bacteriophage λ by Wu and his colleagues (Wu and Kaiser, 1968; Wu and Taylor, 1971). This bacteriophage is a double stranded DNA but each strand extends twelve residues at its 5′ end, as shown in Fig. 13. These two single stranded pieces have complementary sequences and pair together to form the circular duplex. In the linear form they could be filled in using DNA polymerase so that labelled sequences of twelve residues were formed, and by a number of elegant techniques their sequences were determined; these are shown in Fig. 14. These are probably the first sequences to be determined and ascribed a definite position in a DNA molecule.

FIGURE 12

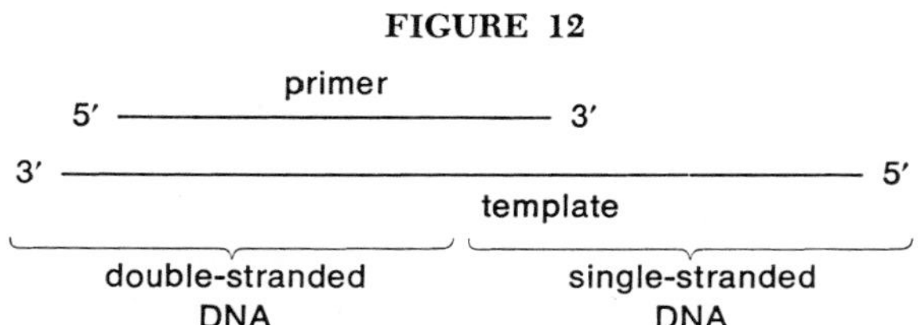

Conditions of DNA necessary for initiation of copying with DNA polymerase.

FIGURE 13

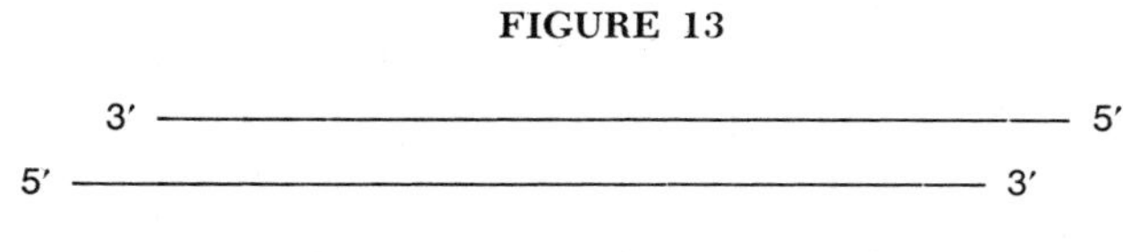

Diagram showing pairing of the two chains of λ DNA.

FIGURE 14

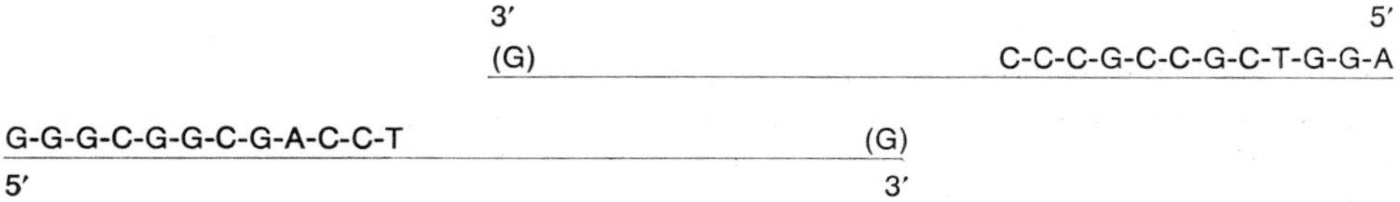

3′ (G) ———— C-C-C-G-C-C-G-C-T-G-G-A 5′

5′ G-G-G-C-G-G-C-G-A-C-C-T ———— (G) 3′

Nucleotide sequences in the "sticky ends" of bacteriophage λ DNA (Wu and Taylor, 1971).

Several other possibilities of achieving a situation as in Fig. 12 can be devised, at least in theory, and it should be possible to apply them as substrates for using DNA polymerase in a pulse-labelling technique.

Another possible approach to DNA sequencing that must be mentioned is the direct visualization of the specifically labelled mononucleotide residues in the electron microscope (Beer, *et al.,* 1971). Although at present there are severe technical difficulties, this method is theoretically possible and would be very rapid and generally applicable to DNA of any size. If the technical problems can be overcome it could be expected ultimately to render any of the above degradation methods for DNA sequencing obsolete and probably also methods for RNA and protein sequencing.

The results of DNA sequencing hitherto have been rather meagre, but the above brief discussion illustrates that a start has been made on the problem. Although there is no general method available at present there are probably more different approaches possible than is the case with RNA, and if these can be successfully applied, it is to be hoped that sequences can be determined and that we can learn more of the way in which the genes carry the information to make an organism.

REFERENCES

Adams, J. M., Jeppesen, P. G. N., Sanger, F. and Barrell, B. G. (1969) *Nature, (London) 223,* 1009

Adams, J. M. and Cory, S. (1970). *Nature, (London) 227,* 570

Barrell, B. G. (1971). in *Procedures in Nucleic Acid Res.* (G. L. Cantoni and D. R. Davies eds.) Vol II, pp 751-779 Harper & Row (New York)

Beer, M., Bartl, P., Koller, T. and Erickson, H. P. (1971). *Methods in Cancer Research, 6,* 286

Billeter, M. A., Dahlberg, J. E., Goodman, H. M., Hindley, J. and Weissmann, C. (1969) *Nature, (London) 224,* 1083

Blattner, F. R., Boettiger, J. K. and Dahlberg, J. E. (1971). *Fed. Proc. Fed. Amer. Soc. Exp. Biol. 30,* part II, 1315 Abs.

Brownlee, G. G. and Sanger, F. (1969). *Eur. J. Biochem. 11,* 395

Burton, K. and Petersen, G. B. (1960). *Biochem. J. 75,* 17

Cory, S., Adams, J. M., Spahr, P-F. and Rensing, U. (1972). *J. Molec. Biol. 63,* 41

Danna, K. and Nathans, D. (1971). *Proc. Nat. Acad. Sci. U.S. 68,* 2913

De Wachter, R., Vandenberghe, A., Merregaert, J., Contreras, R. and Fiers, W. (1971). *Proc. Nat. Acad. Sci. U.S. 68,* 585

Fiers, W., Contreras, R., De Wachter, R., Haegeman, G., Merregaert, J., Min Jou, W. and Vandenberghe, A. (1971). *Biochimie, 53,* 495

Goodman, H. M., Billeter, M. A., Hindley, J. and Weissmann, C. (1970) *Proc. Nat. Acad. Sci. U.S. 67,* 921

Jeppesen, P. G. N., Nichols, J. L., Sanger, F. and Barrell, B. G. (1970). *Cold Spring Harb. Symp. quant. Biol. 35,* 13

Jeppesen, P. G. N. (1971). *Biochem. J. 124,* 357

Jeppesen, P. G. N., Barrell, B. G., Sanger, F. and Coulson, A. (1972). *Biochem. J.* In press

Kelly, T. J. and Smith, H. O. (1970). *J. Molec. Biol. 51,* 393

Ling, V. (1971). *FEBS Lett. 19,* 50

Ling, V. (1972a). *J. Molec. Biol. 64,* 87

Ling, V. (1972b). *Proc. Nat. Acad. Sci. U.S. 69,* 742

Lodish, H. F. (1970). *J. Molec. Biol. 50,* 689
Lodish, H. F. (1971). *J. Molec. Biol. 56,* 627
Murray, K. (1970). *Biochem. J. 118,* 831
Nichols, J. L. (1970). *Nature,* (*London*) *225,* 147
Nichols, J. L. and Robertson, H. D. (1971). *Biochim. biophys. Acta, 228,* 676
Robertson, H. D. and Jeppesen, P. G. N. (1972). In press *J. Molec. Biol.*
Sadowski, P. D. and Hurwitz, J. (1969). *J. Biol. Chem. 244,* 6192
Sanger, F. (1971). *Biochem. J. 124,* 833
Sanger, F., Brownlee, G. G. and Barrell, B. G. (1965). *J. Molec. Biol. 13,* 373
Steitz, J. A. (1969). *Nature,* (*London*) *224,* 957
Wu, R. and Kaiser, A. D. (1968). *J. Molec. Biol. 35,* 523
Wu, R. and Taylor, E. (1971). *J. Molec. Biol. 57,* 491

Mutations at the End of the Iso-l-cytochrome *c* Gene of Yeast

FRED SHERMAN AND JOHN W. STEWART

Department of Radiation Biology and Biophysics, University of Rochester School of Medicine and Dentistry, Rochester, New York

The *cy*1 gene in the bakers' yeast *Saccharomyces cerevisiae* is by far the most characterized gene from any eukaryotic organism. Genetic studies of this gene, along with protein analysis of its gene product, iso-1-cytochrome *c*, has led to the elucidation of the DNA changes associated with various mutational events and to the determination of AUG as a chain initiation codon (Stewart *et al.*, 1971) and to UAA (ochre) and UAG (amber) as chain terminating codons (Stewart *et al.*, 1972; Stewart and Sherman, 1972). Recently, the nucleotide sequence of a portion of the end of the *cy*1 gene was deduced from altered forms of iso-l-cytochrome *c* (Stewart and Sherman, unpublished).

The *cy*1 mutants are, by definition, deficient in activity or amount of iso-l-cytochrome *c*. On the average, approximately one half of intragenic revertants derived from these *cy*1 mutants contain functional iso-l-cytochromes *c* with altered primary structures. It is these altered proteins that have been used to deduce the corresponding nucleotide changes in the gene. We are in the process of characterizing all of our *cy*1 mutants that are associated with the first ten amino acid residues at the amino terminus. This paper summarizes the genetic studies that revealed 21 *cy*1 mutants which were eventually shown to have alterations within the first 33 nucleotides. Included in this paper are tabulations of the structures of 241 iso-1-cytochromes *c* that were obtained from revertants of 20 of these *cy*1 mutants.

ISOLATION OF *cy*1 MUTANTS

Numerous procedures have been developed in our laboratories for detecting yeast mutants whose iso-1-cytochrome *c* is either absent or nonfunctional. The first systematic search consisted of scanning clones of yeast with a spectroscope in order to isolate mutants deficient in the α-band of cytochrome *c* (Sherman,

1964). A large number of mutants with a variety of patterns of cytochrome deficiencies were obtained with this technique after examining approximately 14,800 clones. These included mutants of the unlinked genes *cy*1, *cy*2, *cy*3, *cy*4 and *cy*6 that specifically control cytochrome *c* content. While only one *cy*1 mutant was isolated with the spectroscopic scanning procedure (Table 1), this mutant, *cy*1-2,

TABLE 1

The different selection procedures used to obtain *cy*1 mutants

Method	Number of mutants	Alleles	Reference
Spectroscopic scanning	1	*cy*1-2	Sherman, 1964
Benzidine staining	14	*cy*1-3 → *cy*1-16	Sherman *et al.* 1968
Chlorolactate resistence	195	*cy*1-17 → *cy*1-211	Sherman *et al.* unpublished
Total	210		

was used to establish that the primary structure of iso-1-cytochrome *c* is determined by the *CY*1 locus (Sherman *et al.*, 1966).

A second procedure relied on the haem catalysis of H_2O_2 oxidation of benzidine to stain colonies according to their level of haemoproteins (Sherman *et al.*, 1968). Since the benzidine staining reaction is more diagnostic for cytochrome *c* levels when other haemoproteins are absent, the cytochrome *c* deficient mutants were prepared from ρ^- cytoplasmic strains, which lack cytochromes a, a_3, b and c_1. The ρ^- strains were treated with various mutagens and approximately one million colonies were tested with the benzidine reagents. The mutants that gave a weak or negative benzidine reaction were tested for their ability to complement known *cy*1 mutants; *cy*1 ρ^- tester strains were crossed to the presumptive mutants and the diploids were subjected to the benzidine test. The 14 mutants, *cy*1-3 through *cy*1-16 (Table 1), were selected in this way.

The two methods described above reveal mutants that are deficient in amount of cytochrome *c*, but cannot detect mutants having normal levels of inactive protein. More recently, a method was devised to select mutants deficient in either activity or amount of cytochrome *c* (Sherman, Stewart, Jackson, Gilmore and Parker, unpublished). It relies on the observations that a substantial proportion of mutants that resist the toxic action of chlorolactate cannot utilize lactate for growth, and that nearly all chlorolactate-resistant mutants that cannot utilize lactate are substantially deficient in amount or activity of cytochrome *c*. Apparently the inability to utilize lactate, in contrast to other nonfermentable carbon sources, constitutes a significant mechanism for chlorolactate resistance and is associated with the mutational loss or inactivation of cytochromes *c*. In any case, hundreds of *cy*1 mutants were obtained by crossing the chlorolactate resistant mutants to the deletion mutant, *cy*1-1, and testing the hemizygous diploids for growth on DL-lactate medium. The 210 *cy*1 mutants obtained with the three selection procedures are listed in Table 1, and the various mutagens used to induce them are listed in Table 2.

TABLE 2

Number of *cy*1 mutants obtained with different mutagens
(Sherman *et al.* in preparation)

Mutagen[a]	Complete set	Mutants in NH_2-terminal region[b]
None	41	3
UV	129	17
ICR-170	30	0
NIL	5	1
NA	5	0
Totals	210	21

[a]The mutagen abbreviations are: UV, ultraviolet light; ICR-170, 2-methoxy-6-chloro-9-[3-(*N*-ethyl-N-1-chloroethyl)-aminopropyl]-amino-acridine dihydrochloride; NIL, 1-nitro-soimidazolidone-2; NA, nitrous acid.

[b]Mutants in the region preceding glycine 11.

THE *cy*1 MUTANTS ASSOCIATED WITH THE AMINO-TERMINAL REGION OF ISO-1-CYTOCHROME *c*

An attempt has been made to map all 210 *cy*1 mutants by the frequencies of x-ray induced recombination of various heteroallelic diploids (Sherman *et al.*, unpublished). Recombination frequencies are approximately related to corresponding distances between altered sites in iso-1-cytochrome *c* (Sherman, *et al.*, 1970), but the genetic maps are sufficiently ambiguous as to make precise predictions of the spacings between mutational sites somewhat unreliable (Parker and Sherman, 1969). Nevertheless, altered iso-1-cytochromes *c* from revertants always have indicated that lesions lie within adjacent codons when frequencies of recombination between single-site mutants are negligible. All of the 210 *cy*1 mutants and the deletion mutant, *cy*1-1 have been crossed individually to a variety of *cy*1 tester strains, and the frequencies of x-ray induced recombination determined.

A survey of iso-1-cytochromes *c* from intragenic revertants uncovered several *cy*1 mutants that were associated with the amino terminal region of the protein. As will be discussed below, the *cy*1-13, *cy*1-9 and *cy*1-134 mutants, which were used as tester strains, contained alterations corresponding, respectively, to amino acid position -1 (the initiating methionine residue), position 2 and position 10.

There were 15 *cy*1 mutants that exhibited less than 100 milliunit (recombinants/10^{11} survivors/Roentgen) of recombination frequency when tested with the *cy*1-13 and *cy*1-9 strains (Stewart *et al.*, 1971; Sherman *et al.*, unpublished). The results with the *cy*1-13 crosses, shown in Table 3, suggested that

TABLE 3

The 21 *cy1* mutants associated with amino-terminal region of iso-1-cytochrome *c*

	Mutant number	Inducing mutagen[a]	Recombination[b] Crosses with:		Revertibility[c]	
			cy1-13	*cy1*-134	UV	EMS
Class 1	*cy1*-131	UV	0	+++	+	+++
Class 2	*cy1*-13	NIL	0	+++	+	+
	cy1-51	UV	0	+++	+	+
	cy1-74	UV	0	+++	+	+
	cy1-85	UV	0	+++	+	+
	cy1-100	None	0	+++	+	+
	cy1-133	UV	0	+++	+	+
	cy1-163	None	0	+++	+	+
	cy1-181	UV	0	+++	+	+
Class 3	*cy1*-9	UV	+	+++	+++	+
	cy1-91	UV	+	+++	+++	+
	cy1-147	UV	+	+++	+++	+
	cy1-160	None	+	+++	+++	+
	cy1-172	UV	+	+++	+++	+
Class 4	*cy1*-31	UV	++	+++	+	0
Class 5	*cy1*-135	UV	+++	+	+	+
	cy1-138	UV	+++	+	+	+
	cy1-179	UV	+++	+	+	+
Class 6	*cy1*-134	UV	+++	0	+	0
	cy1-183	UV	+++	0	+	0
Class 7	*cy1*-49	UV	+++	0	0	0

[a]The mutagen abbreviations are: UV, ultraviolet light; NIL, 1-nitrosoimidazolidone-2; EMS, ethyl methane-sulphonate.

[b]See Stewart *et al.* (1971) for quantitative recombination frequencies of crosses *cy1*-13 with Classes 1, 2, 3 and 4 strains. The X-ray mapping milliunits (recombinants/10^{11} survivors/ Roentgen) for the various symbols are as follows: 0, less than 5 milliunits; +, 9 to 25 milliunits; ++ approx. 90 milliunits; +++, greater than 100 milliunits.

[c]See Stewart *et al.* (1971) for quantitative reversion frequencies of Class 1, 2, 3 and 4 mutants with UV and of Class 1 and 2 mutants with EMS (ethyl methane-sulphonate). The +++ for UV signifies 3 to 60 times greater reversion frequencies in comparison to +. The +++ for EMS signifies over 100 times greater reversion frequency than +. The 0 for UV and EMS indicates absent or greatly reduced reversion frequency in comparison to + .

15 *cy1* mutants could be grouped according to the distance from the *cy1*-13 site: 9 mutants mapped at the *cy1*-13 site (class 1 and 2); 5 mutants mapped near the *cy1*-13 site (class 3); and the one mutant *cy1*-31, mapped near the *cy1*-13

All of the iso-1-cytochrome *c* samples were compared with normal iso-1-cytochrome *c* by amino acid compositional analysis and by peptide mapping of tryptic and chymotryptic digests. Iso-1-cytochrome *c* is sufficiently small, containing only 108 residues (Fig. 1), and is of such a composition that it is possible to accurately detect single residue changes of all amino acids except cystine and tryptophan by amino acid analysis of acid hydrolysates (see, for example, Stewart *et al.*, 1971; Stewart *et al.*, 1972; Stewart and Sherman 1972). The peptide maps of tryptic and chymotryptic digest of normal iso-1-cytochrome *c* are shown in Fig. 2, along with the positions and sequences of the peptides from the amino

FIGURE 2

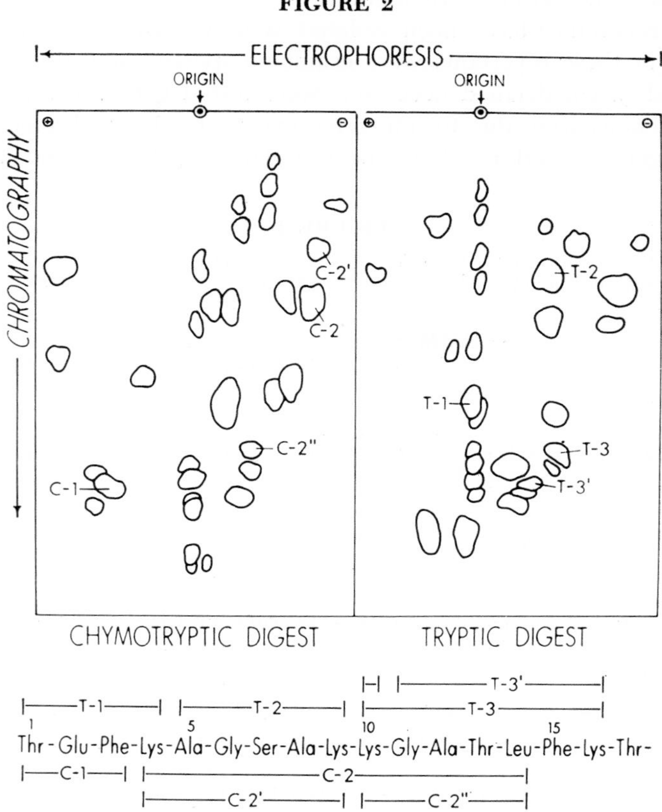

Tracings of peptide maps of normal iso-1-cytochrome *c*. The positions of the chymotryptic and tryptic peptides from the amino terminal region of the protein are indicated on the map and the corresponding sequences are shown below (Narita and Titani, 1969; Stewart *et al.* 1971, 1972; Stewart and Sherman, 1972).

terminal region. The loss or change in position of one or more of the peptide spots that are identified in Fig. 2, is a diagnostic feature of proteins altered in the amino terminal region. The abundance of sites for enzymic cleavages in this region and the small sizes of the resulting fragments make peptide maps sensitive indicators of sequence differences and of the sites of changes. In fact, all the altered sites identified by sequential Edman degradation had been predicted correctly from the changes in peptide maps and amino acid compositions. Sequential Edman degradation of the proteins, from the amino terminus through and beyond the altered region, was used to identify all altered sequences judged to be unique by peptide maps and amino acid compositions, except for many of those from revertants of nonsense mutants; in these cases sequencing was undertaken with only one of the several different proteins that resulted from reversion by single base changes of each of the two nonsense sites. All proteins with identical peptide maps and compositions were assumed to be identical. With a few of the more grossly altered proteins, this assumption was tested and verified by sequential degradation through the altered regions of two or more of the apparently identical proteins.

Relating the genetic code to sequence changes in iso-1-cytochromes *c* from 241 intragenic revertants resulted in the identification of nine different types of mutational lesions in the 21 *cy*1 mutants. In most instances it was possible to deduce the exact nucleotide changes associated with the mutants. In addition, the amino acid changes in the revertant proteins revealed the following nucleotide sequence in m-RNA of a portion of the *CY*1 gene:

−1	1	2	3	4	5	6	7	8	9	10	11	12	13	14
(Met)	**Thr**	**- Glu -**	**Phe**	**- Lys -**	**Ala**	**- Gly -**	**Ser -**	**Ala**	**- Lys -**	**Lys**	**- Gly -**	**Ala**	**- Thr -**	**Leu -**
AUG	ACN	GAA	UUC	AAG	GCC	GGU	UCU	GCU	AAG	AAA	GGU	GCU	ACA	CUN
1	5		10	15		20		25	30		35		40	45

The *cy*1 mutants are conveniently divided into the following types which are discussed below: (a) ochre (UAA) mutants; (b) amber (UAG) mutants; (c) chain initiation mutants; (d) frameshift mutants; (e) the *cy*1-31 mutant; and (f) the *cy*1-49 mutant.

(a) Ochre (UAA) mutants

It is clear that the *cy*1-9 mutant contains a UAA codon at the site corresponding to the normal glutamic acid 2. Amino acid analysis and peptide mapping of all samples and sequencing of four selected samples have established the primary structures of 59 iso-1-cytochromes *c* from intragenic revertants that were obtained either spontaneously or after treatments with the nine different mutagens listed in Table 4 (Stewart *et al.*, 1972; Sherman and Stewart, unpublished). One of the following six amino acids occupies position 2 in 57 out of 59 revertant proteins: glutamine, lysine, glutamic acid, leucine, serine and tyrosine. The codons specifying these six amino acids all differ by one base from UAA (ochre) and UAG (amber). We believe the nonsense codon in *cy*1-9 is UAA, since no tryptophan (UGG) replacement was observed in any of the 59 revertant proteins, and since this replacement can arise by a single base change from UAG but not UAA. In addition, the structural variations observed at and near this site leaves little doubt that tryptophan at position 2 would allow functioning of iso-1-cytochrome *c*. Also, the assignments of amber and ochre, based, respectively, on the presence or absence of tryptophan replacements in revertant proteins from

TABLE 4

Frequencies of amino acid replacements in iso-1-cytochromes *c* from *cy*1-9 revertants and the corresponding base-pair changes

(from Stewart *et al.*, 1972; Sherman and Stewart, unpublished)

Amino acid replacing glutamic acid 2	Codon	Base in UAA affected by reversion	Base change of reversion	Mutagen[a]										
				None	XR	α	EMS	DES	MMS	NIL	NG	NA	UV	Total
Glutamine	*C*AA	first	A·T→G·C	2	4	1	2	1	0	1	2	4	11	28
Lysine	*A*AA	first	A·T→T·A	0	2	1	1	1	0	2	1	0	0	8
Glutamic acid	*G*AA	first	A·T→C·G	1	0	0	0	0	0	0	0	0	0	1
None	U*G*A	second	A·T→G·C	—	—	—	—	—	—	—	—	—	—	—
Leucine	U*U*A	second	A·T→T·A	0	1	2	1	0	1	0	0	1	0	6
Serine	U*C*A	second	A·T→C·G	1	1	1	0	1	0	0	0	0	0	4
None	UA*G*	third	A·T→G·C	—	—	—	—	—	—	—	—	—	—	—
Tyrosine Tyrosine	UA*U* UA*C*	third third	A·T→T·A A·T→C·G }	2	3	2	0	0	2	1	0	0	0	10
Others (two or more base changes)				0	1	1	0	0	0	0	0	0	0	2
			Total	6	12	8	4	3	3	4	3	5	11	59

[a]The mutagen abbreviations are: XR, X-rays; α, polonium-210 alpha particles; EMS, ethyl methane-sulphonate; DES, diethyl sulphate; MMS, methyl methane-sulphonate; NIL, 1-nitrosoimidazolidone-2; NG, N-methyl-N′-nitro-N-nitrosoguanidine; NA, nitrous acid; UV, ultraviolet light.

the two amber mutants, *cy*1-76 and *cy*1-179, and the ochre mutants, *cy*1-2 and *cy*1-9, is completely consistent with the specific action of certain suppressors on these *cy*1 mutants (Gilmore, Stewart and Sherman, 1971; Sherman *et al.*, 1972; Sherman *et al.*, unpublished).

The amino acid replacements in 2 of the 59 proteins from the *cy*1-9 revertants, shown in Table 5, must have occurred by at least two concomitant base-pair

TABLE 5

Altered iso-1-cytochromes *c* from cyv-9 and cyv-179 revertants that occurred by multiple base changes

	Mutagen[a]	
normal		—1 1 2 3 4 5 6 7 8 9 10 11 12 (Met) Thr –*Glu* –Phe –Lys –Ala –Gly –Ser –Ala –*Lys* –Lys –Gly –Ala– AUG ACN GAA UUC AAG GCC GGU UCU GCU AAG AAA GGU GCU 1 5 10 15 20 25 30 35
*cy*1-9		AUG ACN *UAA* UUC AAG GCC GGU UCU GCU AAG AAA GGU GCU
*CY*1-9-AU	XR	Met –*Ile* –*Glu* –Phe –Lys –Ala –Gly –Ser –Ala –Lys –Lys –Gly –Ala – U AUC GAA A
*CY*1-9-BD	α	(Met) Thr –*Leu* –*Leu* –Lys –Ala –Gly –Ser –Ala –Lys –Lys –Gly –Ala – UUA UU$^{A}_{G}$ CUC
*cy*1-179		AUG ACN GAA UUC AAG GCC GGU UCU GCU *UAG* AAA GGU GCU
*CY*1-179-L	NA	(Met) Thr –Glu –Phe –Lys –Ala –Gly –Ser –Ala –*Leu* –*Glu* –Gly –Ala – UUG GAA
*CY*1-179-Z	XR	(Met) Thr –Glu –Phe –Lys –Ala –Gly –Ser –Ala –*His* –Lys –Gly –Ala – CA$^{U}_{C}$
*CY*1-179-AF	XR	(Met) Thr –Glu –Phe –Lys –Ala –*Ala* –Ser –Ala –*Phe* –Lys –Gly –Ala – GCU UU$^{U}_{C}$
*CY*1-179-AW	α	(Met) Thr –Glu –Phe –Lys –Ala –Gly –Ser –Ala ——————Gly –Ala –

Codons shown for the amino acid replacements represent the least possible number of base-changes, and have not been directly verified (from Stewart and Sherman, 1972; Sherman & Stewart, in preparation)

[a]The mutagen abbreviations are: XR, X-rays; α, polonium-210 alpha particles; NA, nitrous acid.

substitutions. The simplest explanation for the x-ray-induced revertant, *CY*1-9-AU, is a C·G → T·A transition at nucleotide position 5, and a T·A → G·C transversion at nucleotide position 7. Similarly, the replacements in the α-particle-induced revertant, *CY*1-9-BD, can be accounted for A·T → T·A transversion at nucleotide position 8 and a C·G → A·T or G·C transversion at nucleotide position 12.

The remaining mutants belonging to class 3 (Table 3) were tested by examining one revertant protein from each of these four *cy*1 mutants. In all instances, glutamic acid 2 was replaced by either glutamine, tyrosine or lysine. Therefore, it is concluded that all five class 3 mutants arose by a G·C → T·A transversion at nucleotide position 7, creating UAA in the m-RNA at amino acid position 2. These results also identify the normal nucleotide positions 7 to 9 as GAA.

(b) Amber (UAG) mutants

Altered iso-1-cytochrome *c* from 52 intragenic revertants established that the defect in *cy*1-179 is due to a UAG codon corresponding to amino acid position 9. Peptide maps, amino acid compositions and five sequences identified the 52 alterations presented in Tables 5 and 6 (Stewart and Sherman, 1972; Sherman and

TABLE 6

Frequencies of amino acid replacements in iso-1-cytochromes *c* from *cy*1-179 revertants and the corresponding single base-pair changes

(from Stewart and Sherman, 1972; Sherman and Stewart, unpublished).

Amino acid replacing lysine 9	Codon	Base in UAG affected by reversion	Base change of reversion	Mutagen[a]						
				UV	NA	DES	XR	α	None	Total
Glutamine	*C*AG	first	A·T→G·C	5	1	0	0	0	0	6
Lysine	*A*AG	first	A·T→T·A	0	0	0	0	0	0	0
Glutamic acid	*G*AG	first	A·T→C·G	0	0	0	0	1	0	1
Tryptophan	U*G*G	second	A·T→G·C	1	4	3	2	5	2	17
Leucine	U*U*G	second	A·T→T·A	4	4	2	2	3	0	15
Serine	U*C*G	second	A·T→C·G	0	0	0	1	0	0	1
None	UA*A*	third	G·C→A·T	–	–	–	–	–	–	–
Tyrosine	UA*U*	third	G·C→T·A }	0	0	1	3	0	4	8
Tyrosine	UA*C*	third	G·C→C·G }							
Others (two or more base changes)				0	1	0	2	1	0	4
			Total	10	10	6	10	10	6	52

[a]Mutagen abbreviations are: UV, ultraviolet light; NA, nitrous acid; DES, diethylsulphate; XR, x-ray; α, polonium-210 alpha particles.

Stewart, unpublished). Lysine 9 was replaced by glutamine, glutamic acid, tryptophan, leucine, serine or tyrosine in 48 out of the 52 revertant proteins. Except for lysine, which is found at this position in normal iso-1-cytochrome *c*, these replacements include all of the amino acids whose codons differ from UAG by one base. The four exceptional revertants, listed in Table 5, are most simply accounted for, respectively, by the concurrent substitutions of 2 or 3 base-pairs, or by the deletion of the 6 base-pairs controlling residues 9 and 10. In several instances, two concomitant base substitutions occurred on either side of an unaltered base (*CY*1-179-L, *CY*1-179-Z and probably *CY*1-9-AU), while in one instance (*CY*1-179-AF) the base substitutions were separated by 7 unaltered nucleotides. The proximity of the concomitant alterations in the *cy*1-9 and *cy*1-179 revertants listed in Table 5, and the lack of alterations at more distant sites, suggests an interdependence in the multiple base changes of each of the revertants. Possibly the initial lesions involved single bases, and modification by inaccurate repair processes produced alterations at distant sites which could be as far away as nine base-pairs.

The action of suppressors on the other two *cy*1 mutants in class 5 suggested that they also contained UAG codons (Sherman *et al.*, unpublished). A tyrosine and a glutamine replacement of lysine 9 in revertant proteins from, respectively, each of these two *cy*1 mutants, confirmed their identity to *cy*1-179. Thus all three class 5 mutants arose from a A·T → T·A mutation at nucleotide position 28. The formation of amber mutations at amino acid position 9 establishes that lysine 9 is encoded by AAG and not AAA. The mutational events describing the formation and reversion of the *cy*1-9 ochre mutant and the *cy*1-179 amber mutant are presented schematically in Fig. 3.

FIGURE 3

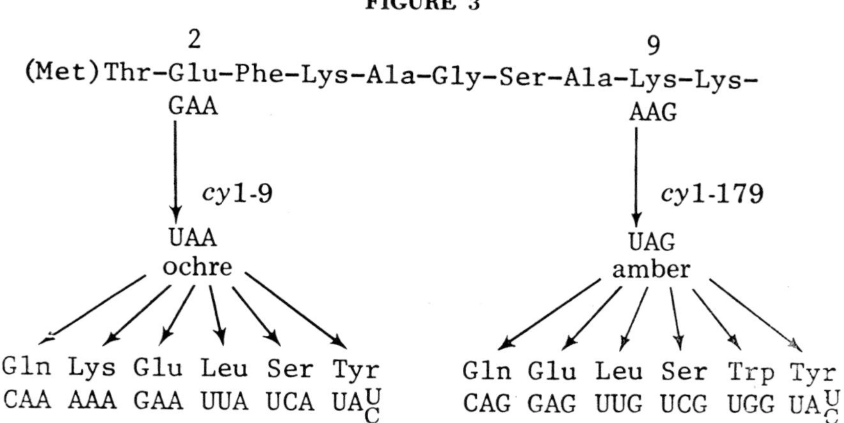

A schematic representation of the mutational events and the amino acid replacements in the iso-1-cytochromes *c* from the revertants of *cy*1-9 and *cy*1-179 that occurred by single-pair changes (Stewart *et al.* 1972; Stewart and Sherman 1972). In order to facilitate the comparison between sequences of amino acids and nucleotides in this figure, as well as in Figs. 4-11, the mutational events are shown as it affects the sequence of bases in m-RNA. The sequence of the 10 amino-terminal residues of normal iso-1-cytochrome *c* is shown at the top of the figure. The codons are presented below the amino acid residues they specify.

(c) Chain initiation mutants

Structural analysis of 45 revertant proteins indicated that the nine *cy*1 mutants from class 1 and class 2 (Table 3) are defective in the initiator codon and that some of the reverse mutations introduce initiator codons at new sites (Stewart

TABLE 7

Frequencies of the various altered iso-1-cytochromes *c* from the revertants of the nine initiation mutants

Types[a]	*cy*1-13	*cy*1-51	*cy*1-74	*cy*1-85	*cy*1-100	*cy*1-131	*cy*1-133	*cy*1-163	*cy*1-181
Normal	10	1	0	0	0	8	1	0	1
Short form	2	1	0	0	0	1	0	0	0
Long forms									
Met-Ile-	10	0	1	1	0	0	0	1	0
Met-Leu-	0	2	0	0	1	0	0	0	1
Met-Arg-	0	0	0	0	0	0	2	0	0
Met-Val- and Val-	0	0	0	0	0	1	0	0	0

The revertants were obtained either after no treatment or after treatments with UV, X-rays, ethyl methane-sulphonate, diethyl sulphate, nitrous acid or 1-nitrosoimidazolidone-2 (from Stewart *et al.* 1971).

[a]The *short form* lacks the four amino terminal residues of the normal protein. The *long forms* carry the indicated appendices at the amino terminus of the normal protein.

et al., 1971). Each of these *cy*l mutants gave rise to some revertants that have longer iso-1-cytochromes *c* containing one of the four following appendages at the amino terminus of the otherwise normal protein (Table 7): Met-Ile-; Met-Leu-; Met-Arg-; and a mixture of approximately 85% Val- and 15% Met-Val-. The long forms were all alike in the revertants from any *cy*l mutants. Also the normal protein occurred in 23 other revertants, while a short form, lacking the four normal amino terminal residues, occurred in four revertants (Table 7).

A model that accommodates all of these results is shown in Figs. 4 and 5. Fig. 4 illustrates the mutations causing the formation and the various reversions

FIGURE 4

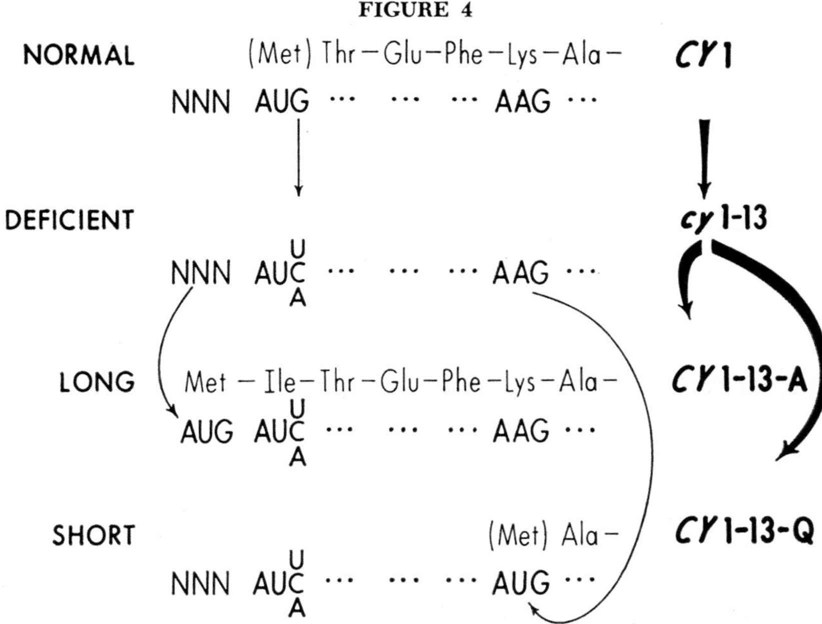

A diagrammatic representation of the mutational events creating the *cy*1-13 mutant, which is deficient in iso-1-cytochrome *c*, and the intragenic revertants, which contain long and short forms of iso-1-cytochrome *c*. *NNN* designates an unknown codon that differs from AUG by a single base. The methionine residues shown in parentheses are excised and are not found in mature proteins (Stewart *et al.* 1971).

of *cy*1-13, the most extensively studied initiation mutant. As shown at the top of the figure, normal iso-1-cytochrome *c* is assumed to have lost an amino-terminal methionine residue through the action of an aminopeptidase. This excised methionine residue is encoded by AUG, which is necessary for chain initiation. No iso-1-cytochrome *c* is produced in the *cy*1-13 mutant, due to the mutation of the initiator codon to any of the three isoleucine codons. The exact reversal of the mutation causes the reappearance of the initiator codon, thus allowing formation of the normal protein. Alternatively, AUG may be formed from an unknown codon, NNN, that precedes the normal initiator, causing the formation of the long protein in *CY*1-13-A. Retention, in this case, of methionine at the amino terminus requires that the activity of the methionine aminopeptidase is dependent on the adjacent residue. An AUG initiator codon at still another location is possible by a single-base substitution in the AAG triplet encoding lysine 4. With this *CY*1-13-Q mutation, a short protein, which lacks the four normal amino terminal residues, is formed after excision of the amino-terminal methionine.

Fig. 5 illustrates the mutational events creating the additional types of long forms of iso-1-cytochrome *c* in revertants of the other *cy*1 mutants. Mutation

FIGURE 5

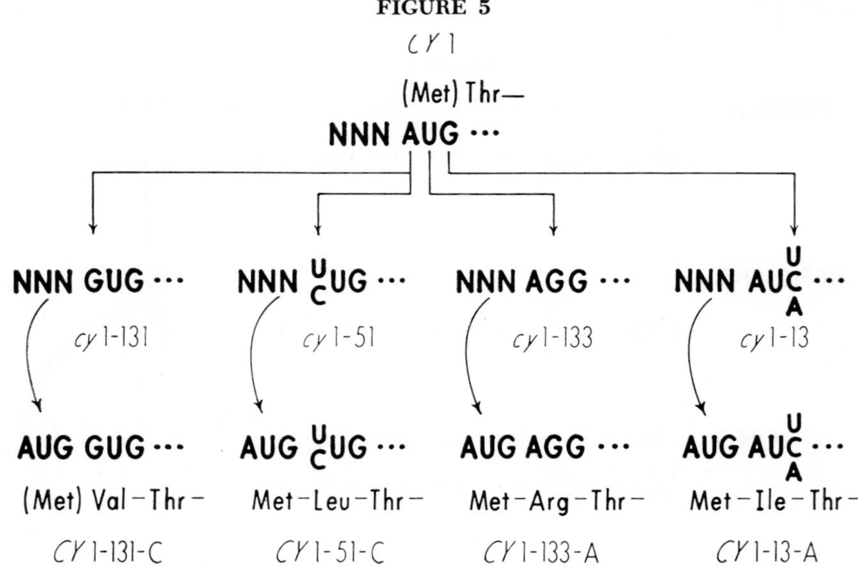

A diagrammatic representation of the mutational events leading to the four types of *cy*1 mutants that are defective in initiation, and of the mutational events creating the intragenic revertants that contain the long forms of iso-1-cytochromes *c*. *NNN* designates an unknown codon that differs from AUG by a single base. The methionine residues, shown in parentheses, are excised from approximately 85% of iso-1-cytochrome *c* in the *CY*1-131-C strain, and from all of the protein in the normal *CY*1 strain (Stewart *et al.* 1971).

of any of the three bases of the AUG initiator codon causes deficiency of iso-1-cytochrome *c* in the resulting *cy*1 mutants. The bottom of Fig. 5 illustrates the reverse mutations that create the new methionine codons from the unknown triplet NNN. It can be seen that all amino acids whose codons differ by one base from AUG are observed in the different long forms, except lysine and threonine. The absence of long proteins that carry either lysine or threonine may plausibly be explained by the improbability of selecting representatives of all nine possible single base changes of AUG in a group of nine *cy*1 mutants. Thus single base-pair substitutions can account for the production of all of these nine *cy*1 mutants and all of their revertants. These results identify AUG as the codon immediately preceding the codon for the amino-terminal threonine of normal iso-1-cytochrome *c* and establish the essential nature of this AUG codon in the beginning of m-RNA.

The high EMS-induced revertibility of the class 1 mutant, *cy*1-131 (Table 3), now can be explained satisfactorily by the selective action of this mutagen for G·C → A·T changes. The *cy*1-131 mutants differ from all of the other initiator mutants by having a codon, GUG, that is able to mutate back to AUG through a G·C → A·T transition. Consistent with this conclusion was finding only normal proteins in EMS-induced revertants of *cy*1-131, while all three types, normal, short and long, occurred in EMS-induced revertants of *cy*1-13.

(d) Frameshift mutants

The sequences of the revertant proteins described below, along with the results of the *cy*1-179 amber mutant, established that the *cy*1-183 mutation was caused by an insertion of a A·T base-pair within or adjacent to the codon of lysine 10. In contrast to the *cy*1 mutants which arose by base substitutions (classes 1, 2, 3 and 5), *cy*1-183 could not be reverted with ethyl methane-sulphonate (EMS) (Table 3), or with N-methyl-N′nitrosoguanidine (NG) and 1-nitrosoimidazolidone-2 (NIL). However a reasonable frequency of *cy*1-183 revertants were obtained with UV, X-rays, α-particles, nitrous acid (NA), nitrogen mustard (HN2), and growth in the presence of methyl methane-sulphonate (MMS). Iso-1-cytochromes *c* from a total of 55 intragenic revertants have been prepared and the sequences of all but two of them have been determined (Stewart and Sherman, unpublished). All proteins that had identical peptide maps and identical amino acid compositions were assumed to have identical sequences. The altered sequence of at least one protein of each type was determined by the Edman degradation technique. These 5 revertant proteins consisted of 17 different sequences, which could be divided into the four groups shown in the four figures, Fig. 6 (types 1-9), Fig. 7 (types 10-13), Fig. 8 (types 14-16), and Fig. 9 (type 17). The distributions of the various types of altered iso-1-cytochrome *c* for each mutagen are tabulated in Table 8.

The bulk of the revertant proteins, 44 of 53 (Table 8), constituted the nine types of the first group. Type 1 is normal and types 2 through 9 contained altered sequences of contiguous amino acids that could be accounted for by simple deletions of single bases from nucleotide positions 8 through 34 (Fig. 6). The five revertant proteins, comprising types 10 through 13, appeared to have arisen by deletions of single bases, with accompanying near-by base substitutions (Fig. 7). The third major group consisted of three kinds of tandem duplications, types 14, 15 and 16, which extended the length of the protein, respectively, by five, four and one amino acid residues (Fig. 8). Type 17 (Fig. 9), the only member of the last group, arose through the deletion of 16 nucleotides, which removed the *cy*1-183 mutation and shortened the protein by five amino acid residues.

After inspection of the amino acid sequences of types 2 through 13 (Figs. 6 and 7), it is concluded that the presumed m-RNA in *cy*1-183 could have either an insertion of A at any of the nucleotide positions 31 through 34, as shown in the figures, or an insertion of G at position 30 or 34. However, the possibility of an insertion of G at position 30 is excluded since the study with *cy*1-179

FIGURE 6

Type	
	−1 1 2 3 4 5 6 7 8 9 10 11
normal	(Met)Thr-Glu-Phe-Lys-Ala-Gly-Ser-Ala-Lys-Lys-Gly-
	AUG ACN GAA UUC AAG GCC GGU UCU GCU AAG AAA GGU
	1 5 10 15 20 25 30 35
	+A
*cy*1-183	AUG ACN GAA UUC AAG GCC GGU UCU GCU AAG AAA AGG U
	—A *or* G
1	AUG ACN GAA UUC AAG GCC GGU UCU GCU AA$^{A}_{G}$ AAA GGU
	(Met)Thr-Glu-Phe-Lys-Ala-Gly-Ser-Ala-Lys-Lys-Gly-
	—U *or* A
-	AUG ACN GAA UUC AAG GCC GGU UCU GC$^{A}_{U}$ AGA AAA GGU
	(Met)Thr-Glu-Phe-Lys-Ala-Gly-Ser-Ala-*Arg*-Lys-Gly-
	—C
2	AUG ACN GAA UUC AAG GCC GGU UCU GUA AGA AAA GGU
	(Met)Thr-Glu-Phe-Lys-Ala-Gly-Ser-*Val*-*Arg*-Lys-Gly-
	—U *or* G
3	AUG ACN GAA UUC AAG GCC GGU UC$^{U}_{G}$ CUA AGA AAA GGU
	(Met)Thr-Glu-Phe-Lys-Ala-Gly-Ser-*Leu*-*Arg*-Lys-Gly-
	—U *or* C
4	AUG ACN GAA UUC AAG GCC GGU $^{U}_{C}$UG CUA AGA AAA GGU
	(Met)Thr-Glu-Phe-Lys-Ala-Gly-*Leu*-*Leu*-*Arg*-Lys-Gly-
	—C *or* G
5	AUG ACN GAA UUC AAG GC$^{C}_{G}$ GUU CUG CUA AGA AAA GGU
	(Met)Thr-Glu-Phe-Lys-Ala-*Val*-*Leu*-*Leu*-*Arg*-Lys-Gly-
	—G
6	AUG ACN GAA UUC AAG CCG GUU CUG CUA AGA AAA GGU
	(Met)Thr-Glu-Phe-Lys-*Pro*-*Val*-*Leu*-*Leu*-*Arg*-Lys-Gly-
	—A
7	AUG ACN GAA UUC AGG CCG GUU CUG CUA AGA AAA GGU
	(Met)Thr-Glu-Phe-*Arg*-*Pro*-*Val*-*Leu*-*Leu*-*Arg*-Lys-Gly-
	—C
-	AUG ACN GAA UUA AGG CCG GUU CUG CUA AGA AAA GGU
	(Met)Thr-Glu-*Leu*-*Arg*-*Pro*-*Val*-*Leu*-*Leu*-*Arg*-Lys-Gly-
	—U
8	AUG ACN GAA UCA AGG CCG GUU CUG CUA AGA AAA GGU
	(Met)Thr-Glu-*Ser*-*Arg*-*Pro*-*Val*-*Leu*-*Leu*-*Arg*-Lys-Gly-
	—A
9	AUG ACN GAU UCA AGG CCG GUU CUG CUA AGA AAA GGU
	(Met)Thr-*Asp*-*Ser*-*Arg*-*Pro*-*Val*-*Leu*-*Leu*-*Arg*-Lys-Gly-

Representation of the formation of the m-RNA sequence in the *cy*1-183 mutant and of the m-RNA sequences and protein sequences that would be expected by deletion of single bases at nucleotide positions 8 through 34 from the *cy*1-183 sequence. The amino acid residues that differ from the normal are shown in *italics*. The methionine residues shown in parentheses are excised and are not found in mature proteins. The types of proteins that have been obtained are indicated by a number at the left of the sequence; the frequencies of each type and their corresponding mutagens are presented in Table 8 (Stewart and Sherman, unpublished).

revertants established that this nucleotide position is normally the G of the AAG triplet encoding lysine 4. In addition, the insertion of G at position 34 is excluded by the results of the two duplications, types 14 and 15 (Fig. 8), which

FIGURE 7

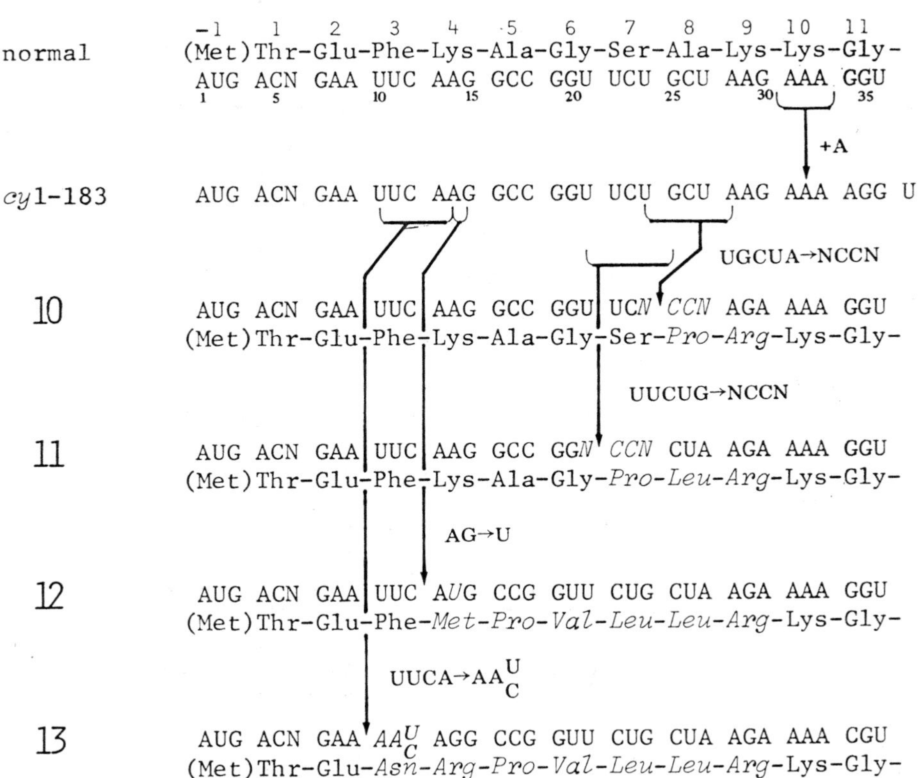

The types of altered iso-1-cytochromes *c* from *cy*1-183 revertants that have occurred by deletions of single bases and by concomitant substitution of bases adjacent to the deletions. The mutational changes of the reversion is indicated above each altered sequence. Other notations are presented in the legend of Figure 6. (Stewart and Sherman, unpublished).

identifies nucleotide position 34 as an A. Thus it is concluded that the *cy*1-183 mutation was due to an insertion of an A·T base pair within or adjacent to the stretch of three A·T base pairs at positions 31 to 33.

Type 9 (Fig. 6), containing the longest stretch of amino acid replacements, unambiguously identifies the normal nucleotides at positions 10 through 27. The type 15 duplication (Fig. 8) unambiguously indentifies the normal nucleotides at positions 24 through 33. The type 14 sequence is consistent with a tandem duplication in the mutant of either position 32-45 as shown, or possibly of position 33-46, and identifies the normal nucleotides at positions 31-44. If one includes the results obtained with the initiator mutants, which identified AUG at nucleotide positions 1-3, and the results obtained with the *cy*1-9 mutant, which identified GAA at nucleotide positions 7-9, then we have established the entire sequence from nucleotide position 1-44, except for the base at position 6.

FIGURE 8

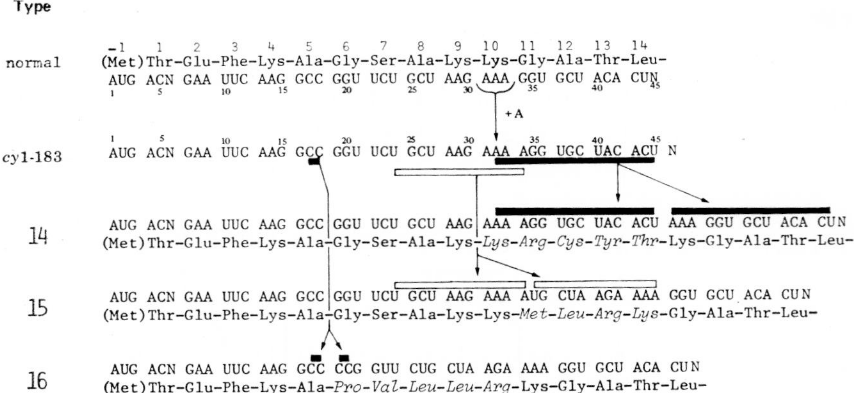

The types of altered iso-1-cytochromes *c* from *cy*1-183 revertants that have occurred by tandem duplications. Other notations are presented in the legend of Figure 6 (Stewart and Sherman, unpublished).

FIGURE 9

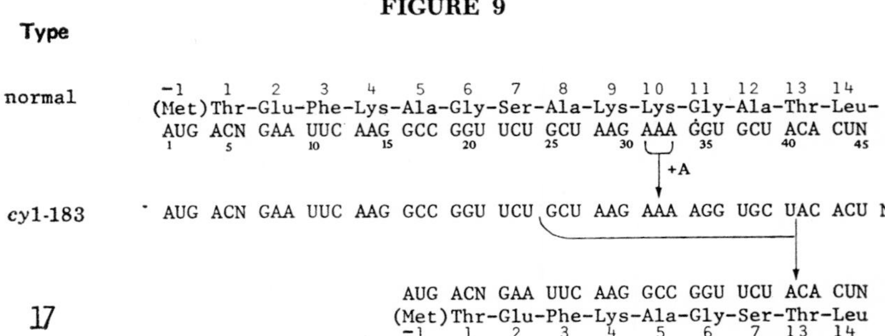

The altered iso-1-cytochrome *c* from the *cy*1-183 revertant that occurred by a deletion. Other notations are presented in the legend of Figure 6 (Stewart and Sherman, unpublished).

The finding of type 4 and type 5 iso-1-cytochromes *c*, respectively, from two *cy*1-134 revertants, indicated that this other class 6 mutant (Table 3) was formed by an insertion of an A at any of the nucleotide positions 31 through 34, or by an insertion of a G at nucleotide position 34. It should be remembered that the exclusion of an insertion of G at nucleotide position 34 in *cy*1-183 depended on the altered amino acid sequence produced by tandem duplications.

(e) The *cy*1-31 mutant

While the lesion in *cy*1-31 could not be unambiguously defined from revertant iso-1-cytochromes *c*, it appears as if the nucleotide alteration is associated with the residue at amino acid position 3. Intragenic revertants containing normal or near normal amounts of iso-1-cytochrome *c* were induced at a reasonable frequency with UV- and X-irradiation and at a low frequency with nitrous acid (NA) treatments. In contrast, we were unable to obtain intragenic revertants

TABLE 8

Frequencies of the various types of altered iso-1-cytochromes *c* from *cy*1-183 revertants and the corresponding changes in mRNA
(Stewart and Sherman, in preparation)

Type[a]	m-RNA changes in revertants	Mutagen[b]							
		UV	XR	α	HN2	MMS	NA	None	Total
1	—A *or* G	0	1	1	1	5	2	0	10
2	—C	1	0	1	0	0	0	0	2
3	—U *or* G	3	0	0	0	0	0	0	3
4	—U *or* C	3	1	1	1	0	0	1	7
5	—C *or* G	0	3	0	3	0	1	0	7
6	—G	0	2	1	3	0	3	1	10
7	—A	0	0	1	0	0	0	0	1
8	—U	1	0	1	0	0	0	0	2
9	—A	0	2	0	0	0	0	0	2
10	UGCUA→NCCN	0	0	0	0	0	1	0	1
11	UUCUG→NCCN	0	1	0	0	0	0	0	1
12	AG→U	0	0	0	1	0	0	1	2
13	UUCA→AA$^{U}_{C}$ . .	0	1	0	0	0	0	0	1
14	Duplication #32–45	0	1	0	0	0	0	0	1
15	Duplication #24–34	0	0	0	0	0	0	1	1
16	Duplication #17–18	0	0	0	1	0	0	0	1
17	Deletion #25–40 . .	1	0	0	0	0	0	0	1
	Total	9	12	6	10	5	7	4	53

[a]See Fig. 6 for types 1-9; Fig. 7 for types 10-13; Fig. 8 for types 14-16; Fig. 9 for type 17.
[b]The mutagen abbreviations are: UV, ultraviolet light; XR, x-rays; α, polonium-210 alpha particles: HN2, nitrogen mustard; MMS, methyl methane-sulphonate; NA, nitrous acid.

from *cy*1-31 either spontaneously or after treatments with a variety of alkylating agents, including ethyl methane-sulphonate (EMS), methyl methane-sulphonate (MMS), diethylsulphate (DES), N-methyl-N′-nitro-N-nitrosoguanidine (NG), 1-nitrosoimidazolidone-2 (NIL), 4-nitroquinoline-1-oxide (NQO), and DL-diepoxybutane (DEB).

Amino acid compositions and peptide maps of 24 revertant iso-1-cytochromes *c* and analysis by Edman degradation of 13 of these proteins established the nine different sequences shown in Fig. 10 (Stewart and Sherman, unpublished). Also tabulated in Fig. 10 are the frequencies of the various types that were induced by each of the three mutagens. The majority of the revertant proteins, 18 out

FIGURE 10

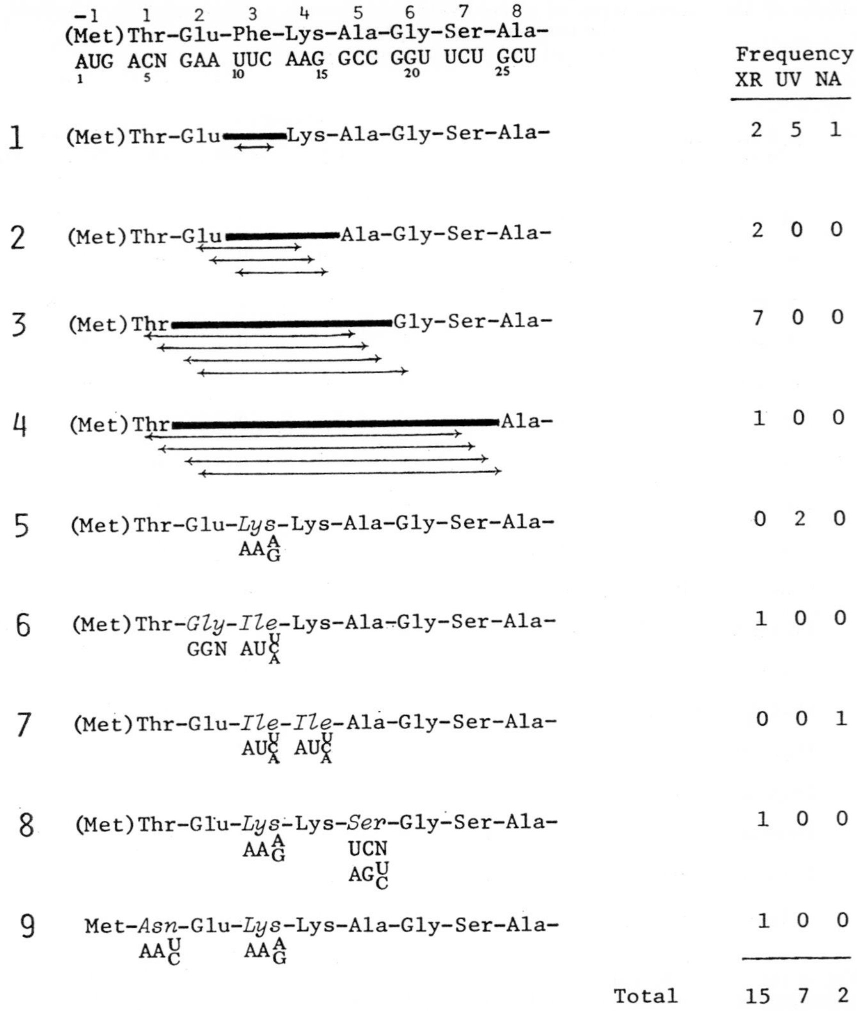

	Sequence	Codons	Frequency XR	UV	NA
	−1 1 2 3 4 5 6 7 8 (Met)Thr-Glu-Phe-Lys-Ala-Gly-Ser-Ala-	AUG ACN GAA UUC AAG GCC GGU UCU GCU 1 5 10 15 20 25			
1	(Met)Thr-Glu———Lys-Ala-Gly-Ser-Ala-		2	5	1
2	(Met)Thr-Glu————Ala-Gly-Ser-Ala-		2	0	0
3	(Met)Thr——————Gly-Ser-Ala-		7	0	0
4	(Met)Thr————————Ala-		1	0	0
5	(Met)Thr-Glu-*Lys*-Lys-Ala-Gly-Ser-Ala-	AA^{A}_{G}	0	2	0
6	(Met)Thr-*Gly*-*Ile*-Lys-Ala-Gly-Ser-Ala-	GGN AU(U/C/A)	1	0	0
7	(Met)Thr-Glu-*Ile*-*Ile*-Ala-Gly-Ser-Ala-	AU(U/C/A) AU(U/C/A)	0	0	1
8	(Met)Thr-Glu-*Lys*-Lys-*Ser*-Gly-Ser-Ala-	AA^{A}_{G} UCN / AG^{U}_{C}	1	0	0
9	Met-*Asn*-Glu-*Lys*-Lys-Ala-Gly-Ser-Ala-	AA^{U}_{C} AA^{A}_{G}	1	0	0
		Total	15	7	2

The altered iso-1-cytochromes *c* from revertants of *cy*1-31. The amino acid and nucleotide sequences of the amino-terminal region of normal iso-1-cytochrome *c* are shown on top. The deleted amino acids (types 1 through 4) can occur by deletion of any one of the corresponding nucleotide segments that lie directly above the arrows. The amino acid replacements are shown in *italics* along with their codons. The methionine residues in parentheses are excised from the normal protein and from the altered types 1 through 8; amino terminal methionine is not removed from the altered type 9. X-irradiation (XR), ultraviolet light (UV), and nitrous acid (NA) were used to induce the various revertants as indicated at the right of the figure (Stewart and Sherman, unpublished).

of 24, contained deletions which included phenylalanine 3 either alone (type 1) or in combination with adjacent residues (types 2-4). In addition, some revertants had single (type 5) and multiple (types 6-8) amino acid replacements that in all cases involved at least position 3.

So far we have been unable to simply explain all of the altered iso-1-cytochromes *c* from the *cy*1-31 revertants. However, it appears clear that *cy*1-31 is altered in or near the region controlling phenylalanine 3, since all revertants had alterations that included this residue and since the genetic mapping results indicated that the defect lies near *cy*1-9. Our simplest explanation is that *cy*1-31 contains a frameshift mutation which generated a nonsense codon and that many of the reverse mutations occurred by deletions which corrected the reading frame and excluded the nonsense codon. The deletion of nucleotides 11 and 12 from the phenylalanine 3 codon would give rise to such sequence. There would always be a UAG or UAA codon if a subsequent correction occurred by a single base deletion at any point to the right of position 11. This explains the amino acid deletions, types 1-4 (Fig. 10) and the lack of reversion of *cy*1-31 by the mutagens that do not appear to cause single-base deletions (EMS, NG and NIL; see [d] Frameshift mutants). However, it is not clear why none of the revertants were formed by single-base deletions at any of the nucleotide positions 4 through 9. Perhaps the nucleotides at positions 4 through 9 are resistant to mutation, for base deletions in this region occurred only twice in the 53 *cy*1-183 revertants (type 9 in Table 8). The lack of *cy*1-31 reversion with MMS and HN2 likewise may be explained by their inability to act on the nucleotides in this restricted region, and by their general inability to cause extended deletions. However, we have no explanation for the single (type 5) and double (types 6-9) amino acid replacements, although finding only lysine or isoleucine at residue position 3 does produce a conspicuous pattern.

(f) The *cy*1-49 mutant

The sole mutant of class 7 (Table 3), *cy*1-49, is currently under investigation, and at the time of this writing we are unable to report the nature of its defect. The mutant *cy*1-49 reverts only with extremely low frequencies after UV- and X-irradiation, and genetic mapping experiments indicate that it may be associated with lysine 10 or an adjacent residue. Iso-1-cytochromes *c* from several intragenic revertants have been prepared and are being analyzed. Although structures have not been determined, preliminary examinations suggested that several residues near lysine 10 are altered.

SUMMARY OF THE NUCLEOTIDE ALTERATIONS IN THE *cy*1 MUTANTS

The nucleotide alterations that were deduced from the abnormal iso-1-cytochromes *c* are summarized in Table 9 and Fig. 11. These results establish at least nine distinct nucleotide changes in the 21 *cy*1 mutants. Some act by preventing translation, either through the formation of the chain terminating codons UAA and UAG, or through modifications of the initiator codon AUG. The complete deficiency in still other *cy*1 mutants are attributed to frameshift mutations which cause extended protein changes beyond the site of the mutations. While the nature of the lesion in the *cy*1-31 mutant has not been definitely estab-

TABLE 9

The alterations in the 21 mutants associated with the first 33 nucleotides of the *CY1* gene

Mutant number	Amino acid position	Nucleotide position	Mutational change in m-RNA	Type of mutant
*cy*1-131	—1	1	AUG→*G*UG	initiator
*cy*1-51 *cy*1-100 *cy*1-181	—1	1	AUG→ *U*/*C* UG	initiator
*cy*1-133	—1	2	AUG→A*G*G	initiator
*cy*1-13 *cy*1-74 *cy*1-85 *cy*1-163	—1	3	AUG→AU*U*/*C*/A	initiator
*cy*1-9 *cy*1-91 *cy*1-147 *cy*1-160 *cy*1-172	2	7	GAA→*U*AA	ochre
*cy*1-31	3	unknown		
*cy*1-135 *cy*1-138 *cy*1-179	9	28	AAG→*U*AG	amber
*cy*1-134 *cy*1-183	10	within 30-34	insertion of A	frameshift
*cy*1-49	10?	unknown		

lished, it is undoubtedly also a frameshift mutant, and may lack the two bases at positions 11 and 12. Indications are that the *cy*1-49 mutant differs from all of the other 20 mutants; work in progress should result in the elucidation of its lesion in the near future.

FIGURE 11

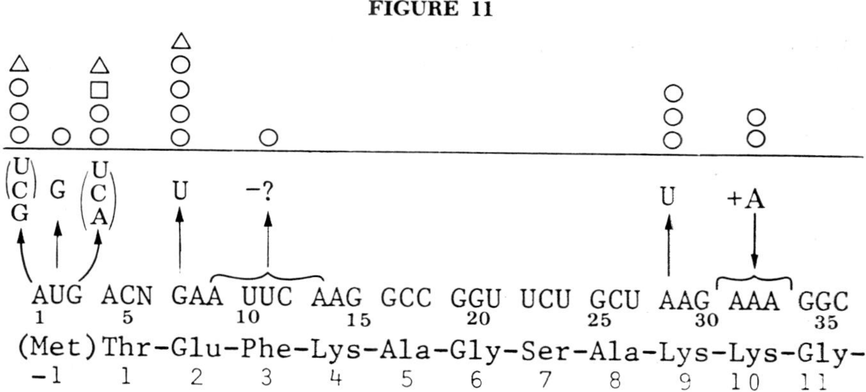

The frequency of various types of *cy*1 mutants that affect the 10 amino-terminal residues. The mutational changes of the m-RNA are indicated by arrows. The symbols represent independent mutants that were obtained either spontaneously (Δ) or after treatments with ultraviolet light (○), or 1-nitrosoimidazolidone-2 (□). See Tables 3 and 9 for a complete description of these 20 *cy*1 mutants.

MUTAGENESIS

Identification of the lesions in the *cy*1 mutants and the development of selection techniques provides a precise system for investigating mutagenesis in both forward and reverse direction. Although there are too few *cy*1 mutants for an extensive consideration of the preferential occurrence of forward mutations, there appears to be indications of "hot spots". While appropriate single-base substitutions at any of the nucleotide positions 7, 13, 28 and 31 could give rise to nonsense codons, only mutations of two of these sites were observed (Table 9 and Fig. 11). The highest number of repeats were the five instances of G·C → T·A transversions at position 7, which constitutes approximately 2.5% of the total number of mutational events. Since frameshift mutants at any point along the chain would be detected, it is significant that the two mutants *cy*1-183 and *cy*1-134 contained insertions at the same site. This "hot spot" may be a reflection of the stretch of three A·T base-pairs.

The amino acid replacements resulting from various single base-pair substitutions of ochre (UAA) and amber (UAG) mutants are tabulated, respectively, in Tables 4 and 6. The amino acid replacements not only specify the type of nucleotide change but also the position of the base. However, it should be noted that no amino acid replacement can occur from G·C → A·T transitions of either triplet, and that tyrosine replacements can occur by both A·T → T·A and A·T → C·G transversions.

The results presented here (Tables 4 and 6) and elsewhere for other sites (Stewart and Sherman, 1968; Sherman *et al.*, 1969), strongly suggest that the nucleotide environment influences spontaneous and induced mutations. This marked dependence on the base position can be seen with the UAG mutant (Table 6), where X-rays and α-particles, nitrous acid, diethyl sulphate and spontaneous events produce more mutations of the second A·T base-pair than of the first A·T base-pair. Also amino acid replacements in *cy*1-9 revertants (Table 4) indicate a strong specificity for A·T → G·C mutations with UV, although this is not the case for the UAG mutant (Table 6) nor for another ochre mutant (Sherman *et al.*, 1969). In addition, the UV specificity is changed in UV-sensitive strains (Lawrence *et al.*, 1970). The base changes associated with spontaneous and induced reversion of amber and ochre mutants so far either appears random, or exhibits a specificity that is not seen at other sites within the same codon or in a similar codon at a different location.

As mentioned above in (c) *Chain initiation mutants,* the *cy*1-131 mutant can revert by a G·C → A·T transition, a change that is not normally detected in reversion of amber and ochre mutants. While all nine of the initiator mutants, including *cy*1-131, are about equally revertible with UV, *cy*1-131 is over 100 times more revertible with EMS (Stewart *et al.*, 1971). However, it is still unknown whether or not the specific action of EMS for G·C → A·T changes occurs at all sites or if it is restricted to certain sites. In any case, this *cy*1-131 mutant, along with the other initiator mutants, the *cy*1-9 ochre mutant and the *cy*1-179 amber mutant are being used to test the mutagenic specificity of numerous agents (Prakash and Sherman, 1972).

The identification of bonafide frameshift mutants provides the opportunity to test mutagens for their ability to add or delete bases in yeast. The reversion studies with *cy*1-183, described above in detail, established that the mutagens UV, X-rays, α-particles, nitrogen mustard (HN2), nitrous acid (NA) and methyl methane-sulphonate (MMS) (Table 8) are capable of deleting bases, while this was not demonstrated for mutagens ethyl methane-sulphonate (EMS), N-methyl-N′-nitro-N-nitrosoguanidine (NG), 1-nitrosoimidazolidone-2 (NIL), and, interesting enough, ICR-170. Although the types of deleted base-pair can not always be identified from the altered sequence, none of the chemical agents (HN2, MMS and NA) have been shown unambiguously to delete an A·T base-pair (types 7-9 in Table 8). Only MMS exhibited a marked site-specific action.

The high frequency of X-ray-induced deletion in *cy*1-131 revertants is worthy of emphasis (Fig. 10). With one exception, only ionizing radiations were capable of producing deletions which clearly encompassed more than two base-pairs (Table 5 and Fig. 10). The one UV-induced deletion (Fig. 9 and Table 8) may be a special case having redundant ends.

GENETIC CODE

So far all indications are that the genetic codes for both yeast and *E. coli* are identical. The altered iso-1-cytochromes *c* described in this paper have led not only to the demonstration of the UAA and UAG chain terminating codons and the AUG chain initiator codon but also have verified the identity of the codon assignments for 28 out of 61 amino acids. The amino acid codons could be inferred from frameshift mutations (Figs. 6 and 8), and single-base mutations of amber, ochre, (Fig. 3) and some initiation mutants (Fig. 5). We are preparing a compilation of the codons that could be derived and are consistent with the alterations reported here and with other altered forms of iso-1-cytochrome *c*.

ALTERED FORMS OF ISO-1-CYTOCHROME *c*

Since the strains were selected for their ability to utilize lactate, all of the revertant iso-1-cytochromes *c* should be at least partially functional, although maybe not to the same degree as the normal protein. A study with suppressed nonsense mutants (Gilmore, Stewart and Sherman, 1971) indicated that growth is only slightly lower in strains having approximately 10% of the normal level of iso-1-cytochrome *c*. Since the revertant strains exhibited normal or near normal growth on lactate medium, and never contained above-normal amounts of cytochrome *c*, it is concluded that the specific activities of the altered forms of iso-1-cytochrome *c* are at least 10% of normal.

Forty-eight different sequences of iso-1-cytochrome *c* occurred in the 241 revertants from the 20 *cy*1 mutants. These 48 sequences, along with two from *cy*1-6

FIGURE 12

−5	−4	−3	−2	−1	1	2	3	4	5	6	7	8	9	10	11	12
												Glu				
					Glx							Ile	Thr	Thr		
					Asx		Tyr	Glu	Lys			Val	Ala	Ser		
			Thr	Pro	Ala	Pro	Val	Ser	Gln		Asn	Ser	Glu	Asn		Glu
		Ala	Ser	Phe	Ser	Gly	Ala	Pro	Pro		Asp	Pro	Asp	Ala		Lys
		|→		|→	|→	|→				|→						
					Thr-	Glu-	Phe-	Lys-	Ala-	Gly-	Ser-	Ala-	Lys-	Lys-	Gly-	Ala-
|→	|→		|→	|→	|→	|→	|→		|→	|→	|→					
Thr	Glu	Phe	Lys	Ala	Gly	Ser	Ala	Ile	Lys	Arg	Cys	Tyr	Thr	Ala		Ser
	Thr	Glu	Phe	Lys	Ala	Gly	Ser	Ala	Ser	Lys	Met	Leu	Arg	Glu		Thr
			Met	Thr	Glu	Phe	Lys	Arg	Pro	Val	Leu	Gly	Leu			
				Met	Asn	Thr	Glu	Met	Thr	Thr	Glu	Phe	His			
				Ile	Ile	Leu	Thr	Glu	Phe	Ala	Thr	Val	Phe			
				Leu		Gln	Ile				Ala	Pro	Ser			
				Arg		Lys	Leu				Pro		Gln			
				Val		Tyr	Asn						Glu			
						Asp							Trp			
													Tyr			

A composite of amino acid sequences of the amino terminal region of 43 different cytochromes *c* from various eukaryotic species (for original references see Dayhoff, 1969; Thompson, *et al.*, 1971; Sugeno, *et al.*, 1971; Nakayama, *et al.*, 1971; Gürtler and Horstmann, 1970, 1971, Ramshaw, *et al.*, 1971; Augusteyn, *et al.*, 1972; Sokolovsky and Moldovan, 1972), and of 50 different altered iso-1-cytochromes *c* from the intragenic revertants. The normal iso-1-cytochrome *c* is presented as a continuous sequence. Other residues found at each position in other species are listed above the normal iso-1-cytochrome *c*, and other residues in the mutants are listed below it. The positions of amino termini in the phylogenetic series are indicated by |⟶ above the normal sequence and in the mutants by |⟶ below it. All of the altered iso-1-cytochromes *c* are described in this paper (from Stewart *et al.* 1971; Stewart *et al.* 1972; Stewart and Sherman, 1972; Sherman and Stewart, unpublished; Stewart and Sherman, unpublished) except for the altered proteins from *cy*1-6 revertants which contain replacements at position 12 (Putterman *et al.* unpublished, cited in Sherman *et al.* 1970).

revertants (Putterman *et al.*, unpublished, cited in Sherman *et al.*, 1970), are presented as a composite in Fig. 12. Also shown in Fig. 12 is a composite of all the known cytochromes *c* from various eukaryotic species, including vertebrates, flowering plants, insects and *Ascomycetes* fungi. If one considers only the amino-terminal region extending to position 12 of iso-1-cytochrome *c* as shown, then the phylogenetic series constitutes 36 different sequences. The positions of the amino termini of both the mutants and the species are also indicated in the figure. Flowering plants contain the longest of the naturally occurring cytochromes *c*, beginning with an acetylated alanine residue at position -3. Vertebrate cytochromes *c*, the shortest in the phylogenetic series, begin with an acetylated glycine at position 6. Amino termini of cytochromes *c* of intermediate length are found at position 2 for insects and at positions -1, 1, and 2, respectively, for various fungi.

While earlier composites, published by Sherman *et al.* (1970) and Sherman and Stewart (1971) are less detailed, they include information of the region extended to position 31. For the present discussion, we will be concerned only with the amino terminal sequences extending to position 12, the region that includes all the alterations reported in this paper.

An examination of Fig. 12 clearly points out the unessential nature of the region preceding serine 7, since it can be varied widely and deleted altogether. Furthermore, the acceptable variability of position 7 through 10 seems to indicate that the side chains of these next 4 residues play no specific structured or functional role. A superficial inspection of the deletions, shown in Fig. 13, appears to indi-

FIGURE 13

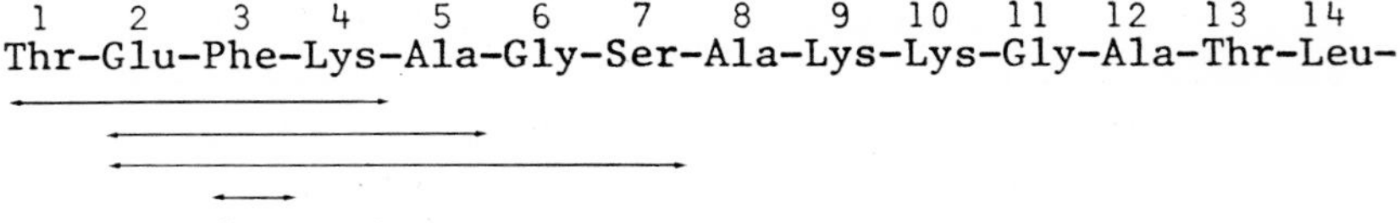

The amino acid deletions that occur in iso-1-cytochrome *c* from revertants of *cy*1-13 and etc. (Table 7), *cy*1-31 (Fig. 10), *cy*1-179 (Table 5) and *cy*1-183 (Fig. 9).

cate that no single amino acid among the 12 amino-terminal residues is necessary for function. However, a close examination of these results does indeed emphasize the essential nature of glycine 11. Most convincing is the preservation of glycine 11 in all of the revertants from *cy*1-183 (Figs. 6-9).

This is spectacularly illustrated with the deletion mutant that encompassed glycine —1 (Fig. 9) and caused glycine 6 to be aligned at position 11. A glycine residue at position 11 is evidently necessary for function. Thus of the two glycine residues that are invariant in the amino terminal region of all eukaryotic cytochromes *c* (Fig. 1), glycine 11 was not mutationally altered in any of the functional iso-1-cytochromes *c*, while glycine 6 was replaced by a variety of residues (Fig. 12), and was deleted in one instance (Fig. 10). The preservation of a glycine residue at position 11 in iso-1-cytochrome *c* (or in the homologous position 6 in vertebrate cytochromes *c*), is understandable from the three-dimensional structure of horse-heart cytochrome *c* (Dickerson *et al.*, 1971), since this position lies in a crowded region inside the molecule and cannot accommodate any side chains. In contrast, glycine 1 in vertebrate cytochromes *c*, which corresponds to glycine 6 in iso-1-cytochrome *c*, is located at the surface of the molecule, and there is no known reason for its evolutionary invariance.

METHIONINE AMINOPEPTIDASE

An exopeptidase, which removes amino terminal residues of methionine when they preceded certain amino acids, was postulated to exist in yeast in order to account for the amino termini of iso-1-cytochromes *c* from initiator revertants (Stewart *et al.*, 1971, see Figs. 4 and 5). The study with the initiator mutants indicated that this enzyme excised methionine completely from residues of threonine and alanine, less efficiently from residues of valine, and not at all from residues of leucine, isoleucine or arginine. A verification and extension of its specificity was provided by two additional revertant proteins having altered amino termini, one with Met-Ile- and the other with Met-Asn-. All of the iso-1-cytochromes *c* with altered amino termini are tabulated in Table 10.

TABLE 10

Iso-1-cyochromes c altered at the amino terminus

(from Stewart *et al.* 1971; Stewart *et al.* unpublished; Sherman & Stewart, unpublished)

Thr-Glu-Phe-	*CY1* (normal)
Ala-Gly-Ser-	*CY1-13-Q* etc.
Val-Thr-Glu- *Met-Val*-Thr-Glu-	*CY1-131-C*[a]
Met-Leu-Thr-Glu-	*CY1-51-C* etc.
Met-Ile-Thr-Glu-	*CY1-13-A* etc.
Met-Ile-Glu-Phe-	*CY1-9-AU*
Met-Asn-Glu-Lys-	*CY1-31-N*
Met-Arg-Thr-Glu-	*CY1-133-A*

The appended residues and replacements are shown in *italics*.

[a]The *CY1-131-C* strain contained a mixture of approximately 85% and 15%, respectively, of the Val- and Met-Val- proteins.

GENETIC CONTROL OF THE AMOUNT OF ISO-1-CYTOCHROME *c*

In addition to mutations at other loci, *cy2* through *cy6*, which cause various levels of deficiency of iso-1 and iso-2-cytochromes *c* (Sherman, 1964; Sherman *et al.*, 1965), mutations within the structural gene can cause diminution of iso-1-cytochrome *c*. Diminished levels of nonfunctional iso-1-cytochrome *c* are increased after incubation of some *cy1* mutants at lower temperature; in these cases it appears as if the diminution could be due to instability of the proteins (Sherman *et al.*, 1968; 1970). In all mutants that were examined, these non-functional proteins proved to be extremely labile during extraction and could not be recovered in pure form.

However, there are certain intragenic revertants that contained slightly decreased levels of iso-1-cytochrome *c* which may not be attributable to instability. The most documented study concerns the initiator revertants which lack the four amino terminal residues (Stewart *et al.*, 1971). One-half of the normal amount of iso-1-cytochrome *c* occurs in all of the independent mutants that contain this short protein. Extensive genetic analysis established that the 50% reduction in amount was due to the same mutation which caused this short form. Since the reverse mutation was associated with a shift of the initiator codon to the right by 12 nucleotides, into a different nucleotide sequence, it was suggested that the diminution of iso-1-cytochrome *c* was due to a decreased efficiency of initiation of translation; the mechanism for such an action is not understood. Normal amounts of iso-1-cytochrome *c* do occur in revertants that contain deletions to the right of the original initiator codon; the most significant example is the type 3 *cy*1-31 revertant (Fig. 10) which lacks the four normal residues that succeed the normal initiator codon.

In addition, low-temperature spectroscopic examinations of intact cells uncovered numerous other intragenic revertants with slightly lower levels of cytochrome *c*. In these instances the heights of the c_α-band were intermediate to those found in normal strains and in the initiator revertants containing the short form of iso-1-cytochrome *c*. Thus it appears that these strains contain approximately 75% of the normal amount of iso-1-cytochrome *c*. So far we have not completed our genetic studies that would test whether these slight deficiencies are due to the reverse mutations within the *cy*1 gene or by chance mutations of other genes. Nevertheless in some instances there is a striking pattern that leaves little doubt of the relationship between the reversion and the slight deficiencies. For example, all 10 of the *cy*1-183 revertants that contained normal iso-1-cytochrome *c* (type 1 in Table 8) also had normal amounts, while all seven type 4 and all seven type 5 revertants contained slight deficiencies. While one would be tempted to relate the deficiencies simply to the -Leu-Leu-Arg- sequence at amino acid positions 7-9, other *cy*1-183 revertants with longer stretches of replacements did not always have such deficiencies; both type 9 revertants (Fig. 6) clearly contained normal amounts of iso-1-cytochrome *c*. Comparable examples are the slight deficiencies in only *CY*1-179-AF and *CY*1-179-AW (Table 5) out of all of the 52 revertants from *cy*1-179 (Table 6). Both of these revertant iso-1-cytochromes *c* contain alterations or losses of two residues. So far we have no explanation for these slight deficiencies.

SYNOPSIS

Twenty-one out of 211 *cy*1 mutants were shown to have alterations within the first 33 out of a total of 330 base-pairs of the gene that determines the structure of iso-1-cytochrome *c*. Nine different mutational lesions were identified by relating the genetic code to the structural variations in iso-1-cytochromes *c* from 241 intragenic revertants of 20 of these *cy*1 mutants. The mutants were either frameshifts, amber (UAG), ochre (UAA) or altered in the AUG initiator codon. From the amino acid changes in the revertant proteins it was possible to deduce the nucleotide sequence for all but one of the first 44 base-pairs of the gene.

ACKNOWLEDGEMENTS

For much of the work cited as unpublished, we wish to acknowledge Mrs. E. Risen for the amino acid analyses; Mrs. N. Brockman for the peptide maps and amino acid sequences; Miss S. Dennis for preparation of the iso-1-cytochromes *c*; and Mrs. M. Jackson for genetic results.

This research was supported in part by the U.S. Public Health Services research grant GM 12702 from the National Institute of Health, and in part by the U.S. Atomic Energy Commission at the University of Rochester Atomic Energy Project, Rochester, New York, and has been assigned USAEC Report No. UR-3490-114.

REFERENCES

Augusteyn, R. C., McDowall, M. A., Webb, E. C. and Zerner, B. (1972) *Biochim. Biophys. Acta, 257,* 264

Dayhoff, M. O. (1969) *Atlas of Protein Sequence and Structure,* vol. 4 Silver Spring, Md.: National Biomedical Research Foundation

DeLange, R. J., Glazer, A. N. and Smith, E. L. (1970) *J. biol. Chem. 245,* 325

Dickerson, R. E., Takano, T., Eisenberg, D., Kallai, O. B., Samson, L., Cooper, A. and Margoliash, E. (1971) *J. biol. Chem. 246,* 1511

Gilmore, R. A., Stewart, J. W. and Sherman, F. (1971) *J. Mol. Biol. 61,* 157

Gürtler, L. and Horstmann, H. J. (1970) *Eur. J. Biochem, 12,* 48

Gürtler, L. and Horstmann, H. J. (1971) *FEBS Lett. 18,* 106

Lawrence, C. W., Stewart, J. W., Sherman, F. and Thomas, F. L. X. (1970) *Genetics, 64,* s36

Nakayama, T., Titani, K. and Narita, K. (1971) *J. Biochem., (Tokyo) 70,* 311

Narita, K. and Titani, K. (1969) *J. Biochem., (Tokyo), 65,* 259

Parker, J. H. and Sherman, F. (1969) *Genetics, 62,* 9

Prakash, L. and Sherman, F. (1972) *Genetics, 71,* s 48

Ramshaw, J. A. M., Richardson, M. and Boulter, D. (1971) *Eur. J. Biochem. 23,* 475

Sherman, F. (1964) *Genetics, 49,* 39

Sherman, F. and Stewart, J. W. (1971) *Annu. Rev. Genetics, 5,* 257

Sherman, F., Taber, H. and Campbell, W. (1965) *J. Mol. Biol. 13,* 21

Sherman, F., Stewart, J. W., Margoliash, E., Parker, J. and Campbell, W. (1966) *Proc. Nat. Acad. Sci. U.S. 55,* 1498

Sherman, F., Stewart, J. W., Parker, J. H., Inhaber, E., Shipman, N. A., Putterman, G. J., Gardisky, R. L. and Margoliash, E. (1968) *J. biol. Chem. 243,* 5446

Sherman, F., Stewart, J. W., Cravens, M., Thomas, F. L. X. and Shipman, N. (1969) *Genetics, 61,* s55

Sherman, F., Stewart, J. W., Parker, J. H., Putterman, G. J., Agrawal, B. B. L. and Margoliash, E. (1970) *Symp. Soc. Exp. Biol. 24,* 85

Sherman, F., Liebman, S. W., Stewart, J. W. and Jackson, M. (1972) *Genetics, 71,* s 58

Sokolovsky, M. and Moldovan, M. (1972) *Biochemistry, 11,* 145

Stewart, J. W. and Sherman, F. (1968) *Proc. XII Int. Congr. Genet., Tokyo 1,* 45

Stewart, J. W. and Sherman, F. (1972) *J. Mol. Biol. 68,* 429

Stewart, J. W., Sherman, F., Shipman, N. A. and Jackson, M. (1971) *J. biol. Chem. 246,* 7429

Stewart, J. W., Sherman, F., Jackson, M., Thomas, F. L. X. and Shipman, N. A. (1972) *J. Mol. Biol. 68,* 83

Sugeno, K., Narita, K. and Titani, K. (1971) *J. Biochem., (Tokyo), 70,* 659

Thompson, E. W., Notton, B. A., Richardson, M. and Boulter, D. (1971) *Biochem. J. 124,* 787

Transcriptional and Translational Control Mechanisms

RNA Polymerases and Transcriptive Specificity in Eukaryotic Organisms

W. J. RUTTER, P. W. MORRIS, M. GOLDBERG, M. PAULE AND R. W. MORRIS

Department of Biochemistry, San Francisco Medical Center, University of California, San Francisco, California, U.S.A.

Selective gene transcription implies an intriguing modality in specificity of the transcriptive system which is uncommon in most enzymatically catalyzed reactions (Fig. 1). For specific gene selection, recognition of initiator DNA

FIGURE 1

Requirements for transcriptive specificity

Recognition of specific DNA sequence	:	Initiating mode
Indiscriminate reading of DNA sequence	:	Transcribing mode
Dissociation of template/RNA polymerase complex	:	Termination mode

Hypotheses:

1. Template restricts transcription	≡	Indiscriminate polymerase (transcribing mode only)
2. Multiple RNA polymerases with inherent specificity	≡	(Polymerases have initiating, transcribing, terminating modes)
3. Specific initiating and terminating factors	≡	Indiscriminate polymerase (transcribing mode only)

sequences by the transcribing unit is required. On the other hand, transcription itself involves an indiscriminate copying of the DNA template. Finally a mechanism for termination probably involves some sensing of DNA sequence. These initiating, transcribing, and terminating modes of the transcriptive system require specificity-determining elements either in the chromosomal template, the RNA polymerase, or in another component of the transcribing system. Three models to explain the specificity are immediately apparent. In the template restriction model, the non-transcribed portions of the DNA are essentially sequestered from an indiscriminate RNA polymerase. Alternatively, one can envision multiple RNA polymerases with inherent specificity for initiating and terminating sequences. In this instance, the polymerases could exist in initiating ,transcribing, and terminating modes. This would probably require at least two conformational states (the initiating and terminating modes could be equivalent). Finally, the specificity of initiation and termination could be provided by specific external factors which would aid an indiscriminate polymerase in selecting sites for transcription.

There is some experimental support for each of these possible mechanisms. The quality of the RNA produced by a prokaryotic polymerase and reconstituted chromatin is at least partially dependent upon the chromosomal proteins (Paul and Gilmour, 1968; Smith, *et al.*, 1969). These experimental results, however, do not necessarily imply that all the specificity resides in the chromatin since initiator sites might be similar in prokaryotic and eukaryotic organisms and hence the polymerases from both sources might react to some extent similarly in both systems.

The possibility that polymerase-associated factors may confer initiation and termination specificity on an indiscriminate polymerase is mechanistically appealing. The discovery of the structure of the core *E. coli* polymerase and associated σ factor (Burgess *et al.*, 1969) dramatically demonstrated the concept, and subsequent discovery of ψ (Travers *et al.*, 1970), M (Davidson *et al.*, 1969) and CAP (deCrombrugghe *et al.*, 1970; Zubay and Chambers, 1969) regulatory proteins as well as the termination factor ρ, further emphasized the principle of positive control of transcription. However, further attempts to demonstrate the general validity of this concept, or in fact, to elucidate the role of the factors in determining specificity have been less successful than one at first hoped; the possibility remains that these factors modulate rather than determine initiation specificity.

In addition to the specific factors which affect transcription, it seems clear that alterations in the structure of the polymerase also play a role in determining specificity. Shorenstein and Losick, 1970 and Leighton *et al.*, 1971 have shown that modification of the polymerase occurs during bacterial sporulation. Similar modulations in structure may occur in other systems. This implies that the polymerase is to some degree a determinant of specificity.

Finally, multiple RNA polymerases have been described in the nuclei of eukaryotic organisms. It is the nature of these enzymes, their properties, structure, specificity, the role they play in transcriptive regulation, which is the subject of this paper. The experimental results to date indicate that there are at least two, and probably three, but not more than half a dozen such polymerases; certainly not enough to prescribe the entire gamut of transcriptive regulation.

Thus these enzymes can only be partial determinants of specificity. All three modes of regulation may be utilized by eukaryotic organisms.

Three cognate nuclear polymerases in eukaryotes

Three years ago, Roeder and Rutter reported the resolution of three nuclear DNA-dependent RNA polymerases. Subsequently, these results have been confirmed in the tissues of a number of species spanning the eukaryotic domain. The general properties of the major classes of enzymes are presented in Table 1. Each

TABLE 1

Properties of Nuclear RNA Polymerases[a]

	I (a) (b)	II (a) (b)	III
DEAE Elution $\sim$M$(NH_4)_2SO_4$	.05—.15	$\sim$.25	$\sim$.35
Mn^{2+}/Mg^{2+} activity ratio	1	5-50	2-3
Ionic strength optimum	.05	.10—.15	0-0.2
Rifampicin inhibition %	0	0	0
α-Amanitin inhibition %	0	100	0

[a]Calf Thymus, Rat Liver, Sea Urchin, Yeast

is characterized by distinct DEAE elution profiles, divalent cation activity ratios, ionic strength, optimum and differential sensitivity to inhibition by α-amanitin, the toxin of the mushroom *Amanita phalloides* (Lindell *et al.*, 1970). This molecule specifically inhibits polymerase II. None of the forms of the enzymes are inhibited by rifamycin, an inhibitor of typical prokaryotic enzymes. Two polymerases have been found in extracts of all organisms tested. The third class of polymerase is not as readily detected but has been demonstrated in nuclear extracts of the rat and amphibian liver, sea urchin embryos and yeast. Heterogeneity in polymerase I was apparent in the earliest chromatographic studies (Roeder and Rutter, 1969, 1970a). Two forms of polymerase I have been reported by Chambon and his colleagues (Chambon *et al.*, 1970; Chesterton and Butterworth, 1971a) in the tissues of higher organisms and by Adman *et al.*, (1972) and Benson (1972) in the nuclei of yeast. Two forms of polymerase II have also been independently discovered by ourselves (Weaver *et al.*, 1971), Chesterton and Butterworth (1971) and Chambon and colleagues (Kedinger *et al.*, 1971). With the exception of slight differences in chromatographic behaviour, two forms of I, as well as the two forms of II are indistinguishable. In fact, the class of polymerase I enzymes, and similarly, the class of polymerase II enzymes, regardless of source, have remarkably similar properties. This must mean that the structure and regulatory aspects of the molecules is highly conserved.

Eukaryotic and prokaryotic RNA polymerases have similar subunit structures

Recent studies from this laboratory have elucidated the basic subunit structure of IIa and IIb, isolated from calf thymus and rat liver (Weaver *et al.*, 1971). These enzymes are composed of two large and two small subunits (Table 2) very reminiscent of the structure of the prokaryotic polymerases typified by the *E. coli* enzyme (Travers and Burgess, 1969). The presence of the two large subunits has also been reported by Chesterton and Butterworth (1971b) and Kedinger *et al.*, (1971). Although there is some disagreement about the absolute value of the mass of the large subunits, the values given are probably within 10%. Recently, Dezelee *et al.*, (1972) have reported a study of the subunits of the

TABLE 2

Apparent structural Homologies between Prokaryotic and Eukaryotic RNA Polymerases

Subunit	*E. coli* Core	II		I
		(a)	(b)	(a) (b)
β'	$155{,}000_1$	$(190{,}000)_1$		(180,000)
β''			$(170{,}000)_1$	
β	$(145{,}000)_1$	$(150{,}000)_1$	$(150{,}000)_1$	(105,000)
α	$(40{,}000)_2$	$(35{,}000)_1$	$(35{,}000)_1$	(65,000)
α'		$(25{,}000)_1$	$(25{,}000)_1$	
Molecular Weight	380,000	400,000	400,000	350-400,000
Structure	$\alpha_2\beta\beta'$	$\alpha\alpha'\beta\beta'$	$\alpha\alpha'\beta\beta''$	

RNA polymerase II isolated from yeast. In this instance, the two large polypeptide chains, molecular weight 180,000 and 150,000 respectively were accompanied by three other smaller molecular weight components of 39,000, 28,000 and 19,000 daltons. Though the stoichiometry of the large subunits is consistently observed and no doubt a fundamental aspect of the subunit structure of polymerase II, the number, stoichiometry, and role of the small subunits remains less certain because of the limited amount of enzyme available. Definitive dissociation and reconstruction experiments have not yet been performed because of the limitation of available enzyme.

Highly purified preparations of polymerase I have been obtained from calf thymus, rat liver and sea urchin. Partially completed studies on this molecule show that it has a molecular mass similar to form II as determined by gel filtration and sedimentation analysis (400,000 daltons). Sodium dodecyl sulphate (SDS) gel patterns of these preparations are dramatically different from those of form II. Bands of approximate mass of 180,000, 105,000 and 65,000 daltons are present: other components of smaller mass are also present. Because of stoichiometric uncertainties, we can not now unambiguously assign a subunit structure, but it is apparent that the subunits of this molecule are largely, if not entirely different from those of form II. This suggests but does not prove that the molecules are separate gene products.

Polymerase III has now been highly purified from sea urchin. This molecule is the same general size as polymerases II and I, but its subunit structure has not been determined, thus the question whether this is a unique molecule or a derivative of the other polymerases remains unanswered.

Mitochondria (and probably other plasmids) have a distinct transcriptional system

Available evidence suggests that mitochondrial RNA synthesis is effected by a transcriptional system unique from that of the nucleus. Küntzel and Schäfer (1971) first reported the presence in Neurospora mitochondria of a polymerase radically different from the nuclear polymerases. The molecule apparently has a single subunit of low mass (about 80,000 daltons), is not inhibited by rifampicin and exhibits specificity for the mitochondrial DNA. Similar experiments have now been reported for rat liver (Reid and Parsons, 1971). Wu and Dawid (1972) also find a single polymerase in *Xenopus* mitochondria. These experiments of a single polymerase activity are consistent with studies carried out by Attardi and coworkers (1970) on mitochondrial RNA synthesis in Hela cells. The transcription of Hela mitochondrial DNA is apparently symmetrical, and furthermore, a single transcript equivalent in length to the total mitochondrial DNA may be formed. In sharp contrast to the above reports, studies on the enzymes from yeast mitochondrion, however, by Criddle and colleagues (Tsai *et al.*, 1971) and by Benson (1972) have suggested that a polymerase cohort similar, but not indentical to that present in the nucleus is present in these mitochondria. Distinct similarities in the characteristics of the nuclear and "mitochondrial" enzymes were evident. These polymerases were of high molecular weight, similar to those found in the nucleus. One was inhibited by α-amanitin. Their general characteristics, however, were significantly different from the nuclear enzyme both with respect to divalent metal ion activation and ionic strength optimum, and on the basis of template specificity. Of course, such studies do not rigorously rule out contamination with nuclear enzymes but attempts to wash off the activity or to extract additional mitochondrial activity have been unsuccessful. It is conceivable that in the yeast mitochondrial system where the genome is considerably larger, transcriptive regulation (and multiple enzymes) do occur. Whatever the specific details, it appears that the mitochondrion is a separate transcriptional system, involving polymerases that are different, at least in part, from those in the nucleus.

It is presumed that a similar situation may be obtained for the transcription of the chloroplasts. A single species has been isolated by Bottomly *et al.*, (1971) and his coworkers from the chloroplasts of maize, of wheat leaf (Polya and Jagendorf, 1971) and by Hallick, Richards and Rutter (unpublished) from the chloroplasts of Euglena.

NUCLEAR LOCALIZATION AND TRANSCRIPTIVE FUNCTIONS OF RNA POLYMERASES I, II AND III

The selective inhibition of polymerase II by α-amanitin allows the facile distinction between polymerases II and I + III. These observations may be supplemented by analysis of the polymerase activities by DEAE Sephadex chromatography. Our previous studies (Roeder and Rutter, 1970a) have shown that polymerase I is localized primarily, if not exclusively within the nucleolus, whereas polymerase II and probably polymerase III is localized within the nucleoplasm (Table 3). The majority (greater than 90%) of RNA transcripts are α-amanitin-sensitive and hence are produced by polymerase II. Less than 10% of the total nuclear transcripts are products of polymerase I activity. As judged by DNA-RNA hybridization studies (Ingles, J., in Blatti *et al.*, 1970), ribosomal RNA is the major product in the presence of α-amanitin. Because of the localization of polymerase I in the nucleolus, it is assumed that polymerase I is

TABLE 3

Transcriptive specificity of Nuclear RNA Polymerases

	I III	II
Nuclear Localization	Nucleolus (Nucleoplasm)	Nucleoplasm
Transcriptive Range	<10%	>90%
Transcriptive Products	Ribosomal Precursor	HRNA
	5S-RNA	m-RNA
	t-RNA	

involved in ribosomal RNA synthesis. Wobus, *et al.*, (1971), Beerman (1971), and Egyhazi *et al.*, 1972, have shown that the synthesis of 4S, 5S, 8S and perhaps a few other RNA species is insensitive to α-amanitin in the polytenic chromosomes of Chironomus. Polymerase I or polymerase III may play this transcriptive role. Since polymerase III is probably localized within the nucleoplasm, its role may be the formation of tRNA and perhaps other small RNA species (see, however, a later section for another possible role of this enzyme). The α-amanitin-insensitive activities, polymerase I and III seem to be involved in constructing the translational system. In these instances, of course, the ultimate gene product is RNA, rather than protein.

POLYMERASES EXHIBIT INITIATION SPECIFICITY ON DNA TEMPLATES

Early studies gave no indication that the inherent transcriptive specificity of the enzymes corresponded to their putative transcriptive function (Table 4). The RNAs produced by polymerases I, II and III on nuclear DNA templates are indistinguishable. More extensive studies have been carried out with polymerases I and II from calf thymus and rat liver. In these instances, the rates of reaction on calf thymus DNA templates are about equivalent and the products similar by GC content and nearest neighbor analysis (Roeder and Rutter, 1969). The rates of transcription of a number of synthetic templates are markedly different. The transcriptive pattern of polymerases I and II are qualitatively similar but quantitatively different. Pyrimidines are read more rapidly than purines (I: C>U>G, A; II: C>G>U>A) (Blatti *et al.*, 1970). There are dramatic differences in the readout of homoploymers as compared with mixed polymers. Thus both the dA and dT strands of the dA:dT homopolymer duplex were read substantially slower than the calf thymus DNA standard employed, but the dAT:dAT copolymer was transcribed several times faster than the calf thymus template. These results indicate that certain isostichs may be read very slowly compared with the DNA and could represent regulation (e.g. termination) signals. Studies with viral templates such as SV40 show that this template in the twisted circular conformation was a poor template for either enzyme but the linear form was a better substrate for polymerase II than for I. Because of the structural homology between

TABLE 4

Specificity for Polymerases I and II for Various Templates

DNATEMPLATE		Rat Liver Polymerase I	Rat Liver Polymerase II
Calf Thymus		(1)[a]	1[a]
dA:dT	(A)	.41	.07
	(U)	.10	<.01
dG:dC	(G)	.73	3.1
	(C)	.22	.22
d(AT)	(U, A)	3.0	4.0
SV_{40}		0.1	1-2
T_4		0.02	0.02
$T_4 + \sigma$		<.001	<.001
$\frac{\text{denatured}}{\text{native}}$ active ratio		1.0	2-10

[a] Similar G/C content, nearest neighbor frequency.

the prokaryotic and eukaryotic enzymes, a study of the transcription of the T4 DNA template in the presence and absence of bacterial σ-factor was carried out (Hager, G., Hall, B. D., Lindell, T. and Rutter, W. J., in Blatti *et al.*, 1970). T4 DNA template was transcribed at only a few per cent of the rate of calf thymus template, a rate closely approximating that of the *E. coli* "core" polymerase. Of course, addition of the σ-factor to the *E. coli* core enzyme increased the rate about 50-fold to the level observed with the calf thymus template. In contrast, successive additions of σ-factor to the mamalian polymerase preparations progressively *inhibited* the residual activity on the T4 template. These results suggest that the σ-factor reacts with the polymerase but produces an inactive rather than an activated complex. We consider this result a confirmation of the general homology between the two systems.

The apparent general lack of specificity in these systems on isolated DNA template is also evident in the studies of Roeder *et al.*, (1970) on the transcription of isolated ribosomal DNA from *Xenopus.* In these investigations, the *Xenopus* polymerase I and polymerase II, as well as *E. coli* polymerase, produced similar (partially symmetric) products on purified ribosomal DNA. Nevertheless, experiments using intact nuclei confirmed the translation of the ribosomal genes was carried out by an α-amanitin-insensitive polymerase (presumably I).

These collective experiments on specificity could be used to support the contention that the *in situ* transcriptive specificity was not a function of the polymerases but resided in the template, or in lost specific initiation factors. An alternative explanation stems from the fact that the templates employed are highly "nicked", and hence contain many artificial initiation sites. It is well know that *E. coli* polymerase initiates transcription at non-specific "nicked" sites, as well as at more specific sites. A similar situation is likely with the eukaryotic organisms since the effectiveness of the template increases as the size of the DNA decreases. If the latter postulate were valid, then a greater degree of transcriptive specificity might be obtained with the various polymerases if high molecular weight preparations of DNA were employed. In addition, preblocking non-specific sites with other polymerases should be possible. As illustrated in Fig. 2,

FIGURE 2

Artificial and True Initiation Sites in Determining Transcriptive Specificity

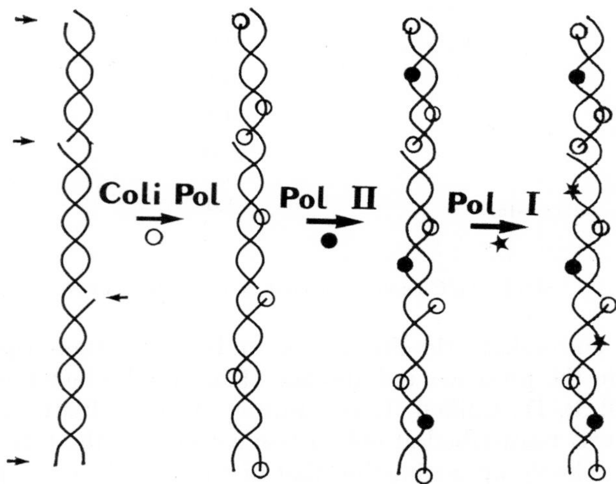

saturating levels of *E. coli* polymerase should bind at both non-specific initiation sites (the nicks) as well as specific sites for that enzyme. Subsequent addition of polymerase II and finally polymerase I might yield incremential increases in transcription sites occupied, and consequently in transcription rate. A number of experimental variations on this general theme have confirmed some degree of transcriptive specificity for the three polymerases Fig. 3. In separate experiments, saturation of high-molecular-weight rat liver DNA by the *E. coli* polymerase, polymerases I and II produced different transcriptive rates. In addition, the products had a significantly different GC content. Sequential additions of the three polymerases indicate progressively increasing rate plateaus, suggesting that additional sites for transcription were available for each of the enzymes. Finally, presaturation with *E. coli* polymerase and/or polymerase II and subsequent blocking by rifamycin and/or α-amanitin respectively should block most non-specific as well as some specific sites. Subsequent addition of the other polymerase or polymerase I showed the presence of specific sites. In particular, presaturation with *E. coli* enzyme or polymerase II and subsequent treatment with polymerase I produced RNA with a high GC content, in some instances approaching that of the ribosomal precursor. Similarly, presaturation with *E. coli* enzyme and rifamycin, and subsequent treatment with polymerase II, yielded an RNA

FIGURE 3

DNA Site-Specific Transcription by RNA Polymerase

		Ecoli Pol		Pol II		Pol I
Single Enzyme Saturation		330 (18 % GC)		75 (24 % GC)		20 (42 % GC)
Stepwise Saturation	Ecoli Pol →	330	+ Pol II →	405	+ Pol I →	425
Sequential Blocking	Ecoli + Rif →	0	+ Pol II →	22 (33 % GC)		
	Ecoli + Rif →	0			+ Pol I →	9 (48 % GC)
			Pol II + α Amanitin →	0	+ Pol I →	18 (50 % GC)
132 (25 % GC) ←	+ Ecoli	0	← Pol II + α Amanitin			

species with intermediate GC content, whereas preblocking with polymerase II and α-amanitin and subsequent transcription with *E. coli* enzyme produced an RNA of relatively low GC content. These experiments indicate a degree of site specificity in each of the polymerases tested. In particular, polymerase I may exhibit specificity for its natural transcriptive sites. A critical test for the enrichment of r-RNA in the product of polymerase I has not yet been made. Further studies have indicated that this specificity is to some extent dependent upon the purification procedures of the enzyme. Thus the transcriptive specificity may be dependent upon specific interacting factors normally fractionating with and perhaps bound to the fundamental enzyme structure.

REGULATION OF RATE AND/OR SPECIFICITY OF THE POLYMERASES BY SPECIFIC "FACTORS"

The search for factors which determine site-specificity for the polymerases or regulate activity is one of the most experimentally active (and frustrating) areas of investigation in this field. It is complicated because increase in rates can be caused artifically by nucleases or by inhibiting the inactivation or interaction of RNA polymerases with surfaces. Furthermore, measurement of specific RNA products by available hybridization procedures is arduous (if not treacherous).

A number of fractions affecting polymerase II and polymerase I activity have been reported. Stein and Hausen (1970) first discovered a protein in thymus extracts which stimulated polymerase II activity on a native but not denatured DNA template at low ionic strength. Seifart (1970) subsequently reported that a similar factor is present in liver tissue. More recently, others have found a similar activity in the extracts of a variety of cells. In addition, there have been a number of reports concerning fractions which stimulate polymerase I activity. In certain instances these have been shown to be due to nuclease activity, but some of these fractions may exert specific effects on the enzyme itself. Dahmus and Lee (1972) have isolated a molecule of 80,000 daltons which stimulates polymerase I and perhaps polymerase II activity (their preparations of polymerase I and II were probably cross-contaminated).

An illustration of factors influencing the rate of transcription by the various polymerases is presented in Table 5. Notice that crystalline albumin (BSA) (presumably free of the nuclease activity recently reported Anai *et al.*, 1972) stimulates the activity, especially of polymerase II, significantly. A heat-stable factor isolated from extracts of calf thymus stimulates activity several-fold. The degree of stimulation is markedly dependent upon the template. It is perhaps significant that little stimulation is achieved on synthetic templates. Substantial and varying stimulations are obtained on various natural DNA templates. This material might stimulate transcription on true, not artificial, initiation sites.

TABLE 5

Specific and Non-Specific Stimulation of RNA Polymerase Activity

DNA TEMPLATE		I	II
		BSA / control activity ratio	
Calf thymus	(Native)	1.1	2.1
	(denatured)		1.0
		BSA + factor[a] / BSA activity ratio	
Calf thymus	(Native)	1.1	2.9
	(denatured)	1.1	1.1
λ Phage	(Native)	0.8	4.9
Poly d (AT)	(U, A)	1.0	1.4
Poly dA:dT	(U)	1.1	1.0
	(A)	1.3	1.2
Poly dI:dC	(C)	1.0	1.0
	(G)	1.0	1.0

[a]Heat stable, macromolecular factor isolated from calf thymus

POSSIBLE REGULATORY MODES OF RNA POLYMERASES: PHOSPHORYLATED POLYMERASE II

Studies of the effects of cyclic AMP have shown it plays a key regulatory role in controlling a number of physiological processes including macromolecular synthesis (Fig. 4). In particular, increases in the levels of cyclic AMP decrease mitosis and DNA synthesis in a variety of cells. Concomitantly, there is perhaps a selective increase in RNA synthesis. Since cyclic AMP stimulates a nuclear protein

FIGURE 4

RNA Polymerases and Reciprocal Regulation of Macromolecular Synthesis?

Phenomena:

↑ cyclicAMP	(Protein phosphorylation) ⟶	↓ DNA synthesis	↑ Specific RNA synthesis

Hypotheses:

1. cyclicAMP affects chromosome structure.
2. cyclicAMP affects transcriptive or replicative enzymes.
3. cylicAMP affects initiation factors for replication or transcription.

FIGURE 5

Phosphorylation of rat liver RNA polymerase with $[\gamma\text{-}^{32}P]$ ATP

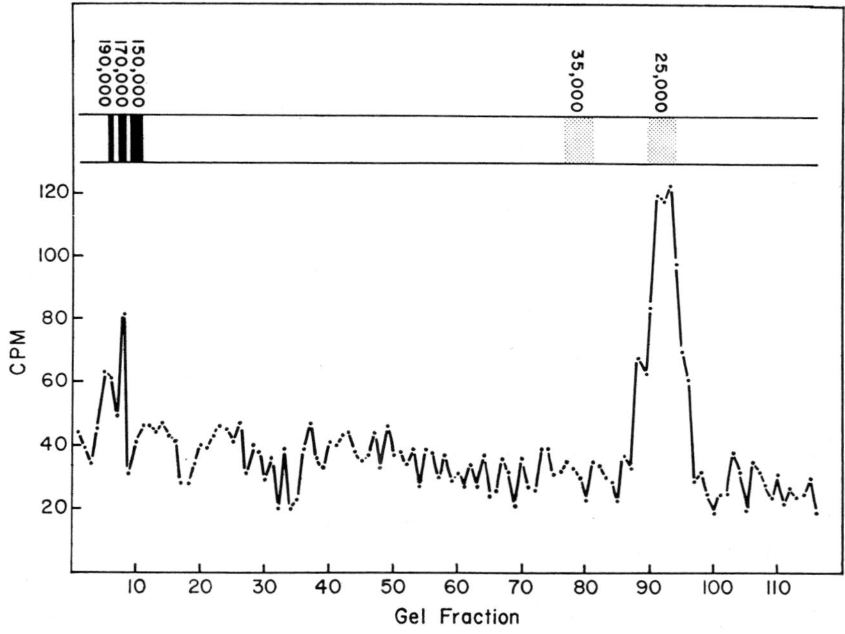

Subsequent to incubation of RNA polymerase II with $[\gamma\text{-}^{32}P]$ ATP the polymerase subunits were resolved by SDS-gel electrophoresis. The co-linear superimposition of the coomassie blue stained proteins with the ^{32}P incorporation profile show the radioactive labelling to be restricted largely to the 25,000 dalton subunit from polymerase II.

kinase (Langan, 1969) it is conceivable that these effects are mediated by phosphorylation of specific nuclear proteins. On the other hand, this compound is known to act directly as an effector of other enzymes (for example, phosphofructokinase). Thus the mode of cyclic AMP action on the overall processes remains unclear. It seemed possible that coordinate effects on the RNA polymerases might occur. We therefore tested for the effects of cyclic AMP itself and of specific phosphorylation of polymerases II. Dr. Marvin Paule found an endogenous cyclic AMP-dependent protein kinase which fractionates with polymerase II until the late stages in the purification procedure. This kinase, or exogenous rabbit muscle protein kinase, phosphorylates rat liver polmerase II. Fig. 5 shows the labelling profile obtained following SDS gel electrophoresis of polymerase II which had been incubated with γ-labelled ATP and protein kinase. Only the 25,000 molecular weight subunit incorporates significant radioactivity. Preliminary experiments suggest a single site is phosphorylated. Experiments to test whether the phosphorylated and non-phosphorylated species of polymerase II have different transcriptive properties are now in progress. Preliminary experiments show no perceptible effect on simple DNA templates but effects on II-specific templates have yet to be tested. The possibility of phosphorylation of polymerase I and/or III is also under investigation.

THE "POLYMERASE I RESPONSE" IN PHYSIOLOGICAL TRANSITIONS

An increase in ribosomal RNA synthesis accompanies many, if not all physiological or developmental transitions (Tata, 1966). The proliferative response after partial hepatectomy, the specific response to hormones such as glucocorticoids in liver, estrogen in the uterus, androgens in the prostate, growth hormone in liver, the recovery of cells from contact inhibition, the differentiation of many organ systems, the reactivation of an inactive nucleus in a cellular heterokaryon, or the stimulation by appropriate cells (for example, lymphoid cell) by a mitogenic agent (for example, phytohaemaglutinin) all include a specific stimulation of RNA synthesis. We have previously shown that there is an increase in polymerase I activity in the liver after treatment with glucocorticoids or after partial hepatectomy (Blatti *et al.*, 1970), or in the uterus after estrogen treatment. This increase in I activity preceeds or is coincident with the increase in ribosomal RNA synthesis observed in these systems. Fig. 6 illustrates the changes in polymerase I and II activity in uterine nuclei after estrogen treatment. An increase in polymerase I activity is apparent soon after estrogen administration, but there is no significant change in polymerase II activity. Similar observations have now been made in a number of other tissues. A selective increase in polymerase I activity has been demonstrated in the prostate gland after testosterone treatment (Mainwaring *et al.*, 1971) in the liver after growth hormone treatment (Smuckler and Tata 1971). In fact, in every instance so examined, an increase in polymerase I accompanies the increased ribosomal RNA synthesis. Thus, the polymerase I-ribosomal RNA response may occur generally in physiological transitions. Whether the increase in polymerase I activity also produces an increased synthesis of 5S and t-RNA and perhaps other RNAs is not yet known.

FIGURE 6

The effect of estrogen treatment on RNA polymerase activity in the immature rat uterus

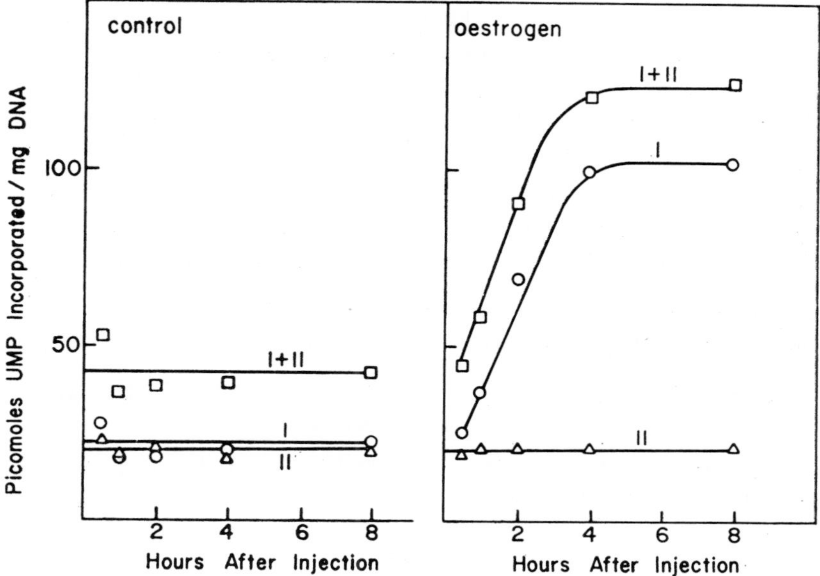

The data is displayed as the nuclear RNA polymerase activity profile following hormone treatment or the corresponding control treatment. Experimental protocol for hormone treatment, nuclear preparation, and assay conditions are as described previously for the rat uterus (Blatti et al., 1970). Polymerase I activity was discriminated from polymerase II and total polymerase activities by α-amanitin.

Recent studies by Yu and Feigelson (1971) have shown that polymerase I turns over rapidly (a half-life of approximately 1½ hours) while polymerase II is relatively stable (a half-life greater than 10 hours). The experiments also emphasize the sensitivity of the ribosomal RNA system to fluctuations in metabolic state. This is consistent with the view that polymerase I is highly regulated (and perhaps regulating).

Several alternative and as yet unexplored hypotheses may be put forward to explain the polymerase I response (Fig. 7): (1) Harris (1968) proposed that ribosomes transport m-RNA from the nucleus to the cytoplasm; (2) there may be multiple classes of ribosomes with different translationsal functions, hence an increase in ribosomal RNA synthesis may in fact signal the production of a new set of ribosomes, and a new specific protein-synthetic system. Different proteins are synthesized on membrane-bound and free polysomes. This segregation of translational function might be mediated by fundamentally different ribosomal populations; (3) ribosomal synthesis may be obligatorily coupled with synthesis of new proteins, for example, specific initiation factors for translation or new classes of m-RNA might stimplate ribosome formation; (4) a product of RNA polymerase I activity may be required for regulation of RNA polymerase II or DNA polymerase function; (5) finally, the polymerase I response may be secondary, and not determinative in the various physiological transition. These various possibilities await experimental test.

FIGURE 7

The *RNA Polymerase I* response in physiological transitions

Inducing influence → → [↑ RNA Pol I activity / ↑ r-RNA Synthesis] → → Effect

Hypothesis:

1. Ribosomes transport m-RNA (Nucleus → Cytoplasm)
2. Multiple classes of functional ribosomes (e.g. membrane bound, free, etc.)
3. Metabolic regulation of ribosomal synthesis (e.g. translational initiation factors stimulate ribosome formation)
4. RNA Pol I product required for RNA Pol II or DNA Pol function
5. Polymerase I response is secondary and not determinative

CHANGES IN POLYMERASE LEVELS DURING SEA URCHIN EMBRYOGENESIS

Some time ago, Roeder and Rutter (1970b) reported selective changes in polymerase levels during the course of the development of sea urchin embryos.

FIGURE 8

Resolution of sea urchin egg RNA polymerase

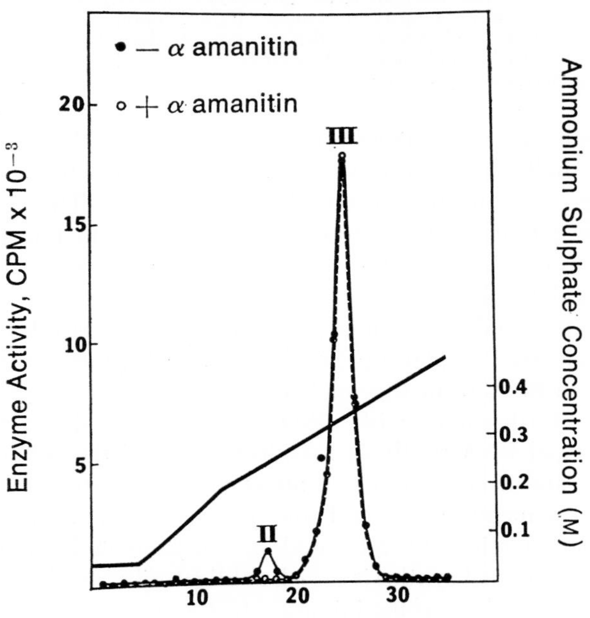

Extracts of unfertilized *Strongylocentrotus purpuratus* eggs were chromatographed over DEAE-sephadex and assayed as previously described (Roeder and Rutter, 1969).

From the mesenchyme blastula to the prism stage, polymerase I activity/cell remained constant while polymerase II and III activity declined. These changes in the relative proportions of the enzymes could be correlated with the progressively increased capacity for ribosomal RNA synthesis. Dr. Paul Morris has recently found a dramatic change in the levels of polymerase during early fertilization periods prior to the mesenchyme blastula. In analogy with amphibian eggs the sea urchin oocyte may synthesize substantial quantities of ribosomal RNA during development. An amplification of the ribosomal genes similar to that demonstrated in *Xenopus* may occur. In the mature egg, however, little RNA synthesis occurs. Early studies showed that in egg extracts there was substantial polymerase activity, but there was little if any inhibition by α-amanitin. Fig. 8 shows that the great majority of activity present is polymerase III. The typical elution profile from a DEAE column gives no evidence for polymerase I, a very low level ($<5\%$) of polymerase II, whereas the dominant activity is polymerase III. The small amount of II activity is probably found in the nucleus since there is some inhibition by α-amanitin in isolated nuclei and no significant inhibition in the cytoplasm. After fertilization, there is a sharp drop in polymerase III and a concommitant rise in II, and later I activity. These studies suggest the interesting possibility that III is a precursor of II or I. These dramatic changes in the RNA polymerase complement are coincident with and perhaps required for the transcriptional events in early embryogenesis.

REFERENCES

Adman, R., Schultz, L. D., and Hall, B. D. (1972) *Proc. Nat. Acad. Sci. U.S.* In press

Anai, M., Hasaguchi, H., and Takagi, Y. (1972) *J. Biol. Chem. 247,* 193

Attardi, G., Aloni, Y., Attardi, B., Ojala, D., Pica-Mattoccia, L., Robberson, D. L., and Storrie, B. (1970) *Cold Spring Harbor Symp. Quant. Biol. 35,* 599

Beerman, W. (1971) *Chromosoma 34,* 152

Benson, R. W. (1972) *Fed. Proc. Fed. Amer. Soc. Exp. Biol. 31,* 478

Blatti, S. P., Ingles, C. J., Lindell, T. J., Morris, P. W., Weaver, R. F., Weinberg, F., and Rutter, W. J. (1970) *Cold Spring Harbor Symp. Quant. Biol. 35,* 649

Bottomly, W., Smith, H. J., and Bogorad, L. (1971) *Proc. Nat. Acad. Sci. U.S. 68,* 2412

Burgess, R. R., Travers, A. A., Dunn, J. J., and Bautz, E. K. F. (1969) *Nature (London) 221,* 43

Chambon, P., Gissinger, F., Mandel, J. L., Kedinger, C., Gniadowski, M., and Meihlac, M. (1970) *Cold Spring Harbor Symp. Quant. Biol. 35* 693

Chesterton, C. J. and Butterworth, P. W. H. (1971) *Eur. J. Biochem. 19,* 232

Chesterton, C. J. and Butterworth, P. W. H. (1971) *FEBS Lett. 15,* 181

Dahmus, M. E., and Lee, S. (1972) *Fed. Proc. Fed. Amer. Soc. Exp. Biol. 31,* 428

Davison, J. Pilarski, L. M. and Echols, H. (1969) *Proc. Nat. Acad. Sci. U.S. 63,* 168

deCrombrugghe, B., Varmus, H. E., Perlman, R. L., and Pastan, I. (1970) *Biochem. Biophys. Res. Commun. 38,* 894

Dezelee, S., Sentenac, A., and Fromageot, P. (1972) *FEBS Letters 21,* 1

Egyhazi, E., D'Monte, B., and Edström, J. E. (1972) *J. Cell. Biol. 53,* 523

Harris, H. (1968) *Nucleus and Cytoplasm* Clarendon Press, Oxford

Kedinger, C., Nuret, P., and Chambon, P. (1971) *FEBS Lett. 15,* 169

Küntzel, H., and Schafer, K. D. (1971) *Nature New Biol. 231,* 265

Langan, T. L. (1969) *J. Biol. Chem. 244,* 5763

Leighton, T. J., Freese, P. K., and Doi, R. H. (1971) *Fed. Proc. Fed. Amer. Soc. Exp. Biol.* *30,* 1069

Lindell, J., Weinberg, F., Morris, P. W., Roeder, R. G., and Rutter, W. J. (1970) *Science,* *170,* 447

Mainwaring, W. I. P., Mangan, F. R., and Peterkin, B. M. (1971) *Biochem. J.* *123,* 619

Morris, P. W., Paule, M., Goldberg, M., Morris, R., Gee-Clough, J. L., and Rutter, W. J. (1972) *Fed. Proc. Fed. Amer. Soc. Exp. Biol.* *31,* 428

Paul, J. and Gilmour, R. S. (1968). *J. Mol. Biol.* *34,* 305

Polya, G. M., and Jagendorf, A. T. (1971) *Arch. Biochem. Biophys.* *146,* 635

Reid, B. D. and Parson, P. (1971) *Proc. Nat. Acad. Sci. U.S.* *68,* 2830

Roeder, R. G. and Rutter, W. J. (1969) *Nature (London)* *224,* 234

Roeder, R. G. and Rutter, W. J. (1970a) *Proc. Nat. Acad. Sci. U.S.* *65,* 675

Roeder, R. G., and Rutter, W. J. (1970b) *Biochemistry* *9,* 2543

Roeder, R. G., Reeder, R. H., and Brown, D. D. (1970) *Cold Spring Harbor Symp. Quant. Biol.* *35,* 727

Seifart, K. H. (1970) *Cold Spring Harbor Symp. Quant. Biol.* *35,* 719

Shorenstein, R. G. and Losick, R (1970) *Nature (London)* *227,* 910

Smuckler, E. A., and Tata, J. R. (1971) *Nature (London)* *234,* 37

Smith, K. D., Church, R. B., and McCarthy, B. J. (1969) *Biochemistry* *8,* 4271

Stein, H., and Hausen, P. (1970) *Eur. J. Biochem.* *14,* 270

Tata, J. R. (1966) *Progr. Nucl. Acid Res. Mol. Biol.* *5,* 191

Tsai, M-J., Michaelis, G., and Criddle, R. S. (1971) *Proc. Nat. Acad. Sci. U.S.* *68*:473

Travers, A. A. and Burgess, R. R. (1969) *Nature (London)* *222,* 537

Travers, A. A., Kamen, R. I. and Schleif, R. F. (1970) *Nature (London)* *228,* 748

Travers, A. (1971) *Nature New Biology* *229,* 69

Weaver, R. F., Blatti, S. P., and Rutter, W. J. (1971) *Proc. Nat. Acad. Sci. U.S.* *68,* 2994

Wobus, U., Panitz, R., and Seafling, E. (1971) *Experientia* *27,* 1202

Wu, and Dawid, I. (1972) *Fed. Proc. Fed. Amer. Soc. Exp. Biol.* *31,* 924

Yu, F. L. and Feigelson, P. (1971) *Proc. Nat. Acad. Sci. (U.S.)* *68,* 2177

Zubay, G. and Chambers, P. (1969) *Cold Spring Harbor Symp. Quant. Biol.* *34,* 753

The control of Gene Expression by Membrane Organization in *Saccharomyces Cerevisiae*

KEVIN A. WARD, S. MARZUKI and J. M. HASLAM
Department of Biochemistry, Monash University, Clayton, Victoria, Australia

INTRODUCTION

The facultative anaerobe, *Saccharomyces cerevisiae,* has considerable advantage as an organism for the study of mitochondrial membranes, as many parameters of mitochondrial membrane composition in this organism are subject to both physiological and genetic manipulation (for review see Linnane and Haslam, 1971). In particular, situations that result in the formation of defective organelles can be experimentally exploited, as the organism does not need respiratory metabolism for growth on fermentable substrates; thus cells of *S. cerevisiae* grown anaerobically on glucose or galactose media contain mitochondrial precursor structures that lack respiratory chain cytochromes and the capacity for oxidative phosphorylation. The ability of *S. cerevisiae* to grow anaerobically also allows the manipulation of the lipid composition of the mitochondrial membranes, since the organism is unable to synthesize unsaturated fatty acids (UFA) or ergosterol in the absence of molecular oxygen (Andreason and Stier, 1953; Bloomfield and Bloch, 1960), but can incorporate into its membranes a wide range of added sterols and UFA's (Bloch *et al.,* 1961; Proudlock *et al.,* 1968). If lipids are present in the anaerobic growth medium in trace amounts, the growth of cells is limited and membrane lipid levels of both UFA and ergosterol fall to as low as one tenth of the values found in the presence of excess lipid supplements (Jollow *et al.,* 1968; Watson *et al.,* 1970; Watson *et al.,* 1971). Recent investigations have shown that the nature of the anaerobic mitochondrial precursor structures is markedly dependent on lipid composition (Watson *et al.,* 1970; Watson *et al.,* 1971). Lipid-depleted precursor structures are deficient in an obvious folded inner membrane, whereas lipid-supplemented structures have a well-developed internal membrane with extensive cristae. Furthermore, the lipid-depleted anaerobic structures are characterised by their having significantly lower levels of tricarboxylate cycle enzymes and ATPase than the lipid-supplemented organelles (Watson *et al.,*

1970; Watson *et al.*, 1971; Vary *et al.*, 1970). The primitive nature of the lipid-depleted structures is most clearly demonstrated however, by the finding that they lack a detectable mitochondrial protein synthetic activity (Watson *et al.*, 1971), in contrast to the active protein synthesizing system found in lipid-supplemented organelles both *in vitro* (Watson *et al.*, 1971) and *in vivo* (Schatz and Salzgeber, 1969). More recently, analysis of mitochondrial RNA by sucrose density gradient sedimentation has shown the loss of mitochondrial ribosomes from lipid-depleted anaerobic mitochondria (Forrester *et al.*, 1971).

The lipid-depleted structures lack detectable amounts of the characteristic mitochondrial r-RNA species (15S and 21S), whereas the latter are present in lipid-supplemented organelles. Instead, a large heterodisperse peak of RNA, sedimenting at 4-10S with a G + C content of 48%, is associated with the lipid-depleted organelle fraction (Forrester *et al.*, 1971). Such a G + C content precludes the possibility that this RNA is a degradation product of mitochondrial r-RNA (G + C content, 27%), although the possibility exists that it is a degradation product of cytoplasmic r-RNA. These results therefore suggest that membrane lipid composition is a key determinant in the maintenance of mitochondrial r-RNA.

Recent studies in our laboratory have led to the suggestion that the mitochondrial ribosome is closely associated with the mitochondrial membrane (Linnane and Haslam, 1970; Bunn *et al.*, 1970). This suggestion was initially based on the study of a number of mitochondrial mutations which result in resistance to the effects of certain antibiotic inhibitors of mitochondrial protein synthesis. Phenotypically, these mutations are expressed as complex cross-resistance patterns involving a number of antibiotic inhibitors, whereas biochemically the mutations appear to be producing a change in the mitochondrial membrane (Bunn *et al.*, 1970). It may be added that one of these mutants initially selected for resistance to mikamycin, an inhibitor of mitochondrial protein synthesis, was also found to be resistant to oligomycin, an inhibitor of the membrane-bound mitochondrial ATPase (Mitchell *et al.*, 1972). Furthermore, the resistance of the whole cells to the antibiotics could be altered by conditions designed to change the mitochondrial membrane composition and organization viz. anaerobic growth.

The necessity for employing detergents in the isolation of mitochondrial ribosomes (Grivell *et al.*, 1971; Morimoto and Halvorson, 1971) supports the concept that mitochondrial ribosomes are membrane bound, and the direct observation by electron microscopy of ribosomes attached to the inner mitochondrial membrane has recently been achieved (Plummer, Green, Watson, Ward and Linnane, unpublished).

Mitochondrial r-RNA is transcribed from mitochondrial DNA in the mitochondria (Wintersberger, 1967), and the mitochondrial DNA-dependent RNA polymerase is very tightly bound to the membrane (Kalf and Faust, 1969; Wintersberger and Wintersberger, 1970; Tsai *et al.*, 1971). On the other hand, mitochondrial ribosomal proteins are apparently synthesized in the cytoplasm by cell sap ribosomes (Davey *et al.*, 1969). Thus the synthesis and assembly of mitochondrial ribosomes are complex processes in which the mitochondrial membrane is clearly an important determinant.

Based on the totality of this data, we have considered the possible role of the mitochondrial membrane in

(1) the function of the mitochondrial RNA polymerase, and:

(2) the assembly and attachment of the mitochondrial ribosome.

This problem has been approached by the study of the events which follow the aeration of anaerobically grown lipid depleted cells. The effects of lipid depletion of mitochondria are fully reversible by aeration of the cells. Initially a rapid synthesis of UFA and ergosterol occurs and this is followed by the formation of an active mitochondrial protein synthesizing system leading eventually to the synthesis of a mitochondrial electron transport chain and the formation of fully functional mitochondria (Watson *et al.*, 1971). The present paper describes some of our current work on the role of membrane lipid composition in the regulation of the activity of the mitochondrial DNA-dependent RNA polymerase; it has in part been previously reported (Watson *et al.*, 1971; Forrester *et al.*, 1971), and some of it is in preparation for publication elsewhere (Forrester, Ward and Linnane, unpublished).

The present paper reports that the absence of mitochondrial r-RNA from lipid-depleted anaerobic structures correlates with a low level of mitochondrial DNA-dependent RNA polymerase activity, whereas organelles isolated from lipid-supplemented cells contain high levels of both r-RNA and RNA polymerase. In addition, the formation of r-RNA upon aeration of lipid-depleted cells is preceded by the appearance of the RNA polymerase.

Apart from the actual level of the RNA polymerase, there exists a second level of control over the synthesis of mitochondrial r-RNA. Thus the antibiotic D(—) chloramphenicol does not affect the synthesis of RNA polymerase during the aeration of lipid-depleted cells, but parallel investigations by Ward, Forrester and Linnane (unpublished) show that the synthesis of mitochondrial r-RNA is inhibited under these conditions. It is suggested that the antibiotic is acting in this instance to perturb mitochondrial membrane organization, with a consequent inhibition of the activity of the membrane-bound RNA polymerase.

MATERIALS AND METHODS

A locally isolated diploid strain of *Saccharomyces cerevisiae,* strain M, was grown on a 4% galactose-1.0% Difco yeast extract-salts medium described previously (Wallace *et al.*, 1968). Tween 80 (4 g/1) which is a source of unsaturated fatty acid, and ergosterol (20 mg/1) were added to the lipid-supplemented anaerobic growth media.

The nomenclature used in earlier publications from this laboratory has been retained (Watson *et al.*, 1971):

(1) An-Gal denotes cells grown anaerobically in 4% galactose-yeast extract medium to yield partially catabolite derepressed anaerobic cells deficient in both unsaturated fatty acid (UFA) and ergosterol (E).

(2) An-Gal+T+E denotes cells grown anaerobically in 4% galactose-yeast extract medium containing Tween 80 and ergosterol to yield lipid-supplemented partially catabolite derepressed anaerobic cells.

(3) Gal-O_2 denotes cells grown aerobically in 4% galactose-yeast extract medium to yield partially catabolite derepressed aerobic cells.

Cells were grown at 28°C and all cultures harvested at early stationary phase when the cell populations of all cultures, including lipid-depleted cells, were shown to be more than 90% viable.

Aerobic Induction

Washed An-Gal cells (4 g dry weight) were suspended in 400 ml of medium consisting of 1% yeast extract-salts and galactose (1%) as carbon source. Chloramphenicol when required was added to the medium at a concentration of 4 mg/ml. Aerobic induction was achieved by vigorously shaking the cell suspension at 28°C in fluted 21 flasks using a rotary shaker. The high cell density (10 mg dry weight/ml) during aerobic induction limited growth to less than 5% increase in cell numbers in 4 h. After induction, all cell suspensions were tested for viability and petite formation, and cultures were routinely found to contain greater than 95% respiratory competent cells.

Preparation of Mitochondrial Fractions

Mitochondria were isolated from yeast protoplasts and purified by discontinuous sorbitol gradient centrifugation as described by Watson *et al.* (1971). Cytoplasmic ribosomes were isolated from the post-mitochondrial supernatant as described by Lamb *et al.* (1968).

Analyses

Mitochondrial RNA Polymerase. Mitochondrial DNA-dependent RNA polymerase activity was assayed in purified mitochondrial fractions by a modification of the method described by Tsai *et al.* (1971). Mitochondria were suspended in tris-HCl (0.01M pH 7.9), $MgCl_2$ (0.01M), EDTA (0.1mM), dithiothreitol (0.1mM), 5% glycerol and disrupted by 50 strokes of a glass-teflon homogenizer rotating at high speed. The fragmented mitochondria (0.5-1.0 mg protein) were then incubated for 20 min. at 37°C in 0.5 ml of incubation medium, containing tris-HCl (0.05M pH 7.9), β-mercaptoethanol (2mM) denatured calf thymus DNA (20 μg); $MgCl_2$ (3mM); ATP, GTP, CTP, (0.8mM each); UTP (0.08mM) and 5μCi of [^{3}H]UTP (13 Ci/m mole). The reaction was terminated by the addition of 5 ml of cold 5% trichloracetic acid containing 0.01M sodium pyrophosphate (to prevent artifactual adsorption of UTP to the precipitate) and the precipitate washed 5 times with this solution, followed by one wash each with ethanol, ethanol-ether (2:1) and ether. The dried precipitate was then dissolved in formamide by heating for 10 minutes at 180°C and radioactivity determined using a liquid scintillation counter. Counting efficiency was approximately 19%.

Lipid Protein and Cytochromes. Lipid analyses were as described previously (Jollow *et al.*, 1968). Protein was estimated by the method of Lowry *et al.* (1951) and cytochromes were identified spectrophotometrically (Clark-Walker and Linnane, 1967).

RNA. The methods and results of RNA analyses will be described elsewhere but the results are included in summary form in certain tables for the purpose of discussing the RNA polymerase results.

RESULTS AND DISCUSSION

Properties of Mitochondrial RNA Polymerase

Mitochondria isolated from aerobically grown cells contain a high level of mitochondrial RNA polymerase activity, which under the present assay conditions corresponds to about 2000 c.p.m. incorporated/20 min incubation/mg mitochondrial protein. The effects of inhibitors on mitochondrial RNA polymerase activity are given in Table 1. The system is inhibited about 90% by high concentrations (50

TABLE 1

The effects of inhibitors on the *in vitro* activity of Mitochondrial DNA-Dependent RNA Polymerase

Inhibitor	Concentration (μg/ml)	RNA Polymerase (% control)
Actinomycin D	50	11
Daunomycin	50	12
α-Amanitin	40	96
Rifampicin	40	100
D(-) Chloramphenicol	4000	98

RNA polymerase activity was measured as described in Methods. In the absence of inhibitors activity was 2100 c.p.m./20 min incubation/mg mitochondrial protein. Results are the average of three experiments.

μg/ml) of either actinomycin D or daunomycin. It is unaffected by α-amanitin, an inhibitor of nuclear polymerase (Dezelee *et al.*, 1970; Ponta et al., 1971; Brogt and Ponta, 1972) or by the prokaryote RNA polymerase inhibitor, rifampicin (Hartmann, 1967). Other authors have reported that the mitochondrial RNA polymerase of *S. cerevisiae* is sensitive (Scragg, 1971) or resistant (Kalf and Faust, 1969; Wintersberger and Wintersberger, 1970; Tsai *et al.*, 1971) to rifampicin. These discrepancies however, may result from the different assay conditions employed. In particular, rifampicin inhibits chain initiation rather than elongation, and the presence of rifampicin resistant incorporation may mean that the RNA polymerase is still attached to mitochondrial DNA and in the process of transcription in our crude extracts. This possibility is in part supported by the observation that radioactive incorporation by our system is not fully dependent on the addition of calf thymus DNA. The enzyme is also resistant to very high levels of the inhibitor of mitochondrial protein synthesis D(—)chloramphenicol (Table 1); the significance of this finding will become apparent on page 000 of this communication, when the effects of this antibiotic on the synthesis of mitochondrial r-RNA are discussed.

The effects of anaerobiosis on mitochondrial RNA polymerase activity

The effects of both anaerobic growth and lipid composition on the RNA polymerase activity of mitochondria are summarized in Table 2. Although the mito-

TABLE 2

The Effect of UFA and Sterol Depletion on Mitochondrial DNA-Dependent RNA Polymerase Activity

Growth Conditions	RNA Polymerase Activities	Mitochondrial Sterol	Mitochondrial Fatty Acids			Mitochondrial r-RNA
			8:0 – 14:0	16:0 + 18:0	16:1 + 18:1	
GAL + O_2	1920	40-60	4	21	75	+++
An-GAL + UFA + E	3140	30-50	10	34	54	++
An-GAL	302	8-14	19	65	15	Not detectable

Lipids were analyzed as described in Methods. Sterol is expressed as μg/mg mitochondrial protein, and fatty acids, denoted by the convention C number of carbon atoms: Number of unsaturated linkages, are expressed as weight % of total fatty acid. RNA polymerase activities are the average of 6-10 experiments. The relative contents of mitochondrial r-RNA are data obtained from Forrester *et al.,* (1971).

chondrial fraction isolated from An-Gal+T+E cells contain slightly lower levels of unsaturated fatty acids and ergosterol than mitochondria from aerobically grown cells, they routinely appear to possess higher levels of RNA polymerase activity. In contrast, the mitochondrial fraction isolated from cells grown anaerobically without lipid supplements contain high levels of short chain saturated fatty acids (8:0 — 14:0) but only approximately 20% of both the ergosterol and UFA levels found in normal mitochondria. These marked changes in lipid composition are accompanied by a large decrease in RNA polymerase activity to about 10% of that found in lipid-supplemented anaerobic mitochondrial structures.

There are a number of possible explanations for the low polymerase activity of the An-Gal mitochondrial fraction. It is known that the enzyme is bound to the membranes of normal and petite mitochondria (Kalf and Faust, 1969; Tsai *et al.*, 1971; Wintersberger, 1970; Wintersberger and Wintersberger, 1970). However, the enzyme is only released from normal membranes by comparatively drastic extraction procedures (Tsai *et al.*, 1971) compared with those conditions required for extraction from the structurally modified membranes of petite mitochondria (Wintersberger, 1970). One obvious explanation for the low An-Gal mitochondrial RNA polymerase activity is that the enzyme is still synthesized *in vivo* but is not firmly bound to the membranes of the lipid-depleted mitochondrial structures, leading to the release of the enzyme into solution during the isolation of these fragile, easily damaged organelles. Alternatively, the RNA polymerase may not be synthesized at all under the growth conditions which lead to lipid-depletion. We have examined these two possibilities by considering the effects of aeration of An-Gal cells on the activity of the mitochondrial RNA polymerase.

The effects of aeration on mitochondrial RNA polymerase activity

Lipid-depleted, anaerobically grown cells were aerated for four hours and the changes which take place in lipid composition and mitochondrial RNA polymerase activity in a typical experiment are shown in Table 3. During aeration both the sterol and UFA levels of the mitochondria increased markedly. The sterol levels changed rapidly during the initial phase of aeration, increasing from 8 μg per mg mitochondrial protein in An-Gal mitochondria to 33 μg per mg mitochondrial protein after 30 minutes aeration. A more gradual increase was then observed throughout the remainder of the aeration period, so that after 240 minutes a sterol level of 51 μg per mg mitochondrial protein was obtained. This latter sterol level is within the range of values found for the sterol content of normal Gal-O_2 mitochondria (Table 2). The amounts of mitochondrial UFA showed a similar time-dependent increase during aeration to that found for the mitochondrial sterol content. Initially, mitochondria isolated from An-Gal cells had a UFA content of 16% of total fatty acids, whereas after 30 minutes aeration, this level increased to 38%. The UFA content continued to rise during the remainder of aeration, so that after 4 hours a value of 63% was attained, this being only slightly lower than the values found in mitochondria from aerobically grown cells.

A rapid appearance of RNA polymerase activity occurs during the aeration and after 30 minutes the RNA polymerase activity returns to the levels found in fully functional mitochondria. Continued aeration for a further 30 minutes results in

TABLE 3

The Effect of Aeration of An-GAL Cells on Lipid Composition and Mitochondrial DNA-Dependent RNA Polymerase Activity

Aeration Time	Mitochondrial RNA Polymerase	Mitochondrial Sterol	Mitochondrial Fatty Acids (Weight %)			Mitochondrial r-RNA
			8:0 – 14:0	16:0 + 18:0	16:1 + 18:1	
0	290	8	19	65	16	Not detectable
30	2200	33	11	51	38	Not detectable
60	4600	42	9	43	48	Not detectable
120	3700	48	9	35	56	+
240	2800	51	9	28	63	+++
240+ Cycloheximide	270	45	10	30	60	Not detectable

Lipids were analysed as described in Methods. Sterol is expressed as μg/mg mitochondrial protein, and fatty acids as weight % total fatty acids. Lipid values represent a single typical experiment. RNA Polymerase Activity (c.p.m./20 min incubation/mg mitochondrial protein) was determined as in methods, and represent the average of three experiments. Cycloheximide was used at a concentration of 20 μg/ml. The relative contents of mitochondrial r-RNA are data obtained by Forrester, Ward and Linnane (unpublished).

a substantial increase in activity above normal levels; thereafter the activity of the enzyme decreases to that found in normal mitochondria, and aeration for an additional 180 minutes has little further effect. This pattern of rapid RNA polymerase synthesis followed by a decrease to normal levels is highly reproducible in different experiments and the changes observed in the polymerase activity closely parallel the increases in the ergosterol and UFA contents of mitochondria during aeration. It is also apparent from Table 3 that there is a considerable range of lipid composition that allows the RNA polymerase to be synthesized and integrated into the mitochondria. That *de novo* synthesis of the polymerase occurs during aeration is suggested by the observation that the increase in RNA polymerase activity but not the synthesis of UFA and ergosterol is prevented by the inhibitor of cytoplasmic protein synthesis, cycloheximide (Table 3). Comparison of the times of aeration required for the formation of RNA polymerase and the detection of mitochondrial r-RNA indicate that there is a substantial lag between the increase in polymerase activity and the synthesis of r-RNA. The RNA polymerase is present in substantial amounts after 30 minutes aeration, whereas r-RNA is not detectable until after 120 minutes of aeration, and does not approximate the amount of r-RNA found in normal mitochondria until after 240 minutes aeration (Forrester, Ward and Linnane, unpublished).

The apparent lag in r-RNA synthesis may merely be a kinetic effect, in that the formation of small amounts of r-RNA is not detectable. Alternatively, it is possible that transcription is controlled by a specific mechanism that requires reorganization of the mitochondrial membrane so as to integrate the RNA polymerase in a manner that allows its function. Indeed, recent studies in our laboratory on the effects of the antibiotic D(—)chloramphenicol on the synthesis of r-RNA during aeration of An-Gal cells, indicate that the control of transcription in mitochondria is also dependent upon some factor other than the level of activity of the enzyme itself (Ward, Forrester and Linnane, unpublished). These results are now discussed.

The effects of chloramphenicol on the formation of mitochondrial RNA polymerase and the synthesis of mitochondrial r-RNA during the aeration of An-Gal cells

The effects of D(—) and L(+)chloramphenicol on the aerobic induction of mitochondrial RNA polymerase, and the synthesis of UFA and ergosterol are shown in Table 4. The action of the antibiotics on r-RNA synthesis (Ward, Forrester and Linnane, unpublished) are also summarized for the purpose of discussion.

The inhibition of RNA synthesis by D(—)chloramphenicol does not appear to be due to a simple inhibition of lipid synthesis. The ergosterol and UFA contents of cells aerated in the presence of the antibiotic approach, but do not quite reach, the levels of normal aerated cells. However, these lipid components do attain the levels found in cells grown anaerobically on lipid supplements, which possess both RNA polymerase and a complete mitochondrial protein synthesizing system. More detailed work designed to correlate more exactly the lipid levels of the cells and mitochondria is currently in progress.

TABLE 4

The Effect of Antibiotics on Mitochondrial RNA Polymerase Activity during Aeration of An-GAL Cells

Aeration Time (min)	Antibiotic	Mitochondrial RNA Polymerase	Cellular Sterol	Cellular UFA Content	Mitochondrial r-RNA
0	—	310	0.4	10	Not detectable
240	—	2900	5.5	72	+++
240	D(−) CAP	2700	3.7	53	Trace
480	D(−) CAP	950	4.2	64	Trace
240	L(+) CAP	4600	6.5	65	+++

Incubations were performed as described in Methods. Mitochondrial RNA polymerase activities (c.p.m./20 min. incubation/mg mitochondrial protein) and the UFA composition (weight % of total fatty acids) and sterol content (mg/g dry weight) of whole cells were determined as described in Methods. D(−) and L(+) chloramphenicol were added at a concentration of 4 mg/ml. Results are the average of three determinations. The relative contents of mitochondrial r-RNA were determined by Ward, Forrester and Linnane (unpublished).

L(+)chloramphenicol does not inhibit the formation of either the RNA polymerase or mitochondrial r-RNA and appreciable synthesis occurs after 2 hours aeration. On the other hand An-Gal cells aerated for 4 or even 8 hours in the presence of D(—)chloramphenicol do not synthesize more than trace amounts of mitochondrial r-RNA, even though the oxygen induced synthesis of the RNA polymerase is normal. Thus despite the presence of high levels of mitochondrial RNA polymerase and of a relatively normal membrane ergosterol and fatty acid composition in aerated cells, D(—)chloramphenicol prevents normal transcription. The inhibition by D(—)chloramphenicol of r-RNA synthesis does not appear to be a kinetic effect in which there is a slowing r-RNA formation, as continued aeration of the cells for as long as 480 min does not result in a significant increase in the amount of mitochondrial r-RNA synthesized (Ward, Forrester and Linnane, unpublished).

D(—)chloramphenicol has been known for some time to have at least two distinct effects on mitochondria. The first obtains with high concentrations of the antibiotic and results in the inhibition of mitochondrial respiration and phosphorylation, but these systems are inhibited both by the L(+) and the (D—) isomers of chloramphenicol (Freeman, 1970). The second effect occurs with low concentrations of the antibiotic, it is specific for the D(—) isomer of chloramphenicol and results in the inhibition of mitochondrial protein synthesis (Lamb *et al.*, 1970). Again, this mechanism of action is apparently ruled out as mitochondrial r-RNA and protein synthesis cannot be detected in mitochondria from An-Gal cells. Thus it appears from the present work that D(—)chloramphenicol is probably acting at a third distinct site.

It has previously been shown that during the growth of Gal-O_2 or An-Gal+T+E cells for many generations, the presence of D(—)chloramphenicol has little effect on the formation of the mitochondrial ribosome (Davey *et al.*, 1969). The proposed third site of action of chloramphenicol is therefore only detected under conditions of lipid-depletion that have initially led to the absence of mitochondrial ribosomes. The consequence of the interaction of the drug with this third site is the *in vivo* inhibition of transcription as both mitochondrial DNA and mitochondrial RNA polymerase are present. We interpret these results to indicate that, when the antibiotic is bound at the proposed membrane site, the nature of the interaction between the polymerase and the membrane is altered, resulting in the inhibition of the synthesis of mitochondrial r-RNA. This could conceivably be the result of the lack of a control factor (e.g. an initiation factor), in the absence of which *in vivo* transcription by the core enzyme is not possible. Alternatively, the various components of the enzyme are all present but not in appropriate juxtaposition to allow *in vivo* transcription. These possibilities and the nature of the mitochondrial membrane-RNA polymerase interactions are presently being investigated both in normal mitochondria and in lipid-depleted organelles.

REFERENCES

Andreason, A. A. and Stier, T. J. B. (1953) *J. Cell. Comp. Physiol. 41,* 23

Bloch, K., Baronowsky, B., Goldfine, H., Lennarz, W. J., Light, R., Norris, A. T. and Scheuerberant, G. (1961) *Fed. Proc. Fed. Amer. Soc. Expl. Bid. 20,* 921

Bloomfield, D. K. and Bloch, K. (1960) *J. Biol. Chem. 235,* 337

Brogt, Th.M. and Planta, R. J. (1972) *FEBS Lett., 20,* 47

Bunn, C., Mitchell, C. H., Lukins, H. B. and Linnane, A. W. (1970), *Proc. Nat. Acad. Sci. U.S. 67,* 1233

Clark-Walker, G. D. and Linnane, A. W. (1967) *J. Cell. Biol. 34,* 1

Davey, P. J., Yu, R. and Linnane, A. W. (1969) *Biophys. Res. Commun. 36,* 30

Dezelee, S., Sentenac, A. and Fromageot, P. (1970) *FEBS Lett., 7,* 220

Forrester, I. T., Watson, K. and Linnane, A. W. (1971). *Biochem. Biophys. Res. Commun. 43,* 409

Freeman, K. B. (1970) *Can. J. Biochem. 48,* 479

Grivell, L. A., Reijnders, L. and Borst, P. (1971) *Biochim. Biophys. Acta. 247,* 91

Hartmann, G., Honikel, K. O., Krusel, F. and Nuesch, J. (1967) *Biochim. Biophys. Acta. 145,* 843

Jollow, D. J., Kellerman, G. M. and Linnane, A. W., (1968) *J. Cell. Biol. 37,* 221

Kalf, G. F. and Faust, A. S. (1969) *Arch. Biochem. Biophys. 134,* 103

Lamb, A. J., Clark-Walker, G. D. and Linnane, A. W. (1968) *Biochim. Biophys. Acta. 161,* 415

Linnane, A. W. and Haslam, J. M. (1970) in *Current Topics in Cellular Regulation,* Vol. 2, (Horecker, B. L. and Stadtman, E. R. eds.) p. 101, Academic Press, New York

Lowry, O. H., Rosebrough, N. J., Farr, A. L. and Randall, R. J. (1951) *J. Biol. Chem. 193,* 165

Mitchell, C. H., Bunn, C., Lukins, H. B. and Linnane, A. W. (1972) *J. Bioenergetics.* In Press

Morimoto, H. and Halvorson, H. O. (1971) *Proc. Nat. Acad. Sci. U.S. 68,* 324

Ponta, H., Ponta, U. and Wintersberger, E. (1971) *FEBS Lett., 18,* 204

Proudlock, J. W., Wheeldon, L. W., Jollow, D. J. and Linnane, A. W. (1968) *Biochim. Biophys. Acta. 152,* 434

Schatz, G. and Salzgaber, J. (1969) *Biochem. Biophys. Res. Commun. 37,* 996

Scragg, A. H. (1971) *Biochem. Biophys. Res. Commun. 45,* 701

Tsai, M. J., Michaelis, G. and Criddle, R. S. (1971) *Proc. Nat. Acad. Sci. U.S. 68,* 473

Vary, M. J., Stewart, P. R. and Linnane, A. W. (1970) *Arch. Biochem. Biophys. 141,* 430

Wallace, P. G., Huang, M. and Linnane, A. W. (1968) *J. Cell. Biol. 37,* 207

Watson, K., Haslam, J. M. and Linnane, A. W. (1970) *J. Cell. Biol. 46,* 88

Watson, K., Haslam, J. M., Veitch, B. and Linnane, A. W. (1971) in *Autonomy and Biogenesis of Mitochondria and Chloroplasts,* (Boardman, N. K., Linnane, A. W. and Smillie, R. M. eds.) p. 163, North Holland Press, Amsterdam

Wintersberger, E. (1967) *Z. Physiol. Chem. 348,* 1701

Wintersberger, E. and Wintersberger, U. (1970) *FEBS Lett., 6,* 58

Wintersberger, E. (1970) *Biochem. Biophys. Res. Commun. 40,* 1179

Transcription and Translation in Mammalian Cells

S. PENMAN, R. PRICE, S. PERLMAN and R. SINGER
Massachusetts Institute of Technology, Department of Biology, Cambridge, Massachusetts, U.S.A.

In this report, we will consider several of the aspects of transcription and translation which are unique to eukaryotic cells. As our knowledge of the macromolecular biochemistry of higher organisms grows, it is apparent that the models of regulatory mechanisms derived from prokaryotes are increasingly unsatisfactory. In this report, we will summarize some of the findings in our laboratory which deal with the unique properties of the systems which transcribe DNA and regulate translation in mammalian cells.

Previously, our laboratory has been concerned with examining the transcription process in nuclei derived from HeLa cells. In this report, we will describe similar experiments performed on cells infected with adenovirus, which appears to constitute a unique and useful system which serves as a model for the processess occurring in uninfected cells. The results obtained with virus-infected cells confirm and extend our previous observations.

TRANSCRIPTION

The macromolecular metabolism specific to adenovirus growth resembles that of the host cell in many ways (Green, 1970). Adenovirus DNA replicates and adeno-specific RNA is transcribed in large molecules in the nucleus of the infected cell (Green *et al.,* 1970). The high molecular weight RNA is apparently cleaved and polyadenylic acid is added. The product is then exported to the cytoplasm where it is translated as messenger RNA by the host cell ribosomes (Green *et al.,* 1970; Raskas, 1971; Raskas and Okubo, 1971; Phillipson *et al.,* 1971; M. Hirsch, unpublished observation). Adenovirus-specific proteins then reassociate with adenovirus DNA in the nucleus to form the mature virus structures (Thomas and Green, 1966; Velicer and Ginsberg, 1968). Thus adenovirus offers a model system in which a relatively small number of genes are expressed through mechanisms which resemble those of the uninfected cell.

In these experiments the transcription of the adenovirus genome in the infected cell is examined. Previous investigators have shown the existence of at least two and possibly three distinct RNA polymerase activities associated with the mammalian cell nucleus (Roeder and Rutter, 1969, 1970a, 1970b). The polymerase activities II and III, as designated by Roeder and Rutter, are located in the nucleoplasm (Roeder and Rutter, 1970a). A previous report suggested that polymerase II appears to be responsible for the major fraction of Hn-RNA synthesis in the uninfected cell (Zylber and Penman, 1971). It will be shown that an activity resembling polymerase II accounts for the major fraction of the transcription of the adenovirus genome when nuclei from infected cells are examined *in vitro*. A similar finding was obtained independently by Wallace and Kates (1972).

Considerable amounts of adenovirus-specific RNA are labelled late in infection in the nuclei of the host cells. However, extensive synthesis still takes place on host cell DNA in the infected cell (Green *et al.*, 1970). The amount of adeno-specific RNA present in the nuclei of infected cells is measured by hybridization of the RNA to an excess of adenovirus DNA bound to cellulose nitrate filters.

The RNA from nuclei of productively infected cells labelled *in vivo* was hybridized to adenovirus DNA filters under conditions of DNA excess. The time course of hybridization is shown in Fig. 1A. The amount of hybridization appears to plateau at approximately 14% of the input RNA, and this is independent of the amount of DNA present on the filters. It appears, therefore, that 14% of the RNA present in infected cell nuclei under the labelling conditions used is adeno-specific, and this agrees well with previous reports (Green *et al.*, 1970).

The RNA labelled *in vitro* in nuclei from infected cells was next examined. A crude nuclear preparation was incubated with three unlabelled nucleoside triphosphates and radioactive UTP. A previous report has shown that high molecular weight heterogeneous RNA is labelled under these conditions, although the amount of RNA elongation amounts to less that a hundred nucelotides (Zylber and Penman, 1971; L. McReynolds, unpublished observation). The principal activity in isolated nuclei appears to be a limited addition to pre-existing molecules. The polymerase activities are inhibited by actinomycin D and the product is completely sensitive to RNAse.

Fig. 1B shows the time course of hybridization of RNA labelled *in vitro* in nuclei from infected cells. The amount of material which hybridizes to adenovirus DNA appears to plateau at approximately 18% of the total labelled RNA. Thus approximately the same fraction of the RNA labelled in the *in vitro* conditions is adenovirus-specific as during labelling *in vivo*.

The activity which transcribes adenovirus DNA is apparently located in the nucleoplasm. In all the experiments described here, the nucleolar activity has been selectively suppressed by pre-treating cells with low levels of actinomycin D (Perry, 1966; Roberts and Newman, 1966; Penman *et al.*, 1968; Zylber and Penman, 1971).

The sensitivity of the activity which transcribes adenovirus DNA to α-amanitin was measured. This cyclic polypeptide has been shown to be a potent and selective inhibitor of polymerase II of the uninfected cell (Roeder and Rutter,

FIGURE 1
The kinetics of hybridization of nuclear RNA

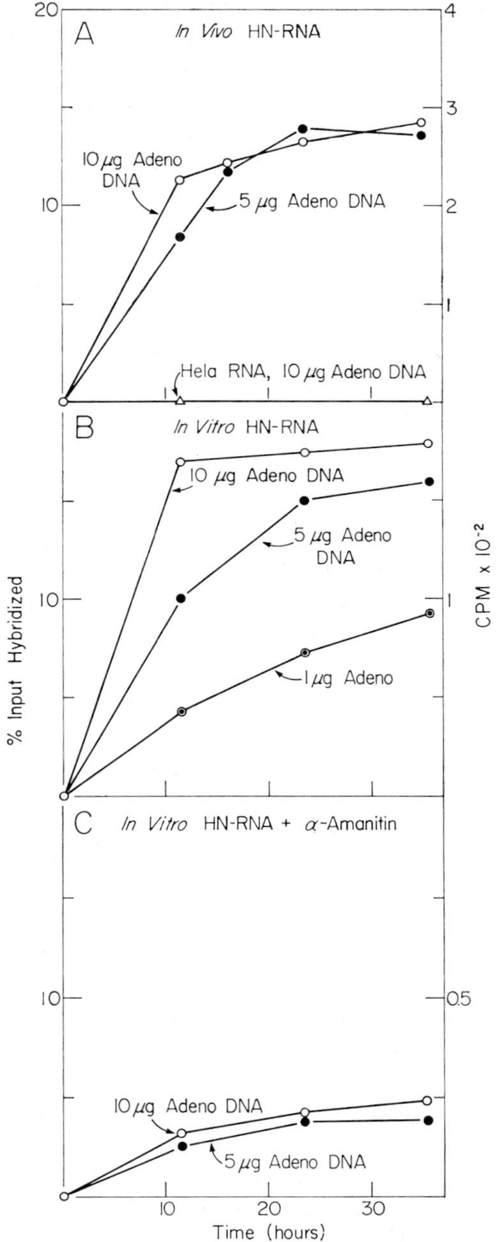

(A) 2 × 10^7 HeLa cells infected with adenovirus were labelled with 200 μci of [^{3}H] uridine for 30 minutes at 16 hours after infection. Nuclei were isolated and RNA extracted as in (Price and Penman, 1972a). Hybridization conditions are given by Price and Penman (1972a).

(B) Nuclei from 6 × 10^6 infected cells at 16 hours after infection were incubated *in vitro* as described by Price and Penman (1972a). RNA was extracted and hybridized to adenovirus DNA as in Fig. 1A.

(C) Same as B, except 0.2 μg/ml of α-amanitin was added to the incubation mixture.

1970; Zylber and Penman, 1971). Nuclei from infected cells are incubated in the presence of α-amanitin. The drug suppresses *in vitro* incorporation to about 15% of the level obtained in the control. The small amount of labelled RNA obtained from α-amanitin-treated nuclei hybridizes to a limited degree with the adenovirus DNA. A plateau of approximately 6% of the input RNA labelled in the presence of α-amanitin is obtained as shown in the data in Fig. 1C. Thus only a small percentage of the small amount of residual synthetic activity observed in the α-amanitin-treated nuclei is adenovirus-specific. Therefore, the bulk of adenovirus DNA is transcribed by an α-amanitin-sensitive polymerase. The α-amanitin resistant adenovirus-specific activity produces RNA which is quite different from the nuclear heterogeneous RNA produced by the α-amanitin-sensitive activity and will be discussed later.

The polymerase activities measured *in vitro* can be further characterized by examining their response to altered ionic strength conditions. In particular, polymerase II is very sensitive to ionic strength, in contrast to polymerase III, which is comparatively indifferent to the ionic conditions of incubation (Roeder and Rutter, 1969; Zylber and Penman, 1971). The nucleoplasmic incorporation in incubated infected cell nuclei is shown in Fig. 2 as a function of varying concen-

FIGURE 2

The effect of ionic strength on total and adenovirus-specific RNA synthesis

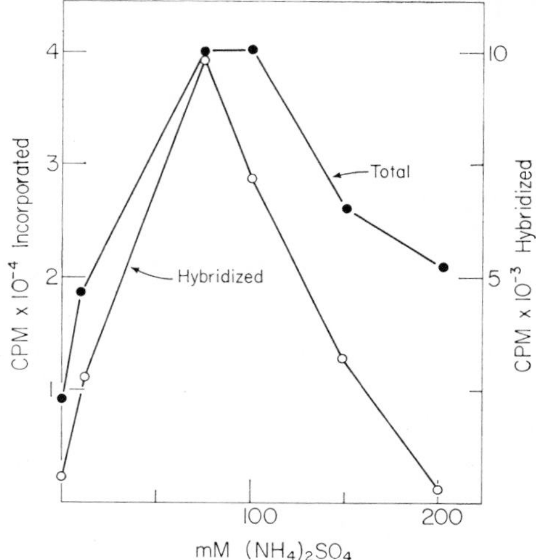

Nuclei were isolated from HeLa cells 16 hours after infection. Cells were treated with 0.04 μg/ml actinomycin D prior to fractionation for 30 minutes to suppress ribosomal RNA synthesis. Crude nuclei were prepared and resuspended in the incubation medium containing the indicated concentrations of ammonium sulphate. Nuclei from 4 × 10^6 cells were used in each sample. [^{3}H] UTP incorporations was allowed to proceed for 5 minutes. Adenovirus-specific RNA synthesis was determined by hybridization as described in Fig. 1.

trations of ammonium sulphate. Both total incorporation and adenovirus-specific synthesis increase significantly with increasing ammonium sulphate up to concentrations of 70 to 100mM. Above this concentration, experimental results become variable possibly due to instability of the nuclear structure at the higher ionic strengths. The response of adenovirus-specific transcription to changes in ionic conditions resembles that of the total incorporation (which under these conditions is due principally to an activity which resembles polymerase II).

The effect of manganese on the polymerase activities measured *in vitro* in nuclei is somewhat different than its effect on purified enzymes. In a previous report it was shown that polymerase II activity is inhibited by manganese, while that resembling polymerase III was relatively indifferent to the ion. Fig. 3 shows

FIGURE 3
The effect of the manganese ion on total and virus-specific RNA synthesis

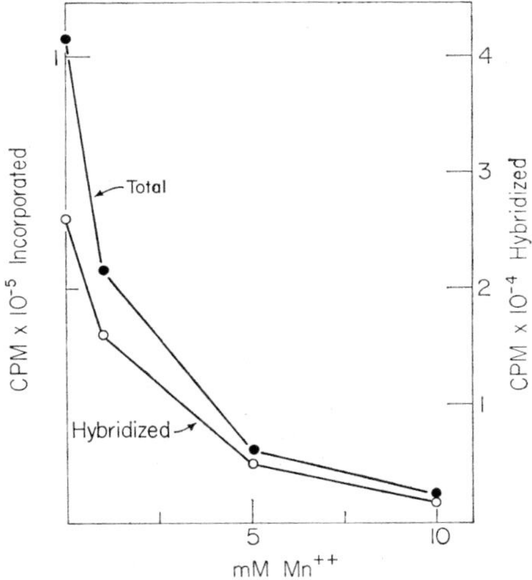

Nuclei were isolated from HeLa cells 16 hours after infection with adenovirus. Labelling and determination of adenovirus-specific synthesis were as described in Fig. 2.

the response of total nucleoplasmic incorporation in nuclei exposed to increasing concentrations of manganese ion in the absence of magnesium. The incorporation decreases as manganese concentration increases in agreement with previous results. The adenovirus-specific RNA, measured by hybridization, also decreases in the presence of manganese and continues to constitute a relatively constant fraction of the total nuclear RNA. The adenovirus transcription activity thus resembles host cell polymerase II by these criteria.

The product of polymerase II in uninfected cells has been shown to be primarily Hn-RNA (Zylber and Penman, 1971). The data in Fig. 4 show the sedimentation profile of RNA labelled by nuclei obtained from adeno-infected cells. The principal product labelled in this case is also Hn-RNA. Hybridization of fractions across the gradient indicate that adeno-specific RNA is heterogeneous in sedimentation values and resembles Hn-RNA. The α-amanitin resistant activity is also shown and this consists primarily of low molecular weight RNA which will be discussed below.

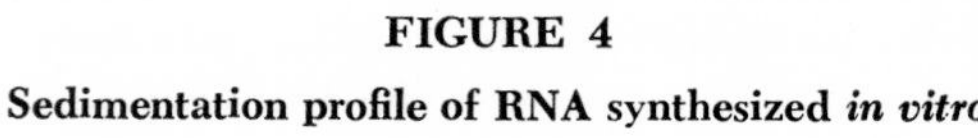

FIGURE 4

Sedimentation profile of RNA synthesized *in vitro*

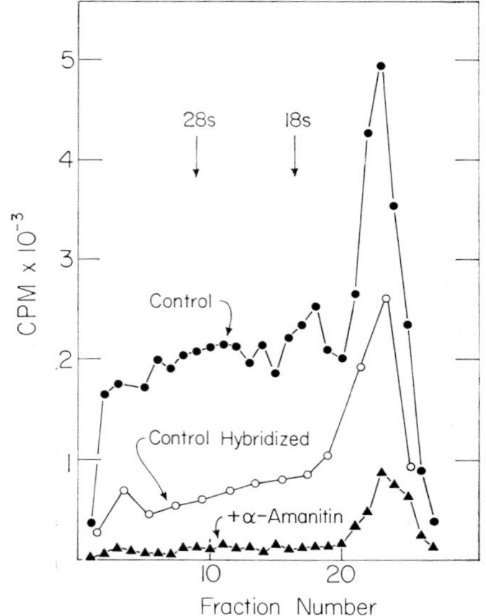

Nuclei were isolated from 1.2×10^7 HeLa cells 16 hours after infection with adenovirus and labelled with [^{3}H] UTP as described in Fig. 2. RNA was extracted by the phenol-SDS method and analyzed by sedimentation through a 15-30% SDS-sucrose gradient. Centrifugation was for 16 hours at 24,000 r.p.m. in the SW 27 rotor. Fractions were collected and radioactivity measured. Aliquots of each fraction were precipitated with ethanol, resuspended in the hybridization solution and hybridized to adenovirus DNA.

A distinct species of small RNA is synthesized in large amounts late in the lytic cycle in adenovirus-infected cells. This RNA, previously designated viral-associated (VA) RNA, corresponds to 5.5S in size, contains a known sequence of 151 nucleotides and appears to be transcribed from the adenovirus genome (Ohe *et al.,* 1969; Ohe and Weissman, 1971). Much more of it is synthesized than of any messenger RNA produced from the adenovirus genome. The next experiments will show that the VA-RNA is transcribed by a distinctly different polymerase system from that producing the large RNA from the adenovirus genome. This RNA polymerase activity, which is shown to reinitiate the synthesis of RNA molecules *in vitro,* appears related to the activity which forms 4S and 5S RNA in the uninfected cell. Thus, during lytic infection, the adenovirus genome is apparently transcribed by two different polymerase activities simultaneously, both of which closely resemble distinct enzymatic activities detectable in the uninfected cell.

Fig. 5B shows the sedimentation distribution of all of the nuclear RNA (host cell plus viral-specific RNA) labelled in the presence of α-amanitin. A small amount of non-viral heterogeneously sedimenting large RNA is labelled, but most of the radioactive RNA sediments in the low molecular weight region of the gradient. The data in Fig. 5B also show that practically all of the low molecular weight product is released from the nuclei during incubation, while the high

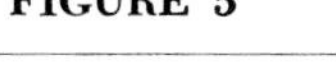

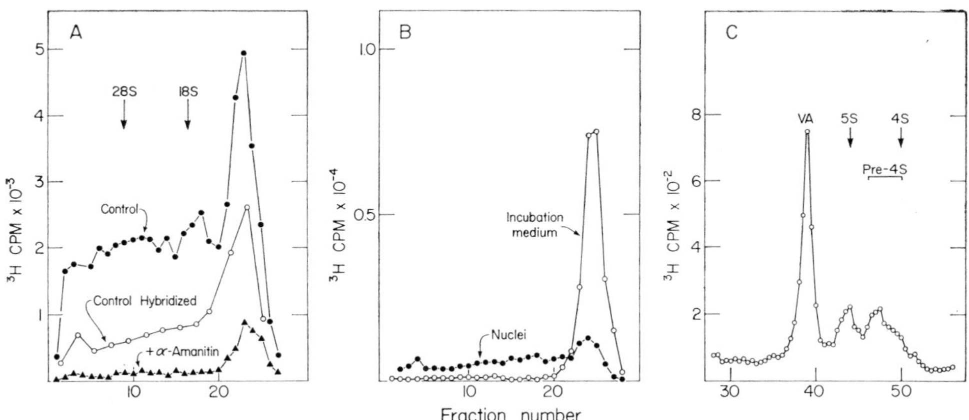

(A) Sedimentation profiles of RNA labelled *in vitro* synthesized in nuclei isolated from adenovirus-infected HeLa cells 16 hours post infection

Nuclei were isolated from 1.2 × 10^7 HeLa cells 16 hours after infection with adenovirus and labelled with [^{3}H] UTP as described by Price and Penman (1972b). RNA was extracted by the phenol-SDS method and analyzed by sedimentation through a 15-30% SDS sucrose gradient. Centrifugation was for 16 hours at 24,000 rev/min in an SW27 rotor. Fractions were collected and radioactivity was measured as described previously. A portion of each fraction was precipitated by ethanol, resuspended in the hybridization buffer, and hybridized to adenovirus DNA as described previously (Price and Penman, 1972a).

(B) Sedimentation profiles of RNA labelled *in vitro* in α-amanitin retained and released from nuclei

Labelling was carried out as described in the legend to Fig. 5, panel A, except that 0.5 μg/ml α-amanitin was added to the incubation mixture. RNA was prepared from the incubation solution and the nuclei. Centrifugation was performed as described in panel A.

(C) Gel electropherogram of RNA labelled *in vitro* in α-amanitin

RNA was prepared from nuclei of infected cells as described in the legend to panel A. Following ethanol precipitation, the RNA was analyzed in a 10% polyacrylamide gel as described previously.

molecular weight RNA remains associated with the nuclei. This same result, the retention of large RNA in the nuclei and the release of small RNA, is also obtained in the absence of α-amanitin when much more Hn-RNA is labelled. Thus it appears that all high molecular weight RNA remains associated with the chromatin of the nuclei, while the low molecular weight product is almost quantitatively released during incubation.

The low molecular weight peak, labelled *in vitro,* is resolved into three distinct RNA, species as shown in Fig. 5C, when analyzed by gel electrophoresis. It seemed likely that one of these low molecular weight RNA species was related to the 5.5S virus-associated RNA produced in adenovirus-infected cells. RNA from the cytoplasm of infected cells labelled *in vivo* is compared to the *in vitro* labelled RNA from nuclei in the gel electropherogram in Fig. 6. The cytoplasmic RNA peak with greatest mobility corresponds to mature 4S RNA, the next slower to 5S RNA. The major peak of radioactivity corresponds to 5.5S RNA and appears only in virus-infected cells. The 5.5S RNA will be designated here as adenovirus-associated RNA (VA-RNA).

FIGURE 6

Co-electrophoresis of low molecular weight RNA synthesized *in vivo* and *in vitro* in HeLa cells 16 hours after infection with adenovirus

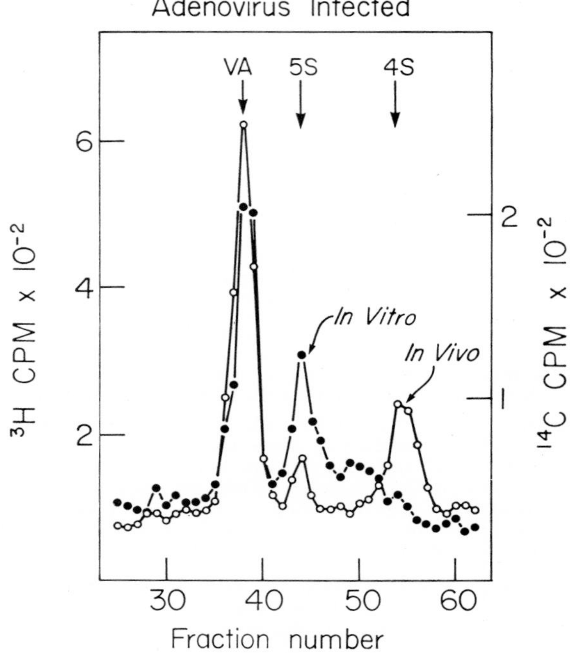

16 hours after infection with adenovirus, 2×10^7 HeLa cells were treated with 0.04 μg/ml actinomycin D for 30 minutes. 0.1 μci/ml [^{14}C] uridine was then added and incubation continued for 60 minutes. The cells were then washed and RNA extracted from the cytoplasm. RNA was labelled with [^{3}H] UTP in isolated nuclei from adenovirus-infected cells as described in Fig. 5. RNA was extracted, mixed with ^{14}C-labelled *in vivo* RNA and analyzed by electrophoresis on 10% polyacrylamide gels. The gels were frozen, sliced and assayed for ^{14}C and ^{3}H as described previously.

RNA labelled *in vitro* shows a considerable similarity to that seen in the cytoplasm of cells labelled *in vivo*. The 4S peak in the RNA labelled *in vivo* is not present in the sample labelled *in vitro*. There is, instead, a somewhat heterogeneous higher molecular weight distribution of more slowly migrating material. Other work has shown that this material has the properties of transfer RNA precursor. Since the primary transcription products would be expected to be labelled when nuclei are incubated *in vitro*, the production of labelled precursor to transfer RNA, rather than the mature species, is not surprising. A monodisperse peak corresponding to 5S RNA is apparent and an RNA species migrating with the VA-RNA is extensively labelled.

The two minor peaks tentatively identified as 5S and pre-4S labelled *in vitro* by nuclei from infected cells are probably the host cell species rather than products of the viral genome. The 5 and pre-4S species are also labelled *in vitro* by nuclei from uninfected cells as shown in the gel electropherogram in Fig. 7, while the VA-RNA is absent. The labelling of both species is again completely resistant to inhibition by α-amanitin.

FIGURE 7

Uninfected HeLa cell low molecular weight RNA synthesized *in vitro*

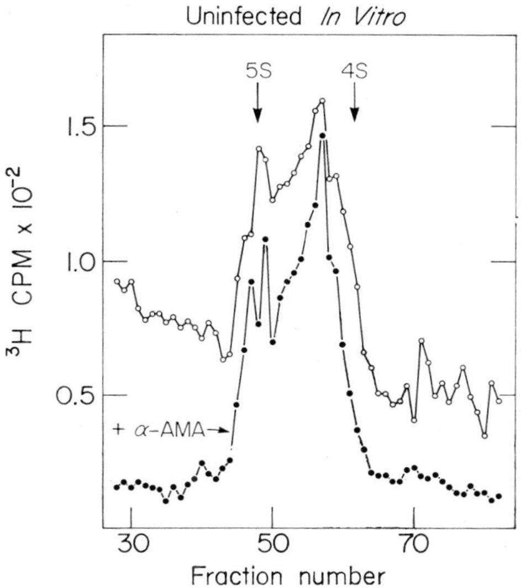

Labelling, RNA extraction and gel electrophoresis were performed as described in Fig. 6.

The following experimental result shows that the VA-RNA labelled either *in vivo* or *in vitro* is indeed complementary to the adenovirus genome. RNA was extracted from the cytoplasm of cells labelled *in vivo* with [^{3}H] uridine at 16 hours after infection. The *in vitro* labelled RNA was prepared as above. The RNA was fractionated by polyacrylamide gel electrophoresis and gel slices corresponding to the VA-RNA were eluted. The RNA obtained was hybridized with both adenovirus and whole cell DNA. The results in Table I show that the VA-RNA hybridizes extensively with adenovirus DNA. In contrast, the amount of hybridization with whole cell DNA is negligible. However, it is not possible to conclude from this latter result whether or not the sequences complementary to the VA-RNA are in fact present in the host genome, since this hybridization would be detected only if the VA-RNA sequence were significantly reiterated.

Surprisingly, the VA-RNA and 5S cellular RNA appear as sharp peaks in the data presented above. The *in vitro* reaction product is not comprised to any observable extent of partially complete RNA molecules which elongate to the final discrete RNA species or of larger molecules which are cleaved to form the discrete species. Nevertheless, the next experiments show that the RNA molecules are indeed products of DNA transcription.

The only model consistent with the data is one in which there is continued re-initiation of transcription coupled with rapid RNA elongation and release of

TABLE 1

Hybridization of VA-RNA to adenovirus and HeLa DNA

	Hybridization to 10 μg adenovirus DNA			Hybridization to 40 μg HeLa DNA		
	^{3}H c.p.m. added	^{3}H c.p.m. hybridized	per cent input hybridized	^{3}H c.p.m. added	^{3}H c.p.m. hybridized	per cent input hybridized
in vivo VA-RNA	728	448	62	3912	19	<1
in vitro VA-RNA	844	419	50	1688	23	1

VA-RNA was purified from the cytoplasm of HeLa cells labelled for 60 minutes with [^{3}H] uridine 16 hours post infection and from isolated nuclei labelled *in vitro* with [^{3}H] UTP as described in Fig. 2. The VA-RNA was purified by acrylamide gel electrophoresis and hybridized to 10 μg adenovirus DNA and 40 μg HeLa DNA attached to nitrocellulose filters, as described previously (Price and Penman, 1972).

the finished molecules. The model predicts the presence of labelled 5′ ends on the *in vitro* product. The products of alkaline hydrolysis of the VA-RNA synthesized *in vitro* were analyzed in order to test this model. Nuclei were incubated in the presence of the three unlabelled nucleoside triphosphates and a [^{3}H]-labelled triphosphate. Following labelling, the nuclei were removed, and the VA-RNA was extracted from the incubation medium and separated on polyacrylamide gels. The VA-RNA peak was eluted from the gel, hydrolysed in 0.3M NaOH and the products analyzed by high voltage electrophoresis on paper. Fig. 8 shows the products obtained for each of the four radioactive nucleotide triphosphates. Cytosine and adenosine are present only as monophosphates, indicating that neither of these labelled bases is found in a terminal position on the VA-RNA synthesized *in vitro*. However, when UTP is the source of radioactivity, some tritium label migrates with the nucleoside, indicating that uridine is found on the 3′ terminal end of the *in vitro* product. Uridine is also the 3′ terminus of VA-RNA labelled *in vivo* (Ohe and Weissman, 1971).

Approximately 1 per cent of the guanine label migrates slightly faster than GTP, the remainder appearing in the monophosphate. This pattern of migration suggests that it is the tetraphosphate, pppGp. The presence of this species in the hydrolysate could only occur if [^{3}H]-labelled guanine is found on the 5′ terminus of the *in vitro* product. The VA-RNA sequence of Ohe and Weissman (1971) also shows a guanine residue in the 5′ position. Thus it appears that a significant fraction of the VA-RNA made *in vitro* contains a labelled 5′ end and must be initiated *de novo*.

Two determinations of the ratio of amount of Gp to presumptive pppGp gave values of 81 and 102. There are 54 G residues in the published sequence of VA-RNA. Thus it appears that at least one-half of the molecules are initiated *in vitro*. This is a lower limit, since any phosphatase activity in the crude preparations would reduce the yield of tetraphosphate.

FIGURE 8

The products of alkaline hydrolysis of VA-RNA labelled with [^{3}H] UTP, ATP, CTP and GTP *in vitro*

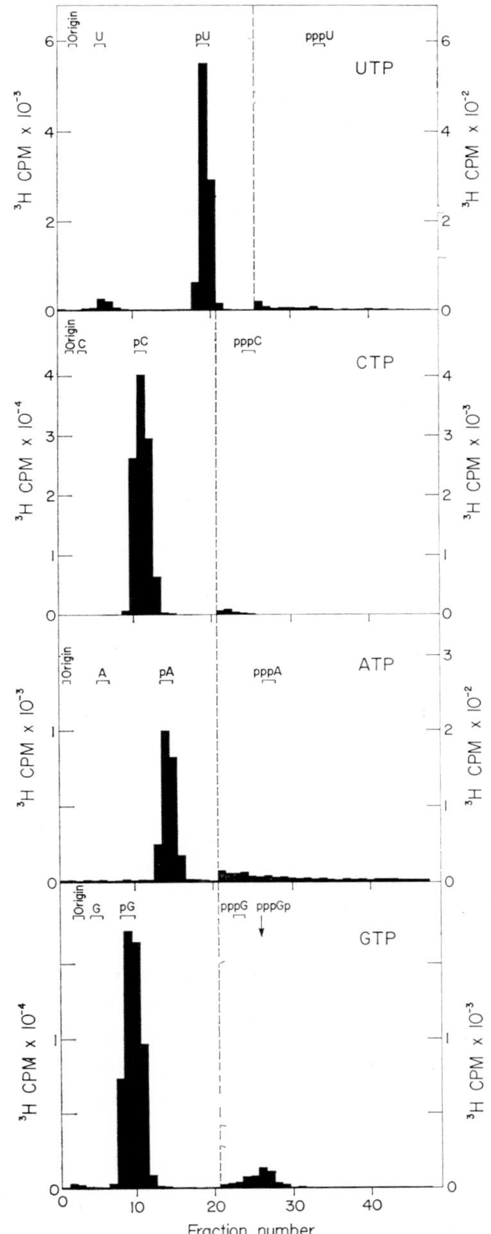

Nuclei were labelled *in vitro* and VA-RNA purified by polyacrylamide gel electrophoresis. After hydrolysis in 0.3 M NaOH, the products were separated by high voltage electrophoresis on paper. The relative migrations of the four samples are not comparable as electrophoresis was performed at different times.

Two determinations of the ratio of Up to U in the *in vitro* labelled VA-RNA gave values of 32 and 23. The published sequence contains 31 U residues. The approximate agreement between the predicted and measured values indicates that the VA-RNA is labelled along its entire length, again supporting the hypothesis of extensive reinitiation.

The labelling reaction described here is distinguishable from the bulk of nuclear activity by its resistance to α-amanitin. The ionic strength dependence of the α-amanitin-resistant activity is shown in Fig. 9. Incorporation of [^{3}H] UTP for a

FIGURE 9

The dependence of *in vitro* labelling of 5S, 4S and VA-RNA on ionic strength

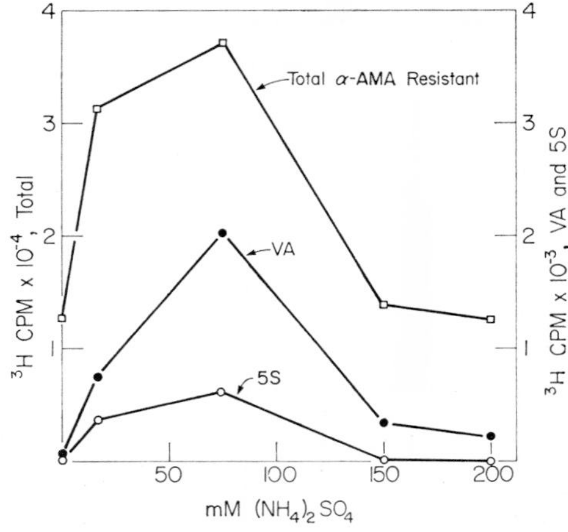

Labelling, RNA extraction and gel electrophoresis were performed as described in Fig. 6. Incubation time was 10 minutes.

fixed labelling time during the period of linear incorporation is measured as a function of ammonium sulphate concentration. The activity synthesizing the VA-RNA is maximal at 75 to 100mM ammonium sulphate and then falls dramatically. The decrease in incorporation of the α-amanitin-resistant activity at supra-optimal ionic strength is significantly greater than for the α-amanitin sensitive activity labelling the heterogeneous RNA as described previously (Price and Penman, 1972a). At lower ionic strengths, labelling of 5 and 4S RNA parallels that of virus-associated 5.5S RNA; at the higher ionic strengths, the inhibition of all incorporation is so drastic that the relatively small 5 and 4S peaks cannot be accurately measured.

The response of the labelling activity described here differs significantly from the major nuclear activities in its response to manganese ion. Fig. 10 shows that increased manganese ion concentration severely inhibits total nuclear incorporation in agreement with previous results. However, in contrast, the labelling of the VA-RNA as well as 5S RNA increases initially with increased manganese concentration in the absence of any added magnesium. Still higher concentrations

FIGURE 10
The effect of Mn^{2+} on the *in vitro* labelling of 5S, VA and total RNA

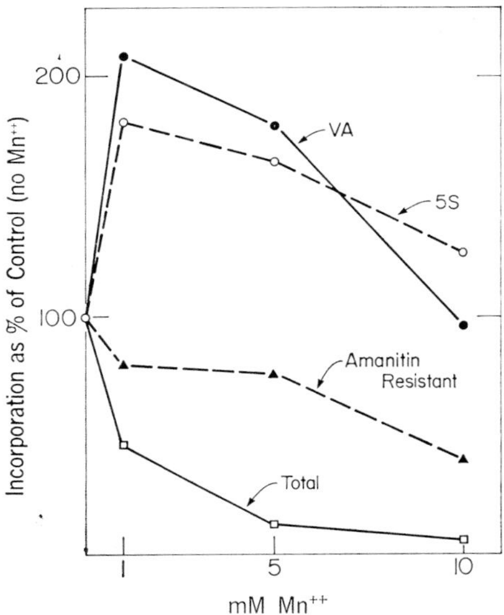

Labelling, RNA extraction and gel electrophoresis were performed as described in Fig. 6. Labelling time was 10 minutes.

of manganese gradually reduce the amount of incorporation. The labelling of VA-RNA is less sensitive to inhibition by Mn^{2+} than the labelling of heterogeneous nuclear RNA (Price and Penman, 1972a). In other work, 5S and 4S synthesis are shown to have the same response to Mn^{2+} as does the one for VA-RNA described here, further suggesting the similarity between the two systems.

The results presented here indicate the simultaneous transcription of the adenovirus genome by two distinct polymerase activities. The activity that labels the large heterogenous RNA containing adenovirus sequences resembles the host cell activity that labels the heterogeneous nuclear RNA. The activity responsible for the synthesis of VA-RNA is distinct from the Hn-RNA polymerase activity in its ionic dependence and its resistance to inhibition by α-amanitin. It closely resembles the activity synthesizing host cell 5S and 4S RNA and is quite different from the nucleolar activity in its response to Mn^{2+}. An RNA polymerase designated polymerase III by Roeder and Rutter is also found associated with the nuclei of eukaryotic cells and this enzyme might conceivably be responsible for the low molecular weight RNA synthesis described here.

These experimental results suggest that the new viral genes present in the nucleus of an adenovirus-infected cell are transcribed by mobilizing two pre-existing host polymerases. One activity, which is probably polymerase II, is specialized for synthesizing high molecular weight RNA, which is precursor to cytoplasmic message. The other activity, which is possibly polymerase III, appears specialized for low molecular weight RNA synthesis.

TRANSLATION

We now consider the problem of the regulation of translation in mammalian cells. The synthesis of proteins from the messenger RNA in eukaryotic cells is different from that of prokaryotes in two important ways. In the mammalian cell, the translation of the m-RNA in the cytoplasm is remote from the transcription in the nucleus, whereas in bacteria these two processes occur almost simultaneously (Byrne *et al.,* 1964; Miller *et al.,* 1970). Bacterial protein synthesis ceases within several minutes after inhibition of RNA transcription with actinomycin D (Levinthal *et al.,* 1962); hence, the m-RNA is short-lived. In metazoan eukaryotic cells, a similar experiment indicates that the m-RNA is much more stable of the order of several hours (Penman *et al.,* 1963). Thus, protein synthesis is not immediately affected by inhibition of transcription.

In a few specialized systems, the lifetime of messenger RNA can be deduced by observing the decay of protein synthesis after RNA synthesis has ceased. Thus the messenger for haemoglobin in the reticulocyte (Rifkind *et al.,* 1964) and for silk fibroin in the silkworm (Suzuki and Brown, 1972) are examples of stable messenger molecules with a lifetime of several days.

The measurement of a messenger lifetime in cells with an active RNA metabolism is more difficult. In one experimental approach, new RNA synthesis was blocked with actinomycin D and the subsequent decay of protein synthesis measured. Under these conditions, protein synthesis generally decreased with a 2½ to 3 hour half-life in a wide variety of systems (Penman *et al.,* 1963; Stachelin *et al.,* 1963; Cheevers and Sheinin, 1970; Craig *et al.,* 1971). However, interpreting the decay of protein synthesis as due to a concomitant degradation of messenger RNA assumes that the availability of messenger molecules is limiting in protein synthesis. In this report, we shall show that messenger RNA decay is not responsible for protein synthesis decay in the presence of actinomycin D. Rather, messenger RNA appears quite stable after the administration of actinomycin D. The subsequent decay of protein synthesis appears to be due to a failure in the initiation of translation.

The decrease in sedimentation pattern of polyribosomes after several hours of treatment with actinomycin D does not yield the expected result assuming a decay in the amount of messenger RNA. Disappearance of messenger RNA should reduce the total amount of polyribosomes but not shift the sedimentation distribution. Yet, polysomes from HeLa cells treated with actinomycin D sediment more slowly than polysomes from normal cells (Fig. 11). This decrease in size implies that messenger RNA becomes more lightly loaded with ribosomes. This could occur should the rate of ribosomal initiation decrease relative to the rate of elongation. The experiment shown in Fig. 11 supports this possibility. Cycloheximide is added to both control and actinomycin D-treated cells at a concentration sufficient to decrease the rate of ribosome movement on message about five-fold (Godchaux *et al.,* 1967; Fan and Penman, 1970; Lodish 1971). The slower initiation is apparently balanced by the diminished elongation rate and a significant recovery of normal-sized polyribosomes occurs in the actinomycin D-treated cells. This resurrection of polyribosomes occurs in the presence of cycloheximide

FIGURE 11

The decay and rebuilding of polysomes in the presence of actinomycin D

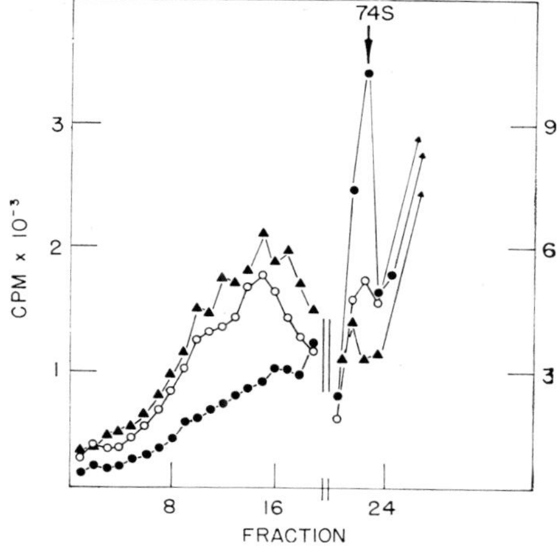

HeLa cells were grown in suspension culture as previously described (Eagle, 1959) One-hundred ml of culture was labelled for 15 hours with 2 μci of [^{14}C] uridine (Schwarz Bio-chemical, 35 mci/mM), so that most of the label was in ribosomal RNA. Polysomes were allowed to decay in the presence of actinomycin D (4 μg/ml, 3 hours) and then 1 μg/ml of cycloheximide was added to $\frac{1}{3}$ the incubation mixture of HeLa cells (50 ml at 4 $\times$ 10^5 cells/ml) for 30 additional minutes. For control cells, the actinomycin D was omitted. Cells were harvested by centrifugation at 2,000 rev./min for 2 minutes and resuspended in RSB buffer (0.01 M NaCl, 0.01 M tris-HCl, pH 7.4, 0.0015 M $MgCl_2$). The cells were then broken with NP40 (Shell Oil Company, 0.5%). Nuclei were removed by centrifugation at 2,000 rev./min for 2 minutes and the cytoplasm layered on an RSB sucrose (15-30% w/v) gradient (12 ml) and centrifuged at 40,000 rev./min in a Spinco SW40 rotor at 4°C for 1 hour. Fractions were collected directly into vials and scintillation fluid (omnifluor, methyoxy ethanol, toluene) added.

Polysomes from normal cells ▲——▲, from actinomycin -D treated cells ●——●, and subsequent cycloheximide treatment O——O. Monomers (74S) particles are on the right-hand scale, polysomes on the left-hand scale.

and suggests that much of the messenger RNA is still present in the actinomycin D-treated cells.

The recent discovery that nearly all messenger RNA contains a length of 3′ terminal polyadenylic acid now allows a quantitative measurement of the amount of messenger RNA in the cell (Darnell *et al.*, 1971; Edmonds *et al.*, 1971; Lee *et al.*, 1971). By adapting a technique developed by Sheldon *et al.* (1972), this RNA containing this poly(A) sequence can be completely and selectively hybridized to polyuridylic acid immobilized on a glass fiber filter. The hybrid can then be melted off the filter and analyzed in a sedimentation gradient. The results shown in Fig. 12 demonstrate the selectivity and efficiency of this hybridizing filter method. The first panel compares the sedimentation profiles of filter-retained and filter-transmitted RNA from the cytoplasm of HeLa cells labelled for 60 minutes in a low level of actinomycin D (0.04 μg/ml) sufficient to differentially suppress synthesis of ribosomal RNA (Perry, 1963; Roberts and Newman, 1966; Penman *et al.*, 1968). The RNA hybridized and then eluted from the filter has a characteristic sedimentation pattern of messenger RNA (Penman *et al.*, 1968); it is hetero-

FIGURE 12

A comparison of RNA hybridized to and washed through the filters under various conditions

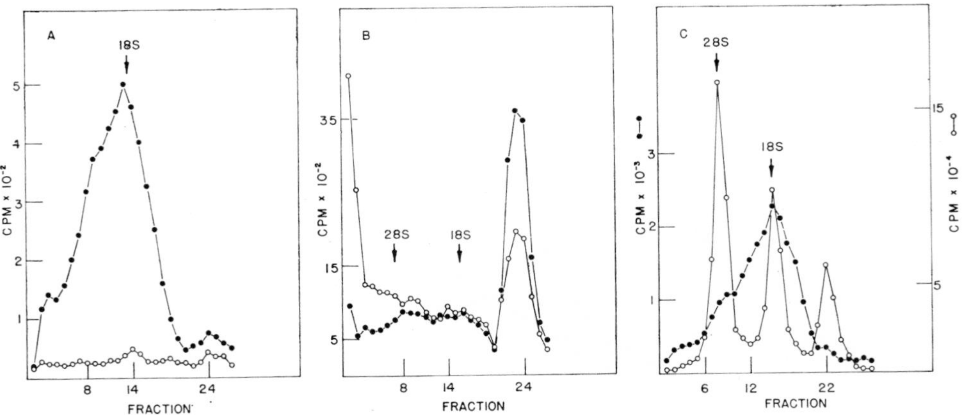

Panels A and B show the effect of cordycepin on the filter-binding and non-binding material. Ten ml of cell culture (2×10^6 cells/ml) were exposed to a low level of actinomycin D (0.04 μg/ml) and ethidium bromide (1 μg/ml). One-half had cordycepin added (20 μg/ml). The cells were then labelled with [^{3}H] uridine (New England Nuclear, 20ci/mM) for 90 minutes. Panel A compares filter-bound material in the presence and absence of cordycepin. Panel B compares the washes from these filters. Panel C represents the wash and bound material from cell cytoplasm labelled for 90 minutes (no drugs present) in 2 μci/ml of [^{3}H] uridine (4×10^5 cells/ml). For all experiments, cells swollen in RSB buffer were lysed with NP40 and the cytoplasm extracted with phenol after the method of Penman (1966), except that the phenol and chloroform were mixed prior to the extraction. The pellet after precipitation with ethanol was resuspended in 2 ml of SDS buffer and bound to a poly(U) filter. Poly(U) filters were made after the method of Sheldon *et al.*, (1972). The sample of total cytoplasm was passed through the poly U filter at about 1 ml/min. The filters were washed with 4 ml of SDS buffer and the material bound was eluted from the filter by passing 4 ml of a formamide solution (50% formamide, 50% elution buffer, 0.01M tris, pH 7.4, 0.5% SDS) at 45°C through the filter. Carrier RNA was added to eluant (made 0.1M NaCl) and wash. Both were precipitated with alcohol and then resuspended in 0.8 ml SDS and layered on an SDS sucrose gradient (15-30% w/v) and centrifuged at 25,000 rev./min in a Spinco SW27 rotor for 18 hours. The gradients were collected, precipitated with 5% trichloracetic acid and assayed for radioactivity. A: Cordycepin cytoplasm ○——○, control (low actinomycin D) cytoplasm ●——●

B: Cordycepin was ○——○, control (low actinomycin D) wash ●——●

C: ●——● eluant, ○——○ wash

geneous and extending from 8S to approximately 30S with a maximum at 18S. The transmitted RNA shown in Fig. 12B consists of 4S RNA and a flat distribution of heterogeneous RNA extending to greater than 60S which has been shown to be unassociated with polyribosomes and is assumed to be unrelated to messenger RNA (Penman *et al.*, 1968). Messenger RNA does not appear in the cytoplasm in the presence of cordycepin (Penman *et al.*, 1970), an adenosine analogue, and, indeed, panel A indicates that this drug inhibits the labelling of nearly all filter-binding material but has little effect on transmitted RNA. Furthermore, the RNA bound to the filter is found associated with polyribosomes and is released by EDTA. It therefore appears to have the behaviour expected of messenger RNA.

Fig. 12C indicates the powerful selectivity of the filters. Total cellular RNA was labelled without suppressing ribosomal RNA synthesis. The total labelled cellular RNA was bound to a filter. Although only a few per cent of this RNA is messenger, this filter technique selectively hybridizes messenger RNA uncontaminated by ribosomal RNA. Less than 0.2% of the input ribosomal RNA is retained on the filter.

The stability of m-RNA in actinomycin D was measured using the poly U filters (Fig. 13). Actinomycin D was added to cells after a 90-minute labelling period,

FIGURE 13

Stability of messenger RNA in the presence of actinomycin D and the effect of polyribosome dissociating agents

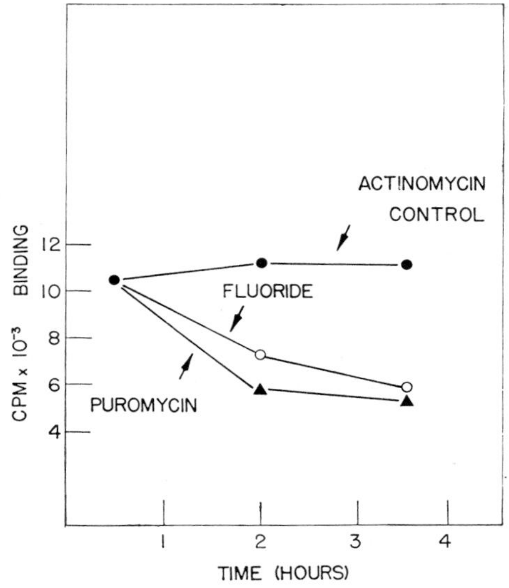

Seventy ml of cells (2×10^6 cells/ml) were incubated for 1 hour in [^{3}H] uridine (10 μci/ml) and then actinomycin D (4 μg/ml) was introduced into the culture.. Twenty minutes after the addition of actinomycin D, the remaining cells were divided into three separate mixtures; one received 200 μg/ml puromycin, one 10 mM NaF, and the last was a control. Aliquots were removed from each mixture at 2 hours and 3½ hours after the addition of actinomycin D. Cytoplasm was prepared as in Fig. 12. The samples were extracted with phenol. All samples were analyzed with the poly-U filters after the procedure as in Fig. 12 and run on gradients along with the wash, in order to correct for variations in recovery. Total counts bound to the filter are plotted.

and measurements made at subsequent intervals taken several hours later. The bound RNA eluted from the filters as well as the transmitted RNA were analyzed on sucrose gradients. The messenger RNA eluted from the filter is identical in all samples in its sedimentation distribution and purity. The data show little decay of the m-RNA eluted from the filter, even after 5 hours in actinomycin D.

Thus messenger RNA in exponentially growing HeLa cells apparently has a half-life much longer than six hours. It is possible, however, that actinomycin D may be interfering with the normal messenger decay process, such as occurs with hormone-induced protein synthesis. We have measured the kinetics of labelling messenger RNA with the filter technique in the absence of any inhibitors. These experiments, which will be described in a future report, indicate a messenger lifetime in the order of 24 hours in exponentially growing HeLa cells.

Mechanisms do exist in the cell which can break down messenger RNA relatively rapidly. In the presence of puromycin (200 μg/ml) or fluoride ion (10mM), the messenger can be seen to decay significantly using the poly-U filter technique (Fig. 13). Similar experiments done with total cytoplasmic RNA pre-labelled in low actinomycin D concentrations to suppress ribosmal RNA yield the same result. Thus, the loss of filter-binding RNA does not result from the loss of the poly(A) sequence from the messenger RNA. Since both puromycin (Joklik and Becker, 1965) and fluoride (Columbo *et al.*, 1968) are polyribosome dis-aggregating agents, the data suggests that messenger RNA stability requires intact polyribosomes and that interruption of normal translation results in rapid messenger decay. However, both drugs can cause serious side reactions so that the conclusion must be viewed as tentative.

These experiments indicate that messenger RNA appears essentially stable in the presence of actinomycin D. We show also that actinomycin D has the additional effect of slowing the initiation of ribosomes onto the messenger RNA. Therefore, the rate of protein synthesis is not a measure of the availability of messenger RNA. Finally, the results using polyribosomal disruptors imply that messenger must be on polyribosomes in order to be stable. Although the demonstration of the existence of m-RNA in the cytoplasm does not prove it functional, the fact that it can be used to build up polyribosomes for as long as 3 hours after exposure to actinomycin D suggests that it probably remains functional for at least this long. Furthermore, a report now in preparation indicates that, after a 24-hour chase, labelled m-RNA is still associated with polyribosomes.

The mechanism of regulating protein synthesis in mammalian cells appears to be distinctly different from that of prokaryotes. The existence of long-lived messenger RNA in mammalian cells suggests that control of protein synthesis occurs partially at the level of translation. In particular, regulation of overall protein synthesis at the level of initiation of translation of messenger RNA has been found under several different conditions in mammalian cells (Adamson *et al.*, 1969; McCormick and Penman, 1969; Tomkins *et al.*, 1969; Fan and Penman, 1970; Liebowitz and Penman, 1971; Lodish, 1971).

The following experiments deal with the question of whether the amount of each messenger RNA present at a given time determines completely, or only partially, the amount of each protein synthesized. Initiation of protein synthesis occurs at different rates for the α and β haemoglobin messages in reticulocytes (Lodish, 1971). No information exists for any other cell as yet. The experiments reported here describe what may be a meaningful model system: the translation of adenovirus-specific messenger RNA late in lytic infection in HeLa cells. It will be shown that messenger RNA molecules coding for different adenovirus proteins are translated at different rates and that this is probably due to differential rates of initiation.

In contrast to the uninfected cells, ribosomes in adenovirus-infected HeLa cells initiate translation at a low rate compared to the rate of peptide elongation. Consequently, the polyribosomes are lightly loaded and sediment slowly as shown in Figs. 14A and 14B. Infected cell polyribosomes are much smaller in sedimentation

FIGURE 14

Effect of cycloheximide on polyribosomes from adenovirus-infected cells

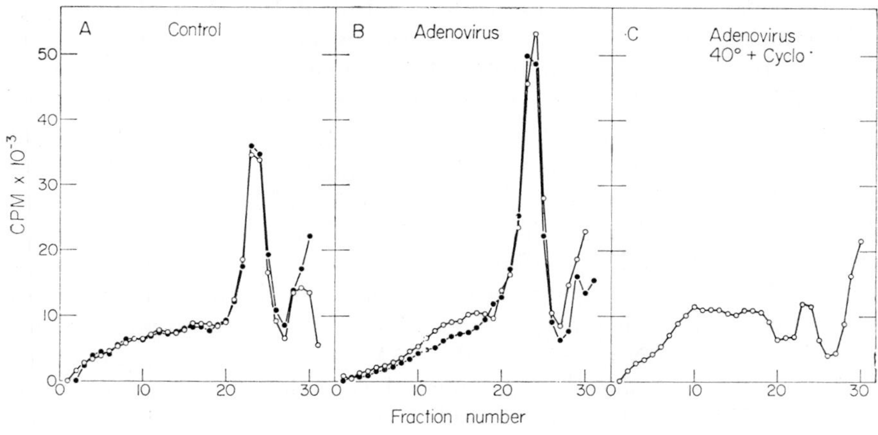

(A) Polyribosomes from uninfected cells

HeLa cells were grown in suspension at a concentration of 4 × 10^5 cells/ml as previously described (Eagle, 1959). 50 ml of cells were labelled with 5 μci [^{3}H] uridine for 40 hours. The cells were incubated at a concentration of 2 × 10^6 cells/ml for 45 minutes. Cytoplasm was prepared as described in Fig. 11. Polyribosomes were analyzed by centrifugation through 15-30% RSB-sucrose (10 mM NaCl, 1.5 mM $MgCl_2$, 10 mM tris, pH 7.4) for 130 minutes at 25,000 rev./min in the Spinco SW27 rotor in the Beckman Ultracentrifuge. Fractions were collected and assayed for radioactivity.

(B) Polyribosomes from infected cells

50 ml of cells were labelled with 20 μci [^{3}H] uridine for 20 hours. The cells were then infected with adenovirus 2 as described in (Price and Penman, 1972a). Twenty hours after infection, cells were incubated at a concentration of 2 × 10^6 cells/ml for 45 minutes and polyribosomes prepared and analyzed as in Fig. 14A.

(C) Effect of cycloheximide on polyribosomes from infected cells

As in B, except that cycloheximide at a concentration of 2 μg/ml was present during the 45-minute incubation period.

velocity, although the size of the polypeptides synthesized in the adenovirus-infected cell are at least comparable to or larger than the average size of uninfected cell proteins (Fan and Penman, 1970). Adenovirus messenger RNA molecules should be on the average as large as those in uninfected cells, and the lower sedimentation velocity suggests a lower density of packing of ribosomes on messenger RNA in the virus-infected cell.

One explanation of the small size of adenovirus polyribosomes is that initiation of protein synthesis relative to peptide elongation occurs at a lower rate in the infected than in the normal cell. If translation is slowed by use of low levels of an inhibitor of elongation, then initiation should no longer be rate limiting. Polyribosomes should become more densely packed and hence more rapid in sedimentation. The results in Fig. 14C show that polyribosomes can indeed be much more heavily loaded in the adenovirus-infected cell if low levels of cycloheximide are added to the medium. The presence of 2 μg/ml cycloheximide reduces the elongation rate by a factor of 5 and the resulting sedimentation distribution of polyribosomes is similar to that of the uninfected cells. Thus, while initiation of translation appears to partially control the relative synthesis of adenovirus-specific proteins, it is possible to reduce control at the level of initiation by severely reducing the rate of translation. In the extreme case where translation is very slow so that initiation is unimportant, the amount of each polypeptide produced should be proportional only to the amount of messenger RNA available.

It is therefore possible to ask whether the different amounts of each viral-specific protein produced are due to different amounts of messenger RNA present in the cytoplasm or differential rates of translation of the available messenger RNA. Low levels of cycloheximide, as described above, render the translation of messenger RNA independent of different rates of initiation or other factors controlling translation. At very low rates of elongation in the presence of the drug, the amount of each polypeptide synthesized should become proportional to the amount of messenger RNA available. Fig. 15A shows a gel electropherogram of adenovirus proteins produced under normal conditions. In a parallel culture shown in Fig. 15B, cycloheximide is added at a concentration of 2 μg/ml. The amount of label in each protein is reduced, but the ratio of label in the different viral proteins is significantly altered. The most dramatic change is seen in the reduction of hexon (II) labelling relative to other proteins. It should be noted that the labelling period in the above experiments was followed by a chase with cold amino acids in the absence of a protein synthesis inhibitor. This procedure insured the completion of polypeptides partially synthesized during the labelling period.

These experiments indicate that adenovirus messenger RNA molecules are translated at different rates. While it seems most likely that this is due to differential initiation of translation of the various adenovirus messages, other possibilities such as varying rates of elongation cannot be ruled out. The observation that virus-infected cell polyribosomes are relatively small but can be greatly increased in size by treatment with low levels of cycloheximide suggests that initiation of translation relative to elongation in general has decreased late in adenovirus infection. In the following experiments, the rate of initiation of translation of adenovirus protein is further depressed by the use of elevated temperatures. Under these conditions the amount of each protein synthesized is greatly reduced, but low levels of cycloheximide return the pattern of translation to that seen at normal temperature in the presence of the drug.

It has been shown previously that the initiation of translation of HeLa cell messenger RNA is greatly depressed at 42°C (McCormick and Penman, 1969). The initiation of translation of adenovirus message fails at a significantly lower temperature and is greatly suppressed at 40°C, a temperature which has little effect on normal cell protein synthesis. The polyribosomes of uninfected cells at

FIGURE 15

Effect of cycloheximide on distribution of adenovirus-specific proteins

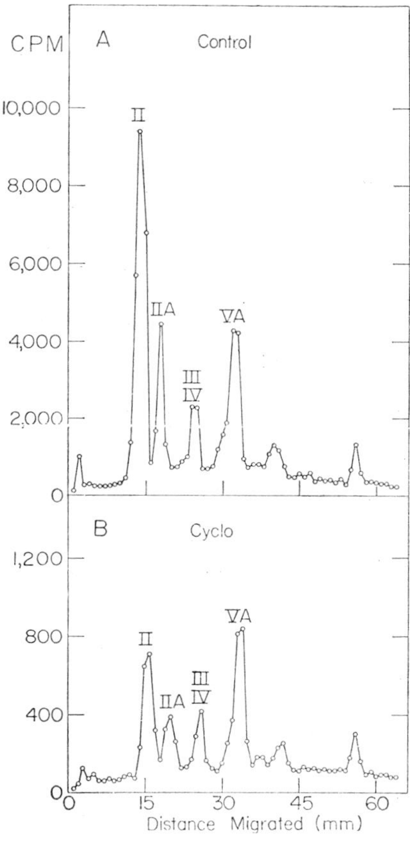

(A) Control

(B) Cycloheximide-treated cells

Twenty ml of HeLa cells were infected with adenovirus as in Fig. 14B. Eighteen hours after infection, the cells were resuspended at a concentration of 4 × 10^5 cells/ml in medium containing 1/10 the normal amount of leucine. One-half was pre-treated with cycloheximide at a concentration of 2 μg/ml. Both aliquots were labelled with 20 μci of [^{3}H] leucine for 45 minutes. At the end of the 45-minute incubation period, the drug was washed out and the cells incubated in complete medium for an additional 10 minutes. Cytoplasm was prepared and the equivalent of 1.6 ml of cytoplasm of each sample was examined by disc gel electrophoresis.

37°C and 40°C are shown in Fig. 16A. Very little change in the sedimentation distribution is seen. Adenovirus polyribosomes at 40°C are even more lightly loaded than at 37°C, as shown in Fig. 16B. Even under these severely depressed rates of initiation, however, a low dose of cycloheximide results in a significant reconstruction of the adenovirus polyribosomes as shown in Fig. 16C. In the next experiment, it will be seen that adenovirus-specific protein synthesis is greatly decreased at 40°C so that disaggregation of adenovirus polyribosomes is not due simply to increased rates of elongation.

FIGURE 16

Effect of elevated temperature on uninfected and adenovirus-specific polyribosomes

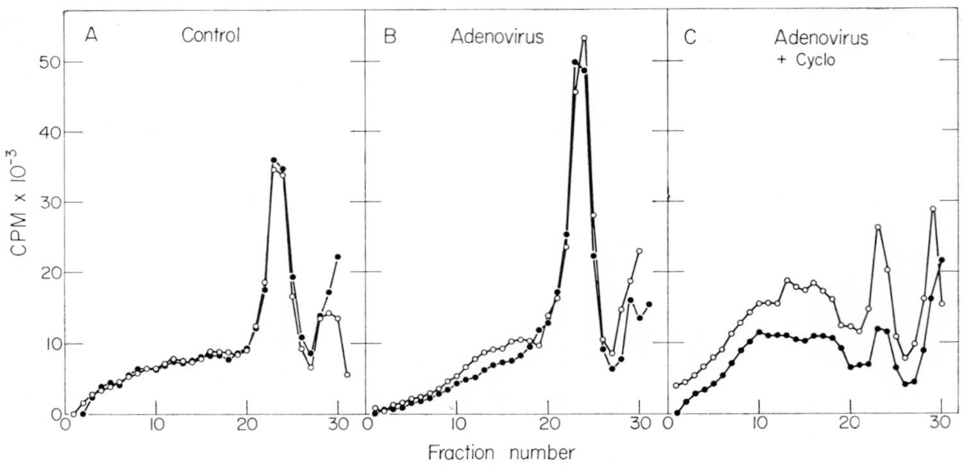

(A) Control polyribosomes

100 ml of cells were labelled with 10 μci [³H] uridine for 40 hours. 50 ml of cells were incubated at a concentration of 2 × 10^6 cells/ml at 37°C for 45 minutes. The other 50 ml were incubated for 45 minutes at 40°C. Cytoplasm was prepared and polyribosomes analyzed as in Fig. 14A.

(B, C) Adenovirus polyribosomes

50 ml of cells were labelled with 10 μci [³H] uridine for 24 hours, at which time they were infected with adenovirus 2 as in Fig. 14A. At 20 hours after infection, the cells were divided into three portions. One-third was incubated at 37°C, one-third was incubated at 40°C and one-third was incubated at 40°C in the presence of 2 μg/ml of cycloheximide, Polyribosomes were prepared and analyzed as in Fig. 14A.

(B) Adenovirus-infected cells incubated at 37°C and 40°C

(C) Adenovirus-infected cells incubated at 40°C in the presence of cycloheximide (A, B)
—○— 37°C —●— 40°C

The effect of increased temperature on the labelling of different adenovirus proteins is shown in Fig. 17. The gel electropherogram in Fig. 17A shows the labelling of proteins at 37°C. At 40°C overall protein synthesis has been reduced and the relative amounts of different adenovirus proteins have changed somewhat. The synthesis of proteins II hexon) and IIA is not as inhibited as is that of proteins III, IV and VA.

Using low levels of cycloheximide to slow translation and consequently rebuild polyribosomes as in Fig 16, the relative rates of labelling of the adenovirus proteins are restored to a pattern which approximates that seen at 37°C in the presence of the drug. This is shown in Fig. 17C. The production of hexon (protein II) is preferentially decreased by treatment with the drug.

The experimental results suggest that adenovirus messenger RNA molecules are translated at different rates and indicate that this may be due to differences in the rate of initiation of translation of the various messenger RNAs. The results resemble those obtained by Lodish (1971) in the reticulocyte, where the mes-

FIGURE 17

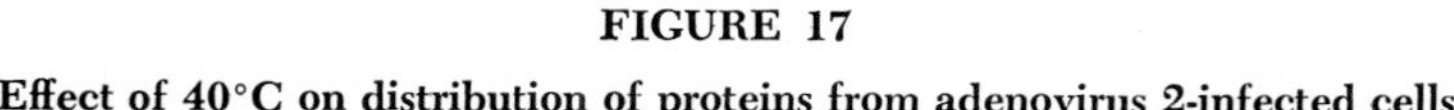

Effect of 40°C on distribution of proteins from adenovirus 2-infected cells

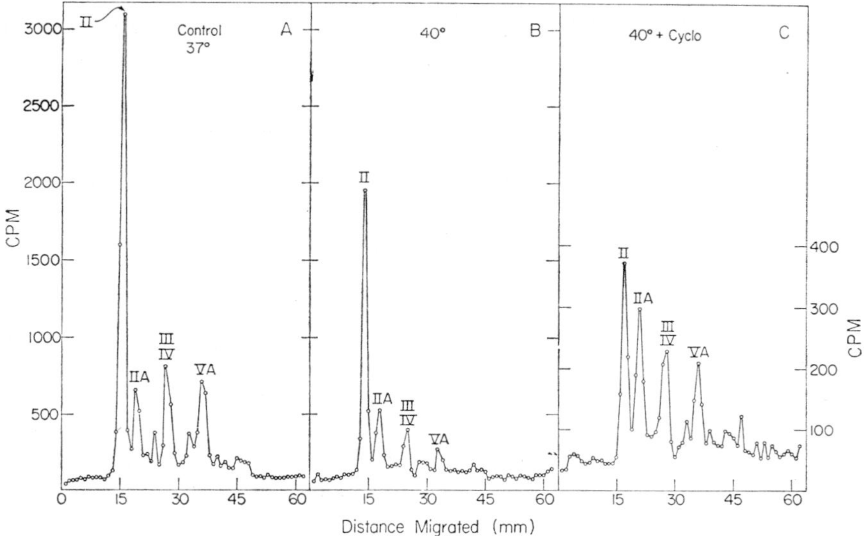

Thirty ml of cells were infected with adenovirus as in Fig. 14A. The cells at 20 hours after infection were resuspended at a concentration of 4×10^6 cells/ml in medium containing 1/10 the normal amount of leucine. One-third was incubated at 37°C, one-third at 40°C and one-third at 40°C in the presence of 2 μg/ml cycloheximide. All three samples were labelled with 25 μci [^{3}H] leucine for 45 minutes. At the end of the incubation, the cells were washed and resuspended in complete medium for an additional 10 minutes. Cytoplasm was prepared and analyzed by gel electrophoresis.

(A) Cells incubated at 37°C

(B) Cells incubated at 40°C

(C) Cells incubated at 40°C in the presence of cycloheximide

senger RNA molecules for α and β haemoglobin are initiated at different rates so that equimolar amounts of the two haemoglobin polypeptides are synthesized.

It is not known at present whether the translation of adenovirus messenger RNA reflects the propensity of pre-existing cellular mechanisms to initiate different messenger RNAs at substantially different rates, or whether this is unique to adenovirus-infected cells. The increased temperature sensitivity of initiation of translation in adenovirus-infected cells is consistent with a viral origin for the temperature-sensitive component of the initiation system. However, it is also possible that the infected cell milieu is sufficiently different from that of the uninfected cell so that protein synthesis initiation is rendered significantly more sensitive to elevated temperature. This latter point may be clarified by studies on adenovirus protein synthesis which occurs before the complete cessation of host cell protein synthesis.

CONCLUSION

We have seen some of the ways in which the mechanisms of transcription and translation in mammalian cells differ from those found in prokaryotes. Transcription is accomplished by several distinct enzymes which appear to be specialized for the type of RNA produced. Thus it appears that ribosomal precursor RNA is synthesized exclusively by an activity resembling polymerase I, while the heterogeneous nuclear RNA, which seems to be precursor to cytoplasmic messenger, is transcribed by a separate enzyme which appears most certainly to correspond to polymerase II. We have seen that low molecular weight RNA, such as 5S and transfer RNA in the uninfected cell and the virus-associated 5.5S RNA in the infected cell, appear to be produced by yet a third polymerase activity which may correspond to polymerase III.

Another aspect of transcription in mammalian cells which has not been dealt with directly in this report is the production of precursor molecules which are much larger than the final products. This is true of both ribosomal RNA and cytoplasmic messenger RNA. The only exceptions known are the low molecular weight species, 5S and the adenovirus-associated 5.5S, which are synthesized, apparently, in a form close to or identical with their final composition. 4S RNA appears be intermediate in that a precursor which is slightly larger than the final product is synthesized. Just why a separate polymerase activity is devoted to different species of RNA is unanswered. It is possible, of course, that the regulatory mechanisms for the different species are quite distinct and that this is reflected in the type of polymerase molecule used as a component for these different systems.

We have also seen that the regulation of protein synthesis is distinct from that in bacteria in that messenger RNA is not the rate-limiting factor in the quantity of protein produced. Rather, messenger RNA is a relatively stable component of the protein synthetic machinery, and short-term regulation, at least, is accomplished at the level of the initiation of translation. In this report, we have shown that, in the adenovirus-infected cell, there is also evidence for the differential utilization of messenger molecules. Thus the amount of protein produced is not a simple function of the amount of messenger RNA coding for a particular protein. We assume that the adenovirus-infected cell is a model system for normal cells and, indeed, Lodish (1971) has shown that the reticulocyte has a mechanism for differential utilization of different messenger molecules. We have not shown whether the intensity of translation of separate messenger molecules can be independently regulated or whether the degree of utilization is a fixed property of each message. Some preliminary results on protein synthesis in metaphase cells suggest that, at least in this unusual condition, the differential rates of messenger utilization can, in fact, be altered by cellular mechanisms.

REFERENCES

Adamson, S. D., Herbert, E., and Kemp, S. F. (1969) *J. Mol. Biol.*, *42*, 247

Byrne, R., Levin, J. G., Bladen, H. A., and Nirenberg, M. W. (1964) *Proc. Nat. Acad. Sci. U.S.*, *52*, 140

Cheevers, W. P., and Sheinin, R. (1970) *Biochim. Biophys. Acta*, *204*, 449

Columbo, B., Vesco, C., and Baslioni, C. (1968) *Proc. Nat. Acad. Sci. US*, *53*, 1437

Craig, N., Kelley, D., and Perry, R. (1971) *Biochim. Biophys. Acta*, *246*, 493

Darnell, J. E., Jr., Wall, R., and Tushinski, R. J. (1971) *Proc. Nat. Acad. Sci. US*, *68*, 1321

Eagle, H. (1959) *Science, 130,* 432

Edmonds, M., Vaughan, M. H., Jr., and Nakagato, H. (1971) *Proc. Nat. Acad. Sci. US, 68,* 1336

Fan, H., and Penman, S. (1970) *J. Mol. Biol., 50,* 655

Godchaux, W., Adamson, S. O., and Herbert, E. (1967) *J. Mol. Biol., 27,* 57

Green, M. (1970) *Annu. Rev. Biochem., 39,* 701

Green, M., Parsons, J. T., Pina, M., Fujinaga, K., Caffier, H., and Landgraf-Leurs, J. (1970) *Cold Spring Harbor Symp., 35,* 803

Joklik, W., and Becker, Y. (1965) *J. Mol. Biol., 13,* 496

Lee, S. Y., Mendecki, J., and Brawerman, G. (1971) *Proc. Nat. Acad. Sci. US, 68,* 1331

Leibowitz, R., and Penman, S. (1971) *J. Virology, 8,* 661

Levinthal, C., Keynan, A., and Higa, A. (1962) *Proc. Nat. Acad. Sci. US, 48,* 1631

Lodish, H. (1971) *J. Biol. Chem., 246,* 7131

McCormick, W., and Penman, S. (1969) *Virology, 31,* 135

Miller, O. L., Jr., Beatty, B. B., Hamkalo, B. A., and Thomas, Jr., C. A. (1970) *Cold Spring Harbor Symp., 35,* 505

Ohe, K., and Weissman, S. (1971) *J. Biol. Chem., 246,* 6991

Ohe, K., Weissman, S., and Cooke, N. (1969) *J. Biol. Chem., 244,* 5320

Penman, S. (1966) *J. Mol. Biol., 17,* 117

Penman, S., Rosbash, M., and Penman, M. (1970) *Proc. Nat. Acad. Sci. US, 67,* 1878

Penman, S., Scherrer, K., Becker, Y., and Darnell, J. E. (1963) *Proc. Nat. Acad. Sci. US, 49,* 654

Penman, S., Vesco, C., and Penman, M. (1968) *J. Mol. Biol., 34,* 49

Perry, R. P. (1963) *Exptl. Cell. Res., 29,* 400

Perry, R. (1966) *Nat. Cancer Inst. Monograph, 23,* 527

Phillipson, L., Wall, R., Glickman, R., and Darnell, J. E. (1971) *Proc. Nat. Acad. Sci. US, 68,* 2806

Price, R., and Penman, S. (1972) *J. Virology, 9,* 621

Price, R., and Penman, S. (1972) *J. Mol. Biol.,* in press

Raskas, H. (1971) *Nature New Biol., 233,* 134

Raskas, H., and Okubo, C. (1971) *J. Cell Biol., 49,* 438

Rifkind, R. A., Danon, D., and Marks, P. A. (1964) *J. Cell. Biol., 22,* 599

Roberts, W. K., and Newman, J. (1966) *J. Mol. Biol., 20,* 63

Roeder, R. G., and Rutter, W. J. (1969) *Nature London, 224,* 234

Roeder, R. G., and Rutter, W. J. (1970a) *Proc. Nat. Acad. Sci. US, 65,* 675

Roeder, R. G., and Rutter, W. J. (1970b) *Biochemistry, 9,* 2543

Sheldon, R., Jorale, C., and Kates, J. (1972) *Proc. Nat. Acad. Sci. US, 69,* 417

Stachelin, T., Wettstein, F. O., and Noll, H. (1963) *Science, 140,* 180

Suzuki, Y., and Brown, D. (1972) *J. Mol. Biol., 63,* 409

Thomas, D. C., and Green, M. (1966) *Proc. Nat. Acad. Sci. US, 56,* 243

Tomkins, G. M., Gelehrter, T. D., Granner D., Martin, D., Samuels, H. S., and Thompson, E. B. (1969) *Science, 166,* 1474

Velicer, L., and Ginsberg, H. (1968) *Proc. Nat. Acad. Sci. US, 61,* 1264

Wallace, R. D., and Kates, J. (1972) *J. Virology, 9,* 627

Zylber, E. A., and Penman, S. (1971) *Proc. Nat. Acad. Sci. US, 68,* 2861

Isolated Chromatin in the Study of Gene Expression

A Personal and Biased Account of Chromatin as of May, 1972

SARAH C. R. ELGIN[a] AND JAMES BONNER

Division of Biology, California Institute of Technology Pasadena, California, U.S.A.

INTRODUCTION TO CHROMATIN

One of the salient features of eukaryotic chromatin is that its transcription by RNA polymerase is limited. Only a small percentage of the DNA sequences in the genome is transcribed in a given tissue at a given time. Although there can be considerable overlap of sequences transcribed, the pattern of RNA transcription is tissue specific (Paul and Gilmour, 1966, 1968; Bekhor *et al.*, 1969; Smith *et al.*, 1969; McConaughy and McCarthy, 1972). We would like to understand the mechanism by which this tissue specific transcription is dictated, and the means by which the program of transcription is changed. One effective method used in studying these problems during the last several years has been the analysis of the macromolecular components and the reactions of isolated chromatin *in vitro*. Chromatin, the interphase form of the cell's chromosomes, is a complex of DNA, RNA, histone, and non-histone chromosomal protein (NHC protein or hertones).[b] We will discuss here some of the more important recent findings concerning the chemical and physical characteristics of the latter three components and some experiments aimed at elucidating their association with the DNA and/or their possible biological roles.

Chromatin can be isolated for *in vitro* study by chemically gentle techniques based on differential centrifugation. In all experiments we will be discussing, the chromatin was prepared by the following technique. Frozen tissue was ground in a Waring blender or alternatively the cells were otherwise lysed in

---

[a] Fellow of the Jane Coffin Childs Memorial Fund for Medical Research.

[b] The term "hertone" has been suggested for the nonhistone chromosomal protein class by R. D. Cole. The word is derived from the notion that this class of proteins may be involved in the more delicate aspects of gene regulation.

saline-EDTA (0.075 M-NaCl plus 0.024 M-EDTA, pH 8). The crude nuclear pellet was isolated by low speed centrifugation. The nuclei were washed and then lysed in 0.01 M-tris-HCl; the chromatin was recovered by centrifugation. This crude chromatin was further purified by centrifugation through a sucrose gradient (maximum 1.7 M-sucrose). The purified chromatin was resuspended and used directly, or sheared to increase solubility and uniformity of physical characteristics (Bonner *et al.*, 1968b). Isolated chromatin is a good model for chromatin as it exists *in vivo*. This has been established by experiments which show that the properties of chromatin as a template for DNA-dependent RNA polymerase are not drastically altered by isolation. As far as can be determined by present analytical techniques, the same regions of the genome are transcribed *in vitro* as *in vivo* (Marushige and Bonner, 1966; Paul and Gilmour, 1966, 1968; Bekhor *et al.*, 1969; Smith *et al.*, 1969; Tan and Miyaga, 1970; Chetsanga *et al.*, 1970). Note that this conclusion must still be tempered by the knowledge that the above-cited *in vitro* transcription studies were carried out using prokaryotic RNA polymerase and that the hybridization competition experiments used to evaluate *in vitro* transcription examine most effectively transcription from repetitious DNA (see below). While it seems likely that much fine control (such as the relative amounts of transcription at various genes) could be lost on isolation, the basic characteristics of chromatin as a template appear to be maintained *in vitro*, and isolated chromatin has proved to be an excellent system for study.

The chemical composition of chromatin in terms of the mass ratios of the macromolecular components has been determined for many tissues. Typical data are

TABLE 1

Chemical Composition of Chromatin[a]

Source of chromatin	No. of prepns	Content relative to DNA, of			
		DNA	RNA	Histone	Hertone
Rat liver	7	1	0.04	1.15	0.95
Newborn rat liver	3	1	0.05	1.07	0.63
Rat Novikoff ascites[b]	3	1	0.13	1.16	1.00
Rat kidney	4	1	0.06	0.95	0.70
Mouse kidney	2	1	0.02	1.11	0.93
Chicken liver	3	1	0.03	1.17	0.88
Chicken erythrocyte	5	1	0.02	1.08	0.54
Pea bud	2	1	0.05	1.10	0.41

[a]Data from Elgin (1971) unless otherwise noted.
[b]Data from Bonner *et al.* (1968a).

given in Table 1. Such mass ratio data usually have a standard deviation of 10% (Elgin and Bonner, 1970). The histone:DNA ratio is generally on the order of 1:1, while the hertone:DNA ratio shows considerably greater variation. The RNA is a relatively minor component.

Recent studies have demonstrated that the DNA of eukaryotes is complex and repetitious. In addition to a single copy (unique) DNA, presumably analogous to that of prokaryotes, 10-40% of the DNA of most eukaryotes is made up of sequences repeated 10^2-10^4 times each (middle repetitious DNA), and 10% of the DNA is made up of sequences repeated 10^6 times (highly repetitious DNA) (Britten and Kohne, 1967; Britten and Davidson, 1971). *In situ* hybridization studies have shown that the centromeric DNA is highly repetitious (Pardue and Gall, 1970); structural roles are generally evoked for this class of DNA. Middle repetitious DNA includes sequences complementary to the "working" RNA's—r-RNA, t-RNA and 5S RNA. A regulatory role has also been suggested for RNA transcribed from middle repetitious DNA (Britten and Davidson, 1969). All the evidence to date suggest that m-RNA is transcribed from unique DNA (haemoglobin, Bishop *et al.*, 1972; silk fibroin, Gage, 1971). Kedes and Birnstiel (1971) have reported that a m-RNA fraction, primarily involved in synthesizing nuclear proteins, hybridizes with the kinetics of middle repetitious DNA. However, the protein synthesized from this m-RNA probably includes hertones as well as histones; thus the derived proteins are of unknown heterogeneity, and this makes the observation difficult to interpret. In evaluating data on the analysis and transcription of chromatin, one must keep in mind that many hybridization studies are carried out under conditions in which most of the DNA or RNA transcripts involved in the analysis represent repetitious sequences. This difficulty can be overcome by using isolated unique DNA in hybridization studies. We now know that both unique and middle repetitious DNA are transcribed in a tissue-specific program *in vivo* (for references see page 161 and McConaughy and McCarthy, 1972), and the fidelity, at least of middle repetitious DNA transcription, is maintained *in vitro*. It should be noted that the complexity of eukaryotic DNA may have unexpected effects on the attempts to dissociate and reconstitute DNA-protein complexes. It is possible that repetitious DNA is critically involved in protein-DNA complexes that dictate the tertiary struture of chromosomes; such a structure would be difficult to reconstitute correctly using typical preparations of sheared chromatin.

THE HISTONES

One can generally assume that the information required for the construction of an organism must be contained in the DNA complement of the genome. However, the timing of the utilization of this information is critical for the orderly development of the organism. Our interest has focused on the association of other macromolecules with the DNA, on the effect this has on the availability of the DNA for transcription, and on the correlation of changes in these associations with changes in transcription. The first protein component of chromatin to be analyzed and studied in detail was the fraction referred to as histones. The histones are small, basic proteins found in association with the DNA of all eukaryotes (Elgin *et al.*, 1971). The fungi are a possible exception (Leighton *et al.*, 1971); however, R. D. Cole has recently isolated histones from *Neurospora* in low yield (personal communication). There are about a dozen histones, described in four classes according to their lysine/arginine ratio, as seen in Table 2. All the histones are very basic, having isoelectric points above 10. They vary in molecular weight from 11,300 to 21,000. Several lines of experimental evidence, summarized below,

TABLE 2.

Properties of the Chromosomal Proteins

The Histones[a]

Class	Lys/Arg ratio	Molecular weight	Asx+Glx/Arg+His+Lys ratio	Number of subfractions
I or f1	22	21,000	0.20	3-8
f2a2 AL		15 000		
II or or . .	~2.5		0.52	2-3
f2b —		13,770		
III or f 3	0.8	14,900	0.66	1
IV or f2al or GAR . .	0.7	11,300	0.41	1

The Hertones[b]

Class	SDS gel bands	pI	Molecular weight	Asx+Glx/ Arg+His+Lys ratio	Est. No. of subfractions
I	θ_{I}	<3.7	49,100	2.7	1
II3-IIIa2	$\pi, \lambda, \kappa, \theta_{II}$	5.4-6.6	49,100-98,200	1.7	4-6
IIIb3′	γ, β	6.4-8.0	16,000-19,000	1.2	4-7
ϵ	ϵ	5.6	31,200	1.4	1

[a]Specific data for calf thymus, Elgin *et al.*, 1971.
[b]Data from rat liver. Elgin and Bonner, 1972.

have shown that histones can cause gene repression, i.e., when histones are associated with DNA, that portion of the genome is not available to RNA polymerase for transcription. The fractionation of chromatin into template-active and template-inactive portions yields fractions whose histone:DNA ratio shows an inverse correlation with template activity and whose hertone:DNA ratio shows an even stronger direct correlation with template activity (Frenster, 1965; Marushige and Bonner, 1971). Selective *in vitro* removal of histones from chromatin by a variety of reagents ($NaClO_4$, NaCl, dilute acid, sodium deoxycholate (DOC), etc.) results in a concomitant increase in template activity (for example, see Marushige and Bonner, 1966); likewise, the addition of histone to DNA in solution causes a proportional decrease in template activity (Shih and Bonner, 1970). It should be noted that the evidence to date does not indicate the extent to which histones are involved in transient gene repression/derepression (e.g., hormone response). There is evidence from *Drosophila* to suggest that the histone-DNA association may be altered in some subtler way than histone removal during gene activation (see Ashburner, 1970, for a review of the *Drosophila* system).

Many years ago it was suggested that the presence of tissue-specific histones might be the mechanism whereby tissue-specific transcription was dictated (Stedman and Stedman, 1950). However, further study has shown that in the preponderance of cases all of the histones are present in all of the tissues of a given organism (Fambrough *et al.,* 1968; Panyim and Chalkley, 1969 and Fig. 1).

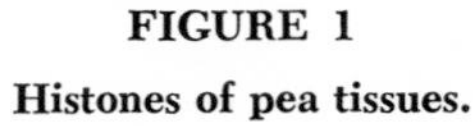

FIGURE 1

Histones of pea tissues.

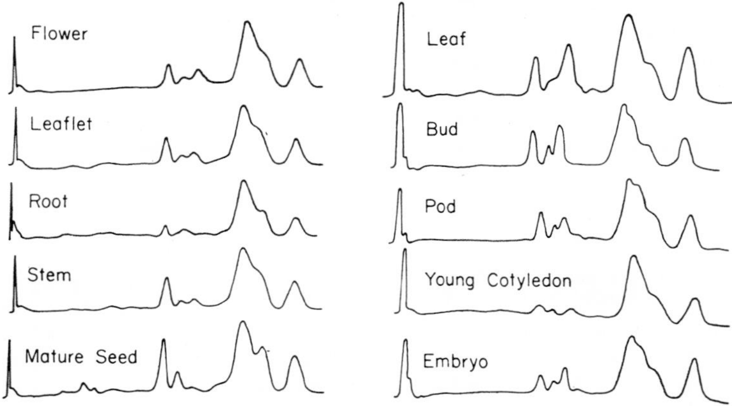

Densitometric tracings of the polyacrylamide gel patterns of pea histones from different tissues. Gels run in 6.25 M-urea, pH 4.3, left to right. Peaks at far left indicate origins of gels (Fambrough *et al.,* 1968).

As indicated in Table 2, only a small number of unique histones have been found; only histone I shows significant variability, where the presence of 3-8 subfractions in different amounts in different tissues and species has been demonstrated (Bustin and Cole, 1968, 1969). There is no good evidence for code sequence specificity in the binding of histones to DNA, although some class preferences can be demonstrated under appropriate conditions. (For example, in 1 M-NaCl polylysine binds preferentially to A-T rich DNA [Leng and Felsenfeld, 1966]; see Elgin *et al.,* 1971 for review.) At present, the best interpretation of the evidence is that other factors (macromolecular chromatin components?) direct or modulate the histone-DNA interaction to obtain tissue-specific gene repression.

Next it is of interest to inquire whether any of the particular histones have particular roles in the repression mechanism, or whether all act equivalently and in concert. To examine this question, Dr. John Smart has looked at the alterations in several biological and physical properties of chromatin during the sequential extraction of histones by two different reagents which removed the histone classes in different order. One reagent, sodium chloride, removes the histones in the order I, II, III-IV, while the second reagent, DOC, removes the histones in the order

II, III-IV, I. This type of experiment allows one to differentiate between the effect of the proportion of total histones, and particular classes of histones, left in association with DNA. In most cases Dr. Smart observed the type of behaviour shown in Fig. 2. Most of the physical characteristics of chromatin reflect

FIGURE 2

Histones and template activity of chromatin.

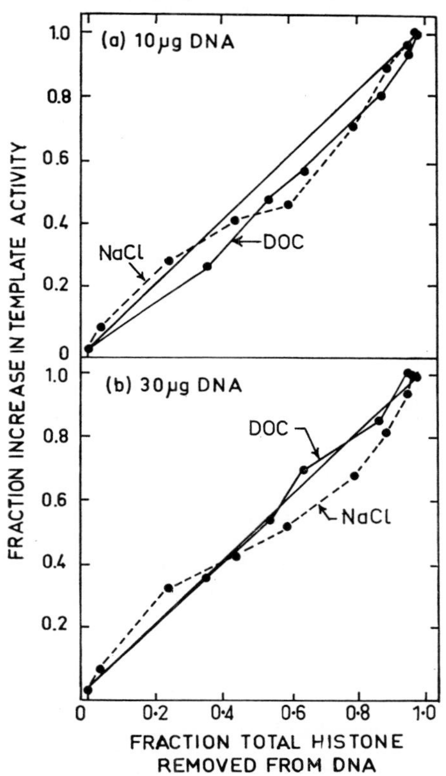

(a) The fraction increase in template activity (10 μg DNA/0.25 ml assay solution) versus the fraction total histone removed from DNA after extraction of chromatin with increasing concentrations of DOC or NaCl. The reactions were catalyzed by *E. coli* RNA polymerase in a high ionic strength medium (0.2 M-KCl). The template activity of native chromatin is assigned the value of 0, while that of chromatin completely dehistonized by 3.0 M-NaCl or 0.15 M-DOC is assigned the value of 1.00. (b) Same as (a) except that the template concentration was 30 μg/0.25 ml assay solution (Smart and Bonner, 1971b).

only the percentage of total histone in association with the DNA and not the class of histone in association with the DNA. This proved true for thermal denaturation, sedimentation velocity and flow dichroism studies of the chromatin, as well as for template activity, the data for which are given in Fig. 2 (Smart and Bonner, 1971a,b). However, the solubility of the partially dehistonized chromatin in 0.15 M-sodium chloride provided the one exception. It is characteristic of chromatin (DNA in association with histone) that it will precipitate at this concentration of salt, while DNA itself will remain in solution. As shown in Fig. 3, the solubility is critically sensitive to the presence of histone I (Smart and Bonner, 1971b). Other observations, particularly circular dichroism studies, have pointed

FIGURE 3

Histones and salt precipitability of chromatin.

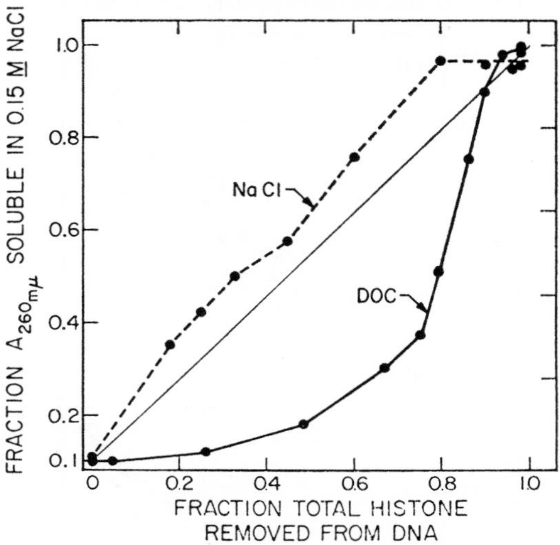

The fraction of E_{260nm} material contained in chromatin which is soluble in 0.15 M-NaCl-0.0025 M-tris, pH 8, versus the fraction of total histone removed from DNA after extraction of chromatin with increasing concentrations of DOC or NaCl. The curves are not significantly altered if the data are plotted using the fraction of total histone net positive charge removed from DNA on the x-axis (Smart and Bonner 1971b).

out that histone I does have a very different influence on the state of aggregation of the DNA in chromatin (Fasman *et al.*, 1970; Shih and Fasman, 1972). The reasons for this are not understood at this time. Two points are of interest. (1) Histone I is the most basic of the histones. Therefore, one might assume that its binding to DNA is most likely to be totally ionic. However, it is the first histone dissociated from chromatin by sodium chloride and other ionic agents. (2) The lack of correlation between solubility and template activity illustrates that these two characteristics of chromatin are not tightly coupled; therefore one can make no extrapolation from the apparent state of aggregation of chromatin to its template activity. It should be noted also that at present we have no consistent and detailed model of the structure of a histone-DNA complex; thus we cannot predict exactly how, or to what extent, histone will limit the accessibility of DNA for various reagents or probes. This, as well as differences in results with different experimental conditions, makes it difficult to interpret experiments aimed at looking at protected or repressed DNA, and free DNA in chromatin, via DNase susceptibility (Clark and Felsenfeld, 1971; Billing and Bonner, personal communication). See DeLange and Smith (1971) for a review of the problem.

FIGURE 4
Primary structures of protamines and histones.

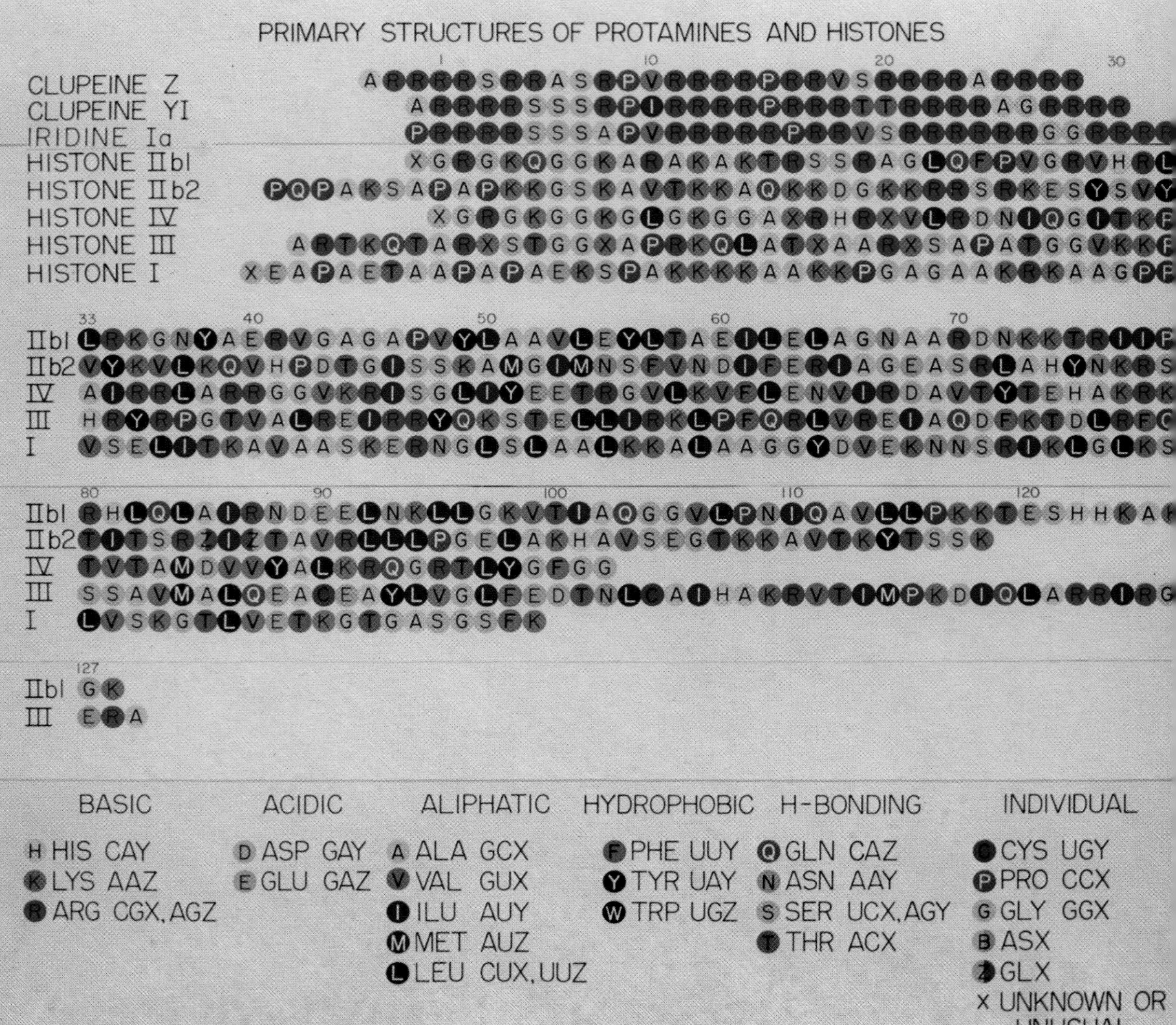

The sequences are given using the standard IUPAC one letter codes for the amino acids. All sequences are for calf thymus histones, except I which is for rabbit thymus, subfraction 3. In histone IIb_1, the first residue is acetylserine. In histone IV, residue 1 is acetylserine, residue 16 is ϵ-N-acetyllysine (50% acetylated), and residue 20 is methyllysine. In histone III, residues 3 and 21 are methyllysine and residue 8 is acetyllysine (our numbering). In histone I, the first residue is acetylserine. Sequence data from the following sources: protamines (clupeine Z, clupeine Y1, and iridine Ia) from T. Ando and co-workers as cited in Phillips (1971); histone IIb_1, from Soutiere *et al.*, (1972) and Yeoman *et al.* (1972); histone IIb_2 from Iwai *et al.* (1970); histone IV from DeLange *et al.* (1969a); histone III from DeLange *et al.* (1972); histone I (incomplete) from Rall and Cole (1971) and R. D. Cole, personal communication.

In an attempt to understand better the role of the histones in chromatin and to gain further insight into the evolution of chromatin, all the histones have now been isolated and at least partially sequenced. These sequences are presented in Fig. 4. The chart has been prepared using a colour code such that functionally related amino acids have similar colours and amino acids whose codons are related by a single base change have similar colours. The data make it clear that while there are extensive functional homologies between the histones, the sequences are so different that one class of histones could not have recently evolved from another. Within a given class, however, the subfractions are very closely related; note the three protamines in Fig. 4 and histone I subfractions as reported by Rall and Cole (1971). The basic amino acids of the histones occur in clusters. Generally there is a cluster at both terminals. In histones II, III and IV there is a dominant cluster at the N-terminal end and a lesser cluster at the C-terminal end. In the case of histone I this appears to be the reverse (note that the C-terminal end of histone I has yet to be sequenced and is not shown in Fig. 4). One also sees particular spacings of the basic amino acids, namely doublets and arrangements where every fourth amino acid is basic, predominate over other possibilities. However, no particular requirement of this sort appears to be absolute. While the protamines have a core common sequence (9-14), which is seen again in histone I, they do show variations on the sequence of one aliphatic amino acid followed by four basic amino acids. There are groupings of two, three, four, five and six arginines. Nothing appears to be sacred —there is nothing to which one can point and say a protein must have this to bind effectively to DNA.

The amount of sequence conservation during evolution varies considerably among the histone classes. Histone IV has been almost completely conserved. Only two amino acid changes are noted between the sequences of IV from calf thymus and pea bud; in pea bud isoleucine replaces valine at 60 and arginine replaces lysine at 77 (DeLange *et al.*, 1969b). Histone III appears to have been also highly conserved throughout evolution (Fambrough and Bonner, 1968; Panyim *et al.*, 1970b; DeLange, personal communication) although a second cysteine residue is found in the histone III of more highly evolved mammals (Panyim *et al.*, 1971a). (III is the only histone that contains any cysteine.) The other classes of histones show greater divergence in different species (for example, see Panyim *et al.*, 1971a). This situation has led to speculation that histones IV and perhaps III are involved in stringent complex formation with the other element of the system that is highly conserved, namely the sugar-phosphate backbone of the DNA molecule. It also suggests that these histones are involved in a complex structure such that the conformation of the protein is rigidly dictated in all regions. No part of the protein is at all times in contact with water or in loose associations that would allow the replacement of a valine by a leucine, for example (Elgin *et al.*, 1971).

The data at hand lead one to conclude that the histones are general and nonspecific chromosomal structural proteins and play a general role in gene repression. A few qualifications and caveats remain in the field; we will discuss three.

(1) The histones are known to undergo minor modifications, phosphorylation, acetylation and methylation. Such modifications could, of course, greatly increase the potential number of different histones. However, there is no evidence at the present time that these mechanisms are involved in providing gene specific binding of histones to DNA. These modifications more likely play a role in chromosome metabolism, e.g., in the transport of histones into the nucleus, in their binding to and removal from DNA, etc. (Louie and Dixon, 1972; Balhorn *et al.*, 1972). This will be discussed further below. (2) It has recently been suggested that there are up to 400 copies of the genes coding for the histones (Kedes and Birnstiel, 1971), although the evidence presented is open to alternative interpretations. Variation in the order of one substitution at a given locus out of 400 potential genes could of course not be picked up by present sequencing techniques. There is no reason to believe that such micro-heterogeneity exists nor that it is relevant to the role of histones; nonetheless, the possibility cannot be ruled out at the present time. A third possibility is that the histone repression is not relevant to the problem of *reversible* gene repression, but that it represents an efficient way to package large amounts of DNA whose function is unknown. This follows from the idea that most of the eukaryotic genome is not transcribed and not used in a sequence-specific fashion, a notion that stems from theories of population genetics which suggest that the rate of mutation is too great for vertebrates to maintain and use their huge amounts of DNA (Ohno, 1972). The arguments in favour of such a drastic genetic load effect depend on a number of parameters whose values are uncertain. For example, the theory requires that mutation rates be constant for all organisms, which is certainly not established and even seems unlikely. The theory states that approximately 2% of the DNA sequences are used to make protein. There is now evidence that at least 20% of the unique DNA sequences (10% of the total DNA) of vertebrates are transcribed tissue specifically (Grouse *et al.*, 1972). McConaughy and McCarthy (1972) showed that the DNA in chromatin which melts at a low temperature is greatly enriched in sequences that are transcribed with tissue specificity and such DNA which melts at low temperatures is free of histone (Smart and Bonner, 1971a; Li and Bonner, 1970). This suggests that histones are responsible for tissue-specific gene repression patterns during development. Since the programming of nuclear activity is reversible (see Gurdon, 1970, for a summary of experiments showing that differentiated nuclei when transplanted into fertilized enucleated eggs can be reprogrammed for major changes in gene activity, resulting in the normal development of eggs) it follows that histones are involved in at least some types of reversible gene repression. Thus, the hypothesis that histones play the role of general, nonspecific gene repressors in chromatin is the best interpretation of present data. One should, however, remember that all this evidence is either circumstantial or results from *in vitro* experimentation. Direct demonstration of the role of histones awaits the development of a eukaryotic test system, equivalent to the prokaryotic *lac* system, in which regulated *in vitro* experiments are possible, and where their validity can be established by biological controls (mutant components, etc.).

HERTONES: INITIAL OBSERVATIONS

The conclusion that the histones represent a general class of gene repressors has led us to explore the other macromolecular components of chromatin to try to discern both the mechanism of specificity and the mechanics of gene activation/inactivation. Our first studies of the hertones were directed at determining the relative heterogeneity and possible tissue specificity of this fraction. These questions can be answered in a preliminary fashion by examining the isolated hertone fraction by sodium dodecyl sulphate (SDS) gel electrophoresis (molecular weight sieving) (Shapiro *et al.*, 1967). The following method is used. Histones are extracted from chromatin with 0.4 M-H_2SO_4; the hertone-DNA pellet is solubilized in SDS, the DNA removed by centrifugation, and the hertones used directly. The results of such a fractionation are shown in Fig. 5. It is clear

FIGURE 5

Gel electrophoresis of chromosomal proteins.

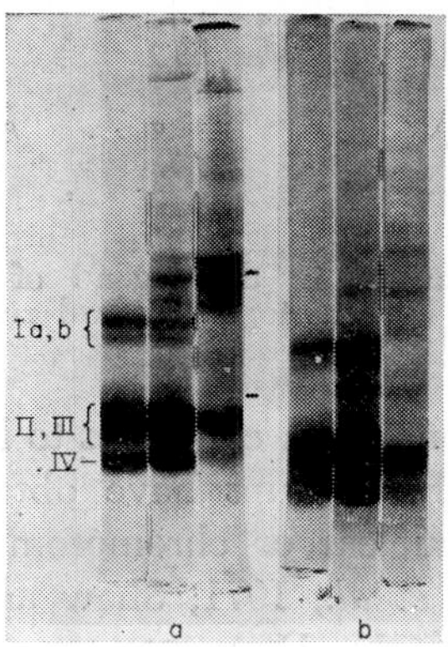

SDS gel electrophoresis of histone, total chromosomal protein, and hertone. Sample size 10-50 μg, 5% acrylamide gels, pH 7.1, run top to bottom. Molecular weight markers at left of 5a indicate position of γ-globulin marker (heavy and light polypeptide chains) run simultaneously; note that the log of the distance of migration is directly proportional to the molecular weight of the polypeptide chain (Shapiro *et al.,* 1967). (a) Rat liver and (b) rat kidney (Elgin and Bonner, 1970).

that this method has achieved a good separation of the histones from the hertones, and that the sum of these two fractions represents the total of the proteins present in the chromatin sample. Note that the total chromosomal protein sample was prepared simply by dialyzing chromatin against the SDS gel sample buffer; thus the hertones cannot be a consequence of acid treatment of the chromatin, as has been suggested (Sonnenbichler and Nobis, 1970). In comparing the hertones of two tissues, namely rat liver and rat kidney (Fig. 6A) one observes that many of the protein bands are the same in the two tissues. However, tissue specific bands are observed (Elgin and Bonner, 1970). The rat kidney gel lacks the bands lambda (λ) and kappa (κ) and has a new one marked omega (ω). We have recently completed an extensive survey of the hertones of

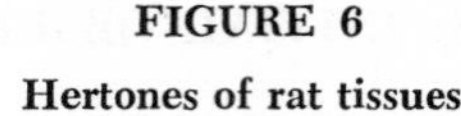

FIGURE 6

Hertones of rat tissues.

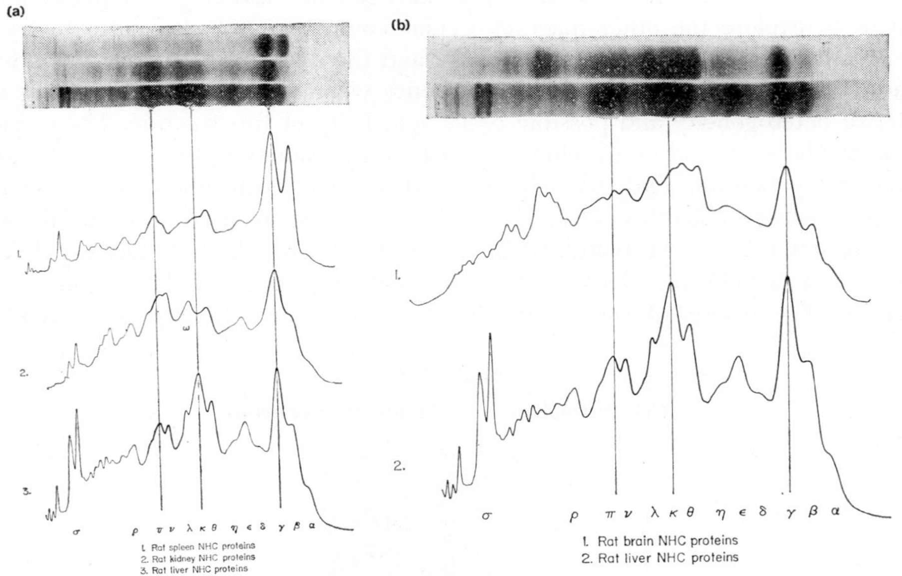

SDS gel electrophoresis as described in the legend of Fig. 5; gels run left to right. Hertones are compared to a rat liver hertone standard run simultaneously (Wu and Elgin, unpublished).

various rat tissues and the conclusion continues to hold (Wu and Elgin, personal communication). Several other groups have obtained similar results using a variety of preparations of nuclear or chromosomal proteins, nuclear phosphoproteins, etc. (MacGillivray *et al.*, 1971; Shaw and Huang, 1970; Platz *et al.*, 1970; Richter and Sekeris, 1972). However, as one examines tissues more dissimilar to liver in terms of embryogenesis, potential hormone response, etc., one obtains increasingly different patterns. In the case of rat brain versus rat liver (Fig. 6B) the band patterns are so different (no more than 50% homology) that one hesitates to suggest the proper homology.

To date, observations in this and other laboratories, using SDS gel electrophoresis, indicate significant quantitative but no qualitative changes in the hertone pattern during hormone response (Elgin, 1971) and development (Hill *et al.*, 1972). However, we have observed a hertone which may be characteristic of a rapid rate of cell division in rat liver cells. When comparing the hertones of regenerating rat liver (cell cycle time *ca.* 21-22h), rat Novikoff ascites cells (a liver-derived tumour cell with cell cycle time *ca.* 24h), and normal rat liver (cell cycle time *ca.* 10^2 days) an additional hertone A is observed in the two former (Elgin, 1971). Our initial observation of the hertones of a given tissue in two different species indicates relatively little variation. In a comparison of the hertones of rat liver and chicken liver, the same molecular weight bands are present in each case (Elgin and Bonner, 1970).

The tissue-specific patterns found for the hertones have suggested that these proteins may be involved in maintaining or altering tissue-specific programs of gene transcription. Several other lines of experimental evidence have led others to suggest such a role for the hertones. In many cases where one has observed a change in template activity of the chromatin of a given tissue, as in development, one observes a correlation between the amount of hertone and the template activity (Marushige and Ozaki, 1967; Marushige and Dixon, 1969). As previously mentioned, experiments in which chromatin has been fractionated into template active and template inactive protein show that *ca.* 80 or 90% of the hertone remains in the template active portion (Frenster, 1965; Marushige and Bonner, 1971). The hertones, in contrast to the histones, show a much more rapid turnover, a characteristic suitable for molecules involved in the variable readout of chromatin (Holoubek and Crocker, 1968). The pattern of hertone turnover may be correlated with the type of RNA synthesis (ribosomal versus DNA-like RNA) in rat liver. The most salient evidence has been the observation of increased hertone as an early response of chromatin to hormone stimulation. Perhaps the best example of this is the observation of an accumulation of nonhistone protein at a locus in the polytene chromosomes of *Drosophila* before that locus begins increased RNA synthesis in response to hormone stimulation (Berendes, 1968). Indeed, in many hormone systems a pattern is emerging which may be referred to as priming. This includes the existence of a protein in the cytoplasm of a target cell which will bind the hormone and help transport it to the nucleus. This complex, or a new one consisting of the hormone and a nuclear protein, will then bind to the chromatin, presumably causing a change in template activity. In this case we will observe an apparent new hertone in response to hormone stimulation. Despite all the positive clues, there is yet to be a definitive demonstration of specific gene activation by a hertone in a eukaryotic system. It is a difficult proposition. The ideal situation would be where the observed activation effect occurs at the normal pH and salt concentration of the nucleus (i.e. approximately 0.2M), where it is clearly an effect on the chromatin or on homologous (eukaryotic) RNA polymerase, where it is gene specific, but is not due to simple co-precipitation of the histone or the accidental addition of histone protease. No experiment to date has met all these criteria. This is not to say that the possibilities mentioned should be considered trivial, but rather that they can lead to incorrect interpretation of results and can mask the desired specificity of gene activation. To work toward such experiments we are extending our studies on both the chemistry and biology of the hertone proteins.

CHROMOSOMAL RNA

Another macromolecule which might be involved in determining the specificity of gene activation is the RNA associated with chromatin (Bonner *et al.*, 1968a). The RNA component of chromatin may include nascent m-RNA, nascent r-RNA, etc., and also includes a number of small RNAs. One of the small RNAs termed chromosomal or c-RNA has been characterized extensively. Chromosomal

RNA is generally of small size (*ca.* 50 nucleotides) with fairly uniform 5′ and 3′ terminal bases and a high content of dihydropyrimidine (7-10%) (Jacobson and Bonner, 1971). This RNA hybridizes, mostly to repetitive DNA sequences, to a very high percentage of the genome (*ca.* 2-5%) (Sivolap and Bonner, 1971; Mayfield and Bonner, 1971). While it is not a degradation product of r-RNA or t-RNA (Dahmus and McConnell, 1969), its includes many of the same sequences as nuclear, high molecular weight, rapidly labelled RNA and may be a product of the processing of these molecules (Holmes, Mayfield, Murthy and Bonner, personal communication). To date its biological role is not well understood. One possibility mentioned above is that this class of RNA molecules is involved in the co-ordinated gene activation seen in response to hormones or to changes in developmental pathways (Britten and Davidson, 1969). The idea that c-RNA is involved in gene expression is supported by observations on the sequence of events following partial hepatectomy in the rat which lead to a major increase in the proportion of chromosomal DNA available for transcription. This increase is preceded by the appearance of new sequences of c-RNA, and this in turn is preceded by the appearance of the same new sequences in nuclear, rapidly labelled, high molecular weight RNA (Mayfield and Bonner, 1972). A second possibility is that c-RNA represents a processing product analogous to the spacer of r-RNA genes and is involved in dictating the processing or packaging of m-RNA from high molecular weight heterogenous nuclear RNA. Such a spacer might have a defined size as in the r-DNA case, with a highly variable internal sequence. The question is under continuing investigation.

HERTONES: CHEMICAL CHARACTERISTICS AND SOME ENZYME ACTIVITIES

While the method of preparing hertones described previously together with anaylsis by SDS gel electrophoresis has provided us with a good deal of comparative data, the technique has several limitations. We have therefore developed a second method for isolating the hertones based on ion exchange chromatography. This second method allows the preparation of hertones for chemical analysis. The chromatin is dissociated by 25% formic acid, 8 M urea, 0.2 M sodium chloride and the DNA removed by centrifugation. The protein can then be quantitatively adsorbed onto Sephadex SE C-25 after decreasing the salt concentration by a factor of ten. The chromosomal proteins are subsequently eluted by a salt gradient. (Steps of 0.1 and 0.2 M-NaCl are followed by a gradient from 0.2 to 0.8 M-NaCl, all in 25% formic acid, 10 M-urea, pH 2.5.).

The technique allows for the separation of the histones from the hertones and for the fractionation of the latter as shown in Fig. 7. Fractions I, II, IIIa and IIIb have been shown to be hertone fractions; fraction IV contains the histones and one hertone, ϵ. Fraction V, eluted from the column with 100% formic acid, contains a residual 10% of the protein, apparently hertone. Scans of SDS gels of each fraction are presented below the elution diagram. Following further purifications using molecular weight sieving and phosphocellulose

FIGURE 7

Fractionation of chromosomal proteins by ion exchange chromatography.

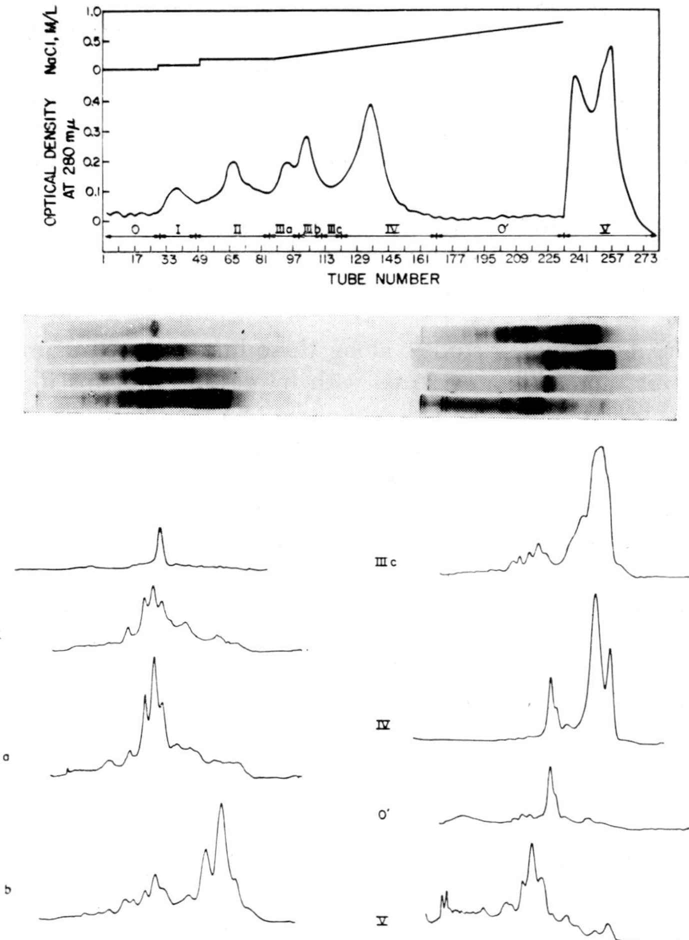

(A) Elution profile. (B) SDS gel electrophoresis of pooled fractions. Gels run left to right (Elgin and Bonner, 1972).

chromatography, we have obtained four fractions of hertones, two of which represent pure proteins and the other two of which are groups of proteins. Assuming that our isolated fractions are representative of the same molecular weight bands in the total hertones, the proteins characterized constitute *ca.* 70% of the hertone protein fraction.

The characteristics of these proteins are shown in Table 2. As can be seen, the hertones cover a much wider range of molecular weight than do the histones, from *ca.* 12,000 to several hundred thousand. The isoelectric points of the hertones range from acidic through neutral to very slightly basic. In all cases the ratio of acidic to basic amino acids is greater than 1, although this calculation does not take into account the extent of amidation. Two classes contain several (4-7) hertones each: a class of middle molecular weight proteins with isoelectric points between 5.4 and 6.6, and a class of relatively low molecular weight proteins with isoelectric points from 6.4 to about 8. The protein ϵ is of particular interest from a chemical point of view in that almost half of its amino acids are acidic or basic (Elgin and Bonner, 1972). Further characterization of this protein is in progress. It is worth noting that the individual hertones and the subfractions listed here are much more soluble than the hertone fraction as a whole. The tendency of the hertones to aggregate with DNA, with histone, and with each other has frequently impeded progress in their study. Once an initial fractionation on the basis of isoelectric point has been accomplished, these problems lessen.

We hope that continuing studies along these lines will tell us more about the hertones *per se,* and will provide us with interesting proteins for the study of evolution as well as providing a source of protein for the production of specific antibodies. Such antibodies could be used both to determine, by inhibition assays, the enzyme activities of various protein fractions and also to localize the molecules *in situ,* thus resolving the perennial question of which proteins really are chromosomal and which proteins may be adventitiously absorbed. However, these isolation techniques involve the exposure of hertones to a very low pH, 2.5, and as such their suitability for the isolation of enzymically active molecules can be questioned. In preparing hertone fractions for the study of enzyme activities and for the study of their effects on template activity, we have had to turn to other, gentler isolation methods such as the extraction of the chromatin with salt. While total quantitative extraction is never achieved by such gentle techniques, certain enzyme activities of interest can be extracted from the chromatin. We are currently studying two such chromosomal enzymes, RNA polymerase and histone protease.

Stanley C. Froehner has been studying the RNA polymerases from rat Novikoff ascites tumour. The enzymes are extracted from the nuclei by strong salt solutions and purified by phosphocellulose and DEAE cellulose chromatography and by sucrose density centrifugation. Three activities, Ia, Ib and II are resolved on chromatography and can also be distinguished by their template preferences (native DNA for Ia-Ib, single-stranded template for II), divalent cation requirements, and ionic strength preferences. Correlation of activity and subunit structure during the final sucrose density gradient purification of Ia suggests that the enzyme has six major polypeptide chains present in stoichiometric amounts as shown by SDS electrophoresis (see Fig. 8). It has been observed that all three enzymes show a considerable increase in activity on native DNA template on addition of a heat-labile factor(s) present in the run-off of the DEAE cellulose column (see Fig. 9). It may be pointed out that this stimulation shows a strong polymerase concentration effect, at least for polymerases Ia and Ib (Froehner, 1972). There is good evidence that it will be necessary to use eukaryotic polymerases to obtain precisely controlled transcription of chromatin

FIGURE 8

Analysis of subunits of rat ascites tumour RNA polymerase Ia.

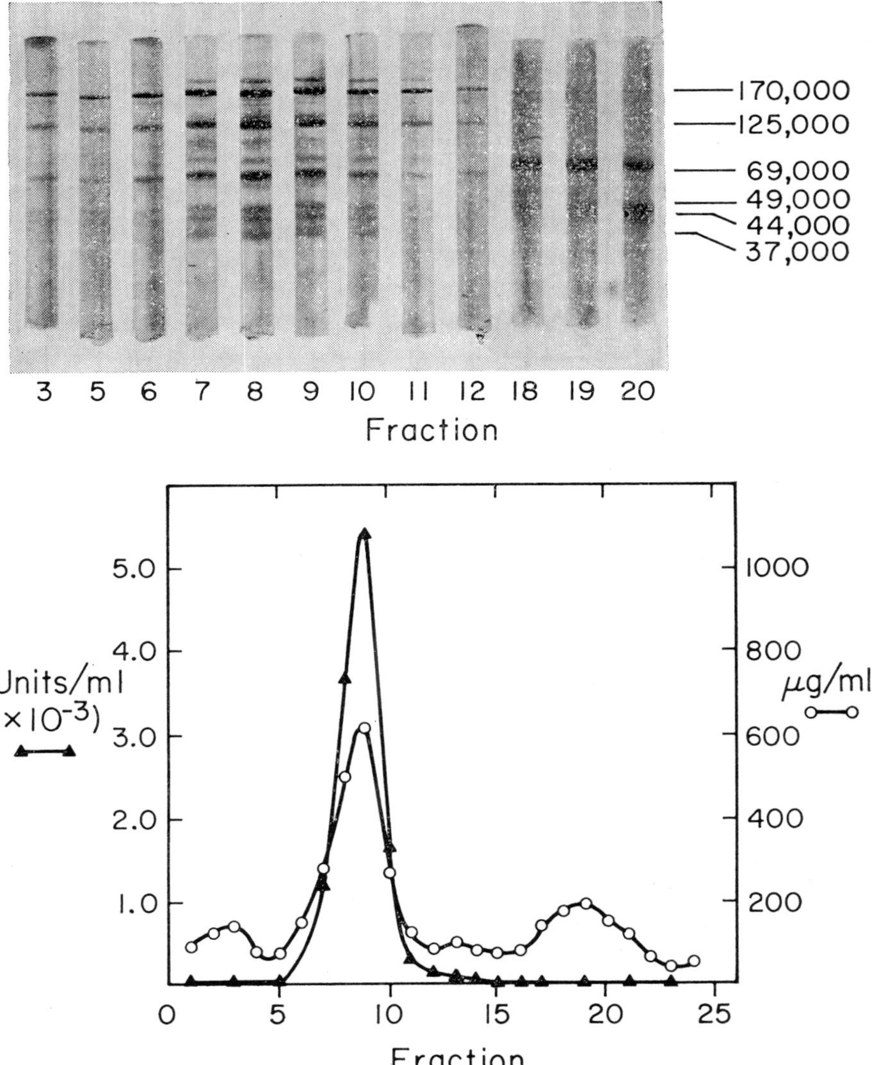

Top: S.D.S. gel electrophoresis of the protein fractions indicated. Bottom: sucrose density gradient sedimentation in 0.2 M-ammonium sulphate of the enzyme preparation showing units of activity and μg of protein per ml for each fraction. Enzyme assay carried out using native calf thymus DNA as template. (One unit of enzyme incorporates one pmole of GTP into acid-precipitable form in 10 min at 37°C.) (Froehner, 1972).

(Butterworth *et al.*, 1971). It is to be hoped that by the use of such enzymes the correct transcription of ribosomal genes from isolated eukaryotic chromatin will soon be demonstrated (D. Brown, personal communication). The work on

FIGURE 9

Stimulatory effect of run-off factor on rat ascites tumour RNA polymerase Ia activity.

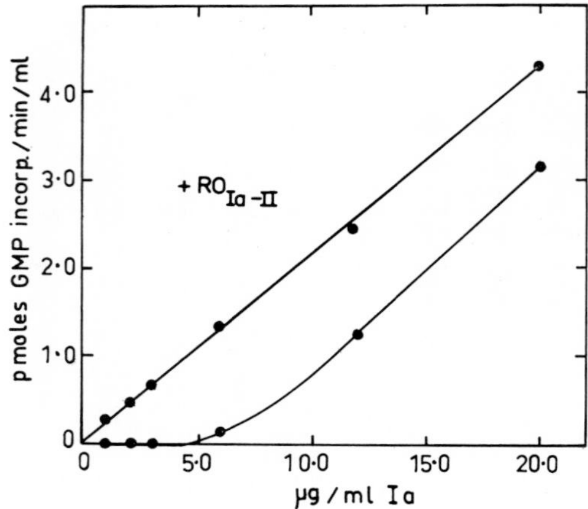

Enzyme assay carried out using native rat liver DNA as template with 138 μg/ml run-off factor in the presence of 100 μg/ml BSA. Note that the run-off factor has no activity alone (Froehner, 1972).

the RNA polymerases and the DEAE cellulose run-off factor is a good example of the classical problem confronting us in the study of chromatin. To understand this system we must simplify it, by purifying it to the essential components. However, in our attempts to so purify we frequently find ourselves purified out of business. We accidentally remove some critical factor, component, or subunit and lose the activity of interest or its specificity. Caution and controls whereby the significance of results can be checked *in vivo* are necessary.

Histone protease from rat liver is another chromosomal enzyme which we are studying. This enzyme was first observed in chromatin as a nuisance (Phillips and Johns, 1959; Panyim *et al.*, 1968). Indeed, its presence calls for extreme care when isolating chromatin for study *in vitro* and for consideration in interpretation of results. The enzyme may be purified from rat liver chromatin 30-fold by extraction with 0.7 M-sodium chloride followed by purification on Sephadex columns. This enzyme greatly prefers histone as a substrate. Taking the activity on histones as 100%, this enzyme exhibits only 18% activity on casein, 9% activity on protamine, and 5% activity or less on RNase, lysozyme, and denatured haemoglobin as substrates. The basis of substrate specificity is unknown; histones are good substrates for many other proteases. An apparently similar histone protease has been studied in calf thymus by Furlan and Jericijo (1967a,b), Furlan *et al.* (1968), and Bartley and Chalkley (1970). Rat liver histone protease appears to cleave the histone molecules at a few specific sites, as judged by the appearance of specific large breakdown products on gel electrophoresis analysis during the reaction. Fig. 10 illustrates the salt and pH effects on enzyme activity (Garrels *et al.*, 1972). The very sharp pH curve is of special interest in considering nuclear events and is useful in controlling the enzyme.

FIGURE 10

Salt and pH effects on histone protease activity.

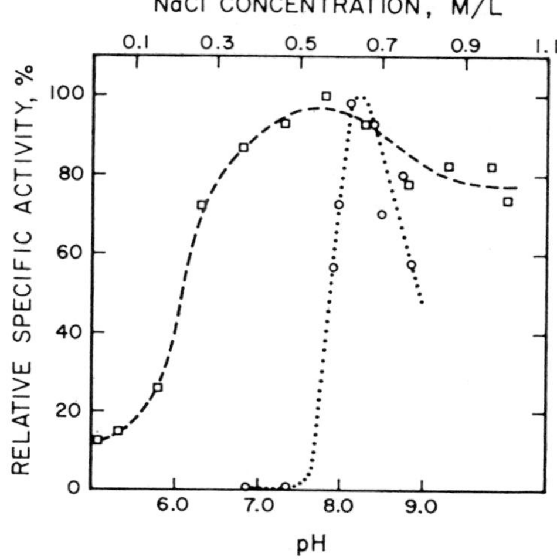

pH effects . . . . at 0.35 M-NaCl and salt effects - - - - at pH 8 (0.005 M-tris) using whole rat liver histone as substrate for partially purified rat liver histone protease. Incubation for 4 h at 37°C at 0.5 mg/ml histone. Histone breakdown measured as the production of new N-terminals using a ninhydrin assay (Garrels *et al.*, 1972).

The presence of histone protease in the chromatin of rat liver, a tissue with very slow cell turnover but capable of hormone responses, regeneration, etc., has suggested to us that histone protease may play an interesting role in chromatin metabolism as opposed to being merely an autodegradative enzyme. If one assumes on the basis of present evidence that the association of histone with DNA results in repression of that DNA as a template, and further that this gene repression is reversible, then one must consider the mechanism by which the histone is removed from the DNA. Classical procedures used *in vitro* include extraction with high salt concentrations (e.g., 0.6-3 M-sodium chloride) or the use of very low pH such as that achieved by 0.2 M-sulphuric acid. Probably neither of these procedures are available to the cell. One real possibility is that histone may be removed by virtue of never having been put on, i.e., that gene activation must await a round of cell division, during which the critical nascent DNA will be protected from deposition of histone and the gene consequently become active. It has been noted in some systems that hormones require cell division to occur, so that hormone-directed gene activation may take place (Stockdale and Topper, 1966). However, this is not the universal case. It has been suggested that histones are removed by competition from another cationic molecule. This indeed happens in the replacement of histone I by protamine during trout spermatogenesis (Marushige and Dixon, 1971). However, the association of a more basic protein with the DNA hardly seems a state of affairs conducive to gene activation. Another suggestion is that the histone could be removed by competition from a strong anion, for instance a double-stranded RNA or a similar oligonucleotide. However, present evidence suggests that under the salt concentrations found in the nucleus, only histone I is free to move from one nucleic acid to another (Clark and Felsenfeld, 1971). In the light of these difficulties the suggestion that histones might be removed by proteolytic degradation *in situ* is attractive. Such degradation of the histones on the DNA has been found in the case of trout spermatogenesis. Here the histones are phosphorylated prior to their degradation into large pieces. The phosphorylation *per se* does not seem to significantly affect their binding to DNA as shown by salt and DOC extraction curves (Marushige *et al.,* 1969); rather the phosphorylation may be a signal marking the histone for degradation by the protease. The process of trout spermatogenesis, however, is a special case where all histones are removed, and the relevance of protease activity to specific gene activation has yet to be conclusively demonstrated.

CONCLUSIONS

Let us conclude by summing up what we now know about chromatin and the role of its various components. That isolated chromatin will maintain its tissue-specific transcription *in vitro* seems to be established. Sufficient data are now in hand to re-evaluate the role of the histones. The number of histone species is very small. The extreme amino acid sequence conservation of some of these molecules during evolution is unprecedented, and suggests that they must participate in a multimolecular complex in direct association with the conserved DNA helix backbone. Although the definitive experiments are lacking, there is nothing to suggest at this time that histones recognize specific base sequences or are involved in determining the specificity of gene repression or gene activation. The histones are well constructed to interact with DNA and at the same time interact with other macromolecular components. They do, however, apparently play a role as general structural proteins and general elements of the mechanism of gene repression. This suggests that other molecules lend specificity to these mechanisms. At the present time it appears that the hertones, while including many homologous molecules, also include tissue-specific proteins. These proteins may be involved in regulating tissue-specific transcription in development and in response. However, it is reasonable to suppose that proteins involved in the activation of a few genes will be present in concentrations much too low to be observed by our present crude techniques. It is reasonable to suppose that the many homologous hertones could include common structural proteins, nuclear membrane elements, stoichiometric enzyme proteins (analogous to T4 gene 32 protein described by Alberts and Frey, 1970) as well as the enzymes of chromosomal metabolism, two of which have been discussed. Studies are progressing on the isolation and characterization of eukaryotic RNA polymerase. Hopefully this work will allow us to find instances of specific initiation of transcription that will allow the process to be studied in some detail. There is no doubt that the use of prokaryotic RNA polymerase may have obscured many specific characteristics of chromatin transcription. The accumulation of nonhistone protein at specific chromosomal loci in *Drosophila* prior to gene activation ("puffing") has suggested that the hertones include enzymes and other molecules involved in the general mechanism of gene activation. The presence of histone protease in chromatin has suggested one way in which histone removal (gene activation?) might occur. Definitive experiments have yet to be done. However, we believe that the systems necessary to do the experiments—the purified chromatin, the knowledge of the components of the chromatin, the isolation and analysis of the enzymes necessary to do the job—are being accumulated, and that progress in this field over the next few years will be quite exciting.

ACKNOWLEDGEMENTS

We would like to thank our colleagues Stanley C. Froehner and Flora Wu for the use of experimental data prior to publication. This work was supported in part by U.S. Public Health Service Grant No. GM-13762, The Jane Coffin Childs Memorial Fund for Medical Research, and the National Science Foundation (grant GB 34160 to Dr. Leroy E. Hood). We would also like to thank the following copyright holders for permission to reprint figures: Figures 1, 5 and 7, the American Chemical Society; Figures 2 and 3, Academic Press, London; Figure 10, Academic Press, New York.

REFERENCES

Alberts, B. M. and Frey, L. (1970) *Nature (London) 227,* 1313

Ashburner, M. (1970) *Adv. Insect Physiol. 7,* 1

Balhorn, R., Chalkley, R. and Granner, D. (1972) *Biochemistry 11,* 1094

Bartley, J. and Chalkley, R. (1970) *J. Biol. Chem. 245,* 4286

Bekhor, I., Kung, G. M. and Bonner, J. (1969) *J. Mol. Biol. 39,* 351

Berendes, H. D. (1968) *Chromosoma 24,* 418

Bishop, J. O., Pemberton, R. and Baglioni, C. (1972) *Nature New Biology 235,* 231

Bonner, J., Dahmus, M. E., Fambrough, D., Huang, R. C., Marushige, K., and Tuan, D. Y. H. (1968a) *Science 159,* 47

Bonner, J., Chalkley, G. R., Dahmus, M., Fujimura, F., Huang, R. C., Huberman, J., Jensen, R., Marushige, K., Ohlenbusch, H., Olivera, B., and Widholm, J. (1968b) in *Methods in Enzymology* (Grossman, L. and Moldave, K., eds.), vol. 12B, p. 3, Academic Press, New York

Britten, R. J. and Kohne, D. E. (1967) *Carnegie Inst. of Washington Yearbook 65,* 78

Britten, R. J. and Davidson, E. H. (1969) *Science 165,* 349

Britten, R. J. and Davidson, E. H. (1971) *Quart. Rev. Biol. 46,* 111

Bustin, M. and Cole, R. D. (1968) *J. Biol. Chem. 243,* 4501

Bustin, M. and Cole, R. D. (1969) *J. Biol. Chem. 244,* 5286

Butterworth, P. H. W., Cox, R. F. and Chesterton, C. J. (1971) *Eur. J. Biochem. 23,* 229

Chetsanga, C. J., Poccia, D. L., Hill, R. J. and Doty, P. (1970) *Cold Spring Harbor Symp. Quant. Biol. 35,* 629

Clark, R. J. and Felsenfeld, G. (1971) *Nature New Biology 229,* 101

Dahmus, M. E. and McConnell, D. J. (1969) *Biochemistry 8,* 1524

DeLange, R. J., Fambrough, D. M., Smith, E. L. and Bonner, J. (1969a) *J. Biol. Chem. 244,* 319

DeLange, R. J. and Smith, E. L. (1971) *Ann. Rev. Biochem. 40,* 279

DeLange, R. J., Fambrough, D. M., Smith, E. and Bonner, J. (1969b) *J. Biol. Chem. 244,* 5669

DeLange, R. J., Hooper, J. A. and Smith, E. L. (1972) *Proc. Nat. Acad. Sci. U.S. 69,* 882

Elgin, S. C. R. and Bonner, J. (1970) *Biochemistry 9,* 4440

Elgin, S. C. R. (1971) Ph.D. Thesis, California Institute of Technology

Elgin, S. C. R. and Bonner, J. (1972) *Biochemistry 11,* 772

Elgin, S. C. R., Froehner, S. C., Smart, J. E. and Bonner, J. (1971) *Adv. Cell Mol. Biol. 1,* 1.

Fambrough, D. M. and Bonner, J. (1968) *J. Biol. Chem. 243,* 4434

Fambrough, D. M., Fujimura, F. and Bonner, J. (1968) *Biochemistry 7,* 575

Fasman, G. D., Schaffhausen B., Goldsmith, L. and Alder, A. (1970) *Biochemistry 9,* 2814

Frenster, J. H. (1965) *Nature (London) 206,* 680

Froehner, S. C. (1972) Ph.D. Thesis, California Institute of Technology

Furlan, M. and Jericijo, M. (1967a) *Biochim. Biophys. Acta 147,* 135

Furlan, M. and Jericijo, M. (1967b) *Biochim, Biophys. Acta 147,* 145

Furlan, M., Jericijo, M. and Suhar, A. (1968) *Biochim. Biophys. Acta 167,* 154

Gage, L. P. (1971) *Carnegie Inst. of Washington Yearbook 70,* 39

Garrels, J. I., Elgin, S. C. R. and Bonner, J. (1972) *Biochem. Biophys. Res. Commun. 46,* 545

Grouse, L., Chilton, M. D. and McCarthy, B. J. (1972) *Biochemistry 11,* 798

Gurdon, J. B. (1970) *Proc. Roy. Soc. Lond. B 176,* 303

Hill, R. J., Poccia, D. L. and Doty, P. (1972) *J. Mol. Biol. 61,* 445

Holoubek, V. and Crocker, T. T. (1968) *Biochim. Biophys. Acta 157,* 352

Iwai, K., Ishikawa, K. and Hayashi, H. (1970) *Nature (London) 226,* 1056

Jacobsen, R. A. and Bonner, J. (1971) *Arch. Biochem. Biophys. 146,* 557

Kedes, L. H. and Birnstiel, M. (1971) *Nature New Biology 230,* 165

Leighton, T. J., Dill, B. C., Stock, J. J. and Phillips, C. (1971) *Proc. Nat. Acad. Sci. U.S. 68,* 677

Leng, M. and Felsenfeld, G. (1966) *Proc. Nat. Acad. Sci. U.S. 56,* 1325

Li, H.-J. and Bonner, J. (1971) *Biochemistry 10,* 1461

Louie, A. J. and Dixon, G. H. (1972) *Proc. Nat. Acad. Sci. U.S. 69,* 1972

MacGillivray, A. J., Carroll, D. and Paul, J. (1971) *FEBS Letters 13,* 204

McConaughy, B. L. and McCarthy, B. J. (1972) *Biochemistry 11,* 998

Marushige, K. and Bonner, J. (1966) *J. Mol. Biol. 15,* 160

Marushige, K. and Ozaki, H. (1967) *Develop. Biol. 16,* 474

Marushige, K. and Dixon, G. H. (1969) *Develop. Biol. 19,* 397

Marushige, K. and Bonner, J. (1971) *Proc. Nat. Acad. Sci. U.S. 68,* 2941

Marushige, K. and Dixon, G. H. (1971) *J. Biol. Chem. 246,* 5799

Marushige, K., Ling, V. and Dixon, G. H. (1969) *J. Biol. Chem. 244,* 5953

Mayfield, J. E. and Bonner, J. (1971) *Proc. Nat. Acad. Sci. U.S. 68,* 2652

Mayfield, J. E. and Bonner, J. (1972) *Proc. Nat. Acad. Sci. U.S. 69,* 7

Ohno, S. (1972) *Develop. Biol. 27,* 131

Panyim, S. and Chalkley, R. (1969) *Biochemistry 8,* 3972

Panyim, S., Jensen, R. H. and Chalkley, R. (1968) *Biochim. Biophys. Acta 160,* 252

Panyim, S., Chalkley, R., Spiker, S. and Oliver, D. (1970) *Biochim. Biophys. Acta 214,* 216

Panyim, S., Bilek, D. and Chalkley, R. (1971a) *J. Biol. Chem. 246,* 4206

Panyim, S., Sommer, K. R. and Chalkley, R. (1971b) *Biochemistry 10,* 3911

Pardue, M. L. and Gall, J. G. (1970) *Science 168,* 1356

Paul, J. and Gilmour, R. S. (1966) *Nature (London) 210,* 992

Paul, J. and Gilmour, R. S. (1968) *J. Mol. Biol. 34,* 305

Phillips, D. M. P. and Johns, E. W. (1959) *Biochem. J. 72,* 538

Phillips, D. M. P. (1971) in *Histones and Nucleohistones* (Phillips, D. M. P., ed.), p. 47, Plenum Press, London and New York

Platz, R. D., Kish, V. M. and Kleinsmith, L. J. (1970) *FEBS Letters 12,* 38

Rall, S. C. and Cole, R. D. (1971) *J. Biol. Chem. 246,* 7175

Richter, K. H. and Sekeris, C. E. (1972) *Arch. Biochem. Biophys. 148,* 44

Sautiere, P., Tyrou, D., Laine, B., Mizoo, J., Lambellin-Breynaert, M-D., Raffin, P., and Biserte, G. (1972) *C. R. Acad. Sci. Paris 274,* 1422

Shapiro, A. L., Viñuela, E. and Maizel, J. V. (1967) *Biochem. Biophys. Res. Commun. 28,* 815

Shaw, L. M. J. and Huang, R. C. (1970) *Biochemistry 9,* 4530

Shih, T. Y. and Bonner, J. (1970) *J. Mol. Biol. 48,* 469

Shih, T. Y. and Fasman, G. D. (1972) *Biochemistry 11,* 398

Sivolap, Y. and Bonner, J. (1971) *Proc. Nat. Acad. Sci. U.S. 68,* 387

Smart, J. E. and Bonner, J. (1971a) *J. Mol. Biol. 58,* 661

Smart, J. E. and Bonner, J. (1971b) *J. Mol. Biol. 58,* 675

Smith, K. D., Church, R. B. and McCarthy, B. J. (1969) *Biochemistry 8,* 4271

Sonnenbichler, J. and Nobis, P. (1970) *Eur. J. Biochem. 16,* 60

Stedman, E. and Stedman, E. (1950) *Nature (London) 166,* 780

Stockdale, F. E. and Topper, Y. J. (1966) *Proc. Nat. Acad. Sci. U.S. 56,* 1283

Tan, C. H. and Miyagi, M. (1970) *J. Mol. Biol. 50,* 641

Yeoman, L. C., Olson, M. O. J., Suzano, N., Jordan, J. J., Taylor, C. W., Starbruck, W. C. and Busch, H. (1972) *J. Biol. Chem.* (in press)

Structural Modifications of Histones in Cultured Mammalian Cells

GEORGE R. SHEPHERD, BILLIE J. NOLAND, JULIA M. HARDIN
Biomedical Research Group, Los Alamos Scientific Laboratory University of California, Los Alamos, New Mexico, U.S.A.
PAUL BYVOET
Department of Pathology, School of Medicine, University of Florida, Gainesville, Florida, U.S.A.

The widespread association of histones with nucleic acids and the ability of histones to influence both *in vivo* and *in vitro* transcription reactions have led to the suggestion that histones may function as direct regulators or as intermediate moderators of gene expression (for reviews see Phillips, 1964; Busch and Steele, 1964; Murray, 1965; Hnilica, 1967; Stellwagen and Cole, 1969). Modification of the nucleoprotein complex, brought about through modification of functional groups of histone molecules, might bring about alterations of chromatin structure and eventually result in gene repression or derepression (Allfrey, 1966; Ord and Stocken, 1968; Allfrey *et al.*, 1964). Histones may also participate in the geometric structuring of the nucleoprotein complex, perhaps in preparation for division, and histone structural alterations may be related to this function (Tidwell *et al.*, 1968). In either case, there should exist a temporal relationship between histone structural alterations and related events. The purpose of this study was to search for such relationships within the cell cycle.

MATERIALS AND METHODS

Chinese hamster ovary (CHO) cells were grown in suspension culture in supplemented F-10 medium, as previously described (Shepherd and Noland, 1968) and were synchronized, where indicated, by the isoleucine-depletion method of Tobey and Ley (1970) and Ley and Tobey (1970). Cell numbers were enumerated by electronic cell counting methods. Cultures were labelled as indicated in the text and were harvested by centrifugation at suitable intervals after labelling or release. Washed cells were frozen immediately upon harvest and preserved at —20°C until needed. Histone fractions were prepared from frozen cell pellets by the method of Johns (1964), as modified for cell cultures by Gurley and Hardin (1968). Electrophoretic analysis of fractions prepared by this method showed them to be of the degree of purity described by Johns (1967). Protein was determined by the method of Lowry *et al.* (1951), while DNA biosynthesis was determined by incorporation of radiothymidine in parallel cultures. Isotope contents of cells and preparations were determined by scintillation counting. Intracellular acetyltransferase, methyltransferase and phosphokinase levels were determined by slightly modified versions (Noland *et al.*, 1971; Hardin *et al.*, 1971; Shepherd *et al.*, 1971a) of existing techniques (Nohara *et al.*, 1968; Kaye and Sheratzky, 1969; Langan, 1968).

FIGURE 1

Temporal patterns of histone acetylation, methylation and phosphorylation in synchronized cultures of Chinese hamster ovary cells

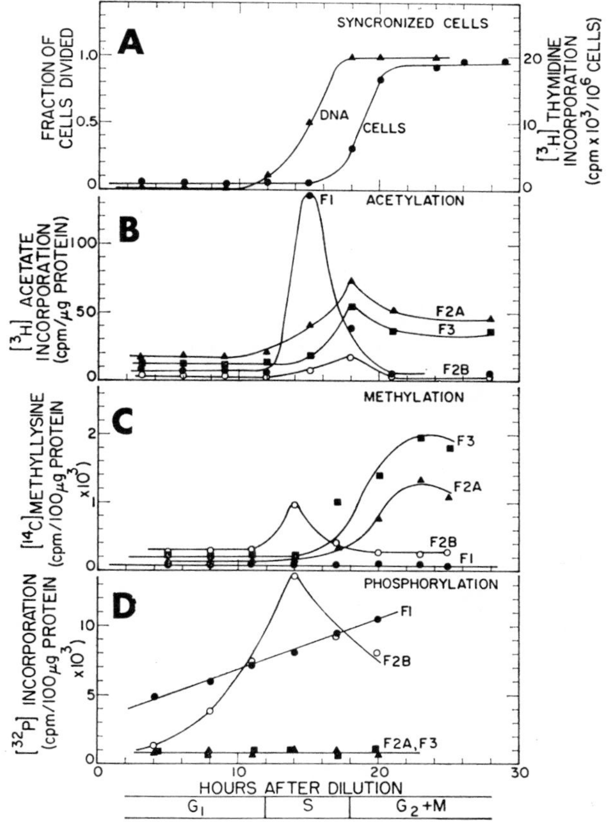

(A) Synchrony pattern illustrated by incorporation of labelled thymidine (▲) and by cell count (●). (B-D) Accumulation and retention of labelled acetate (B), methyl groups (C) and phosphorus (D) by histone fractions F1 (●), F2a (▲), F2b (O) and F3 (■).

RESULTS

Acetylation

Synchronized cultures were labelled continuously with [^{3}H]acetate at levels of 600 μCi/1 of medium. Cells were harvested at appropriate intervals after release during progression of the cell population through one division cycle. The specific activities of individual histone fractions during this period were determined with the results seen in Fig. 1B. The specific activities of histone fractions F2a, F2b and F3 rose rapidly to a maximum in S, coincident with termination of DNA synthesis (Fig. 1A), and declined rapidly thereafter. The specific activity of histone fraction F1 appeared to rise to a maximum in mid-S, coincident with

FIGURE 2

Temporal patterns of transferase and kinase intracellular levels in synchronized cultures of Chinese hamster ovary cells

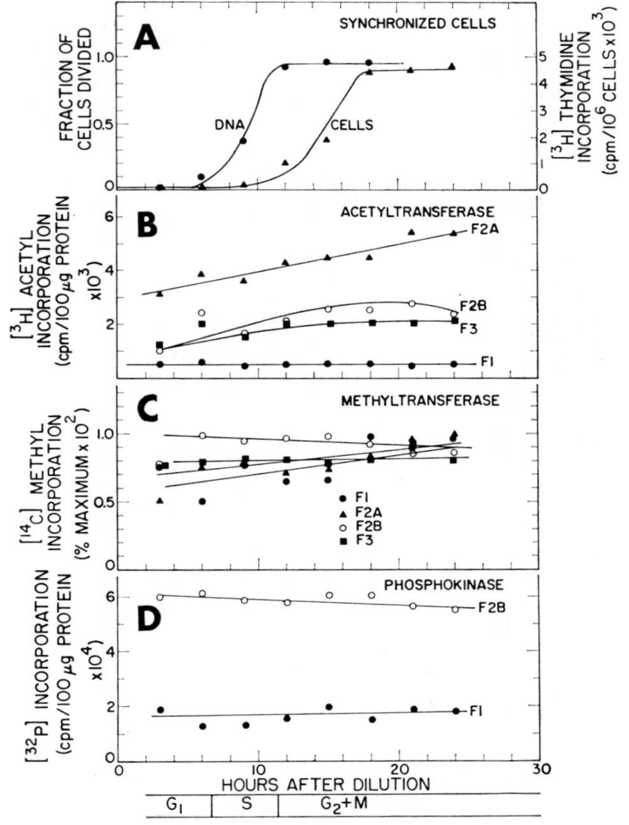

(A) Synchrony pattern illustrated by incorporation of labelled thymidine (●) and by cell count (▲). (B-D) Intracellular acetyltransferase (B), methyltransferase (C) and phosphokinase (D) levels using histone fractions F1 (●), F2a (▲), F2b (O) and F3 (■) as substrates.

maximum DNA synthesis, and then declined. The specific activities of histones F1 and F2b declined to near background levels, while those of histones F2a and F3 declined to approximately half their maximal values (Shepherd *et al.*, 1970).

The acetyltransferase activities of parallel synchronous cultures may be seen in Fig. 2B. No temporal activity maxima were observed which could be correlated with previously observed histone acetylation maxima (Noland *et al.*, 1971).

The metabolic stability of incorporated acetate in histone fractions of exponentially-growing cultures may be seen in Fig 3. Cells labelled with [^{3}H]acetate were harvested under sterile conditions, washed once with unlabelled medium and resuspended in unlabelled medium. Cultures were harvested at suitable intervals, and the specific activities of their histone fractions were determined. It may be seen that the decay slopes for histone acetyl contents are parallel to the theoretical dilution plot. The individual half-lives of the fractions approximate the generation time of the culture, the period in which specific activities of structures exhibiting no turnover would be diluted to half their former values through biosynthesis of new materials from unlabelled precursors. Cells doubly-labelled

FIGURE 3

Turnover of histone acetyl groups in exponentially-growing culture of Chinese hamster ovary cells in unlabelled medium

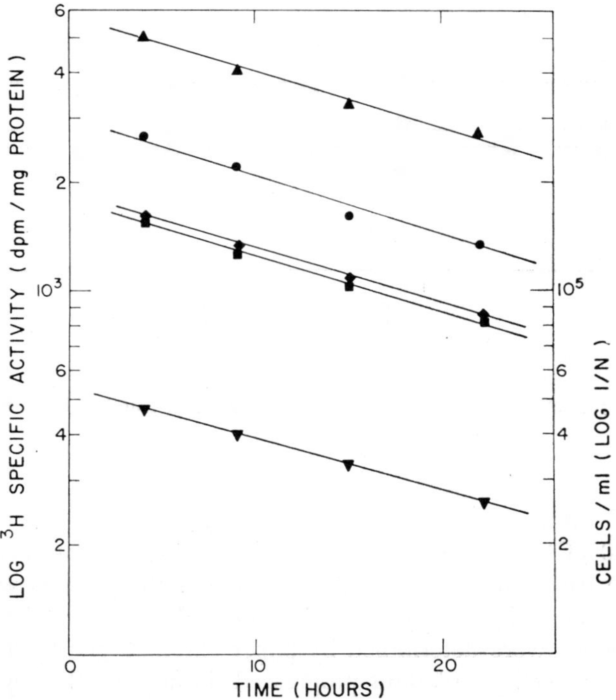

Fraction F1 (●); fraction F2a (▲); fraction F2b (■); fraction F3 (◄►) and theoretical dilution curve (▼) derived from cell growth kinetics.

in culture with [^{3}H]acetate and [^{14}C]arginine were harvested, washed and resuspended as before. Cultures were again harvested at suitable intervals, and the isotopic contents of their histone fractions were determined. The results in Fig. 4 indicate that the ratios of isotope specific activities remained the same throughout the culture period, demonstrating that histone acetyl groups exhibit the same turnover (nil) as their parent histone species (Shepherd *et al.*, 1972a).

FIGURE 4

Ratios of specific activities of [^{14}C]arginine and [^{3}H]acetate of histone fractions of exponentially-growing culture of Chinese hamster ovary cells in unlabelled medium

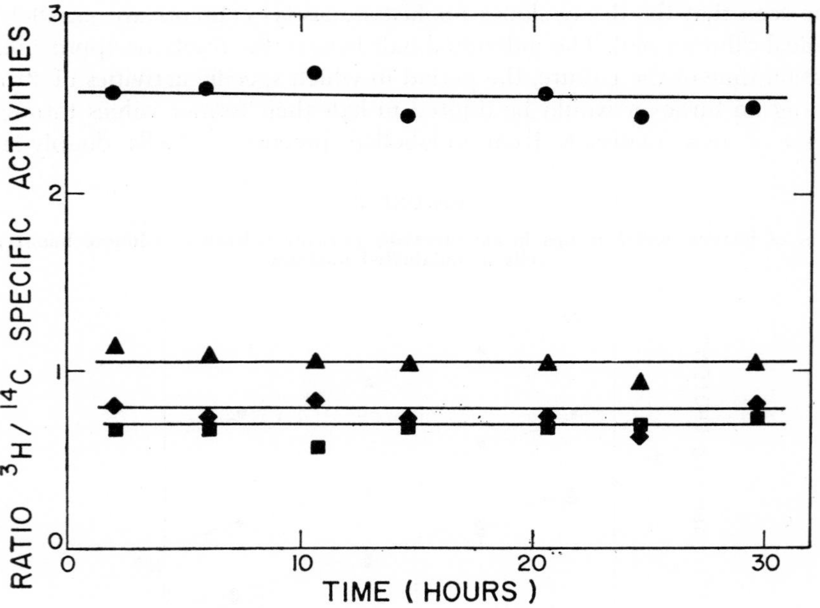

Fraction F1 (●); fraction F2a (▲); fraction F2b (■) and fraction F3 (◀▶).

Methylation

Synchronized cultures were labelled continuously with [methyl ^{14}C]methionine at a level of 40 μCi/1 of medium. Labelled methionine was incorporated into histone peptide backbones, and methionine radiomethyl groups were incorporated into histones as methylated lysine and arginine derivatives. The levels of these labelled compounds were determined by paper chromatography and counting in histone fractions as cell populations progressed through the life cycle (Shepherd *et al.*, 1971b). The results may be seen in Fig. 1C. The [methyl ^{14}C]lysine contents of fractions F2a and F3 began to rise in S, reached maxima after termination of DNA and histone synthesis and began to fall by mid-M. The [methyl ^{14}C]lysine content of fraction F2b rose to a maximum early in S, coincident with initiation of DNA synthesis, and rapidly decreased to its original unmethylated level by late S. Fraction F1 remained unmethylated throughout (Shepherd *et al.*, 1971a).

The methyltransferase activities of parallel synchronized cultures may be seen in Fig. 2C. No alterations of methyltransferase levels were observed within the cell cycle which could be correlated with histone methylation maxima (Hardin *et al.*, 1971).

The metabolic stabilities of histone methyl lysine groups were next investigated. Random cultures were labelled with [methyl ^{14}C]methionine at levels of 50 μCi/1. Cells were harvested at suitable intervals after continuous labelling, histone fractions were isolated, and the distribution of radiomethyl activities among the amino acids of each fraction was determined by ion exchange chromatography. These results may be seen in Tables 1 and 2. Fraction F1 was devoid

TABLE 1

Percentage Distribution of Radioactivity in Each Histone Fraction After 8 h in Unlabelled Medium

Fraction	Methionine	Total methyl lysine	Methyl arginine
F1	91.3	0	0
F2a	48.1	47.3	0
F2b	75.7	11.3	10.8
F3	49.2	46.2	0

TABLE 2

Percentage Distribution of Radioactivity Over *N*-Methyl Derivatives of Lysine in Each Histone Fraction After 8 h in Unlabelled Medium

Fraction	Mono-	Di-	Tri-
F1	0	0	0
F2a	8.0	84.5	7.5
F2b	17.0	57.0	26.0
F3	14.5	68.0	17.5

of methyl lysine or methyl arginine derivatives. Fraction F2b was the only fraction found to contain methyl arginine. It is also evident that the dimethyl lysine derivative contained most of the methyl lysine radioactivity and that significant differences may be observed in the distribution of radiomethyl derivatives among the histone factions. The metabolic stabilities of the histone fractions, based on their methionine contents, and of their methyl lysine and methyl arginine contents may be seen in Fig. 5. The calculated half-lives of these components may be

FIGURE 5

Turnover of histone fractions and their methyl lysine and methyl arginine components

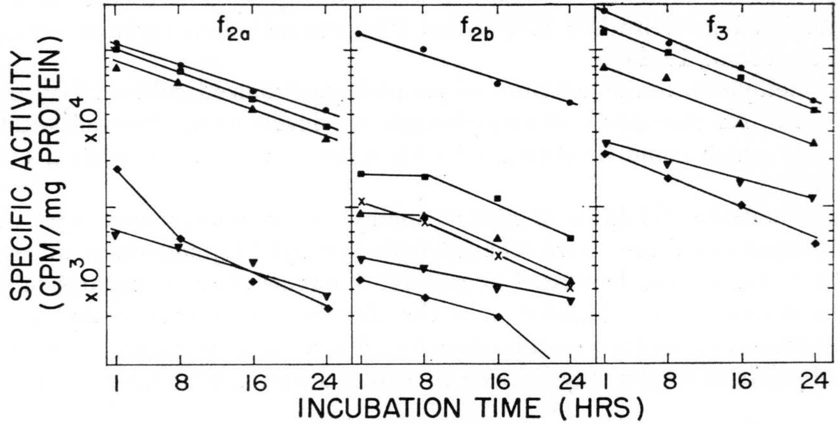

Methionine (●); total methyl lysines (■); monomethyl lysine (◀▶); dimethyl lysine (▲); trimethyl lysine (▼) and methyl arginine (×).

seen in Table 3. These data clearly indicate that histone methylation is fraction-specific and that gross incorporated histone methyl groups are metabolically stable (Byvoet *et al.*, 1972).

TABLE 3

Half-lives (in hours) of Histones and their *N*-Methyl Groups in Random Cultures of CHO Cells

Fraction	Histone methionine	Methyl lysine	Methyl arginine
F1	12.1	0	0
F2a	16.4	14.1	0
F2b	15.6	15.1	12.2
F3	12.3	14.9	0

Phosphorylation

Synchronized cultures were labelled continuously with inorganic radiophosphorous [^{32}P] at a level of 1 mCi/1. Incorporation of label into histone fractions as alkali-labile [^{32}P]phosphate was followed through the cell cycle (Fig. 1D). No evidence was obtained for incorporation of significant quantities of radiophosphorus into fractions F2a or F3. The radiophosphorus content of fraction F1 rose steadily during this period, while the radiophosphorus content of F2b rose to a maximal value in mid-S and declined thereafter (Shepherd *et al.*, 1971b).

Phosphokinase levels in parallel cultures are displayed in Fig. 2D. No direct correlations were observed between kinase levels within the cell cycle and the period of maximum specific activity for histone fraction F2b (Shepherd *et al.*, 1971c).

The metabolic stabilities of histone phosphate groups were next determined. Random cultures were pulse-labelled with inorganic radiophosphorus as before, washed once and released into unlabelled medium. The decay of radiophosphorus contents of histone fractions F1 and F2b were determined with the results seen in Fig. 6. The calculated half-lives of histone phosphate were found to be 7 h

FIGURE 6

Turnover of histone radiophosphorus in exponentially-growing cultures of Chinese hamster ovary cells

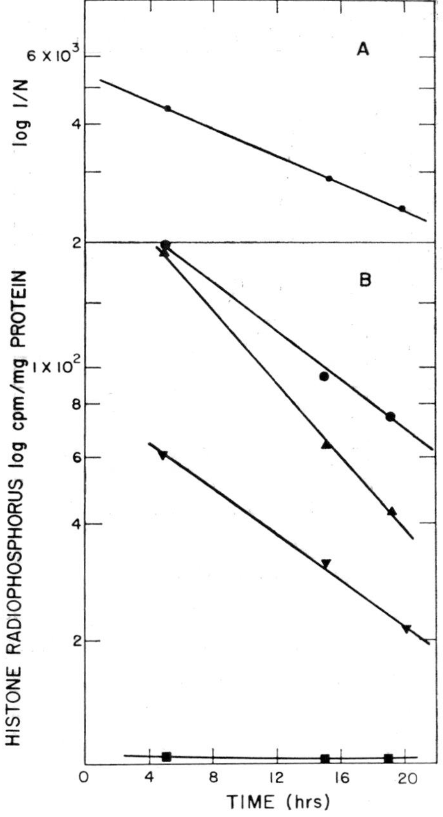

(A) Reciprocal plot of cell growth curve. (B) Radiophosphorus specific activities of histone fractions F1 (▲), F2a (■), F2b (●) and F3 (▼).

for fraction F1 and 13 h for fraction F2b when using a culture with a 16-h division (theoretical dilution) time. To investigate the turnover of these phosphate moieties within the cell cycle, isoleucine-depleted cells were labelled with inorganic radiophosphorus [^{32}P] in stationary-phase and released, after washing, into normal unlabelled medium. The variation of phosphate specific activities

FIGURE 7

Turnover of histone radiophosphorus in synchronized cultures of Chinese hamster ovary cells

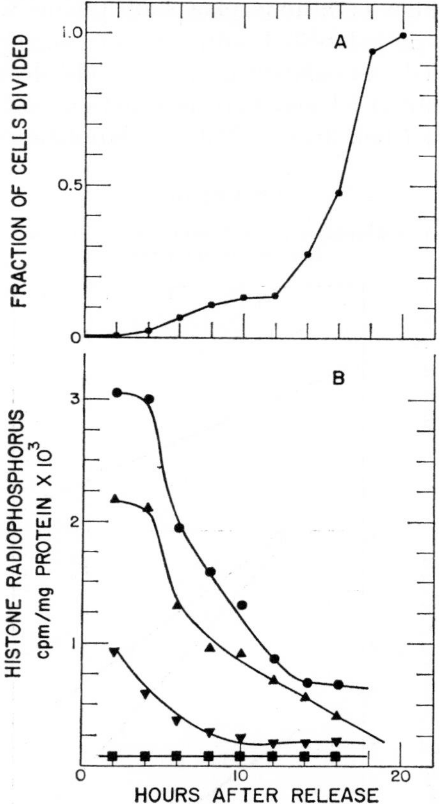

(A) Growth pattern of synchronized cells. (B) Radiophosphorus specific activities of histone fractions F1 (▲), F2a (■), F2b (●) and F3 (▼).

for histone fractions F1 and F2b during the life cycle may be seen in Fig. 7. Little or no alteration of radiophosphorous specific activity was seen in G_1 for either fraction. The specific activities of both fractions dropped to approximately half their G_1 values during S. The specific activity of fraction F1 phosphate continued to drop during G_2-M, while that of histone F2b remained steady during this period (Shepherd *et al.*, 1972b).

DISCUSSION

The purpose of this study was to determine the temporal patterns of histone structural alterations within the life cycle of cultured mammalian cells in the eventual hope of correlating these patterns in a causal manner with other events within the life cycle. Before discussing these correlations, it would be well first to consider the significance and limitations of the data presented in the Results section.

All experiments involving the measurement of rates of incorporation of labelled precursors into histones of cell cultures were performed using continuous labelling techniques. This method of labelling provided us with a continuous cumulative record of the level of labelled precursors within histone molecules, as well as partial data on relative incorporation rates. If we consider the case of accumulation of labelled acetyl, methyl and phosphoryl groups in histones (Fig. 1B-D) in this light, it is apparent that rising specific activity values reflect processes of acetylation, methylation and phosphorylation or the balance between these processes and those of deacetylation, demethylation and dephosphorylation. On the other hand, decreasing specific activity values may be interpreted as (1) a cessation of alteration processes followed by biosynthesis of unaltered molecules, or (2) a shift in the balance, for example, of the relative rates of acetylation and deacetylation in favor of the catabolic process.

A comparison of the patterns of histone acetylation, methylation and phosphorylation within the cell cycle reveals both similarities and differences among these patterns. The processes of acetylation and methylation of histone fractions appear to be discontinuous, as does the phosphorylation of fraction F2b. In general, acetylation appears to occur earlier within the cell cycle than does methylation. Thus, it is tempting to relate acetylation to events occurring in S or early G_2, and indeed, on this basis, histone acetylation may not be dismissed as a possible causal mechanism for events in S (Allfrey, 1966). In contrast, methylation of histone fractions F2a and F3 peaks in late G_2 or M and, thus, may occur too late to participate in events in S. However, by its timing, methylation of histones may represent the process of preparing and condensing chromatin for division (Tidwell *et al.*, 1968). It is far more difficult to interpret histone phosphorylation patterns in this manner. Some disagreement exists in this field as to the identity and purity of phosphorylated histone fractions (Balhorn *et al.*, 1972; Gurley *et al.*, 1971), and these points of disagreement must be settled before we may interpret these metabolic patterns with confidence. If we assume, for the sake of discussion, that our findings are correct, we observe that histone fraction F1 continuously accumulates radiophosphorous during the life cycle, while fraction F2b phosphate specific activity reaches a maximum early in S and declines. If phosphorylation of histone F1 represents a control or transport function, extrapolation would suggest that this function is continuous throughout the life cycle. Phosphorylation of both F1 and F2b appears to occur prior to DNA synthesis. On this basis, phosphorylation may not be excluded as a possible participant in events occurring in G_1 and S. The alteration of fractions F1 (acetylation), F2b (methylation) and F2b (phosphorylation) early in S may prove to be especially interesting in this regard.

We may now address ourselves to the means by which these alterations are accomplished. Our assumption in this portion of the investigation was that, using histone fractions as substrates in suitable *in vitro* reactions, increased capability of cells to alter histones would be reflected in increased transferase and kinase activities in clarified homogenates of these cells. On this basis, the data in

Fig. 2B-D suggest that these cells do not demonstrate discrete increases in acitivity which can be correlated with variations in histone acetylation, methylation and phosphorylation. If so, then it is necessary to postulate alternative mechanisms by which structural alterations are regulated.

In this regard, we are presently uncertain whether all histones are altered equally or if only selected portions of each population are altered. For example, if histone acetylation were the causal event for triggering DNA and histone synthesis in S, then one might expect to find the "old" histones synthesized in a previous cycle to be preferentially altered. If histone acetylation were a necessary event occurring prior to formation of new chromatin, one might expect to find newly-synthesized histones preferentially altered. One may deduce, from the phosphorylation of histones prior to S (Fig. 1D), that "old" histones would, of necessity, be phosphorylated. It is impossible to deduce further information of this sort with regard to acetylation and methylation other than to suggest that either or both types may be altered. However, it is tempting to speculate that the controlling factor in histone structural alterations is the availability of histone substrate, regulated by the removal of chromatin blocking groups or by biosynthesis of unaltered histones. That such regulation exists is suggested by the fact that histone fraction F1, which serves as an excellent substrate in *in vitro* methylation reactions, is not methylated *in vivo.*

We stated earlier in this discussion that observed decreases in specific activity of radiolabelled acetylated, methylated and phosphorylated histones (Fig. 1B-D) might be due either to a cessation of alteration processes followed by biosynthesis of unaltered histones or to a removal of labelled groups by appropriate enzyme systems. In order to decide between these alternatives it was necessary to establish the metabolic stabilities of these groups, with the results seen in Figs. 3-7. It appeared that histone acetyl and methyl groups and histone F2b phosphate are relatively permanent modifications in cultured CHO cells. If true, this would suggest that the rate of accumulation of such groups in histones is dependent upon the availability of suitable intramolecular sites. If several sites within the molecule are altered in the same manner at different times, it would be necessary to postulate a series of control events. If alterations of one type occur at all suitable sites within the molecule at the same time, then it is apparent that incorporation of metabolically-stable groups into histones should be dependent to a very great extent upon biosynthesis of new histones. This, in turn, would lead us to predict that "new" histones would be methylated or acetylated in these cells under normal culture conditions. This prediction is being investigated currently. The stability of F2b phosphate would lead us to the same predicition for this fraction. On the other hand, the turnover rate of histone F1 phosphate suggests that a continuous uncovering of previously phosphorylated sites occurs—an observation agreeing with the continued accumulation of radiophosphorous in this fraction during the cell cycle. This, in turn, would lead to the prediction that both old and new histone F1 may be phosphorylated in normal culture.

In summary, the temporal patterns of histone acetylation and phosphorylation and the early methylation of fraction F2b do not exclude these processes for consideration as control or participating phenomena for events in S and G_2. On this basis, the histones may still be considered candidates for roles as gene regulators or as gene moderators in higher organisms.

SUMMARY

The processes of histone acetylation, methylation and phosphorylation have been studied in random and synchronized cultures of mammalian cells.

The temporal profiles within the life cycle for each of these events have been determined, as have the corresponding levels of the kinase systems which effect these changes.

Turnover rates of histone acetyl, methyl and phosphoryl groups in culture have been determined.

These data are presented together with arguments supporting the hypothesis of multiple roles of histones within the nucleoprotein complex.

ACKNOWLEDGEMENTS

This work was performed under the auspices of the U.S. Atomic Energy Commission and the Associated Western Universities, Inc.

REFERENCES

Allfrey, V. G. (1966). In *Proceedings of the 6th Canadian Cancer Research Conference* (R. W. Begg, C. P. LeBlond, R. L. Noble, R. J. Rossiter, R. M. Taylor and A. C. Wallace, eds.). Vol *6*, p. 313, Pergamon Press, London

Allfrey, V. G., Faulkner, R. and Mirsky, A. E. (1964). *Proc. Nat. Acad. Sci., U.S., 51,* 786

Balhorn, R., Bordwell, J., Sellers, L., Granner, D. and Chalkley, R. (1972). *Biochem. Biophys. Res. Commun. 46,* 1326

Busch, H. and Steele, W. J. (1964). *Adv. Cancer Res. 8,* 41

Byvoet, P., Shepherd, G. R., Hardin, J. M. and Noland, B. J. (1972). *Arch. Biochem. Biophys. 148,* 558

Gurley, L. R. and Hardin, J. M. (1968). *Arch. Biochem. Biophys. 128,* 285

Gurley, L. R. and Walters, R. A. (1971). *Biochemistry 10,* 1588

Hardin, J. M., Noland, B. J. and Shepherd, G. R. (1971). *Exp. Cell Res. 68,* 459

Hnilica, L. S. (1967). *Progr. Nucleic Acid Res. Mol. Biol. 7,* 25

Johns, E. W. (1964). *Biochem. J. 92,* 55

Johns, E. W. (1967). *Biochem. J. 104,* 78

Kaye, A. M. and Sheratzky, D. (1969). *Biochim. Biophys. Acta 190,* 527

Langan, T. A. (1968). *Science 162,* 579

Ley, K. D. and Tobey, R. A. (1970). *J. Cell Biol. 47,* 453

Lowry, O. H., Rosebrough, N. J., Farr, A. L. and Randall, R. J. (1951). *J. Biol. Chem. 193,* 265

Murray, K. (1965). *Annu. Rev. Biochem. 34,* 209

Nohara, H., Takahashi, T. and Ogata, K. (1968). *Biochim. Biophys. Acta 154,* 529

Noland, B. J., Hardin, J. M. and Shepherd, G. R. (1971). *Biochim. Biophys. Acta 246,* 263

Ord, M. J. and Stocken, L. A. (1968). *Biochem. J. 107,* 403

Phillips, D. M. P. (1964). *Progr. Biophys. Biophys. Chem. 12,* 211

Shepherd, G. R., Hardin, J. M. and Noland, B. J. (1971a). *Arch. Biochem. Biophys. 143,* 1

Shepherd, G. R., Hardin, J. M. and Noland, B. J. (1972b), in the press
Shepherd, G. R. and Noland, B. J. (1968). *Exp. Cell Res. 49,* 238
Shepherd, G. R., Noland, B. J. and Hardin, J. M. (1970). *Biochim. Biophys. Acta 228,* 544
Shepherd, G. R., Noland, B. J. and Hardin, J. M. (1971b). *Arch. Biochem. Biophys. 142,* 299
Shepherd, G. R., Noland, B. J. and Hardin, J. M. (1971c). *Exp. Cell Res. 67,* 474
Shepherd, G. R., Noland, B. J. and Hardin, J. M. (1972a). *Exp. Cell Res.* (in press)
Stellwagen, R. H. and Cole, R. D. (1969). *Annu. Rev. Biochem. 38,* 951
Tidwell, T., Allfrey, V. G. and Mirsky, A. E. (1968). *J. Biol. Chem. 243,* 707
Tobey, R. A. and Ley, K. D. (1970). *J. Cell Biol. 46,* 151

The Role of Histones in Avian Erythropoiesis[a]

V. L. SELIGY, G. H. M. ADAMS AND J. M. NEELIN

Biochemistry Laboratory, National Research Council, Ottawa, Canada and Department of Biology, Carleton University, Ottawa, Canada

The maturation process in the nucleated avian erythrocyte has been shown to involve complete cessation of DNA synthesis and cell division (Cameron and Kastberg, 1969; Attardi *et al.,* 1970), pronounced reduction in RNA and protein synthesis (Cameron and Kastberg, 1969; Attardi *et al.,* 1970; Cameron and Prescott, 1963; Scherrer *et al.,* 1966; Kabat and Attardi, 1967), marked nuclear condensation (Kernell *et al.,* 1971; Brasch *et al.,* 1971) increased haemoglobin content and decreased cell buoyancy (Kabat and Attardi, 1967; Mathias *et al.,* 1969; Adams *et al.,* 1971). Furthermore, the erythrocyte nucleus contains major proportions of a characteristic histone V or f2c (Neelin *et al.,* 1964; Hnilica, 1964). Although the actual role *in vivo* of this highly basic histone is unknown, recent studies indicate that it may function both as a repressor of the genome (Seligy and Neelin, 1970) and as an agent supporting condensation of the chromatin in the mature erythrocyte (Brasch *et al.,* 1972).

Further clarification of the role of this cell-specific histone has required a good separation of erythroid cells of the different maturation stages. Such a separation has been approached recently (Adams *et al.,* 1971) by isopycnic centrifugation of blood cells from anaemic geese. Utilizing such a system, we have assessed the relation of histone metabolism to rates of synthesis of DNA, RNA and haemoglobin of erythroid cells at sequential stages of maturation.

EXPERIMENTAL

To obtain blood containing 12-20% erythroblasts (Table 1) adult ganders were injected three times intraperitoneally with 1% (w/v) phenylhydrazine and exsanguinated from a jugular vein 1 day after the last injection. To augment the enrichment of immature cells, washed red blood cells were centrifuged (IEC 947 rotor, 5000 r.p.m. for 30 minutes) into a barrier of 30% (w/w) Ficoll (Pharm-

[a]N.R.C.C. No. 12399.

TABLE 1

Populations of cells at different stages of maturity in peripheral blood of geese recovering from phenylhydrazine-induced anaemia

Type of blood cell population	No. of phenylhydrazine injections	Days between injections	Days of recovery	Percentage of cells						
				LBE	SBE	PE	OE	R	E	WC
Normal	0	—	—	0	0	0	0	6	92	2
Reticulocyte-rich	2	8	3	0	0	2	3	46	47	2
Erythroblast-rich	3	3	1	0.2	2.2	5.8	8.4	19.0	61.8	2.0
(standard deviation, 5 experiments)				(0.1)	(1.8)	(2.2)	(4.3)	(7.8)	(7.1)	(1.0)

acia) in Seligmann's Balanced Salt Solution (SBSS) supplemented with 5% dialyzed chicken plasma. The supernatant and loose pellet were decanted and washed with SBSS supplemented with plasma to lower the Ficoll concentration to less than 10%. Cell separation was performed by layering 85 ml cell suspension (35% packed cells) above a 10% Ficoll overlay on a 15-30% (w/w in SBSS) Ficoll gradient in an IEC Z-15 rotor turning at 1,000 r.p.m. and subsequently centrifuging at 5,000 r.p.m. for 30 minutes. Cells were collected in 10 ml fractions and surveyed by phase microscopy. Adjacent fractions of similar cell compositions were pooled and washed free of Ficoll by centrifugation in SBSS. Differential counts were made after staining slides of 10% neutral formalin-fixed smears with May-Grunwald, Giemsa stain (Lucas and Jamroz, 1961).

For incorporation studies, cells from unfractionated blood or enriched cell fractions from isopycnic separations were washed and prepared for short-term suspension culture in lysine-free, phosphate-free Minimum Essential Medium (MEM), (Eagle spinner-modified), as previously described (Seligy and Neelin, 1971). DNA synthesis was determined by measuring the incorporation of [^{3}H] thymidine into DNA isolated after incubating 0.5 ml cell suspensions (0.2 ml packed cells) with 5 μCi [^{3}H]methylthymidine (Seligy and Neelin, 1970). Measurements of haemoglobin synthesis, histone synthesis and phosphorylation, and RNA synthesis were obtained from cell suspension cultures (0.4 ml packed cells/ml suspension) containing 15 μCi/ml [^{3}H]lysine and 200 μCi/ml carrier-free [^{32}P] orthophosphate. Cells washed once in cold medium were lyzed in saponin-saline (0.05% w/v saponin-0.9% saline) and nuclei were prepared by repeated centrifugation from isotonic saline (Seligy and Neelin, 1970). Haemoglobin was separated from the bulk of the postnuclear cell lysate by electrophoresis for 2 hours at 300V on cellulose polyacetate Sepraphore III strips in Barbital buffer ($I = 0.052$, pH 8.6). Haemoglobin concentrations were determined by a micro-biuret procedure (Goa, 1953) after elution from the polyacetate strips with 0.1 M-NaOH. RNA was purified by the method of Scherrer *et al.* (1966).

For template assays, chromatin was prepared from saline-washed nuclei in dilute Tris-HCl buffer at pH 8.0 (Seligy and Neelin, 1970). Between 5 and 20μg. chromatin-DNA or deproteinized DNA were incubated with 6.3 units of Micrococcal RNA polymerase for 10 minutes at 30°; incorporation into acid-insoluble precipitate was measured after ultrafiltration (Seligy and Neelin, 1970).

Histones were extracted from saline-washed nuclei by repeated extraction with 0.20 M-HCl (Seligy and Neelin, 1971), and precipitated with 9 volumes of acetone. Precipitates were washed with acetone, but not dried, in order to facilitate redissolution for chromatography. Concentrations for specific activities or for chromatographic profiles were determined by turbidity at 400 nm in 18% trichloroacetic acid (Vidali and Neelin, 1968b).

RESULTS AND DISCUSSION

By appropriate injection regimes with phenylhydrazine, populations of goose blood cells are readily obtained in large quantities and are relatively enriched in

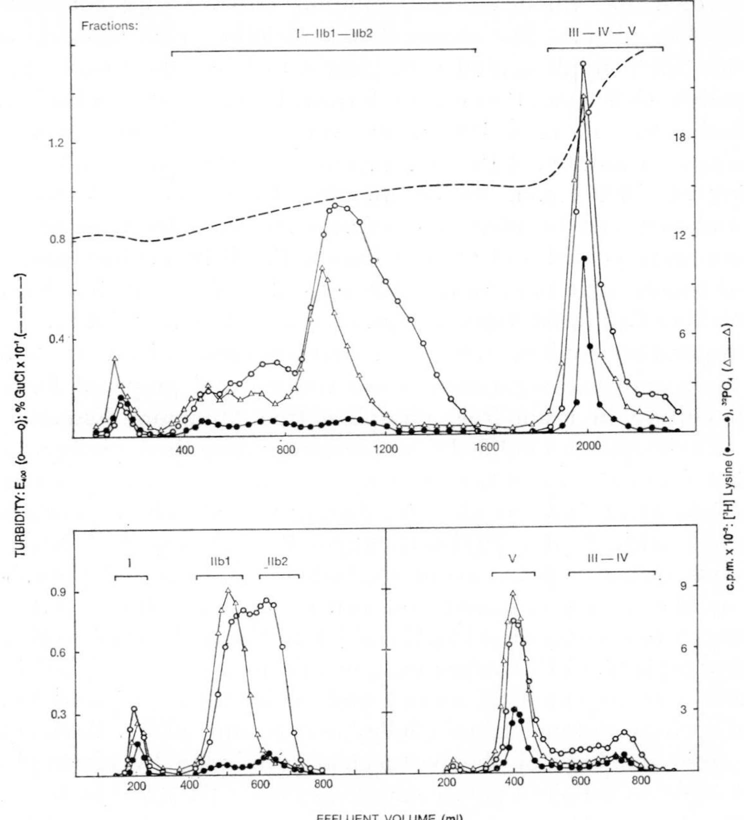

FIGURE 1A (*Top*)

Distribution of [^{32}P]phosphate and [^{3}H]lysine in normal goose erythrocyte histones, fractionated by cation-exchange chromatography. Histones were obtained by repeated acid extraction of nuclei isolated from 120 ml of normal gander erythrocytes after incubation for 1 h (Seligy and Neelin, 1971) with carrier-free [^{32}P]orthophosphate (200 μCi/ml) and [^{3}H]lysine (17 μCi/ml) in phosphate-free MEM. Eagle spinner modified (0.42 ml. packed cells per 0.55 ml medium). Histone, precipitated from the acid extract by addition of 9 volumes of acetone, was redissolved in 7.5% guanidinium chloride (GuCl) in 0.1 M-PO_4 buffer and applied to an Amberlite CG-50 column (2.5 × 60 cm). Histones were eluted from the column using a non-linear gradient of GuCl in 0.1 M-PO_4 buffer, pH 6.8 (Vidali and Neelin, 1968a).

Protein concentration was measured by turbidity in trichloracetic acid (Vidali and Neelin, 1968b). Radioactivity was measured by collecting the protein from the turbidity assays on millipore filters by low vacuum filtration before counting in a liquid scintillation counter. The broken line (——) indicates GuCl concentration from 7.8% to 21%; open circles (O — O), turbidity at 400 nm, i.e. histone concentration; closed circles (● — ●), c.p.m. [^{3}H]lysine; triangles (Δ — — Δ), c.p.m. [^{32}P]phosphate.

FIGURE 1B (*Bottom*)

Separation of partially fractionated histones I-IIb_1-IIb_2 (left) and III-IV-V (right) from cation-exchange chromatography (Fig. 1A) by exclusion chromatography on Bio-Gel P-60. Pooled histone fractions were desalted, concentrated and dialyzed against 0.005 M-H_2SO_4 (Adams and Neelin, 1969) before applying to the column of P-60 (1.6 × 82 cm) which was equilibrated and eluted with 0.01 M-HCl at a flow rate of 30 ml/h.

mature erythrocytes, reticulocytes or erythroblasts (Table 1). It should be noted that various nomenclatures based on staining properties of avian erythroid cells may not be compatible with metabolic characteristics. In the nomenclature which we have used (Scherrer *et al.*, 1966), large basophilic erythroblasts (LBE) correspond to "basophilic erythroblasts" according to Lucas and Jamroz (1961), small basophilic erythroblasts (SBE) and polychromatic erythroblasts (PE) correspond to "polychromatic erythrocytes" (Lucas and Jamroz, 1961), and orthochromatic erythroblasts (OE) or "orthochromatic erythrocytes" (Lucas and Jamroz, 1961) probably do not represent a distinguishable cell type but rather are part of a maturation continuum through reticulocytes (R) to mature erythrocytes (E). The response of individual ganders to phenylhydrazine is highly varied despite comparable sizes, ages and dosages, and mature erythrocytes and reticulocytes still predominate. Nonetheless remarkable metabolic differences among blood cell populations were evident because of the synthetic inertia of the mature cells.

All six major histone fractions may be recovered from avian erythrocytes by a combination of cation-exchange and exclusion chromatography (Vidali and Neelin, 1968a; Adams and Neelin, 1969). If normal red blood cells were incubated simultaneously with [^{3}H]lysine and [^{32}P]phosphate, incorporation of lysine was significant only in the arginine-rich components (Fig. 1A) and chiefly in the erythrocyte-specific histone V (Fig. 1B right). On the other hand phosphate incorporation was marked not only in histone V but also in histone I and especially in IIb1 (Figs. 1A and B left). This phosphate is firmly bound in the form of phosphoserine (Seligy and Neelin, 1971). Of course it cannot be said whether the incorporations represent the activities of relatively mature cells which predominate, or traces of erythroblasts, especially later stages, which may escape enumeration in differential cell counts.

In parallel experiments with regenerating blood cells, [^{3}H]lysine was incorporated into all histones (Fig. 2), about in proportion to the relative lysine contents of the different histone fractions. This is to be expected in a dividing cell population in which histone must be made in concert with new DNA. The distribution of incorporated [^{32}P]phosphate was not much different from that in normal blood, with little or no incorporation into histone IIb2 and large amounts into histone IIb1, but the specific activities of both [^{3}H]lysine and [^{32}P]phosphate were severalfold greater in histones of regenerating erythroid cells than in normal blood cells.

It was clearly desirable to carry the enrichment of erythroid cell types further to relate metabolic changes with stages in differentiation. We have found isopycnic centrifugation through gradients of Ficoll (Noble *et al.*, 1968) most effective for our purposes (Adams *et al.*, 1971). The least mature cells are the least dense, but represent also the smallest proportion, even in regenerating blood of geese. Therefore successive stages of enrichment, discarding the densest cells at each stage, are necessary to obtain a significant number of erythroblasts. The yield of basophilic erythroblasts was still marginal, and the distributions of cells

FIGURE 2A (*Top*)

Distribution of [^{32}P]phosphate and [^{3}H]lysine in histones of regenerating goose blood cells after cation-exchange chromatography. Blood enriched in immature cell forms was obtained from phenylhydrazine-injected adult ganders (Table 1). The methods for labelling of the cells, fractionation of histone on Amberlite CG-50 and measurement of protein and radioactivity of the fractions are described in Fig. 1, except that histones were obtained from nuclei isolated from a 105 ml cell suspension culture (0.42 ml packed cells/ml cell suspension).

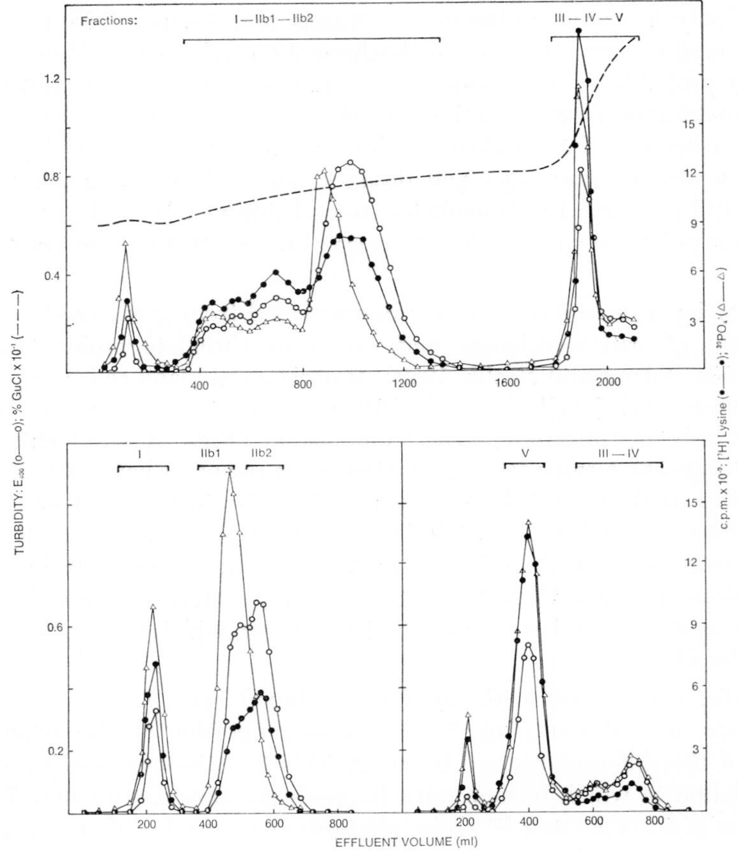

FIGURE 2B (*Bottom*)

Separation of partially fractionated histones I-IIb$_1$-IIb$_2$ (left) and III-IV-V (right) from the cation-exchange chromatogram (top) by exclusion chromatography on Bio-Gel P-60. Analytical procedures and symbols are described in Fig. 1.

through the gradient were broad and overlapping, so that pooling of fractions was somewhat arbitrary, leading to considerable variation in replicate experiments (Table 2). Nonetheless significant enrichments were obtained, and experimental variation could be minimized by performing all metabolic comparisons simultaneously on fractions cut from a single population of known composition. This internal consistency was followed in the ensuing experiments, except where noted (Figure 5 and Table 3).

TABLE 2

Differential counts of regenerating blood cell fractions prepared by zonal centrifugation in Ficoll density gradients (*three different geese and separation experiments*).

Enriched Cell fraction	Density of Ficoll	Total nuclei		Percentage of whole cells (standard deviations in brackets)					
		Free nuclei	Whole cells	LBE	SBE	PE	OE	R + E	WC
Large & small basophilic erythroblast	1.07	32(15)	68(15)	22(11)	59(11)	9(6)	2(2)	5(2)	3(4)
Small basophilic erythroblast	1.08	33(20)	67(19)	6(3)	55(7)	28(7)	8(7)	4(2)	1(1)
Polychromatic erythroblast	1.09	10(1)	90(1)	0.1(0.2)	8(8)	53(13)	13(9)	27(8)	0.5(0.5)
Reticulocyte and mature erythrocyte	1.10-1.12	4(1)	96(1)	0	1(2)	25(10)	15(4)	58(11)	2(1)

The relative composition of four cell fractions used in the following biochemical studies is shown in Fig. 3. Fraction 1 had the highest number of LBE, although SBE was the predominant cell type, approximately the same as in

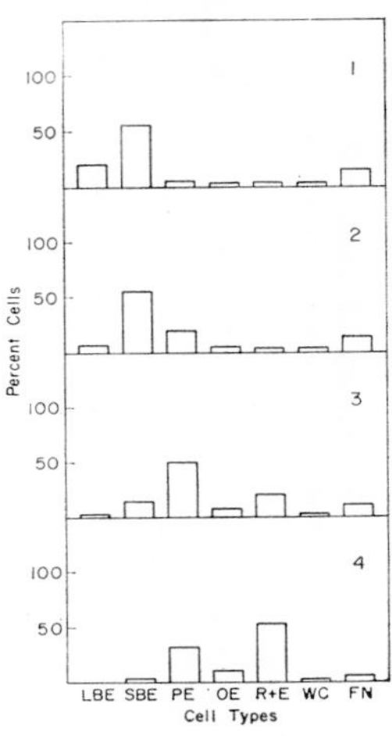

Distribution of cell types in cell fractions 1-4 (increasing in density) obtained from a separation of goose blood enriched in immature erythroid cells (Methods) by centrifugation in a 15-30% Ficoll gradient. Cells were identified according to Scherrer *et al.*, (1966): LBE, large basophilic erythoblast; SBE, small basophilic erythroblast; PE, polychromatic erythroblast; OE, orthochromatic erythroblast; R, reticulocyte; E, mature erythrocyte; WC, white blood cell; FN, free nuclei.

fraction 2. In addition to SBE, fraction 2 also contained considerable PE which occurred in highest numbers in fraction 3. Although fraction 4 had relatively more erythrocytes, this fraction was still quite rich in PE.

Consistent with autoradiographic studies (Cameron and Kastberg, 1969; Cameron and Prescott, 1963) on regenerating avian blood, rates of incorporation of radio-active precursors of nucleic acids and proteins differed in the four fractions studied (Fig. 4). The marked drop in DNA synthetic capacity of cell fractions 2, 3 and 4 in comparison to fraction 1 followed the successive decreases in LBE and SBE. After the initial drop in fraction 2, RNA and haemoglobin synthesis decreased more gradually with the proportion of total erythroblasts in the population. There is no evidence that the amino acid supply is limiting histone synthesis, even in the young cells, (d'Amelio and Costantino-Ceccarini, 1969) under these conditions. Since each cell fraction contained approximately the same numbers of white blood cells (WC) and free nuclei (FN) (Fig. 3), and since similar incorporation rates have also been obtained from cells which were labelled prior to cell separation, it is evident that the differences in synthetic rates of nucleic acids and proteins in the fractionated cells are a direct reflection of the state of differentiation of the erythroid cells rather than of damage or contamination incurred during the separation process.

FIGURE 4

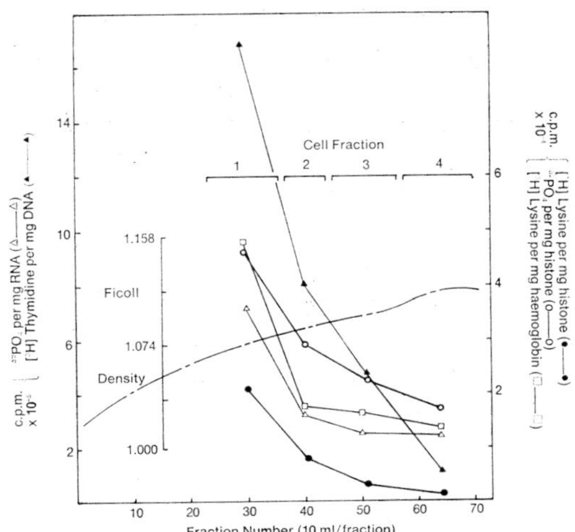

DNA, RNA and haemoglobin synthesis, and histone synthesis and phosphorylation, in relation to degree of maturity of erythroid cells separated by isopycnic centrifugation in a zonal rotor (Fig. 3 and Methods). Cell density and maturity increases from left to right on the abscissa.
Closed circles (● — ●), c.p.m. [^{3}H]lysine per mg histone; open circles (O — O), $^{32}PO_4$ per mg histone; squares (□ — □), [^{3}H]lysine per mg haemoglobin; open triangles (Δ — Δ), $^{32}PO_4$ per mg RNA; closed triangles (▲ — ▲), [^{3}H]thymidine per mg DNA.

Template capacity of isolated chromatin with added bacterial RNA polymerase and [^{3}H]uridylic acid also reflected the state of cell maturation (the assays in

FIGURE 5

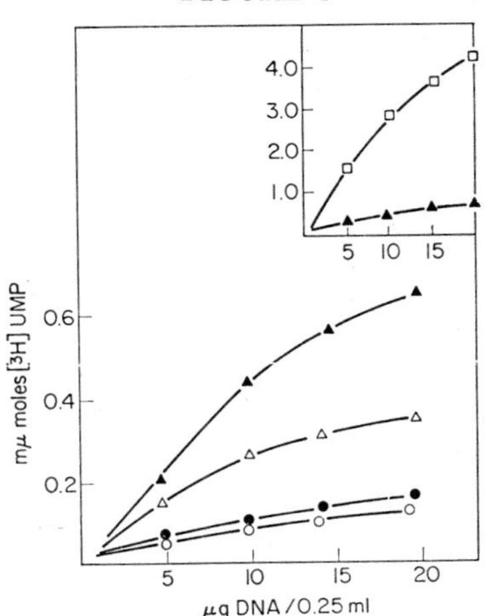

Template activities of goose erythrocyte chromatin and DNA in DNA-dependent RNA synthesis with bacterial RNA polymerase.
Chromatin was prepared from cell fractions 1 (▲ — ▲), 2 (Δ — Δ), 3 (● — ●), and 4 (O — O) after zonal centrifugation through Ficoll (Experiment A, Table 3; similar populations to those in Figures 3 and 4). Assay conditions are described in Methods.
Inset: Comparison of template activities of deproteinized DNA (□ — □) and chromatin from cell fraction 1 (▲ — ▲).

Fig. 5 were performed on the fractions in Experiment A, Table 3, whereas the data in Figures 3, 4, 6 and 7 correspond to Experiment B). The ability of chromatin from mature erythrocytes to act as template in DNA-dependent RNA synthesis was less than 2% of that of deproteinized DNA (Seligy and Neelin, 1970). The corresponding activity of chromatin from the least mature cell fraction was nearly one-seventh that of deproteinized DNA (Fig. 5, inset), and the template capacity declined to about the same level as in mature erythrocytes by the PE stage (Fig. 5).

Earlier biochemical studies of avian erythrocyte histones indicated that both synthesis (Sadgopal and Kabat, 1969; Freedman *et al.*, 1966) and phosphorylation may occur in peripheral blood (Seligy and Neelin, 1971; Gershey and Kleinsmith, 1969). The present study demonstrates that while histone synthesis and phosphorylation both diminished with increased cell maturity, lysine incorporation

TABLE 3

Comparison of radioactive lysine and phosphate incorporation into histones isolated from goose blood cells in suspension culture (A) before and (B) after zonal sedimentation and separation of erythroid cells according to stages of maturation. (*Figs. 3 and 4.*)

Experiment	Zonal Fraction	Specific Activity of Histone Fractions (cpm × 10^{-4}/mg)[a]			
		I-IIb		III-IV-V	
		[^{3}H]Lysine	$^{32}PO_4$	[^{3}H]Lysine	$^{32}PO_4$
A.	1	2.40	3.61	5.66	6.62
A.	2	1.16	1.73	2.69	3.33
A.	3	0.94	1.11	2.09	2.41
A.	4	0.42	0.89	1.07	1.92
B.	1	2.02	2.13	4.60	6.31
B.	2	0.86	1.30	1.53	2.91
B.	3	0.23	0.61	0.66	1.83
B.	4	0.11	0.84	0.25	1.31

[a]Whole histone, extracted from isolated nuclei (Methods) after incubation of cells with isotopic tracers (Fig. 4), was fractionated into two main components (I-IIb and III-IV-V) by Amberlite CG-50 column chromatography (Fig. 6)

paralleled DNA synthesis while phosphorylation declined more slowly (Fig. 4). Further analysis of nuclear histone from each of the cell fractions by miniaturized cation-exchange chromatography (Fig. 6, Table 3) indicated that histone synthesis and phosphorylation occurred differentially, with the highest incorporation of both isotopes in the arginine-rich histone fraction which contained the erythrocyte-specific histone V as the dominant component. It is assumed that synthesis of all histones draws on the same amino acid pool. The observation that lysine incorporation declined more than did phosphate incorporation between cell fractions 1 (Fig. 6b) and 4 (Fig. 6a) corresponds to the differences between regenerating blood (Fig. 2) and normal blood (Fig. 1). However the difference in lysine incorporation between fractions 1 and 4 appears more marked in frac-

FIGURE 6

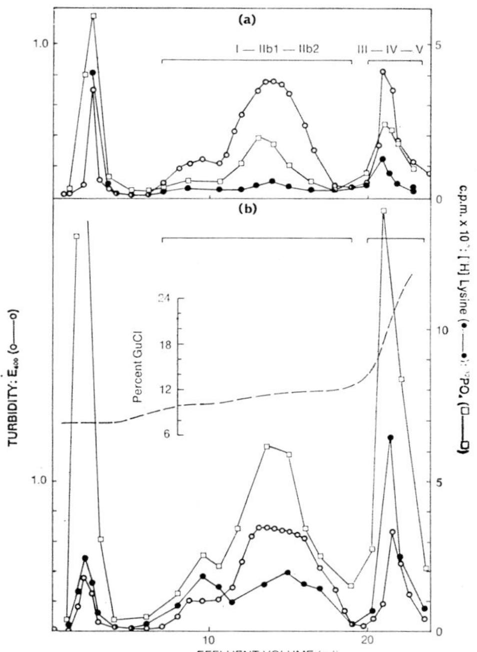

Cation-exchange chromatography of histones from (a) the PE-enriched fraction 4 (Fig. 3, 4) and (b) LBE-enriched cell fraction 1. Cell fractions were pre-labelled with [^{3}H]lysine and ^{32}P-phosphate in suspension culture under standard conditions before isolation of nuclei and extraction of histones. Histones were eluted from an Amberlite CG-50 column (0.5 × 6.0 cm) with a non-linear gradient of GuCl in 0.1 M-sodium phosphate buffer, pH 6.8, and 0.5 ml fractions were collected manually. Paucity of material necessitated the fractionation of histones into only two main CG-50 classes (Table 3). Protein was measured by turbidity in 18% trichloroacetic acid as in Figures 1 and 2. The identity of histones in each pooled fraction was confirmed qualitatively by starch-gel electrophoresis. Radioactivity was measured by collecting the insoluble protein from the turbidity assays on millipore filters by low vacuum filtration before counting in a liquid scintillation counter.

tions incubated after centrifugation in Ficoll (Experiment B, Table 3) than those incubated before separation (Experiment A). These discrepancies in detail may reflect variations in the proportions of mature and immature cells, but there is no evidence that the cells have lost normal synthetic capacities or that phosphorylation has been affected.

Differences among histones in each cell fraction were obscured at this level of resolution, but paucity of material prevented the further fractionation and isolation of the various histones by conventional techniques (Seligy and Neelin, 1971). Therefore, the two main CG-50 histone components from Experiment B (Table 3, Fig. 6) were separated into definitive classes by SDS-polyacrylamide gel electrophoresis (Fig. 7). Again it is evident that all histone classes incorporated labelled lysine but at decreasing rates with increasing cell maturation. Taking into consideration the relative lysine content of each of the histones

FIGURE 7

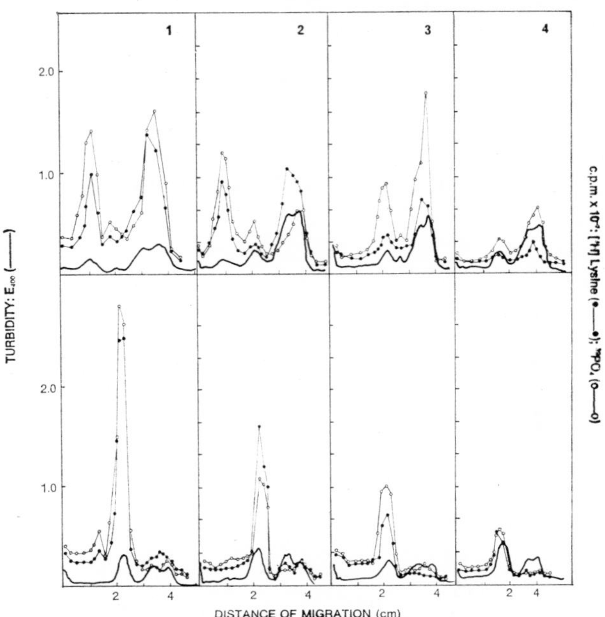

Absorbance and distribution of radioactivity in erythrocyte histones separated by electrophoresis in 12% SDS-polyacrylamide gels. Histones I-IIb and III-IV-V, respectively, were obtained from preliminary fractionation by cation-exchange chromatography (Fig. 6) of labelled whole histone from the four cell fractions described in Fig. 3 and 4. After desalting, the histones were dissolved in 0.1% SDS-0.01 M-PO_4 buffer, pH 6.8, and dialyzed against the same for 15 hours. Protein concentrations were adjusted to approximately 1.0 mg/ml and equal aliquots containing 0.2 mg were loaded on each gel before electrophoresing for 5 hours at 8 mA/gel at 23°C. Gels were stained with Amido Black-ethanol-acetic acid (0.5% Amido Black, 20% ethanol, 7% acetic acid) and destained in 20% ethanol-7% acetic acid before scanning at E_{570} nm. The identification of components contained in each electropherogram was confirmed by electrophoresis of characterized histones by starch-gel (Vidali and Neelin, 1968a)

(Adams and Neelin, 1969; Vidali and Neelin, 1968a), histones I-IV were synthesized at nearly comparable rates, while the synthesis of histone V was relatively elevated, especially in fractions enriched in the least mature cells.

The radioactive profiles in Fig. 7 reveal several significant differences in the relative incorporations of [^{3}H]lysine and $^{32}PO_4$ into the histones of the different cell fractions. In cell fraction 1, c.p.m. of incorporated radiolysine (as plotted) appeared lower than radiophosphate in all histones except III-IV. In fraction 2, phosphate incorporation was decreased relative to lysine incorporation in histones IIb and V. In fraction 3, however, in which lysine incorporation into histones I-IV was markably reduced, phosphorylation of histone IIb was substantially elevated, the bulk of the activity occurring in the region of histone IIb1. Similarly in fraction 3 phosphate incorporation declined less than lysine incorporation into histone V. In fraction 4, the trend of elevated radiophosphate in histone IIb1 was maintained. These results are consistent with the comparison of normal and regenerating blood cells (Figs. 1 and 2), and suggest the possibility that phosphate uptake in histone IIbl may be dissociated from histone synthesis in cell fractions 3 and 4, due to post-synthetic phosphorylation. Increased incorporation of radiophosphate in histone IIbl appears to be correlated with the enrichment of PE (Fig. 3), and phosphorylation in normal blood may reflect a small amount of these late erythroblasts. We have reported (Seligy and Neelin, 1971) that phosphorylation of well-characterized histones in normal goose blood cells occurs mostly in histones IIb1 and V and least in histones IIb2 and III. The favoured phosphorylation of histone IIb1 in absence of amino acid incorporation has been observed in testis (Sung and Dixon, 1970) and regenerating rat liver (Sung *et al.*, 1971), where it has been postulated to be "a specific control event".

There is no evidence in this work of a quantitative loss of any histone during erythrocyte maturation. Other workers have reported a decreasing proportion of histone I (Mazen and Champagne, 1969; Dick and Johns, 1969). The relative amount of histone V in early blood cells appears to vary, and may reflect the fleeting stages early in maturation, where histone V presumably accumulates. Furthermore, the elevated lysine incorporation into histone V in cell fraction 1 is consistent with the erythrocyte-specific histone being synthesized more rapidly than other histones in early erythroblasts. This interpretation corroborates recent unpublished observations (V. Seligy) using metabolic inhibitors, in which synthesis but not phosphorylation of "somatic histones" I-IV is significantly reduced by inhibition of DNA synthesis but not by inhibition of RNA synthesis, while synthesis and phosphorylation of histone V is greatly reduced by inhibition of RNA synthesis. It, therefore, would appear that histone V may be transcribed, synthesized and phosphorylated independently of DNA replication, while the synthesis of the other histones remains coupled to that of DNA. The erythrocyte-specific histone may appear and accumulate in concert with haemoglobin in the erythroid cell, but the histone is maintained in a progressively less phosphorylated state (Adams *et al.*, 1970), as differentiation becomes expressed in cessation of DNA synthesis, ribosome production, chromatin condensation and ultimately RNA synthesis. A similar history may be exemplified by other "shutdown" histones in terminally differentiated cells.

SUMMARY

Erythroid cells from the regenerating blood of experimentally anaemic geese were separated according to successive stages of cell differentiation by isopycnic centrifugation.

Marked differences in synthetic rates of DNA, RNA, haemoglobin and histone, as well as histone phosphorylation were related to the degree of cell maturity. The declining rate of synthesis of total histones in maturing cells paralleled that of DNA, while the rate of haemoglobin synthesis followed RNA synthesis until the reticulocyte stage was reached.

The template capacity of isolated chromatin in DNA-dependent RNA synthesis decreased in a similar manner as differentiation progressed.

Generally, histone phosphorylation was relatively low in histones IIb2, III and IV, and declined more slowly than synthesis in the other histones as cells matured. A marked burst of phosphorylation of histone IIb1 correlated with enrichment in polychromatic erythroblasts. Erythrocyte histone V was synthesized and phosphorylated more actively than the other histones in younger cell populations. Partial displacement of other histones may follow accumulation of this cell-specific histone in basophilic erythroblasts and its subsequent decline in phosphorylation with consequent chromatin condensation and loss in DNA synthesis, template capacity, and RNA synthesis.

ACKNOWLEDGEMENT

The authors are grateful to Mrs. E. Javorsky for unstinting and reliable technical assistance.

REFERENCES

Adams, G. H. M. and Neelin, J. M. (1969), *Can. J. Biochem., 47,* 1121

Adams, G. H. M., Vidali, G. and Neelin, J. M. (1970), *Can. J. Biochem., 48,* 33

Adams, G. H. M., Seligy, V. L. and Neelin, J. M. (1971), *Proc. Can. Fed. Biol. Soc. 14,* 39

Attardi, G., Parnas, H. and Attardi, B. (1970), *Exptl. Cell Res., 62,* 11

Brasch, K., Seligy, V. L. and Setterfield, G. (1971), *Exptl. Cell., Res., 65,* 61

Brasch, K., Setterfield, G. and Neelin, J. M. (1972), *Exptl. Cell Res.,* In the press

Cameron, I. L. and Kastberg, M. L., (1969), *Cytobios., 3,* 229

Cameron, I. L. and Prescott, D. M. (1963), *Exptl. Cell. Res., 30,* 609

d'Amelio, V. and Costantino-Ceccarini E., (1969), *Exptl. Cell Res., 56,* 1

Dick, C. and Johns, E. W. (1969), *Biochim. Biophys. Acta., 175,* 414

Freedman, M. L., Honig, G. R. and Rabinovitz, M. (1966), *Exptl. Cell Res., 44,* 263

Gershey, E. L. and Kleinsmith, L. J., (1969), *Biochim, Biophys. Acta, 194,* 519

Goa, J. (1953), *Scand. J. Clin. and Lab. Invest., 5,* 218

Gutierrez-Cernosek, R. M. and Hnilica, L. S., (1971), *Biochim. Biophys. Acta, 247,* 348

Hnilica, L. S. (1964), *Experientia, 20,* 13

Kabat, D. and Attardi, G., (1967), *Biochim. Biophys. Acta., 138,* 328

Kernell, A. M., Bolund, L. and Ringertz, N. R. (1971), *Exptl. Cell Res., 65,* 1

Lucas, A. M. and Jamroz, C. (1961), in *Atlas of Avian Hematology,* U. S. D. A. Monograph 25

Mathias, A. P., Ridge, D. and Trezona, N. St. G. (1969), Biochem. J., *111,* 583

Mazen, A. and Champagne, M. (1969), *FEBS. Letters, 2,* 248

Neelin, J. M., Callahan, P. X., Lamb, D. C. and Murray, K. (1964) *Can. J. Biochem., 42,* 1743

Noble, P. B., Cutts, J. H. and Carroll, K. K. (1968), Blood *31,* 66

Sadgopal, A. and Kabat, D. (1969), *Biochim. Biophys. Acta., 190,* 486

Scherrer, K., Marchaud, L., Zajdela, F., London, I. M. and Gros, F. (1966), *Proc. Nat. Acad. Sci. U.S., 56,* 1571

Seligy, V. L. and Neelin, J. M., [1970), *Biochim. Biophys. Acta., 213,* 380

Seligy, V. L. and Neelin, J. M., (1971), *Can. J. Biochem., 49,* 1062

Sung, M. T. and Dixon, G. H., (1970), *Proc. Nat. Acad. Sci. U.S., 67,* 1616

Sung, M. T., Dixon, G. H. and Smithies, O., (1971), *J. Biol. Chem., 246,* 1358

Vidali, G. and Neelin, J. M., (1968a), *Eur. J. Biochem., 5,* 330

Vidali, G. and Neelin, J. M., (1968b), *Can. J. Biochem., 46,* 781

Chromosomal Components in Relation to the Differentiation of Avian Red Blood Cells

R. APPELS, R. HARLOW, P. TOLSTOSHEV AND J. R. E. WELLS
Department of Biochemistry, University of Adelaide, Adelaide, South Australia.

During maturation of avian erythroid cells there are striking cytological and biochemical changes. Early, dividing erythroblasts from the bone marrow, active in DNA, RNA and protein synthesis contain within their large nuclei, diffuse interphase chromatin. The early non-dividing progeny cells of the circulation (here simply referred to as reticulocytes, but also classified as early-, mid- or late-polychromatic erythrocytes, Williams, 1972, Appels, *et al.*, 1972) still synthesize RNA and protein. These mature, in turn, with heterochromatinization increasingly obvious, to the mature circulating erythrocyte which contains highly condensed nuclear material and is inactive in macromolecular synthesis. All the major cell types can be obtained as relatively pure populations (Williams, 1972).

In principle at least, the decreasing capacity for transcription and the accompanying heterochromatinization as avian erythroid cells mature, could be effected by increasing levels of negative control elements and/or losses of positive control factors. Histones might fall into the former category and non-histone proteins (also referred to as acidic nuclear proteins) into the latter; chromosomal RNA (c-RNA) has been implicated in both roles (Bekhor, *et al.*, 1969; Huang and Huang, 1969) although positive control is now favoured (Mayfield and Bonner, 1971).

We decided to examine these components from chromatin preparations and in whole cell experiments, in an attempt to correlate the possible contribution of these classical components of eukaryotic chromosomal material to the *in vivo* activities of well defined populations of cells from one cell line. Although histones have been well defined as DNA-associated entities, c-RNA and non-histones have not; a necessary pre-requisite was to define the origins of these two chromatin-associated components.

MATERIALS AND METHODS

Cells

Erythrocytes were obtained from the normal circulation of hens. Reticulocytes (chiefly mid- and late-polychromatic erythrocytes, Appels *et al.*, 1972) and erythroblasts were obtained from the circulation and bone marrow respectively of highly anaemic hens. Where necessary, cells were purified on discontinuous gradients of isotonic bovine serum albumin (BSA), pH 7.4 (Appels *et al.*, 1972). Induction of anaemia (usually over a seven day period) was carefully monitored by haematocrit measurements and by a micro-gradient technique (Williams, 1972; Appels, 1972).

Preparation of nuclei and chromatin

Nuclei were isolated and washed essentially as described by Dingman and Sporn (1964). Chromatin, purified through 1.7M-sucrose, was prepared as described previously (Appels *et al.*, 1972). Crude chromatin refers to the preparation obtained just prior to centrifugation of washed, sheared nuclei through sucrose.

Chromosomal RNA

This was prepared by the method of Dahmus and McConnell (1969) and analysed on 5%-20% sucrose gradients (Beckman-Spinco SW 41 rotor, 41,000 r.p.m., 4 h, 4°C). Analysis of saturated pyrimidine bases in RNA was by the method of McGrath and Shaw, (1967). Direct phenol extractions on chromatin or the 4M-CsCl skin material were carried out by the method of Artman and Roth (1971) at 65°C or at 25°C. Extraction at 4°C (Dahmus and McConnell, 1969) was also used. Essentially the same results were obtained in each case, except that extraction at 65°C resulted in the lowest contamination of RNA preparations by DNA. Solutions for the preparation and sucrose gradient analyses of RNA were treated with diethylpyrocarbonate (Morrison *et al.*, 1970).

Non-histone proteins

(Also referred to as nuclear acidic proteins.) These were isolated by the phenol extraction procedure of Teng *et al.*, (1971) or by hydroxylapatite chromatography essentially as described by McGillivray *et al.*, (1971). Components were resolved on 10% SDS polyacrylamide gels (Weber and Osborn, 1969). Staining was with Coomassie Brilliant Blue (Fairbanks *et al.*, 1971).

Histones

Procedures for the isolation and characterisation of avian erythroid histones have been reported (Appels *et al.*, 1972). Polyacrylamide gels cross-linked with ethylene diacrylate were used to analyse ^{14}C-labelled histones (Cain and Pitney, 1968).

Measurement of transcription *in vitro*

E. coli RNA-polymerase obtained from the low salt glycerol gradient step (Burgess, 1969) was used to transcribe DNA or chromatin. The reaction was carried out at low ionic strength (Bonner *et al.*, 1968).

Incubation of cells

The medium of Schulman (1968) was used, with chicken serum replacing rabbit transferrin.

Other analytical procedures

Protein determinations were carried out by the method of Lowry *et al.* (1951); DNA by the diphenylamine reaction (Burton, 1956); RNA by the orcinol reaction (Dische, 1955) on the perchloric acid-supernatant after alkaline hydrolysis (Fleck and Monro, 1962); phospholipid by the method of Gehrlach and Deuticke (1963) and cholesterol by the method of Clark *et al.*, (1968). Total lipids were extracted from chromatin preparations by the method of Folch *et al.*, (1957).

RESULTS AND DISCUSSION

Template activity of avian erythroid chromatin

Chromatin and DNA preparations from reticulocytes and erythrocytes were assayed for their ability to support RNA synthesis *in vitro* with added *E. coli* RNA polymerase (Fig. 1). Reticulocyte chromatin was a better template than erythrocyte chromatin, suggesting that some degree of the *in vivo* capacity for transcription is preserved (for details of appropriate controls, see Appels, 1972).

Characterisation of RNA associated with avian erythroid chromatin

Chromosomal-RNA (Huang and Bonner, 1965), a heterogeneous 3-4 S RNA species widely distributed in nuclei of different tissues, is claimed to be an essential component for sequence-specific interaction of chromosomal proteins with DNA (Bekhor *et al.*, 1969; Huang and Huang, 1969). It hybridizes to repetitive DNA sequences (Sivolap and Bonner, 1971), and is tissue-specific (Mayfield and Bonner, 1971). These properties are compatible with requirements predicted for control elements in eukaryotic cells (Britten and Davidson, 1969; Crick, 1971). Other properties described for c-RNA include a relatively high percentage of saturated pyrimidines (dihydrouridine, Huang, 1967; dihydroribothymidine, Jacobson and Bonner, 1968), and covalent attachment to chromosomal proteins (Huang, 1967; but see also Artman and Roth, 1971). It has recently been reported that c-RNA may be a degradation product of t-RNA (Heyden and Zachau, 1971) or of r-RNA (Artman and Roth, 1971). Our results (Fig. 2) support the latter contention.

Preparation of c-RNA (Dahmus and McConnell, 1969) essentially involves dispersal of chromatin in 4 M-CsCl (we initially employed sonication to aid solution of the chromatin in CsCl—see discussion below) and centrifugation to isolate the skin material (containing RNA). This is subsequently digested with pronase and phenol extracted. The c-RNA is purified by gradient elution from DEAE-Sephadex. Using this procedure, c-RNA of the type described by Dahmus and McConnell (1969) could be routinely obtained and sucrose gradient analysis of such material (sonicated) is shown in Fig. 2A. It is heterodisperse, but has

FIGURE 1

Template activities of erythroid chromatin preparations and DNA

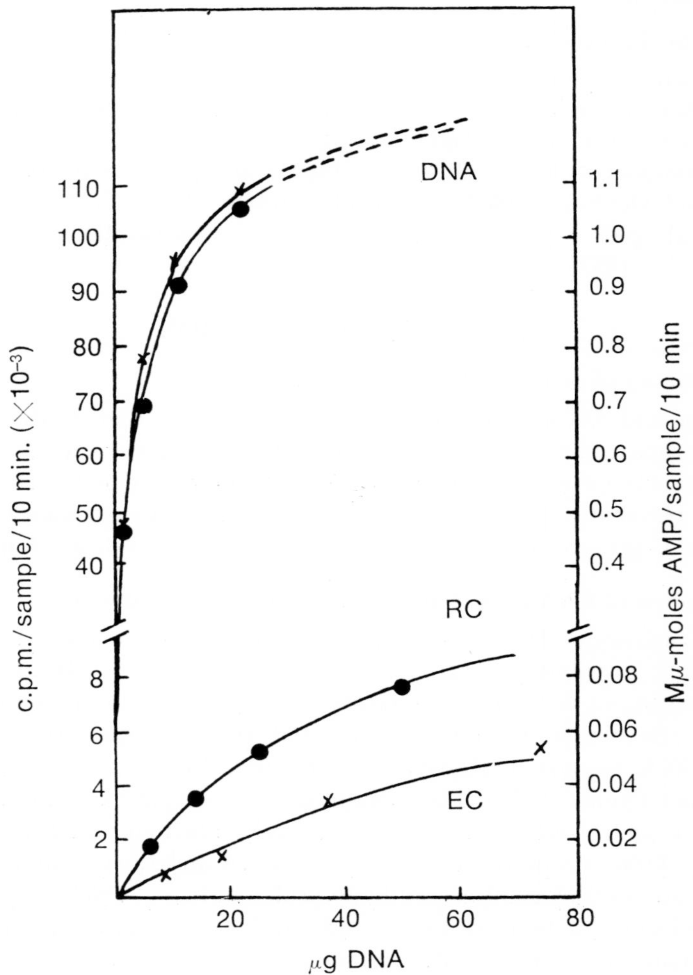

The reaction mixture (0.25 ml final volume) contained tris-HCl 0.04 M, (pH 7.9), $MgCl_2$ 3 mM, EDTA 0.1 mM, potassium phosphate 0.2 mM (pH 7.9), BSA 0.5 mg/ml, UTP, GTP and CTP (in 0.2 mM-EDTA, pH 7), 0.133 mM each, [α-^{32}P] ATP, 0.08 mM (1.0×10^8 c.p.m./μmole) RNA polymerase, 3-5 units, (Burgess, 1969) and DNA (in the form of chromatin or deproteinized DNA), 5-.50μg. Incubations were carried out at 37°C for 10 min. High molecular weight material was dried on GF/C filters and counted in a scintillation spectrometer.

×———×, mature erythrocyte chromatin (or DNA)

●———●, reticulocyte chromatin (or DNA).

a mean S value less than that of chicken t-RNA. Material obtained from crude chromatin was essentially identical (but in greater yield) to that obtained from purified chromatin.

FIGURE 2

Analysis of chicken reticulocyte RNA species

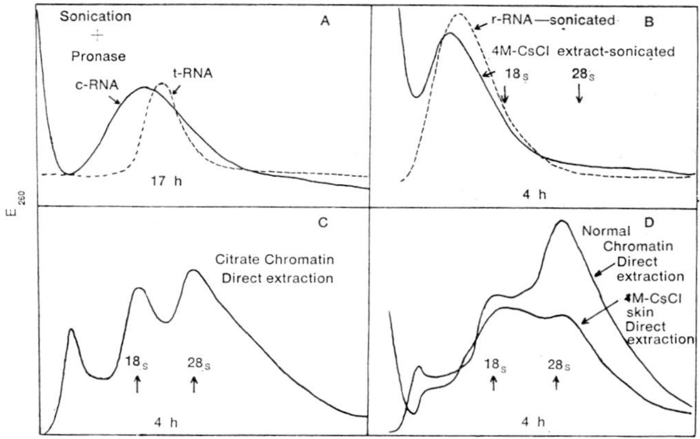

The RNA extracts were analysed on linear 5-20% sucrose gradients, (SW 41 rotor, 41,000 r.p.m., 4°C) for the times shown, and then fractionated through an Optica spectro-photometer fitted with a 2 mm flow cell to measure absorbance at 260 nm.

A. c-RNA, extracted as described in the text. (The dotted line represents reticulocyte t-RNA marker.)

B. Phenol extract of the 4M-CsCl skin of reticulocyte chromatin, prepared as described in the text. (The dotted line represents chicken reticulocyte r-RNA, sonicated for 8 × 15 seconds as described in the text.)

C. Phenol extract of reticulocyte chromatin, prepared by the citric acid method of Busch (1967).

D. Phenol extract of (1) normal reticulocyte chromatin and (2) the 4M-CsCl skin of reticulocyte chromatin, prepared without sonication.

Two lines of evidence suggest that reticulocyte c-RNA is not covalently bound to protein. First, it can be extracted from the 4 M-CsCl skin *without* prior pronase digestion (Fig. 2B) and secondly, the presence of 5 M-urea in the 4 M-CsCl prior to centrifugation, results in sedimentation of most of the RNA away from the protein skin (as previously reported by Artman and Roth, 1971). We also suggest that c-RNA is not a degradation product of t-RNA. Preparations of c-RNA obtained without pronase digestion (cf. Heyden and Zachau, 1971) had no acceptor activity in the presence of avian reticulocyte activating enzymes, whereas activation of chicken t-RNA was readily achieved (unpublished results). In addition, the characteristic alkaline hydrolysis product of dihydrouridine (β-alanine) was found in chicken reticulocyte t-RNA (1.87 moles percent), but not in c-RNA (nor in chicken 18S and 28S r-RNA), under conditions in which 0.3 moles percent was detectable. The alkaline digestion product of dihydroribothymidine (β-amino-isobutyric acid) could not be detected in c-RNA, t-RNA or r-RNA from chicken reticulocytes (limits of detection 0.2 moles per cent). The above results suggest that reticulocyte c-RNA is not related to t-RNA and further that these unusual bases are not of quantitative significance in preparations of this c-RNA.

In Fig. 2B it is shown that sonication (Soniprobe, Type 1130A, Dawe Instruments, U.K., setting 8, max. output, 8 x 15 sec) of chicken r-RNA leads to its partial degradation. If, in preparing c-RNA, sonication is included but pronase digestion prior to phenol extraction omitted, a profile almost identical to sonicated r-RNA is obtained (Fig. 2B). It is likely that ribonuclease activity (both endogenous and as a contaminant of pronase) accounts for the lower S values of c-RNA seen in Fig. 2A compared with that in Fig. 2B. However, sonication itself apparently contributes extensively to degradation of relatively large RNA species (t-RNA was not affected). This latter effect may be due to local heating during sonication.

Sucrose gradient analyses of RNA obtained after direct phenol extraction of chromatin (without 4 M-CsCl treatment or pronase digestion) prepared by the citric acid procedure (Busch, 1967, Fig. 2C) or essentially as described by Dingman and Sporn (1964; Fig. 2D) indicates that the chromatin-associated RNA is predominantly ribosomal. Similarly, omission of sonication or pronase digestion of the 4 M-CsCl skin derived from chromatin, contains relatively high molecular weight RNA (Fig. 2D). We conclude that RNA associated with avian reticulocyte chromatin is predominantly of ribosomal origin.

The non-histone proteins of avian erythroid cells

As a class, non-histone proteins are not well defined. In part this is due to the fact that some investigators use washed nuclei, whereas others start with purified chromatin (membrane material removed through a sucrose barrier, Bonner *et al.*, 1968) for extraction of non-histones. Different preparative procedures have also been employed. Extraction into phenol (after acid extraction to remove histones) is one (Teng *et al.*, 1971) and elution from hydroxylapatite, which avoids acid treatment and the associated generation of possible artefacts (Sonnenbichler and Nobis, 1970), is another (McGillivray *et al.*, 1971). In fact the *in vivo* location of non-histone proteins is not certain. *In vitro* preparations may well contain potential positive control elements such as RNA polymerase and hormone-receptor protein complexes. It is also likely that cytoplasmic proteins (Johns and Forrester, 1969) and membrane-associated proteins will contaminate non-histone preparations. It is perhaps significant also that the acidic nuclear proteins isolated by Teng *et al.* (1971) and a number of ribosomal protein species (Kabat, 1971) are readily phosphorylated *in vivo*. It is not surprising therefore that there are conflicting claims on the one hand for limited heterogeneity of non-histone proteins (Elgin and Bonner, 1970; McGillivray *et al.*, 1971) and on the other that the heterogeneity is extensive (Hill, *et al.*, 1971; Teng *et al.*, 1971).

Despite these complications, there is no shortage of claims made for the role(s) of this class of proteins in the control of eukaryotic gene expression (Gershey and Kleinsmith, 1969; Gilmour and Paul, 1969, 1970; Kamiyama and Wang, 1971; Teng *et al.*, 1971—examples are elaborated in this last reference). To obtain unequivocal results, however, reliable and specific *in vitro* assays for immediate gene products (possibly as part of heavy nuclear-RNA (Hn-RNA) are required;

the technology for this should now be available (Ross, *et al.*, 1972; Melli and Pemberton, 1972).

Here we wish to emphasize two aspects of the non-histone component of avian erythroid cells. Firstly, if washed nuclei are used as a source of non-histone proteins, the SDS gel pattern for all the major cell types is very similar, despite the wide range of activities of the cells investigated. Secondly, quantitative data for crude chromatin (CC), the membrane component (M) removed from CC by 1.7 M sucrose and "pure" chromatin (X), strongly suggest that the major proportion of so-called non-histone proteins can be accounted for quantitatively as being derived from membrane.

Analyses of non-histone proteins of avian erythroid cells. The non-histone component from washed nuclei of the three major avian erythroid cell types was prepared by hydroxylapatite chromatography. It is clear from SDS gel analyses (Fig. 3) that there is a wide distribution of different molecular weight species

FIGURE 3

Comparison of non-histone proteins from washed nuclei of avian erythroid cells

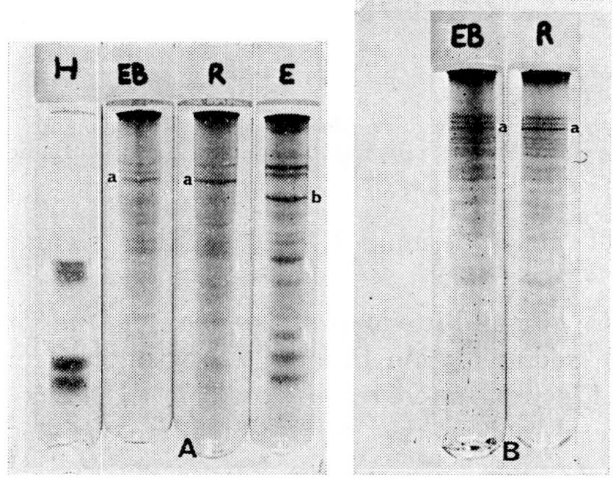

Non-histone proteins (and histones, for comparison), were prepared by hydroxylapatite chromatography and run on 0.1% SDS, 10% polyacrylamide gels (Weber and Osborn, 1969); gels were stained with Coomassie Brilliant Blue (Fairbanks *et al.*, 1971).

A. Non-histone proteins from erythroblasts (EB), reticulocytes (R) and erythrocytes (E). The pattern of histones (H) is also shown.

B. High-resolution gels of non-histone proteins from erythroblasts (EB) and reticulocytes (R).

for each cell type and that the patterns obtained are similar in each case. The erythroblast is a highly active, dividing cell, but the only obvious difference in the non-histone pattern from this cell and the less active, non-dividing reticulocyte is a quantitative difference of the band labelled *a* (Fig. 3A). The similarity of non-histone proteins from these two cell types is even more obvious in Fig. 3B. For the erythrocyte, which is totally inactive in macromolecular synthesis, the pattern of non-histone proteins is similar to that of the other two cell types with the exception of the band labelled *b* (Fig. 3A).

Hydroxylapatite chromatography has also been used to prepare the non-histone component from reticulocyte pure chromatin and the membrane removed from it. Fig. 4 again shows a wide distribution of protein species in each case and

FIGURE 4

Comparison of reticulocyte non-histone proteins from pure chromatin and membrane

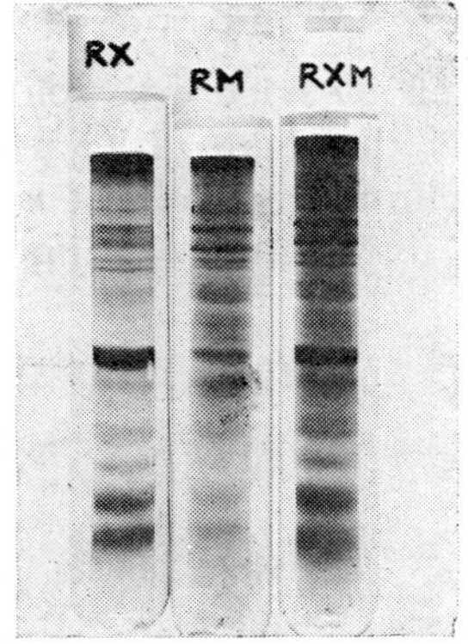

Proteins were prepared and analysed as described in Fig. 3. RX, pure chromatin, RM, membrane, RXM, co-electrophoresis of pure chromatin and membrane non-histone proteins from reticulocytes.

co-electrophoresis of the two samples indicates that there are many common species.

Membrane as origin of non-histone proteins. During the preparation of pure chromatin (Fig. 5), sedimentation through 1.7 M-sucrose removes much membranous material. Electron micrographs confirm that this is chiefly nuclear membrane but they also show that similar membrane material sediments with purified chromatin (Appels, 1972). We have compared the composition of crude chromatin, pure chromatin and membrane for reticulocytes and erythrocytes to determine what proportion of the non-histone protein in pure chromatin could be accounted for by membrane-associated protein.

The details of DNA, protein, cholesterol and phospholipid determinations will be reported elsewhere (Harlow *et al.*, unpublished results). Based on these detailed results we present condensed data for 10 mg of total protein (an arbitrary figure) as a reference (Table I). Determinations of histone were made on material obtained from acid extraction procedures (Appels *et al.*, 1972) or from hydroxylapatite chromatography. Non-histone protein was determined by difference (total protein minus histone protein) or by hydroxylapatite chromatography. The results in Table I show that the membrane and the membrane component of pure chromatin have the same character as judged by their content of phospholipid and cholesterol, which suggests that membrane-

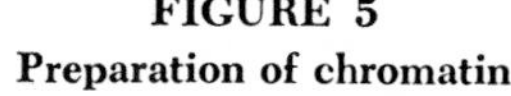

FIGURE 5
Preparation of chromatin

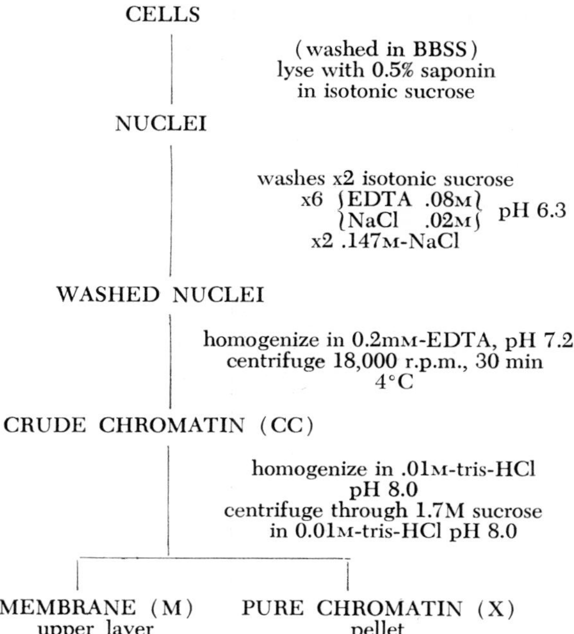

The method is essentially that of Dingman and Sporn (1964). The fractions CC, M and X were assayed for DNA, protein and total lipids (see Table I). For composition of BBSS (Buffered Balanced Salt Solution) and Isotonic sucrose, see Appels *et al.* (1971).

associated protein is also distributed in the same fashion. We show that the ratio of total non-histone protein to lipid components (cholesterol or phospholipid) in membrane and pure chromatin is very similar. Although the absolute values for reticulocytes and erythrocyte fractions are different (Table I) the same relationship is true for both cell types. We conclude from the two sets of data that a major proportion of the non-histone protein in purified chromatin from avian erythroid cells can be accounted for by membrane-associated protein. In contrast, histone is always accounted for as a DNA-associated entity. That is, the ratio of histone/DNA is constant in membrane and pure chromatin.

Lipid material in purified chromatin has been reported before (Rose and Frenster, 1965; Jackson *et al.*, 1968), and it is likely that the membrane is important in replication (Mizuno *et al.*, 1971) and possibly in transcription (Rouvière, *et al.*, 1969). Notwithstanding this, we wish to question the classification of non-histones as chromosomal proteins and their proposed role as specific control elements *in vivo*.

Histones of the avian erythroid cell series

Chicken erythrocytes contain a unique histone (called FV [Neelin, 1964] or f2c [Hnilica, 1964] as used here), a fact which contributed to the initial postu-

TABLE 1

The Distribution of Non-Histone Proteins and Membrane Components on a 1.7 M-Sucrose Barrier

Fraction	mg Total protein	mg Histone	mg Non-histone	% Non-histone	% Cholesterol	% Phospholipid	% DNA
	Reticulocyte						
CC	10	7.20	2.80	100	100	100	100
M	3.14	1.23	1.91	68.2 (±1.49)	76.9 (±3.8)	73.7 (±4.1)	17.7
X	6.86	5.97	0.89	31.8	23.1	26.3	82.3
	Erythrocyte						
CC	10	7.89	2.11	100	100	100	100
M	1.78	0.74	1.04	49.3 (±1.8)	53.7 (±6.3)	53.0 (±6.7)	8.0
X	8.22	7.15	1.07	51.7	46.3	47.0	92.0

The figures show the distribution of crude chromatin (CC) components containing 10 mg of total protein (an arbitrary figure). They are derived from quantitative analyses (to be reported in detail elsewhere) of the components shown in the Table. M, membrane; X, pure chromatin.

lation of histones as specific gene repressors (Stedman and Stedman, 1950). In contrast to this suggestion, histones are now chiefly regarded as structural elements in the organization of nuclear chromosomal material (Georgiev, 1969; Stellwagen and Cole, 1969; Paul, 1970), despite the fact that tissue-specific differences have been found (see above reviews). Crick (1971) suggests that histones interact with eukaryotic DNA in such a way that single-stranded recognition stretches of DNA are exposed. Further understanding of the interaction of basic proteins with DNA may well come from studies with simpler but highly specific systems, such as the interaction of so-called internal proteins with T4 'phage DNA (Stone and Cummings, 1972).

With regard to avian erythroid histones, it has been suggested (Purkayastha and Neelin, 1966) that avian erythroblasts do not contain the tissue-specific histone f2c, while it is agreed that it is present in the non-dividing progeny cells (Purkayastha and Neelin, 1966; Dick and Johns, 1969). We find histone f2c to be present in all cells of the series, including erythroblasts (Appels *et al.*, 1972—see also Stevely, 1971) and have therefore argued against the notion of a simple relationship between the presence of this histone and the cessation of macromolecular synthesis during maturation of the cells. Furthermore, we found no significant difference in total histone to DNA ratios for cells of this series. Although this latter result ruled out gross accumulation of histone on chromatin during maturation it was observed that the relative amount of tissue-specific f2c histone in erythroblasts was lower than that in the non-dividing cells (f2c histone in erythroblasts/f2c histone in non-dividing cells = 0.8). Thus, the possibility of f2c histone accumulation on erythroid chromatin as cell maturation

FIGURE 6

Analysis of [^{14}C]histones isolated from reticulocytes

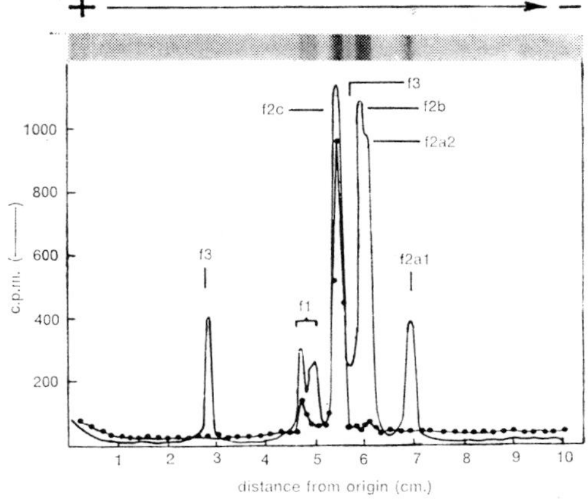

Reticulocytes were incubated with [^{14}C]lysine (20μC) plus [^{14}C]arginine (10μC) *in vitro* for 30 min at 40°. Isolated histones were analysed on polyacrylamide gels (40μg) per gel.) ●———●, distribution of radioactivity in counts/min/1 mm gel slice; ———, densitometer trace of the gel used to obtain the distribution of radioactivity.

proceeded could not be excluded. We have examined this point further by studying the metabolism of histones in these cells.

Metabolism of histones in avian erythroid cells. As expected for dividing cells, all histone species incorporated label when erythroblasts were incubated in the presence of ^{14}C-amino acids (Appels, 1972). However, for reticulocytes (mid- and late-polychromatic erythrocytes) the only histone into which label was incorporated was the f2c histone (Fig. 6). This result was reproducible and was checked by additional histone fractionation procedures (Nelson and Yunis,

FIGURE 7

Pulse-chase experiment with reticulocytes

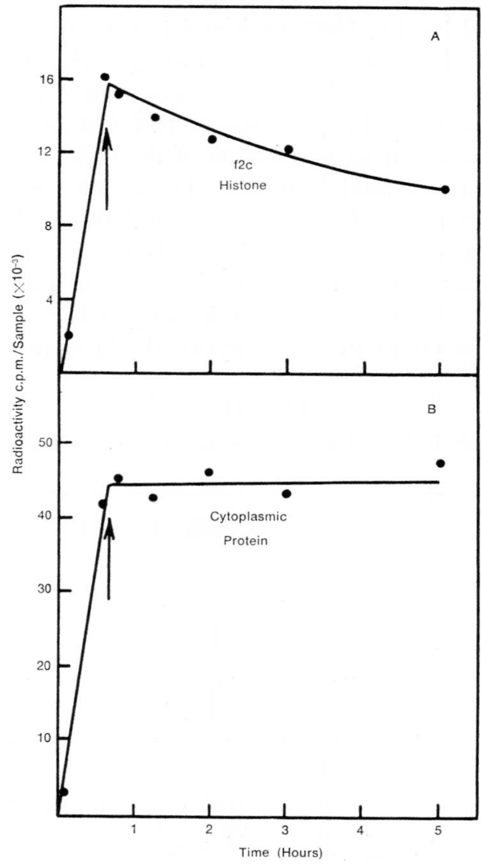

Reticulocytes were incubated in the same conditions as described for Fig. 6 but for 40 min. The arrow indicates termination of the pulse of radioactivity with either puromycin (150 μg/ml) or an excess of lysine plus arginine (0.5 mg/ml each).

A. Radioactivity in histone samples.

B. Radioactivity in cytoplasmic protein from the same cells.

1969; Johns and Diggle, 1969). The incorporation of ^{14}C-amino acids into the f2c histone has allowed us to study the metabolism in cells which are not synthesizing DNA (Williams, 1972).

Metabolic flux of the f2c histone. Reticulocytes were pulse-labelled *in vitro* with [^{14}C]lysine plus [^{14}C]arginine for 40 min (for conditions, see Figs. 6 and 7). The pulse was terminated by addition of puromycin or excess amino acids and samples were taken at intervals during the chase period to determine the radioactivity in cytoplasmic protein (mainly haemoglobin) and f2c histone. In all cases label incorporated into cytoplasmic protein was stable whereas the label in f2c histone was metabolically unstable during the chase period (Fig. 7). From measurements of the specific activity of [^{14}C]lysine plus [^{14}C]arginine in the pool of free amino acids (beginning of the pulse) and in f2c histone (beginning of the chase) we have calculated that the rate of addition of newly synthesized f2c to DNA is of the same order as its rate of release (Appels and Wells, 1972). The metabolic flux of this histone provides a rational explanation for its presence in both active and inactive cells of the series. In the former, machinery for its synthesis *and turnover* would be present and as cells mature both of these functions would decline resulting in their complete absence in the mature erythrocyte.

The metabolism of f2c histone, studied in a situation uncomplicated by requirements for newly synthesized DNA, may also apply to other histones, particularly the f1 histones which show some degree of tissue specificity (Stellwagen and Cole, 1969) and which are possibly related to f2c histone (Greenaway and Murray, 1971). Whether events such as chemical modification (Candido and Dixon, 1971) are involved in histone turnover is not known. We propose that the important point, relating histone to the potential for transcription, may be its rate of synthesis and turnover. This dynamic state could then allow fine control elements to effect specific transcription.

SUMMARY

We have studied chromosomal RNA (c-RNA), non-histone proteins and histones in purified populations of cells (erythroblasts, reticulocytes and erythrocytes) from the avian erythroid cell series, in an attempt to correlate the possible contribution of these components to the *in vivo* activity of the cells of one cell line.

Chromosomal-RNA can be prepared from reticulocyte chromatin, but it appears to arise predominantly from degradation of r-RNA.

Using washed nuclei as a source, the pattern of non-histone proteins from all the major cell types is very similar. A wide spectrum of such proteins as well as membrane material can be prepared from crude or purified chromatin. Analyses of these fractions from reticulocytes and erythrocytes, show that membrane-associated proteins can, in quantitative terms, account for the so-called non-histone proteins.

In vitro incubation of reticulocytes with [^{14}C]lysine plus [^{14}C]arginine shows that the only histone species labelled in these non-dividing cells is the tissue-specific f2c histone. Pulse-chase experiments show that this histone is metabolically unstable (cytoplasmic protein is stable in the chase period). It is proposed that the dynamic state of histone rather than its static level is important in determining the potential of chromatin for transcription.

ACKNOWLEDGEMENTS

This work was supported by the Australian Research Grants Committee.

REFERENCES

Appels, R. (1972). *Ph.D. Thesis,* University of Adelaide.

Appels, R. and Wells, J. R. E. (1972) *J. Mol. Biol.,* in press.

Appels, R., Wells, J. R. E. and Williams, A. F. (1972). *J. Cell. Sci. 10,* 47.

Artman, M. and Roth, J. S. (1971). *J. Mol. Biol. 60,* 291.

Bekhor, I., Kung, G. M. and Bonner, J. (1969). *J. Mol. Biol. 39,* 351.

Bonner, J., Chalkley, R. C., Dahmus, M. E., Famborough, D., Fujimura, F., Huang, R. C. C. Huberman. J., Jensen, R., Marushige, K., Ohlenbusch, H., Olivera, B. and Widholm, J. (1968), in *Methods in Enzymology* (Grossman, L. and Moldave, K., eds.) vol. *12,* p. 3. Academic Press Inc., New York and London.

Britten, R. J. and Davidson, E. H. (1969). *Science. 165,* 349.

Burgess, R. R. (1969). *J. Biol. Chem. 244,* 6160.

Burton, K. (1956). *Biochem. J. 62,* 315.

Busch, H. (1967), in *Methods in Enzymology* (Grossman, L. and Moldave, K., eds.) vol. *12*A, p. 448. Academic Press Inc., New York and London.

Cain, D. F. and Pitney, R. E. (1968). *Anal. Biochem. 22,* 11.

Candido, E. P. M. and Dixon, G. H. (1971). *J. Biol. Chem. 246,* 3182.

Clark, B. R., Rubin, R. T. and Arthur, R. J. (1968). *Anal. Biochem. 24,* 27.

Crick, F. H. C. (1971). *Nature (London), 234,* 25.

Dahmus, M. E. and McConnell, D. J. (1969). *Biochemistry, 9,* 1524.

Dick, C. and Johns, E. W. (1969). *Biochim. Biophys. Acta, 175,* 414.

Dingman, C. W. and Sporn, M. B. (1964). *J. Biol. Chem. 239,* 3483.

Dische, Z. (1955) in *The Nucleic Acids* (Chargaff, E. and Davidson, J. N., eds.), vol. *1,* p. 258, Academic Press, New York.

Elgin, S. C. R. and Bonner, J. (1970). *Biochemistry, 9,* 4440.

Fairbanks, G., Steck, T. L. and Wallach, G. F. H. (1971). *Biochemistry, 10,* 2606.

Fleck, A. and Monro, H. N. (1962). *Biochim. Biophys. Acta, 55,* 571.

Folch, J., Lees, M. and Sloane-Stanley, G. H. (1957). *J. Biol. Chem. 226,* 497.

Gehrlach, E. and Deuticke, B. (1963). *Biochem. Z. 337,* 477.

Georgiev, G. P. (1969). *Ann. Rev. Genet. 3,* 155.

Gershey, E. L. and Kleinsmith, L. J. (1969). *Biochim. Biophys. Acta, 194,* 519.

Gilmour, R. S. and Paul, J. (1969). *J. Mol. Biol. 40,* 137.

Gilmour, R. S. and Paul, J. (1970) *FEBS Lett. 9,* 242.

Greenaway, P. J. and Murray, K. (1971). *Nature, (London) 229,* 233.

Heyden, H. W. and Zachau, H. G. (1971). *Biochim. Biophys. Acta, 232,* 651.

Hill, R. J., Poccia, D. L. and Doty, P. (1971). *J. Mol. Biol. 61,* 445.

Hnilica, L. S. (1964). *Experientia, 20,* 13.

Huang, R. C. C. (1967). *Fed. Proc. 26,* 1933.

Huang, R. C. C. and Bonner, J. (1965). *Proc. Nat. Acad. Sci. U.S. 54,* 960.

Huang, R. C. C. and Huang, P. C. (1969). *J. Mol. Biol. 39,* 365.

Jackson, V., Earnhardt, J. and Chalkley, R. (1968). *Biochem. Biophys. Res. Commun.* 33, 253.

Jacobson, R. A. and Bonner, J. (1968). *Biochem. Biophys. Res. Commun.* 33, 716.

Johns, E. W. and Diggle, J. A. (1969). *Eur. J. Biochem.* 11, 495.

Johns, E. W. and Forrester, S. (1969). *Eur. J. Biochem.* 8, 547.

Kabat, D. (1971). *Biochemistry,* 10, 197.

Kamiyama, M. and Wang, T. Y. (1971). *Biochim. Biophys. Acta.* 228, 563.

Lowry, O. H., Rosebrough, N. J., Farr, A. L. and Randall, R. T. (1951). *J. Biol. Chem.* 193, 265.

McGillivray, A. J., Carroll, D. and Paul, J. (1971). *FEBS. Letts.* 13, 204.

McGrath, D. I. and Shaw, D. C. (1967). *Biochem. Biophys. Res. Commun.* 26, 32.

Mayfield, J. E. and Bonner, J. (1971). *Proc. Nat. Acad. Sci. U.S.* 68, 2652.

Melli, M. and Pemberton, R. E. (1972). *Nature New Biol.* 236, 1972.

Mizuno, N. S., Stoops, C. E. and Peiffer, R. L. (1971). *J. Mol. Biol.* 59, 517.

Morrison, M., Williamson, R., Lanyon, G. and Paul, J. (1970). *Biochem, J.* 119, 59P.

Neelin, J. M. (1964) in *Nucleohistones* (Bonner, J. and T'so, P., eds.) p. 66, Holden. Day Inc., London and Amsterdam.

Nelson, R. D. and Yunis, J. J. (1969). *Expl. Cell. Res.* 57, 311.

Paul, J. (1970). *Current Topics devl. Biol.* 5, 317.

Purkayastha, R. and Neelin, J. M. (1966) *Biochim. Biophys. Acta,* 117, 468.

Rose, H. G. and Frenster, J. H. (1965). *Biochim. Biophys. Acta,* 106, 577.

Ross, J., Aviv, H., Scolnick, E. and Leder, P. (1972). *Proc. Nat. Acad. Sci. U.S.* 69, 264.

Rouvière, J., Lederberg, S., Granboulan, P. and Gros, F. (1969). *J. Mol. Biol.* 46, 413.

Schulman, H. M. (1968). *Biochim. Biophys. Acta,* 155, 253.

Sivolap, Y. M. and Bonner, J. (1971). *Proc. Nat. Acad. Sci. U.S.* 68, 387.

Sonnenbichler, J. and Nobis, P. (1970). *Eur. J. Biochem.* 16, 60.

Stedman, E. and Stedman, E. (1950). *Nature, (London)* 166, 780.

Stellwagen, R. H. and Cole, R. D. (1969). *Annu. Rev. Biochem.* 38, 951.

Stevely, W. S. (1971). *Biochem. J.* 124, 48P.

Stone, K. R. and Cummings, D. J. (1972). *J. Mol. Biol.* 64, 651.

Teng, C. T., Teng, C. S. and Allfrey, V. G. (1971). *J. Biol. Chem.* 246, 3597.

Weber, K. and Osborn, M. (1969). *J. Biol. Chem.* 244, 4426.

Williams, A. F. (1972). *J. Cell. Sci.* 10, 27.

Mechanism of Glucocorticoid Hormone Action and of Regulation of Gene Expression in Cultured Mammalian Cells[a]

JOHN D. BAXTER[b], GUY G. ROUSSEAU[c], STEPHEN J. HIGGINS[d], GORDON M. TOMKINS

To understand the mechanism of gene expression in higher organisms we have been studying the hormonal regulation of specific protein synthesis in mammalian cells. Most of our work has involved cultured rat hepatoma (HTC) cells (Thompson *et al.*, 1966) in which glucocorticoid hormones influence, as far as we know, only five functions (Table 1). Of these, the induction of tyrosine aminotransferase

TABLE 1
Glucocorticoid effects on HTC cells

Function	Influence
Tyrosine aminotransferase synthesis[a]	Increased
Glutamine synthetase[b]	Increased
Cell adhesiveness[c]	Increased
Adenosine-3,5′-cyclic monophosphate (cAMP) diesterase[d]	Decreased
Phenylalanine t-RNA[e]	Increased

[a]From Granner *et al.*, (1970).
[b]From Kulka *et al.*, (1972).
[c]From Ballard and Tomkins (1969, 1970).
[d]From Manganiello and Vaughan (1972).
[e]From Lippman, Yang and Thompson (1972).

[a]*Metabolic Research Unit, Department of Medicine, and Department of Biochemistry and Biophysics, University of California, San Francisco, California.* *Supported by Grant No. 17329 of the National Institute of General Medical Sciences of the National Institutes of Health*
[b]*Supported by a Dernham Senior Fellowship of the American Cancer Society, No. D-177, California Division*
[c]*Chargé de Recherches du Fonds National de la Recherches Scientifique (Belgium) and recipient of a US Public Health Service International Postdoctoral Research Fellowship (1-F05-TW-1725-02)*
[d]*Supported by the Damon Runyon Memorial Fund for Cancer Research No. DRF-630*

(TAT) synthesis has been the most extensively studied. A factor which promotes cell adhesiveness (Ballard and Tomkins, 1969, 1970) is also induced, as well as under special circumstances, glutamine synthetase (Kulka *et al.*, 1972). Finally, preliminary reports suggest that glucocorticoids increase the amount of phenylalanine t-RNA (Yang, Lippman and Thompson, unpublished) and decrease the activity of the phosphodiesterase which degrades adenosine-3′, 5′cyclic monophosphate (cAMP) (Manganiello and Vaughan, 1972).

The effects of glucocorticoids appear to be specific, since the rates of total protein or RNA synthesis (Tomkins *et al.*, 1966), the cell protein and RNA content (Gelehrter and Tomkins, 1967) and their rates of degradation (Aurrichio *et al.*, 1969; Hershko and Tomkins, 1971; Tomkins *et al.*, 1972) are unaffected (Table 2). A number of cytoplasmic enzymes are likewise not influenced by the

TABLE 2

Functions not affected by glucocorticoids in HTC cells

Tyrosine aminotransferase degradation[a]
Cell growth or generation cycle[b]
Total protein or RNA synthesis or degradation[c]
Total protein or RNA levels[d]
Activity of about 30 cytoplasmic or nuclear enzymes[e]

[a]From Auricchio *et al.*, (1969); Hershko and Tomkins (1971); Tomkins *et al.*, (1972).
[b]From Tomkins *et al.*, (1966); Martin and Tomkins (1970).
[c]From Gelehrter and Tomkins (1967); Auricchio *et al.*, (1969); Hershko and Tomkins (1971); Tomkins *et al.*, (1972).
[d]From Gelehrter and Tomkins (1967); Tomkins *et al.*, (1966).
[e]From Tomkins (unpublished).

hormone (Tomkins, unpublished). Therefore, the HTC cell system seems to be a good one for studying the regulation of gene expression.

Until recently much of the information regarding hormonal regulation in this system came from experiments with intact cells in which effects of various steroids, of inhibitors or macromolecular syntheses, and of the cell cycle on TAT synthesis were studied. Both protein and RNA synthesis are necessary for enzyme induction (Thompson *et al.*, 1966). Most interestingly, RNA synthesis is also required for the decline in the rate of TAT synthesis following removal of the inducer steroid (deinduction) (Thompson *et al.*, 1966). Further, there are certain phases in the cell cycle during which deinduction cannot occur (Martin and Tomkins, 1970; Martin *et al.*, 1969a, 1969b). From these observations, a model involving post-transcriptional regulation of TAT synthesis was proposed (Tomkins *et al.*, 1969). Yet, the role of glucocorticoids remains unclear. They appear to promote an accumulation of m-RNA for TAT in the absence of protein synthesis (Peterkofsky and Tomkins, 1968). This could result from a direct effect of steroids on transcription or from an indirect influence on the post-transcriptional regulatory system.

Lately, our efforts have been directed towards reproducing the entire sequence of events for the hormonal induction of TAT in cell-free systems. To this end, we have studied the interactions of inducer steroids with specific receptor proteins which are believed to mediate the hormone effect. We have also investigated the events which immediately follow the formation of the receptor-steroid complex. Further, we have been successful in synthesizing TAT in cell-free systems. Some of these recent studies are discussed below.

HTC CELLS CONTAIN SPECIFIC RECEPTORS FOR GLUCOCORTICOIDS

In common with other investigators (Raspé, 1970; Gammeltoft and Schaumburg, 1972; Beato *et al.*, 1972; Hackney *et al.*, 1970), we find present in target tissues for glucocorticoid hormones, including HTC cells (Baxter and Tomkins, 1971), specific cytoplasmic receptor proteins. In spite of their wide distribution in glucocorticoid-responsive tissues (for review, see Baxter and Forsham, 1972), the receptors are not found in all tissues, (e.g. immature uterus or prostate (Ballard, Baxter, Higgins and Rousseau, unpublished)). The receptors differ from other glucocorticoid-binding proteins such as plasma transcortin and from cellular "non-specific" binding components (Baxter and Tomkins, 1971; Rousseau *et al.*, 1972; Baxter and Tomkins, 1970).

Several lines of evidence implicate the specific receptors as necessary for a glucocorticoid response. First, the rates of steroid association with and dissociation from receptors in intact HTC cells are rapid enough to account for the kinetics of enzyme induction and deinduction (Baxter and Tomkins, 1970). Second, there is a good correlation between the biologic potency of a glucocorticoid (e.g., dexamethasone) and its capacity to bind to the specific receptors (Baxter and Tomkins, 1970, 1971; Rousseau *et al.*, 1972) (Fig. 1). In addition, other

FIGURE 1

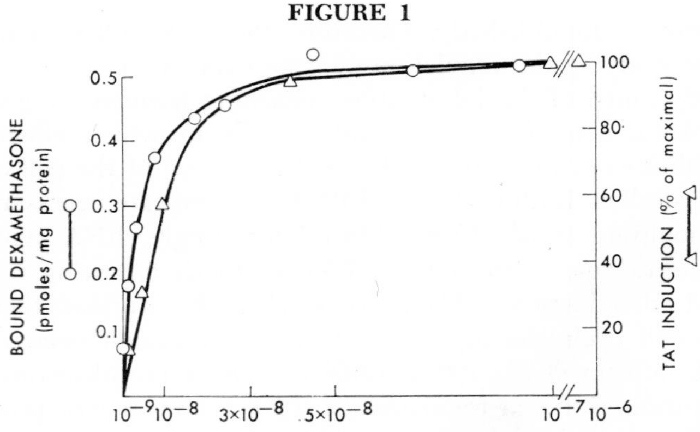

Correlation between the biologic potency of dexamethasone and its capacity to bind to the specific glucocorticoids receptors in HTC cells. The induction data are taken from Samuels and Tomkins (1970). The binding data are taken from Baxter and Tomkins (1971).

inducers such as corticosterone, cortisol, and aldosterone bind to the specific cytoplasmic receptors with an affinity which is related to their potency (Rousseau *et al.,* 1972). These correlations also extend to the induction of cell adhesiveness (Baxter and Tomkins, 1971). Third, a close relationship between binding characteristics and biological actions has also been established in the case of steroids with anti-glucocorticoid activity (Baxter and Tomkins, 1970, 1971; Rousseau *et al.,* 1972). Such steroids inhibit both the binding and actions of glucocorticoids, by themselves binding to the same specific receptor molecule (Rousseau *et al.,* 1972). As shown in Table 3, two of these steroids progestrone and 17 alpha-methyl testo-

TABLE 3

Inhibition by anti-inducers of specific binding or enzyme induction by dexamethasone[a]

Anti-inducer	Induction of TAT by dexamethasone in the presence of anti-inducer (% of control)	[^{3}H] Dexamethasone bound in the presence of anti-inducer (% of control)
None	100	100
Progesterone		
10^{-8}M	113	100
10^{-7}M	93	70
10^{-6}M	44	9
10^{-5}M	12	2
17 α-methyl testosterone		
10^{-8}M	97	105
10^{-7}M	100	103
10^{-6}M	44	64
10^{-5}M	18	22

[a]From Rousseau *et al.,* (1972).

sterone, inhibit induction by dexamethasone to an extent roughly similar to that to which they inhibit binding.

As a fourth line of evidence implicating these specific receptors as necessary for a glucocorticoid response, loss of the receptors is accompanied by a loss of glucocorticoid responsiveness (Hackney *et al.,* 1970; Hollander and Chiv, 1966; Kirkpatrick *et al.,* 1971; Baxter *et al.,* 1971; Rosenau *et al.,* 1972). We have studied this in cultured lymphoma cells which are killed by glucocorticoids (Baxter *et al.,* 1971; Rosenau *et al.,* 1972). The lines are established in continuous culture by Horibata and Harris (1970). These workers, and later ourselves, derived several steroid-resistant lines from the steroid-responsive cells.

The steroid-sensitive cells contain specific glucocorticoid receptors with properties similar to those of the HTC cells (Baxter *et al.,* 1971; Rosenau *et al.,* 1972). In sharp contrast steroid-resistant cells show a marked decrease in the concentration of specific glucocorticoid receptors (Fig. 2) (Rosenau *et al.,* 1972). Further, what little binding activity is present in these cells may have physical characteristics which differ from the principal receptor in the steroid-sensitive cells (Rosenau *et al.,* 1972). Thus, loss of glucocorticoid responsiveness in these

FIGURE 2

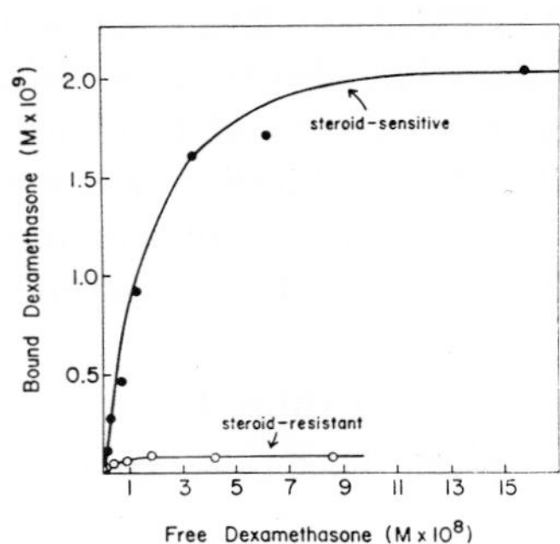

Specific binding of dexamethasone by cytosol of steroid-sensitive (●) and steroid resistant (○) lymphoma cells. Data reprinted from Rosenau *et al.*, (1972). S = steroid (glucocorticoid or anti-glucocorticoid)

cells can be attributed either to a loss of or alteration in the specific glucocorticoid receptors. It should be emphasized that the steroid-sensitive and resistant cell lines are not known to differ in any other way; they have the same light microscopic appearance, karyotype, growth rate, subcellular distribution and content of protein (Baxter *et al.*, 1971).

STEROIDS ACT AS ALLOSTERIC EFFECTORS

What happens upon formation of the receptor-steroid complex? Several lines of evidence suggest that steroids act as allosteric ligands influencing the conformation of the receptors (Rousseau *et al.*, 1972; Samuels and Tomkins, 1970). This model was proposed (Rousseau *et al.*, 1972) to explain the biologic action of various classes of steroids. Subsequently, additional support for it has been obtained following detection and characterization of the specific glucocorticoid receptors (Rousseau *et al.*, 1972).

This model (Fig. 3) assumes that two conformational states of the receptor, active and inactive, are in an equilibrium that is influenced by the steroid. In the

FIGURE 3

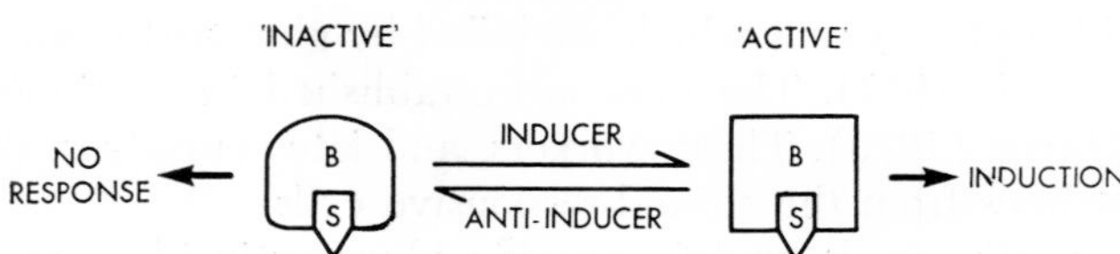

Allosteric model for steroid hormone action. B denotes the glucocorticoid receptor, and the different shapes indicate different conformational states of B.

absence of steroid, the predominant form of the receptor is inactive. An "*optimal*" inducer steroid such as dexamethasone binds predominantly to the active form of the receptor, and thus increases the proportion of molecules in this conformation, resulting in the glucocorticoid response. Conversely, when the *anti-glucocorticoid,* progesterone, binds to the specific receptors no change in the conformational equilibrium occurs, and therefore there is no biological response.

This model provides in addition an interpretation of the effects of other classes of steroids (Fig. 4). Besides optimally active and anti-glucocorticoids,

FIGURE 4

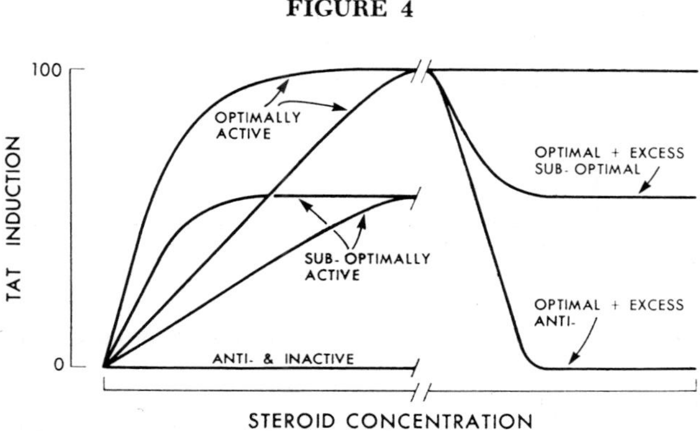

Classes of steroid action. The TAT induction (%) at various concentrations of the different steroids is shown.

there exists a third class of "*sub-optimal*" inducers (Samuels and Tomkins, 1970). These steroids give rise to an intermediate biologic response, even at concentrations high enough to fully saturate the specific receptors. Further, sub-optimal inducers competively inhibit the binding and action of inducers, resulting in the response achieved by the sub-optimal inducer alone. If it is assumed that sub-optimal inducers bind to both active and inactive conformations of the receptor ("nonexclusive binding") there would be only a partial shift of the equilibrium, and therefore only an intermediate biologic response. As illustrated, *inactive* steroids do not influence the induction (Samuels and Tomkins, 1970), and also lack binding activity (Baxter and Tomkins, 1970, 1971).

Further support for the allosteric model is derived from experiments in cell-free cytoplasmic extracts in which the kinetics of inducer and anti-inducer binding and the thermal stabilities of the resulting complexes were compared (Rousseau *et al.,* 1972). Dexamethasone binds quite slowly at 0°C. in spite of its high affinity for the receptors. These kinetics could reflect the postulated conformational changes associated with inducer binding, i.e., the conversion of receptors from "inactive" to "active" forms. If so, it would be expected that anti-inducers should bind faster than inducers since anti-inducers would bind to the "inactive" form of the receptor which predominates in the uncomplexed state. This is in fact the case; the anti-inducer progesterone binds much faster than dexamethasone, even though progesterone binding is slightly weaker (Fig. 5). That the conformations of inducer and anti-inducer-receptor complexes are distinct is further suggested by the difference in their thermostability (Rousseau *et al.,* 1972).

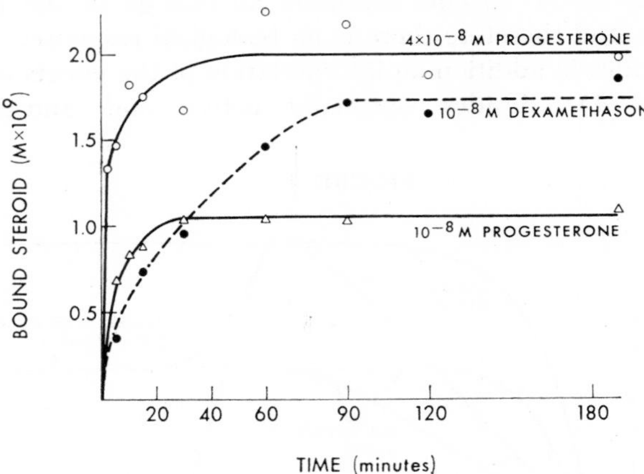

Kinetics of binding of dexamethasone and progesterone. Data reprinted from Rousseau *et. al.* (1972)

RECEPTOR-STEROID COMPLEXES BIND TO NUCLEI

The uncomplexed receptors are found in the cytoplasmic but not the nuclear fraction of HTC cells (Baxter and Tomkins, 1971) (Fig. 6). In contrast, in

FIGURE 6

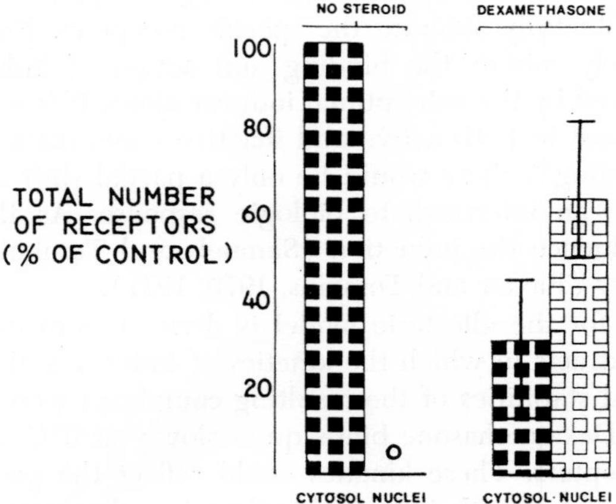

Steroid effect on the location of receptors in HTC cells. The total number of receptors (either bound or uncomplexed) is shown in cultures of control or dexamethasone-treated HTC cells. Data taken from Baxter *et al.,* (1971, 1972); Baxter, Rousseau, Higgins and Tomkins, unpublished.

extracts from cells exposed to dexamethasone, most of the receptors are found in the nuclear fraction and correspondingly fewer are found in the cytoplasmic fraction (Baxter *et al.,* 1971, 1972; Baxter, Rousseau, Higgins and Tomkins, unpublished). Thus, inducers allow a redistribution of the receptor to favour its nuclear localization.

When HTC cells are equilibrated with dexamethasone and the inducers subsequently removed from the incubation medium, there is a rapid dissociation of steroid from the receptors (Baxter and Tomkins, 1970; Baxter *et al.,* 1971, 1972; Baxter, Rousseau, Higgins and Tomkins, unpublished). There also follows a disappearance of specifically bound nuclear steroid, a reappearance of the receptors in the cytoplasm, and a rapid decline in the induced rate of TAT synthesis (Baxter and Tomkins, 1970; Baxter *et al.,* 1971, 1972; Baxter, Rousseau, Higgins and Tomkins, unpublished). The intracellular redistribution which occurs upon the addition or withdrawal of inducer steroids occurs even when protein or RNA synthesis are inhibited (Baxter, *et al.,* 1971, 1972; Baxter, Rousseau, Higgins and Tomkins, unpublished). These findings strongly suggest that in the induced state the cytoplasmic receptor itself remains in the nucleus with the steroid rather than acting catalytically as a transport molecule for the steroid. If the receptors transfered the steroid onto other binding sites, one would have to explain why the receptors do not then return to the cytosol. Their failure to return could of course be due to degradation in the nucleus; however, they do in fact reappear in the cytoplasm when the steroid is removed from the medium. Since the return occurs even when macromolecular synthesis is inhibited, the original receptors and not newly synthesized copies must be involved in the reappearance.

Nuclear binding of receptor-steroid complexes has also been studied in a cell-free system (Baxter *et al.,* 1972a, 1972b; Higgins, Rousseau, Baxter and Tomkins, unpublished). As might have been predicted from the whole cell experiments, isolated nuclei specifically bind the receptor-steroid complex. This interaction is observed only if the cytosol has been "activated" beforehand, for example, by raising the temperature. As suggested from our studies with intact cells, specific

FIGURE 7

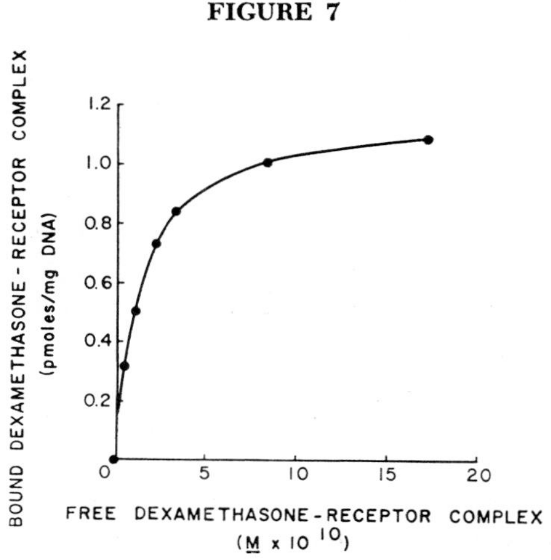

Nuclear binding of receptor-dexamethasone complexes in a cell-free system at various concentrations of complex. The experiment was performed as described by Baxter *et al.,* (1972a).

steroid binding to isolated nuclei requires the cytoplasmic receptors. No specific nuclear binding is observed either in the absence of cytoplasmic macromolecules, or with free steroid, with inducer-bound-transcortin, or with steroid-containing cytosol from the receptor-deficient resistant lymphoma cells (Baxter *et al.*, 1972a). These findings reinforce our hypothesis that the receptor-steroid complex itself associates with the nucleus. Furthermore, the amount of steroid bound to isolated nuclei is accounted for by an equivalent loss of receptor-steroid-complex from the cytosol (Baxter *et al.*, 1972a). Evidently, this cell-free reaction strongly resembles the situation in the intact cell. The cell-free nuclear binding is of high affinity (K_d 2—3 x 10^{-10}M) to a limited number of sites (Fig. 7) (Baxter *et al.*, 1972a; Higgins, Rousseau, Baxter and Tomkins, unpublished). The same number of nuclear sites is found in the intact cell (Baxter, Rousseau, Higgins and Tomkins, unpublished).

DO THE NUCLEAR BINDING SITES CONTAIN DNA?

Obviously then, the nature of the nuclear binding sites becomes a critical issue. Nuclear binding activity is destroyed by deoxyribonuclease (Baxter *et al.*, 1972a), suggesting that DNA is involved. As shown in Fig. 8 (Baxter *et al.*,

FIGURE 8

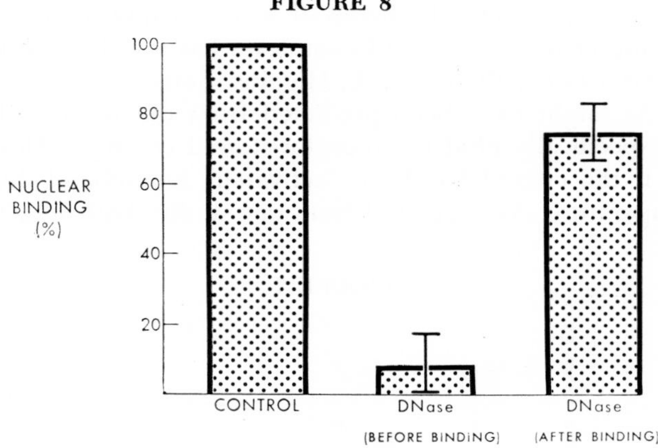

DNA dependency of nuclear binding. The amount of nuclear binding in the cell-free system (Baxter *et al.*, 1972a) following exposure of nuclei to DNase is shown as percent of the binding in nuclei not exposed to the enzyme. Also shown is the effect of DNase on nuclei previously bound with receptor-dexamethasone complex. Data taken from Baxter *et al.*, (1972a).

1972a), isolated nuclei were treated with DNase so that they lost about 40% of their DNA and little or no protein, but retained their microscopic appearance. However, they lost almost all of their capacity to bind receptor-steroid complex. That DNase did not destroy the nuclear architecture nonspecifically was suggested by the finding that receptor-steroid complexes previously bound to the nuclei were not released by subsequent DNase treatment even though as much DNA was released under these conditions (Baxter *et al.*, 1972a).

As a second suggestion that the nuclear sites may contain DNA, we found that receptor-steroid complexes bind with high affinity to purified HTC cell DNA (Baxter *et al.,* 1972a). Cytosol-bound dexamethasone was "activated" and then incubated with pure HTC cell DNA. The DNA and the DNA-bound receptor-steroid complexes were separated from free steroid and receptor-steroid complexes not bound by the DNA using agarose gel filtration (Fig. 9). Under these

FIGURE 9

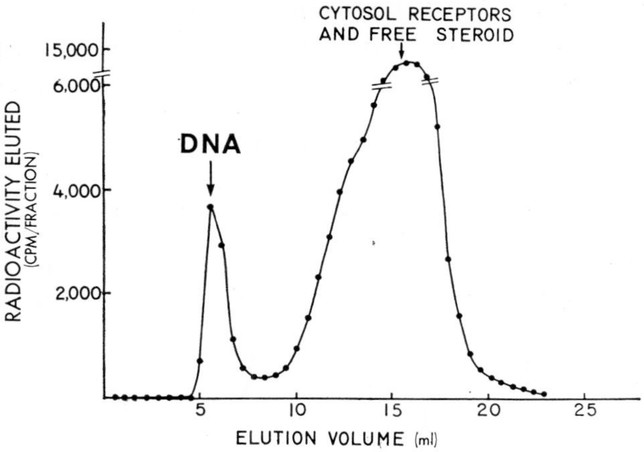

Binding of receptor-steroid complexes to DNA. Agarose-gel filtration of receptor-steroid complex bound to HTC cell DNA (Baxter *et al.,* 1972a).

conditions free steroid does not bind to DNA (Baxter *et al.,* 1972a). The DNA-receptor complex can be refiltered through the gel with less than 30% loss of bound radioactivity, suggesting that the binding to the DNA is of high affinity (Rousseau, unpublished).

Thus, nuclear binding, either in the cell-free system or in the intact cell, resembles the DNA binding reaction, in that: the receptor-steroid complex is required; the "activation" step seems to enhance the binding; and the interactions appear to be of high affinity (Baxter *et al.,* 1972a; Rousseau, unpublished).

The data suggest that the nuclear binding sites contain DNA, but whether DNA is the sole constituent is at present unclear. However, it should be emphasized that the amount of complex bound by pure DNA is greater than that bound by isolated nuclei or nuclei in intact cells (expressed on a DNA basis) (Baxter *et al.,* 1972a). This suggests that if chromatin proteins influence nuclear binding (O'Malley *et al.,* 1972), they do so by restricting rather than enhancing the binding. If the receptor-steroid complex binds to DNA in the intact cell, it seems reasonable to think that the receptor-steroid complex acts on DNA and thereby influences in some way gene transcription. If a comparison is made with the induction of β-galactosidase in *Escherichia coli,* steroids clearly do not act in the same way as β-galactosides by binding to a repressor and removing it from the DNA (Jacob and Monod, 1961). Rather, they would act more analogously to cAMP which, in bacterial systems, combines with a receptor (CAP-protein). This complex then binds to DNA and influences the transcription of specific genes (de Crombrugge *et al.*, 1971; Zubay and Chambers, 1971).

The nature of the receptor-steroid binding sites on the HTC cell DNA remains unknown, although other types of DNA (e.g., prokaryotic and denatured HTC cell DNA) do bind the receptor-steroid complex (Rousseau, unpublished). The specificity of these interactions is presently being studied.

SUMMARY OF THE EARLY EVENTS IN GLUCOCORTICOID ACTION

Our current formulation of the early events in glucocorticoid action is shown in Fig. 10. The steroid freely penetrates the cell membrane (Levinson *et al.*, 1972)

FIGURE 10

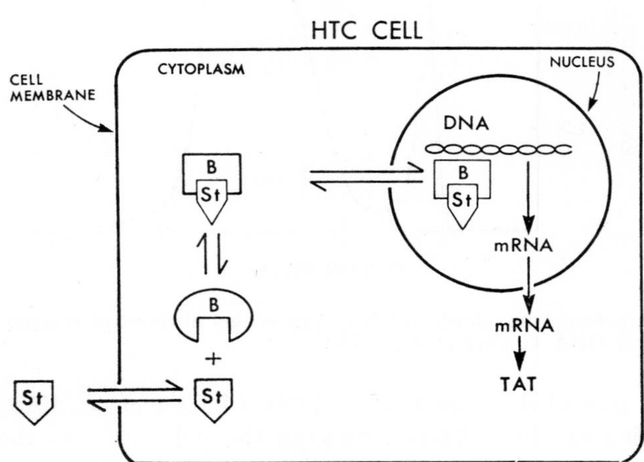

Early events in glucocorticoid action. St = steroid. B = specific glucocorticoid receptor. Different shapes of B indicate different conformations. m-RNA = messenger RNA.

and reversibly binds to receptor protein. Conformational changes occur in the receptor and the resultant complex binds reversibily and with high affinity to a fixed number of DNA-containing sites in the nucleus. In some way this binding influences the transcription of genes involved in the expression of the hormonal effect. Obviously, glucocorticoid unresponsiveness can result from a defect at any point in this sequence of events. In the case of the steroid-resistant lymphoma cells, there appears to be a loss of or alteration in the cytoplasmic receptor.

GLUCOCORTICOIDS INFLUENCE THE LEVEL OF ACTIVE MESSENGER RNA FOR TYROSINE AMINOTRANSFERASE

Several lines of evidence suggest that during the course of induction the level of active messenger RNA for TAT increases.

1. The induction of TAT is not observed if RNA synthesis is blocked by actinomycin D (Fig. 11) (Thompson *et al.*, 1966, Peterkofsky and Tomkins, 1967).

FIGURE 11

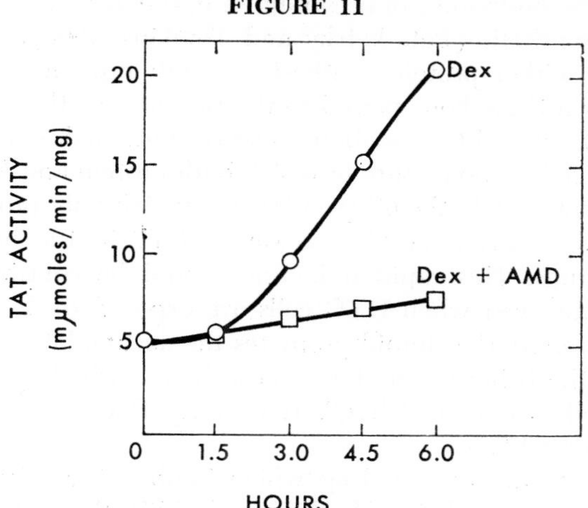

Effect of actinomycin D on induction of TAT in HTC cells. The enzyme was measured at various intervals following exposure of cells to 10^{-5} M dexamethasone or to the same concentration of dexamethasone plus 1μg/ml actinomycin D. Data reprinted from Peterkofsky and Tomkins (1967).

FIGURE 12

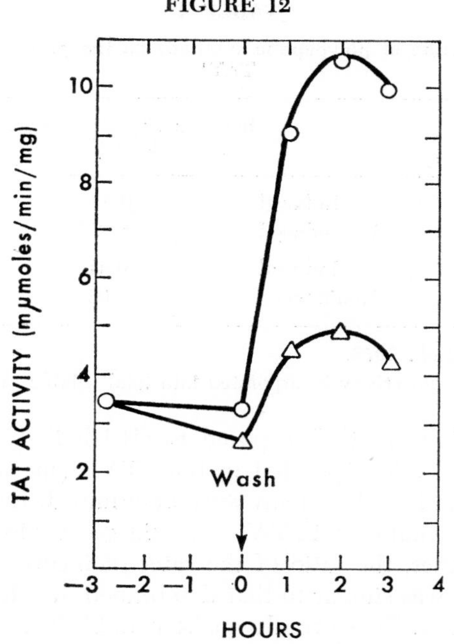

Effect of dexamethasone in the absence of protein synthesis. Two cultures of HTC cells were exposed for 3 hours to (1) dexamethasone and cycloheximide (O), or to (2) dexamethasone plus cycloheximide and actinomycin D (Δ). At zero time the drugs were removed from the cultures by centrifugation and resuspension of the cells in fresh medium. The levels of TAT were then measured at various time thereafter. The small increase in TAT activity in culture (2) probably represents the stimulation of translation by fresh serum (Hershko and Tomkins, 1971; Gelehrter and Tomkins, 1969). Data reprinted from Peterkofsky and Tomkins (1968).

2. An effect of the hormone can be observed in the absence of protein synthesis but not of RNA synthesis (Peterkofsky and Tomkins, 1968). In the experiment shown in Fig. 12, either protein synthesis or both protein and RNA syntheses were inhibited for a three hour period in the presence of the inducer. The drugs were then washed out and the incubations were continued. Only a small increase in TAT was observed in the culture in which both protein and RNA synthesis had been inhibited. However, in the other culture in which only protein synthesis had been inhibited there was a rapid increase in TAT in the hour following the removal of the drugs. This rapid induction should be contrasted with the lag period normally observed when HTC cells are exposed to dexamethasone alone (see Fig. 11). Therefore, the steroid promotes the accumulation of a factor which requires RNA synthesis but not protein synthesis, and which is responsible for the observed increased synthesis of TAT. It is likely that this factor is the active messenger RNA for TAT.

3. Studies with pactamycin, a drug which inhibits the initiation but not the completion of polypeptide chains (Lodish *et al.*, 1971; Stewart-Blair *et al.*, 1971), suggest that cells making TAT at the induced rate contain more polyribosomes engaged in TAT synthesis than do uninduced cells (Scott *et al.*, 1972). In the experiment shown (Table 4), the initiation of protein synthesis was inhibited by

TABLE 4

Effect of inhibiting initiation of polypeptide synthesis on the proportion of nascent chains of TAT[a]

		Radioactivity incorporated into TAT[b]	Ratio (Induced/uninduced)
Pactamycin-treated cells	Induced	0.34	13.6
	Noninduced	0.025	
Control cells	Induced	0.41	13.7
	Noninduced	0.03	

[a]Data taken from Scott *et al.*, 1972.

[b]Expressed as % of the radioactivity incorporated into total soluble protein.

pactamycin. The cells were then exposed to ^{3}H-labelled amino acids, and the amounts of radioactivity incorporated in both TAT (measured by an immunoprecipitation assay) and total proteins were determined. Even in the presence of pactamycin, the incorporation into TAT was still greater in induced cells than in uninduced cells. Further, the ratio of the rates of incorporation in the induced and uninduced states was similar to that determined in cells which had not been exposed to the inhibitor. These results make it unlikely that the effect of glucocorticoids on TAT synthesis is at the level of polypeptide chain elongation or termination. Thus, steroids must influence some prior step in protein synthesis e.g., chain initiation or the number of active messenger RNA molecules.

4. Studies in a cell-free system indicate that the polysome fraction from induced cells has an increased capacity to synthesize TAT when compared with

that from uninduced cells (Klein, Levinson and Tomkins, unpublished; Beck, Beck and Tomkins, unpublished).

On the basis of all these experiments, it seems most likely that in response to an inducer steroid, there is an increase in the level of active TAT messenger RNA. Whether this increase results from an activation of pre-existing messenger RNA, a decrease in its rate of degradation, or an increase in its rate of synthesis remains to be established. Further, we cannot presently exclude the additional possibility (suggested by preliminary observations in a cell-free system (Beck, Beck and Tomkins, unpublished)) that the translation machinery for TAT is somehow specifically stimulated.

THERE IS POST-TRANSCRIPTIONAL REGULATION OF TYROSINE AMINOTRANSFERASE SYNTHESIS

The foregoing data suggest that the steroid, in combination with its receptor, affects gene transcription, resulting in an increased amount of active TAT messenger RNA. This conclusion must be reconciled with the finding that RNA synthesis is also necessary for the deinduction which follows removal of the inducer (Thompson *et al.,* 1966, 1970). As shown in Fig. 13, when the steroid is removed

FIGURE 13

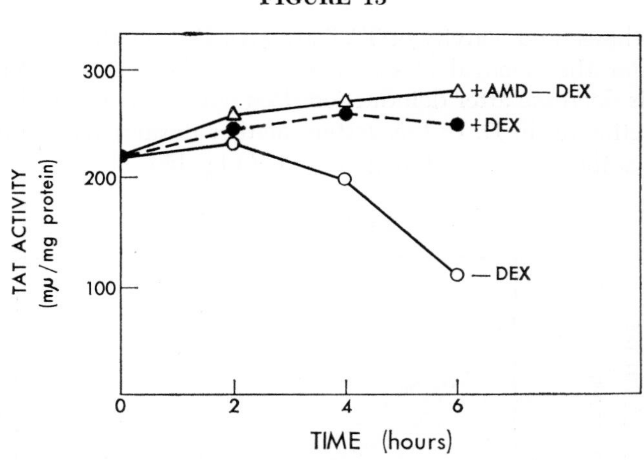

Deinduction. Induced cells were resuspended in fresh medium: without steroid (○); with actinomycin D without steroid (Δ); or with dexamethasone (●). The level of TAT activity was then measured at various time periods thereafter. Data reprinted from Thompson *et al.,* (1970).

from an induced culture, the activity of TAT declines (Thompson *et al.,* 1966, 1970). In striking contrast, if RNA synthesis is inhibited by actinomycin D at the time the inducer is removed, deinduction is prevented (Thompson *et al.,* 1966, 1970). Under the latter conditions TAT activity sometimes even increases ("super induction") (Thompson *et al.,* 1966, 1970). These data led to the suggestion that there is post-transcriptional regulation of TAT synthesis (Tomkins *et al.,* 1969).

The changes in TAT activity during the various conditions of induction and deinduction are due to changes in the rate of TAT synthesis as measured by immuno-precipitation (Tomkins *et al.,* 1972) (Fig. 14). In the uninduced state,

FIGURE 14

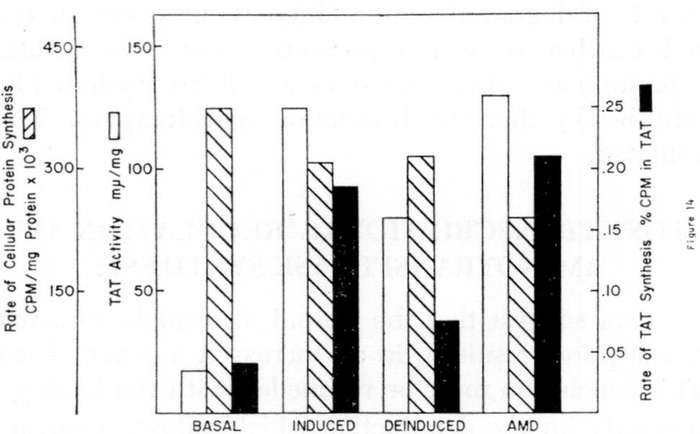

The effect of induction, deinduction and deinduction plus actinomycin D on the activity and rate of synthesis of TAT and on the rate of synthesis of total cell protein (Data reprinted from Tomkins *et al.,* 1972).

the rate of synthesis and activity of TAT are low. Dexamethasone increases both. Four hours after the removal of dexamethasone, the rate of TAT synthesis is decreased. This decrease after deinduction does not occur when RNA synthesis is inhibited by actinomycin D. In this system neither dexamethasone nor actinomycin D influences the rate of TAT degradation (Fig. 15).

FIGURE 15

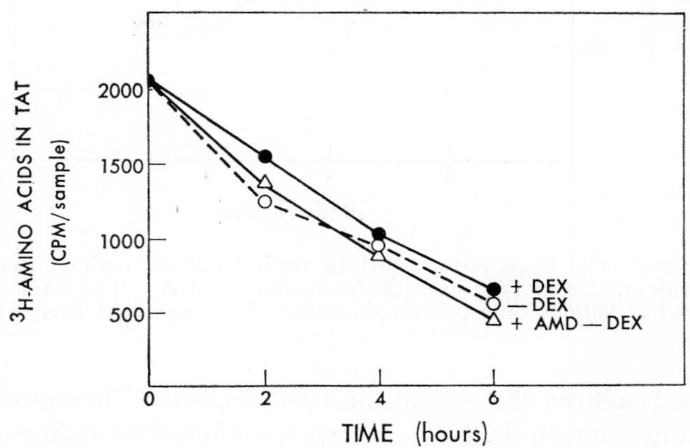

Lack of influence of dexamethasone or actinomycin D on TAT degradation. An induced culture of HTC cells was exposed to ^{3}H-labelled amino acids. The medium was removed and the cells were resuspended in fresh medium alone or medium containing dexamethasone, with or without actinomycin D. At various times thereafter the amount of radioactivity remaining in TAT, as measured by immunoassay, was determined (Data reprinted from Tomkins *et al.,* 1972).

These experiments suggest that there is a labile factor, sensitive to inhibitors of RNA synthesis, which decreases the rate of synthesis of TAT and enhances the rate of degradation of TAT m-RNA. This factor has been termed the "post-transcriptional repressor" (Tomkins *et al.*, 1969).

The scheme shown in Fig. 16 collates all these results. It is assumed that the regulation of TAT involves at least three genes—the structural gene for TAT, a

FIGURE 16

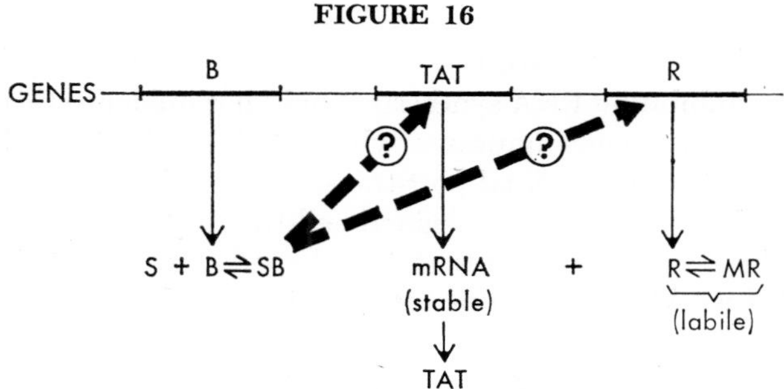

Regulation of TAT synthesis in HTC cells. B = specific glucocorticoid receptor. S = glucocorticoid. R = labile post-transcriptional repressor. SB = steroid—receptor—complex. The ? refers to the fact that the site of action of SB is unknown.

gene coding for the post-transcriptional repressor, and a gene coding for the glucocorticoid receptor. The labile repressor inhibits the activity of the TAT messenger RNA and enhances its degradation. When RNA synthesis is inhibited during deinduction, the repressor rapidly disappears because of its lability. This results in a decrease in the rate of degradation of the existing TAT messenger RNA. Thus, due to increased m-RNA stability and decreased translational repression the induced rate of TAT synthesis is maintained. The occasional increased TAT synthesis following inhibition of RNA synthesis can be also explained. Since the repressor rapidly disappears, it could dissociate from "repressed" messenger RNA and therefore increase the quantity of active messenger RNA.

A completely different experimental approach using synchronized cells lends further support to the model (Martin and Tomkins, 1970; Martin *et al.*, 1969a, 1969b). During the G2, mitotic, and early G1 phases of the cell cycle, TAT is neither inducible nor, if previously induced, deinducible. These results can be explained if it is assumed that neither the repressor nor the TAT messenger RNA are made during these periods. Pre-existing messenger RNA would thus be stabilized and its concentration would be maintained at the level achieved at the commencement of G2. Thus, the rate of TAT synthesis at G2 would also be maintained through to early G1.

There is an increasing body of evidence that this type of post-transcriptional regulation operates in a variety of prokaryotic and eukaryotic systems (for references see Tomkins *et al.*, 1972). This is indicated by the fact that inhibitors of macromolecular synthesis can stimulate several functions, or prevent their deinduction. In certain cases, alternatives to the post-transcriptional hypothesis have been proposed. However, many of these can be discounted as general possibilities, since they cannot explain why a variety of inhibitors produce the same effects, or they cannot account for the findings in all the systems (Tomkins *et al.*, 1972). For example, actinomycin D could eliminate the short-lived messenger RNA's allowing more efficient translation of stable messenger RNA's like that coding for ovalbumin (Rhoads, McKnight and Schimke, unpublished). Clearly this model cannot explain the inhibition of TAT deinduction by actinomycin D. Assuming that there were only transcriptional control, the rapid decline in TAT synthesis following inducer removal would imply that the TAT messenger RNA under these conditions is relatively short-lived ($t_{\frac{1}{2}}$ = 1—3 h (Tomkins *et al.*, 1972)). Therefore, inhibiting RNA synthesis should not influence this decline unless the stability of the messenger improves when RNA synthesis is blocked. This is in fact a restatement of the original post-transcriptional hypothesis. Clearly, other types of data, probably from better defined cell-free systems, are needed in order to determine whether the post-transcriptional model is correct. Nevertheless, it provides a useful working hypothesis for studying regulation of gene expression.

DO THE STEROIDS ACT INDEPENDENTLY OF THE POST-TRANSCRIPTIONAL REPRESSOR OR DO THEY CONTROL IT?

To date, the evidence suggests that receptor-steroid complexes act on DNA and therefore presumably modify transcription. Other experiments point to an accumulation of TAT messenger RNA in the induced state. Furthermore, equally convincing data support the idea that there is post-transcriptional repression of TAT synthesis. These considerations can be reconciled in several ways. The receptor-steroid complexes could stimulate (directly or indirectly) the transcription of the TAT structural gene and the post-transcriptional regulation could operate independently of the hormone. Alternatively, the receptor-steroid complex could inhibit the transcription of the repressor gene which would allow TAT messenger RNA to accumulate. Finally, both the repressor and TAT gene transcription could be under steroid control. These possibilities are illustrated in Fig. 16. Further work will be required to distinguish between them.

SUMMARY

Glucocorticoid hormones regulate gene expression in cultured (HTC) cells by inducing the synthesis of a few specific proteins, of which tyrosine aminotransferase has been studied in the greatest detail. In inducing this enzyme, steroids penetrate the cell membrane and bind reversibly to specific receptor proteins in the cell cytoplasm. Steroids influence the conformation of the receptor so that the receptor-steroid complex binds reversibly to sites in the cell nucleus which may contain DNA. The latter interaction results in an increase in the level of active messenger RNA for tyrosine aminotransferase presumably by influencing the transcription of structural or regulatory genes. A post-transcriptional repressor inhibits the translation and enhances the rate of degradation of the messenger RNA for tyrosine aminotransferase. It is presently unknown whether the transcriptional action of the receptor-steroid complex is to inhibit the formation or action of the repressor or whether post-transcriptional control operates independently of the hormone.

REFERENCES

Auricchio, F., Martin, D. W., Jr. and Tomkins, G. M. (1969). *Nature (London) 224,* 806

Ballard, P. L. and Tomkins, G. M. (1969). *Nature (London) 224,* 344

Ballard, P. L. and Tomkins, G. M. (1970). *J. Cell. Biol. 47,* 222

Baxter, J. D. and Forsham, P. H. (1972). *Am. J. Med.;* in press

Baxter, J. D., Harris, A. W., Tomkins, G. M. and Cohn, M. (1971). *Science 171,* 189

Baxter, J. D., Rousseau, G. G., Benson, M. C., Garcea, R. L., Ito, J. and Tomkins, G. M. (1972a). *Proc. Nat. Acad. Sci., U.S. 69,* 1892

Baxter, J. D., Rousseau, G. G., Higgins, S. J. and Tomkins, G. M. (1972b). *J. Clin. Invest. 51,* 9a

Baxter, J. D., Rousseau, G. G. and Tomkins, G. M. (1971). *Fed Proc. 30,* 1048

Baxter, J. D. and Tomkins, G. M. (1970). *Proc. Nat. Acad. Sci., U.S. 65,* 709

Baxter, J. D. and Tomkins, G. M. (1971). *Proc. Nat. Acad. Sci., U.S. 68,* 932.

Beato, M., Kalimi, M. and Feigelson, P. (1972). *Biochem. Biophys. Res. Commun. 47,* 1464

de Crombrugghe, B., Chen, B., Anderson, W. B., Gottesman, M. E., Perlman, P. L. and Pastan, I. (1971). *J. Biol. Chem. 246,* 7343

Gammeltoft, S. and Schaumburg, B. (1972). *IV Int. Congress of Endocrinology, 256,* 147 Excerpta Medica Foundation, Amsterdam

Gelehrter, T. D. and Tomkins, G. M. (1967). *J. Mol. Biol. 29,* 59

Gelehrter, T. D. and Tomkins, G. M. (1969). *Proc., Nat. Acad. Sci., U.S. 64,* 723

Granner, D. K., Hayaski, S., Thompson, E. B. and Tomkins, G. M. (1970). *J. Mol. Biol. 245,* 1472

Hackney, J. F., Gross, S. R., Aronow, L. and Pratt, W. B. (1970). *Molec. Pharmacol. 6,* 500

Hershko, A. and Tomkins, G. M. (1971). *J. Biol. Chem. 246,* 710

Hollander, N. and Chiu, Y. W. (1966). *Biochem. Biophys. Res. Commun. 25,* 291

Horibata, K. and Harris, A. W. (1970). *Expl. Cell. Res. 60,* 61

Jacob, F. and Monod, J. (1961). *J. Mol. Biol. 3,* 318

Kirkpatrick, A. F., Milholland, R. J. and Rosen, F. (1971). *Nature New Biol. 232,* 216

Kulka, R. G., Tomkins, G. M. and Crook, R. B. (1972). *J. Cell. Biol. 54,* 175

Levinson, B. B., Baxter, J. D., Rousseau, G. G. and Tomkins, G. M. (1972). *Science 175,* 189

Lippman, H., Yang, S. and Thompson, E. B. (1972). *IV Int. Congress of Endocrinology 256,* 5, Excerpta Medica Foundation, Amsterdam

Lodish, H. F., Housman, D. and Jacobsen, M. (1971). *Biochemistry 10,* 2348

Manganiello, V. and Vaughan, M. (1972). *J. Clin. Invest., 51,* 2763

Martin, D. W., Jr. and Tomkins, G. M. (1970). *Proc. Nat. Acad. Sci., U.S. 65,* 1064

Martin, D. W., Jr., Tomkins, G. M. and Bresler, M.A. (1969a). *Proc., Nat. Acad. Sci. U.S. 63,* 842

Martin, D. W., Jr., Tomkins, G. M. and Granner, D. (1969b). *Nat. Acad. Sci., Proc., U.S. 62,* 248

O'Malley, B. W., Spelsberg, T. C., Schrader, W. T., Chytil, F. and Steggles, A. W. (1972). *Nature (London) 235,* 141

Peterkofsky, B. and Tomkins, G. M. (1967). *J. Mol. Biol. 30,* 49

Peterkofsky, B. and Tomkins, G. M. (1968). *Proc., Nat. Acad. Sci. U.S. 60,* 222

Raspé, G. (Ed.) (1970). *Advances in the Biosciences 7. Schering Workshop on Steroid Hormone Receptors.* Pergamon Press, Oxford

Rosenau, W., Baxter, J. D., Rousseau, G. G. and Tomkins, G. M. (1972). *Nature New Biol. 237,* 20

Rousseau, G. G., Baxter, J. D. and Tomkins, G. M. (1972). *J. Mol. Biol. 67,* 99

Samuels, H. H. and Tomkins, G. M. (1970). *J. Mol. Biol. 52,* 57

Scott, W. A., Shields, R. and Tomkins, G. M. (1972). *Proc. Nat. Acad. Sci., U.S.,* in press

Stewart-Blair, M. L., Yanowitz, I. S. and Goldberg, I. H. (1971). *Biochemistry 10,* 4198

Thompson, E. B., Granner, D. K. and Tomkins, G. M. (1970). *J. Mol. Biol. 54,* 159

Thompson, E. B., Tomkins, G. M. and Curran, J. F. (1966). *Proc. Nat. Acad. Sci., U.S. 56,* 296

Tomkins, G. M., Gelehrter, T. D., Granner, D., Martin, D. W., Jr., Samuels, H. H. and Thompson, E. B. (1969). *Science 166,* 1474

Tomkins, G. M., Levinson, B. B., Baxter, J. D. and Dethlefsen, L. L. (1972). *Nature New Biol. 239,* 9

Tomkins, G. M., Thompson, E. B., Hayashi, S., Gelehrter, T. D., Granner, D. and Peterkofsky, B. (1966). *Cold Spr. Harb. Symp. Quant. Biol. 31,* 349

Zubay, G. and Chambers, D. A. (1971). *Regulating the lac Operon* in *Metabolic Pathways, Metabolic Regulation* (Vogel, H. J., Ed.) *Vol. 5* p. 297. Academic Press, New York

Regulation of transcription by glucocorticosteroids

C. E. SEKERIS and W. SCHMID

Institut für Physiologische Chemie der Universität Marburg, Marburg/L., Lahnberge, Western Germany

The effect of glucocorticosteroids on ribonucleic acid biosynthesis in target cells has been amply documented in the past ten years. Glucocorticosteroids stimulate RNA metabolism in the liver (Feigelson *et al.*, 1962, Kenney and Kull, 1963, Jervell, 1963, Sekeris and Lang, 1964) and dramatically inhibit RNA synthesis in the thymus (Nakagawa and White, 1966) and lymph cells (Makman *et al.*, 1968). The action on RNA metabolism is regarded to be causally related to the physiological effects of the hormones. This is obvious in such cases as in the thymus where the hormones lead to cell death: the inhibition of RNA synthesis is one of the manifestation of general catabolic processes. On the other hand the significance of newly synthesized RNA on the gluconeogenic action of cortisol in the liver has been up to now indirectly established e.g. with the help of inhibitors of RNA synthesis whose administration considerably impairs the induction of enzymes involved in gluconeogenesis and the gluconeogenic process itself.

CORTISOL AND RNA SYNTHESIS IN RAT LIVER

The administration of cortisol to rats leads to increase in the rate of incorporation of precursors into liver RNA, first in the nucleus and then in the cytoplasm. The corticosteroids also affect nucleotide metabolism and nucleotide pool size (Feigelson *et al.*, 1969) and therefore it was initially postulated that the effect of the hormones on RNA synthesis may be an indirect one. By isolating nuclei from the liver of control and hormone treated rats and performing RNA synthesis in the presence of an RNA synthesizing mixture under conditions where neither membrane permeability nor nucleotide pools could influence the reaction, we could show that the effect of the hormone is exerted at the level of the cell nucleus (Lang and Sekeris, 1964). This assumption was strengthened by the observation that isolated rat liver nuclei responded to the presence of cortisol with increased synthesis of RNA (Lukacs and Sekeris, 1967). We postulated that either the amount of genetic information available for transcription is increased, or that the amount or activity of the DNA-dependent RNA polymerase is stimulated.

In the first case we would expect to find qualitative differences in the RNA synthesized whereas in the second case only quantitative effects of the hormone should be seen. Therefore experiments were designed a) to quantitate the amount of active chromatin and b) to analyze and characterize the newly synthesized RNA.

Dahmus and Bonner (1965) were the first to show that the amount of chromatin active in RNA synthesis increases three hours after cortisol administration. They transcribed chromatin with *E. coli* polymerase under conditions of enzyme saturation, where the template was the restricting factor for transcription. Under these conditions they observed that liver chromatin from cortisol treated rats was more active as template than control chromatin. We repeated the experiments using much shorter time periods of cortisol administration assuming that the effect on chromatin must be a very rapid one, preceding the stimulation of precursor incorporation into RNA. Thus we were able to observe increase in the template activity of chromatin within 30-60 minutes after administration of cortisol (see Table 1). In fact using the *in vitro* system mentioned above we could

TABLE 1

***In vivo* effect of cortisol on the template activity of isolated rat liver chromatin.**

Chromatin prepared from:	[^{14}C] UTP incorporated into RNA		
	c.p.m.	%	P
Control animals	815	100.0	—
Cortisol injected 2 mg/100g (30 min) ..	988	121.5	0.01
Cortisol injected 2 mg/100g (60 min) ..	912	112.0	0.05

Cortisol suspended in 0.5 ml physiological saline solution was injected i.p. Liver chromatin was prepared from the purified nuclei by a modification of the procedure of Bonner *et al.*, 1968. The template activity of the chromatin was tested with *E. coli* RNA polymerase. Usually chromatin containing 15 μg DNA was used in the presence of an excess of enzyme (5 U). Incubation was for 20 min at 37°C; incorporation without enzyme (115 c.p.m.) has been subtracted. The values represent the mean of three experiments. P has been calculated by Fischer's t-test.

observe that within minutes after addition of cortisol to isolated liver nuclei, template activity of chromatin was increased significantly above that of controls (Beato *et al.*, 1969) (see Fig. 1).

To characterize the newly synthesized RNA a variety of techniques were used. It was early shown that all species of RNA are stimulated. However DNA/RNA hybridization techniques (Drews and Brawerman, 1967; Doenecke and Sekeris, 1970) as well as the assay of messenger activity of RNA on an *in vitro* protein synthesizing system (Lang *et al.*, 1968; Sekeris, 1966) suggested qualitative differences in the RNA synthesized. Due to the limitations of the experimental approaches however, a clear cut demonstration of the *de novo* appearance of new RNA species was not possible at that time.

FIGURE 1

***In vitro* effect of cortisol on the template activity of rat liver nuclear sediment**

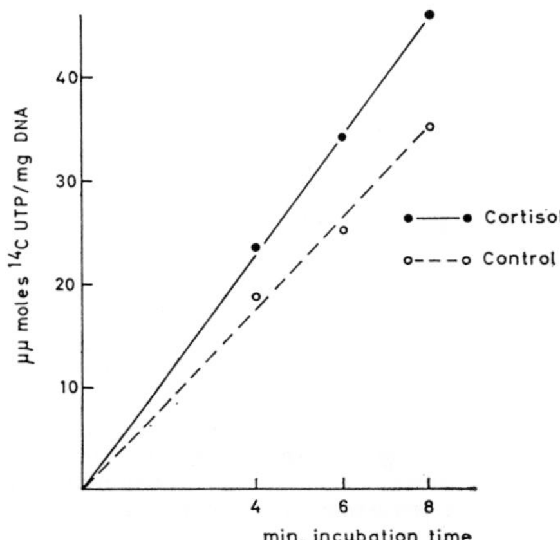

Rat liver nuclei were incubated in the presence of cortisol (10^{-5}M) dissolved in ethanol or with ethanol alone, for 10 min at 37°C and subsequently lysed in hypotonic buffer. The sediment obtained after centrifugation at 10,000 g for 10 min was used as template for *M. lysodeikticus* RNA polymerase. The assay contained 11 μg of DNA as nuclear sediment and 5 units of enzyme. Incorporation in the absence of enzyme has been substracted. Each point is the mean of 6 experiments. P values at 8 min were 0.01 according to Fischer's t-test.

SEQUENTIAL EFFECTS OF CORTISOL ON SYNTHESIS OF EXTRANUCLEOLAR AND NUCLEOLAR RNA

In a time sequence experiment and using the specific inhibitor of polymerase B, α-amanitin (see Fig. 2) we could demonstrate that cortisol administration leads first to the stimulation of amanitin sensitive RNA (DNA-like RNA) and to an inhibition of amanitin resistant RNA, followed in a later time period by stimulation of amanitin resistant RNA. This suggested that cortisol first affects extranucleolar and then nucleolar RNA synthesis. By further fractionating nuclei and assaying for RNA synthesis in the individual fractions (see Fig. 3) we could corroborate these findings and show that in early time periods the synthesis of extranucleolar RNA is stimulated and that of the nucleoli inhibited, whereas in later time periods nucleolar RNA synthesis is highly increased. Similar conclusions had been reached on the basis of base analysis of the synthesized RNA. The extranucleolar fraction most active in RNA synthesis contains template, not precipitable with salt, having a very high activity in RNA synthesis and very probably consisting of the bulk of the active chromatin. The salt precipitable fraction is much less active in RNA synthesis and as seen from Table 2 is only partially inhibited by α-amanitin. This can be either due to the presence of nucleolar contaminants in the fraction or of polymerase which is amanitin sensitive, probably involved in t-RNA biosynthesis. The nucleolar fraction is completely

FIGURE 2

Effect of cortisol on RNA synthesis in the absence and presence of α-amanitin.

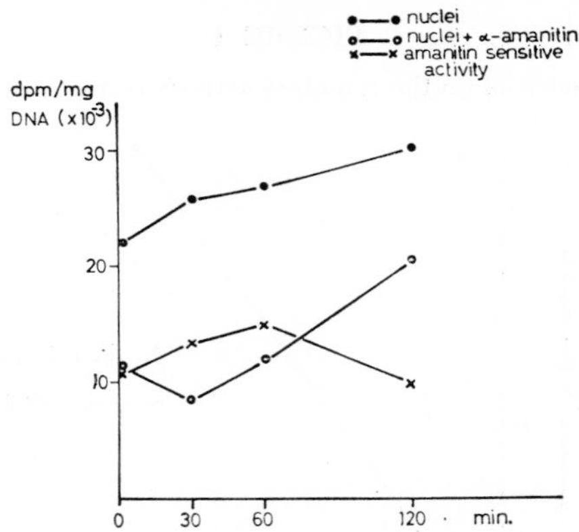

Rats received 2 mg cortisol/100 g body weight i.p. and were sacrificed after 30 and 60 minutes. Nuclei were incubated in the absence ●——● and presence Δ – – Δ of α-amanitin (4μg/ml). The amanitin sensitive RNA synthesis (O – – O) was calculated as the difference between RNA synthesis in the absence and presence of α-amanitin.

FIGURE 3

Effect of cortisol on RNA synthesis in rat liver nuclei and nuclear fractions.

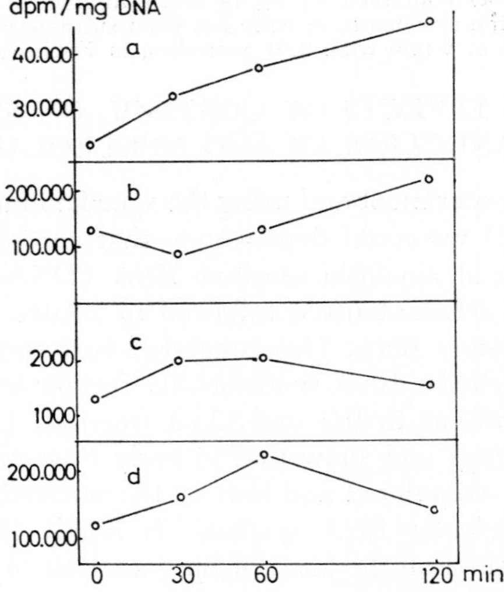

Rats received 2 mg cortisol/100 g body weight i.p. and were sacrificed after various time intervals. Nuclei and clear fractions were then prepared. Values are given as d.p.m./mg DNA. (a) nuclei, (b) nucleoli, (c) 0.14 M NaCl precipitable chromatin, (d) 0.14 M NaCl Supernatant. From Schmid and Sekeris, 1972.

resistant to the toxin, which reflects the purity of the preparation. These results suggested that the stimulatory effect of cortisol of ribosomal RNA is a direct consequence of the stimulation of DNA-like RNA, which may either directly acti-

TABLE 2

Effect of α-amanitin on RNA synthesis by isolated rat liver nuclei and nuclear fractions.

	RNA synthesis (d.p.m./mg DNA)		
	without α-amanitin	with α-amanitin (0.2 μg/ml)	% inhibition
nuclei	22,000	11,200	49
nucleoli	494,000	496,000	0
0.14 M NaCl precipitable fraction	2,200	690	69
0.14 M NaCl soluble fraction	110,000	9,600	91

Nuclei and nuclear fractions were isolated as described in Methods and tested for RNA synthetic capacity in the presence or absence of α-amanitin.

vate ribosomal RNA synthesis or indirectly affect it by acting as messenger for the synthesis of nucleolar proteins regulating synthesis, processing or transport of ribosomal RNA. That normal functioning of the nucleolus is dependent on a continuous synthesis of DNA-like RNA has been shown from experiments with α-amanitin which *in vitro* inhibits RNA polymerase B, having no effect whatsoever on RNA polymerase A (see Fiume and Wieland, 1970). However *in vivo* both DNA-like and ribosomal RNA synthesis is impaired. For a discussion of the effects of α-amanitin *in vivo* and *in vitro* see Sekeris and Schmid, 1972b. The effects are summarized in Table 3 and our current concepts of how DNA-like

TABLE 3

Some effects of α-amanitin on RNA and protein synthesis of the liver cell. (From Sekeris and Schmid, 1972a)

1. Isolated RNA polymerase B inhibited, enzyme A not affected.
2. *In vivo* administration to rats inhibits labelling by radioactive precursors of both DNA-like and ribosomal RNA. Low molecular weight species less affected.
3. Amount of polymerase A extracted from liver nuclei of α-amanitin treated rats (for one or two hours) not affected. No active polymerase B can be isolated.
4. ^{14}C-labelled α-amanitin incubated with rat liver slice-binds to enzyme B but very slightly, if at all, to A.
5. Nuclei isolated from the liver of α-amanitin treated rats do not synthesize DNA-like RNA but do synthesize, in fact much more, of α-amanitin resistant RNA. The α-amanitin resistant RNA which is increased is synthesized in the extranucleolar fraction.
6. Nucleoli isolated from in vivo α-amanitin treated rats show impaired synthesis in vitro.
7. No flow of ribosomal RNA to cytoplasm.
8. Nucleolus segregates within 30-60 min.
9. Bulk of protein synthesis on cytoplasmic microsomes not impaired.
10. Labelling of certain nucleolar proteins impaired by α-amanitin.

RNA affects ribosomal RNA metabolism shown in Fig. 4. The effects of the glucocorticosteroids on both DNA-like and ribosomal RNA were independent of an

FIGURE 4

Control of ribosomal RNA formation by DNA-like RNA.

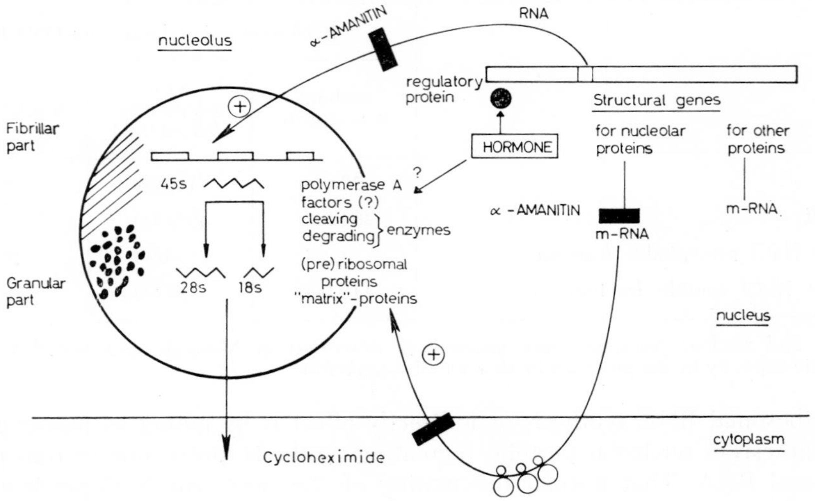

increase in the amount of the respective RNA polymerases (K. H. Seifart, unpublished experiments). The increased RNA synthesis initially reflected the increased template activity of chromatin to support RNA synthesis, although other factors affecting transcription may also be involved.

EFFECTS OF CORTISOL ON CHROMATIN

Turning our attention to chromatin we were interested in following the initial changes leading to increase in its template activity. In designing our experiments we considered two possibilities of activation of chromatin. One was based on the concepts developed by Allfrey and his colleagues (Allfrey, 1966; Allfrey *et al.*, 1964; Kleinsmith *et al.*, 1966) that modifications of the histones by acetylation, methylation or phosphorylation were directly involved in transcriptional control. The second possibility and more difficult to tackle experimentally was that the hormone allosterically modifies proteins negatively controlling gene transcription (Karlson and Sekeris, 1966 a, b) in analogy to the interaction of bacterial inducers with repressor proteins. Due to the fact that possible changes in chromatin connected to its activation should be very rapid we used in our experiments the *in vitro* system of isolated nuclei. In these experiments we incubated rat liver nuclei with the immediate donors of the various groups i.e. acetyl-CoA, S-adenosyl-methionine or ATP and measured the uptake of acetate, methyl and phosphate groups in total histones and in the various histones fractions under normal conditions and after stimulation with cortisol. No significant differences in the rates of incorporation of the labelled groups in histones between control and hormone treated nuclei could be observed (Gallwitz and Sekeris, 1969 a, b; Sekeris *et al.*, 1967; Schiltz and Sekeris, 1969; Schiltz, 1969).

What we did observe was a significant increase of the thiol content of histones as well as of non-histone nuclear proteins (see Table 4, Fig. 5) due to cleavage

TABLE 4

***In vitro* effect of cortisol on the incorporation of [2—^{14}C] iodoacetate ([^{14}C]IA) into proteins of isolated rat liver nuclei.**

Nuclei from	Incorporation of [^{14}C] IA (pmol/mg protein)
Control	1073
Cortisol treated	1203
P	0.01

Rat liver nuclei were incubated with 5 μg cortisol/ml of incubation medium, at 37°C for 10 min. The nuclear sediment obtained after hypotonic lysis of the nuclei, containing between 0.5 and 2.8 mg protein, was incubated with 0.033 μCi [^{14}C] IA 120 min in a final volume of 0.5 ml. Incorporation of [^{14}C] IA into proteins was determined on filter paper discs after heating at 95°C for 20 min. in 12% TCA and repeated washing in 5% PCA. The values are the mean of 15 experiments. P values calculated by the Fischer's t-test. (From Sekeris *et al.*, 1968).

FIGURE 5

Dose dependence of the cortisol effect *in vivo* on RNA polymerase activity of rat liver nuclei.

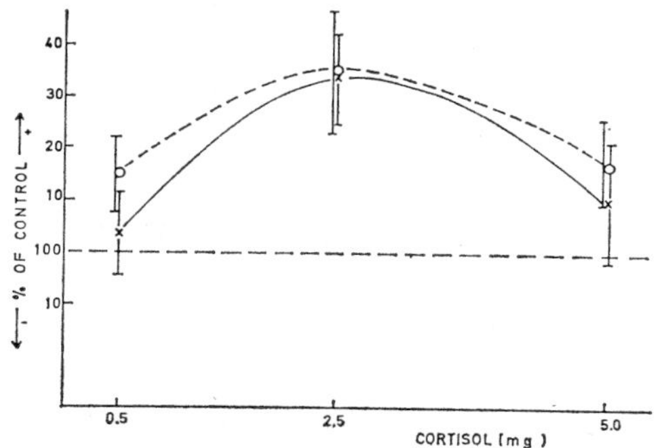

Intact rats (125 g) were injected i.p. with different doses of cortisol suspended in 0.5 ml of a physiological saline solution and rat liver nuclei were prepared. RNA polymerase activity was tested in the sediment obtained after hypotonic lysis of the nuclei in the presence of 3.6 mM $MnCl_2$. The values are expressed as percentage of the controls, which were injected with physiological saline alone. X — X = RNA synthesis. The dotted line (O – – O) corresponds to the incorporation of [^{3}H] iodoacetate into the nuclear proteins. Each point represents the mean and standard deviation of triplicate determinations from five experiments.

of disulphide bonds (Sekeris *et al.*, 1968; Table 5). These experiments suggested that the F_3 histones could *in vivo* either exist as F_3 dimers or could be covalently bound to non-histone proteins by disulphide bonds and that enzymatic oxido-reduction processes triggered by hormones could play a substantial role in gene activation. The magnitude of the effect (10-15% increase in the total thiol content) as well as the number of proteins affected as shown by polyacrylamide gel electrophoresis (Doenecke *et al.*, 1972), demonstrate that the hormone leads to general drastic change in nuclear protein structure. Whether this is a result of

TABLE 5

Effect of cortisol on the thiol content of rat liver nuclear proteins

(a) Ratio of thiol to thiol plus disulphide groups (determined by the dithiopyridine reagent).

	Thiol content (nmol/mg of nuclear protein)	
	Control animals	Cortisol-treated animals
Before $NaBH_4$ treatment	0.280	0.500
After $NaBH_4$ treatment	0.447	0.560
Ratio (—SH)/(—SH) + (S—S) . .	0.628	0.895

(b) Titration of nuclear protein preparations from control and cortisol treated rats with labelled iodoacetate.

	Iodo [^{3}H] acetate incorporated into nuclear sediment (c.p.m./mg of protein)		
	Control nuclei	Cortisol-treated nuclei	Percentage effect
Complete system	29450	32950	+ 12
—8 M-Urea	11820	13225	+ 12
+ 3.5 M-NaCl	62050	67850	+ 9
+ 0.1% Sodium dodecyl sulphate . .	26810	29600	+ 11
+ Sodium thioglycollate	91020	91800	+ 1
+ Sodium borohydride	86100	87910	+ 2

Cortisol (2mg/100g rat) was injected 30 min before sacrifice (from Doenecke *et al.*, 1972).

an amplification of an initially limited hormone protein interaction, or, whether this is the primary result of the hormone action on multiple protein sites remains to be seen. It must be kept in mind that cortisol affects the synthesis of a number of proteins, not only those involved in gluconeogenesis *per se,* but also those of nucleolar components (Schmid and Sekeris, unpublished observations) among others. Considering the magnitude of the stimulation of DNA-like RNA and of the other species of RNA one is led to the conclusion, that the amount of gene copies transcribed must be increased.

Effects of cortisol on nuclear proteins were also detected by the technique of actinomycin D binding. As known, binding of actinomycin D to DNA is restricted by the presence of nuclear proteins reacting with the DNA, so that the degree of binding can be used to titrate the amount of proteins interacting with DNA. Using this method we could show that after action of cortisol either *in vivo* (Homoki *et al.,* 1968) or *in vitro* with isolated nuclei (Homoki *et al.,* 1968) or chromatin (Beato *et al.,* 1970a) more [^{3}H] actinomycin D is bound in cortisol treated than in control preparations (see Table 6, 7). We interpret this finding as removal of chromosomal proteins from the DNA as consequence of their interaction with cortisol.

TABLE 6

Effect of different steroids on the uptake of [³H] actinomycin D by isolated rat liver nuclei

	[³H] actinomycin D incorporated c.p.m. per mg DNA ($\times 10^{-3}$)	
	10 min	20 min
Control	128.40 ± 2.18	156.44 ± 4.02
Cortisol	148.60 ± 5.66	170.72 ± 7.12
Testosterone	151.88 ± 9.73	168.60 ± 8.33
Androstendione	135.33 ± 9.30	148.80 ± 12.51
Pregnenolone	129.61 ± 4.26	147.14 ± 7.55
Control sonicated	143.60 ± 4.07	156.34 ± 3.82
Cortisol sonicated	176.34 ± 11.90	181.96 ± 13.15

Rat liver nuclei were incubated in TSS (2 mg protein/ml) in the presence of different steroids (in concentrations of 3×10^{-5}M) for 10 min at 37°C and then [³H] actinomycin D (20 μg/ml) was added. The incorporated radioactivity was measured in aliquots taken at 10 and 20 min after washing the nuclei three times with TSS. The final pellet was taken in 5 vol. of Nuclear Chicago Solubilizer (NCS) and shaken at 37°C for 4h. Scintillation counting was performed in Bray's solution using a Mark I liquid scintillation counter with a final efficiency of 18% as calculated by the channel ratio method. Each value represents the means and standard deviation of triplicate determination from two experiments. (From Homoki *et al.,* 1970).

TABLE 7

***In vitro* effect of different steroids on the binding of [³H] actinomycin to and on the template capacity of rat liver chromatin**

Chromatin incubated with	[³H] actinomycin bound		Template capacity	
	d.p.m./A260	%	c.p.m./0.1 ml	%
Tris-HC1 buffer	9,950 ± 315	100	2,500 ± 150	100
Cortisol phosphate	11,050 ± 420	110	2,880 ± 150	115
Cortisol	11,120 —	112	2,790 —	112
Tetrahydrocortisol	9,830 —	99	2,450 —	98
Pregnenolone	9,910 —	99	2,550 —	102

(From Beato *et al.,* 1970a)

EFFECT OF CORTISOL ON TRANSCRIPTION: DIRECT ACTION ON CHROMOSOMAL PROTEINS OR MEDIATED THROUGH CYTOPLASMIC FACTORS?

On the basis of our experiments with isolated nuclei we have concluded that cortisol initiates the effects on transcription by directly interacting with the nuclear proteins.

However work on other systems such as the prostate (Liao and Fang, 1969) uterus (Jensen *et al.,* 1967) and chick oviduct (O'Malley, 1972) has failed to show direct effect of the respective hormones on nuclear RNA synthesis. This was ascribed to the absence of cytoplasmic receptor proteins which either help to bring the hormone into the nucleus or are themselves involved in the regulation of transcription (see Raspé, 1970).

Liver cytosol does contain proteins which bind corticosteroids (Beato *et al.*, 1969, 1970b, 1972a) and which facilitate the transport of the hormones into the nucleus (see Fig. 6). The facilitating effect of the supernatant proteins is observed

FIGURE 6

Effect of cytosol macromolecules or bovine albumin on the uptake of [^{3}H] cortisol by isolated rat liver nuclei.

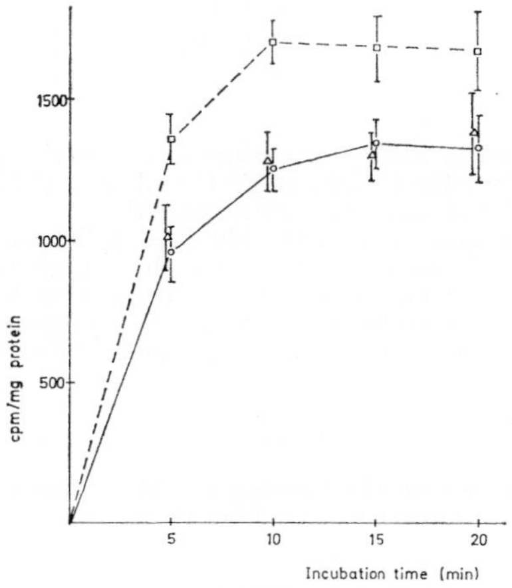

The macromolecular fraction of rat liver cytosol was prepared 20 min after i.p. injection of [^{3}H] cortisol (50 μCi/100 g body weight). The purified rat liver nuclei were incubated at 37°C with this fraction, with [^{3}H] cortisol alone or with [^{3}H] cortisol plus bovine serum albumin, and the radioactivity incorporated into the nuclei was measured after different time intervals.

□ - - □, nuclei incubated with the labelled cytosol macromolecules,

o——o, nuclei incubated with free [^{3}H] cortisol,

△——△, nuclei incubated with free [^{3}H] cortisol plus bovine serum albumin.

Each value represents the mean and standard deviation of four experiments.

if nuclei are isolated with the help of the detergent Triton X-100 but not if prepared with hypertonic sucrose. The effect is dependent on the dose of the hormone, on the pH and on the presence of nucleoside triphosphates. In the presence of the cytosol, the nuclei respond with increased RNA synthesis to doses of cortisol in the range of 10^{-8}-10^{-7}M, i.e. far lower than the amounts directly needed to effect the nuclei (10^{-5}M) (Fig. 7) (Beato *et al.*, 1970b). In order to isolate the macromolecule responsible for the effect we have separated the cytosol proteins on DEAE-cellulose columns into several peaks binding labelled hormones (Figs. 8, 9) and further a) tested their capacity to transport cortisol into the nucleus and b) their effects on RNA synthesis of isolated nuclei in the presence of doses of cortisol (10^{-8} to $^{-}10^{7}$M) which alone are inactive.

FIGURE 7

Effect of the cytosol macromolecules on the response of isolated rat liver nuclei to cortisol

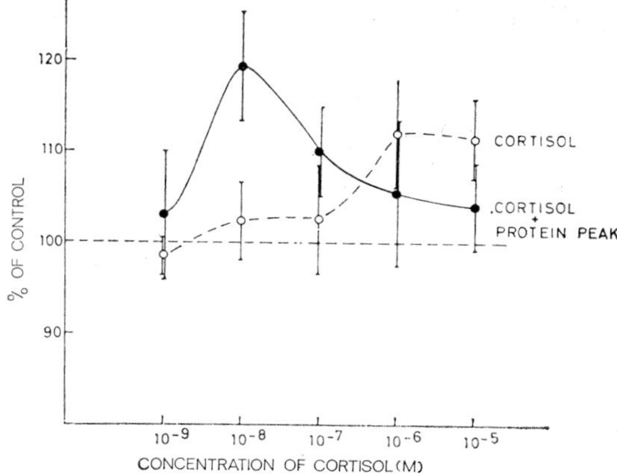

Different concentrations of cortisol were preincubated for 5 min at 37°C, in the presence or absence of cytosol macromolecules for 10 min. The nuclei were lysed in hypotonic buffer and RNA synthesis measured in the nuclear sediment (Beato *et al.*, 1970b).

As shown in Fig. 8 and 9 several proteins binding glucocorticosteroids are resolved. We have tentatively named them I, A, X, B and IV. Peak X is clearly

FIGURE 8

DEAE-cellulose chromatography of rat liver cytosol after injection of [^{3}H] cortisol.

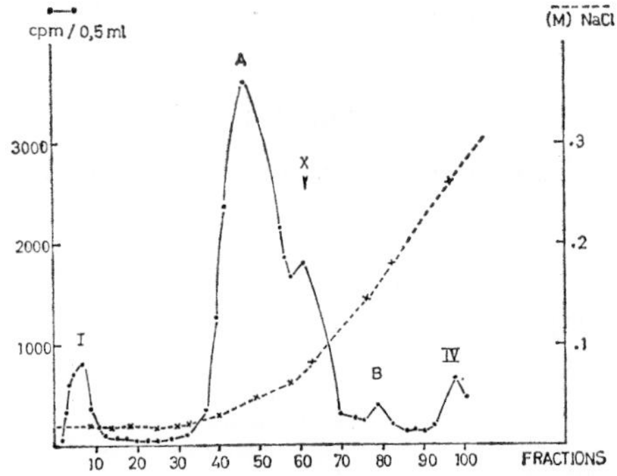

Cytosol was prepared 20 minutes after injection of [^{3}H] cortisol (50 μCi/100 g body weight). Column chromatography as described by Beato *et al.*, (1972a). The individual fractions were checked for absorbance at 280 nm and radioactivity.

separated from peak A only in shallow salt gradients (see Fig. 9). Only peak X can bind [^{3}H] dexamethasone (see Beato *et al.*, 1972b) and seems to be the glucocorticosteroid receptor involved in the intranuclear transport of cortisol. Work on the functional role of these receptor proteins, on the lines mentioned above, is actively pursued in our laboratory.

FIGURE 9

DEAE-cellulose chromatography of rat liver cytosol after injection of [^{3}H] corticosterone.

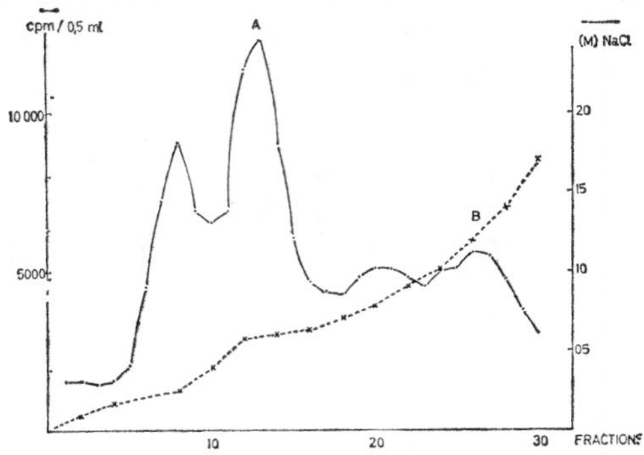

Cytosol was prepared 20 minutes after injection of [^{3}H] corticosterone (50 μCi/100 g body weight). Column chromatography as described by Beato *et al.* (1972a). The individual fractions were checked for radioactivity.

Independent of the presence of cytosol receptors, isolated rat liver nuclei respond, *per se,* to high dose of cortisol with increased RNA synthesis. This has been observed also with nuclei from target organs responding to their respective hormones (Sekeris *et al.*, 1965; Congote *et al.*, 1969; Abraham and Sekeris, 1971). We therefore ascribe to the cytosol receptor proteins a permissive role in the cortisol mediated effect, the exact nature of which still remains to be elucidated.

I have summarized in Table 8 some possible ways by which hormone, receptor proteins and genes could interact. The action of cortisol on isolated rat liver nuclei on the basis of the reported data would fall under heading Ia of the

TABLE 8

Some possible ways by which hormones and recepor proteins may control gene activity

	Effect on RNA synthesis	
	Stimulation	Inhibition
I. HORMONE IS EFFECTOR MOLECULE (cytoplasmic receptor could play a role in the transport of the hormone to the nucleus and in the stabilization of hormone-nuclear protein complexes)	(a) negative control: Hormone acts as derepressor by binding and removing repressor protein from gene (b) positive control: Hormone-nuclear protein complex binds to gene and stimulates transcription	(c) Hormone-nuclear protein complex binds to gene and inhibits transscription
II. CYTOPLASMIC RECEPTOR IS EFFECTOR MOLECULE (hormone needed to bring receptor into nucleus)	(a) receptor binds to chromatin and stimulates RNA synthesis	(b) receptor binds to chromatin and inhibits RNA synthesis.

Table. Although the nuclear repressor proteins with which cortisol interacts are still not identified, we have reasons to believe that they are not related to the cytosol cortisol-binding proteins described above which enter into the nucleus under appropriate conditions. We had already reported (Lukacs and Sekeris, 1967) that nuclei isolated from adrenalectomized rats are more sensitive to cortisol than nuclei from normal rats. However nuclei from normal rats have a much larger cortisol receptor content than the nuclei of adrenalectomized animals. (Beato and Feigelson, 1972). These findings suggest the participation of additional molecules in the regulation of nuclear RNA synthesis by cortisol, at least in the initial stages.

Recent studies from various laboratories demonstrate in a strikingly uniform manner that cytosol receptor proteins when bound to the respective hormones enter the nucleus and bind to the chromatin. Although in none of these studies has binding been correlated to the effects on RNA synthesis it is conceivable that the hormone-receptor complex could act as a positive effector in transcription. In this case either the hormone, the receptor protein or the complex itself could be needed for the effect (see Table 8).

We already know, from the extensive studies concerning bacterial control mechanisms, a variety of examples of positive and negative controls and should strive to extrapolate these findings to higher organisms, taking into account the particular structure of the eukaryotic cells.

In recent experiments we have shown that cortisol can inhibit ribosomal RNA synthesis of isolated rat thymus nuclei (Abraham and Sekeris, 1971). From these nuclei we could extract a receptor protein for glucocorticosteroids (Abraham and Sekeris, in press) which is instrumental for the cortisol effect (see Fig. 10a).

FIGURE 10

Effect of fractions from sucrose gradients on RNA synthesis by thymocyte nuclear sediment.

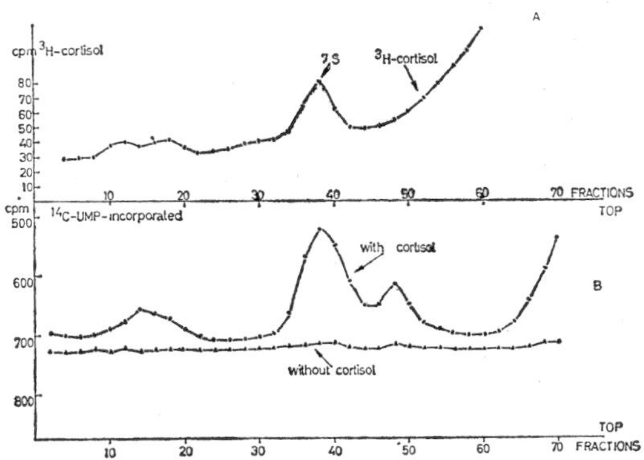

(A) Linear sucrose gradient centrifugation of the protein peak obtained by Sephadex G-25 chromatography of [^{3}H] cortisol labelled thymocyte nuclear extract. ● —— ● ^{3}H-radioactivity.

(B) Fractions from the sucrose gradient were incubated in the absence ▲——▲ or presence ● —— ● of 10^{-5}M cortisol with 0.1 M NaCl-nuclear sediment (0.1 ml containing 150-200 μg of DNA) at 0°C for 60 min. The preparations were subsequently tested for RNA synthesizing capacity. (From van der Meulen *et al.*, 1972).

Nuclei deprived of the receptor protein do not react to cortisol with decreased RNA synthesis. The addition of the receptor to the system restores its sensitivity to cortisol (see Fig. 10b) (van der Meulen *et al.*, 1972).

We have shown that cortisol-receptor complex binds to chromatin (unpublished experiments). In this case we are dealing with a control mechanism as described in heading Ic of Table 8.

CORTISOL AND NEGATIVE CONTROL OF TRANSCRIPTION

The extrapolation of the negative control concept of bacterial regulation to higher organisms (Karlson, 1961, 1963; Karlson and Sekeris, 1966a) poses many specific questions which must be experimentally verified. The complexity of the genome of higher organisms and the higher degree of organization makes this task very difficult. The central postulate is the existence of repressor proteins interacting directly with the DNA in a similar way as the bacterial repressors. From the experiments of Paul and Gilmour (1968) we know that the main factors conferring specificity to transcription of chromatin by exogenous RNA polymerase are the non-histone chromosomal proteins. We therefore turn our attention to these proteins as a potential source of regulatory factors. Due to the difficulty in obtaining non-histone proteins in a soluble form and free of histone contamination we devised a method of isolation (Richter and Sekeris, 1972) starting from chromatin prepared according to Bonner *et al.*, 1968. The method for the preparation of the non-histone proteins is shown in Fig. 11.

FIGURE 11

Isolation of non-histone proteins from chromatin.

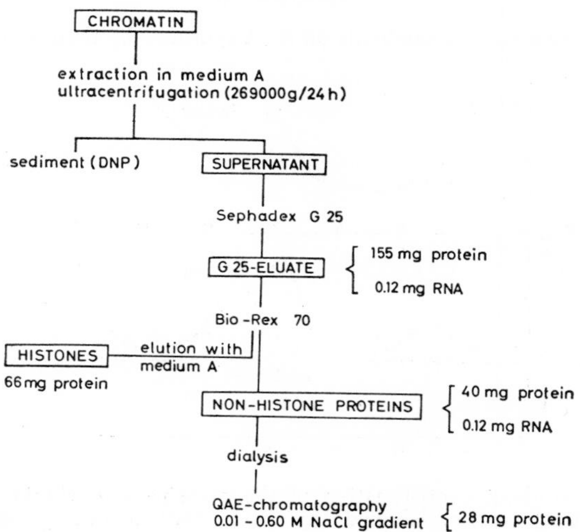

The main characteristics of a repressor protein are 1) its ability to bind to DNA; 2) this binding should lead to inhibition of transcription which 3) should be overcome in the presence of the respective inducer molecule. In bacteria all these criteria have been fullfilled. As a first step in this direction we have tested the ability of the non-histone proteins to a) bind to DNA and b) to affect transcription. Under various conditions only a very small part of the proteins binds to the DNA. However no specificity of the binding could be observed as the proteins recombined with bacterial DNA as well (Roewekamp and Sekeris, unpublished results). On the other hand the addition of non-histone proteins to DNA in solution makes the DNA a less effective template in RNA synthesis using exogenous rat liver RNA polymerase B. Transcription with *E. coli* RNA polymerase is not inhibited, but is in fact slightly stimulated. It is still early to draw conclusions on the basis of these experiments as to the nature and specificity of the inhibitory effect of the proteins. Work on these lines is now actively pursued in our laboratory.

REQUIREMENTS OF AN *IN VITRO* SYSTEM FOR THE STUDY OF TRANSCRIPTION AND ITS CONTROL

In higher organisms the genetic material is present in form of the nucleoprotein complex, chromatin, composed of DNA, RNA and the various chromosomal proteins. The arrangement of these components in chromatin as well as the role of the various components in transcription are still unsolved problems. There are two main approaches towards an understanding of the significance of the chromosomal components. One is the resolution of chromatin in its individual parts, and the systematic analysis of each component in respect to structure and function. The other approach is to sequentially remove and recombine under defined conditions chromosomal components and observe resulting effects on transcription. At the moment several groups are engaged in research on these lines. One of the main obstacles till now, the identification of specific m-RNA's synthesized, has been solved with the introduction as routine methods of *in vitro* protein synthesizing systems directed by exogenous template (Evans and Lingrel, 1969; Aviv *et al.*, 1971; O'Malley *et al.*, 1972).

The first studies on transcription of chromatin with homologous polymerase have appeared (Butterworth *et al.*, 1971; Sekeris, 1971; Sekeris *et al.*, 1972), factors affecting transcription have been identified and isolated (Stein and Hausen, 1970; Seifart, 1970; Sekeris *et al.*, 1972), the RNA polymerases have been purified to homogeneity (Chambon *et al.*, 1972) and judging from the rapid progress and interest of molecular biologists in this field much should be expected soon concerning the mechanism of the activation of genes by hormones.

SUMMARY

Glucocorticosteroids initially stimulate DNA-like RNA synthesis and inhibit ribosomal RNA formation in rat liver. In later time periods ribosomal RNA is highly stimulated. The increased formation of DNA-like RNA is mainly due to the increased capacity of chromatin to support RNA synthesis and not to the increase in the amount of RNA polymerase. The increased production of ribosomal

RNA is very probably triggered by the newly synthesized DNA-like RNA either acting directly or by way of formation of specific proteins.

The increase in template capacity of chromatin is not due to modification of histones by acetylation, methylation or phosphorylation but to a direct interaction of the hormone with chromosomal proteins. The increased thiol content of histones and non-histone proteins suggest the important role of oxido-reduction processes in gene activation.

The role of cytosol receptor proteins in gene activation is seen as permissive: induction is elicited by the steroid molecule itself.

An *in vitro* system composed of chromatin, RNA polymerase and various factors which will permit transcription under physiological conditions and in which the products of transcription will be identified and quantitated is the next step towards a deeper understanding of hormonal control of gene activity.

ACKNOWLEDGEMENTS

We thank Drs. A. D. Abraham, N. van der Meulen, D. Doenecke, M. Beato, K. H. Richter, W. Roewekamp and Prof. K. H. Seifart and D. Gallwitz for collaborative work and stimulating discussions. This work was generously supported by the Deutsche Forschungsgemeinschaft. The competent technical assistance by Mrs. Ch. Pfeiffer, F. Seifart and H. Paul is gratefully acknowledged.

REFERENCES

Abraham, A. D. and Sekeris, C. E. (1971) *Biochim. Biophys. Acta 247,* 562

Allfrey, V. G. (1966) *Cancer Res. 26,* 2026

Allfrey, V. G., Faulkner, R. and Mirsky, A. E. (1964) *Proc. Nat. Acad. Sci. U.S. 51,* 286

Aviv, H., Boime, I. and Leder, P. (1971) *Proc. Nat. Acad. Sci. U.S. 68,* 2303

Beato, M. and Feigelson, P. (1972) *Abstr. Proc. IV. Intern. Congr. Endocrin.,* No. 13

Beato, M., Homoki, J., Lucács, I. and Sekeris, C.E. (1968) *Z. Physiol. Chem. 349,* 1099

Beato, M., Biesewig, D., Braendle, W. and Sekeris, C. E. (1969) *Biochim. Biophys. Acta 192,* 494

Beato, M., Seifart, K. H. and Sekeris, C. E. (1970a) *Arch. Biochem. Biophys. 138,* 272

Beato, M., Braendle, W., Biesewig, D. and Sekeris, C. E. (1970b) *Biochim. Biophys. Acta 208,* 125

Beato, M., Schmid, W. and Sekeris, C. E. (1972a) *Biochim. Biophys. Acta 263,* 764

Beato, M., Koblinski, M. and Feigelson, P. (1972b) *J. Biol. Chem.* in press

Bonner, J., Dahmus, M. E., Fambrough, D., Huang, R. C., Marushige, K. and Tuan, D. Y. H. (1968) *Science 159,* 47

Butterworth, P. H. W., Cox, R. F. and Chesterton, C. J. (1971) *Europ. J. Biochem. 23,* 229

Congote, L. F., Sekeris, C. E. and Karlson, P. (1969) *Exptl. Cell Res. 56,* 338

Dahmus, M. E. and Bonner, J. (1965) *Proc. Nat. Acad. Sci. U.S. 54,* 370

Doenecke, D. and Sekeris, C. E. (1970) *FEBS Letters 8,* 61

Doenecke, D., Beato, M., Congote, L. F. and Sekeris, C. E. (1972) *Biochem. J. 126,* 1171

Drews, J. and Brawerman, G. (1967) *J. Biol. Chem. 242,* 801

Evans, M. J. and Lingrel, J. B. (1969) *Biochemistry 8,* 829

Feigelson, P., Fu-Li-Yu and Hanoune, I. (1969), in *Progress in Endocrinology,* (C. Gual ed.), p. 33, Excerpta Med. Foundation

Feigelson, P., Gros, P. R. and Feigelson, M. (1962) *Biochim. Biophys. Acta 55,* 485

Fiume, L. and Wieland, Th. (1970) *FEBS Letters 8,* 1

Gallwitz, D. and Sekeris, C. E. (1969a) *Z. Physiol. Chem. 350,* 150
Gallwitz, D. and Sekeris, C. E. (1969b) *FEBS Letters 3,* 99
Homoki, J., Beato, M. and Sekeris, C. E. (1968) *FEBS Letters, 1,* 275
Jensen, E. V., Hurst, D. I., De Sombre, E. R. and Jungblut, P. W. (1967) *Science 158,* 385
Jervell, K. F. (1963) *Acta Endocrin Cpnh. 44,* suppl. 88
Karlson, P. (1961) *Dtsch.med.Wschr. 86,* 668
Karlson, P. (1963) *Persp. Biol. Med. 6,* 203
Karlson, P. and Sekeris, C. E. (1966a) *Acta Endocrin. Cpnh. 53,* 505
Karlson, P. and Sekeris, C. E. (1966b) *Rec. Progr. Horm. Res. 22,* 473
Kenney, F. T. and Kull, F. J. (1963) *Proc. Nat. Acad. Sci. U.S. 50,* 493
Kleinsmith, L. J., Allfrey, V. G. and Mirsky, A. E. (1966) *Science, 154,* 780
Lang, N. and Sekeris, C. E. (1964) *Life Sciences 3,* 391
Lang, N., Herrlich, P. and Sekeris, C. E. (1968) *Acta Endocrin. Cpnh. 57,* 33
Liao, S. and Fang, S. (1969) *Vitam. a. Horm. 160,* 272
Lucács, I. and Sekeris, C. E. (1967) *Biochim. Biophys. Acta 134,* 85
Makman, M. H., Dvorkin, B. and White, A. (1968) *J. Biol. Chem. 243,* 1485
Nakagawa, S. and White, A. (1966) *Proc. Nat. Acad. Sci. U.S. 55,* 900
O'Malley, B. (1972) *Proc. IV. Intern. Congr. Endocrin.,* in press
Paul, J. and Gilmour, R. S. (1968) *J. Mol. Biol. 34,* 305
Raspé, G. (1970) ed. *Adv. Biosci.* 7, Pergamon Press, Vieweg
Richter, K. H. and Sekeris, C. E. (1972) *Arch. Biochem. Biophys. 148,* 44
Schiltz, E., 1969 Dissertation, Marburg
Schiltz, E. and Sekeris, C. E. (1969) *Z. Physiol. Chem. 350,* 317
Schmid, W. and Sekeris, C. E. (1972) *FEBS Letters, 26,* 109
Seifart, K. H. (1970), *Cold Spring Harbor Symp. Quant. Biol., 35,* 719
Sekeris, C. E. (1966) in *Molec. Basis of Some Aspects of Mental Activity,* (O. Walaas, ed.) Vol. I, p. 363 Acad. Press
Sekeris, C. E. (1971) *Biochem. J. 124,* 43 p
Sekeris, C. E. and Lang, N. (1964) *Life Sciences 3,* 169
Sekeris, C. E. and Schmid, W. (1972a) *FEBS Letters, 27,* 41
Sekeris, C. E. and Schmid, W. (1972b) *Proc. IV. Intern. Congr. Endocrin.,* in press
Sekeris, C. E., Dukes, P. P. and Schmid, W. (1965) *Z. Physiol. Chem. 341,* 152
Sekeris, C. E. Sekeri, K. and Gallwitz, D. (1967) *Z. Physiol. Chem. 348,* 1660
Sekeris, C. E., Beato, M., Homoki, J. and Congote, L. F. (1968) *Z. Physiol. Chem. 349,* 857
Sekeris, C. E., Schmid, W. and Roewekamp, W. (1972) *FEBS Letters, 24,* 27
Stein, H. and Hausen, P. (1970) *Europ. J. Biochem. 14,* 272
van der Meulen, N. Abraham, A. D. and Sekeris, C. E. (1972) *FEBS Letters, 25,* 116

A Control Mechanism in Gene Expression of Higher Cells Operating at the Termination Step in Protein Synthesis

I. T. OLIVER

Department of Biochemistry, University of Western Australia, Nedlands, Western Australia

Tyrosine aminotransferase (EC 2.6.1.5) is an enzyme whose synthesis is induced in rat liver within a few hours of birth (Sereni *et al.*, 1959; Holt and Oliver, 1968). In the adult animal *in vivo* and the isolated perfused liver, a number of hormones have been implicated as inducers, including insulin, hydrocortisone and glucagon (Kenney, 1963; Civen *et al.*, 1967; Csányi *et al.*, 1967; Holten and Kenney, 1967; Hager and Kenney, 1968). Adenosine 3′,5′-cyclic monophosphate (cyclic AMP) seems to be the intracellular effector for induction by glucagon and adrenalin (Holt and Oliver, 1969a; Wicks, 1968a; 1969).

In various cultured cell lines derived from hepatomas (e.g. Granner *et al.*, 1968; Butcher, *et al.*, 1971) and in foetal liver organ cultures (Wicks, 1968a, b; 1969) some or all of these agents have also been shown to be effective in stimulating the rate of enzyme synthesis.

In the neonatal animal the appearance of hepatic enzyme activity can be blocked with inhibitors of DNA transcription and of RNA translation (Greengard, *et al.*, 1963; Holt and Oliver, 1968) and the process therefore appears to be due to *de novo* enzyme synthesis resulting from the initiation of gene transcription, but other explanations are also tenable. In adult animals and tissue culture systems, the rate of labelling of the specific enzyme protein is increased by the inducers. Glucocorticoids have been closely implicated in postnatal production of the enzyme, since adrenalectomy at birth prevents the accumulation of enzyme activity in the liver (Sereni *et al.*, 1959; Holt and Oliver, 1969a; 1971).

However, some evidence has accumulated which shows that the apparent multiplicity of inductive mechanisms can be explained by enzyme multiplicity and the induction of specific forms by specific hormones. The postnatal accumulation of enzyme activity is shown in Fig. 1. After delivery of late foetal rats by uterine section, there is a reproducible lag phase and then activity begins to accumulate.

FIGURE 1

Postnatal development of tyrosine aminotransferase in neonatal rats

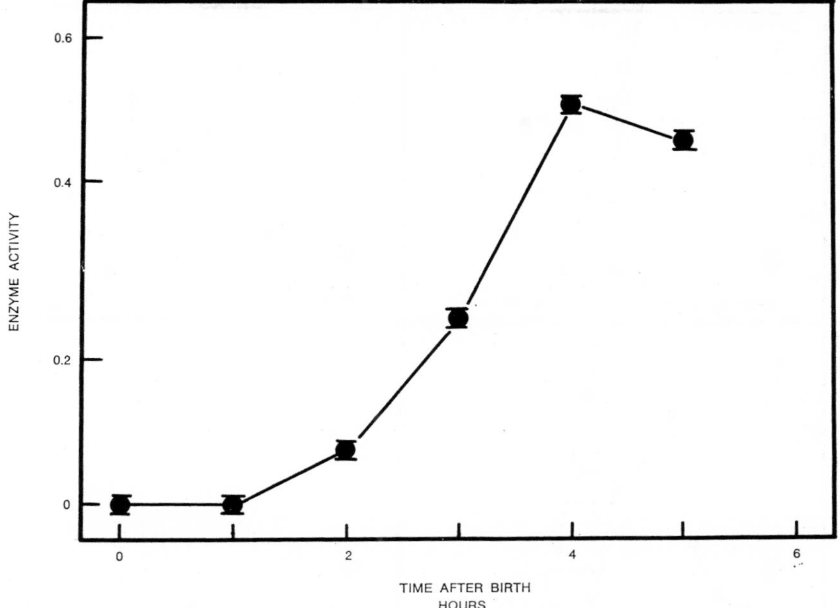

Each litter was delivered by uterine section and placed in a humidicrib at 37°C. At the times shown, at least one animal from each litter was killed and the liver tyrosine aminotransferase activity determined. The points on the figure represent the mean enzymatic activity ±1S.E.M. of the determinations on 8 litters. Enzyme activity is expressed as μmoles p-hydroxyphenylpyruvate (PHPP) produced/h/mg protein at 37°C.

When the pattern of enzyme activity is examined by electrophoresis in polyacrylamide gel, it is seen to change in a fashion which strongly suggests the occurrence of multiple forms as shown in Fig. 2 (Holt and Oliver, 1971). Previous work has also demonstrated that the patterns of activity after gel electrophoresis can be altered in a systematic way by treatment of animals with various hormonal regimes (Holt and Oliver, 1969b). A summary of these findings is that glucagon, adrenalin or cyclic AMP each result in the stimulation of the first postnatal form (Form C), glucocorticoids stimulate both forms C and B, the first and second postnatal forms respectively, while insulin stimulates only Form A, the last postnatal form to appear. A recent paper of Iwasaki and Pitot (1971) has also indicated the occurrence of 3 different forms of the enzyme which are differentially inducible by different hormones. Hydroxylapatite chromatography was used to separate the various forms.

The major results to be described in this paper apply only to Form C of the enzyme which is apparently inducible by glucagon and adrenalin via cyclic AMP as the intracellular effector of these hormones.

In the postnatal production of tyrosine aminotransferase activity in intact animals, the use of puromycin produces anomalous results. When given at

FIGURE 2

Gel electrophoretic profiles of hepatic tyrosine aminotransferase from early postnatal rats

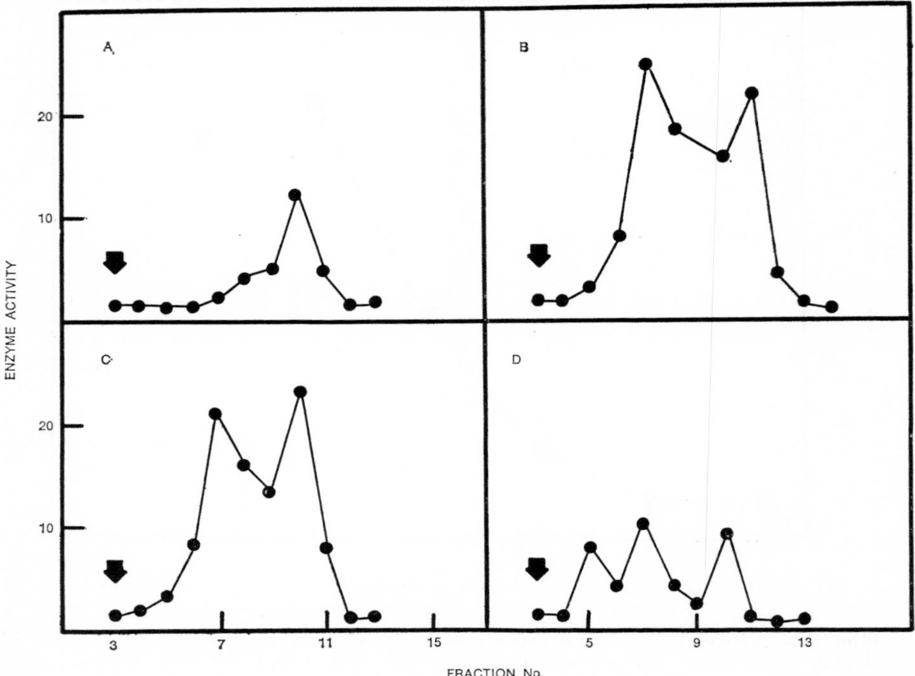

Animals of gestational age, 21-22 days, were delivered by uterine section and placed in a humidicrib at 37°C. Five hours later, half the litter was sacrificed and pooled liver extracts prepared for electrophoresis (A). Ten hours after delivery, the remaining litter-mates were sacrificed and gel electrophoresis of their liver extracts carried out in the same gel slab as above (B). Twentyfour hours after normal vaginal delivery pooled liver extracts were prepared from 4 animals from each of 6 litters. The electrophoretic profile of tyrosine aminotransferase activity is shown in C. For purposes of comparison, the profile obtained from a normal, untreated adult is shown in D. Enzyme activities are in μmoles PHPP/hr/2 mm gel section.

(Reproduced by permission from Holt and Oliver (1971), Int. J. Biochem. *2*, p 214, Fig. 1.)

delivery, puromycin blockades the subsequent production of enzyme. However, if drug administration is delayed until enzyme accumulation has begun, the drug produces an apparent inductive effect. These results are demonstrated in Fig. 3 (Chuah, *et al.*, 1971).

Zone centrifugation of the enzyme from puromycin treated animals sediments as a single peak and the profile is identical with that from control animals (Chuah, *et al.*, 1971). Thus the increased activity is not due to the abortive release of incomplete polypeptides which possess enzymatic activity and another explanation must be sought.

When experiments, similar in design to the puromycin experiment are carried out using cyclic AMP, a similar result is obtained but the stimulation of enzyme accumulation by cyclic AMP is not at all inhibited by simultaneous injection of

FIGURE 3

The effect of puromycin on post-natal development of tyrosine aminotransferase

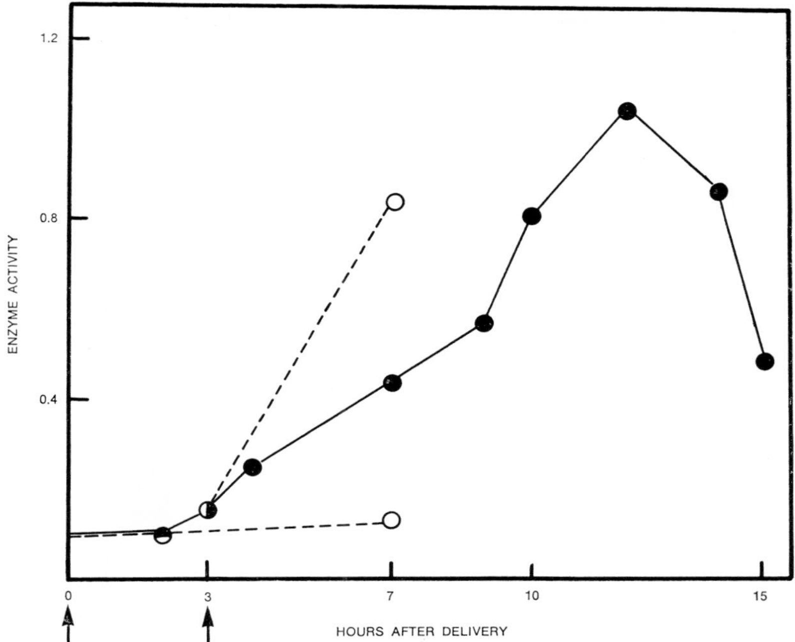

Animals were delivered by uterine section and maintained in the humidicrib at 37°C. Injections of puromycin (750 μg per animal in 50 μl 0.145 M NaCl) were made at delivery or at 3 hours after delivery. Experimental and control animals from the same litter were maintained in the humidicrib at 37°C until sacrificed. Tyrosine aminotransferase activity was assayed in liver extracts from each animal and results are expressed as μmoles PHPP produced/h/mg protein at 37°C. Arrows indicate the time of puromycin injection. - - - - - - -, Puromycin treated. ———, Control.

(Reproduced by permission from Chuah, Holt and Oliver (1971), *Int. J. Biochem.* 2, p 194, Fig. 1.)

actinomycin D even though actinomycin D alone prevents further accumulation of activity in otherwise untreated animals (See Fig. 4.) (Holt and Oliver, 1969a). These results suggest that cyclic AMP functions at a postranscriptional level in enzyme synthesis. Further experiments in 2 day postnatal animals also show that certain litters of animals which show low basal levels of tyrosine aminotransferase, increase markedly in hepatic enzyme activity after injection with adrenalin or puromycin and there is a clear correlation of effect between the two agents (Holt and Oliver, 1969a; Chuah *et al.*, 1971).

These findings suggest a basic working hypothesis. Both the puromycin and cyclic AMP (or adrenalin) effects are due to release of completed enzyme polypeptides from the distal end of the tyrosine aminotransferase polysome and in new born animals and in a proportion of older postnatal animals, polysomes occur in a preloaded condition which contain enzyme peptides at various stages of

FIGURE 4

The effect of actinomycin D on the increase in tyrosine aminotransferase activity mediated by cyclic AMP in neonatal rat liver *in vivo*

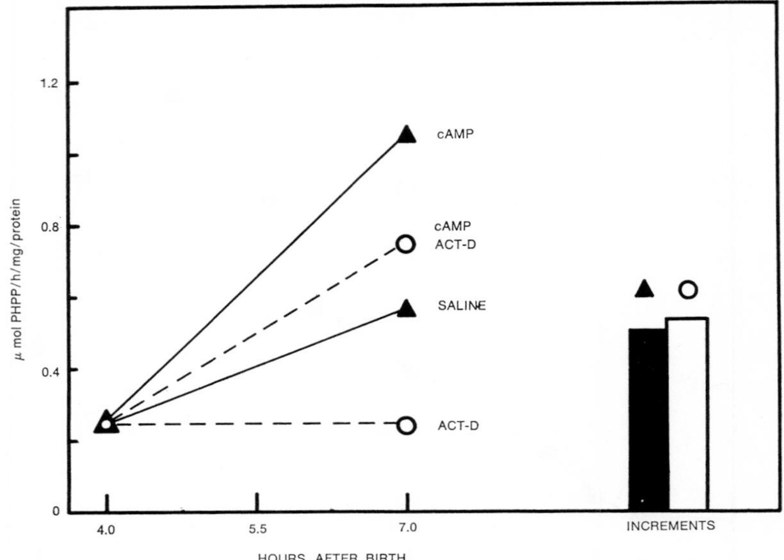

In these experiments dibutyryl cyclic AMP was used. A litter of foetal rats was delivered by uterine section and maintained in a humidicrib at 37°C. At 4 hours after delivery, duplicate rats were injected with 0.145 M NaCl, actinomycin D (3.5 μg), dibutyryl cyclic AMP (50 μg) or actinomycin D together with dibutyryl cyclic AMP. Another pair were killed for enzyme assay. The injected animals were maintained for a further 3 hours and killed for enzyme assay. Tyrosine aminotransferase activity is expressed as μmoles PHPP produced/h/mg protein.

assembly up to and including unreleased peptides complete to the C-terminal amino acid. The system would remain in this condition until the operation of a release signal or factor. Puromycin is apparently an effective but unnatural such factor while cyclic AMP might be the natural release signal.

Microsomes isolated from the liver of adrenalin-inducible animals and incubated with cyclic AMP *in vitro* show an increase in enzyme activity (Fig. 5 and Table 1).

The data and the following results are from Chuah and Oliver (1971). The effect can be obtained at concentrations as low as 10^{-5}M cyclic AMP (Fig. 5). When the microsomes are extensively washed, the occluded enzyme activity is completely removed but the cyclic AMP effect can still be elicited. This makes it unlikely that enzyme activation occurs and in Figs. 6 and 7 it is shown that the enzyme activity released from the microsomes has the same mol.wt. as the normal, cytoplasmic tyrosine aminotransferase. Table 1 shows that dibutyryl cyclic AMP is also effective and that theophylline added to the incubation medium, increases the effectivity of cyclic AMP. This is undoubtedly due to inhibition of cyclic AMP phosphodiesterase which is occluded in unwashed micro-

FIGURE 5

Effect of c-AMP on tyrosine aminotransferase activity of the rat liver microsome-polysome fraction

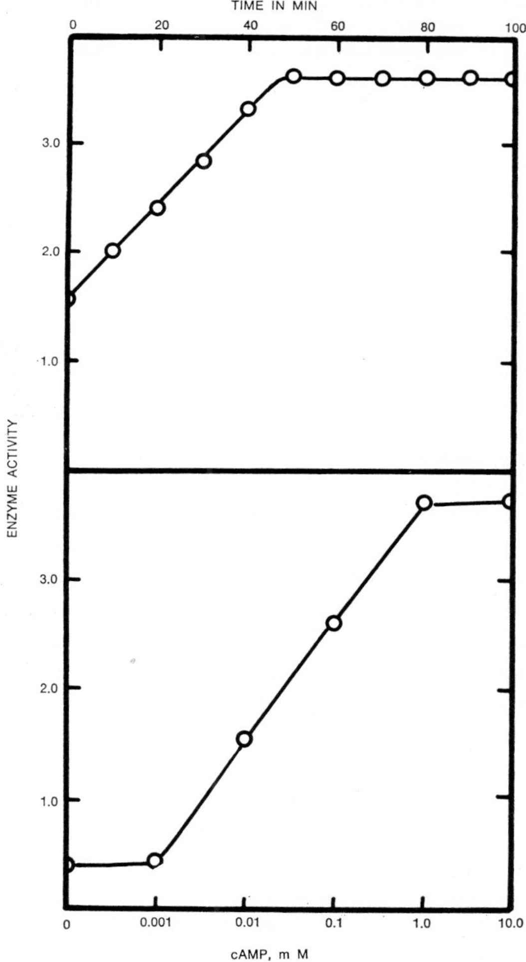

The liver microsome-polysome fraction was prepared from 2-day post-natal rats by procedure 1 and incubated (0.5 ml, 5 mg of protein) in TKM buffer, 5 mM theophylline, and 1 mM c-AMP for various times at 37°C before enzyme assay (A). The liver microsome-polysome fraction was incubated at 37°C for 60 min in TKM buffer, 5mM theophylline, and various concentrations of c-AMP (B). Tyrosine aminotransferase activity was assayed on 100μl aliquots after incubation (micromoles of PHPP per hour).

(Reproduced by permission from Chuah and Oliver, 1971; *Biochemistry 10,* p 2991, Fig. 1.)

some preparations. Cycloheximide at a concentration of 3.5 mM does not prevent the response and puromycin at 1 mM concentration in the presence of NH_4^+ gives the same enzyme yield as cyclic AMP. ATP also promotes release in this system but 5′-AMP is without effect. The ATP stimulation appears to be due to the presence of adenyl cyclase in the microsome fraction, since the inclusion

TABLE 1

Effect of c-AMP on tyrosine aminotransferase of liver microsome/polysome fractions *in vitro*

Additions to Medium (1mM)	Increase in Enzyme Activity
c-AMP	0.91
+ Theophylline (5 mM)	1.83
Theophylline alone	0.0
Dibutyryl c-AMP	1.69
5′ AMP	0.0
ATP	0.87
+ Theophylline	1.97

The liver microsome-polysome fraction (0.5 ml, 5.6 mg protein) was prepared from 2-day postnatal rats by procedure 1 of Chuah and Oliver (1971) and incubated at 37°C for 60 min in a final volume of 1 ml in TKM buffer + 5 mM GSH, 5 mM PEP, 10 i.u. of PK, pH 8.0. Tyrosine aminotransferase was subsequently assayed. Various additions were made at the final concentrations given in the table. The increase in enzyme activity was obtained by subtracting the control value from the value obtained after additions. Activities were calculated as μmoles of PHPP formed/h.

(From Chuah and Oliver, 1971.)

of cyclic AMP phosphodiesterase in the medium prevents the effect and polysomes prepared in deoxycholate are not active in release with ATP although release can be elicited after addition of cyclic AMP and a factor extracted from microsomes with 0.5 M NH_4Cl. Microsomes which are washed at high ionic strength lose the capacity to respond to cyclic AMP but this capacity is restored by addition of the microsomal factor. The microsomal factor binds cyclic AMP in equilibrium dialysis experiments. Table 2 summarizes some further properties of the release system.

TABLE 2

Characteristics of the Release System

1. Localized in rough-surfaced microsomes.
2. Requiries c-AMP but no cofactors for protein synthesis.
3. ATP effect due to microsomal adenyl cyclase.
4. Puromycin promotes enzyme release *in vitro.*
5. Cycloheximide does not block system.
6. Requires factor extractable from microsomes at high ionic strength.
7. Microsome factor binds cyclic AMP.
8. System demonstrable only in adrenalin inducible animals.

The effect is specific for cyclic AMP or its esterified derivative and Table 3 gives a qualitative summary of the results obtained using cognate cyclic nucleotides and some other compounds. The inhibition experiments were carried out with the test compound at 10 times the concentration of cyclic AMP and it is interesting to note that both cyclic IMP and GMP produce a 50% inhibition of release and this point will be further discussed later.

TABLE 3

Specificity of Nucleotides in release function

Nucleotide	Activity in release	Inhibition
c-AMP	+	—
Dibutyryl c-AMP	+	—
c-IMP	—	+
c-GMP	—	+
c-UMP	—	—
c-CMP	—	—
c-TMP	—	—
Deoxy c-AMP	—	—
2′, 3′ c-AMP	—	—
5′ AMP	—	not tested

Data taken from Chuah and Oliver (1971, 1972).

FIGURE 6

Zone centrifugation of tyrosine aminotransferase released from the rat liver microsome-polysome fraction by c-AMP *in vitro*

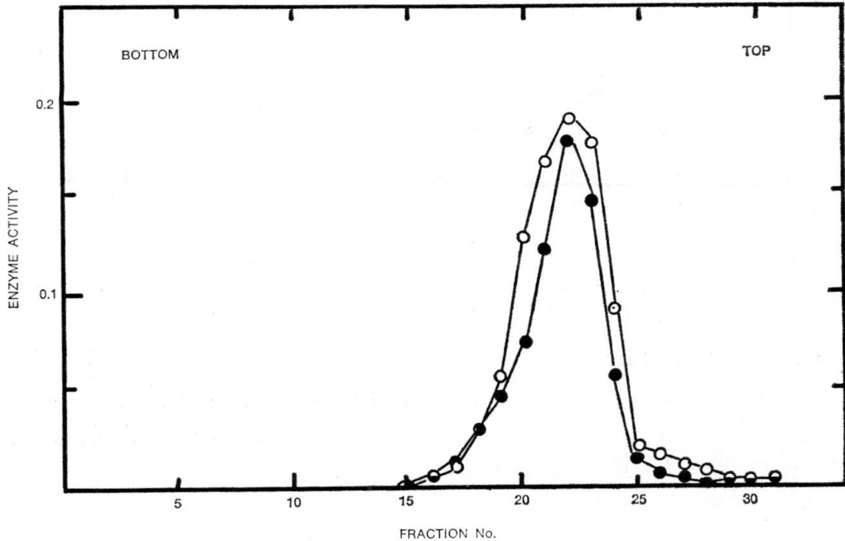

The liver microsome-polysome fraction was prepared from 2-day inducible post-natal rats by procedure 1 and washed three times with 0.25 M sucrose. An aliquot (0.5 ml 10 mg of protein) was incubated in TKM buffer, 5 mM theophylline, and 1 mM c-AMP for 1 h at 37°C. The incubate was then centrifuged at 400,000g for 60 min and the supernatant collected. This supernatant (100 μl) was layered on 4.2 ml of a linear sucrose concentration gradient (20-30% sucrose) in potassium phosphate (5 mM)-EDTA (5 mM) (pH 7.4). The tubes were centrifuged for 12 h at 2-4°C in the SW56 rotor of the Spinco L2-65 ultracentrifuge at 56,000. A preparation of tyrosine aminotransferase from the 400,000g for 60 min supernatant from epinephrine-induced 2-day rat liver was also zone centrifuged in the same way. Fractions were collected through the tube bottom (8 drops each) and incubated for enzyme assay for 1 h (micromoles of PHPP per hour). (●) Enzyme from microsomal incubate; (○) enzyme from epinephrine induction. Points were plotted only for tubes which showed enzyme activity.

(Reproduced by permission from Chuah and Oliver, 1971; *Biochemistry 10,* p 2996, Fig. 3.)

FIGURE 7

Release of tyrosine aminotransferase from rat liver microsomes by c-AMP *in vitro*

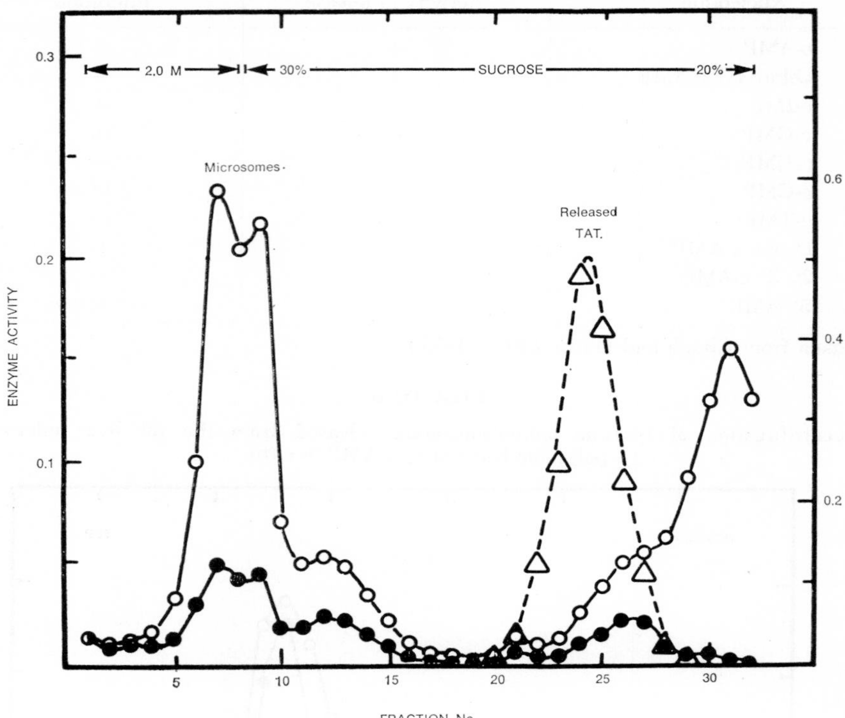

Liver microsomes were prepared by procedure 2 from 2-day inducible rats and washed three times with 0.25 M sucrose. An aliquot of 0.5 ml (10 mg of protein) was incubated in a final volume of 1 ml in TKM buffer, 5 mM theophylline, and in the presence or absence of 1 mM c-AMP at 37°C for 60 min. Sucrose concentration gradients consisted of 1.2 ml of 2.0M sucrose followed by a linear gradient (3 ml) of 30-20% sucrose. Aliquots (100 μl) of the incubates above were layered on the top of the gradients and the tubes centrifuged in the SW56 rotor of the Spinco L2-65 ultracentrifuge for 12 h at 56,000 at 2-4°C. Fractions were collected by puncturing the bottom of the tubes (8 drops each) and assayed from duplicate tubes for enzyme activity (Δ) (micromoles of PHPP per hour) (60-min incubation), or after dilution with 1.0 ml of 0.1 M KOH, for absorbance at 260 nm (○) and at 290 nm (●). Points for tyrosine aminotransferase were plotted only for tubes which showed activity. No activity was found after incubation in the absence of c-AMP. The absorbance profiles are for the preparation incubated with c-AMP.

(Reproduced by permission from Chuah and Oliver, 1971; *Biochemistry 10,* p 2996, Fig. 4.)

All of this data can be interpreted to mean that there is a control point in the synthesis of at least one form of tyrosine aminotransferase (Form C) which occurs at the release step in enzyme synthesis and is controlled by the small molecule cyclic AMP, working through a binding factor which is normally found in liver microsomes. In the system described, no cofactors or substrates of protein synthesis are required for the effect which suggests that peptide chain elongation does not take place. The results with cycloheximide support this suggestion.

The *in vitro* experiments give evidence for the existence of enzyme-specific polysomes which are loaded with unreleased but completed enzyme molecules or subunits. In such a system it is predictable that the release of the C-terminal peptide will lead to a cascade phenomenon in which the other nascent, but incomplete peptides, if they are present, will complete their synthesis and in turn be released by the cyclic AMP mechanism provided that all the cofactor and substrates required for protein synthesis are present to allow peptide chain elongation. The following data from Chuah and Oliver (1972) relate to this question.

Tables 4 and 5 show that such an effect can be demonstrated in both sodium deoxycholate-prepared-(DOC) polysomes and microsomes in the presence of all the cofactors and substrates for protein synthesis. The amount of enzyme released

TABLE 4

Effect of Addition of cofactors and substrates for protein synthesis to DOC polysomes

Liver Fraction	Conditions	Enzyme Activity
DOC	Release	2.75
Polysomes	Cascade (Complete)	7.77
	pyridoxal 5′ phosphate (PALP)	2.68
	—cAMP	0
	+ Cycloheximide (3.5 mM)	2.75
	+ Puromycin (3.5 mM)	2.75

DOC-polysomes were prepared from 2-day old inducible rats and aliquots incubated under different conditions with additions noted above for 60 min at 37°C in a final volume of 1 ml. Aliquots were then taken for tyrosine aminotransferase determination. Enzyme activity is expressed as μmoles PHPP/h. Cascade conditions were as follows: incubation was in a final volume of 1 ml containing pH 5 fraction (1.9 mg protein), 0.85 mM ATP, 0.20 mM GTP, 5 mM PEP, 250 μg of PK, 2.7 mM $MgCl_2$, 35 mM NH_4Cl, 47mM KCl, 4 mM mercaptoethanol, 44 mM sucrose, 5 mM theophylline, 1 mg/100 ml of each of the most common L-amino acids, 35mM Tris-HCl (pH 7.6) 1 or 5 mM cyclic AMP and 1 mM pyridoxal 5′ phosphate. Release conditions were identical except that the ATP regenerator, pH 5 fraction, GTP and free amino acids were omitted. (Data from Chuah and Oliver, 1972, Table 2.)

TABLE 5

Effect of addition of cofactors and substrates for protein synthesis to microsomes

Liver Fraction	Conditions	Enzyme Activity
Microsomes	Release	2.72
	Cascade (Complete)	16.95
	+ Cycloheximide (3.5 mM)	2.72
	+ Puromycin (3.5 mM)	2.72

Microsomes were prepared from 2-day old inducible rats and aliquots incubated under the conditions noted with additions shown for 3h at 37°C in a final volume of 1 ml. Aliquots were then taken for tyrosine aminotransferase determinations. Enzyme activity is expressed as μmoles PHPP/h. Cascade and release conditions were as in Table 4. (Data from Chuah and Oliver, 1972, Table 2.)

under cascade conditions is considerably increased over that released under the release-only conditions, and the increase in yield of enzyme is prevented by cycloheximide and by puromycin. The deletion of t-RNA, aminoacyl synthetases, GTP or a single amino acid from the reaction mixture also reduces the yield to that of release and therefore peptide chain elongation appears to take place in the complete system. The inclusion of pyridoxal phosphate in the reaction is also required.

Fig. 8 demonstrates the kinetics of the system in microsomes and also gives some information relating to the action of pyridoxal phosphate. Microsomes were incubated in the presence and absence of pyridoxal phosphate and the yield of enzyme measured. Microsomes in the latter experiment were removed by centrifugation and the supernatant then incubated with pyridoxal phosphate. As can be seen, the accumulation of enzyme activity, although it reaches the same final value as in microsomes, occurs at a much faster rate.

FIGURE 8

The kinetics and effect of pyridoxal phosphate on cascade synthesis of tyrosine aminotransferase by liver microsomes *in vitro*

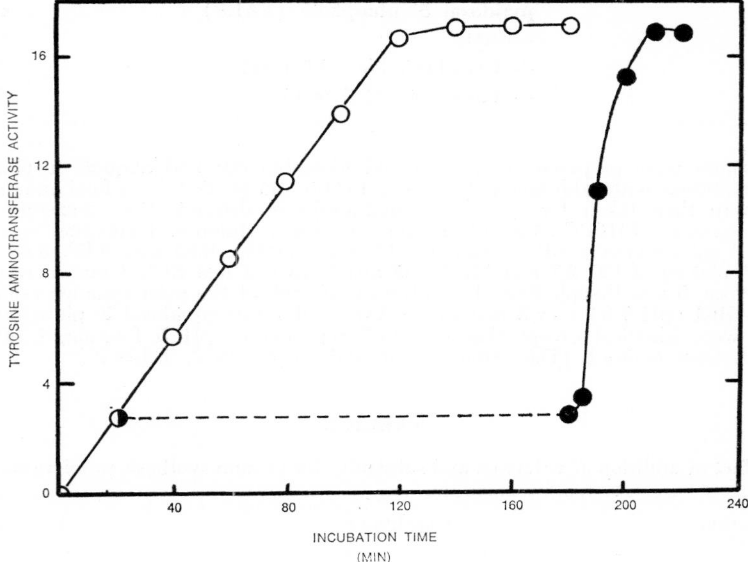

Liver microsomes were prepared from 2-day old inducible rats. Aliquots (0.5 ml, 5.6 mg protein) were incubated under cascade conditions (given in Table 4) for various times at 37°C before enzyme assay. In a second run, PALP was omitted from the incubation which lasted for 180 min. The incubate was chilled in ice and centrifuged at 150,000g for 42 min. The total microsomal-free supernatant was incubated with 1 mM PALP for various times at 37°C before enzyme assay. The 180 min point in the second curve thus represents 0 min of incubation with PALP. Enzyme activity is expressed as μmoles PHPP/h.

○— complete cascade system.

● microsomal-free supernatant incubated with PALP.

(Reproduced by permission from Chuah and Oliver, 1972; *Biochemistry 11,* p 2550, Fig. 1.)

This indicates that pyridoxal phosphate cannot be involved in the translational process since all microsomes are removed by centrifugation and only the supernatant is incubated with the cofactor. It is thus more likely that pyridoxal phosphate combines with newly completed apoenzyme, converting it to the active holoenzyme. It has been further shown (Chuah and Oliver, 1972) that the mol.wt. of the inactive enzyme present after incubation of microsomes in the absence of pyridoxal phosphate is the same as that of the active enzyme completed and released in the presence of the cofactor.

In the experiment illustrated in Fig. 9, microsomes were incubated under cascade condition but in the absence of pyridoxal phosphate (PALP). The

FIGURE 9

Zone sedimentation behaviour of tyrosine aminotransferase released and synthesized in presence and absence of PALP

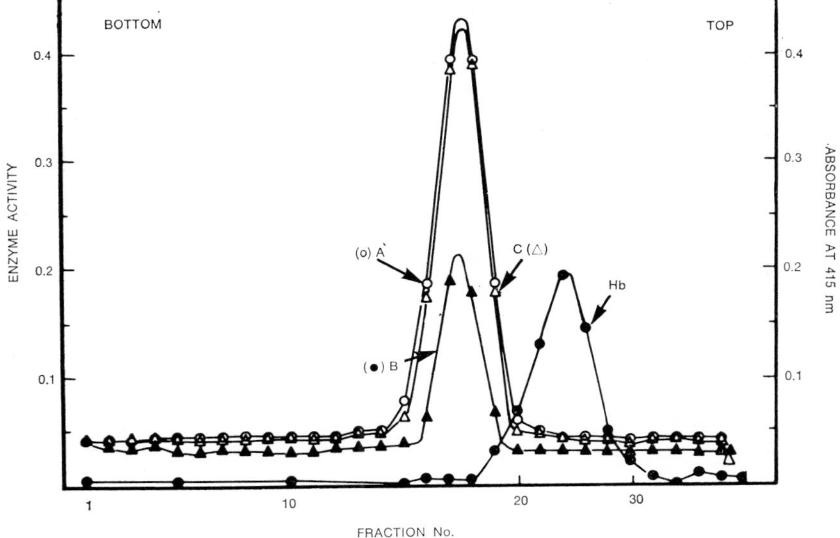

Liver microsomes were prepared from 2-day old inducible rats. Aliquots (0.5 ml, 10 mg protein) were incubated under cascade conditions (as in Table 4), minus PALP for 180 min at 37°C. The incubates were chilled, centrifuged at 150,000g for 60 min and the supernatants harvested. The supernatant was divided into 3 portions, A. B. and C. A was incubated in 1 mM PALP for 30 min at 37°C. Supernatants A, B and C (100 μl each) were layered on 4.2 ml of a linear sucrose gradient (20% to 30%, in 34 mM tris-HCl-47mM KCl-2.7 mM $MgCl_2$ pH 7.6). The tubes were centrifuged at 2-4° for 12 h at 56,000 rpm in the SW56 rotor of the L2-65 Spinco ultracentrifuge. Fractions were collected through the tube bottom (8 drops each). The fractions from tube C, were then incubated in 1 mM PALP for 30 min. All fractions from A, B and C were then incubated for enzyme assay for 60 min. Fractions from a fourth tube were diluted with 1.0 ml of 0.01 M NH_4OH for determination of absorbance at 415 nm to locate haemoglobin (Hb) which contaminates the preparations.

microsomes were removed by centrifugation and the supernatant divided into 3 portions called A, B and C. Portion A was incubated after addition of 1 mM PALP for 30 min at 37°C. Aliquots of all three portions were then zone centrifuged and fractionated. Each fraction from C was incubated with PALP and enzyme activity then determined in all fractions from each of A, B and C. The peak height of B corresponds to the release amount of enzyme since no cascade effect on activity is apparent in absence of PALP (see Fig. 8). If PALP promoted association of enzyme subunits, the peak from C should have a much lower sedimentation rate than the A-peak. In fact it is coincident with both the A and B peaks. Thus the cofactor does not promote the formation of quarternary structure in the enzyme if subunits exist, and the mol.wt. of the enzyme produced in the absence of PALP is the same as that produced in its presence.

When the enzyme activity produced under cascade conditions, is expressed as a ratio to that produced under release conditions, it gives a rough measure of the size of the tyrosine aminotransferase polysome. The maximum value of the ratio obtained in microsomes is six (6.2). In the preparation of microsomes, polysomes are less likely to be broken by shearing forces, or other factors. Independent and

FIGURE 10

Zone centrifugation of DOC polysomes active in release and cascade

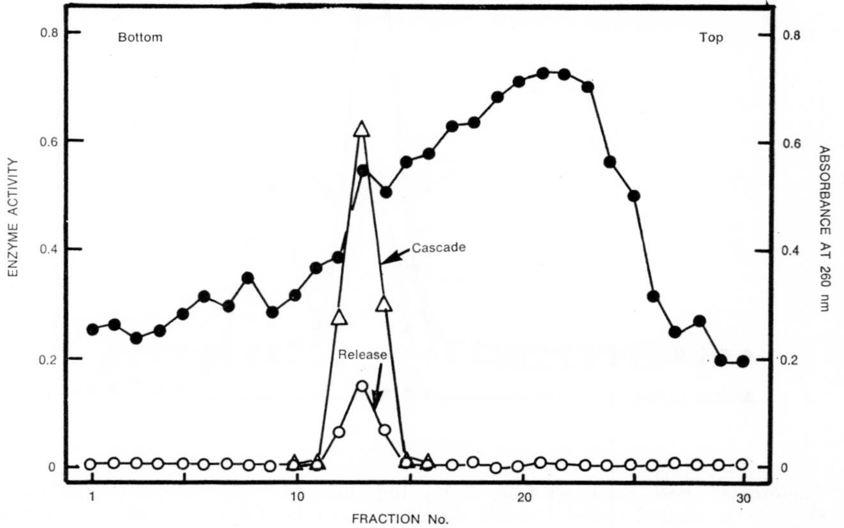

Polysomes were prepared by the deoxycholate procedure from 2-day old inducible rats. The polysome pellets were gently resuspended in 0.2 M sucrose — 2.7 mM $MgCl_2$ — 34 mM Tris-HCl, pH 7.6 with a stirring rod and 200 μl aliquots layered over 4.0 ml of a linear sucrose gradient (20% — 30%) prepared in 34 mM Tris-HCl — 47 mM KCl — 2.7 mM $MgCl_2$, pH 7.6. Tubes were centrifuged at 2-4° for 45 min at 37,000 in the SW56 rotor.
Fractions were collected by bottom puncture (8 drops each). Fractions from one tube were incubated under release conditions in the presence of the microsomal release factor while a parallel series were incubated under cascade conditions including also the microsomal release factor. Fractions from a triplicate tube were diluted with 1.0 ml of 0.1 M NaOH for determination of absorbance at 260 nm.

○ Release

Δ Cascade

● 260 nm absorbance

(Reproduced by permission from Chuah and Oliver, 1972; Biochemistry, in press, Fig. 2.)

direct measurements of the size of tyrosine aminotransferase polysomes can be made by zonal centrifugation since both the enzyme release and cascade phenomenon can be used to locate the polysomes in the gradients. However, it is necessary to use deoxycholate to prepare the free polysomes.

Fig. 10 shows the results of such an experiment using deoxycholate-prepared polysomes. In two such experiments, using the mildest mechanical conditions to prepare the polysomes, the cascade/release ratio was 4.4 (Fig. 10) and 5.8 (not shown). The sedimentation coefficients calculated by the methods of McEwen (1967) were 270 and 290 S. The number of microsomes in the polysomes can also be calculated from the sedimentation coefficient by assigning an arbitrary value of 1 to the 80 S monosome and the value for the tyrosine aminotransferase polysome which shows a maximum cascade/release ratio comes out at 6—7.

FIGURE 11

Disc polyacrylamide gel electrophoresis of the protein-cyclic [^{3}H] AMP complex prepared from the crude microsomal fraction

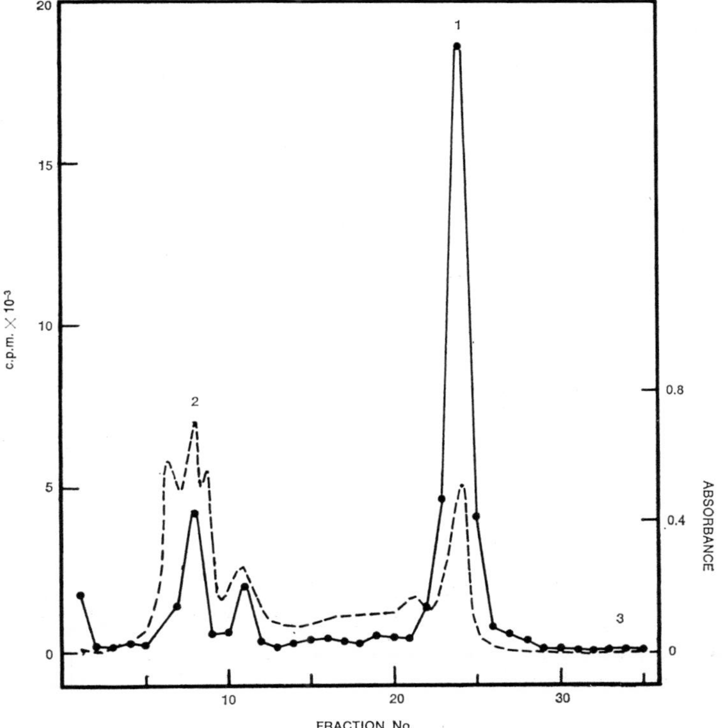

A highly concentrated protein sample was treated with a saturating level of cyclic [^{3}H] AMP and dialysed for 15 hours against two changes of 200 volumes of TM buffer to remove free radioactivity. After electrophoresis the gel was scanned at 280 nm (------) fractionated into counting vials and counted in 10 ml of Diotol (●). The sections numbered 1, 2 and 3 represent areas from which 3 mm sections of gel were taken to be tested for tyrosine aminotransferase release activity.

(Reproduced by permission from Donovan and Oliver, 1972; *Biochemistry*, in press, Fig. 1.)

This independent value is in remarkable agreement with the maximum cascade/release ratio of 5.8—6.2 and indicates a probable value of six ribosomes. Such a polysome is far too small to code for a protein of mol.wt. 115,000 and could code only for a protein of mol.wt. 25,000—30,000. This would indicate that the enzyme has 4 subunits. This is in agreement with the number of pyridoxal phosphate binding sites, and other workers have also suggested a tetrameric structure for this enzyme (Valeriote *et al.*, 1969; Auricchio *et al.*, 1970).

The microsomal release factor which strongly binds cyclic AMP has more recently been partially purified (Donovan and Oliver, 1972). Fig. 11 shows that the binding factor has a high electrophoretic mobility in polyacrylamide gel. The fast running binding peak was also shown to correspond to the release factor by testing various portions of the sectioned gel for release activity. Accordingly, preparative gel electrophoresis was used to purify the factor.

FIGURE 12

Zone ultracentrifugation of the protein-cyclic [^{3}H] AMP complex

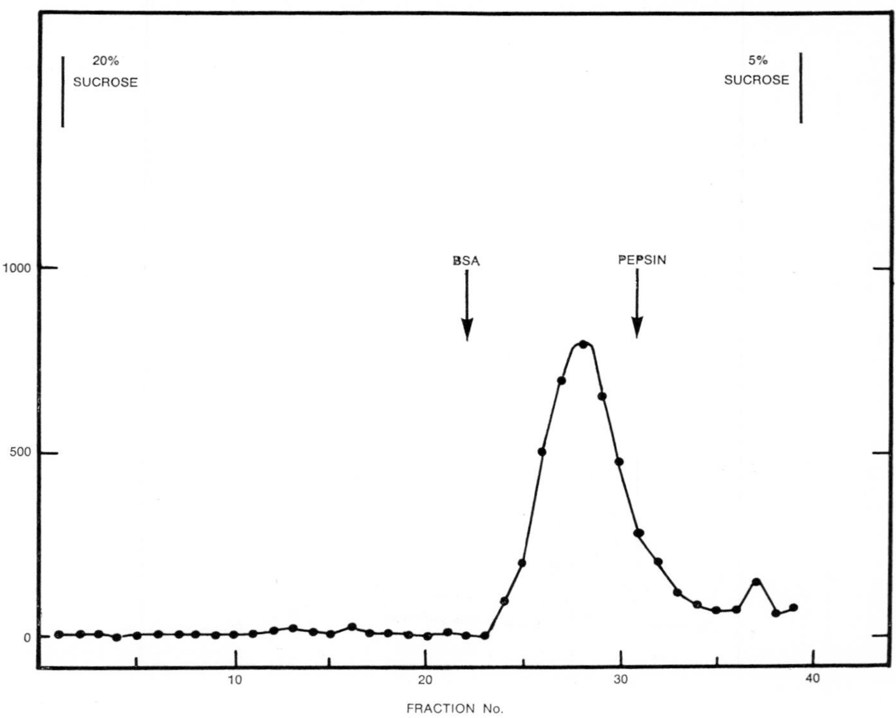

The electrophoretically purified protein was exposed to a saturating concentration of cyclic [^{3}H] AMP and dialysed for 16 hours against 2 changes of 200 volumes of TM buffer to remove free radioactivity. Centrifugation was for 12 hours at 56,000. After fractionation into counting vials the fractions were counted in 10 ml of diotol. Six drop fractions were collected. Bovine serum albumin and pepsin were used as markers.

(Reproduced by permission from Donovan and Oliver, 1972; *Biochemistry*, in press, Fig. 4.)

The binding capacity of the purified factor is completely destroyed by mild heat treatment (10 min at 50°C) or incubation with bacterial pronase. It is not at all affected by incubation with high concentrations of RNAase or DNAase.

The binding protein can be readily followed in zone centrifugation by the radioactivity of bound [^{3}H] cyclic AMP and Fig 12 show the centrifugal profile of the purified protein. The mol.wt. is 40,000. Fig. 13 shows a Scatchard plot of binding determined by equilibrium dialysis. The intrinsic association constant is 3.5×10^{-9} M.

FIGURE 13

Inset: The interaction of cyclic [^{3}H] AMP with the release protein as a function of the equilibrium concentration of the nucleotide

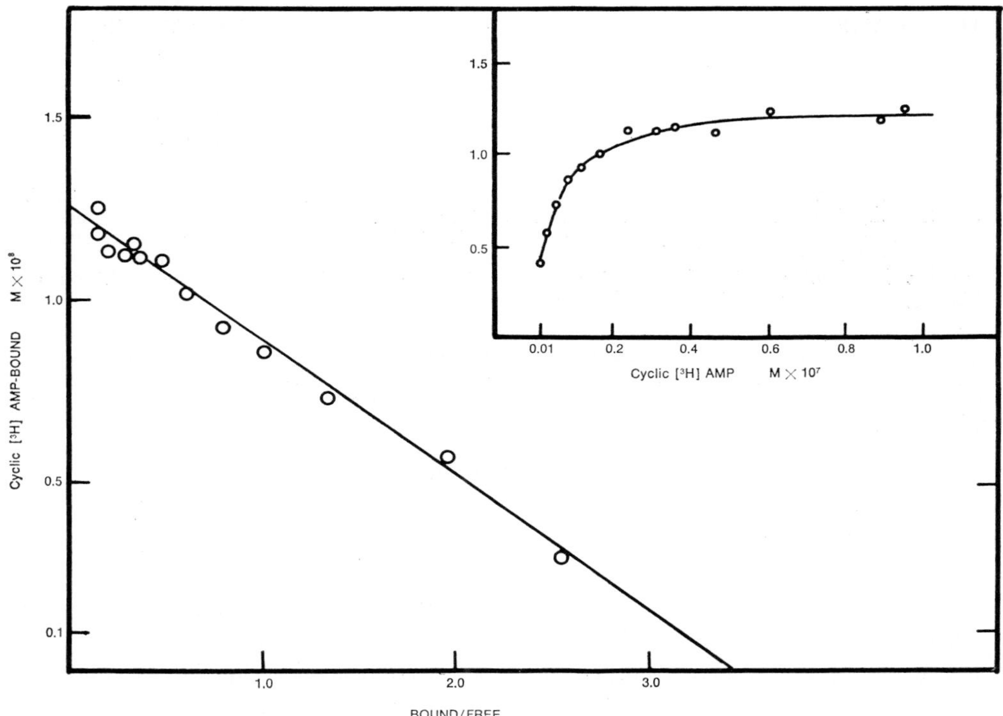

Concentrations of bound nucleotides were standardized to equal protein concentrations (180 mg/l). A Scatchard plot of the same data using a linear regression yielded an intrinsic association constant of 3.6×10^{-9}M.

(Reproduced by permission from Donovan and Oliver, 1972; *Biochemistry,* in press, Fig. 3.)

Table 6 gives a summary of results using a range of nucleotides as competitors of cyclic AMP binding in equilibrium dialysis. It can be seen that the protein recognizes the purine ring, the 3′-5′ cyclic phosphate group and the 2′-0 atom. It is important to note that both cyclic GMP and IMP have inhibitory effects on the release of tyrosine aminotransferase which requires cyclic AMP and in

TABLE 6

Competitors of binding of cyclic AMP to the microsomal release protein

Competitor 100/1	Inhibition %	Ineffective Competitors (< 3%)
c-AMP	99	c-UMP
Monobutyryl-c-AMP	82	c-CMP
c-IMP	46	c-TMP
Dibutyryl-c-AMP	39	ADP
c-GMP	19	ATP
Deoxy c-AMP	9	2′, 3′ c-AMP
		3′ AMP
		Deoxy 5′ AMP
		Deoxy ATP
		Deoxy GMP
		5′ AMP
		5′ GMP

1 ml of the microsomal factor was dialysed against 40 ml of buffer (50 mM Tris-HCl, 8 mM theophylline, 6 mM mercaptoethanol, pH 7.4) for 16 h in the presence of the competitor nucleotide (10^{-6} M) and [^{3}H]cyclic AMP (10^{-8} M). Duplicate aliquots were dialysed against [^{3}H]cyclic AMP only. Binding was determined by the difference in radioactivity between equal volumes of solution from inside and outside the dialysis sac and the inhibition of binding by other nucleotides thus determined.

addition, that deoxy cyclic AMP is inactive in the release function. This constitutes further evidence that the binding protein and the release factors are identical.

CONCLUSIONS

The characteristics of a control mechanism in protein synthesis which operates at the final release step are rather unique and require some emphasis. Firstly, with the demonstration of the cascade phenomenon, the mechanism qualifies as a translational control, since translation is brought to a halt in the absence of cyclic AMP. Secondly, the intracellular location of all elements of the system show that it is a cytoplasmic mechanism. Thirdly, such a control operating *in vivo* may be readily mistaken for an initiation control of the type recently suggested by Tomkins and his colleagues for the control of tyrosine aminotransferase synthesis in HTC cells (Tomkins *et al.*, 1969).

Cycloheximide will blockade most, but not all of the effect since it cannot prevent the release of the finished polypeptide although the drug blocks translocation reactions on the polysome. The completeness of the blockade is determined clearly by the size of the polysome and with large polysomes a small amount of released protein might go undetected. Tyrosine aminotransferase "induction" by cyclic AMP in H-35 cells is not completely blockaded by cycloheximide (Butcher *et al.*, 1971) and this system may represent a termination control.

Fourthly, the release mechanism will result in the production of labelled protein if radioactive amino acids are made available, but only a sequential analysis of the specific protein will detect the difference between an initiation control and the release mechanism.

Cytoplasmic controls over protein synthesis at the post-transcriptional or translational level have been suggested for some time (see Harris, 1970; for review) and the release mechanism described here appears to constitute such a control device, at least for the synthesis of one specific enzyme. Further work is necessary to determine its control significance in the synthesis of other proteins.

ACKNOWLEDGEMENTS

The author is indebted to his colleagues P. G. Holt, C. C. Chuah, G. Donovan and Miss A. Armstrong, without whom much of this work could not have been done. Additional thanks are also due to Miss A. Armstrong for her willing help with figure preparation. The work was financed by grants from The Australian Research Grants Committee and the Medical Research Grants Committee of the University of Western Australia.

REFERENCES

Auricchio, F., Valeriote, F. A., Tomkins, G. M. and Riley, W. D. (1970) *Biochim. Biophys. Acta. 221,* 307

Butcher, F. B., Becker, J. E. and Potter, V. R. (1971) *Exp. Cell Res. 66,* 321

Chuah, C. C., Holt, P. G. and Oliver, I. T. (1971) *Int. J. Biochem. 2,* 193

Chuah, C. C. and Oliver, I. T. (1971) *Biochemistry 10,* 2990

Chuah, C. C. and Oliver, I. T. (1972) *Biochemistry,* in press

Civen, M., Trimmer, B. M. and Brown, C. B. (1967) *Life Sci. 6,* 1331

Csányi, V., Greengard, O. and Knox, W. E. (1967) *J. Biol. Chem. 242,* 2688

Donovan, G. and Oliver, I. T. (1972) *Biochemistry,* in press

Granner, D., Chase, L. R., Aurbach, G. D., and Tomkins, G. M. (1968) *Science, 162,* 1018

Greengard, O., Smith, M. A. and Acs, G. (1963) *J. Biol. Chem. 238,* 1548

Hager, C. B. and Kenney, F. T. (1968) *J. Biol. Chem. 243,* 3296

Harris, H. (1970) in *Nucleus and Cytoplasm,* p. 10, Clarendon Press, Oxford

Holt, P. G. and Oliver, I. T. (1968) *Biochem. J. 108,* 333

Holt, P. G. and Oliver, I. T. (1969a) *Biochemistry 8,* 1429

Holt, P. G. and Oliver, I. T. (1969b) *FEBS Lett. 5,* 89

Holt, P. G. and Oliver, I. T. (1971) *Int. J. Biochem. 2,* 212

Holten, D. D. and Kenney, F. T. (1967) *J. Biol. Chem. 242,* 4372

Iwasaki, Y. and Pitot, H. C. (1971) *Life Sci. 10,* 1071

Kenney, F. T. (1963) Advan. *Enzyme,* Reg. *1,* 137

McEwen, C. R. (1967) *Anal. Biochem. 20,* 114

Sereni, F., Kenney, F. T. and Kretchmer, N. (1959) *J. Biol. Chem. 234,* 609

Tomkins, G. M., Gelehrter, T. D., Granner, D., Martin, D. Jr., Samuels, H. H. and Thompson, E. B. (1969) *Science, 166,* 1474

Valeriote, F. A., Auricchio, F., Tomkins, G. M. and Riley, W. D. (1969) *J. Biol. Chem. 244,* 3618

Wicks, W. W. (1968a) *Science 160,* 997

Wicks, W. W. (1968b) *J. Biol. Chem. 243,* 900

Wicks, W. W. (1969) *J. Biol. Chem. 244,* 3941

"The Mechanism and Control of the Biosynthesis of α-Lactalbumin by the Mammary Gland".

P. N. CAMPBELL
Department of Biochemistry, University of Leeds, England

The mammary gland is potentially an excellent system for the study of the control of protein synthesis in eukaryotic cells. Not only can one study the way in which the synthesis of specific protein is switched on but also the events which occur at the cessation of milk secretion. The influence of hormones on these activities is important and for this purpose several excellent systems employing explants have been devised. Our aim has been to study the mechanism of control of protein synthesis at the level of the ribosome and for this purpose we have chosen to study the synthesis of a specific milk protein, namely α-lactalbumin (αLA), in extracts of guinea-pig mammary gland.

PROPERTIES AND BIOLOGICAL ACTIVITY OF α-LACTALBUMIN.

The original choice of this protein was something of an accident. Our object was to study the synthesis of a specific protein in extracts of mammary gland. We chose the guinea-pig because of its convenient size, the fact that you can fairly easily collect milk from it and because Fraser and Gutfreund (1958) had already shown that active subcellular fractions could be prepared from the lactating gland. We decided on a whey protein and planned to follow the synthesis of β-lactoglobulin. In fact we soon found that guinea-pig whey did not contain this protein but contained only the protein that occurs in most milks as a minor constituent, α-lactalbumin (Brew and Campbell, 1967a). The protein was purified and its synthesis both in slices and crude subcellular extracts of the lactating gland was studied. At that time the physiological role of αLA was not known. Brew and Campbell (1967b) from a knowledge of its general properties and amino acid composition suggested that αLA was a modified lysozyme. This suggestion has stimulated a considerable amount of work but I do not wish to pursue that particular aspect to-day. It was Brodbeck and Ebner (1966) and Brodbeck *et al.* (1967) who first showed that αLA had a role in the synthesis of lactose from UDP galactose and glucose. The basic scheme as now understood is as follows:

(1) UDP- galactose + N-acetylglucosamine $\xrightarrow{\text{'A' protein}}$ N-acetyllactosamine + UDP

(2) UDP- galactose + glucose $\xrightarrow{\text{'A' + 'B' proteins}}$ lactose + UDP.

αLA was designated the 'B' protein of lactose synthetase. Brew *et al.* (1968) showed that the 'A' protein had enzymic activity in its own right and was N-acetylglucosamine galactosyl-transferase.

It should in principle, therefore, be possible to assay for the presence of small quantities of αLA in subcellular fractions such as are used for the synthesis of protein under *in vitro* conditions. One of our aims is to test for the synthesis of αLA by the post-mitochondrial supernatant from Krebs II ascites tumour cells in the presence of RNA isolated from mammary gland. Dr. P. J. Davey has been attempting to work out the conditions for such an assay in our laboratory.

'A' protein has been prepared from cow's milk by methods involving affinity chromatography devised by Brew (personal communication). The assay for αLA is then done in the presence of glucose, UDP- [^{14}C] galactose and 'A' protein

FIGURE 1

The relationship between amount of guinea-pig α-lactalbumin and the amount of lactose synthesized from UDP-galactose and glucose in the presence of bovine 'A' protein

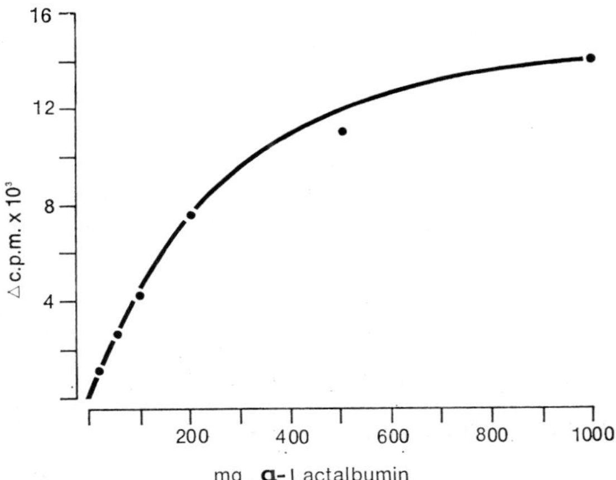

The assay was modified from the method of Brew *et al.* 1968 as suggested by Brew (Personal communication).

The assay system contained the following in a final volume of 50 μl:— 20 μl, of sample containing 'A' protein: 40 mM-N-acetylglucosamine; 11 mM-$MnCl_2$; 36 mM-MOPS [3-(*N*-Morpholino) propanesulphonic acid] pH 7.4; 0.6 mM-UDP-Galactose; 0.1% BSA, and 12,000 c.p.m. as UDP-[^{14}C]Galactose. Incubation was at 37°C for 10 min and was stopped by the addition of 50 μl of 200 mM-EDTA.

The entire sample was transferred to a Dowex 50 column in the H^+ form and washed through with water. The eluate was collected into scintillation vials and counted in 10 ml of a scintillation mix consisting of Toluene 54% v/v, Triton X100 46% v/v, 5.4 g/l PPO.

and the amount of [^{14}C] lactose formed is determined essentially according to the method of Babad and Hassid (1966). The relationship between the amount of αLA present and the amount of ^{14}C recovered in lactose is shown in Fig. 1. The assay at the level of αLA that we wish to determine is difficult for the following reasons. The 'A' protein tends to be unstable, the 'A' protein preparation must be free of αLA, the 'A' protein preparation must also be free of N-acetylglucosamine for the assay does not differentiate between N-acetyllactosamine and lactose and the transferase reaction is very much more active than the synthetase. There is also an added complication in the use of bovine 'A' protein since its affinity for guinea-pig αLA is much less than for bovine αLA. In view of these problems the control reaction in the absence of added αLA is much more significant than is desirable. For the measurement of αLA added to or present in the ascites tumour cell extract it is first necessary to separate the protein containing αLA from the other constituents which would interfere with the assay. This was done by the method described by Michael (1961) for the isolation of serum albumin by precipitation with HCl-methanol.

Table 1 shows some typical results for the assay of αLA added to the post-mitochondrial supernatant of ascites tumour cells. While the recovery of activity

TABLE 1

Assay of Guinea-pig αLA from Extract of Krebs II Ascites Tumour Cells to Which it Had Been Added

Amount of αLA added per ml of PMS	Activity measured in protein extracted from PMS		Total αLA activity recovered	Percent recovery of αLA
(ng)	(c.p.m.)	(ng αLA)		
1000	1490	35	525	53
500	406	10	150	30
100	336	7.5	56	56

Known quantities of guinea-pig αLA were added to 1 ml samples of Krebs II ascites post-mitochondrial supernatant (PMS). The αLA was recovered from the PMS according to the method described for serum albumin by Michael (1961).

To 1 ml of the PMS were added 10 ml of 0.02 M-HCl in methanol at 0°C. Precipitation was allowed to occur for 30 min with occasional stirring. The precipitate was collected by centrifugation and washed once with 2 ml of 0.02 M-HCl in methanol. The combined supernatants were brought to pH 5.9-6.0 with 0.2M-diethylamine in methanol. The precipitate was recovered and washed once with 100% methanol and once with ether. When dry the precipitate was dissolved in 0.2 ml 1mM-NH_4HCO_3 and aliquots were assayed by measuring the stimulation of lactose synthesis.

due to αLA varies we consider that the method should be useful as a qualitative test for the presence of αLA in incubation mixtures effecting the synthesis of protein. We are now planning to improve the methods by the use of affinity columns containing antibody to αLA and expect to be able to assay amounts of αLA of the order of 50 ng.

THE ROLE OF RIBOSOMES IN THE MAMMARY GLAND IN THE SYNTHESIS OF αLACTALBUMIN.

Brew (1969) has suggested how two proteins may interact within the cells in the pregnant and lactating mammary glands for the synthesis of αLA. This is illustrated in Fig. 2. This shows that the transferase, 'A' protein, is located in the

FIGURE 2

The role of membrane-bound polyribosomes in the synthesis of α-lactalbumin and the control of lactose synthesis in the mammary gland. Based on the work of Brew (1969).

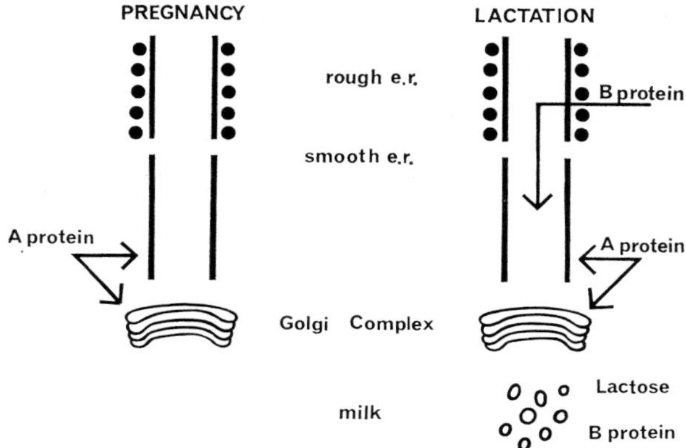

smooth-surfaced endoplasmic reticulum including the Golgi Complex and is active in the pregnant gland for the synthesis of glycoprotein. In the lactating gland the membrane-bound ribosomes in the rough-surfaced endoplasmic reticulum synthesize αLA and this passes down through the smooth-surfaced endoplasmic reticulum where it interacts with 'A' protein and glucose to cause the formation of lactose. At the cessation of lactation the synthesis of αLA ceases and the remaining protein is removed by passing into the milk. Brew suggests that the release of the αLA into the milk is a useful way of rapidly switching off the synthesis of lactose.

THE ACTIVITY OF POLYRIBOSOMES ISOLATED FROM LACTATING MAMMARY GLAND.

It is not difficult to obtain a satisfactory preparation of polyribosomes from lactating gland. Such preparations are active for the incorporation of radioactive amino acids into protein under the usual incubation conditions.

We have tested the ability of polyribosomes from lactating glands to synthesize αLA (Fairhurst *et al.*, 1971). The soluble supernatant, obtained after incubation after addition of carrier αLA, was fractionated on columns of Sephadex G-100, CM-cellulose and DEAE-cellulose. The results are shown in Fig. 3. As a final procedure for the identification of radioactive peptides characteristic of αLA the protein was cleaved with cyanogen bromide. The single polypeptide chain of

FIGURE 3

Purification of α-lactalbumin from the soluble medium after incubation of polyribosomes from mammary glands of lactating guinea-pigs in the presence of ATP

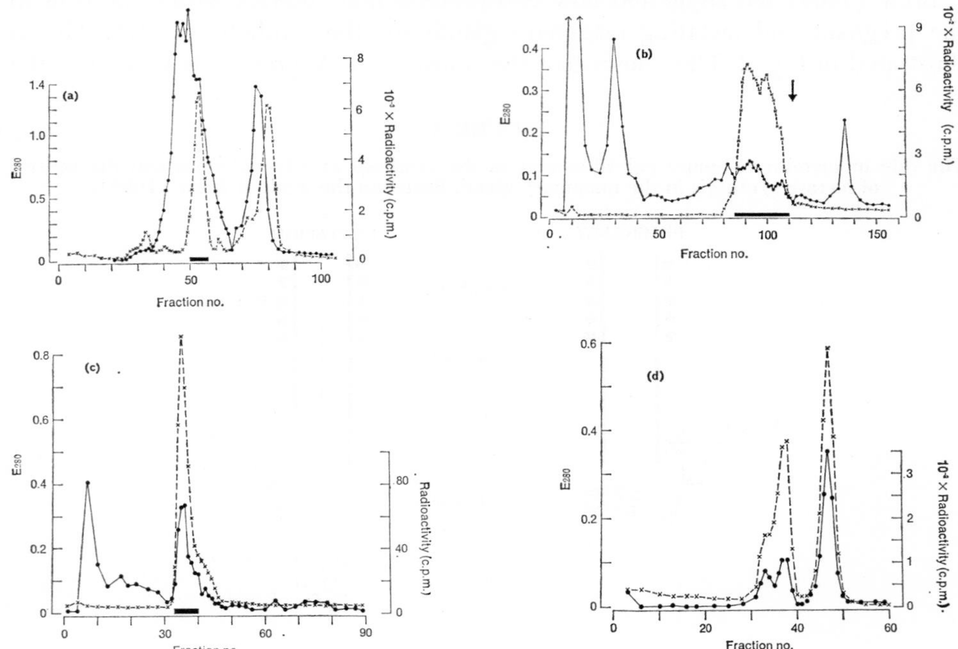

×, E_{280}: •, radioactivity (c.p.m.). In all cases fractions (indicated by the bar) were pooled for further purification. (a) Elution from a column of Sephadex G-100. Samples (0.2ml) of each fraction (8.8ml) were taken for the determination of radioactivity. (b) Elution from a column of CM-cellulose; samples (0.2 ml) of each fraction (6.0 ml) were taken for determination of radioactivity. The arrow indicates the point at which 0.5M-NaCl in 15 mM-sodium acetate buffer, pH 5.0 was applied as eluent. (c) Elution from a column of DEAE-cellulose. Samples (0.5ml) of each fraction (5.8ml) were taken for determination of radioactivity. (d) Fractionation of cyanogen bromide fragments of α-lactalbumin on a column of Bio-Gel P10. The peak at fraction 47 is the C-terminal fragment, that a fraction 38 the N-terminal fragment and that at fraction 33 undegraded α-lactalbumin. Samples (1.0 ml) of each fraction (3.5ml) were taken for the determination of radioactivity.

123 amino acids contains only a single methionine residue at position 90 so that cleavage gives two fragments, an N-terminal of 90 amino acids and a C-terminal of 33 amino acids. The two fragments can be separated on a column of Bio-Gel P10, or G-50 Sephadex as shown in Fig. 3. There is a good coincidence of the radioactivity and absorption profiles indicating that the two peptides characteristics of αLA contained [^{3}H] leucine. The incorporation was dependent on the presence of an energy source during the incubation. Since the N-terminal fragment contains 10 leucines and the C-terminal fragment 4 leucines (Brew personal communication) it was possible to calculate the specific radioactivity of the leucines in the two fragments and show that the C-terminal fragment leucines were on average nine times as active as those of the N-terminal fragment. This is the expected result if protein is synthesized from the N to C terminus.

As a check on the procedures αLA was also synthesized in a system consisting of slices of mammary gland and the results of the analysis of the cyanogen bromide fragments are shown in Fig. 4. In this case the ratio of the specific radioactivity of the leucines in the residues was 1:1.09.

We conclude from these results that the polyribosomes isolated from lactating mammary gland were able to effect the synthesis of at least part of the polypeptide chain of αLA.

FIGURE 4

Separation of cyanogen bromide fragments of α-lactalbumin isolated from slices of mammary gland of lactating guinea-pigs after incubation with [³H] leucine

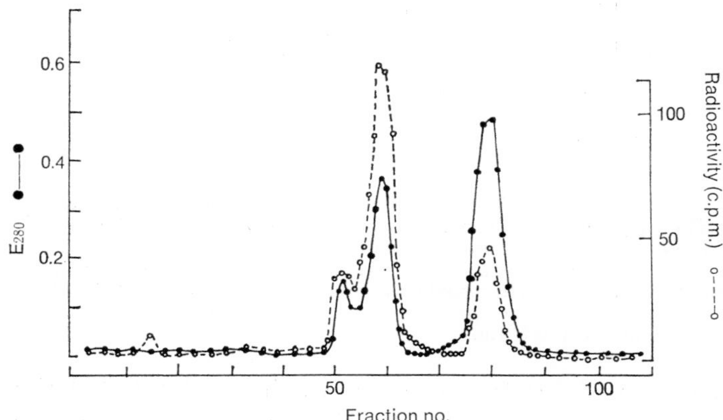

Slices from mammary gland were incubated in a bicarbonate medium for 4h. The tissue was then frozen and thawed, carrier αLA added and the radioactive protein purified from the soluble supernatant. After treatment with cyanogen bromide the fragments were fractionated on a column of Sephadex G-50. The peak at fraction 80 is the C-terminal fragment, that at fraction 57 the N-terminal fragment and that at fraction 52 undegraded αLA. Samples (1.0 ml) of each fraction (3.5 ml) were taken for determination of radioactivity.

A COMPARISON OF THE ACTIVITY OF POLYRIBOSOMES FROM PREGNANT AND LACTATING MAMMARY GLAND.

It has not proved possible to make preparations from pregnant gland that contain large polyribosomes. A comparison of the polyribosome profiles from pregnant and lactating gland is shown in Fig. 5. This result could either mean that the polyribosomes originally present in the cell were degraded during the isolation procedure or were never present. We have tried many different methods for the isolation of polyribosomes which are designed to inhibit any nucleases if they be present but without effect on the polyribosome profile. I will refer to this question again later.

Table 2 shows a comparison of the activity of the polyribosome preparations from pregnant and lactating glands. It will be seen that the ribosomes in the absence of poly(U) are considerably less active when isolated from the pregnant gland than from the lactating glands. Just in case this result might have been due

FIGURE 5

A comparison of the profiles on sucrose density gradients of polyribosome preparations from the mammary glands of pregnant and lactating guinea-pigs

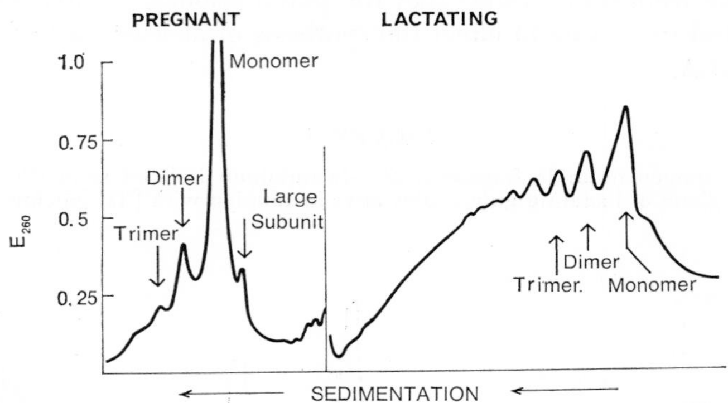

Polyribosomes were prepared from the mammary gland of guinea-pigs as described by Fairhurst *et al.* (1971). The polyribosomes in a buffer containing 50mM-tris, 25mM-KCl and 5mM-Mg acetate were layered onto a sucrose density gradient (15 — 50%) made up in the same buffer. Centrifugation from right to left was in the SW 25 rotor of the Beckman L2/65B centrifuge for 4h at 25,000 r.p.m. at 2°C. The profile was obtained by measuring the E_{260} in a recording spectrophotometer by pumping the gradients from the bottom of the tube through a flow cell at a constant rate of about 1 ml/min.

to the procedure used for the isolation, we have also tested the activity of unfractionated post-nuclear supernatant. The supernatant from pregnant glands contained much less ribosomal-RNA than that from lactating glands but on an RNA basis the pregnant preparation was less active than that from lactating glands. In the presence of poly(U) it will be seen that the ribosomes from pregnant glands were stimulated much more than were those from lactating glands, suggesting a difference in the availability of endogenous m-RNA.

We have examined further the possible differences in the ribosomes from the two types of gland in two ways. Firstly, we have prepared subunits and then

TABLE 2

Comparison of Protein-synthesizing Activity of Polyribosomes from the Mammary Gland of Lactating and Pregnant Guinea-pigs

	No poly (U)	poly (U)	Ratio
Lactating Gland	700	2732	3.9
Pregnant Gland	340	7120	21

Polyribosome preparations (0.22mg) were incubated with [^{14}C] phenylalanine, ATP (2mM), GTP (0.25mM), PEP (10mM), $MgSO_4$ (14mM), cell sap (0.2ml) in a total volume of 1 ml. Poly (U)(100 μg) was added as indicated. Incubation was for 60 min at 37°C. Results are expressed as c.p.m. of protein 100 μg of ribosomes. The radioactivity of the protein extracted at zero-time incubations has been subtracted in all cases.

made hybrids of ribosomes using subunits from the two different sources. The subunits were prepared essentially according to the method of Lawford (1969) which involves "running off" the ribosomes in the presence of puromycin and then separating the subunits by centrifugation on sucrose density gradients in the presence of 0.15M-KCl and 1 mM-$MgSO_4$. The activity of such hybrid ribosomes is shown in Table 3. The results indicate that so far as the synthesis of poly-

TABLE 3

Protein-synthesizing Activity of Ribosomal Subunits from Mammary Glands of Lactating and Pregnant Guinea-pigs

Source of ribosomes		Contents of incubation mixture							
Mammary gland of lactating guinea-pig	Small subunit	+		+				+	
	Large subunit		+	+					+
Mammary gland of pregnant guinea-pig	Small subunit				+		+		+
	Large subunit					+	+	+	
Total radioactivity (c.p.m.)		132	237	1580 (369)	87	194	1583 (281)	1437 (326)	1553 (324)

The ribosomal subunits were prepared and incubated as described by Lawford (1969). The incubation was in the presence of [^{14}C] phenylalanine at a Mg^{2+} concentration of 14mM. The ratio of the amount of large subunit to small subunit incubated was always 2.3:1 (i.e. 0.172 mg of large subunit and 0.075 mg of small subunit). The radioactivity of the protein obtained from the incubation mixture in the absence of ribosomes was subtracted from the results. The numbers in parentheses indicate the straight addition of the activity for the various combinations of subunits incubated independently. There was a complete dependence of activity on the presence of poly (U) (100 μg).

phenylalanine in the presence of poly(U) is concerned there is no defect in the ribosomal subunits from the pregnant gland. We realise that this result must be interpreted with caution and that we cannot conclude that there is necessarily no defect in respect to the translation of endogenous m-RNA.

FIGURE 6

The eukaryotic ribosome cycle.

(*Based on the work of Falvey and Staehelin*, 1970).

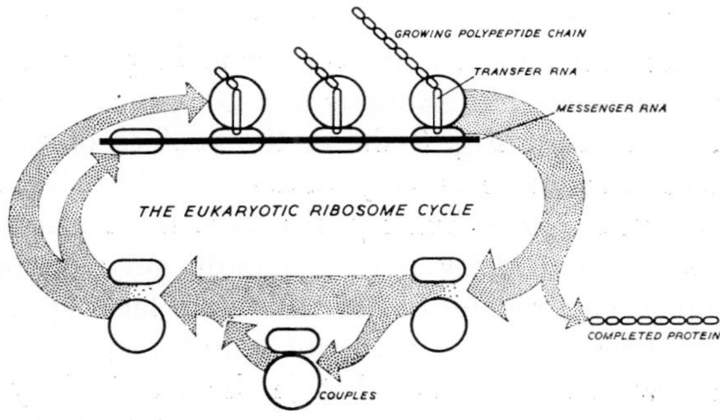

Secondly, Peter Ashby in our laboratory has used the method devised by Falvey and Staehelin (1970) for studying the reversible dissociation of ribosomes into their subunits. One of the ideas arising from their work and that of Hogan and Korner (1968) on liver ribosomes is that some of the 80S ribosomes in a cell represent a pool of subunits that can be drawn on if the rate of protein synthesis increases. We may call the ribosomes in this pool "couples", see Fig. 6. One can differentiate between "couples" and monomeric ribosomes with nascent protein chains attached to them, by the conditions under which they undergo dissociation to subunits. Thus, couples will dissociate in a buffer containing 300mM-KCl and 3mM-$MgSO_4$ whereas monomeric ribosomes with nascent chains attached are stable under these conditions. Fig. 7. shows the behaviour of the

FIGURE 7

The effect of salt concentration on the dissociation of a polyribosome preparation from the mammary gland of pregnant guinea-pigs

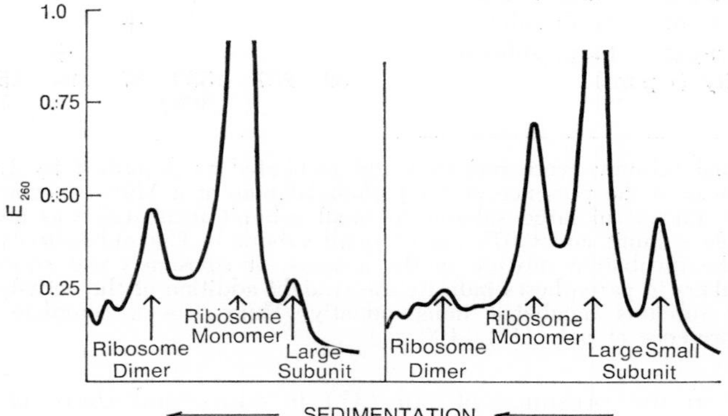

Polyribosomes were prepared from the mammary gland of pregnant guinea-pigs as described by Fairhurst *et al.* (1971). The polyribosomes in buffer were layered onto a sucrose density gradient (15-35%) made up in the same buffer. Centrifugation was from right to left and was in the SW 27 rotor of the Beckman L2/65B centrifuge for 8h at 26,000 r.p.m. The profile was obtained as in Fig. 5. The buffer used for the profile on the *left* was 50mM-tris, 25mM-KCl and 5mM-Mg acetate and for that on the *right* 50mM-tris, 300mM-KCl and 3mM-Mg acetate.

polyribosome preparation (C-ribosomes) of pregnant glands. This profile at the usual salt concentration of 24mM-KCl and 5mM-$MgSO_4$ shows a rather high proportion of ribosome monomers and even some large subunits. Many of the ribosome monomers dissociate into subunits at a salt concentration of 300mM-KCl and 3mM-$MgSO_4$. In contrast Fig. 8 shows the profiles of polyribosomes from lactating glands and here there is very much less dissociation when the salt concentration is changed. The subunit peaks have been characterized by their content of RNA as determined by polyacrylamide gel electrophoresis. On the basis that the dimers are aggregates of 80S ribosomes lacking m-RNA then about 60% of the 80S ribosomes from the pregnant gland dissociate at the high KCl concentration. Thus we can conclude that in the terminology of the ribosome cycle virtually all the single ribosomes in the lactating gland have nascent protein associated with them and are participating in protein synthesis. In the pregnant gland about 60% of the single ribosomes are couples. Whether these

FIGURE 8

The effect of salt concentration on the dissociation of a polyribosome preparation from the mammary glands of lactating guinea-pigs

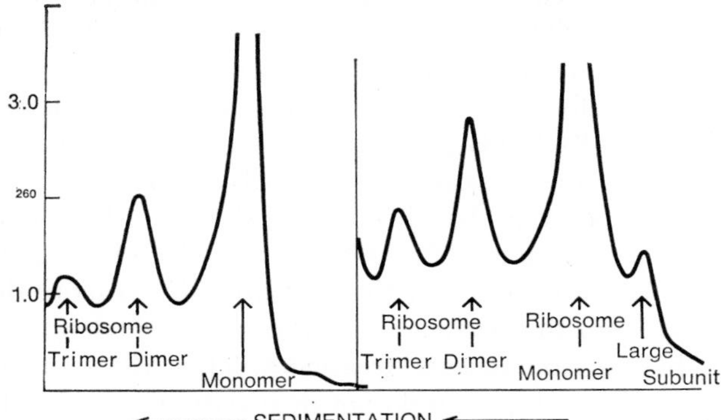

The profiles were obtained as described for polyribosomes from pregnant animals in Fig. 7 except that centrifugation was at 21,000 r.p.m. for 12h.

FIGURE 9

The effect of salt concentration on the dissociation of a polyribosome preparation from guinea-pig liver after treatment with ribonuclease

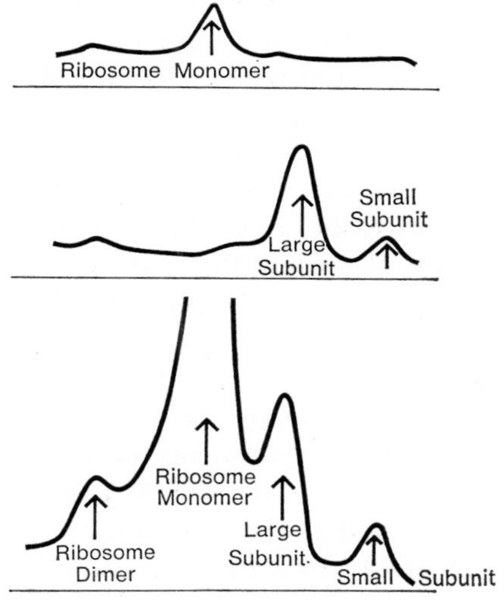

Polyribosomes were prepared from guinea-pig liver as described by Fairhurst *et al.* (1971). The polyribosomes were suspended in buffer and treated as described below. In all cases the sucrose density gradient was 15-35% and centrifugation was from right to left at 26,000 r.p.m. for 7.5h. The profiles were obtained as described in Fig. 5.

Top profile. The poly ribosomes and the density gradient contained a buffer of 50mM-tris, 25mM-KCl and 5mM-Mg acetate.

Middle profile. The polyribosomes and the density gradient contained a buffer of 50mM-tris, 300mM-KCl and 3mM-Mg acetate.

Bottom profile. The polyribosome suspension (1ml) was treated with 5μl of a solution of pancreatic ribonuclease at 1 mg/ml before layering onto a sucrose density gradient. The buffer for the polyribosomes and gradient was as for the Middle profile.

couples cannot associate with m-RNA because of its absence or whether they are defective in that they are lacking a factor is not at present clear. Our previous results suggest that they can certainly associate with poly(U).

Fig. 9 shows the behaviour of liver ribosomes under various conditions. In the top profile a polyribosome preparation has been centrifuged until all the polyribosomes have been pelleted and only the monomer is present in the gradient at 25mM-KCl. At 300mM-KCl this monomer dissociates completely into subunits showing that the monomeric peak consisted of couples. In the bottom profile the polyribosome preparation has been treated with ribonuclease to produce monomers with small fragments of m-RNA attached and no doubt also nascent protein. At 300mM-KCl one sees that there is virtually no more dissociation than would be expected from the presence of the couples which would be present in the preparation before ribonuclease treatment. Thus the monomeric ribosomes produced by ribonuclease digestion do not dissociate in the high KCl buffer. This suggests that the failure to obtain polyribosomes from the mammary gland of pregnant animals is not due to their degradation by nucleases but because they are not present in the gland.

ATTEMPTS TO DETECT m-RNA FOR αLACTALBUMIN.

We postulate that the S value of m-RNA for αlactalbumin with a molecular weight of 14,800 would be about 8. In a search for such an RNA, pregnant and lactating animals have been injected with [^{32}P] phosphate and the RNA isolated from the glands. The various species of RNA have been separated on polyacrylamide gel. Labelled RNA in the 10S region has been consistently found but it has appeared in extracts from the glands of both the pregnant and lactating animals. This does not seem to be a profitable procedure for the detection of m-RNA in this case. We are also using the post-mitochondrial supernatant from Krebs II ascites tumour cells to test for the presence of m-RNA (see e.g. Mathews and Korner, 1970). Although various fractions have stimulated the incorporation of radioactivity into protein the product has not yet been characterized.

Another approach is to inject *Xenopus* oocytes with either polysomes or RNA. I wish to report on a series of experiments performed in collaboration with Dr. Lane in Dr. Gurdon's laboratory at Oxford. Together with Marbaix (Lane, *et al.*, 1971) they have shown that the injection of either polyribosomes from rabbit reticulocytes or 9S RNA from the same source causes the synthesis of rabbit haemoglobin in the oocytes. Moreover they have also shown that a 9-13S fraction of RNA from K-41 mouse myeloma cells is also translated by the frog oocytes (Gurdon, *et al.*, 1971). In view of the ability of polyribosomes from lactating mammary gland to synthesize αLA we have prepared such polyribosomes and Dr. Lane has injected them into oocytes and then incubated them in a medium containing [^{3}H] leucine. We have also prepared an RNA fraction from lactating polyribosomes by extraction with SDS and phenol. This too has been injected into oocytes. After incubation for various times the oocytes were disrupted, the supernatant collected and carrier αLA added. The proteins were then fractionated on Sephadex G-100, CM-cellulose and in some cases the αLA was

FIGURE 10

The separation of αLA from the soluble supernatant of disrupted oocytes after they had been injected with either buffer, polyribosomes from the mammary gland of lactating guinea-pigs or RNA extracted from these polyribosomes.

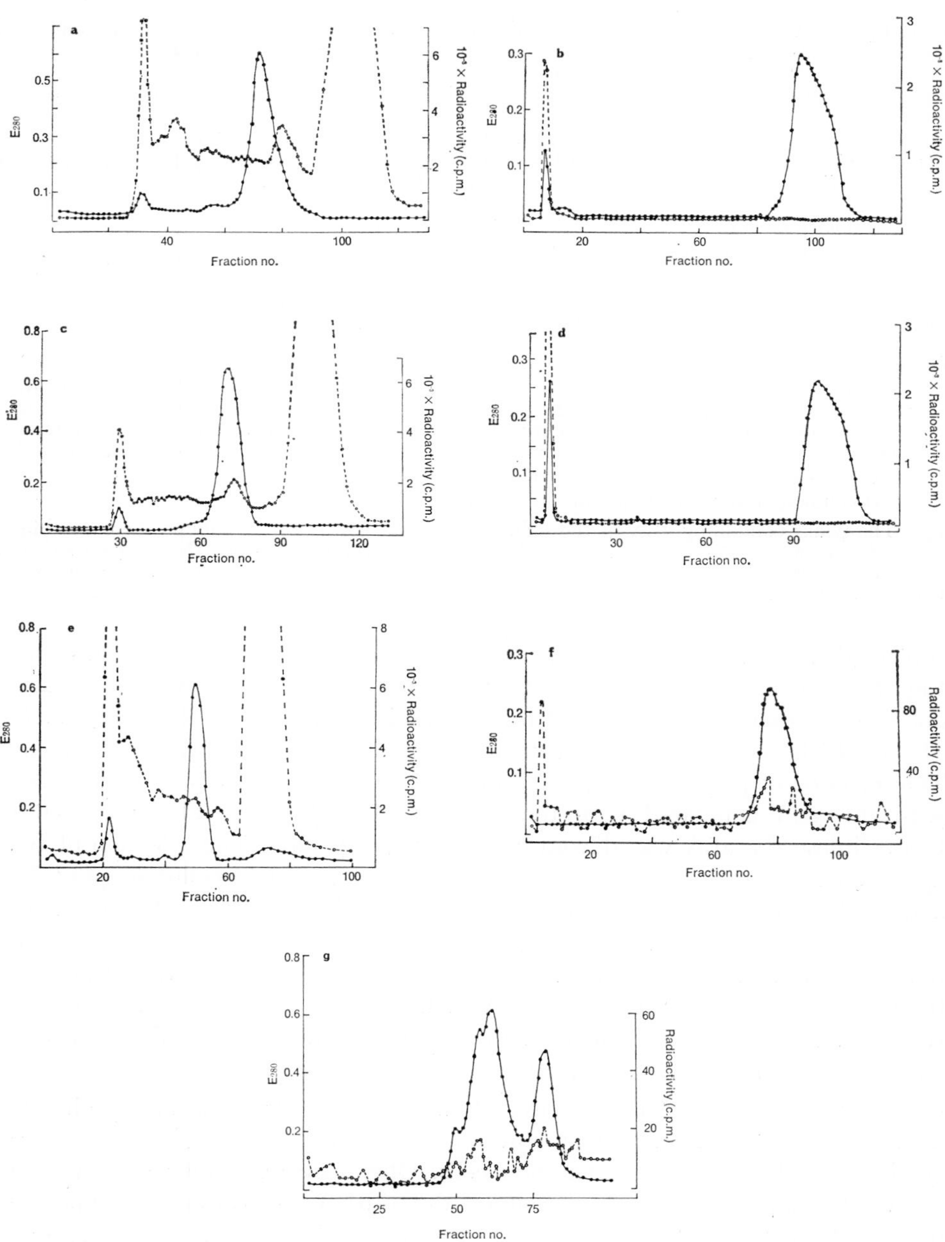

O, radioactivity c.p.m. ●, absorption E_{280}.

Buffer Injection (Fig 10)

(a) About 30 oocytes were injected with 1 μl of buffer as described by Lane *et al.* (1971). After incubation in the presence of [^{3}H] leucine for periods varying from 4 to 16 h at 25°C the cells were frozen.

Later they were thawed, suspended in 1 ml of 0.05 M-NH_4HCO_3, disrupted in a homogeniser, centrifuged at 3000 *g* for 15 min and the supernatant carefully removed. Carrier guinea-pig αLA (25 mg) was added and the radioactive proteins fractionated on a column of Sephadex G-100 in 0.05M-NH_4HCO_3. Fractions were collected, the E_{280} measured and the radioactivity determined as described for Fig. 3. The major peak of absorption at E_{280} represented αLA and the fractions under the peak were collected and subjected to further purification.

(b) Purification of αLA on a column of CM-cellulose in 0.015M sodium acetate buffer at pH 5.0. A gradient of 0→0.05M-NaCl in 0.015M-sodium acetate buffer pH 5.0 was applied. Again fractions were collected for determination of E_{280} and radioactivity.

Polyribosome injection

(c) The oocytes were injected with 1 μl of a solution containing 23 mg/ml of polyribosomes from the mammary gland of lactating guinea-pigs. The oocytes were treated and the radioactive proteins fractionated as described in (a) above.

(d) CM-cellulose fractionation as for (b) above.

RNA injection

(e) The oocytes were injected with 1 μl of a solution containing 6.7 mg RNA/ml. The RNA was extracted from the polyribosomes from the mammary gland of a lactating guinea-pig which were active for protein synthesis *in vitro*. The extraction buffer contained SDS and phenol. All glassware was treated with diethylpyrocarbonate. The oocytes were treated and the radioactive proteints fractionated as described in (a) above.

(f) CM-cellulose fractionation as for (b) above.

(g) The peaks containing αLA in the effluent from the CM-cellulose column were combined, freeze-dried and treated with cyanogen bromide. The fragments were then separated on a column of Sephadex G-50 run in 5% formic acid. The peak at fraction 77 represents the C-terminal fragment and that at 63 the N-terminal fragment with the undegraded αLA behind.

subjected to CNBr degradation. The results obtained to date are shown in Fig. 10. The first peak of radioactivity on the Sephadex columns is no doubt due to the presence of oocyte proteins and the last peak due to free [^{3}H] leucine. The methods seem to separate the αLA well clear of both these contaminants. The injection of polysomes appeared to cause some radioactivity to be associated with the αLA compared with the controls (Fig. 10a and Fig. 10c) but on CM-cellulose the αLA was devoid of activity (Fig. 10d). The injection of RNA appeared to cause some activity in the αLA even after CM-cellulose (Fig. 10f) and so CNBr degradation was applied. The result (Fig. 10g) is rather equivocal. There appears to be some radioactivity associated with the fragments but the radioactivity does not coincide well with the absorption profile. We suspect that there has been some synthesis of αLA in the oocyte as a result of injection of RNA but that it is very inefficient. Further efforts will now be directed to the subfractionation of the RNA with the hope of obtaining a clearer result.

In conclusion I hope that I have shown that the study of the synthesis of αLA provides an interesting system for the study of the control of protein synthesis in differentiated eukaryotic cells. Obviously we want to try and find out whether the pregnant gland contains the m-RNA for αLA and if it does why it is not translated until the onset of lactation. Many problems also remain concerning the cessation of lactation.

ACKNOWLEDGEMENTS

I would like to acknowledge the help of my assistant Mrs Diana McIlreavy who performed in our laboratory many of the experiments described. The work was supported by a grant from The Medical Research Council.

REFERENCES

Babad, H. and Hassid, W. Z. (1966). *J. Biol. Chem. 241,* 2672

Brodbeck, U., Denton, W. L., Tanahasi, N. and Ebner, K. E. (1967). *J. Biol. Chem. 242,* 1391

Brodbeck, U. and Ebner, K. E. (1966). *J. Biol. Chem. 241,* 762

Brew, K. (1969). *Nature (London) 222,* 671

Brew, K. and Campbell, P. N. (1967a). *Biochem. J. 102,* 258

Brew, K. and Campbell, P. N. (1967b). *Biochem. J. 102,* 265

Brew, K., Vanaman, T. C. and Hill, R. L. (1968). *Proc. Nat. Acad. Sci., U.S., 59,* 491

Fairhurst, E., McIlreavy, D. and Campbell, P. N. (1971). *Biochem. J. 123,* 865

Falvey, A. K. and Staehelin, T. (1970). *J. Mol. Biol. 53,* 21

Fraser, M. J. and Gutfreund, H. (1958). *Proc. Roy. Soc. Ser. B. 149,* 392

Gurdon, J. B., Lane, C. D., Woodland, H. R. and Marbaix, G. (1971). *Nature (London) 233,* 177

Hogan, B. L. M. and Korner, A. (1968). *Biochim. Biophys. Acta. 169,* 129

Lane, C. D., Marbaix, G. and Gurdon, J. B. (1971). *J. Mol. Biol. 61,* 73

Lawford, G. R. (1969). *Biochem. Biophys. Res. Commun. 37,* 143

Mathews, M. B. and Korner, A. (1970). *Eur. J. Biochem. 17,* 328

Michael, S. E. (1961). *Biochem. J. 82,* 212

Changes in Protein Synthesis and Degradation Involved in Enzyme Accumulation in Differentiating Liver

F. J. BALLARD[a], M. F. HOPGOOD[a], LEA RESHEF[b], AND R. W. HANSON[c]
CSIRO, Division of Nutritional Biochemistry[a], Adelaide, Australia. Hadassah Medical School[b], Jerusalem, Israel, and Temple University Medical School[c], Philadelphia, U.S.A.

The growth of a mammal from embryo to adult is accompanied by differentiation so that non-specific cells eventually give rise to specific tissues. As part of this developmental process a series of sequential changes occur with the gradual attainment of the mature function characteristic of a tissue. The foetal rat liver, for example, although somewhat similar histologically to adult liver, has very few of the specific functions carried out by the mature organ. Gluconeogenesis, urea formation, amino acid degradation, glucuronide formation, sugar phosphate transformations and drug detoxication reactions are absent or essentially so in the foetal rat liver, and appear during development at birth or at weaning (Schimke and Doyle, 1970; Ballard, 1971). Although little is known about the natural stimuli for the appearance of these pathways, it is a general finding that the pathways become functional when rate-limiting enzymes appear for the first time. Thus an essential problem in understanding development is to resolve the factors responsible for specific gene expression: how are the respective genes repressed during the early growth of foetal liver and what changes initiate gene expression at a particular developmental stage?

One experimental approach to these questions has been to identify the physiological stimuli that initiate the formation of enzymes during development. For example, it has been shown that certain hormones cause increases in the activities of some enzymes in adult liver, and these activity changes are due to the appearance of new enzyme protein (Schimke *et al.*, 1965a; Kenney, 1962). Generally the hormones increased enzyme activities by stimulating the rate of enzyme synthesis; i.e. protein synthesis was accelerated and probably the rate of messenger RNA (m-RNA) formation. There have been several attempts to extend this work on hormonal induction to developmental problems, but hormones which increase the rate of enzyme synthesis in adult tissues did not cause the premature appearance of these enzymes in the foetus (Sereni *et al.*, 1959; Nemeth, 1959; Walker and Holland, 1965). Greengard (1969) recognised that a different hormone may induce the initial appearance of an enzyme as compared to activity changes in the adult. Thus glucagon or cyclic 3′5′ AMP, the latter compound administered as the dibutyryl derivative (DBcAMP), initiated tyrosine aminotransferase formation in foetal rat liver, while these agents had no effect on activity in the adult. Conversely, glucocorticoids increased the activity of tyrosine aminotransferase in adult liver but were not effective in the foetus (Greengard and Dewey, 1967).

Studies in our laboratories have centred on establishing the role of phosphoenolpyruvate carboxykinase during the initial appearance of gluconeogenesis in rat liver in the period immediately after birth. The very rapid development of this enzyme is not only of physiological importance to the newborn animal, but it also provides a model for the study of mechanisms by which the appearance of a new protein may be controlled. Some experiments with phosphoenolpyruvate carboxykinase induction are shown in Fig. 1, with the normal postnatal changes

FIGURE 1

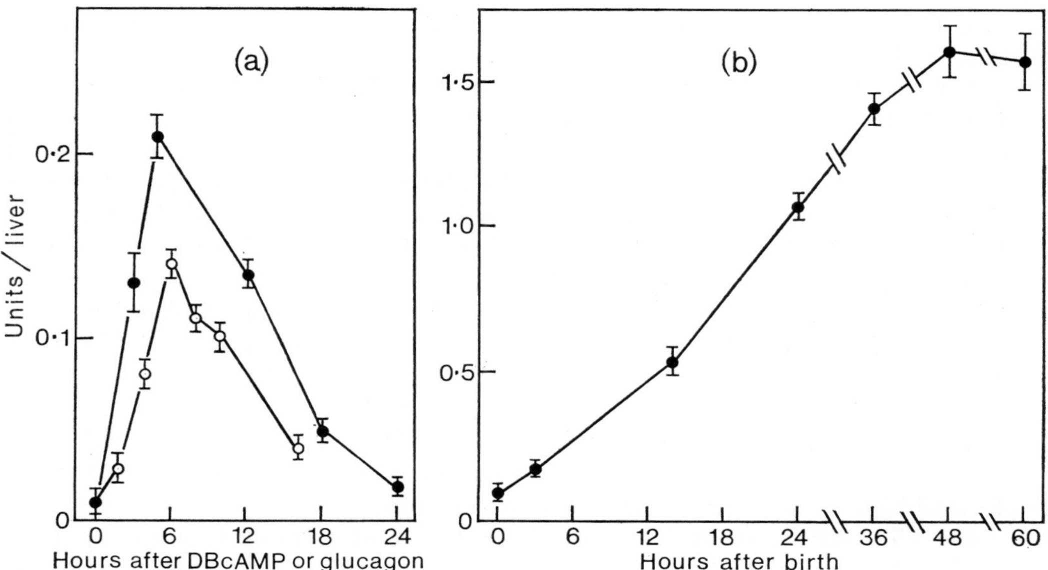

Changes in phosphoenolpyruvate carboxykinase activity in rat liver. (a) Activity in foetal liver is shown after the injection of either 1 μmol DBcAMP (●) or 100 μg glucagon (○). (b) The normal developmental pattern of enzyme. Values are means ± S.E.M. Phosphoenolpyruvate carboxykinase was measured according to Chang and Lane (1966).

in activity given as a comparison. Glucagon or DBcAMP increase the hepatic enzyme activity in 20 day foetuses but in contrast to normal development, this activity is not sustained. It is probable that the fall in activity in foetal liver is a result of hormone inactivation, because larger injection doses or multiple injections do maintain the activity at a high level. We consider that the premature induction of phosphoenolpyruvate carboxykinase caused by glucagon in foetal liver may be similar operationally to the normal postnatal development of the enzyme, because in the latter case blood glucagon levels do increase substantially (J. Girard, personal communication) while injection of glucose into the neonatal rat retards the increase in enzyme activity (Yeung and Oliver, 1968) and would probably reduce the glucagon concentration.

How do glucagon and DBcAMP act to increase the activity of phosphoenolpyruvate carboxykinase, and is the mechanism similar to that operating during the postnatal appearance phase? Theoretically it is possible for the increase in enzyme activity to be produced by either one or a combination of the following events:

1. A precursor of the enzyme may be present in foetal liver. The stimuli would activate this precursor in some undefined way to form phosphoenolpyruvate carboxykinase.
2. The rate of enzyme synthesis may be increased by an increase in the transcription rate of the relevant gene. There would thus be additional amounts of m-RNA specific for phosphoenolpyruvate carboxykinase.
3. There may be a stimulation in the translation rate or efficiency of specific m-RNA that is already present in foetal liver.
4. If the rate constant for phosphoenolpyruvate carboxykinase degradation was large in foetal liver, a reduction in this rate constant would result in more enzyme activity.

There is evidence that mechanisms 1, 2 and 4 can participate in the activity changes of various enzymes in adult liver (Schimke and Doyle, 1970), but translational control is less well established. Studies with developing liver indicate the requirement for synthesis of new RNA, since actinomycin D is an effective inhibitor of tyrosine aminotransferase, serine dehydratase, glucose 6-phosphatase (Greengard, 1969) and phosphoenolpyruvate carboxykinase (Yeung and Oliver, 1968) appearance. These experiments argue against the possibility of enzyme activation, and imply that a change in translation rate is not an important factor, at least with the enzymes mentioned. However, Chuah and Oliver (1971) find that a fraction of the postnatal increase in tyrosine animotransferase is insensitive to inhibition by actinomycin D. These authors offer evidence that part of the cyclic 3′5′ AMP stimulation is caused by a release of preformed enzyme from loaded polysomes, so that protein synthesis is not involved. This mechanism for the appearance of enzymes during development does not fit into the classification listed above, but is obviously a form of translational control.

Possibility of a phosphoenolpyruvate carboxykinase precursor in foetal rat liver

The most satisfactory method that can be used to resolve the control of phosphoenolpyruvate carboxykinase synthesis requires specific antibodies against the enzyme. With antibodies the enzyme can be quantitatively precipitated so that radioactive amino acids may be used to measure the rates of enzyme synthesis and degradation. The antibody prepared against liver cytosol phosphoenolpyruvate carboxykinase is inactive when tested with mitochondrial phosphoenolpyruvate carboxykinase (Ballard and Hanson, 1969) but has an equivalent specificity against the cytosol enzyme from liver, adipose tissue (Ballard and Hanson, 1969) and kidney (Longshaw and Pogson, 1972). To test whether a preformed antigen is present prior to the formation of enzyme activity, we have measured the amount of antigen in foetal liver with the specific antibody. The techniques used have been described in detail (Philippidis *et al.*, 1972) and are simplified by the finding that a constant volume of antibody will inactivate a fixed activity of enzyme, regardless of the enzyme concentration. This linear response is not shared by all antibody-antigen systems (Majerus and Kilburn, 1969; Knox, *et al* 1970). The experimental results (Table 1) show that only with

TABLE 1

Phosphoenolpyruvate carboxykinase antigen and activity in liver

Source of cytosol enzyme	Phosphoenolpyruvate carboxykinase	
	Activity (units/g)	Antigen (units/g)
20 day foetus	0.04 ± 0.01	0.07 ± 0.01
20 day foetus, 3 h after 50 μg glucagon	0.40 ± 0.07	0.37 ± 0.03
1 day young	3.85 ± 0.25	4.02 ± 0.36

Enzyme was assayed for activity and amount of antigen as described previously (Philippidis *et al.*, 1972). One unit of activity is the amount of enzyme that catalyses the fixation of 1 μmol of ^{14}C-labelled bicarbonate per min at 37°C. One unit of antibody inactivates 1 unit of phosphoenolpyruvate carboxykinase in the liver cytosol fraction from adult rats. Values are the mean ± S.E.M.

the uninduced state in foetal liver is there an excess of phosphoenolpyruvate carboxykinase antigen over activity, and this additional 'enzyme' only amounts to 0.03 units per g of tissue. As this is less than 10% of the enzyme activity induced 3 hours after injection of glucagon into foetal rats, the increase in phosphoenolpyruvate carboxykinase activity cannot be due to an activation of pre-existing, enzymatically inactive, antigen. This conclusion could not be extended to an inactive 'enzyme' if the material in foetal liver possessed an antigenic activity very much less than the native enzyme. Such a possibility is unlikely since the antigenic determinants would need to be effectively masked in the precursor enzyme.

Changes in the synthesis rate of phosphoenolpyruvate carboxykinase

A second possibility considered for the increase in phosphoenolpyruvate carboxykinase activity was an increase in the rate of enzyme synthesis. The rate of enzyme synthesis relative to other cytosol proteins was measured by determining the incorporation into phosphoenolpyruvate carboxykinase and total protein 40 min after an injection of [^{14}C] leucine. We note that the rate of

TABLE 2

Rates of synthesis of phosphoenolpyruvate carboxykinase in liver

	Per cent incorporation into phosphoenolpyruvate carboxykinase
20 day foetus, untreated	0.09 ± 0.06
20 day foetus, 3 h after 0.2 μmol DBcAMP	1.26 ± 0.08
20 day foetus, 3 h after 50 μg glucagon	0.80 ± 0.08
22 day foetus, untreated	0.18 ± 0.03
Newborn, 2 h	0.92 ± 0.23
Young, 2 days	2.02 ± 0.16

The relative rate of enzyme synthesis was measured after intraperitoneal injection of [^{14}C] leucine (Philippidis *et al.*, 1972). Values are expressed as the percent of radioactivity in phosphoenolpyruvate carboxykinase as compared to the radioactivity in cytosol proteins, and are given as the means ± S.E.M.

enzyme synthesis is extremely low in foetal liver (Table 2) and that glucagon or DBcAMP injected into the foetus produce a ten fold increase in synthesis rate. A similarly large increase in enzyme synthesis occurs during the postnatal period. These experiments do not distinguish between a transcriptional or translational effect on enzyme synthesis, but preliminary experiments show that the changes in the rate of synthesis do not occur when actinomycin D is injected before or with the hormones. Actinomycin D also abolishes the increase in phosphoenolpyruvate carboxykinase activity associated with birth (Yeung and Oliver, 1968). If we assume that actinomycin D is specific in its effect on RNA synthesis and it does not affect translation of m-RNA, it is likely that glucagon, DBcAMP and birth all produce an increase in the transcription rate of the gene specific for phosphoenolpyruvate carboxykinase. Complete proof would require an effective assay for the specific m-RNA concentration in liver, an experimental approach that is technically feasible (Gurdon *et al.*, 1971; Rhoads *et al.*, 1971). In this context it is significant that phosphoenolpyruvate carboxykinase makes up 2% of the proteins being formed in the liver cytosol of 2 day rats. Since the specific acitivity of this enzyme at purity is 16 units per mg of protein and it has a molecular weight of 74,000 (Ballard and Hanson, 1969), the activity of 6.5 units per g of liver in the 2 day rats represents a phosphoenolpyruvate carboxykinase concentration of 400 μg per g of liver or approximately 1 x 10^{-5}M.

Changes in the degradation rate of phosphoenolpyruvate carboxykinase

Enzyme degradation has been measured as the rate at which radioactivity disappears from the phosphoenolpyruvate carboxykinase pool after labelling with [^{14}C] or [^{3}H] leucine. We express this degradation or decay rate relative to the radioactivity in cytosol proteins, because with young animals one cannot be certain that all the radioactive leucine injected does in fact remain in the animal.

Degradation rates of phosphoenolpyruvate carboxykinase can vary with dietary or hormonal conditions (Table 3). When the activity is at a steady state, either

TABLE 3

Half times of hepatic phosphoenolpyruvate carboxykinase during steady state conditions and when the activity is falling

	Units enzyme per g	t½ (h)
Steady States:		
2 day young	5.21 ± 0.41	13
Adult refed	1.24 ± 0.09	12
Adult fasted	5.77 ± 0.31	13
Declining Activities:		
DBcAMP treated foetuses		6
Adult, fasted during refeeding		6
Adult, insulin-treated diabetics		5.5

Enzyme degradation was measured as the rate of disappearance of isotope from the phosphoenolpyruvate carboxykinase pool, except for insulin treated diabetics and DBcAMP foetuses where the half time is the rate of activity loss. Values with 2 day rats are from Philippidis *et al.* (1972) and those with insulin treated diabetics are from Reshef *et al.* (1970).

with high or low enzyme content, the t½ for degradation is approximately 13 h. During a period when the activity is falling, such as caused by refeeding a fasted adult animal, the injection of insulin to diabetics, or after the peak of activity

FIGURE 2

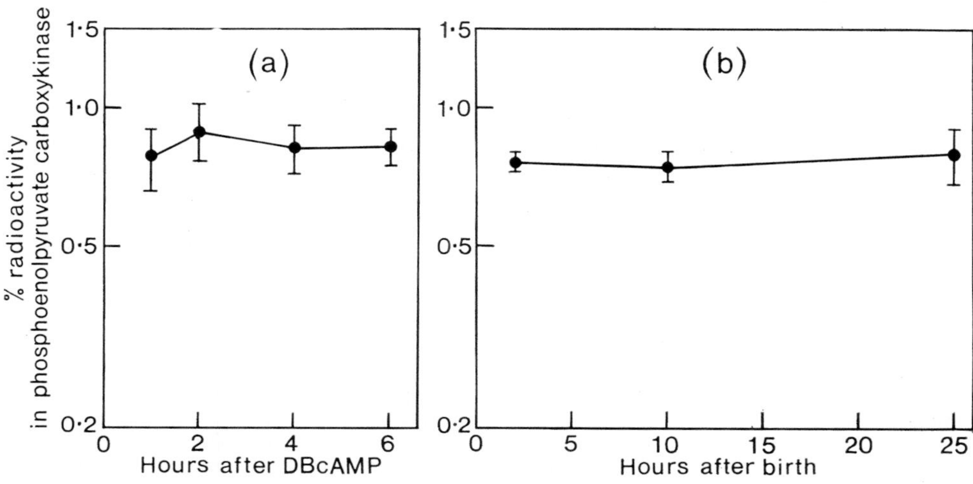

Radioactivity remaining in phosphoenolpyruvate carboxykinase at various periods after (a) injection of 0.2 μmol DBcAMP into 20 day foetal rats or (b) birth. Methods have been described in detail by Philippidis *et al.* (1972). Values are expressed as means ± S.E.M.

caused by DBcAMP injection into foetuses *in utero*, phosphoenolpyruvate carboxykinase is degraded with a t½ of 6 h. However, no radioactivity is lost when the enzyme pool is labelled with [^{14}C] leucine during conditions under which the activity is increasing (Fig. 2). Thus phosphoenolpyruvate carboxykinase is not degraded during the appearance phases caused by birth or by DBcAMP injection *in utero*.

A degrading system for phosphoenolpyruvate carboxykinase

The experiments outlined above show that degradation of enzyme occurs several hours after injection of glucagon or DBcAMP into foetal rats, but they do not establish that degradation is caused by the hormones directly, or by the fact that phosphoenolpyruvate carboxykinase begins to attain a high activity. Possibly the appearance of enzyme causes a specific or semispecific degrading system to become functional. A problem with this concept of specific degrading enzymes is that each proteinase would have to be degraded by another proteinase, etc., etc., an idea that is absurd (see Schimke and Doyle, 1970). Nevertheless, there is evidence that proteinases or cathepsins show some substrate specificity. For example, (Katunuma *et al.*, 1971a, 1971b) have isolated an enzyme that specifically inactivates pyridoxal enzymes and another enzyme specific for NAD-dependent dehydrogenases. In another study, Otto (1971) demonstrated the inactivation of glucokinase, aldolase and pyruvate kinase by cathepsin B1 from liver, while some 15 other enzymes were not affected.

TABLE 4

Protein degradation in newborn rat liver *in vitro*

Sample	Measurement	Incubation time 0	4 h
Liver slice			
	cytosol protein (mg/g)	66	66
	phosphoenolpyruvate carboxykinase	1.11	0.85
	lactate dehydrogenase	321	285
	glucose 6-phosphate dehydrogenase	4.34	3.44
Homogenate			
	cytosol protein (mg/g)	59	47
	phosphoenolpyruvate carboxykinase	1.11	0.57
	lactate dehydrogenase	307	241
	glucose 6-phosphate dehydrogenase	5.63	1.42

Livers from animals 3 h after birth were either sliced and incubated with shaking in Krebs-Ringer-bicarbonate containing 20 mM glucose or homogenized in 0.25 M sucrose and incubated. The gas phase was 95% O_2, 5% CO_2 and the temperature was 37°C. After incubation, slices were homogenized in the incubation fluid and all homogenates centrifuged at 100,000 g for 30 min to obtain cytosol fractions. Assays were carried out according to the following methods: phosphoenolpyruvate carboxykinase (Chang and Lane, 1966); lactate dehydrogenase (Kornberg, 1955); glucose 6-phosphate dehydrogenase (Ballard and Oliver, 1964); protein (Lowry *et al.*, 1951). Values are the means of three determinations and enzyme activities are expressed as international units per g tissue at 37°C.

We have measured phosphoenolpyruvate carboxykinase inactivation in slices and homogenates of liver from newborn rats in an attempt to see whether enzyme degradation can be shown *in vitro* during a period when there is no degradation *in vivo*. We find that phosphoenolpyruvate carboxykinase is inactivated under these conditions (Table 4) as are lactate dehydrogenase and glucose 6-phosphate dehydrogenase. Inactivation of the enzymes in homogenates was more complete than in slices. Although phosphoenolpyruvate carboxykinase is inactivated in the *in vitro* preparations, we cannot be sure than an inactivating system can function *in vivo*, because the tissue has been disrupted. There are several reports that enzyme inactivating systems require ATP for their effectiveness (Steinberg and Vaughan, 1956; Schimke *et al.*, 1965b; Hershko and Tomkins, 1971) but we cannot show an ATP requirement for the inactivation of phosphoenolpyruvate carboxykinase in liver slices.

A further possibility that might account for the absence of enzyme degradation during the induction periods is the complete absence of proteolysis at this time. We have tested this suggestion by using the double label technique described by

FIGURE 3

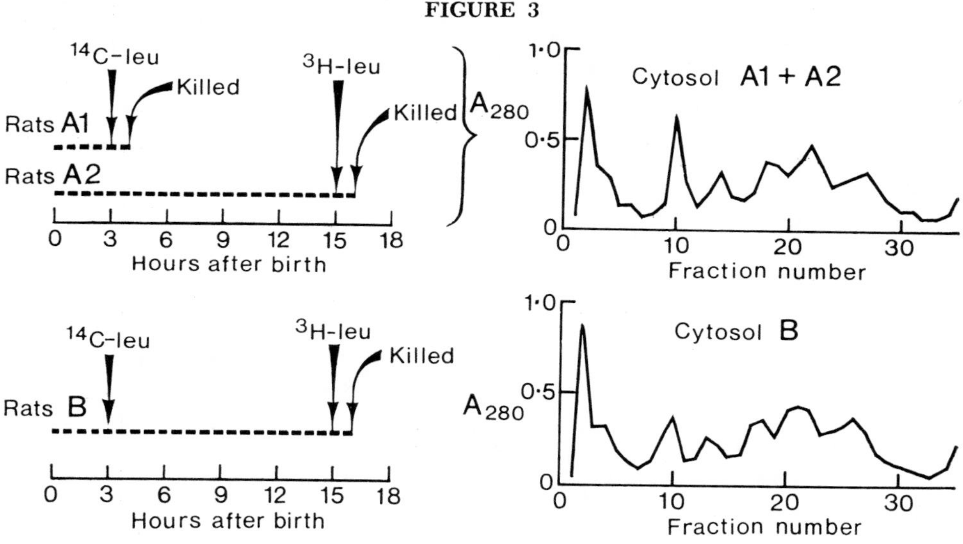

Measurement of relative degradation rates in proteins of the cytosol fraction from newborn rat liver: experimental plan. All rats were in the same litter. Some (A1) were injected with 3 μCi [^{14}C] leucine 3 hours after birth while others (A2) received 20 μCi [^{3}H] leucine 15 hours after birth. A third group (B) were injected with 3 μCi [^{14}C] leucine 3 hours after birth and 20 μCi [^{3}H] leucine 15 hours after birth. All animals were killed 40 min after their injection or 2nd injection and the livers pooled within each group. Livers were homogenized in 4 volumes of 0.25 M sucrose and centrifuged at 100,000 g for 30 min to prepare cytosol fractions. Equal parts of cytosol fractions from rats A1 and A2 were mixed and 2.5 ml chromatographed on a DEAE cellulose column. At the same time 2.5 ml of liver cytosol from rats B was applied to a similar DEAE cellulose column (0.7 cm^2 $\times$ 10 cm) and the two columns eluted with a logarithmic gradient from 0 to 0.4 M NaCl in 0.05 M tris pH 8.0, using a total volume of 360 ml and a stream splitter, so that equal volumes of eluting buffer were applied to each column. The absorbance at 280 nm was measured in each 5 ml fraction.

Schimke *et al.* (1968). Leucine labelled with carbon 14 was injected 3 hours after birth to label the proteins being synthesized in liver at that time. Twelve hours later the same animals were injected with [^{3}H] leucine and were killed after 40 min. Thus the tritium radioactivity in any protein or protein fraction represents the extent of synthesis, while the ratio of tritium to carbon-14 gives a relative measure of degradation, since proteins synthesized in the period after the first injection would not be labelled. This technique presupposes no reutilization of carbon 14 in degraded and resynthesized proteins, and also assumes a steady state with an identical rate of protein synthesis at the times of the two injections.

FIGURE 4

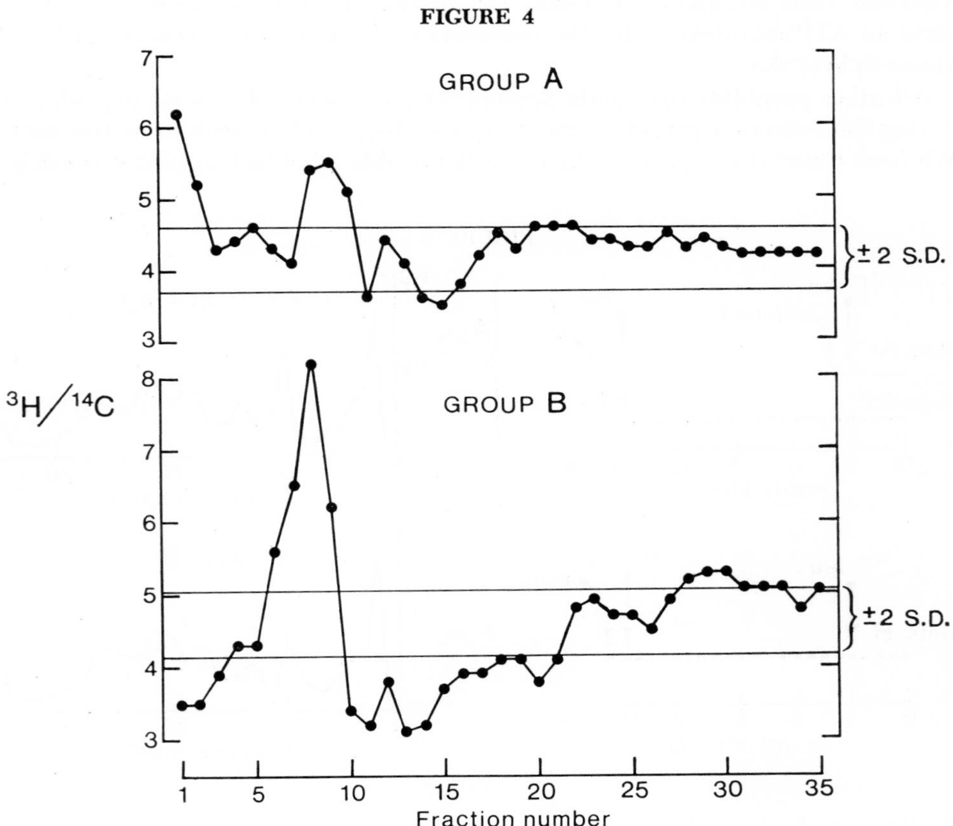

Measurement of relative degradation rates in proteins of the cytosol fraction from newborn rat liver: radioactivity determinations. Protein was precipitated in each fraction from the columns described in Fig. 3, washed with 10% trichloracetic acid and dissolved in 0.2 ml of NCS solubilizer. Dissolved protein was counted in 10 ml of a solution containing 4 g of 2, 5-diphenyloxazole and 100 mg of 1, 4-bis-(4-methyl-S-phenyloxazol-2-yl) benzene per litre of toluene. Radioactivity is expressed as the ratio of disintegrations/min of tritium to disintegrations/min of carbon-14 in each fraction. The variability is indicated (± 2 standard deviations) where 20 μCi [^{3}H] leucine and 3 μCi [^{14}C] leucine were mixed and injected into other animals, cytosol fractions prepared and chromatographed, and the radioactivity in protein fractions determined.

After killing the animals, cytosol proteins were chromatographed on a DEAE cellulose column (Fig. 3) and the ratio of tritium to carbon 14 determined in the proteins of each column fraction (Fig. 4). The variability of the total method is determined by injecting a mixture of [^{3}H] leucine and [^{14}C] leucine into other animals and chromatographing the liver cytosol fractions. This variability is indicated in Fig. 4 as the mean ^{3}H/^{14}C ratio $\pm$ two standard deviations. The ratios shown for group B in Fig. 4 are those determined by the Schimke procedure. It can be seen that fractions 6 to 9 have high ratios, indicating substantial degradation of proteins between 3 and 15 hours of age. Other fractions contain proteins with isotope ratios significantly lower than could be accounted for by technique variability. The degradation rate of these proteins would be slower than the average.

As mentioned above, a problem of this method is the necessity of a steady state with regard to protein synthesis. We cannot assume such a state in the newborn animals and have consequently corrected for any change in the rate of protein synthesis. This was accomplished by measuring rates of synthesis in different animals from the same litter as those used for group B. Cytosol fractions from livers of 3 h-old rats which had been injected with [^{14}C] leucine and 15 h-old rats which received [^{3}H] leucine, were pooled and chromatographed on DEAE cellulose. Ratios of tritum to carbon 14 in protein fractions are shown in Fig. 4. We note that the ratio is higher than can be accounted for by random variation in fractions 1,2,8,9 and 10, indicating a greater relative rate of synthesis for proteins in these fractions at 15 hours after birth. However, it is clear that the ^{3}H/^{14}C ratios are different in the fractions from the two columns so that a heterogeneity of degradation rates must occur. We conclude that the absence of phosphoenolpyruvate carboxykinase inactivation in livers of newborn rats does not reflect a general lack of protein degradation.

General considerations

The experiments with normal, postnatal development and with phosphoenolpyruvate carboxykinase induction by cyclic AMP *in utero* indicate that the increase in activity is produced by an increase in the rate of enzyme synthesis and a lack of degradation. A similar situation occurs with substrate-induced enzyme induction in bacteria where there are large increases in the rate of enzyme synthesis with no protein degradation during the accumulation period (Mandelstam, 1963). Perhaps some of the mechanisms and controls on the *initial* synthesis of proteins are similar in all organisms. This speculation must be tested with a variety of proteins, since the only other study of synthesis and degradation rates of a protein during initial appearance is the work of Silpananta and Goodridge (1971) with chicken liver malic enzyme. These authors found similar results to those reported here, although the change in enzyme degradation was less pronounced. However, a lowered degradation rate would be difficult to detect with malic enzyme because its half life under steady state conditions was long, being about 55 hours (Silpananta and Goodridge, 1971).

REFERENCES

Ballard, F. J. (1971), in *Physiology of the Perinatal Period* (Stave, U., ed.), p. 417, Appleton-Century-Crofts, New York

Ballard, F. J. and Hanson, R. W. (1969). *J. Biol. Chem. 244,* 5625

Ballard, F. J. and Oliver, I. T. (1964). *Biochem. J. 90,* 261

Chang, H. C. and Lane, M. D. (1966). *J. Biol. Chem. 241,* 2413

Chuah, C. C. and Oliver, I. T. (1971) *Biochemistry 10,* 2990

Greengard, O. (1969). *Science, 163,* 891

Greengard, O. and Dewey, H. K. (1967). *J. Biol. Chem. 242,* 2986

Gurdon, J. B., Lane, C. D., Woodland, H. R. and Marbaix, G. (1971). *Nature, (London), 233,* 177

Hershko, A. and Tomkins, G. M. (1971). *J. Biol. Chem. 246,* 710

Katunuma, N., Kito, K. and Kominami, E. (1971a). *Biochem. Biophys. Res. Commun. 45,* 76

Katunuma, N., Kominami, E. and Kominami, S. (1971b). *Biochem. Biophys. Res. Commun. 45,* 70

Kenney, F. T. (1962). *J. Biol. Chem. 237,* 1610

Knox, W. E., Yip, A. and Reshef, L. (1970), in *Methods Enzymol. 17A,* 415

Kornberg, A. (1955). *Methods Enzymol. 1,* 441

Longshaw, I. D. and Pogson, C. I. (1972). *J. Clin. Invest.* in the press

Lowry, O. H., Rosebrough, N. J., Farr, A. L. and Randall, R. J. (1951). *J. Biol. Chem. 193,* 265

Majerus, P. W. and Kilburn, E. (1969). *J. Biol. Chem. 244,* 6254

Mandelstam, J. (1963). *Ann. N.Y. Acad. Sci. 102,* 621

Nemeth, A. M. (1959). *J. Biol. Chem. 234,* 2921

Otto, K. (1971). in *Tissue Proteinases* (Barrett, A. J. and Dingle, J. T., eds.), p. 1, North-Holland, Amsterdam

Philippidis, H., Hanson, R. W., Reshef, L., Hopgood, M. F. and Ballard, F. J. (1972). *Biochem. J. 126,* 1127

Reshef, L., Hanson, R. W. and Ballard, F. J. (1970). *J. Biol. Chem. 235,* 5979

Rhoads, R. E., MacKnight, G. S. and Schimke, R. T. (1971). *J. Biol. Chem. 246,* 7407

Schimke, R. T. and Doyle, D. (1970). *Annu. Rev. Biochem. 39,* 929

Schimke, R. T., Ganschow, R., Doyle, D. and Arias, I. M. (1968). *Fed. Proc. Fed. Amer. Soc. Exp. Biol. 27,* 1223

Schimke, R. T., Sweeney, E. W. and Berlin, C. M. (1965a). *J. Biol. Chem. 240,* 322

Schimke, R. T., Sweeney, E. W. and Berlin, C. M. (1965b). *J. Biol. Chem. 240,* 4609

Sereni, F., Kenney, F. T. and Kretchmer N. (1959). *J. Biol. Chem. 234,* 609

Silpananta, P. and Goodridge, A. G. (1971). *J. Biol. Chem. 246,* 5754

Steinberg, D. and Vaughan, M. (1956). *Arch. Biochem. Biophys. 65,* 93

Walker, D. G. and Holland, G. (1965). *Biochem. J. 97,* 845

Yeung, D. and Oliver, I. T. (1968). *Biochem. J. 108,* 325

Gene Expression and Development

Obligatory Requirement for DNA Synthesis During Myogenesis, Erythrogenesis and Chondrogenesis[a]

H. HOLTZER, R. MAYNE, H. WEINTRAUB AND G. CAMPBELL

Department of Anatomy School of Medicine University of Pennsylvania Philadelphia, Pennsylvania

The purpose of this symposium might best be served by stressing that mechanisms of gene expression underlying cell physiology are likely to be quite different from those initiating cell differentiation. Responses of cells to hormones such as glucocorticoids, or other exogenous molecules such as cyclic AMP, important though they may be from the viewpoint of cell physiology, are of much less importance from the viewpoint of cell differentiation. The origins of cell diversity and the maintenance of the basic phenotypes of metazoan cells cannot be attributed to hormones, embryonic inductions or other schemes predicted on exogenous, didactic molecules which instruct virginal, biochemically undifferentiated cells to differentiate; this is because at no stage in development are there biochemically undifferentiated, genetically uninstructed cells. At most, exogenous molecules allow differentiated cells to express genetic commitments which were determined earlier (Holtzer, 1963, 1968; Holtzer *et al.*, 1972).

The central enigma of cell differentiation is concerned with the legacy, both nuclear and cytoplasmic, which a cell receives directly from its mother and indirectly from earlier ancestral cells, how the cell alters that endowment, and how and what it transmits to its descendants. In brief, the problem of differentiation may be defined as how information which is not readily available in a mother cell is made available in daughter cells. This definition of differentiation excludes many problems of cell physiology of mature cells and cell physiology of embryonic cells, problems which are often confused with problems in developmental biology.

[a]This work was supported by grants from the National Institute of Child Health and Human Development (HD-00189), the National Science Foundation (GB-5047X), the American Cancer Society and Muscular Dystrophy Associations of America.

One of the conventional problems in biochemical embryology has been determining when cells in an emerging lineage first synthesize the terminal luxury molecules appropriate to that lineage, i.e., such molecules as myosin, haemoglobin, chondroitin sulphate, or tryptophane pyrrolase. Efforts are then made to correlate the time of appearance of such molecules with "inducers", kinds of RNAs present, initiating factors, as well as to determine the rates of turnover. This methodology must certainly add to our knowledge of cell biochemistry, but it will not necessarily add to our understanding of the causal events responsible for cell differentiation. The challenging problems of myogenesis, erythrogenesis, or chondrogenesis are largely over by the time a given cell synthesizes and accumulates its first molecules of myosin, haemoglobin, or chondroitin sulphate. In order to understand the step-wise antecedent genetic events leading to a cell with the machinery for assembling myosin, haemoglobin, or chondroitin sulphate, one must understand the history of that cell. More specifically, one must investigate the kinds of molecules synthesized by its mother and grandmother cell so that these obligatory precursors were *compelled* to yield a generation of cells uniquely and exclusively committed to synthesizing myosin *or* haemoglobin *or* chondroitin sulphate.

In this review, experiments will be described which suggest that changes in the synthetic programs of cells are tightly coupled to phases of the mitotic cycle and that movement from one precursor compartment to the next in the myogenic, erythrogenic and chondrogenic lineages depends upon DNA synthesis and a small number of critical cell cycles. Critical cell cycles leading to daughter cells which synthesize molecules that the mother cell did not synthesize differ from the more commonly studied cell cycles which leads to daughters which synthesize the same molecules produced by the mother cells. Cell cycles reprogramming the synthetic repetoire of daughter cells have been termed "quantal cell cycles", whereas cell cycles yielding replicas of the mother cell have been termed "proliferative cell cycles" (Ishikawa *et al.*, 1968; Holtzer and Abbott, 1968; Holtzer, 1970a).

MYOGENESIS

Early somites in a 2-3 day chick embryo consist of 5 x 10^3 to 2 x 10^4 mesenchymal cells. Their progeny differentiate into myogenic, chondrogenic, and fibrogenic cells. The precise time at which somite cells speciate into myogenic, chondrogenic, or fibrogenic lineages is not known. Frank chondroblasts depositing chondroitin sulphate are not observed *in vivo* until late day 4, yet *precursor* chondrogenic cells are present at least from the second day onwards (Holtzer and Matheson, 1970). In marked contrast, several hundred post-mitotic mononucleated myoblasts with cross-striated myofibrils are present in early 2-day embryos (Holtzer *et al.*, 1957; Holtzer and Sanger, 1972). The myofibrils in these mononucleated myoblasts are morphologically and biochemically indistinguishable from those observed in mature animals; there are no "embryonic" myofilaments or myofibrils. At the time of its formation a single somite already harbours between 32 and 128 presumptive myoblasts plus a larger but unknown number of

FIGURE 1

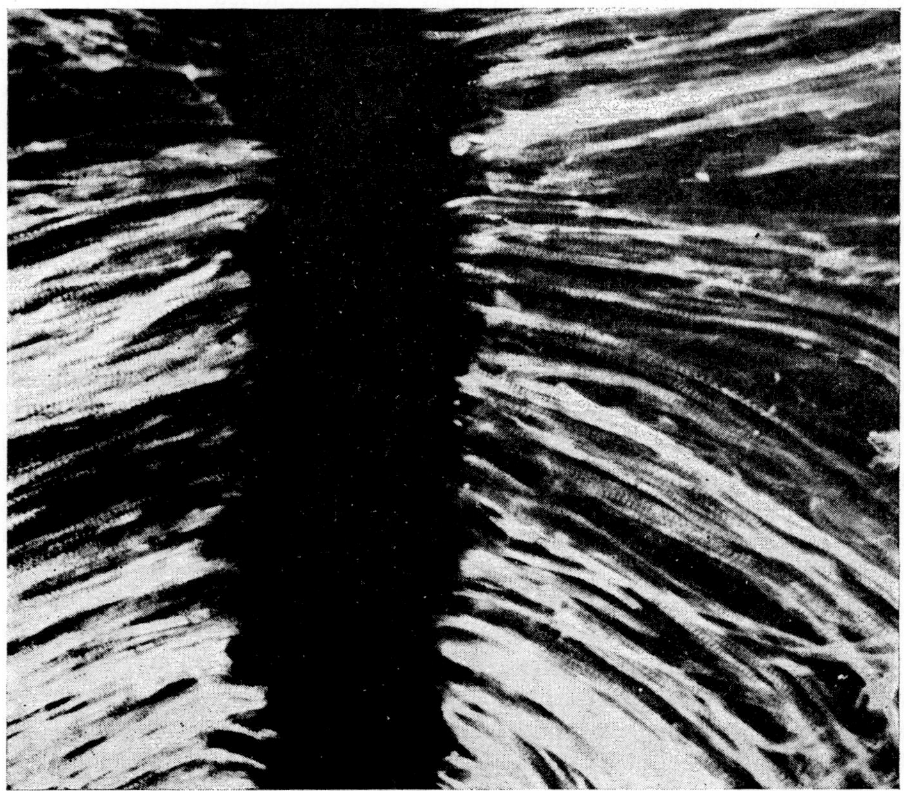

Low power, fluorescence micrograph of a whole-mount of a 3-day chick embryo trunk treated with fluorescein-labelled antimyosin. Though this preparation is over 2mm in thickness and contains innumerable skin, dermal, presumptive chondrogenic and presumptive myogenic cells as well as nerve and notochord cells, only some 1,200 mononucleated myoblasts in each myotome bind the antibody to the A bands in the myofibrils. Two successive somites and the intervening intersomite septum rich in mesenchymal cells are illustrated. Owing to the thickness of the preparation it is impossible to view all the individual myofibrils in the same focal plane.

myogenic beta cells. Myogenic beta cells and presumptive myoblasts are respectively the ante-penultimate and penultimate cells in the myogenic lineage (Holtzer, 1970b; Holtzer and Bischoff, 1970; Holtzer *et al.*, 1972). Presumptive myoblasts do not synthesize myosin, tropomyosin, nor do they have acetylcholine receptors on their surfaces (Frambaugh and Rash, 1971). Replicating presumptive myoblasts may display properties of a relatively stable stemcell population (Konigsberg, 1963; Yaffee, 1969). After one particular round of DNA synthesis and the ensuing mitosis, presumptive myoblasts can yield post-mitotic mononucleated daughter myoblasts. By the time these daughter myoblasts are no more than 4 or 5 hours old, they will have synthesized and assembled into filaments enough myosin, actin and tropomyosin so as to be identified *in situ* with fluorescein-labelled antibodies (Okazaki and Holtzer, 1965) or the electron microscope. Some hours later myofilaments can be identified with the polarizing or phase microscope (Holtzer, 1970b; Fischman, 1970, 1972).

Daughter myoblasts display synthetic programs strikingly different from those of their mother cells, the presumptive myoblasts. Accordingly, the cell cycle between presumptive myoblasts and myoblasts is termed a quantal cell cycle. This particular cell cycle and mitosis allows, or triggers, the initiation of the coordinated synthesis of at least three terminal luxury molecules of muscle-myosin, actin, and tropomyosin (Okazaki and Holtzer, 1965; Holtzer, 1970a; Holtzer and Sanger, 1972). The time or conditions under which the synthesis of other terminal luxury molecules of muscle, such as creatine phosphokinase or myoglobin, begins in these post-mitotic cells has not been determined.

Question: Is there an obligatory requirement for DNA synthesis in order for presumptive myoblasts to yield myoblasts? Alternatively, would presumptive myoblasts held in mitosis, or at the G1-S interface, have the option of coordinately synthesizing myosin, actin and tropomyosin? In brief, is cell maturation a function of the passage of time or is maturation coupled to phases of the mitotic cycle or even to particular mitotic cycles? Replicating presumptive myoblasts were arrested in M with colcimide or vinblastine for up to 15 hours. These M-arrested cells did not bind fluorescein-labelled antibodies to myosin or tropomyosin, nor would they yield antigens that would from precipitin bands in Ouchterlony plates upon extraction (Bischoff and Holtzer, 1969). The myogenic cells in metaphase lacked in their cytoplasm both the thick 160Å, myosin filaments and the thin 60Å, actin filaments; they were rich, however, in the 100Å intermediate-sized filaments found in many cell types (Ishikawa *et al.*, 1968). Clearly, myogenic beta cells and presumptive myoblasts entering M had not synthesized myosin, or tropomyosin in their previous G1, S, or G2. However, within hours after the mitotic inhibitors were washed out, these cells completed mitosis, and their daughter cells were in all respects normal myoblasts (Bischoff and Holtzer, 1968).

The following experiments indicate that replicating presumptive myoblasts held in the G1-S interface do not have the option of synthesizing myosin, or tropomyosin: The brachial segments of 3-day chick embryos were transected in the midline and either the left or right halves incubated for 10 hours in normal medium plus 5-fluorodeoxyuridine (FdU) (10^{-6}M); the contralateral halves grown in normal medium served as controls. The somites were glycerinated, treated with labelled antibodies against myosin or tropomyosin, squashed, and the numbers of individual, mononucleated, post-mitotic myoblasts with striated myofibrils counted under the fluorescence microscope (Fig. 1). In a second series, the brachial segments were first grown in FdU for 10 hours and then either the right or left halves were removed from the inhibitor, washed several times, and then grown for 15 hours in the presence of excess cold 2'-deoxyribosylthymine (dT). These experiments were designed to demonstrate that FdU did not kill presumptive myoblasts. The results are shown in Tables 1 and 2. They demonstrate (1) that the option to synthesize DNA *or* the myofibrillar proteins is not

TABLE 1

Numbers of striated myoblasts/posterior 3-day myotome

	Control	FdU-Treated
A.	960	520
B.	1220	830
C.	810	410
D.	1350	750

Three day trunks were transected into right and left halves and reared as organ cultures for 10 hours in normal medium plus 10^{-6}MFdU and 2×10^{-6}MU for 10 hours. After treatment with fluorescein-labelled antimyosin, the myotomes were squashed and the individual striated mononucleated myoblasts counted.

TABLE 2

Numbers of striated myoblasts/posterior 3-day myotome

	FdU-treated and not reversed	FdU-treated and reversed
A.	380	840
B.	650	1050
C.	430	1010

The trunks were transected into right and left halves and organ cultured in FdU and U for 10 hours. Either the right or left half was removed, washed, and then grown in normal medium with excess cold dT for an additional 15 hours; the other half remained in FdU.

open to the presumptive myoblasts and (2) if the synthesis of the contractile proteins is to occur, the presumptive myoblasts *must* synthesize DNA and form daughter nuclei.

Experiments using Cytochalsin-B suggest that though nuclear divisions are required, cytokinesis is not obligatory (Sanger *et al.*, 1971; Sanger and Holtzer, 1972).

One round of DNA synthesis in 5-bromodeoxyuridine (BdU) allows presumptive myoblasts to continue replication, but their daughters do not differentiate into post-mitotic myoblasts (Stockdale *et al.*, 1964; Okazaki and Holtzer, 1965). BdU-suppressed myogenic cells undergo numerous proliferative cell cycles, providing they are kept in low doses of the analog. Fig. 2 illustrates that the absorption spectra of a variety of cytochromes from BdU-suppressed cells are indistinguishable from untreated controls; similarly the total cytochrome oxidase activity per DNA is the same for both groups of cells. This differential suppression of terminal luxury molecule synthesis over essential molecule synthesis occurs when approximately 20% of the nascent thymine residues of the DNA are substituted by BdU. One round of DNA synthesis at concentrations of BdU that result in less than 20% thymine substitution inhibits fusion, but does not block the synthesis

FIGURE 2

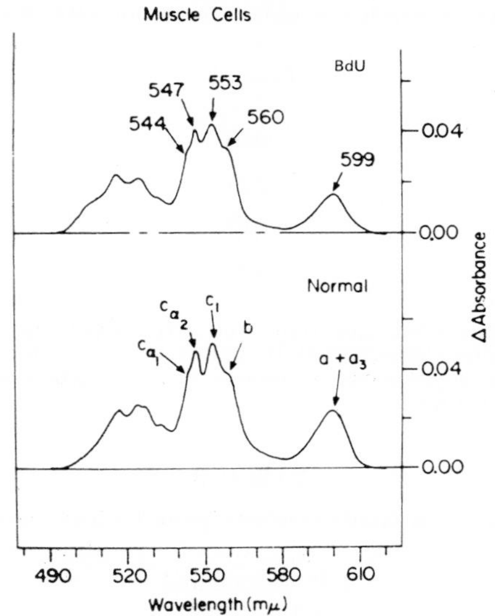

A comparison of the absorption spectra based on equivalent amounts of DNA from control cultures and cultures treated with BdU. One of the paradoxical effects of the BdU-suppressed cells is that they synthesize normal cytochromes, but fail to synthesize myoglobin.

of contractile proteins. There is evidence that higher concentrations of BdU, resulting in greater than 20% thymine substitution disturb the synthesis of essential molecules (Holtzer and Abbott, 1968; Bischoff and Holtzer, 1970). If after three weeks in the analogue, fully suppressed cells are transferred to normal medium and allowed to synthesize normal strands of DNA for 3 to 5 generations, normal myoblasts emerge (Bischoff and Holtzer, 1970). BdU therefore does not act as a conventional mutagen, since (A) 100% of the cells are affected and (B) the covert genetic commitment of the presumptive myoblasts is not disturbed.

In summary: (1) If presumptive myoblasts do not synthesize DNA and do not yield daughter nuclei in a G1 cytoplasm, they will not initiate the synthesis of myosin, tropomyosin, or actin for myofilaments, (2) one round of DNA synthesis in BdU suffices to prevent the initiation of the new genetic program that normally occurs as a consequence of the quantal cell-cycle leading from mother *presumptive* myoblast to daughter myoblasts.

ERYTHROGENESIS

There are approximately 8 x 10^5 haematocytoblasts in a 30-hour chick embryo. These covertly differentiated cells in the erythrogenic lineage do not synthesize haemoglobin (Hb). Between the 30th and 60th hour each haematocytoblast synthesizes DNA and divides, forming daughter cells which are the first generation erythroblasts. By the time it is 5 hours old, each 1st generation erythroblast has synthesized sufficient quantities of Hb to allow microspectrophotometric measurements of the contents of a single cell (Thorell and Raunick, 1966; Campbell, *et al.*, 1971). The 1st generation erythroblasts serve as progenitors for 6 subsequent generations of erythroblasts. Each generation of erythroblast synthesizes Hb throughout most of the mitotic cycle and each generation synthesizes a characteristic amount of Hb. Each generation can be distinguished cytologically and each

FIGURE 3

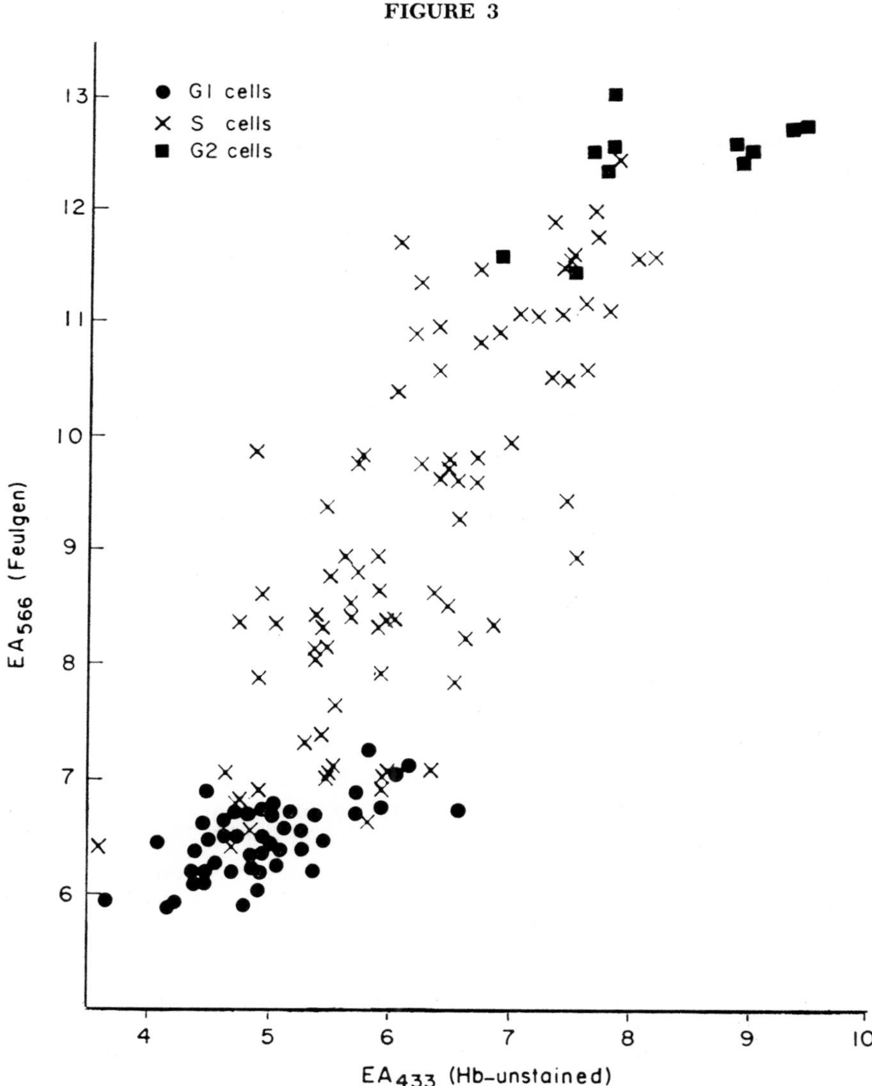

A plot of Hb measurements made on primitive erythroblasts from a day 4 embryo and of DNA measurements made on the same cells. The Hb measurements are expressed as EA_{566} units. The location of each cell in different compartments of the cell cycle was determined by [^{3}H]dT radioautography and by Feulgen microspectrophotometry.

generation displays characteristic G1, S and G2 phases. Earlier generations of erythroblasts synthesize Hb more rapidly than do more mature cells in the lineage; they also have shorter G1 and S periods (Weintraub *et al.*, 1972).

Evidence that Hb is synthesized throughout most of the cell cycle is shown in Fig. 3. Smears were made of 5th generation erythroblasts that had been pulsed with [^{3}H]dT. An appropriate area was selected and mapped photographically. Hb absorbance measurements were made on individual erythroblasts. The slide was then stained for Feulgen microspectrophotometry and estimates of the DNA were obtained for each cell previously measured for its Hb content. Subsequent radioautography indicated which cells were in S. G1 and G2 cells were distinguishable on the basis of their DNA values. For each cell then there were two determinations—relative amounts of Hb and position in the cell cycle. Clearly, erythroblasts synthesize Hb throughout S. Further experiments are required to determine the rates of Hb synthesis in G1, G2 and M.

There are few opportunities for comparing phases of the mitotic cycle from generation to generation *within* an emerging lineage and most studies stress the constancy of the various phases of the mitotic cycle. It was of interest, therefore, that not only did the over-all duration of the mitotic cycle lengthen in successive generations of erythroblasts, but so did the duration of G1, S and G2 as well.

Weintraub *et al.* (1971) also noted that three different generations of erythroblasts co-existed in the same circulating plasma. From this it follows that endogenous rather than exogenous molecules must regulate the characteristic phases of the mitotic cycle that shift from generation to generation. It is unlikely that these cells respond immediately to a specific environmental substance that lengthens their cell cycle. This means that the duration of the G1, S and G2 in any one generation of erythroblast must have been programmed, at a minimum, three generations earlier. Evidence favoring the notion that this programming occurred in the haemotocytoblast for all 7 generations of erythroblasts is discussed elsewhere (Weintraub *et al.*, 1971).

Perturbations of the mitotic cycle similar to experiments performed on myogenic cells, have been carried out on (1) haematocytoblasts and (2) the different generations of erythroblasts. If haematocytoblasts *in vivo* are treated with colcimide, FdU, or BdU, first generation erythroblasts do not emerge. If the effect of FdU is reversed within 5 hours by excess cold dT, the subsequent production of 7 generations of Hb producing erythroblasts is normal. Reversal of the effects of colcemide cannot be observed *in vivo* since the compound cannot be washed out of the egg. Nevertheless, the similarity in response of the haematocytoblasts and presumptive myoblasts to colcimide, FdU and BdU suggests these two particular generations of precursor cells share equivalent roles in their respective lineages. The obvious difference is that the presumptive myoblast produces a post-mitotic myoblast within one generation, whereas the haematocytoblast is programmed to yield 6 generations of replicating erythroblasts before producing the equivalent post-mitotic erythroblast. The fact that erythroblasts undergo a pre-set number of proliferative cell cycles following the terminal quantal cell

FIGURE 4

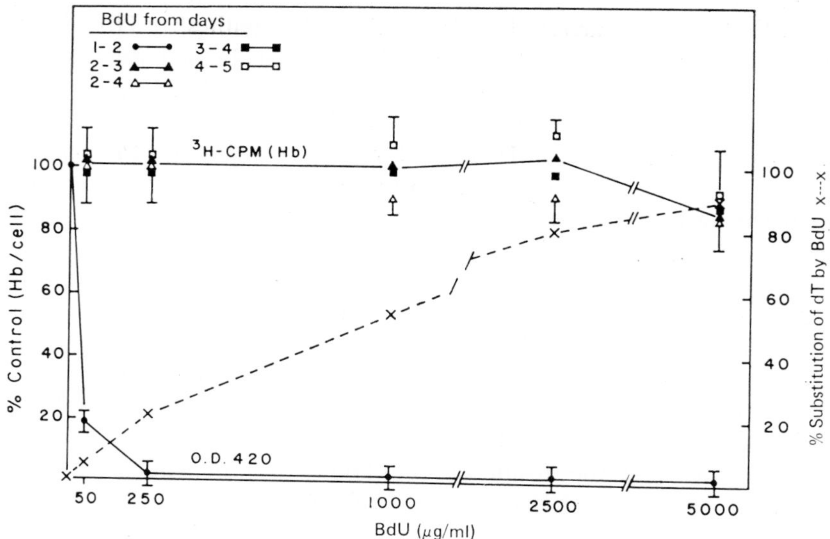

Transition from BdU sensitivity to BdU resistance during normal erythrogenesis. 3ml of a BdU solution was injected into chick eggs of the stated ages at the stated concentrations. Cells were harvested immediately after exposure. For each BdU concentration the percentage substitution of thymine by BdU was determined on CsCl gradients. Hb per cell was monitored in three ways and scored as a percentage of controls: for all experiments except those from day 1 to day 3, [^{3}H] Leucine (10μc/ml) incorporation into CM-cellulose purified Hb was measured on a per cell basis. For the same cells, Hb per cell was measured cytophotometrically. Lastly, inhibition of Hb synthesis from day 1 to day 3 was measured spectrophotometrically as Hb/embryo. The solid line describes the effects of BdU on Hb, the dashed line represents the percentage substitution by BdU

cycle of the haematocytoblast must not obscure the main biological principle that normal cells synthesizing haemaglobin do not yield the enormous numbers of progeny produced by erythroid stem cells in bone marrow. Erythroblasts, like myoblasts, function under stringent restraints of DNA synthesis. These differences between erythroblasts and myoblasts with respect to the limited capacity of DNA synthesis of the former probably represents an adaptation of the erythropoietic system. Given the requirement placed on the erythropoietic system for a specific number of mature red cells per hour, cell division in erythroblasts programmed for only 6 divisions enable the erythropoietic stem cells to divide fewer times during the lifetime of the organism. Thus, the burden of division-related deleterious effects is placed on cells destined to die, the erythroblasts, rather than the more stable stem cells.

If the cell cycle leading from presumptive myoblast to myoblast is a quantal cell cycle, then that cell cycle leading from haematocytoblast to 1st generation erythroblast is also a quantal cell cycle. In contrast, the sequence of cell cycles leading from one generation of erythroblasts to the next involves proliferative cell cycles. Accordingly, it is worth noting that the biological response of proliferating haematocytoblasts to FdU and BdU is quite different from the response of the replicating erythroblasts. If 3rd, 4th, or 5th generation erythroblasts are treated with FdU, the arrested cells continue to synthesize Hb. Apparently, programmed erythroblasts synthesizing Hb are not immediately affected by whether or not the cell engages in DNA synthesis. Erythroblasts arrested by FdU for 10 hours possess 50% more Hb per cell than untreated controls. When these cells are released from their FdU block, they proceed through the remainder of their mitotic program. The amount of Hb per cell remains above normal until the fully-matured, post-mitotic state is reached. These post-mitotic cells cut off the synthesis of Hb when the amount per cell has reached the normal quantity for that cell type. That is to say, the post-mitotic progeny of cells that were blocked by FdU in the 3rd, 4th, or 5th generation, cease synthesizing Hb before the normal controls do (Campbell, unpublished). These findings suggest that the characteristic total number of Hb molecules synthesized from a clone of erythroblasts is set during that quantal cell cycle in which the haematocytoblasts give rise to the 1st generation erythroblasts. The normal constancy of Hb content per cell per generation is the result of coupling two independently determined factors: (A) a given rate of Hb synthesis and (B) cell cycle times programmed to lengthen in a precise manner with each stage in the erythroblast lineage. FdU, by protracting the duration of a given mitotic cycle, simply allows the arrested cells a greater length of time for Hb synthesis to proceed before the subsequent division.

There is an even more conspicuous difference between the responses of haematocytoblasts and erythroblasts to BdU (Fig. 4). Incorporation of BdU into the DNA of erythroblasts even to the extent of 80% substitution of the thymine residues (Weintraub *et al.*, 1972), has no detectable effect on: (1) the parameters of the mitotic cycle (2) the cytology of the cells, or (3) the production of characteristic amounts of Hb per cell per generation. This is all the more remarkable since 20% substitution of BdU for the thymine in the DNA of presumptive myoblasts, haematocytoblasts, chondroblast, fibroblasts or somite cells, is sufficient to produce the BdU syndrome (Abbott and Holtzer, 1968; Schulte-Holthausen, *et al.*, 1969; Bischoff and Holtzer, 1970; Abbott *et al.*, 1972; Mayne, Abbott and Holtzer, unpublished).

CHONDROGENESIS

In vivo and *in vitro* somite cells must be induced by interacting with the embryonic spinal cord or notochord in order to yield progeny that differentiates into chondroblasts which synthesize and deposit chondroitin sulphate (Holtzer and Detwiler, 1953; Watterson *et al.*, 1954; Holtzer, 1968). Extirpation of the inducing tissues *in vivo* or rearing young somites *in vitro* without inducers results in a progeny which does not transform into recognizable chondroblasts. Grafting spinal cord into areas of the somites which normally will yield muscle and fibroblasts causes the progeny of such cells to become frank chondroblasts. These findings are shown in Figs. 5 and 6. Claims of isolating an esoteric cartilage-

FIGURE 5

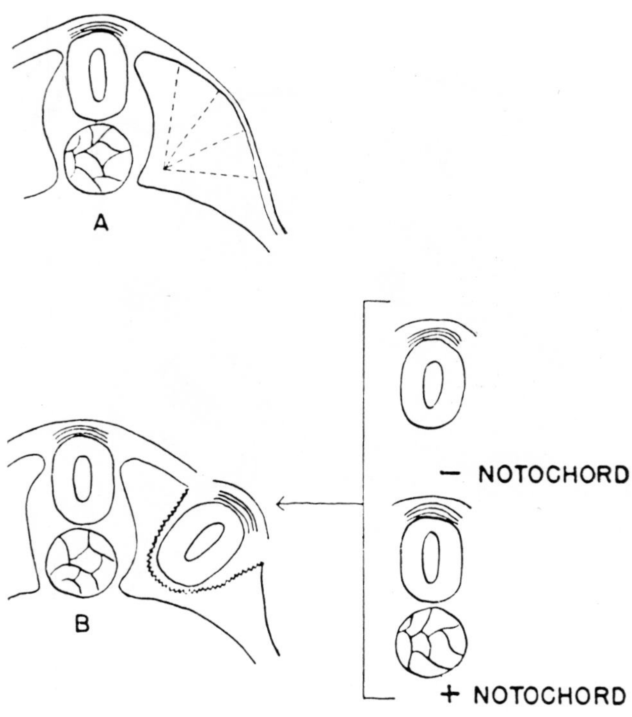

Diagrammatic representation of implanting spinal cord alone or spinal cord plus notochord into the somites of another embryo. The dotted lines in A indicate the different radii followed by the initial incision. B shows the graft implanted into the gap resulting from the incision. (Holtzer and Detwiler, 1953).

inducing molecule have not been confirmed in other laboratories (Thorp and Dorfman, 1967) and the claim of observing "spontaneous" cartilage *in vitro* only confirms that the inductive interaction occurs very early *in vivo* (Holtzer, 1964; Holtzer and Matheson, 1970).

The difference in behavior between (1) a cluster of stage 12-13 somites alone and (2) a similar cluster of somites plus a small piece of notochord is striking. Thousands of typical chondroblasts emerge in the notochord-somite cultures, whereas not a single chondroblast emerges in cultures of somites by themselves (Holtzer, 1964; Holtzer, 1968). Fig. 7 shows the amount of chondroitin sulphate produced by somites by themselves and by somites plus notochord. It is still unclear whether the small amount of sulphate incorporated by somites alone is authentic chondroitin sulphate of the kind produced by frank chondroblasts, whether it is another, unknown sulphated molecule, or whether it is synthesized by the fibroblasts in these cultures (Shubert and Hammerman, 1968).

The following experiments suggest that interference with the normal mitotic schedule during and shortly after the notochord-somite inductive interaction inhibits subsequent chondrogenesis without impairing the ability of the cells to

FIGURE 6

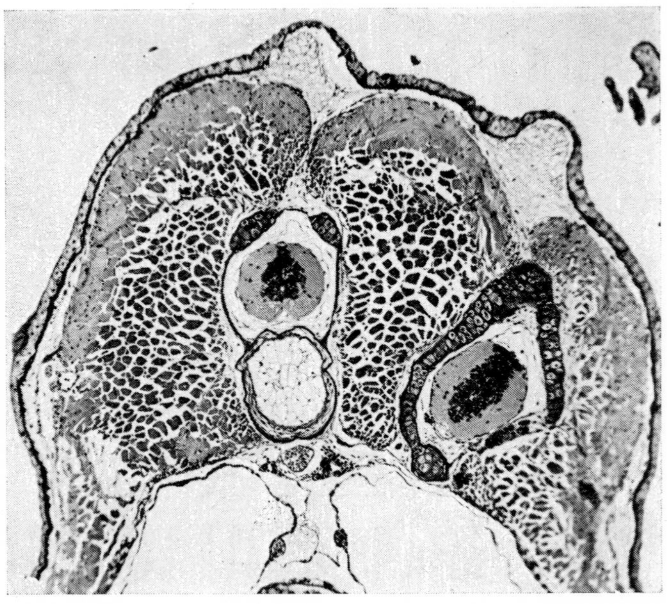

A section through an animal one month after implanting a 3-segment length of spinal cord into the somites as indicated in Figure 5. Note the remarkably normal secondary, cartilaginous vertebral column induced by the grafted spinal cord. It is worth stressing that the first detectable event associated with the induction of chondrogenic cells by the spinal cord is a burst of mitotic activity in the responding somite cells; only days later do some of the progeny of these mitotic cells deposit chondroitin sulphate.

undergo proliferative cell cycles. If notochord-somite cultures are exposed to colcimide or FdU for 12 hours on the first day of culture and then reared in normal medium containing dT for another 9 days, they do not yield chondroblasts. Such cultures do not synthesize amounts of chondroitin sulphate greater than that of cultures of somites alone. If the FdU is added to 3 day old cultures after the cells have been induced and many have withdrawn from the mitotic cycle, chondrogenesis is not blocked.

Fig. 8 summarizes experiments in which BdU was incorporated into the DNA of replicating somite cells during, and immediately after, the notochord-somite inductive interaction. Clearly, the incorporation of BdU into the replicating precursor chondrogenic cells suppresses the emergence of the terminal chondroblasts. If the analogue is added to 3 day old cultures, or after some cells have been induced and have withdrawn from the mitotic cycle, chondrogenesis is not blocked.

The failure of either colcimide, FdU- or BdU-treated somites to recover and yield chondroblasts is not due to generalized cell death or to blocking further cell cycles. Treated somite cells have been grown and sub-cultured for many generations, and though they replicate normally, they do not differentiate into recognizable chondroblasts (Abbott *et al.*, 1972).

FIGURE 7

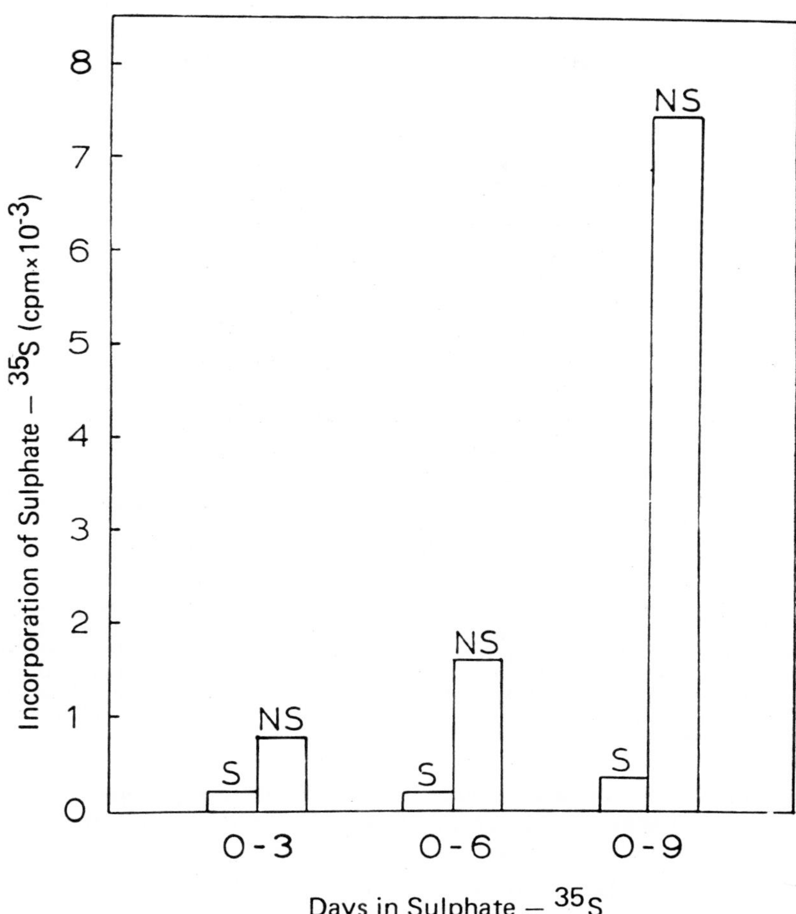

Stage 12-13 somites alone (S cultures) or stage 12-13 somites plus a small piece of notochord (NS cultures) were grown and analyzed for chondroitin sulphate as described in Abbott, *et al.* (1972). From these results it is clear that somite cells by themselves simply do not synthesize the quantities of chondroitin sulphate that characterize a recognizable chondroblast. It is also worth stressing that NS cultures 3 days old when trypsinized will yield individual chondrogenic cells capable of yielding chondrogenic clones; S cultures 3 days old do not yield chondrogenic clones when challenged in the same way.

It is instructive, though difficult, to contrast the effects of interfering with the cell cycle of somite cells evolving into chondroblasts with the effects of interfering with the cell cycle of somite cells evolving into myoblasts. Ten hour exposures to colcimide, FdU, and BdU reversibly block movement from the presumptive myoblast compartment, to the myoblast compartment. The effect of these same inhibitors for greater periods of time on somitic chondrogenesis appears more drastic and suggests that additional factors are involved. One such factor may be the "near neighbor interference effect". Deposition of chondroitin sulphate

FIGURE 8

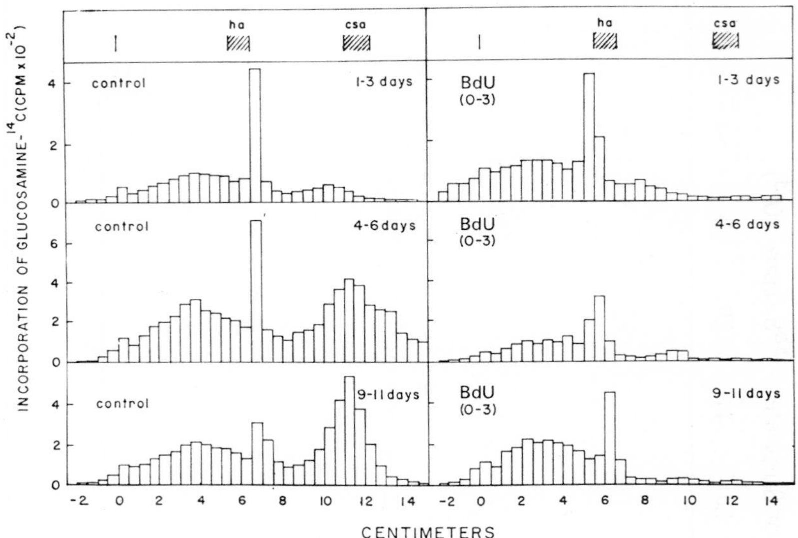

Separation by high voltage electrophoresis of the mucopolysaccharide fraction obtained from notochord-somite cultures exposed to [^{3}H] glucosamine on days 1-3, 4-6 or 9-11. Cultures were exposed to either BdU (10 μg/ml) or dT (10 μg/ml) from days 0-3. Electrophoresis was carried out on cellulose acetate strips in pyridinium formate buffer (pH 3.0, 500 volts, 120 min. 0°C). Equal aliquots from a total of 9 separate cultures were dried into each strip. After electrophoresis each strip was cut into 0.5cm sections, and radioactivity determined in 5ml scintillation fluid.

is readily interfered with when chondroblasts contact inappropriate cell surfaces (Chacko, *et al.*, 1969b; Chacko *et al.*, 1969a). It is also possible that the somite contains large numbers of chondrogenic beta cells and only small numbers of presumptive chondroblasts (Holtzer *et al.*, 1972) and that dispersed chondrogenic beta cells cannot undergo *in vitro* the quantal cell cycle leading to presumptive chondroblasts.

As the culture medium supplies all the nutritional factors required by chondroblasts for the synthesis and deposition of chondroitin sulphate, it follows that the cells in the treated cultures remain arrested in one or other of the compartments antecedent to the frank chondroblast compartment. One possibility is that unless induced presumptive chondroblasts undergo a cell cycle within a limited time period, their commitment to synthesize chondroitin sulphate decays and that they then differentiate into cells which synthesize hyaluronic acid predominantly (Davidson *et al.*, 1972 Mayne *et al.*, 1972).

In summary: Interference with the normal mitotic schedule of somite cells during the time they are responding to the inductive activity of spinal cord and notochord blocks further chondrogenesis and is consistent with the observation made two decades ago that the first signs of induction *in vivo* is a burst of mitotic activity in the responding somite cells (Holtzer and Detwiler, 1953).

DISCUSSION

Most current models of differentiation assume that repressor or derepressor sites on the chromosomes are available throughout the mitotic cycle. The experiments reviewed in this report stress that this is probably not the case. Rather, it is proposed that the transmission of an ongoing synthetic program to daughter cells, associated with proliferative cell cycles, or the reprogramming associated with quantal cell cycles, may be coupled to specific phases of the mitotic cycle and even to specific cell cycles.

Experiments with three developing systems implicate DNA synthesis with the events of subsequent differentiation. In one group of experiments the synthesis of DNA was prevented, for example with FdU; in a second group of experiments, newly synthesized DNA was altered by replacing thymine with BdU. Either of these procedures blocked myogenesis, erythrogenesis and chondrogenesis, providing it was introduced into the appropriate cell in the emerging cell lineage. Clearly any conclusion from these experiments must first deal with the effects of these drugs on cell viability. As a measure of viability, parameters such as multiplication, gross macromolecular synthesis, and reversibility may be used. For BdU-treated cells, most of these parameters testify to the viability of each system; for FdU treated cells, the problems are more difficult since a key factor, cell multiplication, becomes a meaningless measure. The crucial experiment is to show that the inhibition of DNA synthesis for a fixed interval results, upon reversal, only in a delay in the appearance of the differentiated state and that the full potential of that state is eventually achieved. The experimental evidence indicates this to be the case for the terminal stages of myogenesis, erythrogenesis and chondrogenesis. The observation that similar exposure of earlier cells in a lineage to FdU or BdU has a more drastic effect on subsequent differentiation will most certainly be worth probing.

It is clear from a number of studies that although free DNA is readily accessible to RNA polymerases, most of the DNA in a given cell is not. This restriction on the template activity of DNA is presumably imposed by molecules that bind to the DNA. During the process of replication it may be that in order for the replication apparatus to work efficiently the molecules associated with the DNA must be removed. Following duplication of the DNA, the cell must then design a method to reassemble the entire chromosome. Not only must the system reassociate, presumably by a self-assembly mechanism, but it must do this in duplicate. If all the structures in the ensemble are not present in the appropriate concentrations when the pre-replicative DNA doubles there will arise a period of instability between the control molecules and their paired DNA binding sites. Such an instability might provide the appropriate conditions for the institution of new synthetic programs in daughter cells. In this context, FdU prevents the required instabilities from arising; BdU, possibly by altering the way proteins bind to the DNA (Lin and Riggs, 1972) would preclude the events required to establish the new genetic program in the daughter cells.

However, this emphasis on DNA synthesis and replication as the central event for generating cell diversity must always be viewed from the perspective of the cell. The synthetic program of a given cell is the integrated function of two constantly interacting domains—the cytoplasmic, including the cell surface, and the nuclear. The understandable goal of molecular biologists of unravelling the basic machinery of *in vitro* transcription and translation must not obscure the biological fact that the selection of which sets of genes are to be expressed in a given cell is the function of the cytoplasm. It is within this context that the concept of quantal and proliferative cell cycles must be considered.

It is likely that some molecules in the cytoplasm of a presumptive myoblast were synthesized by its mother, the myogenic beta cell. It is even possible that the presumptive myoblast inherited a modest number of unique molecules synthesized by its great grandmother or even earlier ancestral cells. Such molecules would be the molecular equivalents of the polar plasm transmitted through successive generations of "determined" or covertly differentiated germ cells (King, 1970; Mahowald, 1968). In addition, of course, there are those luxury molecules translated by the presumptive myoblast from its own unique m-RNAs. We propose that the accumulation of several of these cytoplasmic molecules in critical concentrations alters the replicating DNA of the presumptive myoblast. This alteration, it is postulated, can occur only while the DNA is being replicated. This alteration, whether it be the binding of protein, or RNA or an induced conformational change, does not occur at the level of the structural genes for myosin, actin, tropomyosin, creatine phosphokinase or myoglobin, but at a higher order genetic unit. Such a unit can be thought of as a region for "terminal myogenesis". This region of the DNA cannot be transcribed or expressed in an S, G2 or M cytoplasm. However, in a G1 cytoplasm this "terminal myogenesis" region can initiate the transcription of all the individual m-RNAs that characterize the terminal phase of myogenesis. A similar scheme with appropriate modifications for inductive interactions could be drawn for erythrogenesis and chondrogenesis.

The concept that a modest number of quantal cell cycles alternating with variable numbers of proliferative cell cycles is the major mechanism in differentiation readily accounts for delayed or premature gene expression. The capacity to respond to a hormone or other exogenous changes such as pH, ionic changes, or substrate shifts, follows a particular quantal cell cycle. If the cell should undergo this particular quantal cell cycle in the absence of, say, a hormone or substrate, the unique phenotypic response would not be observed. Such a covertly differentiated cell might undergo subsequent proliferative replications. If the hormone be added to such covertly differentiated cells, a prompt expression of overt terminal differentiation may be expected. Alternatively, the quantal cell cycle generating the competence to respond to a hormone can normally occur well before the cells are exposed to the agent. In this case early introduction of the hormone or substrate will elicit a "premature" response, one which could be misconstrued as occurring in the absence of a particular cell cycle.

REFERENCES

Abbott, J. and Holtzer, H. (1968) *Proc. Nat. Acad. Sci. U.S.* *59,* 1144

Abbott, J., Mayne, R. and Holtzer, H. (1972) *Develop. Biol.* (in press)

Bischoff, R. and Holtzer, H. (1968) *J. Cell Biol.* *36,* 111

Bischoff, R. and Holtzer, H. (1969) *J. Cell Biol.,* *41,* 188

Bischoff, R. and Holtzer, H. (1970) *J. Cell Biol.,* *44,* 134

Campbell, G., Weintraub, H. and Holtzer, H. (1971) *J. Cell Biol.* *50,* 669

Chacko, S., Abbott, J., Holtzer, S. and Holtzer, H. (1969a) *J. Exp. Med.* *130,* 417

Chacko, S., Holtzer, S. and Holtzer, H. (1969b), *Biochem. Biophys. Res. Commun.* *34,* 183

Davidson, E., Abbott, J. and Holtzer, H. (1972), *Fed. Proc. Fed. Amer. Soc. Exp. Biol.* *31,* 434

Fischman, D. (1970) *Current Topics in Devel. Biol.,* *5,* 235

Fischman, D. (1972) *in*: Research in Muscle Development and the Muscle Spindle (Banker, Przybylski, Meulen and Victor eds.). *Excerpta Medica,* Amsterdam

Frambaugh, D. and Rash, J. (1971) *Develop. Biol.* *26,* 55

Holtzer, H. (1959) *Exptl. Cell Res.,* Suppl. 7, 234

Holtzer, H. (1963) Comments on induction during cell differentiation. XIII Colloq. Ges. Physiol. Chem. Springer-Verlag, Heidelberg

Holtzer, H. (1964) *Biophys. J.,* *4,* 239

Holtzer, H. (1968) Induction of chondrogenesis: a concept in quest of mechanisms. In: *Epithial-Mesenchyme Interactions,* (R. Billingham, ed.) Williams and Wilkins Co., Baltimore

Holtzer, H. (1970a) Proliferative and quantal cell cycles in the differentiation of muscle, cartilage and red blood cells, In: *ISCB Symposium, Control Mechanisms in Tissue Cells,* ed. H. Padykula, Academic Press, New York

Holtzer, H. (1970b) Myogenesis, In: *Cell Differentiation,* (Schjeide, O. and de Vellis, J. eds.) Van Nostrand Reinhold Co.

Holtzer, H. and Abbott, J. 1968, Oscillations of the chondrogenic phenotype in vitro, In: *Stability of the Differentiated Cell,* H. Ursprung, ed. Springer-Verlag, Heidelberg

Holtzer, H. and Bischoff, R. (1970) Mitosis and myogenesis, In: *Physiology and Biochemistry of Muscle,* II, (Briskey, E. and Cassons, L. eds.) University of Wisconsin Press, Madison, Wisconsin

Holtzer, H. and Detwiler, S. 1953, *J. Exp. Zool.,* *123,* 335

Holtzer, H. and Matheson, D. (1970) Induction of chondrogenesis in the embryo, In: *Chemistry and Molecular Biology of the Intercellular Matrix,* ed. E. Balazs, Academic Press, New York

Holtzer, H. and Sanger, J. W. (1972) Myogenesis: Old views rethought. In: Research in muscle development and the muscle spindle, (Banker, B., Przybylski, R., Van der Meulen, J. and Victor, M. eds.) *Excerpta Medica,* Amsterdam

Holtzer, H., Marshall, J. and Fink, H. (1957) *J. Biophys. and Biochem. Cytology,* *3,* 705

Holtzer, H., Weintraub, H., and Mayne, R. (1972) In: *Current topics in developmental biology,* 9 (Moscona, A. and Monray, S. eds.) In Press, Academic Press, New York and London

Ishikawa, H., Bischoff, R., and Holtzer, H. (1968) *J. Cell Biol.* *38,* 538

King, R. (1970) *Ovarian Development in Drosophila melanogaster,* Academic Press, New York

Konigsberg, I. (1963) *Science* *140,* 1273

Lin, S. and Riggs, A. (1971) *Biochem. Biophys. Res. Commun.,* *45,* 1542

Mahowald, S. (1968) *J. Exp. Zool.,* *167,* 237

Mayne, R., Abbott, J. and Holtzer, H. (1972) *Exp. Cell. Res.,* in press

Okazaki, K. and Holtzer, H. (1965) *J. Cyto. Histochem.,* *13,* 726

Sanger, J., Holtzer, S., and Holtzer, H. (1970) *Nature, New Biology,* *229,* 121

Sanger, J. and Holtzer, H. (1972) *Proc. Nat. Acad. Sci. U.S.,* *69,* 253

Schubert, M. and Hammerman, D. (1968) *A Primer on Connective Tissue,* Biochemistry, Lea and Febiger, Philadelphia

Schulte-Holthausen, H., Davidson, E., Chacko, S., Holtzer, H. (1969) *Proc. Nat. Acad. Sci. U.S. 63,* 864

Stockdale, F., Okazaki, K., Nameroff, M. and Holtzer, H. (1964) *Science 146,* 533

Thorell, B. and Raunick, L. (1966) *Ann. Med. Exp. Biol. Fenn., 44,* 131

Thorp, F. and Dorfman, S. (1967) In: *Current Topics in Developmental Biology,* Vol. 2. (Moscona, S. S. and Monray, S. eds.) Academic Press, New York

Watterson, R., Fowler, I. and Fowler, B. (1954) *Amer. J. Anat. 95,* 337

Weintraub, H., Campbell, G. and Holtzer, H. (1971) *J. Cell Biol., 50,* 652

Weintraub, H., Campbell, G. and Holtzer, H. (1972) *J. Mol. Biol.* (In press)

Yaffee, D., 1969, In: *Current Topics in Developmental Biology,* Vol. 4, p. 37 (Moscona, S. S., and Monray, S. ed.) Academic Press, New York

The Dependence of Gene Expression on Membrane Assembly

C. G. DUCK-CHONG AND J. K. POLLAK
Department of Histology and Embryology, University of Sydney, Sydney, N.S.W., Australia

A distinctive feature of eukaryote cells is the presence of extensive intracellular membrane systems, which are organized to give rise to a number of different organelles. In prokaryotes such membrane systems are either absent altogether, or present only in a rudimentary form.

The lipoprotein membranes which compose and surround organelles are endowed with many important functions. Enzymes may be loosely associated with these membranes, or they may form an integral part of the membrane structure. The activity of a membrane enzyme does not necessarily reflect the amount of enzyme present within the cell, as the expression of enzyme activity may be influenced by the interaction of the enzyme with the membrane complex.

We would like to propose a mechanism for the regulation of gene expression at the level of membrane assembly or organisation. Two separate examples will be described to illustrate this post-translational control. The first example deals with the expression of the membrane-bound enzyme, glucose 6-phosphatase (EC 3.1.3.9) (G6Pase), in foetal, neonatal, suckling and adult rat liver, while the second example is concerned with the maturation of rat liver mitochondria.

THE DEPENDENCE OF THE EXPRESSION OF G6PASE ON MEMBRANE STRUCTURE

G6Pase is localized on the membranes of both the rough and the smooth endoplasmic reticulum and remains associated with isolated microsomes (Tice and Barrnett, 1962; Hers *et al.*, 1951). It is a multifunctional enzyme, exhibiting both glucose 6-phosphate phosphohydrolase (hydrolase) activity and inorganic pyrophosphate-glucose phosphotransferase (transferase) activity.

Hydrolase: glucose 6-phosphate (G6P) $+ H_2O \longrightarrow$ glucose $+ P_i$

Transferase: glucose $+ PP_i \longrightarrow$ G6P $+ P_i$

Other activities have been reported (see review, Nordlie, 1969; Lueck and Nordlie, 1970). However, only the hydrolase and transferase activities will be considered in this paper.

In developing rat liver, a stage of extensive morphological and enzymic differentiation occurs during the time immediately surrounding birth (Dawkins, 1966; Greengard, 1971; Dallner *et al.*, 1966a, 1966b; Leskes *et al.*, 1971). Smooth endoplasmic reticulum, absent before birth, proliferates rapidly during the first 24 hours after birth. Many liver enzymes appear for the first time either a few days before or just after birth. Among the "cluster" of enzymes which appear during late foetal development is G6Pase. A second dramatic increase in G6Pase, commencing immediately after birth, is well documented (Kretchmer, 1959; Burch *et al.*, 1963; Dawkins, 1966; Yeung *et al.*, 1967; Greengard, 1969; Leskes *et al.*, 1971). This is illustrated in Fig. 1 (adapted from Yeung *et al.*, 1967). Published values for maximum activity in postnatal rats range from 2 to 3.5 times the adult value, the peak activity being reported variously at 2 to 7 days after birth. It should be noted that in all these investigations G6Pase was measured in terms of the hydrolase activity. While the presence of transferase activity in both late foetal and postnatal rats has been reported (Fisher and Stetten, 1966; Lueck and Nordlie, 1970), no information was available to indicate the rate at which transferase accumulates during this time.

FIGURE 1

Neonatal development of G6Pase, measured as hydrolase activity.

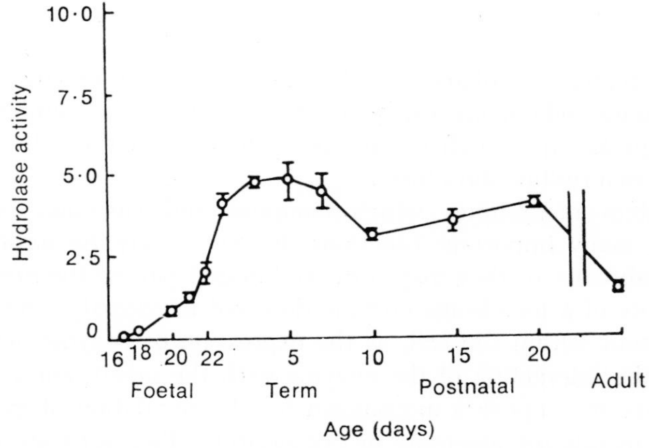

Hydrolase activity is expressed as μmol Pi produced/mg protein/h at 30°C. The vertical bars represent ± 1 S.E.M. (Adapted from Yeung *et al.*, 1967.)

In foetal and postnatal rat livers, both protein synthesis and membrane proliferation occur more rapidly than in adult rat liver. We wanted to know whether the G6Pase produced in the liver during this period of rapid growth has similar properties to the G6Pase in adult liver. This problem was approached in two ways:

1. Hydrolase and transferase activities were measured using whole homogenates of livers from foetal, neonatal, suckling and adult rats, to determine whether changes in hydrolase activity around the time of birth are, in fact, paralleled by changes in transferase activity.

2. The same homogenates were used to measure the response of the hydrolase activity to treatment with Triton-X100 (Triton), a non-ionic detergent. It is known that both activities of G6Pase can be enhanced by treating homogenates or microsomes from adult rat liver with one of a variety of detergents or by pretreating microsomes at pH9.5—9.8 before assay (Nordlie, 1969; Stetten and Burnett, 1967).

The results of these experiments are presented in Table 1. The pH chosen for the assay of each activity was near to the optimal pH for that activity in adult rat liver (Nordlie *et al.*, 1968). It has been shown that, with the possible exception of foetal liver, the stability of the enzyme under the conditions of the assays is comparable in homogenates from all ages, and that the hydrolase results were not significantly affected by non-specific phosphatase, measured using β-glycerophosphate as substrate (Duck-Chong, unpublished).

The expected changes in hydrolase activity were obtained (Table 1), the results being in close agreement with those of other workers who also used Wistar rats (Yeung *et al.*, 1967 and Fig. 1). Increases in transferase activity coincided in time with increases in hydrolase activity. However the transferase/hydrolase ratio was not constant. The ratio increased from 0.8 before birth to a maximum of 2.6 at 4-7 days after birth and then decreased to 1.2 in the adult rat. In addition, the response of the hydrolase activity to Triton treatment was not uniform. The enzyme in 19 day foetal rat liver was inactivated by Triton treatment, whereas at 4-7 days after birth the same treatment increased the hydrolase activity by 270%. Hydrolase activity in adult rat liver was increased by only 45-75%. Similar variations in the response of the hydrolase from neonatal, suckling and adult rat liver have also been observed after treatment with sodium deoxycholate (Duck-Chong, unpublished).

As the above assays were carried out at only one pH for each activity, it was possible that the results reflected changes in pH optima during development and perhaps the presence of isoenzymes. The effect of pH on each activity both before and after Triton treatment was investigated, using either whole homogenates or isolated microsomes from the livers of 5-7 day suckling or adult rats. The results of two experiments are shown in Fig. 2. Similar pH activity curves were obtained whether whole homogenates or microsomes were used, and the response of the hydrolase activity of the microsomes to Triton treatment was much the same as the response of the homogenate used for their isolation. The differences between G6Pase from suckling and adult liver, already described, were again apparent. The following additional points should be noted:

1. In microsomes from both 5 day suckling and adult rat liver the optimal pH for the hydrolase activity was pH5.5-6.0 before Triton treatment and pH6 after Triton treatment.

TABLE 1

G6Pase activities in whole homogenates of rat liver

Age of rats	Enzyme activities[a] Transferase pH4.5	Hydrolase pH6.5	Transferase/ hydrolase ratio	% activation of hydrolase by Triton	Protein/ g liver (mg)
Foetal (19-21 days)	0.4	0.6	0.6	− 30	133
	1.7	2.0	0.8	+ 75	111
Neonatal (0-1 day)	6.9	5.2	1.3	+ 80	139
	11.1	8.0	1.4	+ 60	134
Suckling (4-7 days)	21.8	10.7	2.0	+270	140
	27.4	13.1	2.1	+240	160
	23.5	9.2	2.6	+290	144
(14-16 days)	12.8	7.0	1.8	+170	187
	16.6	9.5	1.8	+140	174
Adult, female (approx. 200 g)	7.5 (8.6)	7.5	1.0 (1.2)	+ 45	186
	7.7 (8.8)	5.6	1.4 (1.6)	+ 75	214
	6.5 (7.4)	6.1	1.1 (1.3)	+ 50	209

All rats were of the Wistar strain. 10% homogenates in 0.25M sucrose were supplemented with 0.25 volumes of either distilled water or 1% Triton (w/v). After 10-15 min. in ice, 0.25M sucrose was added to dilute the homogenate to a final concentration of 2% or 4% for Triton treated and untreated samples respectively. Hydrolase activity was assayed using 20mM G6P. The released P_i was measured after precipitating proteins with $HC1O_4$. Transferase activity was assayed using 180mM glucose and 20mM PP_i. The reaction was stopped by the addition of chloroform and G6P was measured (Duck-Chong and Pollak, 1972). In each case, 0.1ml homogenate was added to 0.4ml substrate in 80mM Na succinate/80mM Na maleate buffer. Incubations were carried out in duplicate at 30°C for 5 min. Protein was assayed by a modified Lowry method (Campbell and Sargent, 1967). The results in parenthesis are corrected for differences in the optimal pH of transferase from adult and postnatal liver, see text.

[a]Activities are expressed as μmol product formed/min/g liver.

FIGURE 2

The effect of pH on the transferase and hydrolase activities of isolated microsomes, before and after treatment with Triton.

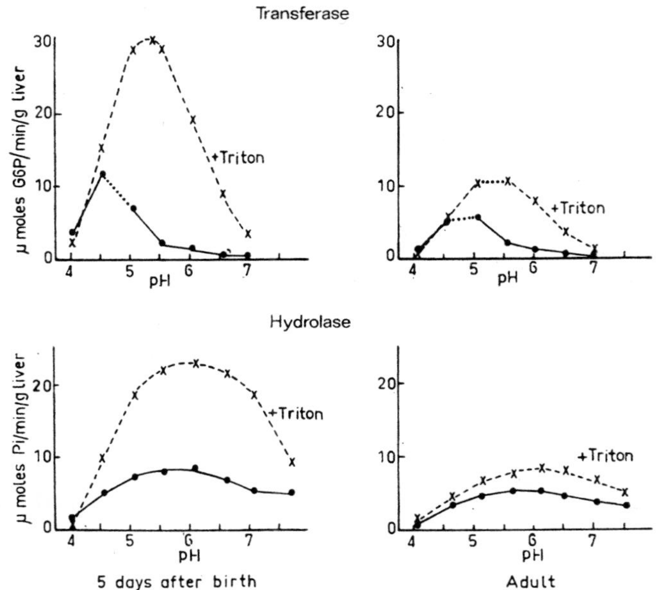

Microsomes were prepared from 10% homogenates in O. 25M sucrose. Nuclei and mitochondria were sedimented and washed twice by centrifuging at 6000g for 10 min. Microsomes were collected from the combined supernatants at 105,000g for 60 min and resuspended in 0.25M sucrose (equivalent to 12-18mg liver/ml). Enzyme activities were assayed after supplementation of the microsomes with either distilled water or Triton and dilution with sucrose, as in Table 1. The recovery of microsomes, determined by relating the hydrolase activity of the microsomes to the hydrolase activity of the original homogenate, was 65% and 61% for 5 day and adult liver respectively. The yields of microsomal protein from 5 day and adult liver were 32mg and 40mg/g liver respectively.

2. In each case, the optimal pH for the transferase activity in the absence of Triton lay between pH4.5 and 5.0 and was shifted to pH5.3 by Triton treatment. Additional experiments have shown that, in the absence of Triton, the peak activities occur at pH4.5 and pH4.8 in 1-12 day postnatal and adult rat liver respectively. As a result, the transferase/hydrolase ratio in adult liver tends to be underestimated by assaying at pH4.5. Corrected transferase activities and transferase/hydrolase ratios for adult rat liver are shown in parenthesis in Table 1. The transferase/hydrolase ratio varied at least two fold throughout postnatal development even after correction of the transferase results.
3. The transferase/hydrolase ratio measured at the optimal pH for each activity after Triton treatment appeared to be the same at both ages.

The last point was investigated further and the results are shown in Table 2. In homogenates and microsomes prepared from neonatal, suckling and adult rat livers, the transferase/hydrolase ratio after Triton treatment was constant at 1.4 ± 0.1 S.D.

TABLE 2

The ratio of transferase activity/hydrolase activity after Triton treatment

Age of rats	Transferase/Hydrolase Ratio	
	Homogenates	Microsomes
Neonatal (0-1 day)	1.4 1.5	
Suckling (5-7 days)	1.2 1.5	1.3 1.3
(12-14 days)	1.4 1.5	
(21 days)	1.4	
Adult	1.3 1.3	1.4

Transferase and hydrolase activities of homogenates or microsomes were measured at pH5.3 and pH6 respectively, after pretreatment with Triton, as described in Table 1.

These findings are in agreement with other results which indicate that the intrinsic differences between the transferase/hydrolase ratios of G6Pase from the livers of fed, fasted, diabetic and hormone-treated rats are removed or reduced by detergent treatment (Fisher and Stetten, 1966; Nordlie and Snoke, 1967; Nordlie *et al.*, 1968).

Thus it appears that changes in the total amount of G6Pase present during development are associated with variations in the expression or suppression of each activity. Qualitative and quantitative changes in the composition of the microsomal membranes are known to occur during the period investigated (Dallner *et al.*, 1966a,b). Hence the micro-environment provided by the membranes is not constant throughout development. Moreover, treatment of the microsomes with detergent, which is known to remove significant amounts of protein and phospholipid from the membranes (Palade and Siekevitz, 1959; Hunter and Korner, 1966) abolished the characteristic differences in the transferase/hydrolase ratios of G6Pase from rat liver at different ages.

It therefore seems likely that the observed variations in the transferase/hydrolase ratios of G6Pase in untreated preparations are in some way related to variations in the interaction of the enzyme with the microsomal membrane. It follows that the amount of each activity expressed as a result of gene activation may depend not only on the number of enzyme molecules synthesized, but also on the nature of the microsomal membranes at that time. Thus, enzyme-membrane interaction, by modifying the activities of G6Pase, may act as an additional regulatory step in the process of gene expression.

THE MATURATION OF THE INNER MITOCHONDRIAL MEMBRANE

The second example illustrating the dependence of gene expression on membrane assembly is based on some of our work on the biogenesis of fully functional mitochondria during the development and differentiation of rat liver.

Rat or chick liver mitochondria have been separated into two well-defined fractions by isopycnic centrifugation (Pollak and Munn, 1970; Pollak and Woog, 1971; Packer *et al.*, 1971). The mitochondria equilibrating in the less dense region of the gradient (d=1.184) are referred to as B2 mitochondria, while those equilibrating in the denser region (d=1.216) are called B3 mitochondria. The ratio of B2/B3 mitochondria depended on the physiological state of the rat (Pollak and Munn, 1970). In the early developmental stages (6 days prenatal) B3 mitochondria predominated, but during development the B2/B3 ratio increased. The B2/B3 ratio was 0.1 six days before birth, 0.7 two days before birth, 2.6 one day after birth and varied between 3 and 8 in adult rats (Pollak and Munn, 1970; Woog and Pollak, unpublished). However, the two fractions (B2 and B3) had the same protein/lipid ratio, protein/phospholipid ratio, phospholipid and total fatty acid composition and virtually the same distribution of marker ensymes for inner mitochondrial membranes, outer mitochondrial membranes and mitochondrial matrix (Pollak and Munn, 1970).

The latter results, as well as data from other experiments not relevant to the present discussion (Pollak and Munn, 1970), led to the concept that the equilibration of B3 mitochondria in the denser part of the isopycnic sucrose gradient was due to the greater permeability of the inner mitochondrial membrane to low-molecular weight compounds such as sucrose.

The relationship between B2 and B3 mitochondria

The predominance of B3 mitochondria during the earlier stages of liver development and their subsequent disappearance concomitant with the appearance of B2 mitochondria suggested a precursor-product relationship between B3 and B2 mitochondria (see Scheme 1, Fig. 3). However, pulse labelling of rat liver slices taken from two developmental ages (6 days prenatal and 2 days prenatal) showed no significant transfer of label from B3 to B2 at either developmental age (Pollak, 1971). In fact from 6 days prenatal to 6 days postnatal, B2 mitochondria always had a slightly greater specific activity than B3 mitochondria, whether isotope labelling was carried out *in vivo* or using tissue slices.

An alternative model was proposed, in which two populations of mitochondria were considered to proliferate at inversely changing rates (see Scheme 2, Fig. 3). Turnover studies were carried out to determine the "apparent" halflife of mitochondria between the 14th and 22nd day of gestation. The experiments do not represent true halflife evaluations, since foetal liver is a rapidly growing tissue and therefore not in the steady state condition required for halflife determinations. Nevertheless it was considered that a non-linear relationship between the log of the counts remaining in mitochondrial proteins in relation to time would indicate at least that the results were compatible with the second scheme.

FIGURE 3

Model systems to account for the changing proportions of B2 and B3 mitochondria during foetal and postnatal development

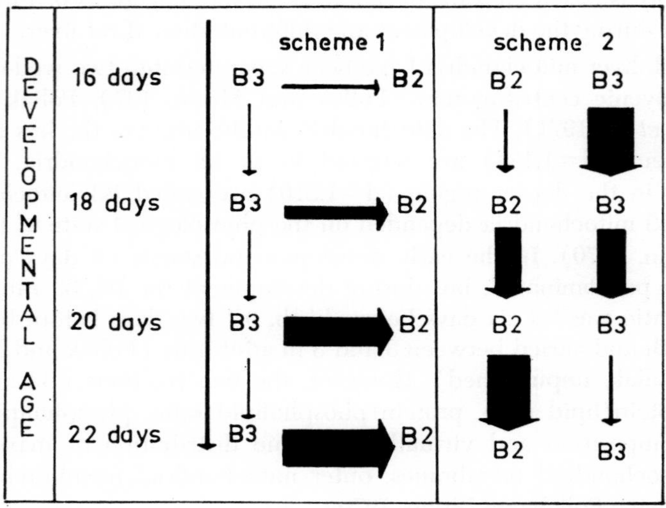

The results, as presented in Fig. 4, show clearly that a straight line relationship held for all the mitochondria that were examined, thus indicating that scheme 2 is unacceptable. In addition, in foetal rat liver, B2 and B3 mitochondria had identical halflives (Table 3). In adult rat liver the halflives of total mitochondria and B2 mitochondria were similar, while that of B3 mitochondria was considerably shorter (Table 3), lending further weight to the suggestion that in adult liver B3 mitochondria may be aged mitochondria getting on towards the end of their life span (Pollak and Munn, 1970).

TABLE 3

Halflives of Rat Liver Mitochondria

Age of rats	Halflife in days		
	Total Mitochondria	B2 Mitochondria	B3 Mitochondria
Foetal	1.75	1.68	1.68
Adult	9.0	9.6	4.8

The halflives of the mitochondria were determined from the plots in Fig. 4, using the relationship: $t_{\frac{1}{2}}$ (halflife) $= \frac{\ln 2}{k}$ where $k = \frac{2.3}{t} \log \frac{Co}{Ct}$. t = time interval at which the measurement of remaining counts took place and Co is the extrapolated value of d.p.m./mg protein at 0 min and Ct the d.p.m./mg protein at the time t.

FIGURE 4

The decay of mitochondrial proteins isolated from pregnant and foetal rat liver

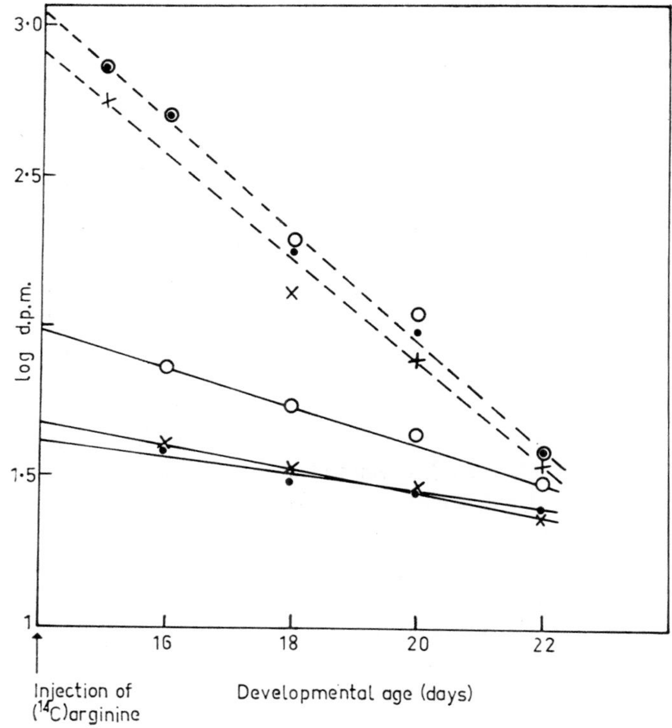

Decay curves were prepared for B2, B3 and total mitochondrial preparations isolated from foetal rat livers and the livers of their mothers as described by Pollak and Munn (1970). On the 14th day of pregnancy, the mother rats, from a Wistar strain, were injected with 10 μCi L-[guanido-^{14}C] arginine in 0.9% NaCl (specific activity = 50 mCi/mmol). At 1, 2, 4, 6 and 8 day intervals the pregnant rat and foetal rats were killed and B2, B3 and total mitochondria were isolated from a portion of the adult rat liver and the pooled livers of the foetal rats. Suitable samples were placed on Whatman 3MM paper disks and treated by the method of Mans and Novelli (1961). The paper disks containing the samples were placed into counting vials with 10ml scintillation fluid (0.4% PPO and 0.01% POPOP in toluene) and counted in a Packard 2111 scintillation spectrometer. All samples were counted for at least 4000 c.p.m. and corrected for counter efficiency, quenching and background. The log of the specific activities (d.p.m. /mg protein) were plotted against the time elapsed from the moment of the isotope injection. Proteins were determined by the method of Lowry *et al.* (1951).

Decay curves of total adult mitochondria, × —— ×; B2 adult mitochondria, • —— •; B3 adult mitochondria, ○ —— ○; total foetal mitochondria, × - - - ×; B2 foetal mitochondria, • - - - •; B3 foetal mitochondria, ○ - - - ○.

Scheme 3 (Fig. 5) is compatible with all the data presented so far. This scheme represents a cyclic pathway of mitochondrial biosynthesis, which includes the alternating production of B2 and B3 mitochondria. Mitochondrial division produces B3 daughter-mitochondria with sucrose-permeable or "leaky" inner membranes; these become converted to "non-leaky" B2 mitochondria, which eventually divide again (Fig. 5). It is proposed that the synthesis of protein required to

FIGURE 5

An alternative model for mitochondrial biosynthesis involving B2 and B3 mitochondria in an alternating cyclic process: Scheme 3

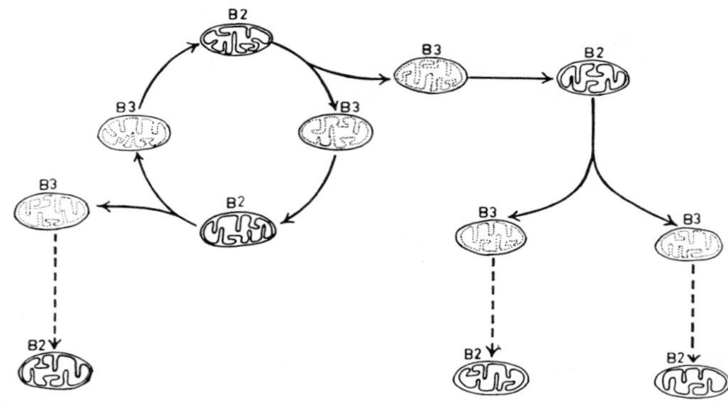

convert B3 to B2 mitochondria is the rate-limiting step in foetal rat liver, while in adult rat liver mitochondrial division is rate-limiting. Scheme 3 (Fig. 5) accounts for the identical halflife of B2 and B3 mitochondria in a rapidly growing tissue, such as foetal rat liver. It is realized that the halflife of a complex mixture of different proteins, such as make up a mitochondrion, cannot be treated as an average halflife of the constituent proteins, particularly in decay experiments (Koch, 1962); therefore the numerical values have limited significance. Nevertheless the very similar decay curves for foetal B2 and B3 mitochondria reinforce the concept that they are alternating stages derived from one mitochondrial population.

The role of cytoplasmic and mitochondrial protein synthesis in the maturation of the inner mitochondrial membrane

In short term *in vivo* labelling experiments, the higher specific activity of proteins from B2 mitochondria of foetal rat liver compared with the specific activity of proteins from B3 mitochondria of the same tissue (Table 4), is consistent with the proposal that additional protein synthesis is required during the conversion of B3 to B2 mitochondria. In foetal rat liver the presence of chloramphenicol (CAP) reduced the ratio of d.p.m. in B2 mitochondria to d.p.m. in B3 mitochondria from 1.32 to 1.04 (Table 4). This selective effect of CAP on B2 mitochondria implies that B3 mitochondria are not capable of mitochondrial protein synthesis. In adult rat liver mitochondria no significant differences in the d.p.m. incorporated into B2 and B3 mitochondria were apparent, either in the presence or absence of CAP (Table 4). The lack of effect of CAP on the mitochondria of non-dividing cells has been previously pointed out by Firkin and Linnane (1969).

In a recent investigation concerned with the incorporation of amino acids into isolated rat liver mitochondria, Coote and Work (1971) found that 93% of the label incorporated into mitochondrial proteins was present in a pH 11.5 insoluble

TABLE 4

Effect of chloramphenicol on *in vivo* labelling of B2 and B3 mitochondrial proteins

	Foetal mitochondria			Adult mitochondria		
	B2 d.p.m.	B3 d.p.m.	B2 d.p.m. / B3 d.p.m.	B2 d.p.m.	B3 d.p.m.	B2 d.p.m. / B3 d.p.m.
Control	423	321	1.32	78	83	0.94
+CAP	370	355	1.04	92	86	1.07

Female Wistar rats that were between 18-20 days pregnant were injected with 0.9% NaCl containing 4 mg CAP/ml. The volume was chosen so that 35mg CAP/kg were injected. In a control pregnant rat a similar volume of 0.9% NaCl was injected. After 30 min 4μCi L-[U-^{14}C] leucine (10mCi/mmol) in 0.2 ml 0.9% NaCl were also injected and the rats killed after a further 60 min. B2 and B3 mitochondria were isolated from the foetal livers and the maternal liver of the control and the CAP-injected rats as described by Pollak and Munn (1970). The preparation of samples for scintillation counting and the counting were carried out as described in Fig. 4.

fraction. This fraction, which was also rich in phospholipids, contained about 15% of the total mitochondrial protein and on polyacrylamide gel electrophoresis the proteins of this fraction separated into at least 20 protein bands (Coote and Work, 1971). Ten of these were labelled and therefore may be regarded as being coded for by mitochondrial DNA, while the other bands were not labelled and therefore are thought to have originated from cytoplasmic protein synthesis. In earlier studies Neupert *et al.*, (1967) as well as Work (1968) showed that virtually all the label incorporated by isolated mitochondria is associated with the proteins of the inner mitochondrial membrane.

In our hands the method of isolating the pH 11.5 insoluble mitochondrial proteins proved very reproducible and in 23 preparations from 5-7 day suckling rat liver we found that this fraction represented 14.1% $\pm$ 0.44 S.E.M. of the total protein. The *in vivo* inhibition studies presented in Table 5 show that the labelling of the pH 11.5 fraction derived from B2 mitochondria was 42% inhibited by CAP and 67% inhibited by cycloheximide (CH). As B2 mitochondria predominate in adult rat liver, these results confirm the findings of Coote and Work (1971), that the mitochondrial proteins of the pH 11.5 insoluble fraction from adult rat liver are partly derived from mitochondrial and partly from cytoplasmic protein synthesis. In contrast, the incorporation of amino acids into the pH 11.5 insoluble fraction of B3 mitochondria was virtually unaffected by CAP, but was completely inhibited by CH (Table 5).

These results thus show clearly that B3 mitochondria differ from B2 mitochondria in that the former are incapable of carrying out protein synthesis. Consequently they lack some of the proteins that are present on the inner membranes of B2 mitochondria and may require cytoplasmic protein synthesis as the first step in the conversion of B3 to B2 mitochondria.

TABLE 5

Inhibition of amino acid incorporation into mitochondrial proteins of 5-7 day suckling rat liver

	% Inhibition by	
	Chloramphenicol	Cycloheximide
B2 mitochondria	14	95
B3 mitochondria	0	100
pH 11.5 insoluble fraction of B2 mitochondria	42	67
pH 11.5 insoluble fraction of B3 mitochondria	4	100

Suckling rats, 5-7 days old and weighing 10-13g were selected from litters with the same birth date. These were divided into three groups (equal numbers from each litter in each group). The groups were injected either with CH (1mg/10g rat) or CAP-succinate (5mg/10g rat). CH and CAP-succinate were dissolved in 0.9% NaCl. The third group was injected with 0.9% NaCl. In all instances the injected volume did not exceed 100μl. The cycloheximide injection was repeated after 10 min, at 20 min each rat was injected with 0.5 μCi of L-[U-^{14}C] leucine in 10 μl (10 mCi/mmol). After another 30 min the rats were killed; B2 and B3 mitochondria were prepared as described in Fig. 4. The pH 11.5 insoluble fractions were prepared as described by Coote and Work (1971). The preparation of samples for scintillation counting and the counting were carried out as described in Fig. 4.

The development of respiratory control

Mitochondria from foetal rat liver exhibited no respiratory control (Mintz *et al.*, 1967 and Table 6). Not even in the presence of bovine serum and Mg^{2+} during the entire isolation procedure and in the assay mixture was respiratory control evident (Pollak, unpublished). It was thought that the relative scarcity of B2 mitochondria in foetal rat liver may have been the cause of this lack of respiratory control. However respiratory control was still lacking when concentrated suspensions of mitochondria were used. The concentrations of these suspensions were calculated to contain sufficient B2 mitochondria to allow the measurement of respiratory control, assuming that the B2 mitochondria from foetal liver behave in the same way as adult rat liver mitochondria.[a] The lack of respiratory control in foetal rat liver mitochondria, even in the presence of a sufficient number of B2 mitochondria, suggests that a two-step process may be involved; the first step results in the conversion of B3 into B2 mitochondria with respect to inner membrane permeability, while the second step is concerned with the establishment of respiratory control.

Preliminary results indicate that liver mitochondria of neonatal rats less than 15 hours old have a respiratory control index of 1.8 (Table 6). Liver mitochondria isolated from 6 day suckling rats show a further increase in respiratory control,

[a]It was found by dilution experiments that an excess of 1 mg adult mitochondrial protein per 2 ml reaction mixture had to be present in the oxygen electrode vessel to carry out measurements of respiratory control. So far it has not been technically possible to isolate enough B2 and B3 mitochondria from foetal rat liver to undertake respiratory control measurements, but such separations will have to be carried out in the future.

which is however still lower than that of adult rat liver mitochondria (Table 6). The ratio of ADP added to oxygen consumed during the return to state 4 respiration (ADP/O ratio) of mitochondria from 6 day suckling rats is also lower than that of adult rat liver mitochondria (Table 6).

Mitochondria from foetal rat liver lack respiratory control but are capable of oxidative phosphorylation when this process is measured in the presence of an ATP-trap such as glucose plus hexokinase (Table 6). The P/O ratio of foetal rat liver mitochondria is significantly lower than that of adult rat liver mitochondria (Table 6). In all these respects the foetal rat liver mitochondria resemble the loosely-coupled mitochondria of skeletal muscle from a hypermetabolic patient (Ernster *et al.*, 1959; Luft *et al.*, 1962; Ernster and Luft, 1963). Ernster and Luft (1963) suggested, that a subtle aberration at the molecular level, such as a change in the lipoprotein responsible for the maintenance of a water-free environment around the sites where energy-rich bonds are formed, might be the reason for the lack of respiratory control. It is considered that in the case of foetal rat liver, where rapid mitochondrial biosynthesis is occurring, the composition or organization of the inner mitochondrial membrane is

TABLE 6

Respiratory control index, ADP/O and P/O ratios of rat liver mitochondria at various developmental ages

Age of rats	Respiratory Control Index (state 3/state 4)	ADP/O	P/O
Foetal	1 ± 0 (8)	—	1.1 ± 0.12 (8)
Neonatal 0—15 h	1.8 (1)	0.5 (1)	—
5—7 day suckling rats	2.7 ± 0.01 (14)	1.2 ± 0.06 (14)	0.7 ± 0.14 (6)
Adult	5.4 ± 0.35 (18)	1.7 ± 0.08 (18)	2.1 ± 0.09 (9)

Substrate: Sodium succinate

Results are presented ± S.E.M. and the number of experiments carried out is given in parenthesis.

Mitochondria were isolated and fractionated as described by Pollak and Munn (1970). Respiratory control and ADP/O (the ration of ADP added to the amount of oxygen taken up during the return to state 4 respiration) were measured as described previously (Pollak and Woog, 1971). P/O ratios were determined by replacing 0.2 ml of water in the reaction mixture with 0.2 ml 0.25M glucose containing an excess of hexokinase to trap the ATP formed in the reaction. Mitochondria, sodium succinate and ADP were added in quick succession and the oxygen uptake was measured for a known time interval; the reaction was stopped by pipetting 0.5 ml of the incubation mixture into 0.5 ml cold chloroform. The G6P formed during the incubation period was measured by a modification of the method of Duck-Chong and Pollak (1972).

such that the action of the respiratory enzymes present on the membrane is significantly modified, resulting in immature cell organelles (either B3 or immature B2 mitochondria) whose metabolism differs from that of the mature organelle. It is only when immature mitochondria predominate, as for example in foetal rat liver, that the membrane effect on gene expression becomes apparent, though it may be observed to a progressively decreasing extent in 0-1 day neonatal and 6 day suckling rat liver mitochondria (Table 6).

CONCLUSIONS AND SUMMARY

Two systems have been described which are dependent on membrane structure for the expression of genetic information.

G6Pase increases with development, but in addition the expressed activity, especially the hydrolase activity, is a variable which depends on the interaction of the enzyme with the microsomal membranes. The loosening of structural relationships by detergent action increases G6Pase activities to variable degrees at different developmental stages, demonstrating the controlling action that a membrane may exercise on its constituent enzymes.

The results of labelling experiments indicate that mitochondrial biosynthesis is a cyclic process involving alternate steps of division and maturation of the inner membrane. During the maturation of foetal rat liver mitochondria, the ability to carry out oxidative phosphorylation precedes the ability to exercise respiratory control. This implies that the essential enzymes involved in oxidative phosphorylation are present already in immature mitochondria and the tight coupling of oxidative phosphorylation appears not to be dependent on the transcription and translation of the appropriate enzymes, but on the maturity and organization of the inner mitochondrial membrane.

ACKNOWLEDGEMENTS

This work was aided by a University of Sydney Research Grant and by a Grant from the University of Sydney Postgraduate Medical Foundation.

A gift of chloramphenicol from Parke-Davis & Co. is also gratefully acknowledged. One of us (C. D-C.) was supported by a Commonwealth Post-Graduate Research Award.

REFERENCES

Burch, H. B., Lowry, O. H., Kuhlman, A. M., Skerjance, J., Diament, E. J., Lowry, S. R. and Von Dippe, P. (1963) *J. Biol. Chem.* *238,* 2267

Campbell, P. N. and Sargent, J. R. (1967) in *Techniques in Protein Biosynthesis* (Campbell, P. N. and Sargent, J. R., eds.), p. 299, Academic Press, London

Coote, J. L. and Work, T. S. (1971) *Eur. J. Biochem.* *23,* 564

Dallner, G., Siekevitz, P. and Palade, G. E. (1966a) *J. Cell. Biol.* *30,* 73

Dallner, G., Siekevitz, P. and Palade, G. E. (1966b) *J. Cell. Biol.* *30,* 97

Dawkins, M. J. R. (1966) *Brit. Med. Bull.* *22,* 27

Duck-Chong, C. G. and Pollak, J. K. (1972) *Anal. Biochem.* *49,* 139

Ernster, L., Ikkos, D. and Luft, R. (1959) *Nature* (*London*) *184,* 1851

Ernster, L. and Luft, R. (1963) *Expl Cell Res. 32,* 26

Firkin, F. C. and Linnane, A. W. (1969) *Expl Cell Res. 55,* 68

Fisher, C. J. and Stetten, M. R. (1966) *Biochim. Biophys. Acta. 121,* 102

Greengard, O. (1969) *Biochem. J. 115,* 19

Greengard, O. (1971) in *Essays in Biochemistry* (Campbell, P. N. and Dickens, F., eds), vol. 7, p. 159, Academic Press, London and New York

Hers, H. G., Berthet, J., Berthet, L. and de Duve, C. (1951) *Bull. Soc. Chim. Biol.,* Paris, *33,* 21

Hunter, A. R. and Korner, A. (1966) *Biochem. J. 100,* 73p

Koch, A. L. (1962) *J. Theor. Biol. 3,* 283

Kretchmer, N. (1959) *Pediatrics 23,* 606

Leskes, A., Siekevitz, P. and Palade, G. E. (1971) *J. Cell. Biol. 49,* 264

Lowry, O. H., Rosebrough, N. J., Farr, A. L. and Randall, R. J. (1951) *J. Biol. Chem. 193,* 265

Lueck, J. D. and Nordlie, R. C. (1970) *Biochem. Biophys. Res. Commun. 39,* 190

Luft, R., Ikkos, D., Palmieri, G., Ernster, L. and Afzelius, B. (1962) *J. Clin. Invest, 41,* 1776

Mans, R. T. and Novelli, G. D. (1961) *Arch. Biochem. Biophys. 94,* 48

Mintz, H. A., Yawn, D. H., Safer, B., Bresnick, E., Liebelt, A. G., Blailock, Z. R., Rabin, E. R. and Schwartz, A. (1967) *J. Cell Biol. 34,* 513

Neupert, W., Brdiczka, D. and Bücher, Th. (1971) *Biochem. Biophys. Res. Commun. 27,* 488

Nordlie, R. C. and Snoke, R. E. (1967) *Biochim. Biophys. Acta. 148,* 222

Nordlie, R. C., Arion, W. J., Hanson, T. L., Gilsdorf, J. R. and Horne, R. N. (1968) *J. Biol. Chem. 243,* 1140

Nordlie, R. C. (1969) *Ann. N.Y. Acad. Sci. 166,* 699

Packer, L., Pollak, J. K., Munn, E. A. and Greville, G. D. (1971) *Bioenergetics 2,* 305

Palade, G. E. and Siekevitz, P. (1959) *J. Biochem. Biophys. Cytol. 2,* 171

Pollak, J. K. (1971) *Proc. Aust. Biochem. Soc. 4,* 83

Pollak, J. K. and Munn, E. A. (1970) *Biochem. J. 117,* 913

Pollak, J. K. and Woog, M. (1971) *Biochem. J. 123,* 347

Stetten, M. R. and Burnett, F. F. (1967) *Biochim. Biophys. Acta. 139,* 138

Tice, L. W. and Barrnett, R. J. (1962) *J. Histochem. Cytochem. 10,* 754

Work, T. S. (1968) in *Biochemical Aspects of the Biogenesis of Mitochondria,* Slater, E. C., Tager, J. M., Papa, S. and Quargliariello, E. eds.), p. 367, Adriatica Editrice, Bari

Yeung, D., Stanley, R. S. and Oliver, I. T. (1967) *Biochem. J. 105,* 1219

Patterns of Gene Activity in Larval Tissues of the Blowfly, *Calliphora*

J. A. THOMSON
Genetics Department, University of Melbourne, Parkville, Victoria, Australia

It is now widely recognised that the giant polytene chromosomes found in terminally differentiated tissues of certain organisms, most notably amongst the Diptera, offer unique opportunities for correlated cytological and biochemical approaches to the study of gene activity and its regulation (Ashburner, 1970). Unfortunately, until the advent of mass isolation techniques (e.g. Boyd *et al.*, 1968; Zweidler and Cohen, 1971), tissues with polytene chromosomes have been better known to cytologists than to biochemists. One consequence has sometimes been the selection of experimental material for studies of gene activity on the basis of favourable chromosome banding and chromosome continuity. Such studies have proved especially helpful in revealing the relation of local chromosome dispersion ("puffing") to gene activity (Beermann, 1952; Ashburner, 1970). On the other hand, the occurrence and functional significance of larger scale chromosome dispersion, involving whole blocks of band-interband sequences, may escape attention if a criterion which ensures a relatively low level of gene read-out is employed in choosing tissues for investigation (Thomson, 1969).

In the blowfly larva, protein synthesis in the body wall (epidermis and associated musculature), the fat body (Price, 1966) and the salivary glands takes place predominantly during the first two-thirds of larval life (Martin *et al.*, 1969; Kinnear *et al.*, 1971). Both the fat body and salivary glands show a high degree of polyteny. The present work arises from an attempt to investigate chromosome structure, and especially the characteristic sequence of contemporaneous nucleolar events, during the main period of protein synthesis.

Apart from its essential role in ribosome formation, the nucleolus clearly has an obligatory role in the transfer of information from nucleus to cytoplasm (Harris, 1970). The nucleolus often shows marked specialization from tissue to tissue and from one developmental stage to another (Busch and Smetana, 1970). An examination of the nuclear ribonucleoprotein inclusions of five larval tissues of the blowfly showed that these nucleoli are of a number, size and form characteristic both of the particular tissue and stage of development (Thomson and Gunson, 1970). Since these tissues have highly polytene interphase nuclei, there is here an excellent opportunity to observe the relationship between the nucleoli and the chromosomes at stages in cell growth and differentiation when markedly different amounts and kinds of protein are being synthesized.

THE LARVAL FAT BODY AND SALIVARY GLANDS IN *CALLIPHORA*

Larvae of *C. stygia* (Melbourne strain) hatch at about 0.12 mg body weight and moult to the second instar 24 h later (0.25 mg) at 21-22°C. The second

FIGURE 1

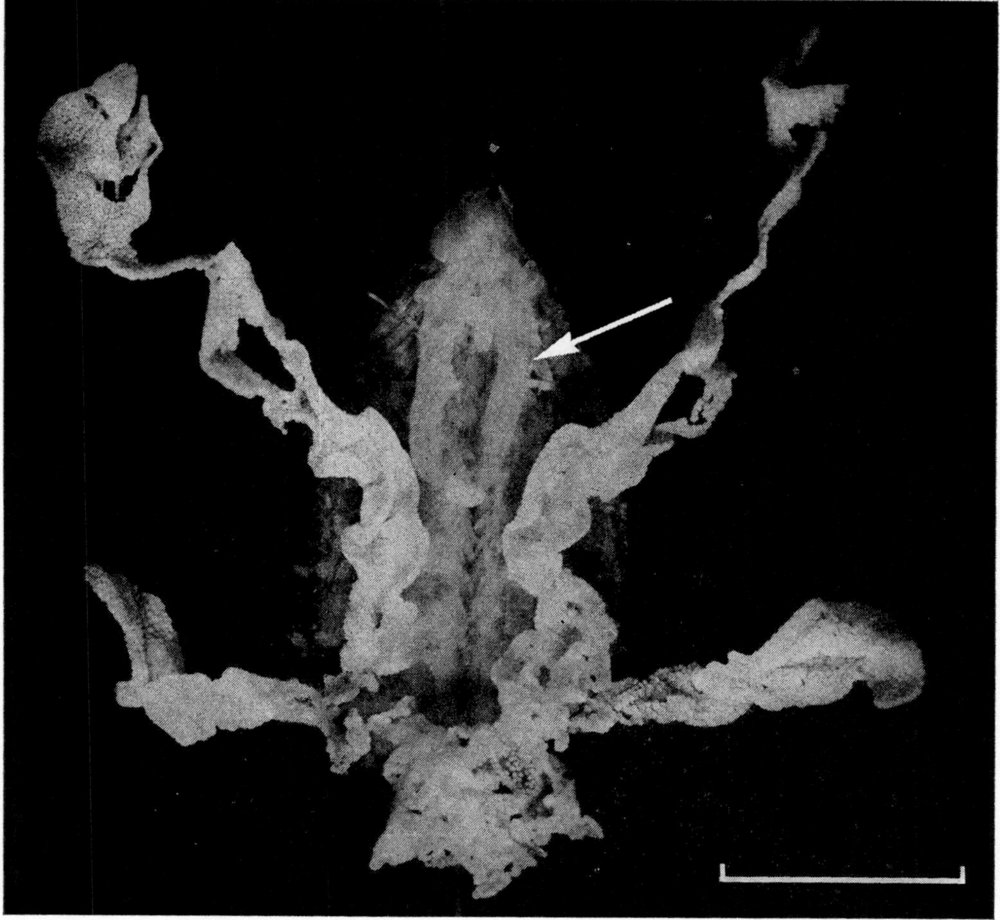

Dissection of *Calliphora* larva at the late wandering stage (day 10) to show the fat body comprising paired anterior and dorso-lateral lobes (extended to either side of the dissection) and a single, folded, highly fenestrated posterior lobe. The salivary glands (arrow) are also paired, being joined posteriorly by a small clump of fat-body cells. The scale bar represents 5 mm.

moult takes place on day 3 (15 mg). The third instar larvae continue to feed actively until day 6, gaining rapidly in weight. By early day 7 the crop is almost empty of food and becomes gas-filled; the larva has now reached a peak weight of 120 mg and enters a wandering phase, moving away from food into the substratum. The larvae become quiescent late on day 10 and pupariate on day 11, when the body weight has fallen to 95-100 mg (Kinnear *et al.*, 1968).

The fat body and salivary glands are already fully formed at the time of hatching, although the cells are extremely small. The fat body contains about 11 500 cells arranged in flat sheets one cell thick in the anterior and dorsolateral lobes (Fig. 1). The cells are irregularly pentagonal in outline and measure about 100 x 150 μm in the mature larva. In the posterior lobe, the fat body cells are more rounded in shape. The salivary glands (Fig. 1) consist of two tubular lobes comprising together 1500-1600 cells, again in a single layer. No cell division occurs in these organs subsequent to hatching. After puparium formation both tissues undergo histolysis, a process which starts in the fat body at the anterior extremity and proceeds progressively towards the posterior lobe.

PROTEIN SYNTHESIS IN FAT BODY AND SALIVARY GLANDS

Cell growth is spectacular during the first half of larval life, and very high rates of protein synthesis characterize the tissues of feeding larvae at this period (Martin *et al.*, 1969; Kinnear *et al.*, 1971). At the peak of their activity, each cell of the fat body synthesizes about 0.4 μg protein *per diem*. Most of this protein is released into the haemolymph (Martin *et al.*, 1969). At the conclusion of the feeding period, the rate of protein synthesis drops sharply. A simultaneous decrease in the rate of protein synthesis is also seen at this time in the salivary gland and in tissues of the body wall. The reduction in the salivary gland (Table 1) is

TABLE 1

Incorporation of [^{3}H] leucine into proteins of salivary gland cells at various larval ages

Age of larvae (days)	Specific activity of protein[a] (c.p.m./mg)
5	480500[b]
7	260500
11	60000

[a]After 5 min incorporation *in vivo*.

[b]Data from Martin, Kinnear and Thomson (1969); values shown are means from 3 treated groups of 5 larvae at each age, after injection of 2 μCi [^{3}H] leucine (23.5 Ci/mmol) per animal.

more gradual than in the case of fat body. Protein synthesis in the fat body of quiescent larvae (just prior to pupariation) increases again slightly, to a level above that seen in wandering stage larvae (Martin *et al.*, 1969).

Studies of the times of synthesis of the principal electrophoretically separable proteins of the fat body and salivary gland provide clear evidence of two distinct phases of synthetic activity in larval life (Kinnear *et al.*, 1971). In the case of fat body, the 22 soluble proteins which are quantitatively most abundant in tissue homogenates fall into 2 groups. One group comprises those proteins synthesized throughout larval life, and the other those synthesized only during the feeding stage, with quite a sharp cut-off at the end of this period.

The pattern of synthesis of the predominant soluble proteins of salivary gland cells and secretion (see Figs. 6-7 in Kinnear *et al.*, 1971) also shows a marked qualitative change at the end of the feeding period. Most of the protein synthesis in salivary gland cells at the wandering stage is attributable to a relative increase in the production of two components at a time when the synthesis of eight others ceases. Both the amount of protein synthesis and the number of species involved decrease in the salivary gland by the quiescent stage.

The pattern of synthesis of the *major* tissue proteins is more consistent with coordinate switching of blocks of genes at a few crucial stages than with progressive smaller scale changes in gene activity during third instar larval life. The gradual appearance of certain proteins, for instance of particular enzymes related to pupariational processes, is of course not precluded. However, any such additional gene activities late in larval development are likely to involve few loci compared with the number implicated during the feeding stage.

It will be apparent from this summary of the cellular activities of fat body and salivary gland that a major phase of gene read-out, which must obviously precede or accompany translation, is to be found in these tissues during the first half of larval life.

DEVELOPMENT OF POLYTENE NUCLEI

Development of the polytene nuclei of the fat body and salivary glands follows a similar pattern. The salivary gland nucleus has been selected for detailed description.

At hatching, the larval salivary gland cells show a strikingly heteropycnotic Feulgen-positive body overlying a prominent nucleolus (seen as a clear area in Fig. 2a). The remainder of the chromatin surrounds this nucleolus in a thin layer. The nuclei undergo rapid enlargement during the next two days of larval life. The heteropycnotic body seen so prominently at hatching elongates, and often now appears as two separate bodies (Figs. 2b, c). By the time the larva has reached a body weight of 2.5 mg, separate irregular and very extended chromosomal threads can be seen. The formation of recognisable polytene chromosomes has occurred by the 4 mg stage (Fig. 2d); here the individual chromosomal strands occupy much of the nuclear volume (except for the nucleolar area, see below). Certain chromosome regions show clear transverse banding, suggesting a drawing together of laterally replicated chromatid elements (compare Bier, 1959), and the homologues then appear paired in some sections and separated in others.

FIGURE 2

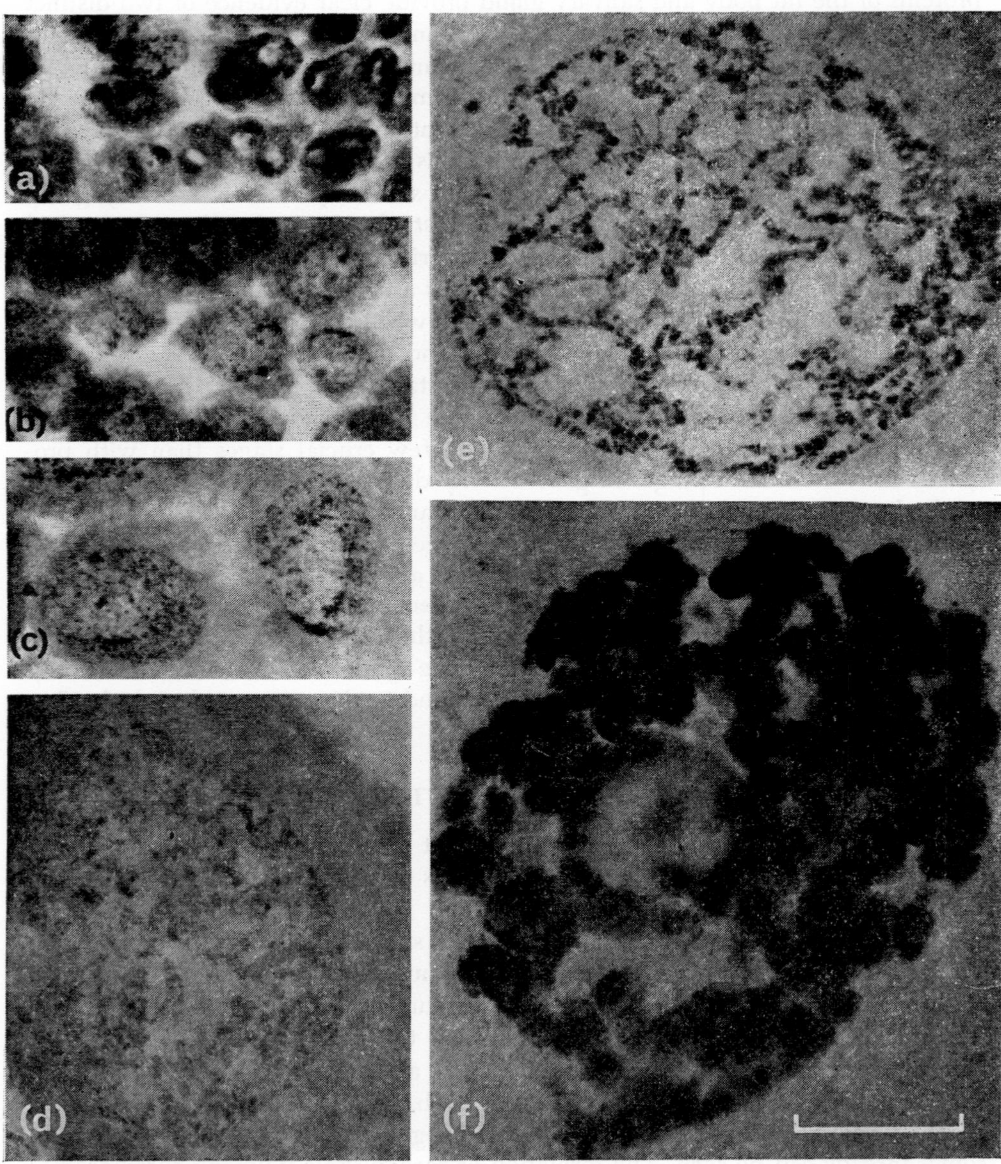

Stages in the development of the polytene chromosomes of salivary-gland nuclei: (a) at hatching, 0.12 mg body weight; (b) at 0.25 mg; (c) at 1.5 mg; (d) at 4 mg; (e) at 12 mg; (f) at late wandering stage, 100 mg. Lactic-orcein stain. The scale bar represents 25 μm.

Two processes become more and more conspicuous as replication continues: the lateral association of chromosomal elements is much tighter and, in many regions, the chromosomes appear to contract lengthwise so that they come progresssively to occupy relatively less of the total nuclear volume (Fig. 2e). The genome now consists of three visibly distinct classes of chromatin: (1) the remnants of the condensed nuclear mass seen so prominently in the early post-hatching stage (this material does not appear to have replicated in proportion to the remainder of the genome); (2) short chromosomal segments of varying length, with the characteristic band-interband elements of the fully developed polytene structure; (3) diffuse chromatin regions between the banded chromosomal segments.

Subsequent development of the polytene chromosomes involves further replication with increase in chromosome width, and progressive longitudinal contraction of banded segments, so that close to the time of puparium formation, individual bands are closely stacked together and some of the more strongly stained ones may appear to fuse. Within these banded segments, local expansion to form puffs can be seen at certain positions especially in the broader chromosomes of wandering and quiescent larvae (Fig. 2f). Chromosomal sections which were fully dispersed at the earlier feeding stage become less conspicuous in these older nuclei, but do not appear to develop banding.

The fat body and salivary gland chromosomes differ in the total number of major bands seen in the nuclei, and both differ from the linearly continuous, fully banded chromosomes of certain pupal trichogen and footpad nuclei (Thomson, 1969), in which 5 of the 6 elements of the haploid genome are represented. Comparison of the length of the banded sequences in salivary gland and fat body nuclei with those of the trichogen nucleus establishes that the average length of these segments is considerably less than that of whole chromosome arms.

Contrary to the vew expressed earlier (Thomson, 1969), the segments of the genome in fat body and salivary gland cells which remain dispersed and are often associated with nucleoli, do not appear to arise from previously banded chromosomal sections. Rather, such regions seem to represent segments which did not undergo contraction and condensation at the early stage when these processes were affecting adjacent portions of the chromosome.

DEVELOPMENT OF NUCLEOLI

The early stages of nucleolar development have been studied in squash preparations and whole cell mounts after staining with Feulgen-light green or toluidine blue-molybdate (Love and Liles, 1959; Love and Suskind, 1961), and after silver impregnation (Estable and Sotelo, 1952) with or without subsequent gold toning (see Figs. 3, 4). Cytochemical identification of ribonucleoprotein has depended in addition on nuclease digestion coupled with gallocyanin-chrome alum staining (Thomson and Gunson, 1970) and on the radioautographic demonstration of uridine uptake. The formation of multiple nucleoli in cells of the feeding stage of the fat body and salivary gland follows the same general pattern, although differing in detail and in timing. The salivary gland has been selected here for detailed description.

Nucleoli of the larval salivary gland

At hatching, most nuclei of the salivary gland show with lactic-orcein (Fig. 2a), or Feulgen-light green, an eccentric lightly stained nucleolar area below a

FIGURE 3

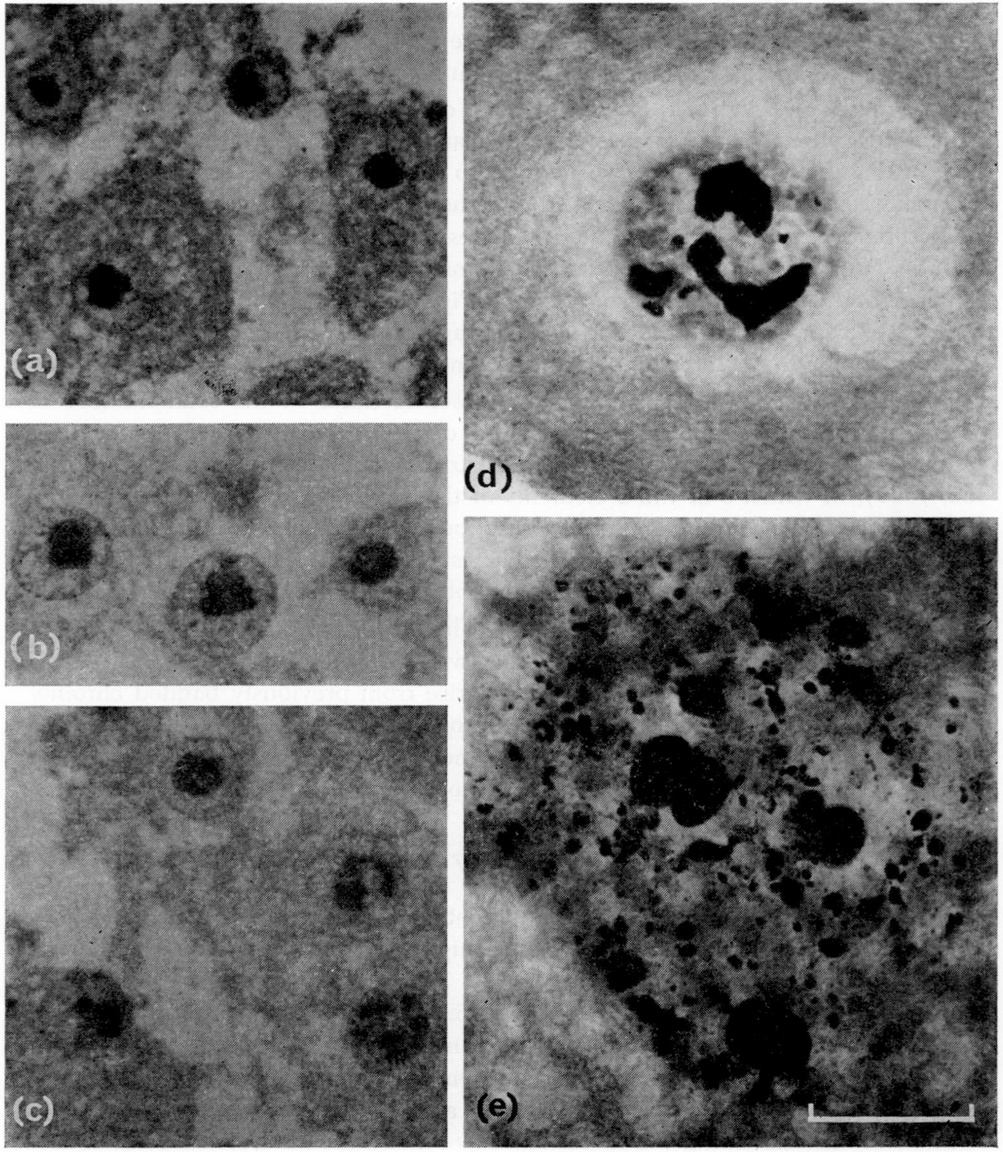

Stages in the development of the nucleoli in salivary-gland nuclei (a-d) and in a fat-body nucleus (e): (a) at 2 mg body weight; (b), (c) at 3 mg; (d), (e) at 85 mg. Silver impregnation; (d), (e) gold toned. The scale bar represents 25 μm.

strongly-stained cap of condensed chromatin. This chromatin cap is always situated closer to the cell membrane than to the bulk of the cytoplasm. The nucleolus rapidly expands and increases in volume, appearing somewhat less regular in outline (Fig. 3a). The ribonucleoprotein becomes concentrated to the periphery of the nucleolus and a light central region is then frequently visible (Figs. 3a, c). As the nucleolar area continues to expand, the structure opens out into a horseshoe-shaped configuration of two to three large lobes between which connecting strands can be seen. The lobes move peripherally and may separate further, so that by this stage two to five main blocks of nucleolar material (Type 1 inclusion bodies; Thomson and Gunson, 1970) are evident in typical nuclei. As the undispersed segments of the polytene chromosomes become more clearly defined, the chromosomal association of these blocks is more pronounced (about 10 mg stage). The nucleolar material increases in bulk after the second moult and the blocks are irregular in number from nucleus to nucleus, suggesting a tendency for the separate nucleolar sections to fuse. The total amount of nucleolar material, however, appears very similar from one nucleus to another in the same gland. Staining of both the nucleoli and chromosomes, as with silver impregnation followed by lactic-orcein treatment, shows the close association of dispersed chromosomal segments with the inclusion material (*e.g.* at arrow, Fig. 4a) in which only a small portion of the nuclear content is in the plane of focus. The nucleolar material reaches its maximum development in the cells of the salivary gland during day 5 of larval life (Fig. 3d); thereafter the inclusion bodies show increasing shrinkage and vacuolation, and finally a tendency to break up.

By the time protein synthesis has decreased (late day 7 to day 8), only small vacuolated remnants of nucleolar material remain. During this wandering stage, a second kind of nucleolar structure commences to form (Type 2 inclusion body; Thomson and Gunson, 1970). This structure has a granular core of ribonucleoprotein surrounded by a lightly staining cortex (Fig 2f). Two of these nucleoli appear to form if the maternal and paternal homologues are not synapsed at the nucleolus organiser region, and are not close enough for the two nucleoli to fuse into a single body. Again, if two of these structures are present, each is about half the volume of the usual single Type 2 nucleolus at a corresponding stage, the total being about 10 percent of the volume of the nucleus at the time of puparium formation. A further increase in this proportion relative to the total nuclear volume occurs as cytolysis approaches after pupariation.

Nucleoli of the larval fat body

Lobulation and the eventual break-up of an initially single, central nucleolar structure occurs in fat body cells as in those of the salivary glands. Separation of the nucleolar material into discrete rounded blocks is, in cells of the fat body, much more complete and less variable from nucleus to nucleus. By the 35 mg stage (Fig. 4b), with very active protein synthesis, the nucleoli are seen as a series of regular globular bodies of fairly uniform size associated with the chromosomes which lie around the nuclear periphery. Subsequently, the nucleoli occupy much of the centre of the nuclei (especially well seen in undehydrated preparations,

FIGURE 4

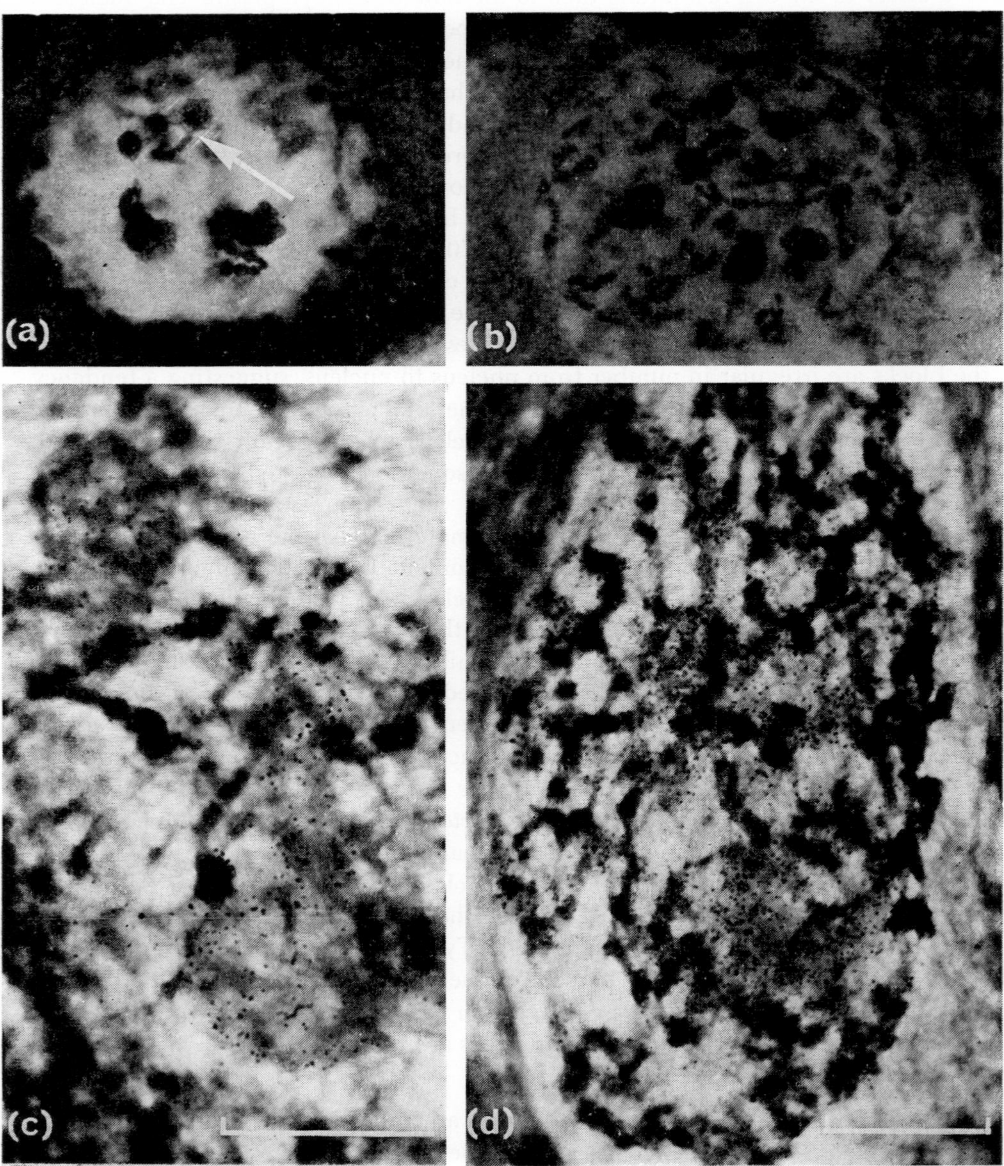

Association of dispersed chromosomal segments with nucleoli: (a) salivary gland, at 30 mg body weight; (b) fat body at 35 mg; (c), (d) radioautographs of salivary gland from 90 mg larva after 15 min incubation *in vitro* with [^{3}H] uridine. The scale bars represent 25 μm; (a), (b), (d) at same magnification.

such as those stained heavily with lactic-orcein; Thomson and Gunson, 1970). The individual bodies then commence to disperse and vacuolate (Fig. 3e) and largely disappear by the end of day 6, when the main period of protein synthesis in this tissue is concluding.

Sites of RNA synthesis in salivary gland nuclei

Chromosomal association of Type 1 nucleolar material in salivary gland cells is seen clearly in both the silver-orcein preparations and in radioautographs (Fig. 4). Radioautographs were made from squash preparations of salivary glands from feeding larvae at day 5, after incubation *in vitro* for 15 min or 45 min in insect saline containing [5-^{3}H] uridine (0.5 μCi/μl; 20 Ci/mmol). Exposures of 2-3 weeks with Ilford K2 emulsion were used. The labelling pattern of typical nuclei was qualitatively similar after both incubation times. Incorporation of [^{3}H] uridine took place fairly generally over all nucleolar masses in any given nucleus (Figs. 4c, d), and over dispersed chromosomal segments, especially those close to the nucleoli. In Fig. 4d, uptake by extensive, lightly stained chromosomal regions is marked. A transverse alignment of silver grains over some individual expanded (puffed) chromosome bands can be seen in other segments. It is most striking that densely stained, compact chromosomal regions do not show incorporation of the labelled nucleoside except peripherally to the condensed chromatin.

Again the radioautographic evidence suggests an intimate relation between whole chromosome segments which are largely dispersed, and the multiple nucleoli of the salivary gland cells in feeding larvae.

Evidence of nucleolar function

The pattern of nucleolar development in the larval tissues of *Calliphora* is consistent with the observations of nucleoli in the salivary gland cells of *Drosophila virilis* made by Painter and Biesele (1966). These authors record the occurrence of multiple nucleoli in the second instar salivary gland and the variable number and size (or even apparent absence) of nucleoli in cells from third instar larvae, but they do not distinguish the large, regular, balloon-like nucleolus which develops in these cells prior to pupariation (Type 2 nucleolus of the present study). Electron microscopy of the nucleoli from salivary glands of second instar larvae (Painter and Biesele, 1966) provided evidence of the role of these structures (comparable to Type 1 nucleoli described here) in ribosome production. Highly involuted surfaces covered in ribosome producing tassels form fringes about the fibrillar internal centres at this stage. The break up of these multiple nucleoli, and their vacuolation is accompanied by increasing release of the ribosomes to the cytoplasm. This process leaves the fibrous core material as small nucleolar residues comparable with those of the wandering stage salivary gland in *Calliphora*.

The nucleoli distinguished here as Types 1 and 2 may ultimately both be related to the nucleolus organiser regions but in different ways and at different times. At least in the nuclei of *Drosophila* salivary gland cells, the nucleolar organiser region contains the r-DNA sequences (Pardue *et al*, 1970). Apart from the morphological and developmental differences between these structures in

Calliphora cells, there appear to be at present two lines of evidence supporting a distinction between them. Firstly, the amount of extractable pre-r-RNA (sedimenting at approximately 41-43S and 38S; Thomson and Schloeffel, unpublished data) parallels the development of the Type 1 nucleoli, but not that of Type 2, in spite of the volume of the latter in salivary gland cells. Secondly, the Type 2 inclusions do not show peripheral association with dispersed segments of the genome but maintain a continuous basal attachment to the nucleolus organiser region. The function of Type 2 nucleoli is speculative at present: it cannot be predominantly storage of pre-r and r-RNAs. More likely, these structures may be related in some way to nucleotide salvage during the cytolysis which follows puparium formation.

Jacob and Danieli (1970) showed that the nucleolus of the midge *Smittia* contains DNA segments replicating autonomously and asynchronously with respect to the remainder of the genome. Clearly it will be important to establish the relation of the nucleolar DNA to the multiple Type 1 nucleoli of the early larval tissues of *Calliphora.* That these are derived from a structure which is single in embryonic cells may indicate an initial association with the nucleolar organizer which is lost after fragmentation.

CONCLUSION: THE CYTOLOGY OF GENE ACTIVITY

By comparison with other nuclei (pupal trichogen and footpad cells) the chromosomes of fat body and salivary gland do not show the full complement of bands (Thomson, 1969). The individual chromosomes cannot be followed as linearly complete structures in these larval tissues. Rather, the genome consists of a series of band-interband-band sequences of varying length joined by dispersed chromatin. The differences in the chromosome banding patterns seen in different tissues reflect not only differences in the appearance of the minor bands of the chromosomes (Beermann, 1956), but also the dispersion of a different set of chromosome segments in each tissue (compare Kosswig and Shengün, 1947). Such dispersed segments appear to remain in a state analogous to the normal interphase condition of nuclei which do not develop giant banded chromosomes. These segments are available for transcription prior to, or during, periods of massive synthetic activity early in the course of differentiation in these cells. Banded chromosomal regions seem analagous to reversibly condensed and inactivated portions of the genome such as, for example, the "chromocentral bodies" of the maturing avian erythrocyte (Cameron and Prescott, 1963). Local puffing of one or a few chromosomal bands is seen here as a smaller scale phenomenon superimposed on a basic coarser pattern of genomic control originating earlier in development. Puffing (Ashburner, 1970) involves activation for read-out of relatively few, specific, loci.

The development, during periods of intense protein synthesis, of a characteristic number of multiple nucleoli with a regular distribution pattern and size range in each tissue, and the association of these bodies with chromosomal segments other than the nucleolar organiser, is particularly interesting from the genetical viewpoint. Such tissue specificity of nucleoli cannot be due merely to differences in the total amount of nucleolar material in cells of different function: the maximum volume of the nucleolar material in the fat body and salivary gland nuclei of feeding *Calliphora* larvae is very similar. It seems probable that the number and size of the Type 1 nucleolar masses in these tissues reflect the spacing and chromosomal distribution of the active gene sequences with which the nucleoli associate. If these active sections of the genome are in close proximity, larger aggregates may form by lack of separation, or by fusion, of nucleoli associated with adjacent regions (as in the salivary gland, but not the fat body, of *Calliphora*).

Whatever the explanation for the diversity of nucleolar patterns, the chromosomal associations and tissue-specificity of nucleoli during periods of intense protein synthesis suggest that here the nucleoli are seen participating in some aspect of the processing, stabilization or transport of gene products. These observations raise also the relation of transport of messenger-RNA to ribosomal precursors, and then the question of ribosomal specificity. Evidence is growing for the view that specific populations of ribosomes are programmed for particular synthetic activities during the development of eukaryotes (*e.g.* Tata, 1968). In the polytene tissues of *Calliphora,* we may be observing directly a series of essential cellular processes coupling ribosome synthesis to messenger synthesis, a view consistent with the chromosomal association of certain ribosomal precursors in the salivary gland nuclei of *Chironomus* (Ringborg and Rydlander, 1971).

ACKNOWLEDGEMENTS

This work has inevitably taken on a perspective gained from more detailed collaborative studies with A. G. Willis, D. C. Rogers and M. M. Gunson, which have yet to be published. M-D. Martin and J. F. Kinnear have contributed much to investigations of patterns of protein synthesis which provided the evidence of distinct phases of gene read-out during larval development.

I am also greatly indebted to H. S. Revell and K. R. Radok for excellent assistance in the laboratory. Financial support has been received through the Australian Research Grants Committee (Grant D65/15167).

REFERENCES

Ashburner, M. (1970) *Adv. Insect Physiol. 7,* 1

Beermann, W. (1952) *Chromosoma 5,* 139

Beermann, W. (1956) *Cold Spring Harbor Symp. Quant. Biol. 21,* 217

Bier, K. (1959) *Chromosoma 10,* 619

Boyd, J. B., Berendes, H. A. and Boyd, H. (1968) *J. Cell Biol. 38,* 369

Busch, H. and Smetana, K. (1970) *The Nucleolus,* 1st edn., p. 35, Academic Press, London and New York

Cameron, I. L. and Prescott, D. M. (1963) *Exptl Cell Res. 30,* 609

Estable, C. and Sotelo, J. R. (1952) *Stain Technol. 27,* 307

Harris, H. (1970) *Nucleus and Cytoplasm,* 2nd edn., p. 120, Oxford University Press, London

Jacob, J. and Danieli, G. A. (1970) *Experientia 26,* 1390

Kinnear, J. F., Martin, M-D., Thomson, J. A. and Neufeld, G. J. (1968) *Aust. J. Biol. Sci. 21,* 1033

Kinnear, J. F., Martin, M-D. and Thomson, J. A. (1971) *Aust. J. Biol. Sci. 24,* 275

Kosswig, C. and Shengün, A. (1947) *J. Hered. 38,* 235

Love, R. and Liles, R. H. (1959) *J. Histochem. Cytochem. 7,* 164

Love, R. and Suskind, R. G. (1961) *Exptl. Cell Res. 22,* 193

Martin, M-D., Kinnear, J. F. and Thomson, J. A. (1969) *Aust. J. Biol. Sci. 22,* 935

Painter, T. S. and Biesele, J. J. (1966) *Proc. Nat. Acad. Sci. U.S. 56,* 1920

Pardue, M. L., Gerbi, S. A., Eckhardt, R. A. and Gall, J. G. (1970) *Chromosoma 29,* 268

Price, G. M. (1966) *J. Insect Physiol. 12,* 731

Ringborg, U. and Rydlander, L. (1971) *J. Cell Biol. 51,* 355

Tata, J. R. (1968) *Nature (London) 219,* 331

Thomson, J. A. (1969) *Currents mod. Biol. 2,* 333

Thomson, J. A. and Gunson, M. M. (1970) *Chromosoma 30,* 193

Zweidler, A. and Cohen, L. H. (1971) *J. Cell Biol. 51,* 240

Hormonal and Environmental Modulation of Gene Expression in Plant Development.

B. KESSLER

Division of Plant Physiology, The Volcani Center, Agricultural Research Organization, Bet Dagan; and Department of Life Sciences, Bar-Ilan University, Ramat Gan, Israel.

Developmental processes are ultimately controlled by the expression of genes, but the phenotypic expression of ontogeny varies with environmental conditions. The state of a cell is described largely by the gene products active at a given time and the development of an organism may be viewed as sequential gene-directed events that are ordered in time with respect to one another. Changes in the state of development are often mediated by simple external signals. In recent years hormones have been assigned transducing functions, linking environmental and genetic control mechanisms (Bonner, 1965; Spirin *et al.*, 1966), the interactions of which are expressed in the phenotypic manifestation of gene-directed activities.

The view emerging at present is that primary hormonal actions are presumably exercised at several sites and levels of organization, but the hypothesis that hormone actions are intimately associated with the metabolism of RNA and proteins in target tissues is still prevalent and supported by evidence from a variety of experiments (Nooden and Thiman, 1963; Key and Shannon, 1964; Galston and Davies, 1969). From the available data it is still difficult to conclude whether hormonal effects on nucleic acids and proteins are an expression of their action on, or close to, the gene, or only remote manifestations of their activity. However, although the precise molecular mechanisms are still unknown, it is conceivable that, where differentiation and the emergence of new growth patterns are concerned, triggering external signals (such as physical stimuli-e.g. photoperiodic conditions, light and temperature, hormones or embryonic inductive agents) are connected with or transduced to control mechanisms regulating the metabolism of nucleic acids. Such a case has been described for one of the principal plant hormones, namely, gibberellin A_7 (GA_7), which binds highly specifically to AT-rich DNAs (Kessler and Snir, 1969; Kessler, 1971a) and, in combination with a DNA ligase, induces the formation of circle-like loops in nuclear DNA (Kessler, 1971b, 1971c).

In the genome of eukaryotes, a sizeable fraction is composed of highly reiterated nucleotide sequences (Britten and Kohne, 1968) and a second fraction which is comprised of nucleotide sequences present in only one or a few copies per haploid genome. It has been suggested that at least some of the highly redundant nucleotide families are concerned with regulation (Britten and Davidson, 1969; Crick, 1971), being able to bind specific inducing agents such as hormones (Britten and Davidson, 1969). No experimental findings have yet been published to substantiate such interactions, but because of the binding of gibberellin (GA) to DNA (Kessler and Snir, 1969; Kessler, 1971a, 1971b, 1971c) this hormone could be a natural candidate for such studies.

In a discussion of control systems, adenosine 3′,5′-cyclic monophosphate (3′,5′-cAMP) occupies at present a central position as a common intracellular mediator of the action of many hormones in animals and the second-messenger model of 3′,5′-cAMP is rather well documented (Robison *et al.*, 1968). In bacteria, also, 3′,5′-cAMP is considered as a positive effector functioning at the level of DNA transcription (Pastan and Parlman, 1970) in enzyme induction phenomena. In plants, however, analogous evidence is very limited and not yet satisfactory. Endogenous 3′, 5′-cyclic mononucleotides have not yet been identified in plants, although exogenous 3′,5′-cAMP at relatively high concentrations mimiced some activities of gibberellins (Kessler, 1969; Duffus and Duffus, 1969; Galsky and Lippincott, 1969, Nickells *et al* 1971; Earle and Galsky, 1971; Kamisaka and Masuda, 1971; Kamisaka *et al.*, 1972) and auxins (Salomon and Mascarenhas, 1971a; Kamisaka, 1971; Kamisaka and Masuda, 1970; Pollard, 1970); also an adenyl cyclase activity has been noted (Pollard, 1970; Salomon and Marcarenhas, 1971b; Wood *et al.*, 1972). For these data to be of biological significance, it would be necessary to show that 3′5′-cyclic mononucleotides exist in plants in general and in responsive tissues in particular. The purpose of this investigation was therefore, to find possible answers to the following questions:

I. Do 3′,5′-cyclic mononucleotides exist in plants?

II. If positive, do they and at what level interact with gibberellin, a phytohormone the action of which can be substituted in some cases by 3′,5′-cAMP (Kessler, 1969; Duffus and Duffus, 1969; Galsky and Lippincott, 1969; Nickells *et al.*, 1971; Earle and Galsky, 1971; Kamisaka and Masuda, 1971; Kamisaka *et al.*, 1972) and guanosine 3′,5′-cycle monophosphate (3′,5′-cGMP) (Kessler, 1969)?

III. Are there any specific gene-related expressions of cyclic mononucleotide-gibberellin interactions at the physiological level?

IV. As gibberellin does bind specifically to DNA (Kessler and Snir, 1969), do gibberellin-cyclic mononucleotide interactions also involve DNA?

MATERIALS AND METHODS

Extraction and assay of 3′,5′-cAMP

The tissues were homogenized rapidly in sodium acetate buffer solution (50 mM, pH 4). The homogenate was boiled for 2-5 min, centrifuged at 18,000 rev./min for 20 min, and the supernatant was used for the assay. The extraction method employed by Gilman (1970) was less effective for plant material.

For the assay of 3′,5′-cAMP we employed Gilman's (1970) method which was slightly modified to suit plant material (R. Levinstein and B. Kessler, unpublished). 3′,5′-cAMP-specific protein kinase was prepared according to Miyamoto *et al* (1969) from bovine muscle. For routine assays the ammonium sulphate fraction was sufficiently sensitive.

Endogenous gibberellins

Endogenous gibberellins were extracted according to Barendse *et al.*, (1968) and quantitatively assayed by the barley endosperm test (Coombe *et al.*, 1967).

α-amylase activity in aleurone layers

The experimental procedures were in principle as described by Chrispeels and Varner (1967). Five aleurone layers were used in 1 ml of incubation medium; each treatment was given in four replicates and each experiment was repeated at least four times.

α-amylase activity in embryonectomized barley seeds

α-amylase activity was measured as reducing sugars released into the tissue and the medium (Chrispeels and Varner, 1967). Two embryo-less half seeds were incubated in 1 ml medium. The reducing sugar was assayed with Summer's (1924) reagent.

Growth of Lemna gibba G3

A stock culture of Lemna gibba G3 was kindly supplied by Dr. W. S. Hillman. were kept under short-day conditions (8 h illumination, 16 h darkness), at 25°C. Most of the experiments were performed under inducing long-day conditions, namely, 16 h illumination and 8 h darkness.

Tissue cultures of soybeans

Soybean var. Acme was cultured on Miller's 1961 medium. The callus was subcultured at least 25 times before use in these experiments.

Preparation of DNA

DNA was prepared as described previously (Kessler, 1971a). In the present experiments the salt precipitation phase was excluded and the isopropanol purification step (Marmur, 1961) was included. The absorption ratios were $E_{260}/E_{230} = 2.08\text{-}2.14$ and $E_{260}/_{280} = 1.86$. The buoyant density of cucumber DNA in CsC1 was $\delta^{20} = 1.693$ g.cm^{-3}.

DNA re-annealing and hydroxylapatite fractionation

DNA was sheared according to Dr. M. D. Chilton (personal communication). DNA was dissolved in 0.1 M acetate buffer (pH 4.2). This solution was heated at 70°C for 35 min, then cooled and adjusted to pH 13 with 1 M NaOH. After an additional heating at 50°C for 10 min, the solution was neutralized with 1 M NaH_2PO_4 and dialyzed against dilute salt solution. The DNA thus obtained was about half the size obtained by shearing at 12,000 psi, *i.e.*, giving a molecular weight of about 500,000.

The solution of sheared DNA (phosphate buffer, 0.12 M) was adjusted to pH 7, heat denatured, and then renatured at 60°C. The kinetics of renaturation is expressed in Cot values (Britten and Kohne, 1968). At different time intervals the DNA solution was loaded onto hydroxylapatite columns (HAP—Bio Rad) at 60°C, the binding capacity of which for double-stranded DNA had been determined previously. The renatured DNA was eluted from the column with 0.4 M phosphate buffer as described by Bernardi (1969).

Quantitative determination of DNA

DNA was isolated quantitatively according to Schmidt and Thannhauser (1954) and determined with the Cerriotti method as modified by Keck (1956).

Thymidine Incorporation

After 18 h of incubation, DNA synthesis was estimated by [^{3}H] thymidine incorporation into acid-insoluble materials. Aleurone layers were incubated in buffer solutions (Chrispeels and Varner, 1967) containing [^{3}H] thymidine (40 μCi/ml, specific activity 17.5 Ci/mmole). Then the aleurone layers were washed with buffer and homogenized with 5% trichloracetic acid containing 10^{-4} M TMP and 10^{-4} M dCMP. Samples were collected on GF/C filters. The filters were washed with 60 ml of cold trichloracetic acid, 10 ml of alcohol-ether (3:1) and 10 ml of ether, and then dried and counted.

RESULTS

I. The occurrence and distribution of endogenous 3′,5′-cAMP in plants

We adapted Gilman's (1970) method for the determination of 3′,5′-cAMP in plant material. Representative assay data are shown in Table 1. The tissue levels vary with the plant species, the organ and the plant age. In mango and avocado seedlings, particularly high levels of 3′,5′-cAMP were detected which reached levels known from animal tissues particularly rich in 3′,5′-cAMP, such as brain (Gilman, 1970). The values are independent of the amount of tissue analyzed and the extent of dilution in the assay (with tissue dilutions of 2 to 20 times).

II. Cyclic mononucleotide-gibberellin interactions (with B. Kaplan)

The findings that endogenous 3′,5′-cAMP exists in plants (Table 1) and that exogenous 3′,5′-cAMP simulates the activity of gibberellin (Kessler, 1969; Duffus and Duffus, 1969; Galsky and Lippincott, 1969; Nickells *et al.*, 1971; Earle and Galsky, 1971; Kamisaka and Masuda, 1971; Kamisaka *et al.*, 1972), made a closer inquiry on the level of their apparent relationships warranted.

The study of hormonal effects is greatly facilitated in cells which (i) contain very low endogenous levels of the hormone involved or (ii) do not form it normally. In this respect, the embryonectomized barley seed is of particular experimental value as its response to an exogenous hormone, namely, gibberellin, is of a "none or all" character (Paleg, 1965). Of particular value is the aleurone

TABLE 1

The Distribution of 3′,5′-cAMP[a] in Different Plant Species, Organs and Age Groups

Species	Age of Plant	Organ	3′,5′-cAMP (pmoles/g fresh weight)
Pea embryo var. Progress No. 9	16 h inhibition	—	176
Cucumber var. Bet Alpha	6 days	Stems + Cotyledons	124
		Roots	123
Lettuce var. Grand Rapids	3 days	Leaves + Roots	265
Tobacco var. Xanthi	11 days	Leaves	55
		Roots	70
	19 days	Leaves	60
		Stems	56
		Roots	114
	25 days	Leaves	100
		Stems	70
		Roots	120
	32 days	Apical Buds	180
Wheat var. CCC	3 days	Leaves[b]	180
		Roots	170
Cleopatra Mandarin	8 months	Leaves	340
		Roots	350
Shamouti Orange	7 months	Growth Flush	380
Mango var. Cantara	12 months	Young Leaves	4800
		Mature Leaves	2800
		Roots	1560
Mango var. Cantara	24 months	Young Growth Flush	3800
		Young Leaves	2000
		Mature Leaves	1500
Avocado var. Fuerte	12 months	Young Leaves	9600
		Mature Leaves	4900
		Roots	1300

[a]cAMP was determined according to Gilman (1970) as modified in our laboratory (R. Levinstein and B. Kessler, unpublished).

[b]In all cases, young leaves and white roots were selected.

layer of the mature barley seed, a tissue of living cells which surrounds the starchy endosperm.

One of the earliest events in germinating barley seeds is the conversion in the endosperm of starch to reducing sugars to supply the needs of the growing embryo. The activity of the relevant hydrolytic enzymes increases markedly, early in the germinating process. The increase of these enzymes requires the presence of the embryo (Paleg, 1965), which can be replaced by gibberellic acid (GA_3) (Paleg, 1960). *I.e.*, when the embryo is removed from the seed prior to germination, the release of reducing sugars in the embryo-less barley half-seeds is promoted by low concentrations of exogenous gibberellins (Paleg, 1960), as a result of an increased formation and secretion of α-amylase by the living aleurone cells surrounding the endosperm (Paleg, 1960, 1964). This increase in α-amylase activity in the embryo-less barley endosperm, due to the *de novo* synthesis of the enzyme (Paleg, 1960, Filner *et al.*, 1969), is wholly dependent upon the added gibberellin.

3′,5′-cAMP (Duffus and Duffus, 1969; Galsky and Lippincott, 1969; Kessler, 1969) and 3′,5′-cGMP (Kessler, 1969) substitute for gibberellin in this system. The time course of the development of α-amylase activity in response to added GA_3, 3′,5′-cAMP and 3′,5′-cGMP is given in Fig. 1. The major appearance of

FIGURE 1

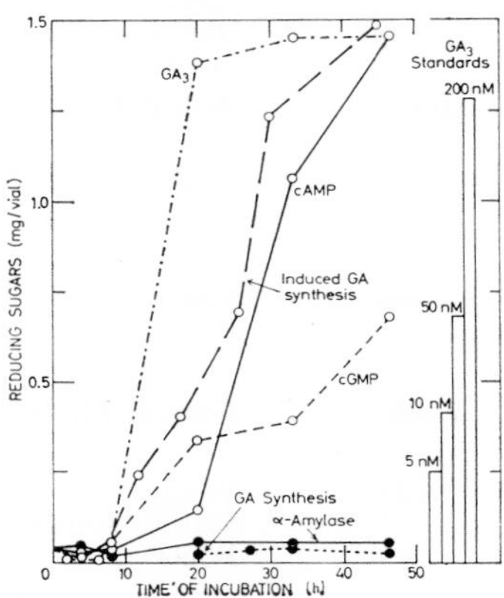

The effects of GA_3, 3′,5′-cAMP and 3′,5′-cGMP on the formation of α-amylase and endogenous GA in embryonectomized barley endosperm. The formation of α-amylase and GA was followed with time. α-amylase activity is expressed as μg reducing sugar (in glucose equivalents). The level of endogenous GA is expressed in GA_3-equivalents.

3′,5′-cAMP- and 3′,5′-cGMP-induced α-amylase activity was delayed by approximately 12 hours relative to GA_3.

The induction of α-amylase by gibberellin, 3′,5′-cAMP and 3′,5′-cGMP was strongly inhibited by inhibitors of RNA (azaguanine, 5-fluorouridine) and protein (puromycin, cycloheximide) syntheses (Kessler, 1969). On the other hand, inhibitors of DNA synthesis (mitomycin C, sarcomycin, 5-fluorodeoxyuridine) differentiated markedly between the enzyme formation induced by gibberellin and that induced by the cyclic mononucleotides.

The inhibitors of DNA synthesis prevented enzyme synthesis induced by the cyclic mononucleotides, but not by GA_3 (Table 2).

TABLE 2

The Effects of Inhibitors on the Induction of α-Amylase Formation in Embryonectomized Barley Endosperm by 3′,5′-cAMP and Gibberellin A_3. The Incubation was Continued for 32 h.

Inhibitor	% Inhibition Concentration of Inhibitor (μM) 1	 10	 100
(1) Gibberellin (70 nM)[a]			
Sarcomycin	0	0	5
Mitomycin C	0	0	3
CCC	0	6	4
AMO-1618	0	4	5
(2) 3′,5′-cAMP (1 mM)[a]			
Sarcomycin	82	94	100
Mitomycin C	94	94	95
CCC	49	93	93
AMO-1618	0	90	93

[a]The absolute values for controls, gibberellin A_3 and 3′,5′-cAMP were 31 ± 7, 753 ± 45 and 681 ± 32, respectively, in μg reducing sugars (glucose equivalents) per vial under standard experimental control. The effects of the inhibitors *per se* were not significantly different from the control values.

Of particular interest were the effects of inhibitors of gibberellin biosynthesis (Kessler, 1969), namely, (2-chloroethyl) trimethylammonium chloride (CCC) and 2-isopropyl-4-dimethylamino-5-methylphenyl-1-piperidinecarboxylate methyl chloride (AMO-1618) (Table 2); with increasing concentrations of these compounds the cyclic mononucleotide-induced formation of α-amylase was progressively inhibited.

These data pointed to the possibility that the induction of α-amylase by 3′,5′-cAMP and 3′,5′-cGMP involves first a stage of DNA synthesis, followed by the formation of gibberellin. The results presented in Fig. 1 and Table 2 indicate that the induction of α-amylase by 3′,5′-cAMP appears indeed to be coupled with the activation of DNA and an endogenous gibberellin synthesis (see also: B. Kessler and Bina Kaplan, *Physiol. Plant* in press).

The apparent involvement of 3′,5′-cAMP-activated DNA synthesis has been further substantiated by thymidine incorporation studies (Table 3). Sarcomycin and mitomycin C inhibited the 3′,5′-cAMP-enhanced incorporation of [^{3}H] thymidine into DNA (Table 3) as well as the formation of endogenous gibberellin (Table 4). The latter was also totally inhibited by CCC and AMO-1618 (Table 4). The hypothesis is, therefore, rather conclusive that 3′,5′-cAMP and 3′,5′-cGMP act at the level of the genome as primary inducers in the formation of at least one hormone, namely, gibberellin. No other cyclic isomer of the 3′,5′- or 2′,3′-configuration, or any of the 5′-mononucleotides or the 2′-(3′)-isomers, had any similar gibberellin-inducing activity.

TABLE 3

Incorporation of [^{3}H] Thymidine into DNA of Barley Aleurone Tissue and its Inhibition by Sarcomycin and CCC

(Concentration of 3′,5′-cAMP was 1 mM and of the Inhibitors 100 μM)

Treatment	Incorporation (as % of control)
Control	100
3′,5′-cAMP	537
3′,5′-cAMP + Sarcomycin	89
3′,5′-cAMP + CCC	127

TABLE 4

The effect of sarcomycin, CCC and AMO-1618 upon the 3′,5′-cAMP-induced formation of gibberellin in embryonectomized barley endosperm.

Treatment	Activity, nM (in gibberellin A_3 equivalents)
Control	0.1
3′,5′-cAMP	10.0
3′,5′-cAMP + Sarcomycin	1.1
3′,5′-cAMP + Mitomycin C	0.8
3′,5′-cAMP + AMO-1618	0.2
3′,5′-cAMP + CCC	0.4

50 g of barley half seeds were incubated for 32 h in solutions containing 3′,5′-cAMP (5 mM) and 3′,5′-cAMP + inhibitors (100 μM), respectively. After incubation, the half seeds were thoroughly washed, gibberellin was extracted (Barendse *et al.*, 1968), chromatographed (Barendse *et al.*, 1968) and assayed in the barley endosperm test (Coombe *et al.*, 1968). Only the fractions located at R_f 0.6-0.7 showed gibberellin-like activities, corresponding to gibberellin A_3.

Aleurone tissue is composed of non-dividing cells which contain large numbers of ribosomes. The aleurone's response to GA is explained by transcriptional control (Filner *et al.*, 1969). The activation of thymidine incorporation into DNA by 3′,5′-cAMP indicates that aleurone cells are capable of showing DNA synthesis or turn-over, similar to that reported for non-dividing elongating tissues (Nitzan and Lang, 1966).

III. Physiological aspects of cyclic mononucleotides-gibberellin interactions

The findings reported in the previous section indicated strongly that at least some activities of 3′,5′-cAMP do involve the activation of DNA synthesis, coupled to an induction of gibberellin biosynthesis. It was, therefore, desirable to investigate effects of cyclic mononucleotides on phenomena which are believed to involve direct gene activation processes. The present hypotheses on differentiation and morphogenesis implicate such gene control and some plant systems offer great advantages for inquiries in this field because of the possibility of experimentally triggering new morphogenic growth patterns at will. In this account we present two examples for demonstration, namely, (i) photoperiodic-dependent floral induction, a photomorphogenetic phenomenon which is believed to involve specific gene activation or de-activation, and (ii) differentiation and organization in callus cultures.

(i) **The effect of cyclic mononucleotides on flowering and multiplication of the duckweed Lemna gibba G_3 (with N. Steinberg).** Lemna gibba is a long-day plant which, under short-day conditions, grows at a greatly reduced rate indeterminately and relatively indefinitely in its vegetative pattern. When transferred to long-day conditions, the multiplication rate of this plant increases markedly, concurrently with the transition from a strictly vegetative growth to reproductive floral induction. Cyclic mononucleotides strongly effected both phenomena (Table 5 and Fig. 2). Under long-day conditions 3′,5′-cAMP promoted flowering

TABLE 5

The Effect of Mononucleotides on the Flowering and Multiplication of Lemna gibba G3 After 19 Days of Growth

Treatment	No. of Fronds		Flowering (No. of flowers as % of plants per replicate)	
	Long Day	Short Day	Long Day	Short Day
Control	107 ± 9	85 ± 8	36 ± 4	0
3′,5′-cAMP	104 ± 5	65[a] ± 3	76[b] ± 3	0
3′,5′-cGMP	114 ± 8	45[b] ± 5	14[a] ± 5	0
2′,3′-cAMP	103 ± 3	73 ± 9	24[a] ± 9	0
2′,3′-cGMP	165[a] ± 6	123[a] ± 3	0	0
5′-AMP	111 ± 4	61 ± 5	16[a] ± 8	0
5′-GMP	111 ± 6	56 ± 2	10[a] ± 5	0
2′-(3′)-AMP	82[a] ± 4	43[b] ± 2	5[b] ± 2	0
2′-(3′)-GMP	101 ± 8	55[a] ± 8	7[b] ± 3	0

[a]Significantly different from control at $p < 0.01$.
[b]Significantly different from control at $p < 0.001$.

but had no effect on multiplication. On the other hand, 2′,3′-cGMP reduced flowering close to zero, but enhanced considerably frond multiplication, both under long-day and short-day conditions. In fact, 2′,3′-cGMP replaced the requirement for long days (Fig. 3), i.e., when grown under short-day conditions in the presence of 2′,3′-cGMP, the plant multiplied as under long photoperiods.

Inhibitors of gibberellin biosynthesis inhibited the multiplication of the fronds. 2′,3′,-cGMP *per se* has no effect on gibberellin biosynthesis, suggesting an interaction between 2′,3′-cGMP an endogenous gibberellin. Indeed, the addition to 2′,3′-cGMP of either gibberellin or the gibberellin-inducing 3′,5′-cAMP, enhanced frond multiplication markedly (Fig. 4). Furthermore, 2′,3′-cGMP induced a net DNA synthesis 3 days earlier than the onset of the 2′,3′-cGMP-related increase

FIGURE 2

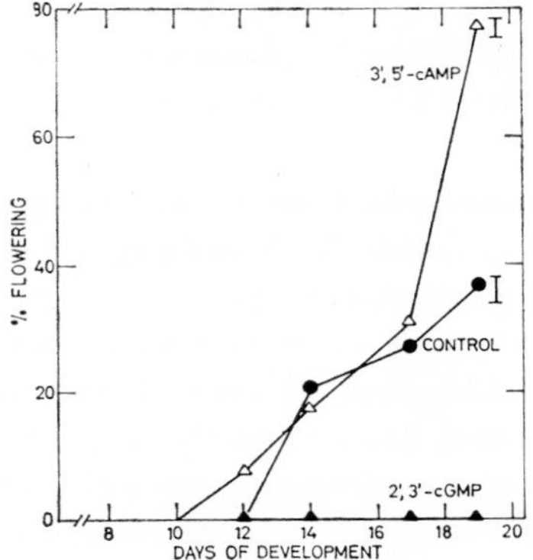

The effect of cyclic mononucleotides on the flowering of Lemna gibba G3. The vertical bars indicate the standard error.

FIGURE 3

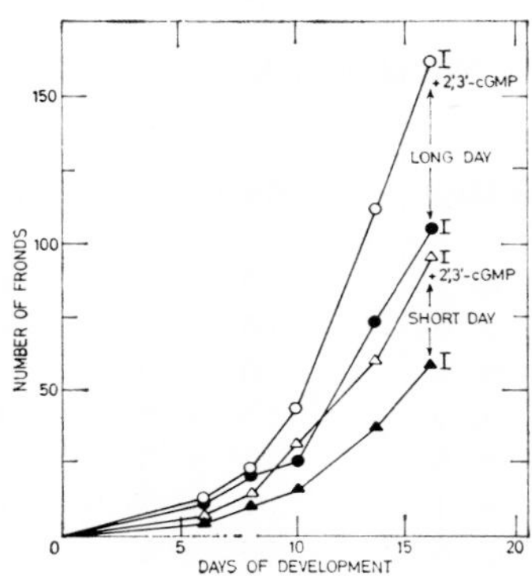

The effect of 2′,3′-cGMP on the multiplication of Lemna gibba G3 under long-day (inducing) and short-day (non-inducing) photoperiodic conditions. The vertical bars indicate the standard error.

FIGURE 4

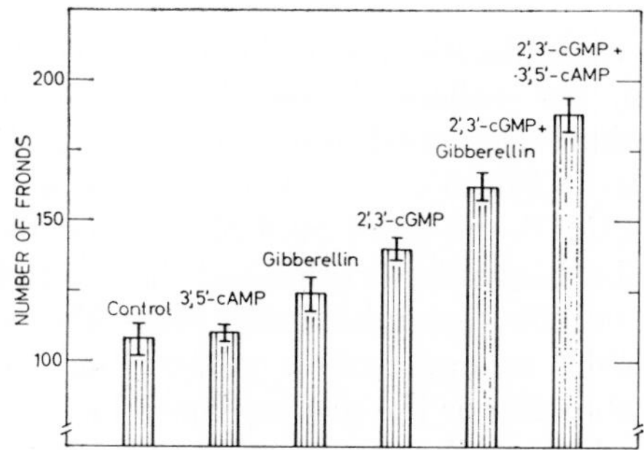

The effect of GA_3 and 3′,5′-cAMP on the 2′,3′-cGMP-enhanced multiplication of Lemna gibba G3.

in multiplication (Fig. 5). We were, therefore, led to assume that, similar to 3′,5′-cAMP and 3′,5′-cGMP, 2′,3′-cGMP also exerts its effect, in cooperation with 3′,5′-cAMP or gibberellin, at the level of the genome.

FIGURE 5

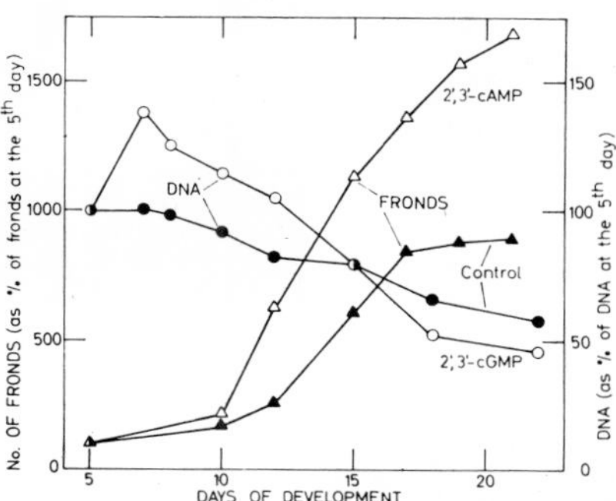

The effect of 2′,3′-cGMP on the synthesis of DNA and the 2′,3′-cGMP-related onset and rate of multiplication of Lemna gibba G3. The data are represented as percent of the respective DNA level and number of fronds on the fifth day of development.

(ii) **The effect of cyclic mononucleotides and gibberellin on differentiation in soybean tissue cultures (with Ilana Karaz).** Tissue cultures of soybean and pine-seedlings were highly responsive to some cyclic mononucleotides but not to others (Karaz, 1971). Of the 2′,3′- and 3′,5′- cyclic isomers as well as the 5′- and 2′-(3′) mononucleotides, only 3′,5′-cAMP and 3′,5′-cGMP induced differentiation (Fig. 6). 3′,5′-cAMP, in the presence of high concentrations of β-indole acetic acid (IAA) and a cytokinin, induced the formation of enlarged cells (Fig. 6 I,B) which developed into isolated small groups (Fig. 6 I, C,D). When the concentrations of IAA and cytokinin were lowered, the newly formed vascular elements became organized into strand-like formations (Fig. 6 I, E,F,G) which in some cases formed a circular structure (Fig. 6 I, H). The differentiation-inducing effect of 3′,5′-cAMP could be replaced by gibberellin (Fig. 6 I, B-D)—except for the formation of circular structures which were induced by 3′,5′-cyclic purine mononucleotides (Fig. 6 I, E-H) only. Fig. 6 II shows a phase contrast photograph of the ring-like structure formed in the presence of 3′,5′-cAMP. Similar to the data presented in Table 2, inhibitors of DNA synthesis prevented the formation of differentiated cells induced by 3′,5′-cAMP but not their induction by gibberellin. It is, therefore, plausible that the 3′,5′-cAMP-induced differentiation is another case involving a DNA-activation, as described in the first section.

IV. Age-and organ-dependent variations in the sequence diversity of cucumber DNA and effects of 3′,5′-cAMP and gibberellin on the nucleotide sequence redundancy.

The information derived so far from our data concerning the interactions between cyclic mononucleotides, gibberellin and DNA indicated the following:

In previous papers we showed that:

FIGURE 6

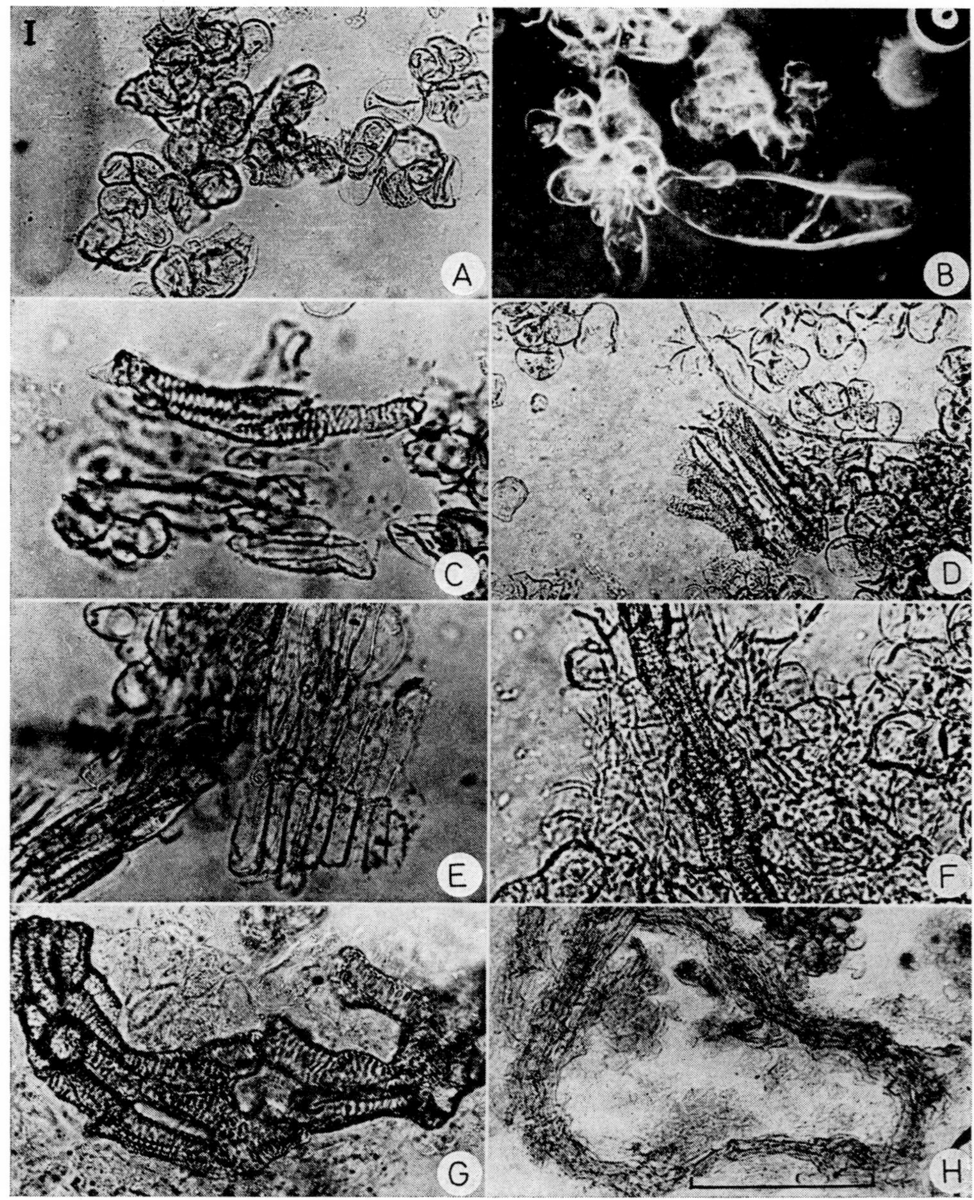

The effect of 3′,5′-cyclic mononucleotides on the differentiation of callus cell cultures.
6I. Development of structure from undifferentiated cells. A: control culture; B-D: 3′,5′-cAMP or GA_3 added, × 740; E-G: 3′,5′-cAMP; H: 3′,5′-cAMP. Scale marking: A-G 50μ; H 25μ.

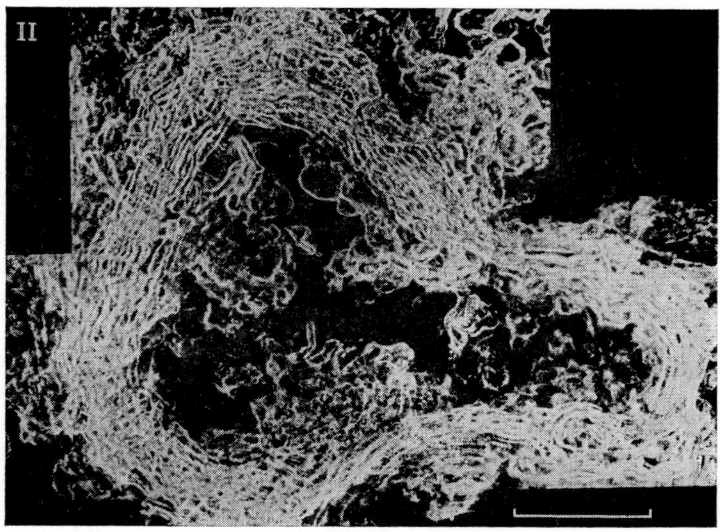

6II. Phase-contrast photograph of picture 6I, H. Scale marking: 50μ.

(i) Gibberellin A_7 bound to AT-rich duplex DNA originating from higher plants (Kessler and Snir, 1969; Kessler, 1971a).

(ii) Gibberellin bound to single nicks, affecting markedly the activity of nick-requiring enzymes, such as DNA ligase (Weiss and Richardson, 1967; Lindahl and Edelman, 1968). In the presence of GA_7, a plant DNA ligase (Kessler 1971b) converted linear duplex plant DNA to conformations consisting in part of internal loops (Kessler, 1971c). Such DNA conformations were also formed in thymal DNA in the presence of growth hormone (Kessler, 1971c). The activity of another nick-requiring enzyme, namely, DNA polymerase (Kornberg, 1969), also responded markedly to varying concentrations of gibberellin (Jona Snir and B. Kessler, unpublished).

(iii) The capability of eukaryotic DNA to form internal loops in response to specific hormones decreased with advancing age (Kessler, 1971c).

(iv) As a probable consequence of the gibberellin-DNA interactions, GC-rich DNA fractions were formed which in some cases could be separated as satellites in CsCl gradients (Kessler, 1971c). These heavy DNA fractions were organ-specific and their rate of formation varied with age, temperature and photoperiod and corresponded to the level of endogenous and applied hormone in an organ-specific fashion (Kessler and Snir, 1971).

In the present account we showed that:

(i) Cyclic mononucleotides induced the biosynthesis of gibberellin at the level of DNA.

(ii) 3′,5′-cAMP promoted flower induction.

(iii) 2′,3′-cGMP, in combination with gibberellin or the gibberellin inducer (3′,5′-cAMP), enhanced multiplication, probably via a stimulation of DNA synthesis. In the process of multiplication, 2′,3′-cGMP also replaced the requirement for long days.

(iv) 3′,5′-cyclic purine mononucleotides and gibberellin induced differentiation in undifferentiated tissue cultures of soybean. 3′,5′-cAMP induced also the formation of morphological ring-like structures.

The above-outlined intimate gibberellin-DNA and cyclic mononucleotide-DNA interactions warranted a closer examination of the genome involved. DNA from many eukaryotes, including plants, differs from prokaryotic DNA in that it exhibits heterogeneous renaturation kinetics (Britten and Kohne, 1968; Britten and Davidson, 1969). The time required for renaturation of separated strands is proportional to the number of different nucleotide sequences and their relative concentration. RNA/DNA hybridization studies indicated that RNA from cortisol-treated and control rats hybridized to DNA of different nucleotide sequence diversity (Doenecke and Sekeris, 1971), indicating (according to these authors) a formation of different species of RNA under the influence of the hormone. On the other hand, a similar hybridization difference could also result from hormone-induced changes in the redundancy of nucleotide sequences, a possibility which could have some bearing on the understanding of the gibberellin (and/or cAMP)-DNA interactions. As a first approach we therefore measured the renaturation kinetics of sheared DNA of roots and the green parts of cucumber seedlings, employing the hydroxylapatite method for the separation of single and double strands (Bernardi, 1969). As expected (Britten and Kohne, 1968) the cucumber DNA can be divided, as shown by the reannealing profiles of Fig. 7,

FIGURE 7

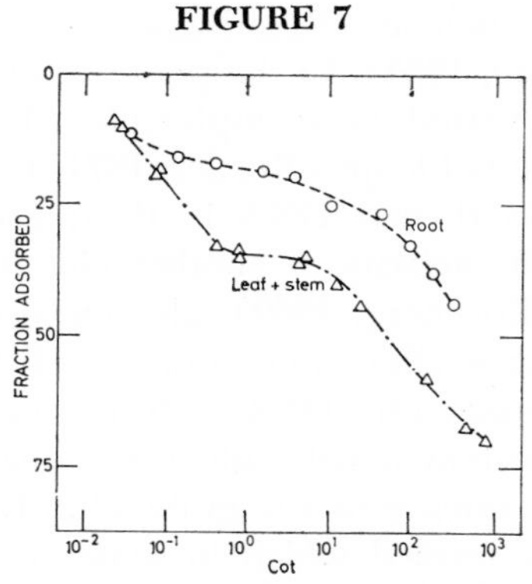

Renaturation kinetics of cucumber DNA from roots and the green plant parts, as measured by hydroxylapatite chromatography. Each DNA had been sheared previously to about 180,000 mol. wt. The mixture was heat denatured at 100°C for 10 min and incubated at 60°C to allow renaturation. At various times samples were removed, diluted in 0.12 M phosphate buffer and loaded on hydroxylapatite columns. The ordinate shows the fraction of DNA reassociated. The Cot value for each sample, presented on the abcissa, is the product of the nucleotide molarity of cucumber and the time (seconds) of incubation at 60°C (Britten and Kohne, 1968).

into three rather distinct categories of nucleotide sequences. *I.e.*, a rapidly reassociating fraction of highly reiterated nucleotide sequences, a second fraction composed of sequences of intermediate redundancy and a major fraction of unique composition, comprising presumably those nucleotide sequences which are present in only one copy per haploid chromosome set. The data of Table 6,

TABLE 6

Composition of Cucumber Nuclear Genome with Regard to Components of Various Degrees of Sequence Redundancy

Organ	Category of DNA	% of total genome	Amount of total genome (g)	$Cot_{1/2}$	Genome size relative to *E. coli*	$Cot_{1/2}$ expected if all sequences unique	Redundancy ($Cot_{1/2}$ expected/ $Cot_{1/2}$ found)
Roots	Rapid	16	8.0×10^{-14}	1.05×10^{-2}	21	1.7×10^{2}	1.5×10^{4}
	Intermediate	12	6.0×10^{-14}	8.0×10^{-1}	16	9.0×10^{1}	1.1×10^{2}
	Slow	72	3.6×10^{-13}	3.8×10^{2}	95	7.6×10^{2}	2.0×10^{0}
Shoots + Leaves	Rapid	34	1.7×10^{-13}	5.0×10^{-2}	45	3.6×10^{2}	7.2×10^{3}
	Intermediate	4	2.0×10^{-14}	6.0×10^{-1}	5	4.0×10^{1}	6.7×10^{1}
	Slow	62	3.1×10^{-13}	3.4×10^{2}	82	6.6×10^{2}	1.9×10^{0}

Calculated from Fig. 7 and based on the following facts: (1) haploid nuclear cucumber genome consists of 5×10^{-13} g (Jona Snir, personal communication); (2) the haploid genome of *E. coli* is 3.8×10^{-15} g; (3) $Cot_{1/2}$ for reannealing *E. coli* is 8.

TABLE 7

Changes with Age of the Nucleotide Sequence Diversity of Cucumber Nuclear DNA

Organ	Age (Days)	Category of DNA	% of total genome	Amount of total genome (g)	$Cot_{1/2}$	Genome size relative to *E. coli*	$Cot_{1/2}$ expected if all sequences unique	Redundancy ($Cot_{1/2}$ expected/ $Cot_{1/2}$ found)
Roots	4	Rapid	20	1.0×10^{-13}	1.05×10^{-2}	26	2.0×10^{2}	1.9×10^{4}
		Intermediate	10	5.0×10^{-14}	8.0×10^{-1}	13	1.0×10^{2}	1.3×10^{2}
		Slow	70	3.5×10^{-13}	5.0×10^{2}	92	7.4×10^{2}	1.5×10^{0}
	12	Rapid	12	6.0×10^{-14}	1.05×10^{-2}	16	1.3×10^{2}	1.2×10^{4}
		Intermediate	7	3.5×10^{-14}	1.05×10^{-1}	9	7.2×10^{1}	6.9×10^{2}
		Slow	81	4.05×10^{-13}	9.2×10^{2}	106	8.5×10^{2}	1.0×10^{0}
Shoots + Leaves	4	Rapid	34	1.7×10^{-13}	5.0×10^{-2}	45	3.6×10^{2}	7.2×10^{3}
		Intermediate	2	1.0×10^{-14}	8.0×10^{-1}	2.7	2.2×10^{1}	2.8×10^{1}
		Slow	64	3.2×10^{-13}	6.1×10^{2}	84	6.7×10^{2}	1.1×10^{0}
	12	Rapid	35	1.75×10^{-13}	5.0×10^{-2}	46	3.7×10^{2}	7.4×10^{3}
		Intermediate	2	1.0×10^{-14}	6.0×10^{-2}	2.6	2.1×10^{1}	3.5×10^{1}
		Slow	63	3.15×10^{-13}	7.0×10^{2}	83	6.6×10^{2}	1.0×10^{0}

Calculated from experimental Cot profiles and based on the following facts: (1) haploid nuclear cucumber genome consists of 5×10^{-13} g (Jona Snir, personal communication); (2) the haploid genome of *E. coli* is 3.8×10^{-15} g; (3) $Cot_{1/2}$ for reannealing *E. coli* is 8.

TABLE 8

The Effect of cAMP and GA_3 on the Sequence Diversity of Cucumber Nuclear DNA Derived from Various Organs

Organ	Treatment	Category of DNA	% of total genome	Amount of total genome (g)	$Cot_{1/2}$	Genome size relative to *E. coli*	$Cot_{1/2}$ expected if all sequences unique	Redundancy ($Cot_{1/2}$ expected/ $Cot_{1/2}$ found)
Roots	Control	Rapid	18	9.0×10^{-14}	1.1×10^{-2}	24	1.9×10^{2}	1.7×10^{4}
		Intermediate	9	4.5×10^{-14}	6.1×10^{-1}	11	8.8×10^{1}	1.4×10^{2}
		Slow	73	3.65×10^{-13}	7.5×10^{2}	96	7.7×10^{2}	1.1×10^{0}
	3′,5′-cAMP	Rapid	27	1.35×10^{-13}	1.0×10^{-2}	36	2.9×10^{2}	2.9×10^{4}
		Intermediate	3	1.5×10^{-14}	4.0×10^{-1}	4	3.2×10^{1}	8.0×10^{1}
		Slow	70	3.5×10^{-13}	5.0×10^{2}	92	7.4×10^{2}	1.5×10^{0}
	GA_3	Rapid	18	9.0×10^{-14}	1.1×10^{-2}	23	1.8×10^{2}	1.6×10^{4}
		Intermediate	9	4.5×10^{-14}	6.1×10^{-1}	11	8.8×10^{1}	1.4×10^{2}
		Slow	73	3.65×10^{-13}	7.5×10^{2}	96	7.7×10^{2}	1.1×10^{0}
Shoots + Leaves	Control	Rapid	34	1.7×10^{-13}	4.5×10^{-2}	45	3.6×10^{2}	8.0×10^{3}
		Intermediate	4	2.0×10^{-14}	6.5×10^{-1}	5	4.0×10^{1}	6.2×10^{1}
		Slow	62	3.1×10^{-13}	5.05×10^{2}	82	6.6×10^{2}	1.3×10^{0}
	3′,5′-cAMP	Rapid	37	1.85×10^{-13}	3.0×10^{-2}	49	3.9×10^{2}	1.3×10^{4}
		Intermediate	4	2.0×10^{-14}	7.0×10^{-2}	5	4.0×10^{1}	5.7×10^{2}
		Slow	59	2.95×10^{-13}	3.0×10^{1}	78	6.2×10^{2}	2.1×10^{1}
	GA_3	Rapid	34	1.7×10^{-13}	5.0×10^{-2}	45	3.6×10^{2}	7.2×10^{3}
		Intermediate	16	8.0×10^{-14}	3.0×10^{-1}	21	1.7×10^{2}	5.7×10^{2}
		Slow	50	2.5×10^{-13}	6.0×10^{2}	66	5.3×10^{2}	0.9×10^{0}

Calculated from experimental Cot profiles and based on the following facts: (1) haploid nuclear cucumber genome consists of 5×10^{-13} g (Jona Snir, personal communication); (2) the haploid genome of *E. coli* is 3.8×10^{-15} g; (3) $Cot_{1/2}$ for reannealing *E. coli* is 8.

derived from Fig. 7, show that the nucleotide sequence redundancy of the DNA is organ-dependent. In the green parts of the cucumber seedlings about 60% of the haploid nuclear genome consists of unique sequences, while in roots 70% of the sequences are unique. Marked differences between roots and the green parts were also noted in the highly-repeated fractions (16% in roots versus 34% in the green parts).

As the plants grew older, the relative sequence diversity of the genome of the green plant parts remained unchanged (Table 7). In the roots, on the other hand, the unique fraction tended to increase, primarily on account of the highly-repeated fraction.

Of particular interest were the differential responses of the plant genome to 3′,5′-cAMP and GA_3 (Table 8). GA_3 had no effect on DNA originating from the roots but, in the shoot and leaves, resulted in an increase of the intermediate DNA fraction from 4% to 16%, with a nine-fold increase in the average sequence reiteration: from 6.2×10^1 to 5.7×10^2 copies (Table 9).

The responses to 3′,5′-cAMP were rather pronounced in the rapid fraction of the root genome (increasing from 18% to 27%) and in the redundancy factor. The average sequence reiteration increased in the rapid root fraction by a factor of 1.7 (from 1.7×10^4 to 2.9×10^4 copies). Particular changes were observed in the redundancy of the various regions of DNA derived from the green parts. The rapid, intermediate and slow regions of this DNA from control plants consisted, respectively, of 8000, 62 and one gene copy per haploid genome, while in the cAMP-treated plants those same sequences were now present: 1.6, 9.2 and 15.0 times per haploid genome, respectively (Table 9).

TABLE 9

Changes in Redundancy with Increasing Age and Under the Influence of cAMP and GA_3

Organ	Category of DNA	Age (Redundancy at age 12 days/redundancy at age 4 days)	Treatments (Redundancy of treatment/ redundancy of respective control)	
			cAMP	GA_3
Root	Rapid	0.5	1.7	1.0
	Intermediate	5.4	0.6	1.0
	Slow	0.7	1.0	1.0
Shoot + Leaf	Rapid	1.0	1.6	1.0
	Intermediate	1.2	9.2	9.0
	Slow	1.0	15.0	1.0

DISCUSSION AND CONCLUSIONS

In this account we provided data which throw some light on relationships between cyclic mononucleotides, a plant hormone (gibberellin) and the plant genome. Also, some physiological processess which are believed to be expressions of primary actions of genes, being triggered via an activation or deactivation of specific genes such as phytochrome-controlled photomorphogenetic processes (Drumm and Mohr, 1967; Durst and Mohr, 1966; Jacobs and Mohr, 1966; Mohr, 1966; Schopfer, 1967) and the multiplication and differentiation of cells, are profoundly influenced by cyclic mononucleotides. The evidence presented above also shows some novel features in their mode of interactions.

3′,5′,-cyclic AMP is widely distributed in the animal world (Robison *et al.*, 1968, 1971) and, while its original role was attributed to its action as "second messenger" transmitting the effect of hormones into their particular target cells, 3′,5′-cAMP has since then proved also necessary for switching on many of the inducible enzymes in bacteria (Pastan and Parlman, 1970). In plants very little is yet known about possible functions of cyclic mononucleotides and their occurrence, though postulated, has not yet been reported. Our work now provides evidence that 3′,5′-cAMP does exist in plants too, both annuals and perennials (Table 1). The level of 3′,5′-cAMP in the leaves is related to the age (or growth rate) of the leaves and of the plants. In contrast to animals, 3′,5′-cAMP functions as a primary inducer which acts on, or very close to, the genome, to induce the production of a hormone, i.e. 3′,5′-cAMP first activates DNA synthesis, followed by GA biosynthesis.

From our present and previously published (Kessler and Snir, 1969; Kessler, 1971a, 1971b, 1971c; Kessler and Snir, 1971) data we may conclude that the cyclic nucleotide-GA system constitutes a regulatory mechanism which includes (i) the induction of hormone synthesis at the genome level, and (ii) a regulation of nucleotide sequence (gene ?) redundancy by the members of this system. In fact, all the above results, taken together with previous reports (Kessler and Snir, 1969; Kessler, 1971a, 1971b, 1971c; Kessler and Snir, 1971), strongly indicate that plant DNA is considerably heterogeneous in nature. The nucleotide sequence redundancy was calculated from renaturation kinetics profiles. The genomic sizes of the various DNA fractions were not adjusted for G + C content as the test plants were relatively low for G + C (32%), and also because contradictory evidence has been presented on its effect (Wetmure and Davidson, 1968; Gillis *et al.*, 1970). The heterologous DNA from cucumber roots and from the green parts renatured independently of *E. coli* DNA, meaning a minimum nucleotide sequence difference of 20-30% (Laird *et al.*, 1969).

This heterogeneity was not constant during the life cycle of the plant, but differed in different organs constituting the plant. Furthermore the "genomic" redundancy changed with age and environmental conditions and was modulated by cyclic mononucleotides and gibberellin in an organ-specific mode. In response to both 3′,5′-cAMP and GA, the sequence redundancy of the intermediate DNA fraction of the green parts increased nine fold. On the basis of our above contention that 3′,5′-cAMP functions as GA-inducer, we postulate that 3′,5′-cAMP does not act directly upon this DNA fraction but induces the synthesis of GA and it is then the hormone that brings about the nine fold increase of the sequence redundancy.

While GA affected only the sequence redundancy of the intermediate DNA fraction of the green plant parts, the effects of 3′,5′-cAMP were more general in the sense that other DNA fractions were affected too, *i.e.*, the sequence redundancy of the unique DNA fraction of the green parts increased 15 times and of the rapid root DNA about twofold.

Britten and Davidson (1969) and Crick (1971) assigned regulating functions to the repetitive sequences of the genome. Repetitive sequences are thought to be hormone-sensitive regulating centres (Britten and Davidson, 1969) and they have also been found to interact with chromosomal RNA (Sivolap and Bonner, 1971), which has been implicated as control element. We postulate that 3′,5′-cAMP, as inducer agent for GA-synthesis, affects (as found—Tables 8 and 9) the repetitive fraction, leading to an amplification of GA-forming genes; the GA thus synthesized will then activate the intermediate DNA fraction in the green parts of the plant. We further postulate that GA-directed processes, in cases which apparently depend on gene activity, such as cell elongation (Nitzan and Lang, 1966) and xylem formation (Fig. 6), are channeled to this particular intermediate GA-sensitive DNA fraction.

In addition to GA-induction, 3′,5′-cAMP has been found to possess also other specific functions in genome-dependent processes such as flower formation (Table 5) and xylem organization (Fig. 6 H). As a working hypothesis we relate these activities not to the regulating repetitive part of the genome but to the coding unique fraction (Sivolap and Bonner, 1971), the redundancy of which increased about 15 times under the influence of 3′,5′-cAMP.

We do not yet have enough data to correlate the remarkable influence of 2′,3′-cAMP on the multiplication of the duckweed Lemna gibba and its capacity to replace the long-day requirement to specific fractions of the genome—apart from the fact that DNA is apparently also involved in this process (Fig. 5).

The various postulated interactions among the regulator or inducer molecules, the sequence diversity of the genome and some probably coupled physiological processes are outlined in the scheme shown in Fig. 8.

This scheme proposes a basis for further detailed studies of long-term major changes in the morphogenetic pattern of plant development such as differentiation and the transition from vegetative to reproductive development, which in many cases are triggered by environmental factors. On the basis of our results we assume that the inducing environmental influences are transduced, via cyclic mononucleotides as primary regulating agents, to the genome, which responds with a rearrangement of its nucleotide sequence diversity, probably reflecting selective gene amplification. Gene amplification is apparently restricted to the cells of higher organisms (Brown and David, 1968) and occurs at sites within the eukaryotic DNA which are genetically redundant (Gall, 1969), as are the ribosomal RNA cistrons. Studies of the relationships between the physiological control of gene amplification and unstable gene redundancy suggest (Tartoff, 1970) that the level of ribosomal genes is physiologically determined during embryonic development. Our data on the changing nucleotide sequence redundancy, in response to regulating agents and inducing environmental conditions,

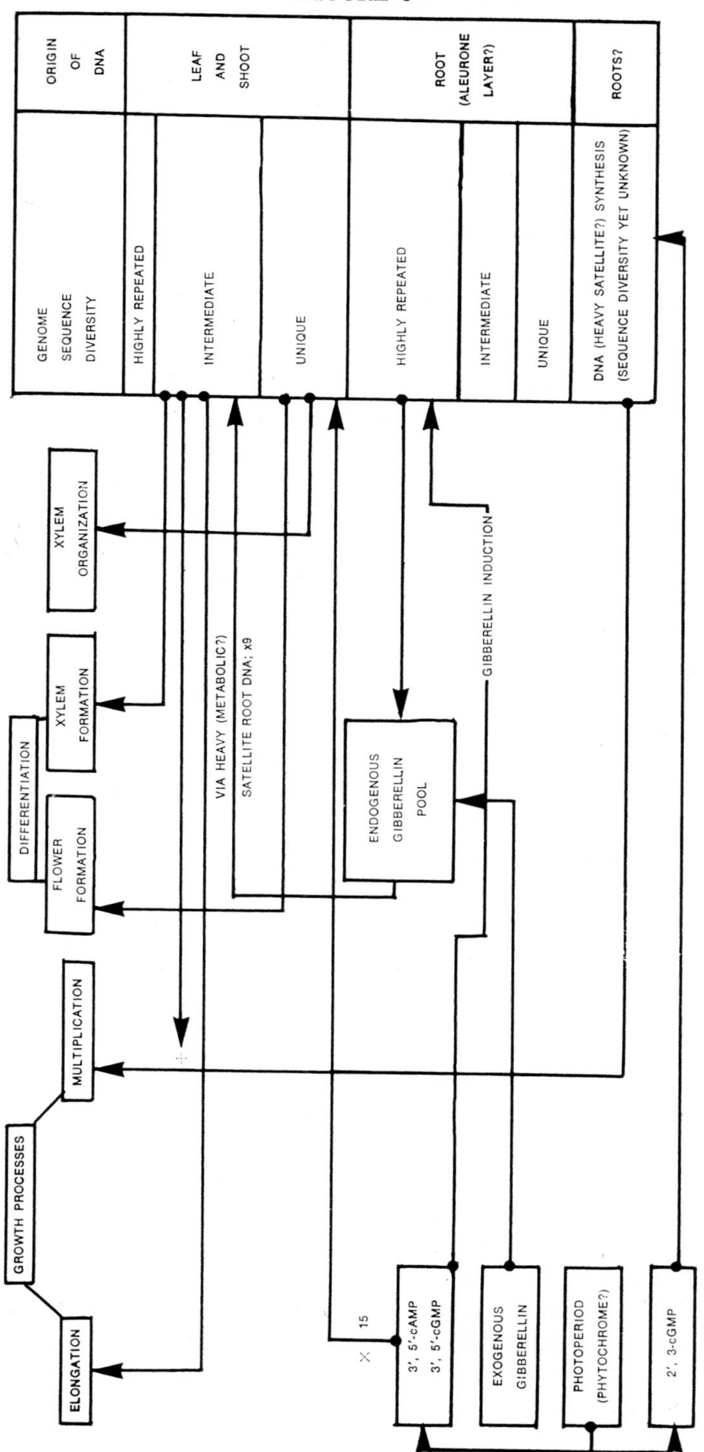

Schematic representation of the possible relationships among regulating agents (cyclic mononucleotides, gibberellin), the nucleotide sequence diversity of the genome and some gene-related physiological processes.

permit and even encourage studies of the significance and role of repeated and non-repeated DNA in plant development and its correlation with gene-activity requiring physiological processes affected by the same regulators (Table 10).

TABLE 10

Summary of the Effects of Cyclic Mononucleotides and GA on Some Physiological Processes and Common Genome Fractions Affected

Inducer or Promoter	Physiological Process	Common Genome Fraction Affected by Promoters of a Respective Process
3′,5′-cAMP	GA Formation	Rapid Root and/or Unique Fractions
3′,5′-cAMP and GA	α-Amylase Formation	Intermediate Shoot Fraction
3′,5′-cAMP	Flowering	Unique Shoot Fraction
3′,5′-cAMP and GA	Differentiation	Intermediate Shoot Fraction
3′,5′-cAMP	Organization	Unique Shoot Fraction
3′,5′-cAMP and GA	Elongation	Intermediate Shoot Fraction
2′,3′-cAMP	Multiplication	Still unknown

A final word about the position of cyclic mononucleotides in plant control mechanisms. We believe that we now have ample evidence for the hypothesis that cyclic purine mononucleotides are key substances which induce hormone biosynthesis and probable other cell activities at the level of genome. In this respect, the involvement of 2′,3′-cyclic mononucleotides is also notable. As these cyclic mononucleotides are derived as intermediates in the catabolic metabolism of RNA, they could couple an enhanced breakdown or turnover of RNA with an induction of morphogenetic events, thus placing RNase as part of a regulatory system functioning in plant development.

We believe that the various parts of this scheme are accessible for detailed experimental exploration, and efforts in this direction are now being made in our laboratory.

ACKNOWLEDGEMENTS

This research was supported in part by grants from the U.S. Dept. of Agriculture (FG-Is-261) and the Bar-Ilan Research Committee (154-1-51).

Contribution from The Volcani Center, Agricultural Research Organization, Bet Dagan, Israel. 1972 Series, No. 1280 -E.

The following students for the M.Sc. degree cooperated in this work: Mrs. Ilana Karaz in the tissue culture experiments, Mrs. Bina Kaplan in the cyclic nucleotide-GA interaction studies, Mr. N. Steinberg in the experiments with duckweed, Mrs. Hadassa Schlesinger in the renaturation studies and Miss Rela Levinstein in the assay of 3′,5′-cyclic AMP in plants and in the thymidine incorporation studies. I am greatly indebted to Mrs. Jael Frank for her competent technical assistance.

REFERENCES

Barendse, W. M., Kende, H. and Lang, A. (1968). *Plant Physiol. 43,* 815

Bernardi, G. (1969). *Biochim. Biophys. Acta 174,* 425

Bonner, J. (1965). In *The Molecular Biology of Development,* p. 155. Clarendon Press, London

Britten, R. J. and Davidson, E. H. (1969). *Science 165,* 345

Britten, R. J. and Kohne, D. E. (1968). *Science 161,* 529

Brown, D. and David, I. (1968). *Science 160,* 272

Chrispeels, M. J. and Varner, J. E. (1967). *Plant Physiol. 42,* 398

Coombe, B. G., Cohen, D. and Paleg, L. G. (1967). *Plant Physiol. 42,* 113

Crick, F. (1971). *Nature (London) 234,* 25

Doenecke, D. and Sekeris, C. E. (1971). *FEBS Lett. 8,* 61

Drumm, H. and Mohr, H. (1967). *Planta 72,* 232

Duffus, C. M. and Duffus, J. H. (1969). *Experientia 25,* 581

Durst, F. and Mohr, H. (1966). *Naturwissenschaften 20,* 531

Earle, R. M. and Galsky, A. G. (1971). *Plant Cell Physiol. 12,* 727

Filner, P., Wray, J. L. and Varner, J. E. (1969). *Science 165,* 358

Gall, J. (1969). *Genetics 61* (Suppl.), 121

Galsky, A. G. and Lippincott, J. A. (1969). *Plant Cell Physiol. 10,* 607

Galston, A. W. and Davies, P. J. (1969). *Science 163,* 1288

Gillis, M., De Ley, J. and De Cleene, M. (1970). *Eur. J. Biochem., 12,* 143

Gilman, A. G. (1970). *Proc. Nat. Acad. Sci., U.S. 67,* 305

Hillman, W. S. (1961). *Am. J. Bot. 48,* 413

Jacobs, M. and Mohr, H. (1966). *Planta 69,* 187

Kamisaka, S. (1972). In *Plant Growth Substances,* 1970 (Carr, D. J., ed.) p. 654, Springer Verl., Berlin

Kamisaka, S. and Masuda, Y. (1970). *Naturwissenschaften 57,* 546

Kamisaka, S. and Masuda, Y. (1971). *Plant Cell Physiol. 12,* 1003

Kamisaka, S., Sano, H. and Masuda, Y. (1972). *Plant Cell Physiol. 13,* 167

Karaz, I. (1971). *M.Sc. Thesis.* Bar-Ilan Univ., Ramat Gan, Israel (Hebrew, with English summary)

Keck, K. (1956). *Arch. Biochem. Biophys. 63,* 446

Kessler, B. (1969). *First Annual Report to the U.S. Dept. Agric.* Project A40-FS-17.

Kessler, B. (1971a). *Biochim. Biophys. Acta 232,* 611

Kessler, B. (1971b). *Biochim. Biophys. Acta 240,* 330

Kessler, B. (1971c). *Biochim. Biophys. Acta 240,* 496

Kessler, B. and Snir, J. (1969). *Biochim. Biophys. Acta 195,* 207

Kessler, B. and Snir, J. (1971). *Third Annual Report to the U.S. Dept. Agric.,* Project A40-FS-17

Key, J. L. and Shannon, J. C. (1964). *Plant Physiol. 39,* 360

Kornberg, A. (1969). *Science 163,* 1410

Lang, A. (1970). *Annu. Rev. Plant Physiol. 21,* 537

Laird, C. D., McConaughy, B. L. and McCarthy, B. J. (1969). *Nature (London) 224,* 149

Lindhal, T. and Edelman, G. M. (1968). *Proc. Nat. Acad Sci., U.S. 61,* 680

Marmur, J. (1961). *J. Mol. Biol. 3,* 208

Miller, C. O. (1961). *Proc. Nat. Acad. Sci., U.S. 47,* 170

Miyamoto, E., Kuo, J. F. and Greengard, J. (1969). *J. Biol. Chem. 244,* 6395

Mohr, H. (1966). *Phytochem. Phytobiol. 5,* 469

Nickells, M. W., Schaefer, G. M. and Galsky, A. G. (1971). *Plant Cell Physiol. 12,* 717

Nitzan, J. and Lang, A. (1966). *Plant Physiol. 41,* 965

Nooden, L. D. and Thiman, K. V. (1963). *Proc. Nat. Acad. Sci., U.S. 50,* 194

Paleg, L. G. (1960). *Plant Physiol. 35,* 293

Paleg, L. G. (1964). *Colloq. Int. Centre Nat. Rech. Sci. 123,* 303

Paleg, L. G. (1965). *Annu. Rev. Plant Physiol. 16,* 291

Pastan, I. and Parlman, R. (1970). *Science 169,* 339

Pollard, C. J. (1970). *Biochim. Biophys. Acta 201,* 511

Robison, G. A., Butcher, R. W. and Sutherland, E. W. (1968). *Annu. Rev. Biochem., 37,* 149

Robison, G. A., Butcher, R. W. and Sutherland, E. W. (1971). *Cyclic AMP.* Academic Press, N.Y.

Salomon, D. and Mascarenhas, J. P. (1971a). *Z. Pflanzenphysiol. 65,* 385

Salomon, D. and Mascarenhas, J. P. (1971b). *Life Sci. 10,* 879

Schmidt, G. and Thannhauser, S. T. (1954). *J. Biol. Chem. 161,* 83

Schopfer, P. (1967). *Planta 72,* 297

Sivolap, Y. M. and Bonner, J. (1971). *Proc. Nat. Acad. Sci., U.S. 68,* 387

Spirin, A. S., Mosana, A. A. and Monroy, A. (1966). *Current Topics in Developmental Biology,* vol. 1, p. 1. Academic Press, New York.

Summer, J. B. (1924). *J. Biol. Chem. 62,* 287

Tartoff, K. (1970). *Science 171,* 294

Weiss, B. and Richardson, C. C. (1967). *Proc. Nat. Acad. Sci., U.S. 57,* 1021

Wetmure, J. G. and Davidson, N. (1968). *J. Mol. Biol. 31,* 349

Wood, H. N., Lin, M. C. and Braun, A. C. (1972). *Proc. Nat. Acad. Sci., U.S. 69,* 403

DNA and RNA Synthesis During Growth by Cell Expansion in *Vicia Faba* Cotyledons

ADÈLE MILLERD
Division of Plant Industry, CSIRO, Canberra, A.C.T., Australia

One approach to the problem of differential gene expression is to study a tissue which makes large amounts of one or a few specific gene products. The developing seeds of legumes, for example broad bean (*Vicia faba* L.) or pea (*Pisum sativum* L.) offer an attractive system for such a study. During development these seeds accumulate large amounts of tissue specific globulins which consist of two high molecular weight proteins, legumin (mol. wt. 320,000; Bailey and Boulter, 1970) and vicilin (mol. wt. 186,000; Danielsson, 1949). The corresponding proteins in the two species are immunologically closely related (Millerd *et al.*, 1971).

A major attempt to study differential gene expression in developing seeds was first made by Dr. James Bonner's group. These workers used immature seeds of pea for analyzing the regulation and transcription of chromatin preparations from cotyledon cells (summarized in Bonner *et al.*, 1968). More recently, Dr. D. Boulter and his colleagues have published extensive studies of cotyledon development in the broad bean, particularly in relation to altered levels of free and membrane-bound ribosomes (Briarty *et al.*, 1969; Payne and Boulter, 1969) and to changes in the protein synthesis machinery (e.g. Payne *et al.*, 1971*a*; Payne *et al.*, 1971*b*).

The work presented in this paper was initiated at La Jolla with Dr. H. Stern and Dr. M. Simon and has continued in Canberra with Dr. P. R. Whitfeld and Dr. W. Dudman. We wish to present our studies of legumin and nucleic acid synthesis during cotyledon development in the broad bean.

Our experimental materials are developing seeds at all stages, from pod formation to the initiation of seed maturation. In the very young seed, the cotyledons are very small and are bathed in liquid endosperm. As the seed develops, the cotyledons increase in size and soon make up the bulk of the seed. The cotyledons apart from layers of epidermal cells, consist of parenchymatous tissue and vascular strands.

During development, the tissue-specific globulins, legumin and vicilin are synthesized and stored in the cytoplasm as protein bodies. These storage proteins are utilized following germination, when they are broken down and supply nitrogenous compounds to the developing seedling. Here, however, we are discussing only their synthesis during seed development on the mature plant.

We have used as a measure of development, the length of the long axis of the cotyledon, and all data are expressed relative to this length. This parameter can be well correlated with fresh weight and days from flowering (Millerd *et al.*, 1971).

In a developing plant organ, there are frequently two phases of growth, an initial one of intensive cell division followed by a period of growth by cell expansion. These two phases were defined in *Vicia* by correlating the stage of development with the number of cells (Fig. 1). During early cotyledon growth, cell number increased rapidly and essentially the full complement of cells, approximately 2.8×10^6, was reached when the cotyledon was 9 to 10 mm long. In cotyledons from 10 mm to 30 mm there was a constant number of cells and growth occured by cell expansion.

FIGURE 1

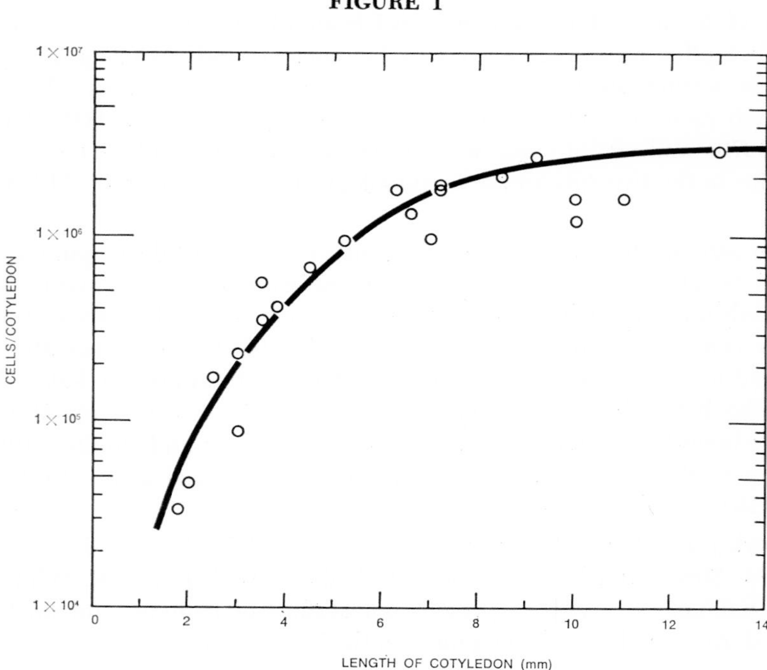

Relationship of stage of development and number of cells per cotyledon in *V. faba*. Cell number was determined according to Rijven and Wardlaw, 1966. Data from Millerd *et al.* (1971) (reproduced with permission).

We wished to describe specifically and quantitatively the pattern of legumin accumulation during cotyledon development.

Globulins were extracted from freshly harvested seeds and legumin was extensively purified. Purified legumin was used as an antigen in rabbits and antiserum of high titre was obtained. When the antiserum was tested by immunodiffusion, a single sharp precipitin band appeared, but with time, a faint second band was visible. This second band showed a reaction of identity with vicilin. This result had two possible explanations. The purified legumin could have contained a trace of vicilin, or legumin and vicilin may have shared antigenic determinants and hence would always show cross-reactivity. We have excluded the second possibility and have shown legumin and vicilin to be separate antigens. Our antiserum did contain both antilegumin and antivicilin but the antivicilin titre was much lower and the antiserum was used at a dilution where antivicilin did not interfere.

Legumin was quantitatively determined by micro-complement fixation. The various checks which were made (Millerd *et al.*, 1971), gave confidence that the assay for legumin was specific and that a single molecular species, rather than a heterogeneous population of molecules, was being measured. In addition, it was shown that large amounts of vicilin did not interfere with the assay of legumin.

The pattern of legumin accumulation during cotyledon development was then determined. Cotyledons at all stages of development were harvested and from pooled samples of each size, protein was exhaustively extracted using 0.2M NaCl in buffer. The total protein and legumin contents of the extracts were then determined.

We could not say that legumin was ever entirely absent from the cotyledons since even very young cotyledons (5 mm) contained 0.2 μg legumin (Table 1)

TABLE 1

Legumin content of cotyledons at various stages of development

Length of cotyledon (mm)	Extractable protein (mg) per cotyledon	Legumin (μg) per cotyledon
5	0.22	0.2
8	1.0	1.2
11	2.75	18.7
14	5.6	140
18	10.0	700
20	14.7	1.91×10^3
25	39.2	10.58×10^3

Samples of each size of cotyledon were exhaustively extracted with 50 mM tris, pH 7.8, containing 0.2 M NaCl. After centrifugation, the protein and legumin contents of the extracts were determined.

and this was the lower limit of our assay. When we could not detect legumin, it was less than 0.05% of the total protein. The data of Table 1 show that when the cotyledon reached about 10 mm in length, and cell division was essentially complete, there was a marked increase in legumin accumulation, and as shown in Fig. 2, legumin ultimately constituted 27% of the extractable protein.

The data of Fig. 2 show that, after the full complement of cells had been reached, there was a marked quantitative change in gene expression. The phase prior to this would be suitable for studying the initiation of specific gene expression. In cotyledons more than 18 mm long, legumin synthesis was very rapid, indicating that many cells were involved.

FIGURE 2

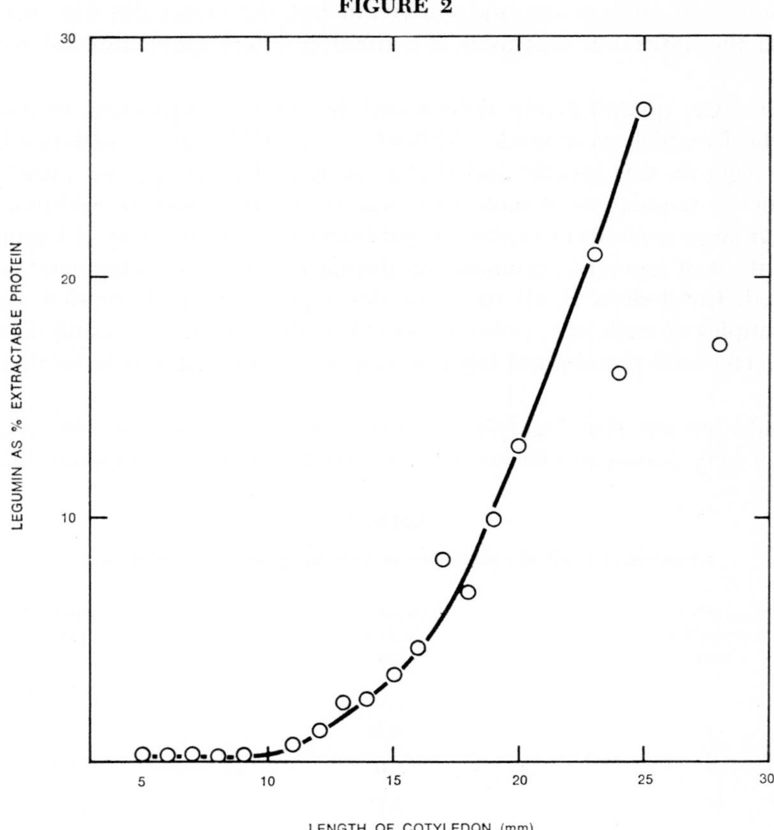

Change in legumin content during development of *V. faba* cotyledons. Samples of each size of cotyledon were exhaustively extracted with 50 mM tris, pH 7.8, containing 0.2 M NaCl. After centrifugation, the protein and legumin contents of the extracts were determined. Data from Millerd *et al* (1971) (reproduced with permission).

We next turned our attention to nucleic acid synthesis during cotyledon development. We inquired if this burst of protein synthesis was correlated with changes in any specific nucleic acid species. Cotyledons were removed from seeds rapidly synthesizing legumin and were exposed to a mixture of [^{3}H] uridine and [^{3}H]adenosine for 22 hours. Nucleic acids were isolated and resolved by electrophoresis on acrylamide gels and the distribution of the label was determined. There was considerable incorporation (Fig. 3) into ribosomal RNA, both 25S and 18S, and into t-RNA, but little indication of marked incorporation into other species.

FIGURE 3

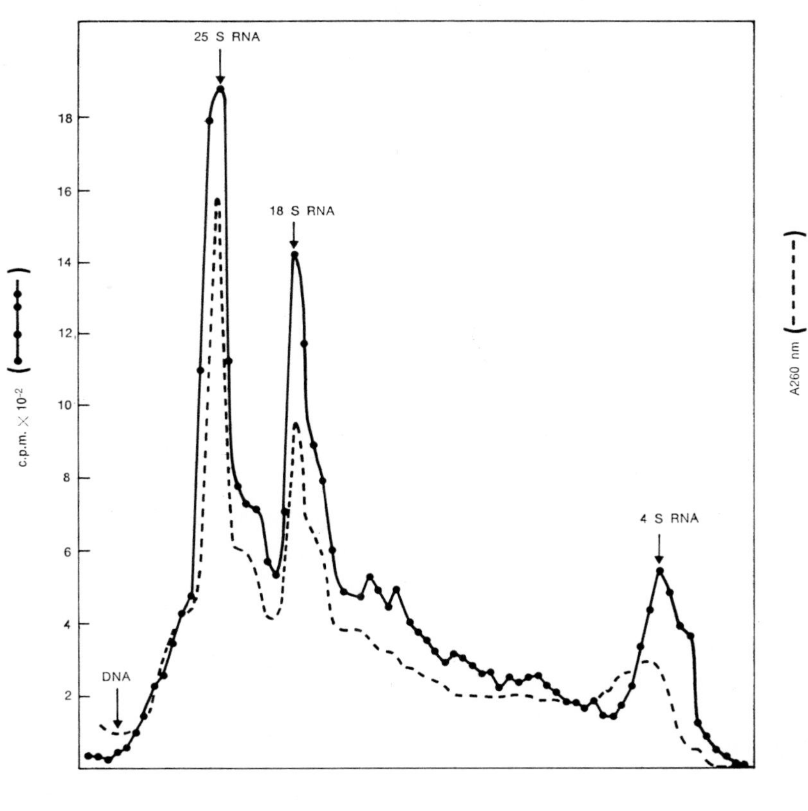

Incorporation of [^{3}H]uridine and [^{3}H]adenosine into nucleic acids. Detached *V. faba* cotyledons (20 mm) were incubated (22 h, 25°C) in nutrient (Linsmaier and Skoog, 1965) containing [^{3}H]uridine (4μci/ml) and [^{3}H]adenosine (4 μci/ml). Nucleic acids were isolated, resolved on acrylamide gels, and the distribution of the label determined.

A less expected observation was made when similar cotyledons, composed of non-mitotic cells, were exposed to [^{3}H]thymidine. There was considerable incorporation into DNA.

The fate of the labelled thymidine was studied by autoradiography of Feulgen-stained squashes. In all cases, the label was restricted to the nucleus. Even after incubation with [^{3}H]thymidine for 15 minutes, some cells were heavily labelled.

After 22 hours incubation, all parenchyma cells and cells of the vascular strands showed incorporation into the nucleus.

We have determined quantitatively the amount of DNA and RNA present in the cotyledons at all stages of development. The data (Table 2) show that during growth by cell expansion, non-mitotic nuclei continue to synthesize DNA.

TABLE 2

Changes in nucleic acid content of cotyledons during growth by cell expansion

Length of cotyledon (mm)	Fresh wt. (g)	RNA (mg) per cotyledon	DNA (μg) per cotyledon
10	0.14	0.35	65
11	0.19	0.44	89
19	0.53	1.3	239
24	1.05	3.3	455
28	1.80	5.0	696

Nucleic acids were extracted according to Williams and Rijven (1965). Total nucleic acids were estimated at 260nm assuming 1 mg nucleic acid/ml had $E^{1cm}_{260\ nm}$ of 30 units. DNA was determined (Burton, 1956) and RNA obtained by difference.

From analyses of DNA content and estimations of the number of cells present, DNA content per cell can be estimated. The DNA content of a cell from a 13 mm cotyledon would be about 4.5×10^{-11}g, but for a 29 mm cotyledon, it may average 33.2×10^{-11}g. Kung and Williams (1969) have reported 5.8×10^{-11}g DNA per nucleus from root tips of *V. faba*.

When nuclei of typical parenchyma cells from the two phases of cotyledon development were compared, photomicrographs showed (Fig. 4) that the marked increase in DNA per cell during growth by cell expansion was reflected in a dramatic increase in nuclear and nucleolar size. Preliminary quantitative microspectrophotometry on this material indicated the DNA content of the large nuclei to be about x10 that of the small nuclei.

To ascertain whether synthesis of DNA beyond the cell division phase was a general feature of legume cotyledons, the incorporation of [^{3}H]thymidine during growth by cell expansion was checked in two other species. Cotyledons were removed from developing seeds of *P. sativum* var. Alaska and from fenugreek (*Trigonella foenum graecum* L.) and incubated in [^{3}H]thymidine. In both types of seed, the developmental stage was well beyond completion of cell division and checks showed no mitotic figures in Feulgen-stained squashes. With both pea and fenugreek it was found that [^{3}H]thymidine was rapidly incorporated. Labelled nucleic acids were isolated and separated on acrylamide gels. The label was present in DNA, was removed by DNase and was unaffected by RNase. Incor-

FIGURE 4

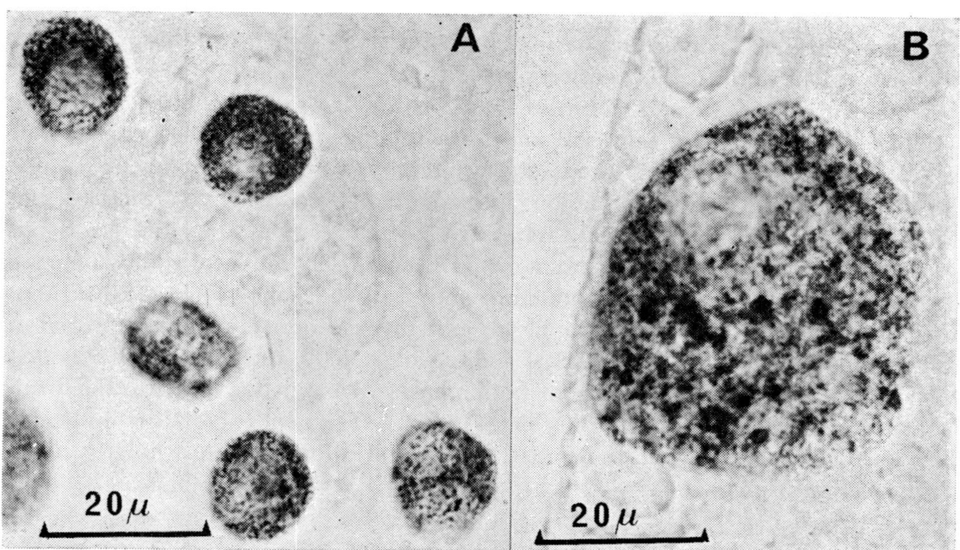

Photomicrographs of nuclei from developing cotyledons of *V. faba*. Material was prepared as a Feulgen-stained squash. A: cotyledon length 8 mm; B: cotyledon 26 mm.

poration of [^{3}H]thymidine does not necessarily imply extensive DNA synthesis, it may be due to repair. However, microscopic examination of pea cotyledons showed a dramatic increase in nuclear size during growth by cell expansion, and this agrees with measurements of nuclear DNA (Smith, 1971).

The increased DNA per nucleus may represent replication of part (selective amplification) or all of the genome (endoreduplication). As a criterion to assist in distinguishing between these alternatives, the buoyant density of DNA isolated from cotyledons in the cell division phase of growth was compared with that of DNA isolated from cotyledons growing by cell expansion. Both DNA preparations gave a single band of density 1.696 g/cm^3, a value in good agreement with those reported for nuclear DNA from roots of *V. faba* (Kung and Williams, 1969) and from leaves (Wells and Birnstiel, 1969). It was concluded that the extra DNA synthesized during cell expansion had the same overall base composition as the 2c DNA and, as there was no evidence of a satellite band in the DNA from the cell expansion phase, it was unlikely that there had been extensive amplification of a specific segment of the genome.

Additional support for the conclusion that the extra DNA synthesized in the cell expansion phase was similar to the pre-existing DNA came from experiments in which DNA labelled with [^{3}H]thymidine at an early stage of cell expansion, was centrifuged to equilibrium in CsCl either with DNA from the cell division phase or with DNA from late in the cell expansion phase. Coincidence of label with absorbance was observed in both cases.

The time required for reassociation of denatured DNA reflects the number of different sequences represented and their relative concentrations (Britten and Kohne, 1968). A comparison was made of the reassociation kinetics of DNA isolated from the cell division phase of cotyledon growth with those of DNA from cotyledons in the cell expansion phase.

As shown in Fig. 5, the kinetics of reassociation are similar for the two types of DNA. Reassociation in high salt and formamide at moderate temperature is a useful method for estimating moderately reiterative and unique sequences since the DNA must be incubated for long periods. However, using this method, the rate of reassociation of the highly reiterative sequences was so rapid, it was not possible to obtain accurate measurements. The small separation of the two curves at low C_0t values was not reproducible and was, we believe, caused by these experimental difficulties. When the highly reiterative sequences were measured at higher temperature (61°C) in 0.12 M sodium phosphate, the kinetics of reassociation were identical for DNA from both phases of cotyledon growth and from shoots of germinated seeds of *V. faba* (Fig. 6). The evidence, then, indicates that in these large nuclei there has been replication of the whole genome (endoreduplication).

FIGURE 5

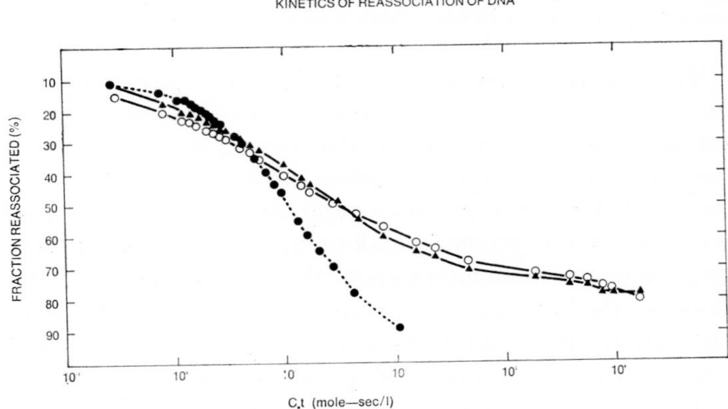

Comparison of reassociation kinetics of *V. faba* DNA from cotyledons in the cell division phase of growth (▲) with that from cotyledons in the cell expansion phase of growth (○). *E. coli* DNA (● —— ●) was used as a standard to control fluctuations in reassociation conditions. The rates of reassociation were measured at 37°C in 5 × SSC, 50% formamide by the optical method. The rates of reassociation have been corrected to those in 0.12M phosphate by the data of Britten and Smith (1970).

The function of this extra DNA is not known. It may play a role in the specialized process of storage protein synthesis, but the failure to demonstrate selective amplification would argue against this. Not only does DNA increase in these cells, but also RNA, starch and protein (both metabolic and storage). The specialized function of these cotyledon cells is the important one of supplying nutrients for the early development of the young seedling. The cytoplasm of these cells is filled with rough endoplasmic reticulum, protein bodies, and starch grains; the

FIGURE 6

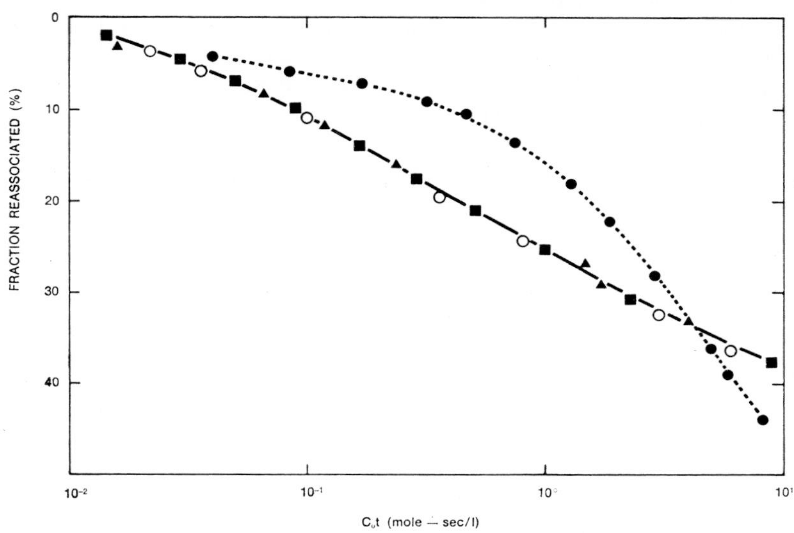

Reassociation profiles of the highly reiterative sequences of *V. faba* DNA from cotyledons in cell division phase of growth (▲); from cotyledons in cell expansion phase of growth (○); from shoots of germinated seeds (■). *E. coli* DNA (● —— ●) was used as a standard to control fluctuations in reassociating conditions. The rates of reassociation were measured at 61°C.

nucleus is cramped within the cytoplasm. There may be no space for the development of a mitotic apparatus. One can speculate that, in evolution towards an efficient storage organ, there has developed the principle of non-mitotic, wholesale replication of cellular contents. In this connection, one notes that storage tissue is frequently more than diploid. In many seeds, from monocotyledons and dicotyledons, a triploid endosperm functions as the storage organ.

ACKNOWLEDGEMENT

DNA was characterized in collaboration with Dr. P. R. Whitfeld.

REFERENCES

Bailey, C. J. and Boulter, D. (1970) *Eur. J. Biochem. 17,* 460

Bonner, J., Dahmus, M. E., Fambrough, D., Huang, R. C., Marushige, K. and Tuan, D. Y. H. (1968) *Science 159,* 47

Briarty, L. G., Coult, D. A., and Boulter, D. (1969) *J. Exp. Botany, 20,* 358

Britten, R. J. and Kohne, D. E. (1968) *Science 161,* 529

Britten, R. J. and Smith, J. (1970) *Yearb. Carnegie Inst. 68,* 378

Burton, K. (1956) *Biochem. J. 62,* 315

Danielsson, C. E. (1949) *Biochem. J. 44,* 387

Kung, S. D. and Williams, J. P. (1969) *Biochem. Biophys. Acta. 195,* 434

Linsmaier, E. M. and Skoog, F. (1965) *Physiol. Plantarum. 18,* 100

Millerd, A., Simon, M., and Stern, H. (1971) *Plant Physiol. 48,* 419

Payne, E. S., Boulter, D., Brownrigg, A., Lonsdale, D., Yarwood, A. and Yarwood, J. N. (1971a) *Phytochem. 10,* 2293

Payne, E. S., Brownrigg, A., Yarwood, A. and Boulter, D. (1971b) *Phytochem. 10,* 2299

Payne, P. I. and Boulter, D. (1969) *Planta* (Berl.) *84,* 263

Rijven, A. H. G. C. and Wardlaw, I. F. (1966) *Expl. Cell Res. 41,* 324

Smith, D. L. (1971) *Ann. Bot. 35,* 511

Wells, R. and Birnstiel, M. (1969) *Biochem. J. 112,* 777

Williams, R. F. and Rijven, A. H. G. C. (1965) *Aust. J. Biol. Sci. 18,* 721

Gene Expression in Differentiated Cells

Induction of δ-Aminolevulinic Acid Synthetase in Perfused Rat Liver by Drugs, Steroids, Lead and Adenosine-3′, 5′-Monophosphate

A. M. EDWARDS AND W. H. ELLIOTT
Department of Biochemistry, University of Adelaide, Adelaide, South Australia

Δ-Aminolevulinic acid (ALA) synthetase catalyses the condensation of succinyl CoA and glycine to yield ALA. This is the first and rate limiting step in the haem biosynthetic pathway and it is now accepted that the rate of haem synthesis in liver is regulated by changes in either the activity or level of ALA synthetase (De Matteis, 1967). This paper is concerned with the factors that determine the level of ALA synthetase in mammalian liver. Study of ALA synthetase control may provide information both on regulation of the expression of a specific gene in liver and on the nature of the genetic defect in the hereditary human diseases, the porphyrias, in which the level of ALA synthetase is abnormally high (Taddeini and Watson, 1968).

In normal liver, the level of ALA synthetase is very low (Irving and Elliott, 1969) but it can be increased dramatically under certain conditions (Granick and Urata, 1963; Granick, 1966). Changes in the enzyme level occur rapidly because of the short half lives (60-70 min) of both the enzyme and the messenger RNA which codes for it (Tschudy *et al.*, 1965a). In this respect ALA synthetase is an attractive system for studies on specific gene regulation in liver.

The current state of knowledge on some of the factors involved in regulation of hepatic ALA synthetase is summarised in Fig. 1. For a more complete review, see Kappas *et al.*, (1968a) and De Matteis (1967). The enzyme may be induced by a variety of drugs of widely differing structure. Inducing drugs are generally lipid soluble and many are substrates for the microsomal mixed function oxidase system. It might be noted that important components of this system, in particular, cytochrome P-450, are haemoproteins: up to 70% of haem synthesis in the liver is devoted to providing the haem prosthetic group of microsomal cytochromes (Schmid *et al.*, 1966; Marver, 1969). There is thus an important link between the level of ALA synthetase and the ability of the liver to detoxify foreign chemicals via the microsomal mixed function oxidase system (Marver, 1969; Tephly *et al.*, 1971).

FIGURE 1

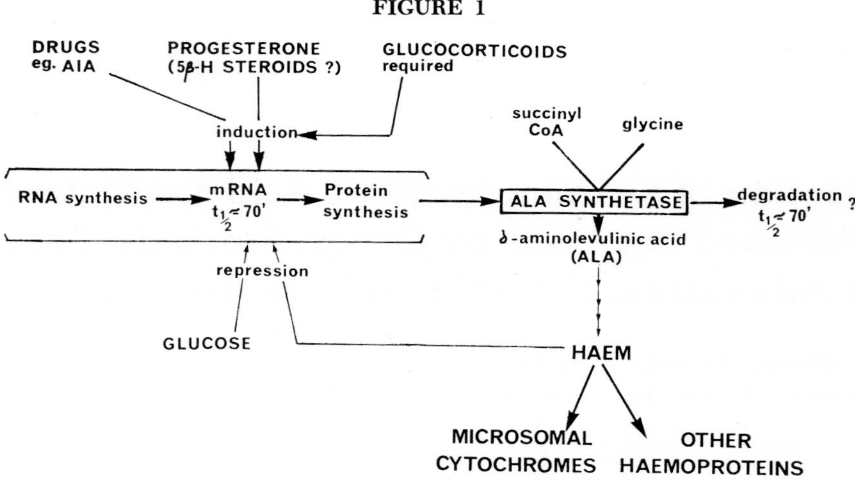

Summary of some important aspects of the regulation of ALA synthetase in mammalian liver.

The drug most frequently used in studies on ALA synthetase induction is allylisopropylacetamide (AIA) which causes a 4-5 fold induction of the enzyme in liver within 3 hours of subcutaneous administration to rats.

Certain steroids also cause induction. In chick embryo liver either *in vivo* or in tissue culture, a number of sex steroid derivatives with a 5β-H structure are potent inducers (Kappas *et al.*, 1968a) but so far *in vivo* studies on rats have failed to provide evidence that this is the case in mammalian liver (Kappas *et al.*, 1968b). However, progesterone, which is a relatively weak inducer in chick embryo liver (Kappas *et al.*, 1968a), has been found to induce hepatic ALA synthetase in mammals (Kaufman *et al.*, 1970; Jones and Emans, 1969).

Glucocorticoids do not themselves induce the enzyme but there is evidence that they exert some "permissive" effect in regulation of ALA synthetase since inducing drugs do not increase the enzyme level in adrenalectomised animals unless hydrocortisone is also administered (Marver, 1966; Matsuoka *et al.*, 1968).

Induction by drugs is repressed by feeding glucose and a substantial induction can be observed only in starved animals (Tschudy *et al.*, 1964; Marver *et al.*, 1966).

Finally perhaps the most important element in the regulation of ALA synthetase may be repression by the end product, haem. Such repression has been shown to occur in cultured chick embryo liver cells (Kappas *et al.*, 1968a; Sassa and Granick, 1970) and in whole animals (Hayashi *et al.*, 1968). Haem repression assumes a central role in most proposed models for ALA synthetase regulation (Kappas *et al.*, 1968a; De Matteis, 1967). In these models it has been suggested that haem acts as a co-repressor in regulating synthesis of ALA synthetase. Recent evidence, however, suggests that the role of haem is more complex than that of simple co-repressor (Schneck *et al.,* 1971; Hayashi *et al.,* 1972).

While it seems likely that many of the important factors in the control of hepatic ALA synthetase are now known, the mechanism of control remains to be established. It is known that induction by drugs and steroids is prevented by administration of cycloheximide or actinomycin D indicating that continuing RNA and protein synthesis are required for an increase in enzyme level (De Matteis, 1967). Existing evidence also suggests that inducers and repressors affect the enzyme level in most cases by changing its rate of synthesis rather than its rate of degradation (Marver *et al.,* 1966; Stein *et al.,* (1970); Sassa and Granick, 1970) but definitive evidence on this point must await studies with a specific antibody to ALA synthetase. Our colleague, Malcolm Whiting, has recently described preparation of an antibody to partially purified ALA synthetase (Whiting, 1972). Initial studies with this antibody showed that no enzymically inactive, immunologically cross-reacting component can be detected in normal liver, a finding consistent with the proposal that induction involves *de novo* synthesis of new enzyme. Nearly all studies to date on control of ALA synthetase in mammalian liver have involved *in vivo* experiments. For further work on this problem it seemed desirable to develop an *in vitro* system in which some of the complexities of *in vivo* experiments could be avoided. Since one object of this work was to provide information about the biochemical defect in the human porphyrias, it was desirable that the *in vitro* system should be derived from mammalian liver. Granick and Kappas and their co-workers have used primary cultures of chick embryo liver for *in vitro* studies on control of ALA synthetase (Kappas *et al.,* 1968a) but comparison of their findings for avian liver with findings from *in vivo* mammalian studies suggest some apparent differences between the two systems (e.g., in the effects of 5βH-steroids as described above). Furthermore, the tissue culture studies have provided little information on the involvement of glucocorticoids and repression by glucose observed in mammals.

We have therefore investigated the possibility of using perfused rat liver as an experimental system for the study of control of ALA synthetase levels.

EXPERIMENTAL METHODS

The perfusion methods used were based on those of other workers: apparatus and surgical procedure were as used in Krebs' laboratory (Hems *et al.,* 1966) and the perfusate was similar, though not identical to that of Hager and Kenney, (1968). The perfusate contained 40% fresh rat blood, glucose (5 mg/ml) and casein hydrolysate (6 mg/ml) in Krebs Phosphate Bicarbonate Ringer. In perfusion experiments, circulation to the liver was interrupted for only about 60 seconds, the perfusate flow rate was 15 ml/min and the liver retained an even pink-brown colour throughout the 3 hour experimental period. Bile production was 0.45 ml/h which is comparable to that reported by other workers. The rats used to provide the liver or bled to provide rat blood were starved for 42 hours and anaesthetized with nembutal before use. Test compounds were added to the perfusate (final volume, 95 ml) and their effects on total ALA synthetase activity in liver assessed by assaying homogenates prepared from small samples of liver removed by ligation (Seglen and Jervell, 1969) at the beginning of the perfusion and at various times over the following 3 hours. ALA synthetase was assayed using a colorimetric method essentially as described by Irving and Elliott (1969). (10 mM EDTA was added to incubations to prevent conversion of ALA to porphobilinogen by cytoplasmic ALA dehydratase).

RESULTS

Induction in control perfusions

Fig. 2 shows the change in total ALA synthetase activity over a 3 hour perfusion period either without added compounds or with the drug, AIA. The results represent pooled data from several experiments and the vertical bars show standard errors of the means.

FIGURE 2

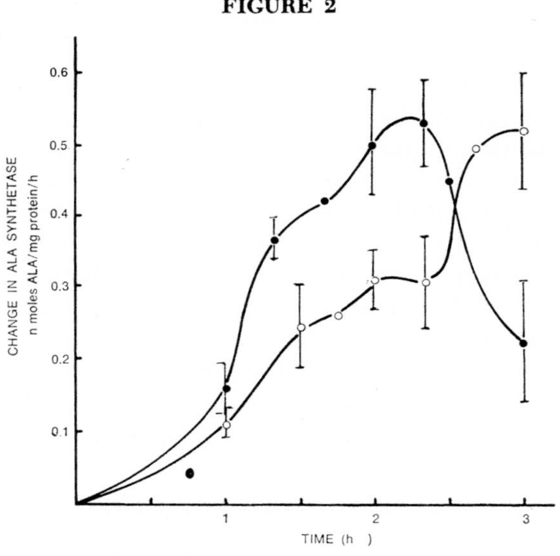

Effects of perfusion without added inducers and with AIA on total ALA synthetase activity in rat liver. Experimental methods are as described in the text. Curves represent mean changes in activity, from the number of experiments shown in brackets. Vertical bars indicate standard errors of means, calculated where 4 or more experiments contributed to the mean. No addition (○) [5]; AIA, 550 μg/ml (●) [4]. Mean zero time ALA synthetase activity was 0.18 nmoles ALA/mg protein/mg protein/h.

In the absence of added inducers there was an increase in enzyme activity corresponding to a 3-fold induction over 3 hours. The increase was prevented by the addition of cycloheximide (20 μg/ml) to the perfusate. At this concentration, cycloheximide inhibits protein synthesis 95% in perfused rat liver (Seglen and

Jervell, 1969). The observation of a substantial induction under "control" conditions was unexpected and the reason(s) for this have not been established with certainty. However, two contributory factors were identified. One is the use of nembutal to anaesthetize the rats. Intraperitoneal injection of nembutal (sodium phenobarbitone) was found to cause a 3-4 fold induction of ALA synthetase over 3 hours *in vivo*. If rats used in perfusion experiments were anaesthetized with ether, which caused no induction *in vivo*, the "control induction" particularly at later times was partly eliminated.

The smaller induction seen in the absence of nembutal or other exogenous inducers seems to be due in part to surgical trauma since sham operations on rats produce a similar 2-fold induction in ether anaesthetized rats.

Effect of AIA

AIA was found to have relatively little effect on the level of ALA synthetase in perfused liver when present in concentrations (550 μg/ml) at which it caused marked induction of ALA synthetase in cultured chick embryo liver cells (Granick, 1966). At early perfusion times, the increase in activity was somewhat greater with AIA than in the control, but at later times activity was lower in the

FIGURE 3

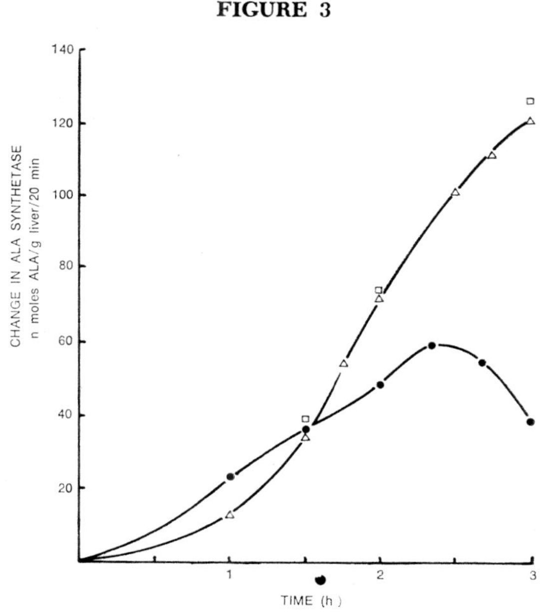

Comparison of the effects of AIA on total ALA synthetase activity *in vivo* and in perfused liver. The results for perfused liver were obtained as in Fig. 2. The *in vivo* results were obtained either by killing rats at various times after subcutaneous injection of AIA or by removing a series of liver samples at various times after AIA injection from anaesthetized, sham operated rats. Curves show means from the number of experiments in brackets. *In vivo*, not operated, AIA, 350 mg/kg (□) [each point, the mean of 4 rats]; *in vivo*, sham operated, AIA 350 mg/kg (△) [4]; perfused, AIA, 550 μg/ml (●) [4]. Mean zero time ALA synthetase activity was 29 nmoles ALA/g liver/20 min.

presence of the drug. The relative effects of AIA in perfused liver and *in vivo* are shown in Fig. 3. Even discounting the contribution of the control in the perfusion experiment, overall induction is less than *in vivo*. Preliminary experiments in which perfusate composition or AIA concentration were varied have failed to give a greater response to AIA.

Effects of steroids

Steroids were next tested as possible inducers of ALA synthetase in the isolated liver. Progesterone has been shown in this laboratory and by other workers (Kaufman *et al.*, 1970; Jones and Emans, 1969) to induce *in vivo* in rats. The effects of perfusion with progesterone are shown in Fig. 4. The increase in enzyme level was significantly greater than in the control; the maximal activity reached was 4-5 times the basal level. The increase could be prevented by addition of cycloheximide (20 μg/ml) to the perfusate. The overall induction in the presence of progesterone was essentially the same as seen for a comparable period of induction by progesterone *in vivo*.

FIGURE 4

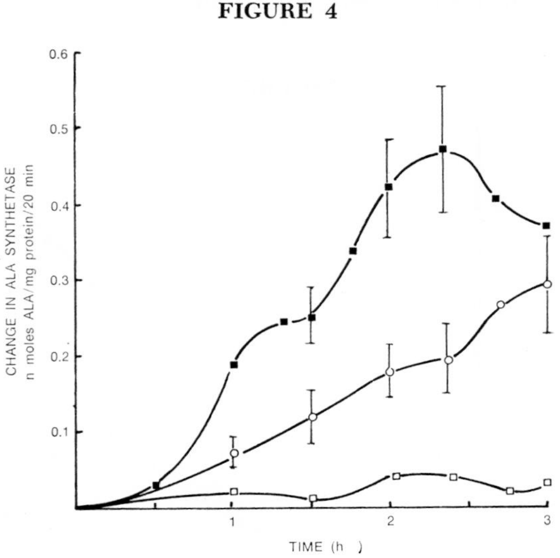

Effect of progesterone on total ALA synthetase activity in perfused rat liver. Procedure was as in Fig. 2. No addition (○) [5]; progesterone, 12 μg/ml (■) [5]; progesterone plus haemin, 2×10^{-5}M (□) [2]. Mean zero time ALA synthetase activity was 0.12 nmoles/mg protein/20 min.

The effects in perfused liver of two steroids with a 5β-H structure known to be potent inducers in chick embryo liver (Kappas *et al.*, 1968a) are shown in Fig. 5. Etiocholanolone, an androgen derivative, had little effect at the concentrations tested. However, pregnanolone, a 5β-H metabolite of progesterone was found to be a potent inducer, causing a 6-fold induction over a 3 hour experiment. This result suggests that at least some 5β-H steroids may be inducers in mammalian

FIGURE 5

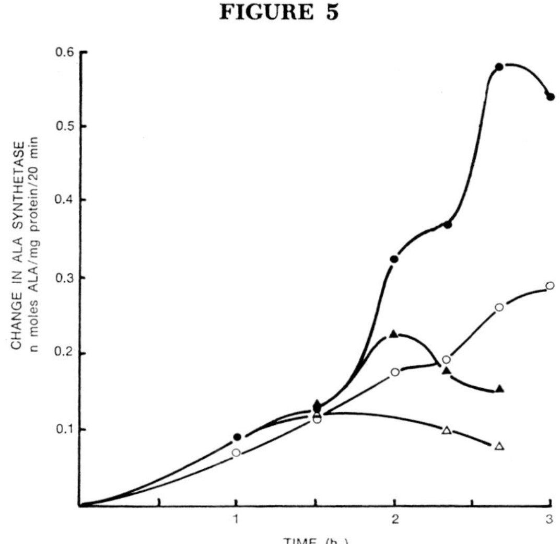

Effect of some 5β-H-steroids on total ALA synthetase in perfused rat liver. Procedure was as in Fig. 2. No addition (○) [5]; pregnanolone 12 μg/ml (●) [1]; etiocholanolone, 12 μg/ml (△) [1]; etiocholanolone, 6 μg/ml (▲) [1].

liver: failure to demonstrate induction in whole animals by these compounds (Kappas *et al.*, 1968b) may reflect the limitations of *in vivo* experiments.

The glucose effect

As described above, the *in vivo* induction of ALA synthetase by drugs is prevented by glucose feeding but little more is known about this effect. In preliminary experiments we found that induction of mitochondrial ALA synthetase was as great or greater with perfusate containing glucose as with perfusate containing no substrate or alternative substrates such as butyrate or lactate. In the perfusion experiments described above, substantial inductions were observed with glucose concentrations in the perfusate, greater than levels of blood glucose which *in vivo* completely block induction by drugs. These observations suggested that glucose repression was an indirect effect, perhaps mediated as is the case for glucose repression of other liver enzymes (Jost and Rickenberg, 1971) by changes in the levels of insulin and glucagon which in turn modulate the concentration of hepatic cyclic AMP. The finding, shown in Fig. 6, that the dibutyryl derivative of cyclic AMP caused a 5-6 fold induction of ALA synthetase in perfused liver is consistent with this proposal. We have also found that in the presence of high levels of insulin, induction by progesterone in perfused liver is reduced by 40%.

Effect of haem and of an inhibitor of haem synthesis

As shown in Fig. 4, induction by progesterone is blocked completely by addition to the perfusate of 2×10^{-5} M haemin. This provides evidence that at this concentration, haem represses by a direct effect on the liver.

FIGURE 6

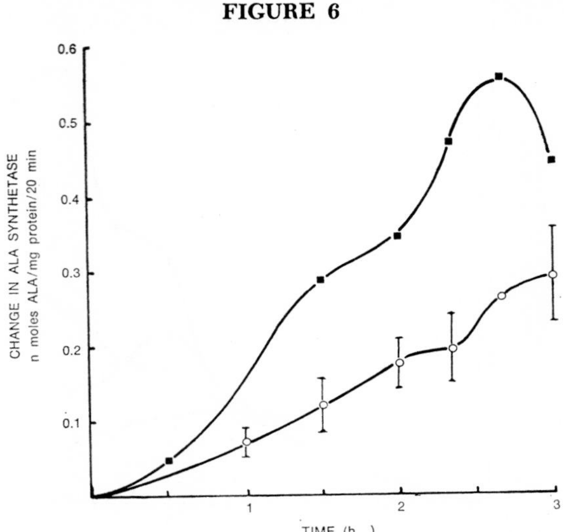

Effect of dibutyryl cyclic AMP on total ALA synthetase activity in perfused rat liver. Procedure was as in Fig. 2. No addition (○) [5]; dibutyryl cyclic AMP (■) [2].

Since it has been proposed by Granick (1966) that in normal liver, ALA synthetase is kept at a low level by haem repression, it might be argued that inhibition of haem synthesis in the presence of continued haem utilization and

FIGURE 7

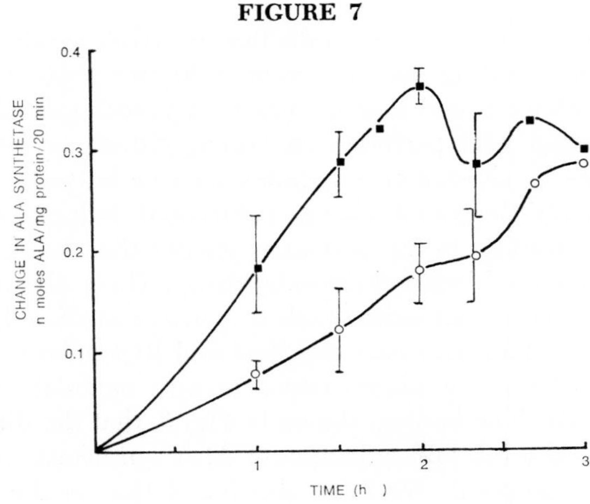

Effect of lead acetate on total ALA synthetase activity in perfused rat liver. Procedure was as in Fig. 2. No addition (○) [5]; lead acetate, 10^{-3}M (■) [5].

breakdown should cause induction by releasing haem repression. Evidence from other workers (Dresel and Falk, 1956) shows that concentrations of lead acetate greater than 10^{-4} M inhibit haem synthesis at least 80%. Fig. 7 shows the effect on ALA synthetase activity of perfusing with 10^{-3} M lead acetate. At early perfusion times lead caused a small increase in enzyme level over that in controls. As in previous experiments, the increase was blocked by cycloheximide. It remains to be established whether this effect of lead does in fact result from inhibition of haem synthesis. It is noteworthy that the clinical and biochemical symptoms of lead poisoning resemble those of acute intermittent porphyria (Goldberg, 1968). A high level of ALA synthetase in both diseases could help explain this observation. However, there is little evidence from *in vivo* experiments in this or other laboratories (Gibson and Goldberg, 1970; Stein *et al.*, 1970; cf. Morse, *et al.*, 1971) that lead causes an increase in ALA synthetase levels.

DISCUSSION AND CONCLUSIONS

Perhaps the most significant aspect of this work is that it has demonstrated the feasibility of using the perfused rat liver as a system for studying ALA synthetase regulation in mammalian liver.

The observed induction by pregnanolone may be significant in relation to the hereditary human disease, acute intermittent porphyria in which the level of ALA synthetase is high (Tschudy, *et al.*, 1965b). Kappas and his co-workers have recently provided evidence for altered steroid metabolism, with increased production of certain 5β-H steroids, during attacks of acute intermittent porphyria (Kappas, *et al.*, 1971). Our present observation that at least one 5β-H steroid is an active inducer in mammalian liver suggests that it may be possible to establish a link between perturbed steroid metabolism and an elevated level of ALA synthetase in the disease.

On the other hand AIA, a potent inducer *in vivo,* was relatively ineffective in the isolated liver. No clear interpretation can yet be attached to this observation although it raises the possibility that certain prerequisites for induction by drugs are lacking in the isolated liver. The possibility that the primary site of drug action is extrahepatic seems unlikely since AIA is a good inducer in cultured chick embryo liver cells (Granick, 1966).

Haem was shown to repress by direct action on the liver. Conversely, lead acetate which might be expected to reduce intracellular haem levels caused a small induction of ALA synthetase.

The repression of induction by high carbohydrate diets appears to be an indirect effect, possibly mediated by changes in liver cyclic AMP levels. It might be noted that the small "control induction" apparently due to surgical trauma may be due to release of adrenaline and a resultant rise in liver cyclic AMP concentration.

Interpretation of results obtained with the perfusion system so far is rendered more difficult by the fact that the perfusion medium contains rat blood with undefined components. With the information on control of ALA synthetase in isolated liver provided by these preliminary studies, it should be possible in further work to determine more completely the requirements for induction by using fully defined perfusion media.

REFERENCES

De Matteis, F. (1967) *Pharmacol. Rev. 19,* 523

Dresel, E. I. B. and Falk, J. E. (1956) *Biochem. J. 63,* 80

Gibson, S. L. M. and Goldberg, A. (1970) *Clin. Sci. 38,* 63

Granick, S. (1966) *J. Biol. Chem. 241,* 1359

Granick, S. and Urata, G. (1963) *J. Biol. Chem. 238,* 821

Goldberg, A. (1968) *Seminars in Hematol. 5,* 424

Hager, C. B. and Kenney, F. T. (1968) *J. Biol. Chem. 243,* 3296

Hayashi, N., Kurashima, Y. and Kikuchi, G. (1972) *Arch. Biochem. Biophys. 148,* 10

Hayashi, N., Yoda, B. and Kikuchi, G. (1968) *J. Biochem. 63,* 446

Hems, R., Ross, B. B., Berry, M. N. and Krebs, H. A. (1966) *Biochem. J. 101,* 284

Irving, E. A. and Elliott, W. H. (1969) *J. Biol. Chem. 244,* 60

Jones, A. L. and Emans, J. B. (1969) in *Metabolic effects of Gonadal Hormones and Contraceptive Steroids,* (H. A. Salhanick, D. M. Kipnis and R. L. Van de Wiele, eds.) p. 68, Plenum Press: New York

Jost, J.-P. and Rickenberg, H. V. (1971) *Annu. Rev. Biochem. 40,* 741

Kappas, A., Bradlow, H. L., Gillette, P. N. and Gallagher, T. P. (1971) *Ann. N.Y. Acad. Sci. 179,* 611

Kappas, A., Levere, R. D. and Granick, S. (1968a) *Seminars in Hematol. 5,* 323

Kappas, A., Song, C. S., Levere, R. D., Sachson, R. A. and Granick, S. (1968b) *Proc. Nat. Acad. Sci., U.S., 61,* 509

Kaufman, L., Swanson, A. L. and Marver, H. S. (1970) *Science, 170,* 320

Marver, H. S. (1966) *Biochem. J., 99,* 31c

Marver, H. S. (1969) in *Microsomes and Drug Oxidations,* (J. R. Gillette, A. H. Conney, G. J. Cosmides, R. W. Estabrook, J. R. Fouts and G. J. Mannering, eds.), p. 495 Academic Press Inc., New York

Marver, H. C., Collins, A., Tschudy, D. P. and Rechcigl, M. (1966) *J. Biol. Chem. 241,* 4323

Matsuoka, T., Yoda, B. and Kikuchi, G. (1968) *Arch. Biochem. Biophys. 126,* 530

Morse, B. S., Guiliani, D. and Rusin, J. (1971) *Clin. Res. 19,* 399

Sassa, S. and Granick, S. (1970) *Proc. Nat. Acad. Sci., U.S., 67,* 517

Schmid, R., Marver, H. S. and Hammaker, L. (1966). *Biochem. Biophys. Res. Commun. 24,* 319

Schneck, D. W., Tyrrell, D. L. J. and Marks, G. S. (1971) *Biochem. Pharmacol. 20,* 2999

Seglen, P. O. and Jervell, K. F. (1969) *Hoppe-Seyl. Z. Physiol. Chem. 350,* 308

Stein, J. A., Tschudy, D. P., Corcoran, P. L. and Collins, A. (1970) *J. Biol. Chem. 245,* 2213

Taddeini, L. and Watson, C. J. (1968) *Seminars in Hematol. 5,* 335

Tephly, T. R., Hasegawa, E. and Baron, J. (1971) *Metabolism, 20,* 200

Tschudy, D. P., Marver, H. S. and Collins, A. (1965a) *Biochem. Biophys. Res. Commun. 21,* 480

Tschudy, D. P., Perloth, M. G., Marver, H. S., Collins, A., Hunter, G. and Rechcigl, M. (1965b) *Proc. Nat. Acad. Sci., U.S., 53,* 841

Tschudy, D. P., Welland, F. H., Collins, A. and Hunter, G. (1964) *Metabolism, 13,* 396

Whiting, M. J. (1972) *Proc. Aust. Biochem. Soc. 5,* 12

The Anaemia-Induced Reversible Switch From Haemoglobin A to Haemoglobin C in Goats and Sheep:

The Two Haemoglobins Are Present in the Same Cell during the Changeover

M. D. GARRICK, R. F. MANNING, M. REICHLIN and M. MATTIOLI

Departments of Pediatrics, Medicine and Biochemistry
State University of New York at Buffalo
Department of Biology, University of Virginia

Over eight years ago, two groups (Blunt and Evans, 1963; Van Vliet and Huisman, 1964) reported that anaemic episodes led to the replacement of sheep haemoglobin A (HbA = $\alpha_2\beta^A{}_2$) by a distinct haemoglobin C (HbC = $\alpha_2\beta^C{}_2$). Huisman *et al.* (1967) found a similar situation in goats three years later. Recently, Wilson *et al.* (1970) reported that the Moufflon and the Barbary Sheep, when anaemic, also switch haemoglobins. Not all species of ruminants are capable of such a replacement; switching has not been found in domestic cattle (Bell and Huisman, 1968).

Three major haemoglobin genotypes (AA, AB, BB) are found in sheep, reflecting two different structural genes for the β-chains, β^A and β^B. During anaemic episodes the β^A-chain of AA or AB animals is replaced by a β^C-chain. The β^B-chain is not replaced in either AB or BB sheep. All three β-chains differ extensively in their amino acid sequences (Boyer *et al.*, 1967; Wilson *et al.*, 1966). The amino acid sequence of the non α-chain of ovine foetal haemoglobin also differs from each of these (Wilson *et al.*, 1966). In nonanaemic goats the most common β-chain type has been designated β^A (Huisman *et al.*, 1967; Garrick and Charlton, 1969). During anaemic episodes, it is replaced by a chain designated β^C. Table 1 summarizes the differences among goat and sheep β-chains. Two variant β-chains have been found in goats: β^D (Adams *et al.*, 1968; Garrick and Charlton, 1969) and β^E (Wrightstone *et al.*, 1970). Both are alleles of goat β^A in the usual sense of the term and in two other properties:

(1) their amino acid sequences differ only slightly, by one and three residues, respectively; (2) they are replaced by β^C-chains during anaemic episodes.

TABLE 1

Difference Matrix for Goat and Sheep β-Chains[a]

	A Sheep	A Goat	B Sheep	C Sheep	C Goat
A Sheep	0	4	7	16	15
A Goat	4	0	7	19	18
B Sheep	7	7	0	21	21
C Sheep	16	19	21	0	1
C Goat	15	18	21	1	0

[a]Data extracted from Dayhoff (1969), expressed as minimal number of amino acid residues difference.

The switch from A to C is mediated by a humoral factor (Boyer *et al.*, 1968a; Gabuzda *et al.*, 1968), probably erythropoietin (Thurmon *et al.*, 1970). As an animal recovers from anaemia, the situation reverses; and HbA replaces HbC (Van Vliet and Huisman, 1964; Boyer *et al.*, 1966; Blunt *et al.*, 1969). Boyer *et al.* (1968b) have shown that hypoxia at high altitudes may induce the A to C switch.

The description above of the A⟷C switch shows that this system has many properties that make it attractive for the study of the biochemical basis of gene expression in higher organisms. Before it could be treated as a paradigm analogous to the *lactose* operon in *E. coli,* however, it is important to know if HbA and HbC are in the same cell. If they are, one cell lineage is probably capable of making either HbA or HbC depending on the erythropoietin level. Biochemical studies of the effects of erythropoietin on erythroid cells cultured *in vitro* would then be in order. If HbA and HbC are found in separate cells, then goats and sheep possess two clones of erythroid cells. The clones would be expected to respond differently to erythropoietin.

This study was undertaken to determine whether or not HbA and HbC can be found in the same cell during the changeover. There are precedents for either situation: both HbF and HbA are present in erythrocytes from infants (Betke and Kleihauer, 1958; Dan and Hagiwara, 1967a, b). Tadpole and frog haemoglobins, however, are in separate cells (Rosenberg, 1970; Moss and Ingram, 1968; Maniatis and Ingram, 1971). The cells containing embryonic vs adult haemoglobin of the mouse originate in different organs, probably representing distinct lines (Fantoni *et al.*, 1969).

To determine whether HbA and HbC are found in the same cell we used two approaches. Incorporation studies detected the intracellular exchange of HbA tetramers with subunits made available by inhibiting β^C synthesis (Manning, 1972). To confirm and extend these observations, anti-bodies specific for either HbA or HbC were prepared and employed for indirect fluorescent staining of erythrocytes from goats and sheep that were switching (Garrick *et al.*, 1973). This technique also demonstrated that the two haemoglobins were present in the same cell.

INCORPORATION STUDIES OF A⟷C SWITCH IN GOATS

The same type of α-chains is common to both HbA and HbC in goats (Huisman *et al.*, 1968). This observation suggests that inhibiting β-chain synthesis for one haemoglobin could affect tetramers of α- and β-subunits for the other via the resulting increment in α-chains. Fig. 1 portrays this rationale. Normally the rates of α- and β-synthesis are approximately equal. Reducing β^C-synthesis augments (or results in) an α-chain pool. If HbA and HbC are present in the same cell, the increased (or new) labelled α-chain pool will exchange with HbA. If the two haemoglobins are in separate cells, HbA is not available for exchange with such a pool.

FIGURE 1

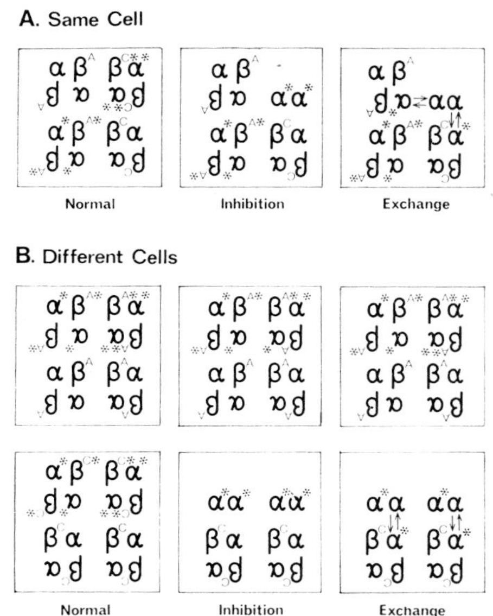

Diagram of anticipated consequences of inhibiting β^C-synthesis. Each square represents a reticulocyte containing unlabelled chains (α, β^A and β^C) and labelled ones (α^*, etc.). The α and β monomers in symmetrical arrays represent haemoglobin tetramers, while α-chains not associated with β's represent the free α-chain pool. (A) Haemoglobins A and C are made in the same cell. (i) Normally, α- and β-chains are made at the same rate. (ii) Inhibiting the β^C-synthesis upsets the balance between α- and β-chain production and leads to the formation of a labelled α-chain pool (iii) α-chains from the pool exchange into haemoglobin already present in the cell, replacing unlabelled α-monomers and increasing the specific activity of α-chains in HbA. (B) Haemoglobins A and C in different cells. (i) As in (A) α-and β-chain production are normally balanced, and (ii) inhibiting β^C-synthesis leads to the formation of a labelled α-chain pool. Nevertheless, (iii) HbA is not affected, since it is not in the same cell as the α-chain pool.

This model depends on being able to (1) inhibit β^C-chain (or β^A-chain) synthesis specifically. (2) detect exchange of α-subunits with haemoglobin tetramers, and (3) distinguish intracellular vs extracellular exchange. Below we demonstrate that each of these three conditions was achieved. The experimental procedures are outlined in the text and legends. Further details may be found in Manning (1972).

Specific inhibition of β^C synthesis

Both α- and β^A-chains of goat haemoglobin contain no isoleucine residues whereas the β^C-chain contains one (Huisman *et al.*, 1967 and 1968). Hence, starvation of goat reticulocytes for isoleucine should reduce only β^C-chain synthesis. Since omitting isoleucine from the amino acid mixture of Borsook *et al.* (1957) did not affect β^C-incorporation, we chose to replace isoleucine in the mixture with L-o-methylthreonine (OMT), an isostere (Smulson and Rabinovitz, 1968). By reducing the level of ile-t-RNA, OMT delays ribosomes at the isoleucine codon(s) of m-RNA (Hori and Rabinovitz, 1968), reducing the synthesis of polypeptides containing isoleucine.

Among human haemoglobin chains, the γ-chain contains 4 isoleucines, while α- and β- have none. Honig *et al.* (1969) used OMT to inhibit γ-chain synthesis in reticulocytes from infants without reducing α- and β-chain synthesis. This inhibition resulted in a labelled α-chain pool which exchanged with the tetramers present in the cells.

The effect of OMT on goat globin synthesis was analyzed in several experiments (Table 2). Phenylhydrazine was used to induce anaemia in goats which had only HbA at the start of the experiment. When the goats had only HbC, as detected on DEAE-Sephadex (Huisman *et al.*, 1967), washed reticulocytes

TABLE 2

Effect of OMT on Amino Acid Incorporation into Goat Globin

Expt.	Label	Chain Separation	OMT Conc. (mM)	α Incorporation as % Control	β^c Incorporation as % Control
1	^{3}H-Leu	8M urea, $NaPO_4$ gradient	0	100	100
			25	103	71
2	^{3}H-Amino acids	Pyridine-formate gradient	0	100	100
			25	97	64

Red cells from an anaemic goat were incubated with either a mixture of ^{3}H-amino acids (133 μc/ml) or ^{3}H-leucine (83 μc/ml) for 4 hours in either a complete mixture of amino acids or a mixture in which 25mM OMT had replaced isoleucine. Globin, prepared from 30 mg of the cell-free haemolysate was chromatographed on carboxymethylcellulose in either 8M urea (Clegg *et al.*, 1966) or pyridineformic acid (Dintzis, 1961). Specific activities were determined from the counts per minute and absorbance at 280 nm for four fractions from the centre of each peak.

were incubated with labelled amino acid(s) at 37°C under 95%O_2/5%CO_2. Haemolysates prepared from these incubations were resolved into globin chains by chromatography on carboxymethylcellulose either in 8M urea using a sodium phosphate gradient, pH 6.7 (Clegg *et al.*, 1966) or in a pyridine-formate gradient (Dintzis, 1961). Both techniques revealed that OMT inhibited β^C-synthesis by one-third without affecting α-synthesis.

Other experiments (not shown), using reticulocytes synthesizing β^A-, β^C- and α-chains, showed that OMT did not alter the proportion of labelled β^A- and α-chains in the haemolysate while inhibiting β^C-synthesis.

Extracellular exchange of labelled subunits with haemoglobin tetramers

Weatherall *et al.* (1969) have detected both intracellular and extracellular exchange between α-chains and haemoglobin tetramers using human haemoglobins. Extracellular exchange was easily demonstrated for goat haemoglobins, too (Table 3). A portion of the lysate from Table 2—experiment 1 was incubated with unlabelled goat HbA. Exchange of labelled α-chains into the HbA was detected by determining the label found in HbA after subsequent separation of the HbA from HbC.

TABLE 3

Extracellular Exchange Experiment

Incubation Temperature °C	Specific Activity		
	Haemoglobin cpm/mg		Ratio HbA/HbC
	HbA	HbC	
—10	18	177	0.102
4	17	161	0.106
37	26	174	0.149

A portion of the lysate in Table 2—expt. 1 was mixed with unlabelled HbA, then divided into three aliquots; one was frozen immediately at —10°C; the other two were incubated at 4° and 37° for 36 hours then kept at −10°C for six days. Haemoglobins A and C were separated on DEAE-Sephadex. Specific activities were determined from the absorbance at 544 nm and the radioactivity of each fraction.

In the model experiment diagrammed in Fig. 1, it is necessary to minimize and quantitate extracellular exchange which could falsely indicate that HbA and HbC are made in the same cell. Since the β^A- and β^D- chains differ by a single amino acid residue (Adams *et al.*, 1968), HbD should exchange with α-subunits in a fashion similar to HbA. Thus HbD from another goat could be used to monitor extracellular exchange during the test for intracellular exchange. Both HbD and HbA acquired label from α-subunits in an extracellular exchange experiment verifying this premise.

Intracellular exchange, showing that HbA and HbC are made in the same cell

To test for intracellular exchange, reticulocytes from an anaemic goat making about three-quarters HbC and one-quarter HbA were incubated with ^{14}C-amino acids. Different times of incubation were used in an effort to promote exchange and OMT was employed to generate a labelled α-chain pool by inhibiting β^{C}-chain synthesis. Myoglobin and HbD were added to several incubations at the time of lysis, HbD to detect extracellular exchange and myoglobin to serve as a monomer marker during gel filtration. After lysis, the haemoglobin tetramers were quickly separated from subunits by gel filtration on Sephadex G-75 to minimize extracellular exchange. The three haemoglobins were then separated on DEAE-Sephadex (Dozy *et al.*, 1968) and the amount of label in each determined. Label was not found in HbD (Table 4) indicating that no detectable extracellular exchange had occurred.

The α- and β-chains of purified HbA were then separated to determine whether intracellular exchange had occurred (Table 4). The mean α/β ratio for the series with 25 mM-OMT was 1.09, while a mean of 0.96 was obtained with OMT absent. These means differ significantly ($P < 0.05$ by Student's t-test); therefore intracellular exchange did occur. The G-75 gel filtration was omitted for one

TABLE 4

Incorporation into α-and β-Chains from Goat HbA

Time of Incubation (hours)	OMT Conc. (mM)	Additions to Lysing Medium	Gel Filtration (G-75)	Specific Radioactivity: Chains[a] α(A)	Chains[a] β^{A}	Ratio $^{a}/\beta$	Ratio[b] HbD/HbA x10^2
2	0	HbD, Mb	+	1880	1850	1.01	0[c]
12	0	None	+	3250	3450	0.95	—
12	0	HbD, Mb	+	2820	3030	0.93	0
2	25	None	+	2490	2220	1.12	—
2	25	HbD, Mb	+	1720	1630	1.05	0
12	25	None	+	3440	3180	1.08	—
12	25	HbD, Mb	+	3070	2760	1.11	0
2	25	HbD, Mb	—	2820	2100	1.34	1.41

Red cells from an anaemic goat were incubated with ^{14}C-amino acids for either 2 or 12 h One half of the cells from each incubation were lysed in an equal volume of 0.1% saponin, the other half was lysed in the presence of 8 mg goat HbD (to detect extracellular exchange) and 5 mg equine heart myoglobin (to serve as a molecular weight marker). Cell walls were removed by centrifugation at 17,000g for 3 minutes. The lysates were then passed through Sephadex G-75 to remove the α-chain pool. The peak of the tetramer region was concentrated and mixed with unlabelled carrier. Haemoglobins were separated on DEAE-Sephadex (Dozy *et al.*, 1968). Purified haemoglobin A was concentrated and chromatographed on carboxymethylcellulose in 8M urea.

[a]Chain specific radioactivity is given as counts per minute per unit of absorbance at 280nm. Differences in specific radioactivity for β^{A} reflect mainly the length of incubation and, to some extent, our failure to add the same amount of carrier to each haemolysate.

[b]Specific radioactivity was calculated as counts per minute per mg of haemoglobin, then the ratio calculated.

[c]0 indicates that the ratio was less than 0.07.

sample and the lysate kept at 4°C for 36 hours to verify that both the precautions for avoiding and the method for monitoring extracellular exchange were of utility. Both the marked increase in the α/β ratio for HbA and the label in HbD indicate that extracellular exchange occurred in this sample.

Since the augmented α-pool created by inhibiting β^C-synthesis exchanged *intracellularly* with HbA, goat HbA and HbC are made in the same cell. We were unable to extend this approach to sheep, since OMT failed to inhibit ovine β^C-chain synthesis. For this reason, because the extent of change in Table 4 was small and also because an exchange-based approach does not permit an estimate of what portion of erythrocytes have both haemoglobins during switching, we developed the alternative approach described below.

IMMUNOCHEMICAL STUDIES OF THE A⟷C SWITCH IN GOATS AND SHEEP

Based on the data summarized in Table 1, goat and sheep HbA are more closely related to each other than either goat HbA and goat HbC or sheep HbA and sheep HbC. Similarly, goat HbC and sheep HbC are closer than the intraspecific comparisons. These relationships suggest that one should be able to obtain sera that will react with goat and sheep HbA but not with the two HbC's and vice-versa. The method used to obtain and apply such sera are outlined in the text and legends; additional details may be found in Garrick *et al.* (1973).

Reaction of antibodies directed against ovine haemoglobins with the homologous and heterologous antigens

Sheep HbA and HbC were purified according to Dozy *et al.* (1968) except that DEAE-Cellulose (Whatman DE-52) replaced DEAE-Sephadex. Three rabbits were immunized with each haemoglobin (Reichlin *et al.*, 1964) and their response determined by quantitative precipitin analysis of the sera (Heidelberger and Kendall, 1935). Two sera were selected for more extensive characterization by macro-complement fixation (Mayer *et al.*, 1948). These sera possessed the desired specificities (Fig. 2). Anti-sheep HbA reacted extensively with sheep HbA, nearly as well with goat HbA, considerably less extensively with sheep HbC and less yet with goat HbC. Anti-sheep HbC reacted best with sheep HbC, less with goat HbC and markedly less with sheep HbA and goat HbA.

The results in Fig. 2 are generally compatible with Table 1. A few anomalies are of some interest: Both sera distinguish goat vs sheep HbC more extensively than goat vs sheep HbA; also, the difference matrix might lead one to expect that anti-sheep HbA could react better with goat HbC than sheep HbC instead of the opposite result which was obtained. The distinction of goat vs sheep HbC by the sera is not readily compatible with the reported sequence difference, β^{45} = *thr* in sheep and *ser* in goats (Dayhoff, 1969). Isolated goat β^C-chains have one more negative charge than sheep β^C-chains (Garrick and Manning, unpublished), which observation supports the immunochemical results rather than the sequence data. The sequences of the goat chains have been less thoroughly determined; hence, it is likely that a few residues deduced for goat β^C-chain must be altered.

FIGURE 2

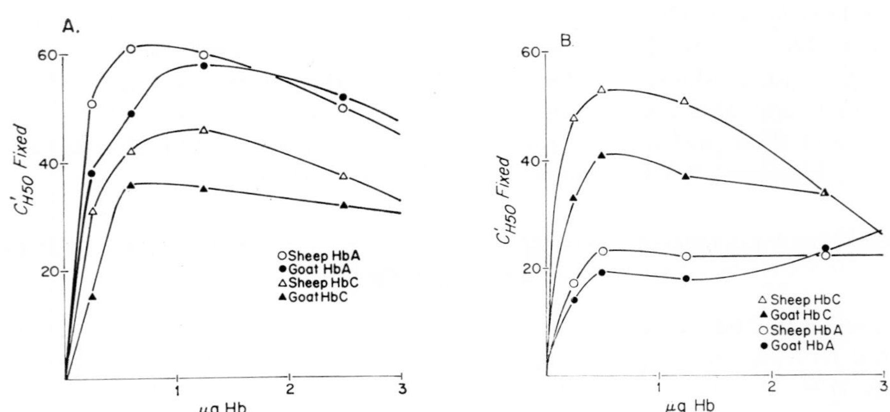

Complement fixation by goat and sheep haemoglobins reacting with homologous and heterologous unabsorbed sera.

A. Reaction of HbA and HbC from both goats and sheep with anti-sheep HbA (diluted 1/200).

B. Reaction of HbA and HbC from both goats and sheep with anti-sheep HbC (diluted 1/80).

Absorbed sera and immunofluorescent staining

Portions of the two sera were absorbed with sufficient heterologous ovine haemoglobin to reach the observed equivalence point, the precipitate removed, and half again as much antigen added with no further precipitate formed. These absorbed sera were specific for the homologous haemoglobin by double diffusion analysis (Ouchterlony, 1949) and macro-complement fixation.

Both of the above tests employed purified haemoglobin in testing the absorbed sera. The best criterion for their specificity is in terms of their utility, namely their reaction with erythrocytes. Erythrocytes with essentially only HbA or HbC are needed to test them in this context. Fig. 3 shows that both goats and sheep can yield erythrocytes with little or no HbA present and that sheep red cells with essentially only HbA can also be obtained.

Smears were prepared according to Dan and Hagiwara (1967b) and reacted with one of the absorbed sera followed by staining with fluorescein-conjugated ovine anti-rabbit immunoglobulin (Coons, 1958) G. Erythrocytes from a sheep with essentially only HbA present stained extensively with anti-A, but fewer than 1% reacted with anti-C (Fig. 4). Similar results were obtained with goat erythrocytes except that about 5% stained with anti-C, probably reflecting the observation that the non-anaemic goat contained a detectable fraction of HbC.

FIGURE 3

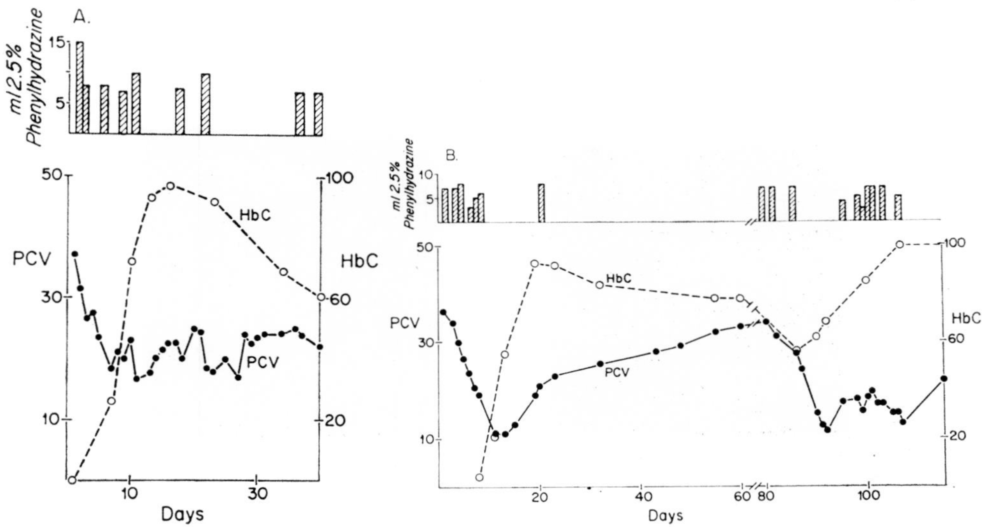

Induction of and recovery from anaemic episodes in goats and sheep. Typical responses are shown for a sheep (A) and a goat (B). Packed cell volume (PCV = haematocrit) is plotted in % as is the amount of HbC. The amount of neutralized phenylhydrazine (2.5% w/v in normal saline) administered is indicated above. When smears from the goat on day 19 (B) were examined by fluorescent staining, the small fraction of HbA remaining was distributed among more than 10% of the cells. Phenlyhydrazine injections were therefore reinitiated at day 78 to stimulate essentially complete replacement of the HbA by day 107.

FIGURE 4

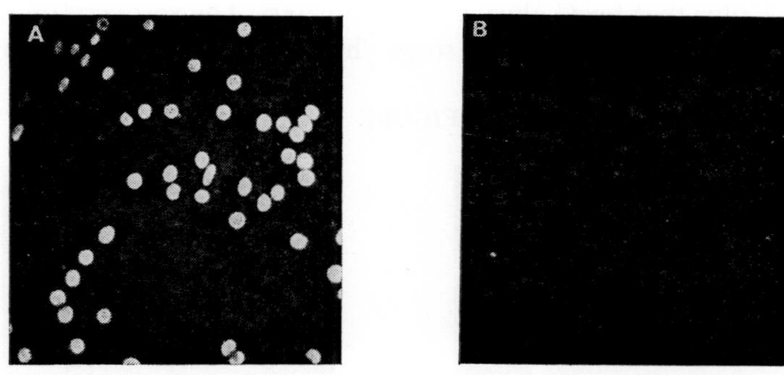

Indirect fluorescent straining of sheep erythrocytes with essentially 100% HbA. Copyright Academic Press (N.Y.).

A. Staining with absorbed anti-sheep HbA (diluted 1/10).

B. Staining with absorbed anti-sheep HbC (diluted 1/10).

Fig. 5 portrays the results with erythrocytes from a goat with at least 99% HbC after being made anaemic by phenylhydrazine injection. Fewer than 1% of the cells stained uniformly with anti-A. A portion exhibited punctate fluorescent bodies, but they fluoresced white rather than green. These probably represent Heinz bodies (haemoglobin precipitates) induced by the phenylhydrazine treatment (Itano, 1970). Nearly all cells stained with anti-C, however, with the punctate bodies staining green.

FIGURE 5

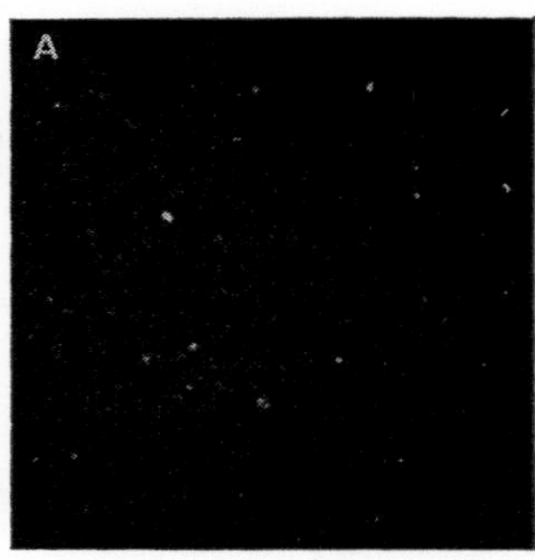

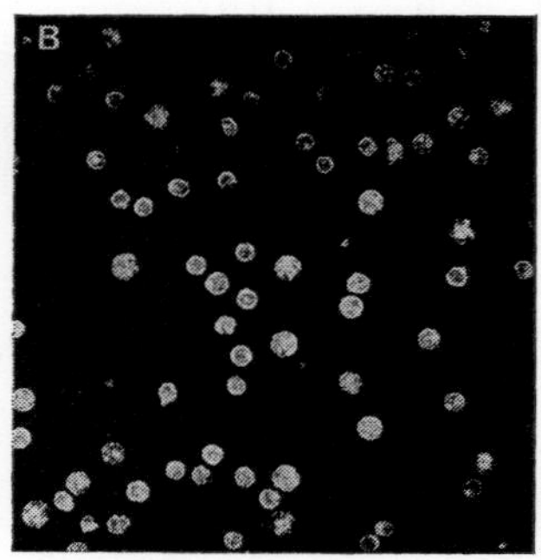

Indirect fluorescent staining of goat erythrocytes with essentially 100% HbA. Copyright Academic Press (N.Y.).

A. Staining with absorbed anti-sheep HbA (diluted 1/10).

B. Staining with absorbed anti-sheep HbC (diluted 1/10).

Immunofluorescent staining during the A⟷C switch

During either the switch from HbA to HbC or its reversal as animals recovered from anaemia, evidence from indirect fluorescent staining indicated that many erythrocytes contained both haemoglobins. In particular, when both represented significant fractions of the haemolysate, the sum of the fraction of red blood cells staining for A plus that for C always exceeded 100%. The recovery phase provided the best example (Fig. 6). At the stage shown, 78% of haemoglobin was HbC

FIGURE 6

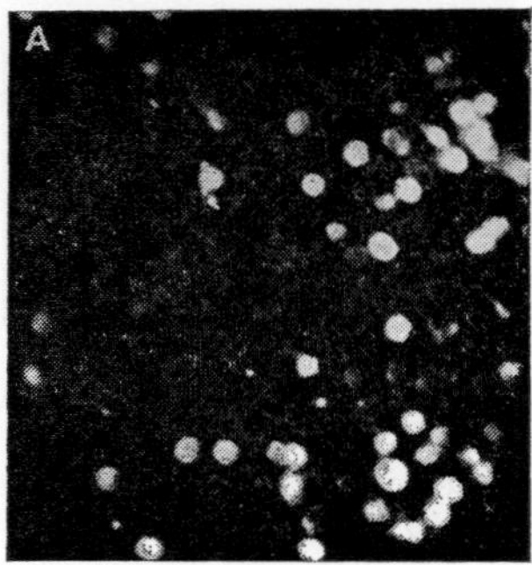

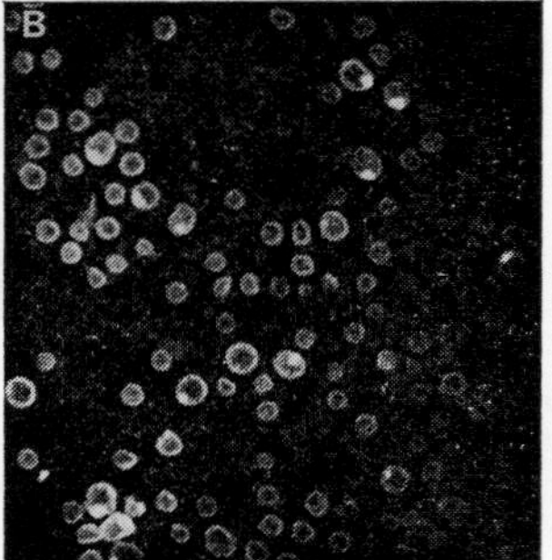

Indirect fluorescent staining of goat erythrocytes switching from HbC to HbA. Copyright Academic Press (N.Y.). Smears were prepared on day 55 for the goat used in Fig. 3, at which time the haemolysate contained 78% HbC and 22% HbA.

A. Staining with absorbed anti-sheep HbA (diluted 1/10).

B. Staining with absorbed anti-sheep HbC (diluted 1/10).

and 22%, HbA. Only about 3% of the cells failed to stain for HbC, while more than 5/6 stained for HbA, with about half the intensity.

One might wonder how well the situation found in Fig. 6 can be distinguished from the opposite result, a clonal distribution of the haemoglobins. To answer this question, a mixture of erythrocytes was prepared and examined (Fig. 7). The mixture contained sheep erythrocytes with at least 99% HbA plus goat erythrocytes with 99% HbC, in the proportions of 2:1 (assuming the mean cellular volume of both types is the same). About 2/3 of the cells stained for HbA with the remainder obviously negative. About 1/3 stained strongly for HbC, with the remainder staining very slightly. The reaction of HbA-containing cells with anti-HbC in Fig. 7B differs from the more obviously negative results obtained in Fig. 5B when smears of the same cell were prepared in the absence of HbC-containing cells. We attribute this difference to adsorption of HbC by HbA-containing cells subsequent to haemolysis of a small portion of the HbC-containing cells, since it is known that phenylhydrazine treatment renders the surviving erythrocytes susceptible to haemolysis. No such haemolysis occurred in the cells stained in Fig. 6, reflecting the 34 day interval after the preceding phenylhydrazine injection. The results obtained in Fig. 6 are clearly distinctive when compared to the "clonal" results modelled by Fig. 7, confirming our conclusion that the two haemoglobins are present in the same cell during A⟷C switching.

FIGURE 7

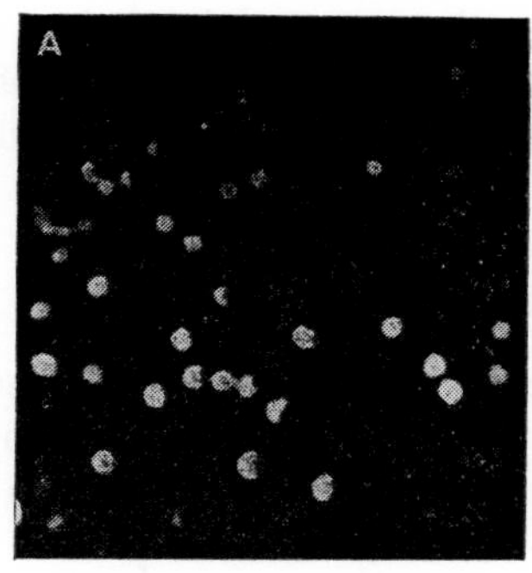

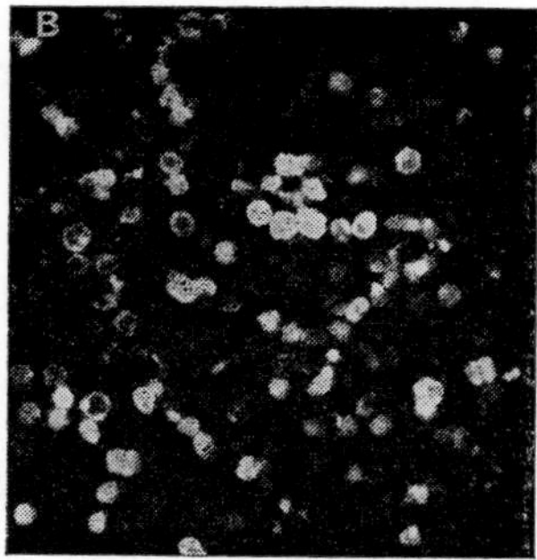

Indirect fluorescent staining of a mixture of erythrocytes containing essentially 100% HbA and erythrocytes containing essentially 100% HbC. Copyright Academic Press (N.Y.). Two volumes of the blood used in Fig. 4 were mixed with three volumes of the blood used in Fig. 5. Packed cell volumes were 40% and 13% respectively, therefore the volumes of erythrocytes in the mixture was in the proportion HbA-containing: HbC-containing = 2:1.

A. Staining with absorbed anti-sheep HbA (diluted 1/10).

B. Staining with absorbed anti-sheep HbC (diluted 1/10).

PERSPECTIVES IN THE A⟷C SWITCH

In the light of these results, it is now meaningful to attempt to answer certain biochemical questions relative to the regulation of gene expression during A⟷C switching. In this connection, Nienhuis and Anderson (1972) have just shown that the type of haemoglobin synthesized by sheep erythroid cells is reflected by the functional m-RNA extracted from them as determined using a cell-free m-RNA-dependent protein synthesizing system from rabbits.

We therefore propose the following model for switching: Erythroid stem cells divide to form either more stem cells or "committed" cells capable of developing into erythrocytes. The latter cell usually goes on to differentiate into an erythrocyte under the influence of its local environment and the typically low levels of erythropoietin in the circulation. These conditions promote transcription (or perhaps proper processing of transcripts) of the HbA genes. Higher levels of erythropoietin stimulate such cells to differentiate more quickly and to transcribe (or process the m-RNA for) HbC genes. Such levels can still stimulate expression of HbC genes after HbA synthesis and slower normal differentiation have already begun. When high levels of erythropoietin are followed by lower levels, a cell already transcribing HbC genes can still transcribe HbA genes. Possibly the expression of either gene is not forbidden until the nucleus becomes pycnotic and is extruded.

A comparison to the foetal to adult haemoglobin switch of humans is also of interest. In humans, switching also involves the presence of both haemoglobins in the same cell and probably involves a mechanism similar to the model above. Normally, however, once HbA has become the main haemoglobin, HbF synthesis does not revive. Hall and Motulsky (1968) were able to obtain significant production of HbF by marrow cultures from adult humans, suggesting that properties of the local cellular environment determine which haemoglobin is made. An understanding of what these properties are and how they determine gene expression is not only of theoretical value, but could also be of practical value in providing therapy for haemoglobinopathies like sickle cell disease and β-thalassemia.

SUMMARY

During anaemic episodes, goats and sheep which normally have haemoglobin A in their peripheral circulation replace it entirely with haemoglobin C. This changeover, which is reversed upon recovery from anaemia, involves the replacement of the β^A-chain with a structurally distinct β^C-chain without any apparent alteration of the α-chain. To analyse further the biochemical basis of this alteration in gene expression, one needs to know whether the change involves the replacement of a clone of erythroid cells that can make only HbA by a separate clone that can make only HbC, or whether the same cell line differentiates to make either or both haemoglobins, depending upon the degree of anaemia. Results from two independent techniques support the latter hypothesis: (1) When goat reticulocytes are incubated *in vitro* with L-o-methylthreonine, an isostere of isoleucine, incorporation into the β^C-chain is inhibited while synthesis of β^A- and α-chains is unaffected. These observations reflect the presence of an isoleucyl residue in the β^C-chain and its absence from the β^A- and α-chains. These incubations and subsequent haemoglobin purifications have been carried out under conditions where *extracellular* exchange is not permitted between the HbA tetramers and the α-subunits which accumulate due to inhibition of β^C-chain synthesis. Nevertheless, an exchange of labelled α-chains into HbA can be detected, indicating that A and C are made within the same cell. (2) Using fluorescent anti-rabbit immunoglobulin G, plus rabbit specific anti-HbA or anti-HbC, the presence of both haemoglobins can be demonstrated in nearly all erythrocytes at times when both are present in haemolysates.

ACKNOWLEDGEMENTS

This research was supported by grants Nos. AM10428, AM14923 (previously 10391), and RR05400 of the National Institutes of Health. M.R. is supported by Career Development Award 5 KO3-AM20729 of the National Institutes of Health and Designated Research funds from the Veterans' Administration. The data in Table 2—Expt. 2 were provided by Dr. L. M. Garrick. Dr. Marco Rabinovitz kindly gave us an initial supply of OMT before it was commercially available. Mr. Kenneth Pembroke allowed us to screen 175 sheep belonging to him to find the AA animals. Fig. 4 through 7 are reproduced from Garrick *et al.* (1973) with permission from the editor and publisher. They are copyrighted by Academic Press (N.Y.).

REFERENCES

Adams, H. R., Boyd, E. M., Wilson, J. B., Miller, A. and Huisman, T. H. J. (1968) *Arch. Biochem. Biophys. 127,* 398-405

Bell, J. T. and Huisman, T. H. J. (1968) *Am. J. Vet. Res. 29,* 479-482

Betke, K. and Kleihauer, E. (1958) *Blut 41,* 241-246

Blunt, M. H. and Evans, J. V. (1963) *Nature (London) 200,* 1215-1216

Blunt, M. H., Huisman, T. H. J. and Lewis, J. P. (1969) *Aust. J. Exp. Biol. Med. Sci. 47,* 601-611

Borsook, H., Fischer, E. H. and Keighley, G. (1957) *J. Biol. Chem. 229,* 1059-1071

Boyer, S. H. IV, Hathaway, P., Pascasio, F., Orton, C., Bordley, J. and Naughton, M. A. (1966) *Science 153,* 1539-1543

Boyer, S. H. IV, Hathaway, P., Pascasio, F., Orton, C., Bordley, J. and Naughton, M. A. (1967) *J. Biol. Chem. 242,* 2211-2232

Boyer, S. H. IV, Crosby, E. F. and Noyes, A. N. (1968a) *Johns Hopkins Med. J. 123,* 85-91

Boyer, S. H. IV, Crosby, E. F., Noyes, A. N., Kaneko, J. J., Keeton, K. and Kinkl, J. (1968b) *Johns Hopkins Med. J. 123,* 92-94

Clegg, J. B., Naughton, M. A. and Weatherall, D. J. (1966) *J. Mol. Biol. 19,* 91-103

Coons, A. H. (1958) Fluorescent antibody methods. In *General Cyto-chemical Methods* (J. F. Danielli, ed.), vol. 1, 399-422. Academic Press, New York

Dan, M. and Hagiwara, A. (1967a) *Exp. Cell Res. 46,* 596-598

Dan, M. and Hagiwara, A. (1967b) *Jap. J. Human Genetics 12,* 55-61

Dayhoff, M. O. (1969) Atlas of Protein Sequence and Structure 1969. National Biomedical Research Foundation, Silver Springs, Md

Dintzis, H. M. (1961) *Proc. Nat. Acad. Sci. U.S. 47,* 247-261

Dozy, A., Kleihauer, E. F. and Huisman, T. H. J. (1968) *J. Chromatog. 32,* 723-727

Fantoni, A., de la Chapelle, A., Chui, D., Rifkind, R. A. and Marks, P. A. (1969) *Ann. N.Y. Acad. Sci. 165,* 194-204

Garrick, M. D. and Charlton, J. P. (1969) *Biochem. Genet. 3,* 393-402

Garrick, M. D., Reichlin, M., Mattioli, M. and Manning R. F. (1973) *Develop. Biol. 30,* 1-9

Gabuzda, T. G., Schuman, M. A., Silver, R. K. and Lewis, H. B. (1968) *J. Clin. Inv. 47,* 1895-1904

Hall, J. G. and Motulsky, A. G. (1968) *Nature (London) 217,* 564-571

Heidelberger, M. and Kendall, F. E. (1935) *J. Exp. Med. 62,* 697

Honig, G. R., Rowan, B. Q. and Mason, R. G. (1969) *J. Biol Chem. 224,* 2027-2032

Hori, M. and Rabinovitz, M. (1968) *Proc. Nat. Acad. Sci. U.S. 59,* 1349-1355

Huisman, T. H. J., Adams, H. R., Dimmock, M. O., Edwards, F. B. and Wilson, J. B. (1967) *J. Biol. Chem. 242,* 2534-2541

Huisman, T. H. J., Brandt, G. and Wilson, J. B. (1968) *J. Biol. Chem. 243,* 3675-3686

Itano, H. (1970) *Proc. Nat. Acad. Sci. U.S. 67,* 485-492

Maniatis, G. M. and Ingram, V. M. (1971) *J. Cell Biol. 49,* 390-404

Manning, R. (1972) M.A. Dissertation. Dept. of Biol., Univ. of Virginia, Charlottesville, Va., U.S.A.

Mayer, M., Osler, A. C., Bier, O. G. and Heidelberger, M. (1948) *J. Immun. 59,* 195

Moss, B. and Ingram, V. M. (1968) *J. Mol. Biol. 32,* 493-504

Nienhuis, A. and Anderson, W. F. (1972) *Proc. Nat. Acad. Sci. U.S.,* in press

Ouchterlony, O. (1949) *Acta Pat. Microbiol. Scand. 26,* 507

Reichlin, M., Hay, M. and Levine, L. (1964) *Immunochem. 1,* 21-30

Rosenberg, M. (1970) *Proc. Nat. Acad. Sci. U.S. 67,* 32-36

Smulson, M. E. and Rabinovitz, M. (1968) *Arch. Biochem. Biophys. 124,* 306-341

Thurmon, T. F., Boyer, S. H., Crosby, E. F., Shepard, M. K., Noyes, A. N. and Stohlman, F. (1970) *Blood 36,* 598-606

Van Vliet, G. and Huisman, T. H. J. (1964) *Biochem. J. 93,* 401-409

Weatherall, D. J., Clegg, J. B., Na-Nakorn, S. and Wasi, P. (1969) *Brit. J. Haematol. 16,* 251-267

Wilson, J. B., Edwards, W. C., McDaniel, M., Dobbs, M. M. and Huisman, T. H. J. (1966) *Arch. Biochem. Biophys. 115,* 385-400

Wilson, J. B., Miller, A. and Huisman, T. H. J. (1970) *Biochem. Genet. 4,* 677-688

Wrightstone, R. N., Wilson, J. B., Miller, A. and Huisman, T. H. J. (1970) *Arch. Biochem. Biophys. 138,* 451-456

Gene Expression in Liver Endoplasmic Reticulum

LARS ERNSTER
Department of Biochemistry, University of Stockholm, Stockholm, Sweden.

The study of gene expression in higher organisms is rendered difficult by the complexity of their cells and tissues. In contrast to prokaryotes, where the translation and transcription of genetic information take place in one compartment, in eukaryotes the two processes are separated topologically by the nuclear membrane. In addition, genetic information in these cells is also present, and is utilized, within certain cytoplasmic organelles, notably chloroplasts and mitochondria, and the operation of the nuclear and extranuclear genomes is coordinated by some yet unknown mechanisms; this topic was the subject of an excellent symposium held here in Australia in late 1969 (cf. Boardman *et al.*, 1971). The association of many enzyme functions with various intracellular membranes also calls for a coordination of protein synthesis within the synthesis of other membrane constituents, primarily phospholipids (Ernster and Orrenius, 1965; Siekevitz *et al.*, 1967; Siekevitz, 1972). A further complexity in multicellular organisms exists in the form of a regulation of gene expression between different cells and tissues by means of environmental, neural and hormonal stimuli (Knox *et al.*, 1956; Tata *et al.*, 1963; Tata, 1967a,b; Greengard, 1970; Knox, 1972).

The purpose of the present survey is to illustrate some of the above problems by studies of the endoplasmic reticulum (ER) of rat liver. The ER (Porter and Kallman, 1952), or its fragments obtained upon tissue fractionation (Palade and Siekevitz, 1956), called microsomes (Claude, 1941, 1943a,b), represent a membranous network extending over the cytoplasm of most eukaryotic cells (Claude, 1948; Palade and Porter, 1954; Palade, 1955b, 1956; Porter, 1961; Palade, 1971). It lacks the complexity of chloroplasts and mitochondria in that it is not known to contain any DNA. On the other hand, in many types of cells the ER membranes serve as a site of attachment of a major or minor part of the cytoplasmic ribosomes (Palade, 1955a; Palade and Siekevitz, 1956; Tashiro, 1958; Tashiro and Siekevitz, 1965; Sabatini *et al.*, 1966; Blobel and Potter, 1967a,b,c), which are engaged in the synthesis of those proteins manufactured by the cell for export (Siekevitz and

Palade, 1960; Palade *et al.*, 1962; Palade, 1966). By an intricate mechanism these proteins are transferred into the lumen of the ER (Redman *et al.*, 1966; Redman and Sabatini, 1966; Redman, 1967; Malkin and Rich, 1967; Blobel and Sabatini, 1970; Sabatini and Blobel, 1970), which thus serves as a transport canal for cellular secretion (Palade *et al.*, 1962; Peters, 1962; Glaumann *et al.*, 1968; Takagi and Ogata, 1968; Redman, 1968, 1969; Hicks *et al.*, 1969; Glaumann, 1970; Glaumann and Ericsson, 1970; Takagi *et al.*, 1970; Peters *et al.*, 1971).

Associated with the membranes of liver microsomes are a variety of enzymes, including transferases, hydrolases, and electron-transport systems (Schneider and Hogeboom, 1951; de Duve *et al.*, 1955, 1962; Ernster *et al.*, 1962; Siekevitz, 1963). The ER plays a major role in cholesterol biosynthesis (Tchen and Bloch, 1955; Chesterton, 1966; Goad, 1970), phospholipid metabolism (Wilgran and Kennedy, 1963; Schneider, 1963; Dawson, 1966; Stoffel and Schiefer, 1968; McMurray and Dawson, 1969; Jungawala and Dawson, 1970; Williams and Bygrave, 1970), and in the biosynthesis of lipoproteins (Marsh, 1963, 1971; Mookerjea, 1971, 1972), glycoproteins (Molnar *et al.*, 1965, 1969; Lawford and Schachter, 1966; Spiro and Spiro, 1966; Molnar and Sy, 1967; Hallinan *et al.*, 1968; Priestley *et al.*, 1969; Schachter *et al.*, 1970; Sierr and Uhr, 1970; Redman and Cherian, 1972) and glycolipids (Caccam *et al.*, 1969; Zatz and Barondes, 1969; Tetas *et al.*, 1970; Behrens and Leloir, 1970; Behrens *et al.*, 1971; Molnar *et al.*, 1971; Richards *et al.*, 1971; Jankowski and Chojnacki, 1972; Lennarz and Scher, 1972; Dallner *et al.*, 1972). The liver ER is the site of glucose-6-phosphatase (Hers and de Duve, 1950), a key enzyme in glycogenolysis. It is also the site of cytochrome P-450 (Klingenberg, 1958; Garfinkel, 1958; Omura and Sato, 1963, 1964a,b), which, together with a flavoprotein known as NADPH-cytochrome *c* reductase (Horecker, 1950; Williams and Kamin, 1962; Phillips and Langdon, 1962; Kamin *et al.*, 1965), constitutes a monooxygenase system active in the oxidation of steroid hormones and fatty acids (ω-oxidation), as well as the oxidative detoxication of a large number of foreign compounds including many drugs (Brodie *et al.*, 1955; Cooper *et al.*, 1965; Ernster and Orrenius, 1965; Orrenius, 1965a; Orrenius and Ernster, 1964, 1971; Gillette, 1966; Gillette *et al.*, 1972). Another electron-transport system associated with the ER, consisting of cytochrome b_5 (Strittmatter and Ball, 1952, 1954; Chance and Williams, 1954; Strittmatter and Velick, 1956a) and the flavoprotein NADH-cytochrome b_5 reductase (Strittmatter and Velick, 1956b, 1957; Strittmatter, 1965), appears to be engaged in fatty acid desaturation (Oshino *et al.*, 1971; Oshino and Sato, 1971). Liver microsomes also contain a nucleoside diphosphatase and a nucleoside triphosphatase (Ernster and Jones, 1962), with yet unknown metabolic functions. All of these enzymes are present in both the ribosome-containing ("rough") and ribosome-free ("smooth") segments of the ER membranes (Dallner, 1963; de Duve, 1971). In general, there seems to exist no definite evidence for a specialization for any given enzymatic function within different segments of the ER, although subfractionation of microsomes does indicate a certain extent of heterogeneity in enzyme composition (Dallner *et al.*, 1968; Dallner and Ernster, 1968; Dallman *et al.*, 1969; Glaumann *et al.*, 1969; Glaumann and Dallner, 1970; de Duve, 1971; Svensson *et al.*, 1972). Turnover studies of ER enzymes and other chemical constituents reveal a highly dynamic steady state, where different components are renewed at different rates (Omura *et al.*, 1967; Siekevitz *et al.*, 1967; Schimke and Doyle, 1970; Siekevitz, 1972).

Characteristic of the liver ER is a high degree of adaptation to varied physiological and experimental conditions. An extensive reorganization of the liver ER takes place, for example, in the newborn rat (Dallner *et al.*, 1966a,b), reflecting an adaptation to extrauterine life. The most striking enzymatic change consists of a several-fold increase in glucose-6-phosphatase, followed by the appearance of

FIGURE 1

Changes in the enzyme pattern of rat-liver microsomes

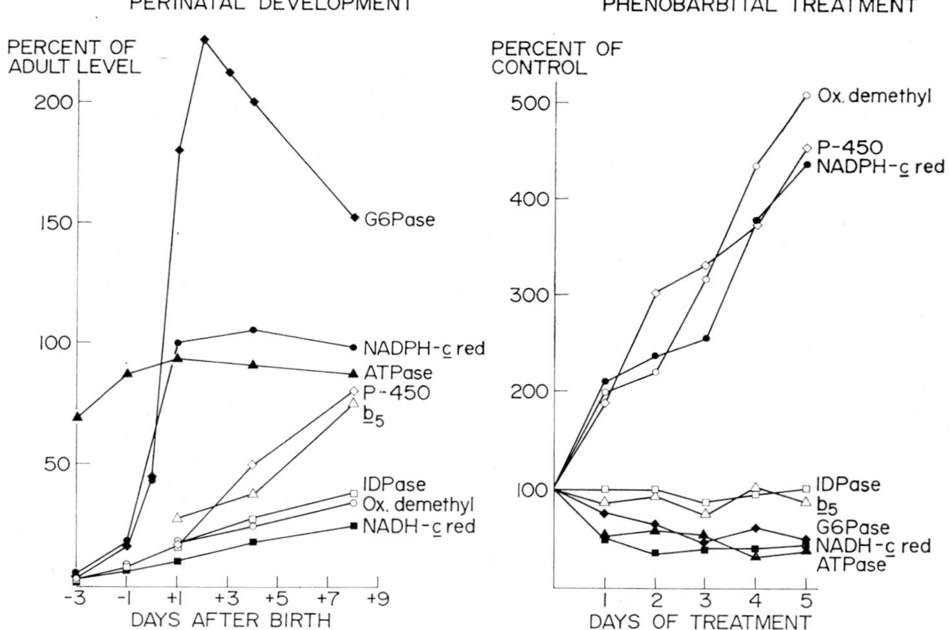

Left: During perinatal development (From Dallner *et al.*, 1966b)
Right: After treatment of rats with phenobarbital (daily intraperitoneal injection of 100 mg phenobarbital per kg body-weight) (From Orrenius *et al.*, 1965).
Enzyme activities are expressed on the basis of total microsomal protein. Abbreviations: G6Pase, glucose-6-phosphatase; NADH-*c* red, NADH-cytochrome *c* reductase; NADPH-*c* red, NADPH-cytochrome *c* reductase; b_5, cytochrome b_5; P-450, cytochrome P-450; ox. demethyl, oxidative demethylation of aminopyrine; ATPase, adenosine triphosphatase; IDPase, inosine diphosphatase.

increased NADPH- and NADH-linked electron-transport activities (Fig. 1, *left*). In all cases, the increase is due to enhanced rates of enzyme synthesis, as indicated by turnover measurements, and is accompanied by a gross reorganization of the ER, as revealed by the appearance of smooth-surfaced profiles resulting probably from an outgrowth of new membranes from the rough-surfaced segments of the ER. As elegantly demonstrated by Greengard (1969; cf. also Greengard *et al.*, 1972), the increase in enzyme levels can be enhanced by the administration of

TABLE 1

Microsomal Enzyme Levels in Rat Liver During Regeneration Following Partial Hepatectomy

(A. Claude, L. Ernster, R. W. Estabrook, C. P. Lee, K. Nordenbrand)
(unpublished data)

Days after hepatectomy	Percentage of values before hepatectomy							Morphology
		Enzyme activities[a]				Enzyme contents[a]		
	Liver weight	NADH-cyt. *c* red.	NADPH-cyt. *c* red.	G-6-P-ase	Oxid. demethyl. of aminopyrine	Cyt. b_5	Cyt. P-450	
2	52	72	96	44	53	71	69	Depletion of glycogen. Abundance of lipid droplets and of non-membrane-bound polysomes.
5	100	95	96	37	67	90	100	Glycogen re-appears. Progressive re-sorption of lipid droplets. Well-developed rough ER, moderate smooth ER.
8	97	109	68	39				Glycogen normally abundant. Lipids, rough and smooth ER normal.
15	97	163	44	47	82	90	100	As after 8 days.
15 + PB[b]	103	128	256	39	305	112	209	Abundance of smooth ER.

[a] Enzyme activities and contents are related to total microsomal protein.
[b] Treatment for 5 days with 100 mg/kg phenobarbital per day.

various hormones, including glucagon, epinephrine and thyroxine, as well as by 3′,5′-cylic AMP. Indications of a reorganization of the ER, although less pronounced and in the opposite direction, are also found in adult rats during liver regeneration following partial hepatectomy (Claude, Ernster, Estabrook, Lee and Nordenbrand, unpublished results; Table 1). In both cases we seem to be dealing with a complex pattern of changes in gene expression in the liver ER, connected with cell growth and differentiation.

More specific effects on the enzyme pattern of liver ER are found in rats under the influence of certain experimental conditions, such as treatment with alloxan, which causes changes in the kinetics of glucose-6-phosphatase (Segal and Washko, 1959; Jakobsson and Dallner, 1968), or dietary deficiency in essential fatty acids, which induces an enhanced activity of fatty acid desaturation (Oshino, 1972a,b). Most studied among this type of effects, however, is the drug-induced synthesis of the liver-microsomal monooxygenase system. This process appears to constitute a striking example of the now widely studied phenomenon of substrate-induced synthesis of a membrane-bound multienzyme complex, and represents, in fact, one of the few instances where this fundamental problem presently is available to experimental approach in a multicellular organism.

FIGURE 2

Effects of puromycin on the phenobarbital-induced increase in monooxygenase activity, rate of phospholipid synthesis, and phospholipid content of rat-liver microsomes

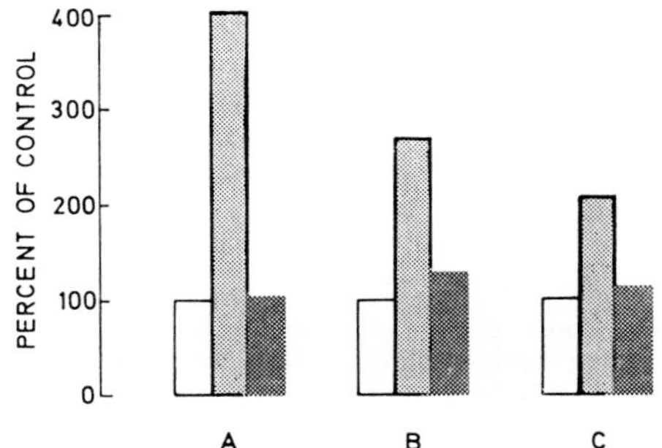

A: Monooxygenase activity (measured with aminopyrine as substrate) (From Orrenius *et al.*, 1965)

B: Rate of phospholipid synthesis (rate of incorporation of $^{32}P_i$) (From Orrenius and Ericsson, 1966)

C: Phospholipid content (expressed on the basis of microsomal protein) (From Orrenius and Ericsson, 1966).

White bars: control.

Gray bars: phenobarbital (100 mg/kg body-weight. A and C: 3 daily injections; B: 1 injection 12 h before sacrifice).

Black bars: phenobarbital + puromycin (12.5 mg/kg body-weight, injected simultaneously with phenobarbital).

Administration of drugs or other substrates of the monooxygenase system is known to induce a greatly increased catalytic capacity of this system in liver microsomes (Remmer, 1959; Conney and Burns, 1959). It is now well established that the induction process leads to a selective and parallel increase in the levels of the flavoprotein and cytochrome components of the monooxygenase system, leaving other microsomal enzyme activities largely unaffected (Orrenius and Ernster, 1964; Orrenius *et al.*, 1965; Ernster and Orrenius, 1965; Remmer and Merker, 1965. Cf. Fig. 1, *right* and Table 1). It involves increased rates of both protein and RNA

FIGURE 3
Effect of actinomycin D on phenobarbital-induced increase in monooxygenase activity of rat liver microsomes.

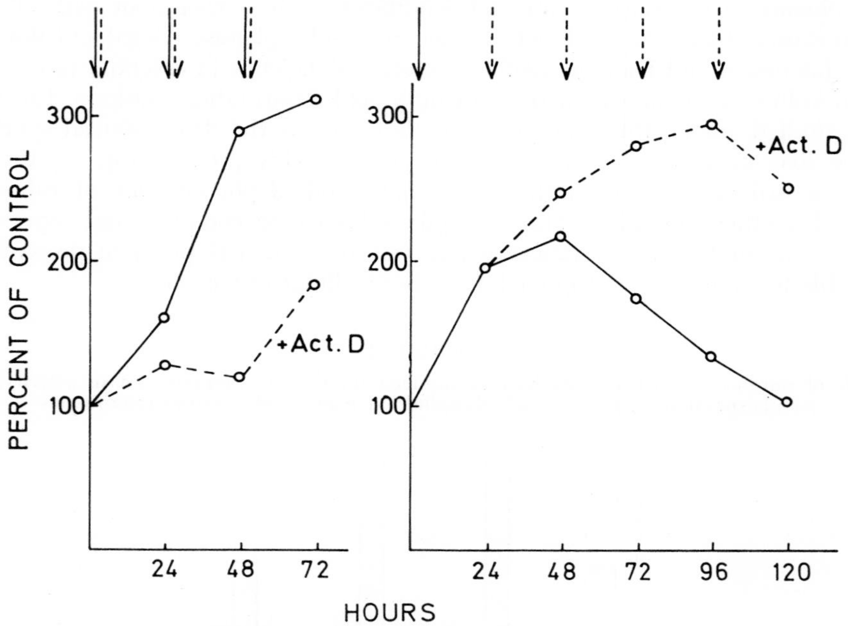

Left: Simultaneous administration of phenobarbital (100 mg/kg body-weight; solid arrows) and actinomycin D (80 μg/kg body-weight; dotted arrows) (From Orrenius *et al.*, 1965) *Right*: Repeated injections of actinomycin D (80 μg/kg body-weight; dotted arrows) after one injection of phenobarbital (100 mg/kg body-weight; solid arrow) (From Orrenius and Ericsson, 1966).

synthesis, as indicated by its sensitivity to puromycin (Conney and Gilman, 1963; Orrenius *et al.*, 1965; Fig. 2) and actinomycin D (Gelboin and Blackburn, 1963; Orrenius *et al.*, 1965; Fig. 3, *left*), as well as increased DNA-dependent RNA polymerase activities (Orrenius *et al.*, 1968; Fig. 4). Moreover, actinomycin D, when administered to rats after prior treatment with phenobarbital, gives rise to a "superinduction", i.e., a further enhancement of the monooxygenase activity, and even a delay of its regress following the cessation of drug treatment (Orrenius and Ericsson, 1966; Fig. 3, *right*). This phenomenon, which was first observed in rat liver by Garren *et al.* (1964) in connection with steroid hormone-induced synthesis of tyrosine transaminase and tryptophane di-oxygenase, is now known to occur in many instances of enzyme induction in both eukaryotes and prokaryotes (cf. Tomkins *et al.*, 1972), and is interpreted as the reflection of a post-

FIGURE 4

Phenobarbital-induced increase in nuclear RNA polymerase activities of rat liver

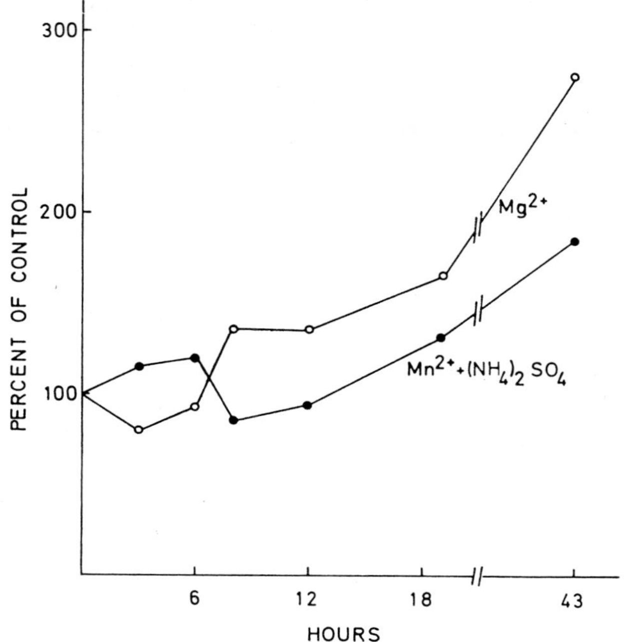

Rats were treated with 80 mg phenobarbital per kg body-weight at 0 and 24 h. Mg^{++}— and Mn^{++}+ $(NH_4)_2SO_4$-activated RNA polymerase activities were assayed as described by Tata and Widnell (1966). (From Orrenius *et al.*, 1968).

transcriptional (Tomkins *et al.*, 1969, 1972) or transcriptional (Nebert and Gelboin, 1970) control of induced enzyme synthesis. It has also been shown (Kato *et al.*, 1965, 1966) that liver microsomes isolated from phenobarbital-treated animals possess, after removal of endogenous messenger-RNA, an increased capacity for polyuridylic acid-dependent incorporation of phenylalanine into protein, indicating an increase in both the endogenous messenger-RNA content and the total number of microsomal binding-sites for messenger-RNA.

Direct evidence for an enhanced rate of enzyme synthesis during phenobarbital induction has been obtained in turnover studies of both the flavoprotein (Jick and Shuster, 1966; Kuriyama *et al.*, 1969) and the cytochrome (Greim, 1970; Greim *et al.*, 1970) components of the liver-microsomal monooxygenase system. Simultaneously, there appears to occur a decrease in the rate of degradation of the flavoprotein (Jick and Shuster, 1966), and under certain conditions also of the cytochrome (Greim, 1970), which may further contribute to the increased enzyme levels. Although some other microsomal enzymes, such as UDP-glucuronyl transferase, may increase as well after prolonged phenobarbital administration (Zeidenberg *et al.*, 1967), the initial phase of the induction is highly selective and restricted to the monooxygenase system. This selectivity is clearly revealed by recent measurements reported by Dehlinger and Schimke (1972), based on an elegant double-labelling technique for turnover studies of proteins *in vivo* (Schimke *et al.*, 1968; Schimke and Doyle, 1970; Dehlinger and Schimke, 1971). On the other hand, preceding the drug-induced increase in the levels of the components of microsomal monooxygenase, there is a rapid increase in mitochondrial δ-aminolevulinic acid synthetase (De Matteis, 1967; Marver, 1969; Kaufman *et al.*, 1970), the rate-limiting enzyme in haem biosynthesis (Granick and Urata, 1963; Granick, 1966). This observation, together with the finding that repressors of haem biosynthesis prevent the induction of the microsomal monooxygenase system by phenobarbital (Marver *et al.*, 1968; Marver, 1969; Kaufman *et al.*, 1970), has led to the suggestion that haem biosynthesis may play an important role in the induction process.

An important but still unresolved problem in connection with the drug-induced synthesis of the liver-microsomal monooxygenase system concerns the mechanism by which the inducing drug stimulates the synthesis of enzyme-specific messenger-RNA. A direct interaction between the drug and a regulator of gene transcription, e.g. a repressor, is a plausible possibility, which has been considered particularly

FIGURE 5

Effect of adrenalectomy and castration of rats on the liver-microsomal monooxygenase activity (measured with aminopyrine or testosterone as substrates) and cytochrome b_5 and P-450 contents

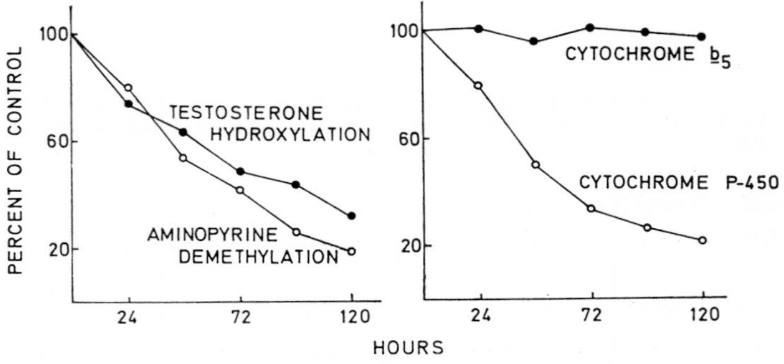

Operation was performed at 0 h. (From Orrenius *et al.*, 1968).

in the case of polycyclic hydrocarbons as inducers (Nebert and Gelboin, 1970). Alternatively, the induction may involve a mediator of cellular origin, such as a hormone and/or cyclic-AMP. There are indications of a need for steroid hormones in both the normal maintenance and maximal inducibility of cytochrome P-450-linked monooxygenase activity in the liver ER (Lotlikar *et al.*, 1964; Orrenius *et al.*, 1968; Nebert and Gelboin, 1969; cf. Figs. 5 and 6), and, although these effects may be secondary, due to changes in nutritional balance (Nebert and Gelboin, 1969), they are quite specific in the sense that other microsomal enzyme levels seem to remain unaffected (cf. Fig. 5, *right*). Since drugs are known to compete with steroid hormones for both their binding to cytochrome P-450 (Leibman *et al.*, 1969; Kupfer and Orrenius, 1970; Orrenius *et al.*, 1970) and their metabolism by way of the cytochrome P-450-linked monooxygenase system (Rubin *et al.*, 1964;

FIGURE 6

Effect of treatment of steroid-deficient rats with phenobarbital and steroid hormones on the liver-microsomal monooxygenase activity

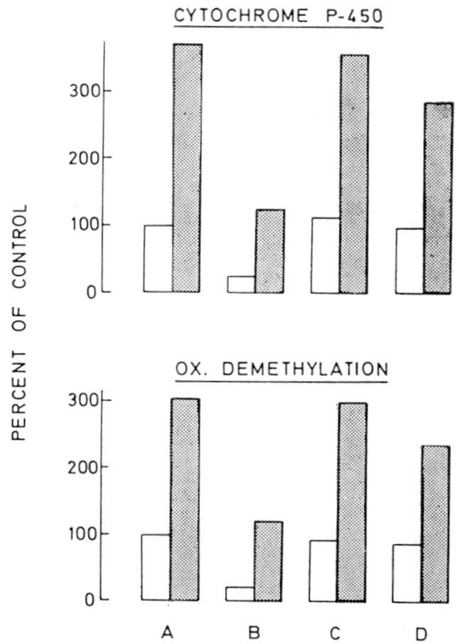

Rats were made steroid-deficient by adrenalectomy and castration 4 days before the experiment. Microsomal monooxygenase activity was measured with aminopyrine as substrate.
White bars: no phenobarbital
Gray bars: 3 daily injections with 80 mg phenobarbital/kg body-weight.
A: Sham-operated controls
B: Steroid-deficient rats
C: Steroid-deficient rats, treated with 3 daily doses of 20 mg prednisolone/kg body-weight
D: Steroid-deficient rats, treated with 3 daily doses of 80 mg testosterone propionate/kg body-weight. (From Orrenius *et al.*, 1968).

Tephly and Mannering, 1968; Orrenius and Thor, 1969), it is conceivable that drug treatment gives rise to increased cellular levels of steroid hormones available for binding to specific receptors involved in the mediation of gene activation. Recent progress in identifying and characterizing the mode of action of such receptors in several instances of steroid-induced enzyme synthesis (cf. Tomkins *et al.*, 1970, 1972; Jensen *et al.*, 1971), and the possibility of studying drug-activated enzyme induction in isolated organs (Juchau *et al.*, 1965) or cells (Holtzman *et al.*, 1972) and in cell cultures (Nebert and Gelboin, 1968a, b, 1969, 1970), may open new ways of approaching this problem. Preliminary experiments by S. Orrenius also suggest a possible involvement of cyclic-AMP in the drug-induced enzyme synthesis, as indicated by an enhancement of the induction by caffeine, an inhibitor of phosphodiesterase, and an inhibition by the β-adrenergic blocking agent, dichloro-isoproterenol (Fig. 7). These observations are now being extended by direct measurements of the cyclic-AMP and adenyl cyclase levels during induction.

A striking phenomenon accompanying drug-induced enzyme synthesis is the proliferation of ER membranes (Remmer and Merker, 1963; Orrenius *et al.*, 1965; cf. Fig. 8). From the available evidence it appears that the induced enzyme synthesis begins in the rough-surfaced areas of the ER, but leads eventually to an accumula-

FIGURE 7

Effects of treatment of rats with caffeine and dichloro-iso-proterenol (DCI) on the phenobarbital (PB)-induced increase in liver-microsomal monooxygenase activity.

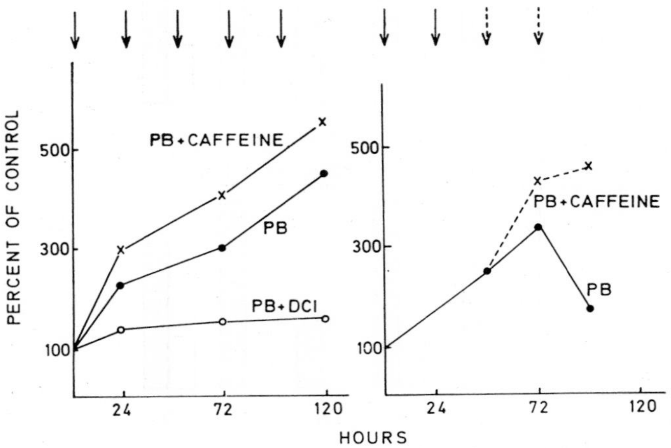

Left: Repeated injections (*arrows*) of caffeine (5 mg/kg body-weight) or DCI (25 mg/kg body-weight) simultaneously with phenobarbital (100 mg/kg body-weight)
Right: 2 injections of phenobarbital (80 mg/kg body-weight; *solid arrows*) followed by 2 injections of caffeine (5 mg/kg body-weight; *dotted arrows*).
Monooxygenase activity was measured with aminopyrine as substrate.
(From S. Orrenius, unpublished results).

tion of smooth-surfaced profiles, rich in components of the monooxygenase system (Orrenius, 1965b; Ernster and Orrenius, 1965; cf. Fig. 9). Thus, the synthesis of new membranes seems to be geared to the formation of new enzymes, by a mechanism that may involve an outgrowth and budding off of smooth-surfaced membrane profiles from the rough-surfaced ER at the sites where the enzyme synthesis takes place. The drug-induced enzyme synthesis is accompanied by an early increase in the rate of phospholipid synthesis, leading to an increased level of membrane phospholipids (Orrenius, 1965b) which, like the membrane proliferation, is sensitive to puromycin (Orrenius and Ericsson, 1966; cf. Fig. 2). Simultaneously, there occurs a decrease in the rate of phospholipid catabolism (Holtzman and Gillette, 1968; Orrenius, 1968), which further contributes to the increased phospholipid level. Significantly, also, an increased phospholipid content of the ER membranes persists for a relatively long period of time after the regress of monooxygenase activity following cessation of drug treatment (Orrenius and Ericsson, 1966; cf. Fig. 10). Administration of actinomycin D, so as to inhibit drug-induced enzyme synthesis, largely prevents the accumulation of smooth-surfaced ER (Orrenius *et al.*, 1965; cf. Fig. 8), but has relatively little effect on the initial

FIGURE 8

Electron micrographs illustrating the effects of treatment of rats with phenobarbital and phenobarbital + actinomycin D on the liver ER.

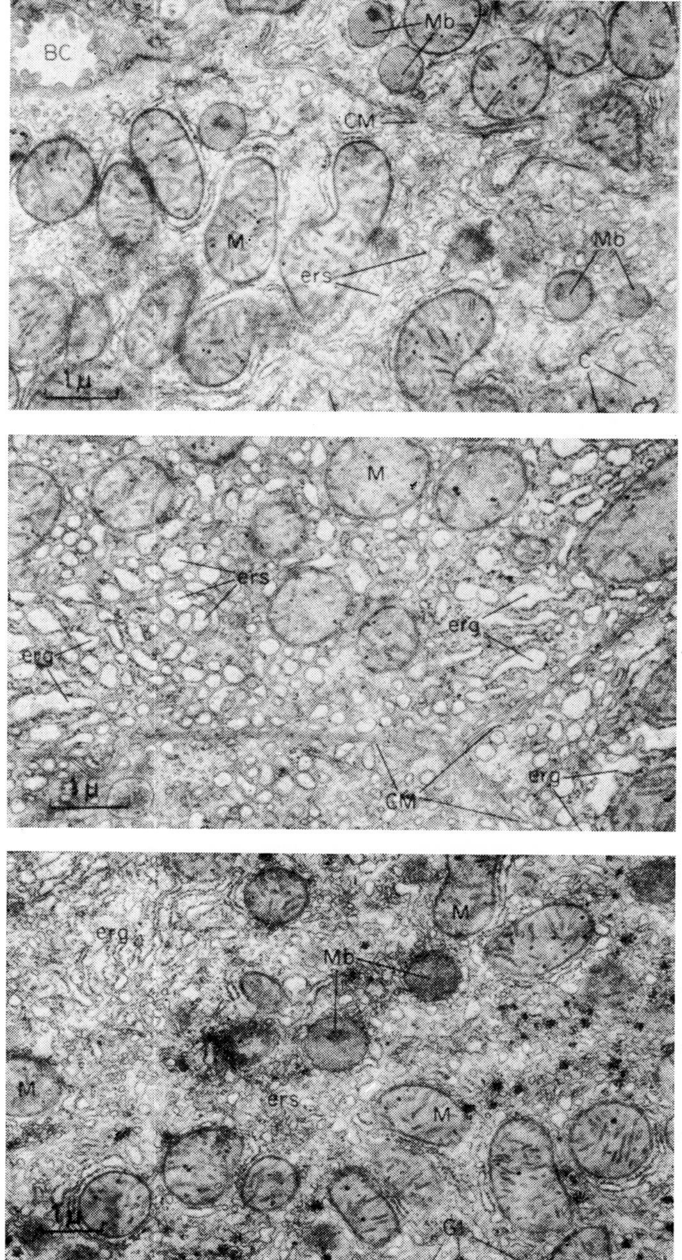

Top: Control (intraperitoneal injection with physiological saline for 5 days).
Middle: Intraperitoneal injection with 5 daily doses of 100 mg phenobarbital/kg body-weight.
Bottom: Intraperitoneal injection with 5 daily doses of 100 mg phenobarbital + 80 μg actinomycin D/kg body-weight.
(From Orrenius *et al.*, 1965).

FIGURE 9

Phenobarbital-induced increase in total protein, monooxygenase activity and its component enzymes in rough—(●) and smooth—(O) surfaced rat-liver microsomes

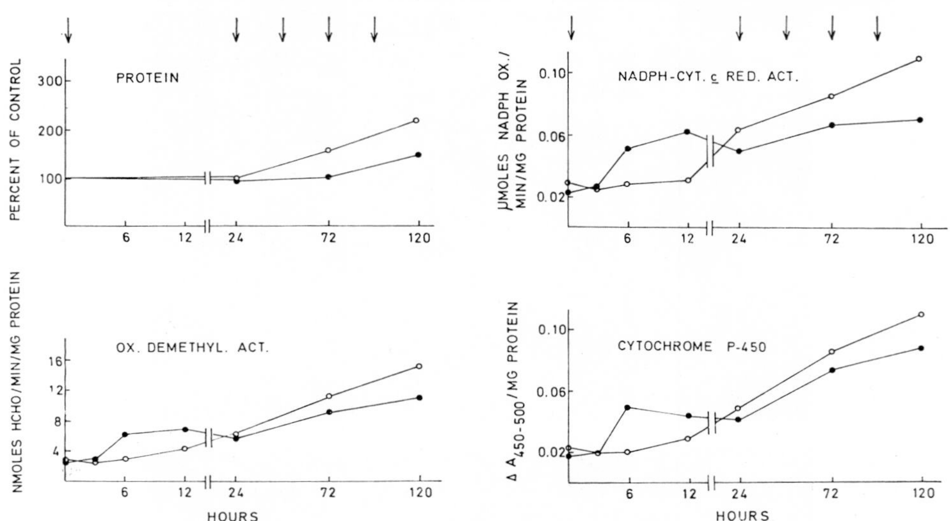

Monooxygenase activity was measured as the oxidative demethylation of aminopyrine. 5 injections (*arrows*) of 100 mg phenobarbital/kg body-weight.
(From Ernster and Orrenius, 1965).

FIGURE 10

Effect of phenobarbital treatment of rats on the monooxygenase activity and phospholipid content of liver microsomes

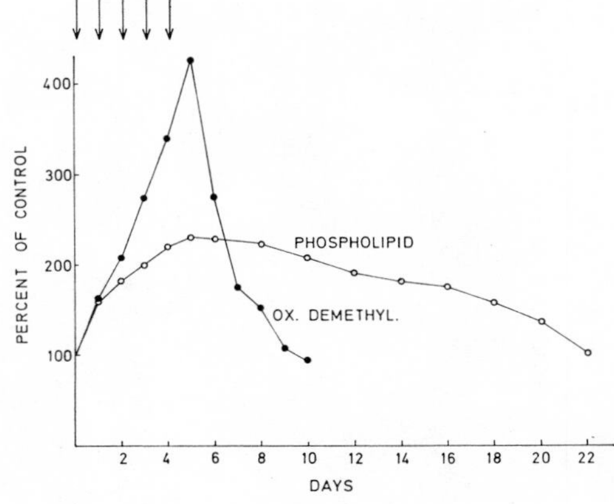

Monooxygenase activity was measured as the oxidative demethylation of aminopyrine. 5 injections (*arrows*) of 100 mg phenobarbital/kg body-weight.
(From Orrenius and Ericsson, 1966).

increase in the rate of phospholipid synthesis (Ernster and Orrenius, 1965; Orrenius and Ericsson, 1966). It thus appears that the latter event, although dependent on protein synthesis (as indicated by its sensitivity to puromycin) does not require a *de novo* synthesis of messenger-RNA. An investigation of the levels and activities of various enzymes involved in microsomal phospholipid metabolism during drug-induced enzyme synthesis may be of interest for gaining further insight into the regulatory mechanisms responsible for the coordination of enzyme induction and membrane biogenesis.

In conclusion, it is evident from this condensed survey that the liver endoplasmic reticulum may serve as a particularly rich source of information concerning various biochemical and physiological aspects of gene expression in higher organisms.

ACKNOWLEDGEMENTS

I wish to thank my colleagues Gustav Dallner and Sten Orrenius for many years of fruitful collaboration and for valuable discussions during the preparation of this manuscript. Work quoted from this laboratory was supported by grants from the Swedish Cancer Society and the Swedish Medical and Natural-Science Research Councils.

REFERENCES

Behrens, N. H. and Leloir, L. F. (1970) *Proc. Nat. Acad. Sci. U.S. 66,* 153

Behrens, N. H., Parodi, A. J., Leloir, L. F. and Krisman, C. R. (1971) *Arch. Biochem. Biophys. 43,* 375

Blobel, G. and Potter, V. R. (1967a) *J. Mol. Biol. 26,* 279

Blobel, G. and Potter, V. R. (1967b) *J. Mol. Biol. 26,* 293

Blobel, G. and Potter, V. R. (1967c) *J. Mol. Biol. 28,* 539

Blobel, G. and Sabatini, D. D. (1970) *J. Cell Biol. 45,* 130

Boardman, N. K., Linnane, A. W. and Smillie, R. M., (Editors) (1971) *Autonomy and Biogenesis of Mitochondria and Chloroplasts.* North Holland Publ. Co., Amsterdam

Brodie, B. B., Axelrod, J., Cooper, J. R., Gaudette, L., La Du, B. N., Mitoma, C. and Udenfriend, S. (1955) *Science 121,* 603

Caccam, J. F., Jackson, J. J. and Eylar, C. H. (1969) *Biochem. Biophys. Res. Commun. 35,* 505

Chance, B. and Williams, G. R. (1954) *J. Biol. Chem. 209,* 945

Chesterton, C. J. (1966) *Biochem. Biophys. Res. Commun. 25,* 205

Claude, A. (1941) *Cold Spring Harbor Symp. Quant. Biol. 9,* 263

Claude, A. (1943a) *Science 97,* 451

Claude, A. (1943b) in *Frontiers in Cytochemistry,* (Hoerr, N. L., ed.) p. 111 Jacques Cattell Press, Lancaster

Claude, A. (1948) *Harvey Lectures 43,* 121

Conney, A. H. and Burns, J. J. (1959) *Nature (London) 184,* 363

Conney, A. H. and Gilman, A. G. (1963) *J. Biol. Chem. 238,* 3682

Cooper, D. Y., Levin, S., Narasimhulu, S., Rosenthal, O. and Estabrook, R. W. (1965) *Science 147,* 400

Dallman, P. R., Dallner, G., Bergstrand, A. and Ernster, L. (1969) *J. Cell Biol. 41,* 357

Dallner, G. (1963) *Acta Pathol. Microbiol. Scand.* Suppl. 166

Dallner, G., Behrens, N. H., Parodi, A. J. and Leloir, L. F. (1972) *FEBS Lett. 24,* 315

Dallner, G., Bergstrand, A. and Nilsson, R. (1968) *J. Cell Biol. 38,* 257

Dallner, G. and Ernster, L. (1968) *J. Histochem. Cytochem. 16,* 611

Dallner, G., Siekevitz, P. and Palade, G. E. (1966a) *J. Cell Biol. 30,* 73

Dallner, G., Siekevitz, P. and Palade, G. E. (1966b) *J. Cell Biol. 30,* 97

Dawson, R. M. C. (1966) in *Essays in Biochemistry,* (Campbell, P. N. and Greville, G. D. eds.) Vol. 2, p 69 Academic Press, New York

de Duve, C. (1971) *J. Cell Biol. 50,* 20D

de Duve, C., Pressman, B. C., Gianetto, R., Wattiaux, R. and Appelmans, F. (1955) *Biochem. J. 60,* 604

de Duve, C., Wattiaux, R. and Baudhuin, P. (1962) *Advan. Enzymol. 24,* 291

Dehlinger, P. J. and Schimke, R. T. (1971) *J. Biol. Chem. 246,* 2574

Dehlinger, P. J. and Schimke, R. T. (1972) *J. Biol. Chem. 247,* 1257

De Matteis, F. (1967) *Pharmacol. Rev. 19,* 523

Ernster, L. and Jones, L. C. (1962) *J. Cell Biol. 15,* 563

Ernster, L. and Orrenius, S. (1965) *Fed. Proc. 24,* 1190

Ernster, L., Siekevitz, P. and Palade, G. E. (1962) *J. Cell Biol. 15,* 541

Garfinkel, D. (1958) *Arch. Biochem. Biophys 77,* 493

Garren, L. D., Howell, R. R., Tomkins, G. M. and Crocco, R. M. (1964) *Proc. Nat. Acad. Sci. U.S. 52,* 1121

Gelboin, H. V. and Blackburn, N. R. (1963) *Biochim. Biophys. Acta 72,* 657

Gillette, J. R. (1966) *Adv. Pharmacol. 4,* 219

Gillette, J. R., Davis, D. C. and Sasame, H. A. (1972) *Annu. Rev. Pharmacol. 12,* 57

Glaumann, H. (1970) *Biochim. Biophys. Acta 224,* 206

Glaumann, H. and Dallner, G. (1970) *J. Cell Biol. 47,* 34

Glaumann, H. and Ericsson, J. L. E. (1970) *J. Cell Biol. 47,* 555

Glaumann, H., Kuylenstierna, B. and Dallner, G. (1969) *Life Sci. 8,* 1309

Glaumann, H., von der Decken, A. and Dallner, G. (1968) *Life Sci. 7,* 905

Goad, L. J. (1970) in *Natural Substances Formed Biologically from Mevalonic Acid,* (Goodwin, T. W., ed.) p. 45 Academic Press, New York

Granick, S. (1966) *J. Biol. Chem. 241,* 1359

Granick, S. and Urata, G. (1963) *J. Biol. Chem. 238,* 821

Greengard, O. (1969) *Science 163,* 891

Greengard, O. (1970) in *Biochemical Actions of Hormones,* (Litwack, G., ed.) p. 53 Academic Press, New York

Greengard, O., Federman, M. and Knox, W. E. (1972) *J. Cell Biol. 52,* 261

Greim, H. (1970) *Naunyn-Schmiedebergs Arch. Pharmakol. 266,* 261

Greim, H., Schenkman, J. B., Klotzbucher, M. and Remmer, H. (1970) *Biochim, Biophys. Acta 201,* 20

Hallinan, T., Murty, C. N. and Grant, J. H. (1968) *Arch Biochem. Biophys. 125,* 715

Hers, H. G. and de Duve, C. (1950) *Bull. Soc. Chim. Biol. 33,* 21

Hicks, S. J., Drysdale, J. W. and Munro, H. N. (1969) *Science 164,* 584

Holtzman, J. L., Rothman, V. and Margolis, S. (1972) *Biochem. Pharmacol. 21,* 581

Holtzman, J. L. and Gillette, J. R. (1968) *J. Biol. Chem. 243,* 3020

Horecker, B. L. (1950) *J. Biol. Chem. 183,* 593

Jakobsson, S. and Dallner, G. (1968) *Biochim. Biophys. Acta 165,* 380

Jankowski, W. and Chojnacki, T. (1972) *Biochim. Biophys. Acta 260,* 93

Jensen, E. V., Numata, M., Brecher, P. I. and DeSombre, E. R. (1971) in *The Biochemistry of Steroid Hormone Action. Biochem. Soc. Symp.* (Smellie, R. M. S., ed.) No. 32, p. 133, Academic Press, New York

Jick, H. and Shuster, L. (1966) *J. Biol. Chem. 241,* 5366

Juchau, M. R., Cram, R. L., Plaa, G. L. and Fouts, J. R. (1965) *Biochem. Pharmacol. 14,* 473

Jungawala, F. B. and Dawson, R. M. C. (1970) *Eur. J. Biochem. 12,* 399

Kamin, H., Masters, B. S. S., Gibson, Q. H. and Williams, C. H. (1965) *Fed. Proc. 24,* 1164

Kato, R., Jondorf, W. R., Loeb, L. A., Ben, T and Gelboin, H. V. (1966) *Mol Pharmacol. 2,* 171

Kato, R., Loeb, L. A. and Gelboin, H. V. (1965) *Biochem. Pharmacol. 14,* 1164

Kaufman, L., Swanson, A. L. and Marver, H. S. (1970) *Science 170,* 320

Klingenberg, M. (1958) *Arch. Biochem. Biophys. 75,* 376

Knox, W. E. (1972) *Enzyme Patterns in Fetal, Adult and Neoplastic Rat Tissues,* S. Krager AG, Basel

Knox, W. E., Auerbach, V. H. and Lin, E. C. C. (1956) *Physiol. Rev. 36,* 164

Kupfer, D. and Orrenius, S. (1970) *Eur. J. Biochem. 14,* 317

Kuriyama, Y., Omura, T., Siekevitz P. and Palade, G. E. (1969) *J. Biol. Chem. 244,* 2017

Lawford, G. R. and Schachter, H. (1966) *J. Biol. Chem. 241,* 5408

Leibman, K. C., Hildebrandt, A. G. and Estabrook, R. W. (1969) *Biochem. Biophys. Res. Commun. 36,* 789

Lennarz, W. J. and Scher, M. G. (1972) *Biochim. Biophys. Acta 265,* 297

Lotlikar, P. D., Enomoto, M., Miller, E. C. and Miller, J. A. (1964) *Cancer Research 24,* 1835

McMurray, W. C. and Dawson, R. M. C. (1969) *Biochem. J. 112,* 91

Malkin, L. I. and Rich, A. (1967) *J. Mol. Biol. 26,* 329

Marsh, J. B. (1963) *J. Biol. Chem. 238,* 1752

Marsh, J. B. (1971) in *Plasma Lipoproteins,* (Smellie, R. M. S., ed.) p. 89 Academic Press, New York

Marver, H. S. (1969) in *Microsomes and Drug Oxidations,* (Gillette, J. R., Conney, A. H., Cosmides, G. J., Estabrook, R. W., Fouts, J. R. and Mannering, G. J. eds.) p. 495 Academic Press, New York

Marver, H. S., Schmid, R. and Stützel, H. (1968) *Biochem. Biophys. Res. Commun. 33,* 969

Molnar, J., Chao, H. and Ikehara, Y. (1971) *Biochim. Biophys. Acta 239,* 401

Molnar, J., Robinson, G. B. and Winzler, R. J. (1965) *J. Biol. Chem. 240,* 1882

Molnar, J. and Sy, D. (1967) *Biochemistry 6,* 1941

Molnar, J., Tetas, M. and Chao, H. (1969) *Biochem. Biophys. Res. Commun. 37,* 684

Mookerjea, S. (1971) *Fed. Proc. 30,* 143

Mookerjea, S. (1972) in *Protides of Biological Fluids, 19th Colloquium,* (Peeters H. ed.) p. 135 Pergamon Press, Oxford

Nebert, D. W. and Gelboin, H. V. (1968a) *J. Biol. Chem. 243,* 6243

Nebert, D. W. and Gelboin, H. V. (1968b) *J. Biol. Chem. 243,* 6250

Nebert, D. W. and Gelboin, H. V. (1969) *Arch. Biochem. Biophys. 134,* 76

Nebert, D. W. and Gelboin, H. V. (1970) *J. Biol. Chem. 245,* 160

Omura, T. and Sato, R. (1963) *Biochim. Biophys. Acta 71,* 224

Omura, T. and Sato, R. (1964a) *J. Biol. Chem. 239,* 2370

Omura, T. and Sato, R. (1964b) *J. Biol. Chem. 239,* 2379

Omura, T. Siekevitz, P. and Palade, G. E. (1967) *J. Biol. Chem. 242,* 2389

Orrenius, S. (1965a) *J. Cell Biol. 26,* 713

Orrenius, S. (1965b) *J. Cell Biol. 26,* 725

Orrenius, S. (1968) in *The Interaction of Drugs and Subcellular Components,* (Campbell, P. N. ed.) p. 97 J. and A. C. Churchill, Ltd., London

Orrenius, S. and Ericsson, J. L. E. (1966) *J. Cell Biol. 28,* 181

Orrenius, S., Ericsson, J. L. E. and Ernster, L. (1965) *J. Cell Biol. 25,* 627

Orrenius, S. and Ernster, L. (1964) *Biochem. Biophys. Res. Commun. 16,* 60

Orrenius, S. and Ernster, L. (1971) in *Cell Membranes: Biological and Pathological Aspects,* (Richter, G. W. and Scarpelli, D. G. eds.) p. 38 Williams & Wilkins Co., Baltimore

Orrenius, S., Gnosspelius, Y., Das, M. L. and Ernster, L. (1968) in *Structure and Function of the Endoplasmic Reticulum in Animal Cells,* p. 81. (Gran, F. C. ed.) p. 81, Universitetsforlaget, Oslo.

Orrenius, S., Kupfer, D. and Ernster, L. (1970) *FEBS Lett. 6,* 249

Orrenius, S. and Thor, H. (1969) *Eur. J. Biochem. 9,* 415

Oshino, N. (1972a) *Arch. Biochem. Biophys. 149,* 369

Oshino, N. (1972b) *Arch. Biochem. Biophys. 149,* 378

Oshino, N., Imai, Y. and Sato, R. (1971) *J. Biochem. (Tokyo) 69,* 155

Oshino, N. and Sato, R. (1971) *J. Biochem. (Tokyo) 69,* 169

Palade, G. E. (1955a) *J. Biophys. Biochem. Cytol. 1,* 59

Palade, G. E. (1955b) *J. Biophys. Biochem. Cytol. 1,* 567

Palade, G. E. (1956) *J. Biophys. Biochem. Cytol. Suppl. 2,* 85

Palade, G. E. (1966) *J. Am. Med. Ass. 198,* 815

Palade, G. E. (1971) *J. Cell Biol. 50,* 5D

Palade, G. E. and Porter, K. R. (1954) *J. Exp. Med. 100,* 641

Palade, G. E. and Siekevitz, P. (1956) *J. Biophys. Biochem. Cytol. 2,* 171

Palade, G. E., Siekevitz, P. and Caro, L. G. (1962). in *Ciba Foundation Symposium on the Exocrine Pancreas,* (de Reuck, V. A. S. and Cameron, M. B. eds.) p. 23, J. & A. Churchill, Ltd., London

Peters, T., Jr. (1962) *J. Biol. Chem. 237,* 1186

Peters, T., Jr., Fleischer, B. and Fleischer, S. (1971) *J. Biol. Chem. 246,* 240

Phillips, A. H. and Langdon, R. G. (1962) *J. Biol. Chem. 237,* 2652

Porter, K. R. (1961) in *Biological Structure and Function,* Vol. I, (Goodwin, T. W. and Lindberg, O. eds.) p. 127, Academic Press, London

Porter, K. R. and Kallman, F. L. (1952) *Ann. N.Y. Acad. Sci. 54,* 882

Priestley, G. C., Pruyn, M. L. and Malt, R. A. (1969) *Biochim. Biophys. Acta 190,* 154

Redman, C. M. (1967) *J. Biol. Chem. 242,* 761

Redman, C. M. (1968) *Biochem. Biophys. Res. Commun. 33,* 55

Redman, C. M. (1969) *J. Biol. Chem. 244,* 4308

Redman, C. M. and Cherian, M. G. (1972) *J. Cell Biol. 52,* 231

Redman, C. M. and Sabatini, D. D. (1966) *Proc. Nat. Acad. Sci. U.S. 56,* 508

Redman, C. M., Siekevitz, P. and Palade, G. E. (1966) *J. Biol. Chem. 241,* 1150

Reid, E. (1967) in *Enzyme Cytology,* (Roodyn, D. B. ed.) p. 321, Academic Press, London

Remmer, H. (1959) *Naunyn-Schmiedeberg Arch. Pharmakol. 235,* 279

Remmer, H. and Merker, H. J. (1963) *Klin. Wschr. 41,* 276

Remmer, H. and Merker, H. J. (1965) *Ann. N.Y. Acad. Sci. 123,* 79

Richards, J. B., Evans, P. J. and Hemming, F. W. (1971) *Biochem. J. 124,* 957

Rubin, A., Tephly, T. R. and Mannering, G. J. (1964) *Biochem. Pharmacol. 13,* 1007

Sabatini, D. D. and Blobel, G. (1970) *J. Cell Biol. 45,* 146

Sabatini, D. D., Tashiro, Y. and Palade, G. E. (1966) *J. Mol. Biol. 19,* 503

Schachter, H., Jabbal, I., Hudgin, R. L., Pinteric, L., McGuire, J. E. and Roseman, S. (1970) *J. Biol. Chem. 245,* 1090

Schimke, R. T. and Doyle, D. (1970) *Annu. Rev. Biochem. 39,* 929

Schimke, R. T., Ganschow, R., Doyle, D. and Arias, I. (1968) *Fed. Proc. 27,* 1223

Schneider, W. C. (1963) *J. Biol. Chem. 238,* 3572

Schneider, W. C. and Hogeboom, G. H. (1951) *Cancer Research 11,* 1

Segal, H. L. and Washko, M. E. (1959) *J. Biol. Chem. 234,* 1937

Siekevitz, P. (1963) *Annu. Rev. Physiol. 25,* 15

Siekevitz, P. (1972) *Annu. Rev. Physiol. 34,* 117

Siekevitz, P. and Palade, G. E. (1960) *J. Biophys. Biochem. Cytol. 7,* 619

Siekevitz, P., Palade, G. E., Dallner, G., Ohad, I. and Omura, T. (1967) in *Organizational Biosynthesis,* (Vogel, H. J., Lampen, J. O. and Bryson, V. eds.) p. 331, Academic Press, New York

Sierr, C. J. and Uhr, J. (1970) *Proc. Nat. Acad. Sci. U.S. 66,* 1183

Spiro, R. G. and Spiro, M. J. (1966) *J. Biol. Chem. 241,* 1271

Stoffel, W. and Schiefer, H.-G. (1968) *Hoppe Seyler's Z. Physiol. Chem. 349,* 1017

Strittmatter, P. (1965) *Fed. Proc. 24,* 1156

Strittmatter, C. F. and Ball, E. G. (1952) *Proc. Nat. Acad. Sci. U.S. 38,* 19

Strittmatter, C. F. and Ball, E. G. (1954) *J. Cell. Comp. Physiol. 43,* 57

Strittmatter, P. and Velick, S. F. (1956a) *J. Biol. Chem. 221,* 253

Strittmatter, P. and Velick, S. F. (1956b) *J. Biol. Chem. 221,* 277

Strittmatter, P. and Velick, S. F. (1957) *J. Biol. Chem. 177,* 847

Svensson, H., Dallner, G. and Ernster, L. (1972) *Biochim. Biophys. Acta 274,* 447

Takagi, M. and Ogata, K. (1968) *Biochem. Biophys. Res. Commun. 31,* 845

Takagi, M., Tanaka, T. and Ogata, K. (1970) *Biochim. Biophys. Acta 217,* 148

Tashiro, Y. (1958) *J. Biochem. 45,* 803

Tashiro, Y. and Siekevitz, P. (1965) *J. Mol. Biol. 11,* 649

Tata, J. R. (1967a) *Nature (London) 213,* 566

Tata, J. R. (1967b) *Biochem. J. 104,* 1

Tata, J. R., Ernster, L., Lindberg, O., Arrhenius, E., Pedersen, S. and Hedman, R. (1963) *Biochem. J. 86,* 408

Tata, J. R. and Widnell, C. C. (1966) *Biochem. J. 98,* 604

Tchen, T. T. and Bloch, K. (1955) *J. Amer. Chem. Soc. 77,* 6085

Tephly, T. R. and Mannering, G. J. (1968) *Mol. Pharmacol. 4,* 10

Tetas, M., Chao, H. and Molnar, J. (1970) *Arch. Biochem. Biophys. 138,* 135

Tomkins, G. M., Gelehrter, T. D., Granner, D., Martin, D., Jr., Samuels, H. and Thompson, E. B. (1969) *Science 166,* 1474

Tomkins, G. M., Levinson, B. B., Baxter, J. D. and Dethlefsen, L. (1972) *Nature New Biol. 239,* 9

Tomkins, G. M. and Martin, D. W., Jr. (1970) *Annu. Rev. Genetics 4,* 91

Wilgran, G. S. and Kennedy, E. P. (1963) *J. Biol. Chem. 238,* 2615

Williams, C. H. and Kamin, H. (1962) *J. Biol. Chem. 237,* 587

Williams, M. L. and Bygrave, F. L. (1970) *Eur. J. Biochem. 17,* 32

Zatz, M. and Barondes, S. H. (1969) *Biochem. Biophys. Res. Commun. 36,* 511

Zeidenberg, P., Orrenius, S. and Ernster, L. (1967) *J. Cell Biol. 32,* 528

The Use of Neurological Mutants as Experimental Models

P. MANDEL, J. L. NUSSBAUM[a], N. NESKOVIC, L. SARLIÈVE and E. FARKAS and O. ROBAIN

Centre de Neurochimie du CNRS, Strasbourg, France
Lab. de Neuropathologie, Hôpital St-Vincent-de-Paul, Paris, France

There is a growing desire among scientists to understand the genetic bases of human development, function and disease as is shown by the increasing number of investigations carried out on eukaryotic cells. Such investigations are the first steps on the way to a complete understanding of gene expression in higher animals. Although most progress has, up until now, been made by studying the relatively simple microorganisms there are many examples of observations made in higher animals which have opened new approaches to understanding the basic mechanisms of heredity in molecular terms. In this respect we can recall the advances made by studying the haemoglobins and haemoglobin synthesis, and the progress in understanding the genetic basis of the inborn errors in complex lipid metabolism in brain.

The advances made in molecular biology by studying the properties of mutant microorganisms are well-known. It seems likely that similar advances can be made by studying mutants of higher organisms, particularly in understanding how complex processes, difficult to reproduce experimentally in simple but valid systems, are controlled. The experiments to be reported here show how, from studies on neurological mutants of mice, some understanding of the genetic control of a complex phenomenon which occurs during brain development, the formation of the myelin sheath, can be gained.

It is known from morphological observations that the myelin sheath is formed by the plasma membranes of glial cells as they wrap around the axon (Luse, 1956). Despite its origin from a plasma membrane, the myelin membrane has a remarkably simple composition. Myelin is almost devoid of enzyme activities with the exception of 2′, 3′-cyclic AMP 3′-phosphohydrolase (CNP) (Kurihara

[a]Attaché de Recherche au CNRS.

et al., 1969, 1970). The protein profile is correspondingly simple. Over 90% of the proteins are accounted for by the proteolipid protein (Folch and Lees, 1951), the basic proteins and the Wolfgram proteins (Wolfgram and Rose, 1961; Waehneldt and Mandel, 1970). The lipid composition is also extremely specific: myelin is extremely rich in lipid having a lipid:protein ratio of 4:1. Cerebrosides and sulphatides are concentrated in this structure. Cholesterol and phospholipids are also found in myelin, but only trace amounts of gangliosides have been detected (Mokrasch, 1969).

Before the formation of the myelin sheath, glial cells proliferate and differentiate. Some of the primitive spongioblasts take on the characteristic form of oligodendrocytes which line up parallel to the axonal processes. Thus myelination could be controlled genetically at a molecular level (the synthesis of myelin specific proteins and lipids), at a supramolecular level (the assembly in the oligodendroglia of myelin consistuents such as cerebrosides, sulphatides, proteolipid proteins, basic proteins and the Wolfgram proteins), or at a morphogenetic level (cell division and differentiation).

Two recessive mutations in the mouse which cause defective myelination in the central nervous system have been described by Sidman *et al.* (1964): the Jimpy and the Quaking mutation. A third mutation, the myelin synthesis deficient (MSD) mutants, was described by Meier and McPike (1970). Two of these, the Jimpy and MSD mutants are sex-linked. Only traces of myelin exist in their brains and spinal cords.

Quaking mice are autosomal recessive mutants. The myelin content in Quaking brain and spinal cord is about 30% of the normal. The phenotype is recognized in all three mutants around the 10-12th day after birth, that is shortly after the period when myelination normally starts. The mutant behaviour is characterized by a marked tremor of the hind quarters and by susceptibility to spontaneous or mechanically stimulated seizure. While the life span of the Jimpy and MSD mutants is 28 to 30 days, it may be nearly normal for Quaking mice.

In contrast to the myelin deficiency in brain and spinal cord, the peripheral nerves of all mutants are normally myelinated. This shows clearly that a process which is impaired in one part of the nervous system can be normal in another part within the same animal suggesting a differential genetic control. Peripheral nerve myelin is not chemically identical to central myelin, and is formed not by oligodendrocytes but by the morphologically distinct Schwann cells.

STUDIES OF MYELIN-SPECIFIC MOLECULES

2′, 3′-cyclic AMP 3′-phosphohydrolase (CNP)

This enzyme described by Drummond *et al.* (1962) was found by Kurihara and Tsukada (1967) to be concentrated in the myelin fractions. Following this enzymic activity in the normal strains we found a rapid increase, whereas in the Jimpy mutants the activity remained extremely low and was slightly higher in Quaking mice, although the values reached were not higher than 30% of the normal. Similarly, low activities were found in the spinal cord (Kurihara *et al.*, 1969,

1970), however the difference between the normal and the mutant mice was less than in brain (Fig. 1).

FIGURE 1

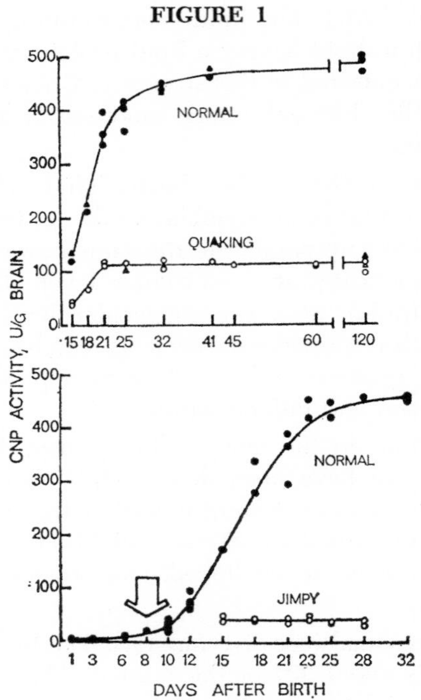

2′, 3′-cyclic AMP 3′-phosphohydrolase in the developing brains of Jimpy, Quaking and respectively normal mice. Each point represents one animal.
○, ● : male
▲, △ : female

This enzyme can be used as a quantitative marker for myelin in brain (Kurihara *et al.*, 1969, 1970) although the enzyme is associated also with glial cells (Zanetta *et al.*, 1972) and at very low activities with other cell membranes (Sudo *et al.*, 1972).

Myelin proteins

Alterations in myelin proteins in the mutants were investigated after isolation of a myelin fraction by polyacrylamide gel electrophoresis in the presence of sodium dodecylsulphate (Waehneldt and Mandel, 1970; Waehneldt, 1971). Alternatively the myelin-specific proteolipids were determined directly on the homogenate, thus avoiding any loss of material during isolation (Mandel *et al.*, 1971; Nussbaum and Neskovic, 1971).

In Jimpy and Quaking mice, the levels of proteolipid proteins were extremely low as shown by densitometry of polyacrylamide gel electrophoresis of chloroform-methanol extracts of total brain (Fig. 2). Peak 7, which corresponds to the myelin proteolipid proteins, is present in the mutants at levels close to those

FIGURE 2

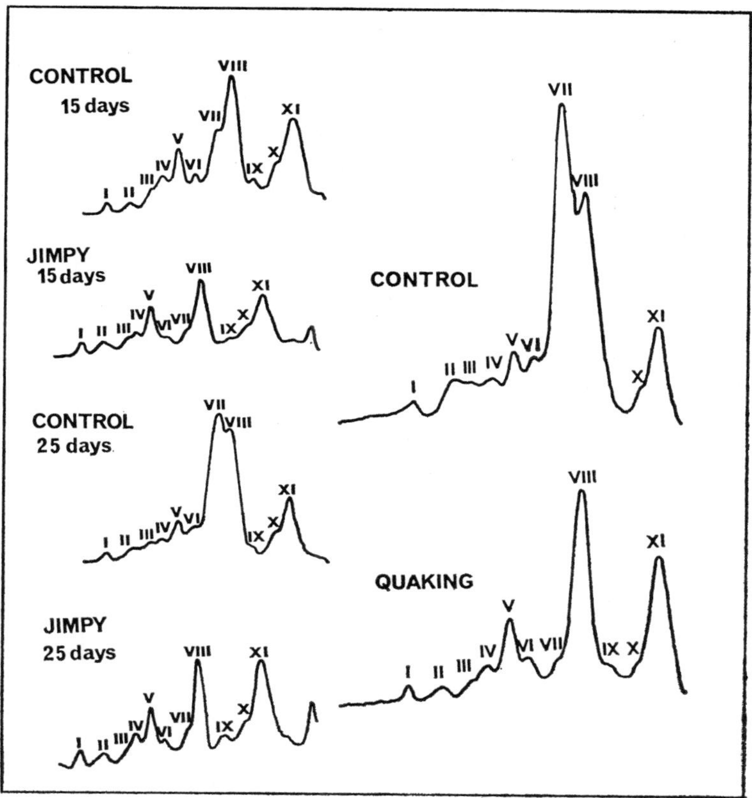

Densitometric scales of whole brain proteolipids from Jimpy and control mice of some age and adult Quaking and controls in 12% polyacrylamide gels. The band VII corresponds to the specific myelin proteolipid.

which exist in mice before or at the early period of myelination. There is no increase of myelin proteolipid proteins in the mutants during myelination. Basic proteins could be detected in the Quaking mice but hardly in Jimpy.

Myelin lipids

The phospholipid composition of normal and Jimpy brains is quite similar for the whole life span of the animals (Tables 1, 2) (Nussbaum *et al.,* 1968; Nussbaum *et al.,* 1969). Slightly lower values of ethanolamine plasmalogens and sphingomyelin were found in Jimpy brain as compared to the normal, but other phospholipids are also myelin-specific. During the myelination period (Fig. 3) the rate of accumulation of phosphatides is lower than normal in the mutants and there is no increase in cholesterol, a quantitatively important constituent of myelin sheath.

The most striking difference in the Jimpy mice concerns the cerebrosides and sulphatides which normally increase markedly during brain development, but

TABLE 1

Percent Distribution of Phospholipid Components in Normal and Jimpy Mouse Brains

Phospholipids	17 days		19 days		23 days		26 days		29 days	
	N	J	N	J	N	J	N	J	N	J
Phosphatidylcholine	44.6	46.6	44.6	46.6	43.3	45.2	41.9	45.7	41.0	45.2
Choline plasmalogens	1.0	0.9	0.8	1.0	1.0	1.0	1.0	0.9	1.2	1.0
Phosphatidylethanolamine	25.2	25.0	26.8	26.1	24.6	25.1	21.7	25.7	23.2	25.4
Ethanolamine plasmalogens	10.3	8.8	9.6	8.1	11.8	8.8	14.4	8.8	14.1	8.4
Phosphatidylserine	8.3	8.0	8.8	7.5	9.0	8.4	9.1	8.0	9.4	8.7
Sphingomyelin	3.9	3.6	4.4	4.1	4.3	3.9	4.8	4.3	5.2	4.2
Phosphatidylinositol	3.7	4.0	4.2	3.8	3.5	3.6	3.9	4.6	3.7	3.9
Cardiolipin	2.4	2.6	1.9	2.5	2.1	2.9	2.4	3.1	2.1	2.8

Values are the means of 2-4 experiments.
N, normal mice
J, Jimpy mice

TABLE 2

Percent distribution of phospholipids and galactolipid/phospholipid ratio in subcellular fractions of normal and Jimpy mice brains

Lipids	SUBCELLULAR FRACTIONS									
	Myelin		Microsomes		Mitochondria		Synaptosomes		Supernatant	
	N	J	N	J	N	J	N	J	N	J
			PHOSPHOLIPIDS Percent of total lipid phosphorus							
PE	43.7	36.3	33.9	32.4	35.2	36.2	33.8	31.6	25.3	28.0
PC	34.4	39.2	41.7	44.1	41.5	40.8	40.1	39.8	48.0	46.4
SPH	4.7	4.5	3.5	3.9	3.7	3.8	4.5	4.5	4.4	3.9
PS	11.0	14.5	12.8	12.4	5.2	5.1	10.0	10.4	10.4	11.7
PI	2.8	3.1	4.5	4.7	3.7	3.8	4.9	5.3	7.2	6.9
PA	1.4	0.6	0.6	0.7	—	—	0.4	0.4	0.7	0.5
CL	—	—	—	—	6.7	6.7	1.6	1.9	0.3	—
LPC	—	—	—	—	2.6	2.3	2.0	2.4	2.4	2.0
X	—	—	—	—	—	—	1.3	1.5	—	—
Origin	2.0	1.8	2.9	1.7	1.2	1.0	1.5	2.2	1.2	0.9
C/PHT mM/m	282	3.02	20.4	1.49	—	—	—	—	—	—
SUL/PHT mM/m	57.6	1.0	6.25	2.32	—	—	—	—	—	—

Abbreviations:

PE	: phosphatidylethanolamine	PA	: phosphatidic acid
PC	: phosphatidylcholine	CL	: cardiolipin
SPH	: sphingomyelin	LPC	: lysophosphatidylcholine
PS	: phosphatidylserine	X	: unknown substance
PI	: phosphatidylinositol	C/PHT	: cerebrosides/phospholipids
J	: Jimpy mice	SUL/PHT	: sulphatides/phospholipids
N	: Normal mice		

FIGURE 3

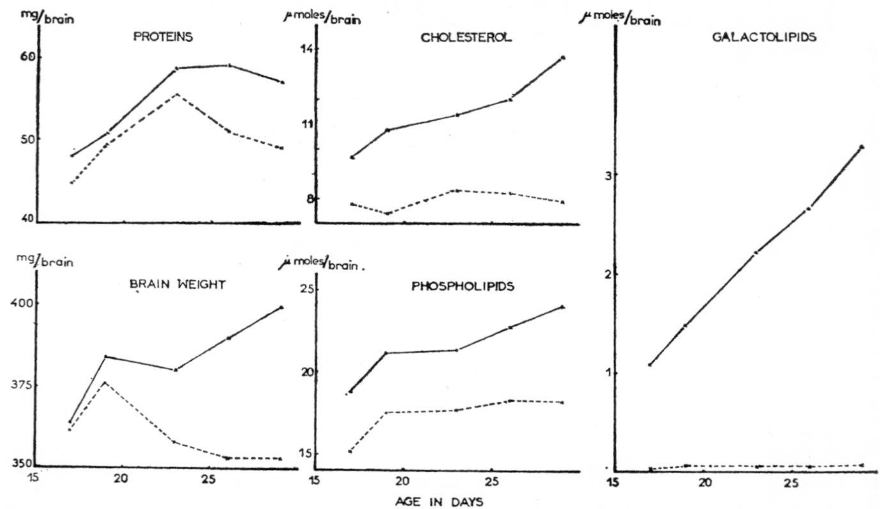

Wet weight, protein and lipid contents of normal (————) and Jimpy (-----) mouse brains as a function of age.

TABLE 3

Ganglioside Distribution in Normal, Quaking and Jimpy Mouse Brains

	25 days		28 days		60 days	
	N	Q	N	J	N	Q
Total NANA µg/wet wt.	650	620	545	575	550	531
GM2	3.7	3.9	4.6	4.2	14.3	10.2
GM1	12.1	7.1[a]	12.1	8.0[a]		
			7.2	7.9		
GD3	37.7	37.1			38.1	37.1
GD1a			30.6	30.3		
GD2	8.5	9.1	10.2	11.4	10.5	10.4
GD1b	12.0	12.2	12.6	12.6	12.8	12.9
GT1	19.0	18.6	15.3	15.4	15.3	14.9
GQ1	5.8	6.6	4.9	4.6	4.9	5.0
Origin	2.2	4.0	2.7	3.4	3.6	4.1

[a]$p < 0.01$. The values (means of 4-6 experiments) are expressed as percent of total NANA chromatographed.

N, normal mice

Q, Quaking mice

J, Jimpy mice

NANA, acetylneuraminic acid.

which are present only at very low levels in the mutants at any stage. Less pronounced alterations in cerebroside and sulphatide content were also found in MSD mutants. Gangliosides which are mainly neuronal compounds are present in quite normal amounts in the mutants except for the ganglioside species GM1 which is slightly decreased (Table 3) (Kostic *et al.*, 1969, 1970; Mandel *et al.*, 1971). It has been suggested by Suzuki *et al.*, (1967) that this ganglioside exists in myelin. The lower amount of this ganglioside in the myelin deficient mutants supports strongly this hypothesis of the presence of GM1 gangliosides in myelin. But since the reduction in this ganglioside is less marked than that in cerebrosides and sulphatides, it suggests that GM1 is not exclusively localized in myelin nor produced exclusively in oligodendroglia in contrast to the galactolipids.

Long chain unsaturated and hydroxy fatty acids are cerebroside constituents. A reduction in the level of these fatty acids has been reported in Quaking mice (Baumann *et al.*, 1968) and we have observed a similar reduction in Jimpy mice (Table 4) (Nussbaum *et al.*, 1971). This suggests that the pathway of fatty acid chain elongation is disturbed in the mutants.

TABLE 4

Non-Hydroxylated Fatty Acid Composition of the Myelin Sphingolipids of Control and Jimpy Mice (percent of total).

Fatty Acids	Cerebrosides		Sulphatides		Sphingomyelin	
	C	J	C	J	C	J
Σ16-18 : 1	10.67	65.00	21.00	91.80	59.00	92.84
Σ19-24 : 1	89.33	35.00	79.00	8.20	41.00	7.16

C = Control Mice J = Jimpy

BIOSYNTHESIS OF MYELIN-SPECIFIC MOLECULES

Since the main effect was on the levels of cerebrosides and sulphatides it was likely that the genetic control of the synthesis of these myelin specific compounds was altered.

Biosynthesis of cerebrosides

From the data concerning the myelin and gangliosides it appears that the enzymic equipment for the synthesis of sphingosine and ceramide is unimpaired in the mutants, although we cannot exclude the possibility that ceramide synthesis in oligodendrocytes is blocked in view of the reduction in the level of GM1. Thus we focused our attention on the transfer of galactose necessary for the synthesis of galactolipids (Fig. 4) (Neskovic *et al.*, 1969, 1970; Mandel *et al.*, 1971).

Two possible acceptors for galactose during cerebroside synthesis have been suggested: sphingosine and ceramide. The UDP-galactose (sphingosine or ceramide) galactosyl-transferase which carries out this reaction, increases

FIGURE 4

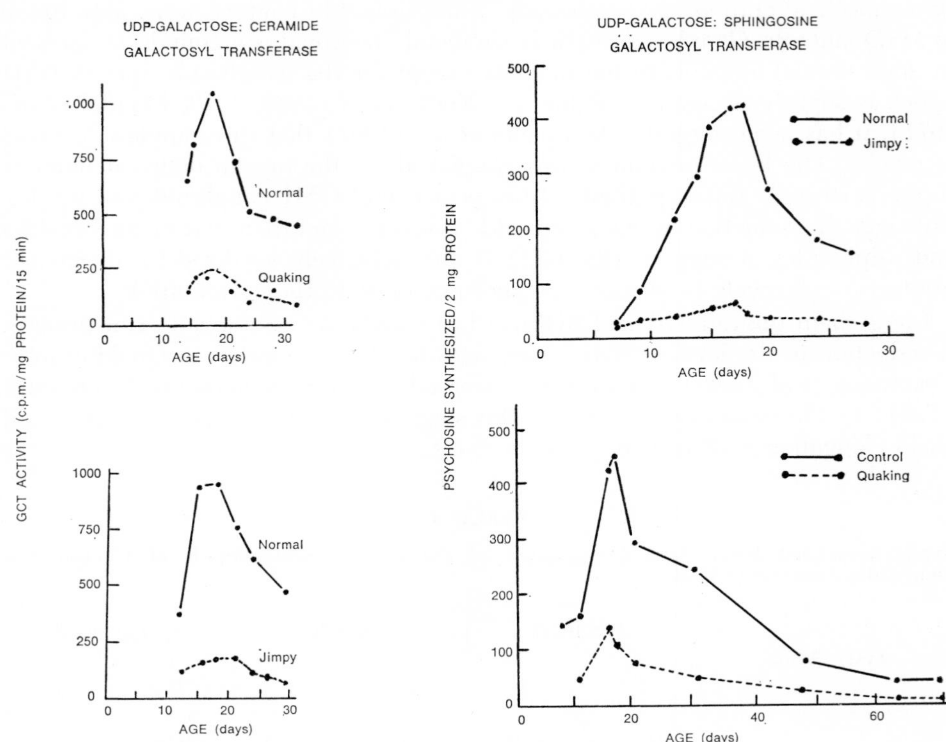

Galactosyl transferases in brain of two neurological mutants of mouse.
UDP-galactose ceramide galactosyl transferase in Quaking (●-----●) with corresponding normal mice (●———●) and in Jimpy (●-----●) with corresponding normal mice (●———●). UDP-galactose sphingosine galactosyl transferase in Jimpy (○-----○) with corresponding normal mice (●———●) and in Quaking (○-----○) with corresponding normal mice (●———●).

sharply during the myelination period then decreases later in control animals. The same pattern was observed in the Jimpy and the Quaking mice although the levels of enzyme activity reached were very much lower. Lower activities of this enzyme were also found in MSD mutants (Sarliève *et al*, 1971).

The alteration of the enzyme activity paralleled the degree of myelin deficiency. By performing crossed incubations the possibility was ruled out that the changes observed during myelination in control animals and the differences between control mutants were due to the effect of activation or inhibition by some effectors. (Mandel *et al.*, 1972). In the peripheral nerves of the mutants, the galactosyl-transferase activities were close to normal (Table 5).

Biosynthesis of sulphatides

Two possibilities for the synthesis of sulphatides in normal and mutant mice were investigated: the transfer of sulphate to psychosine and to cerebrosides (Sarliève *et al.*, 1971; Nussbaum and Mandel, 1972). As for the activity of UDP-

TABLE 5

Glycolipid Synthesizing Enzymes in Peripheral Nerves of Two Myelin Deficient Mutants (Quaking and Jimpy) of mice.

Tissue and Mutant Strain	Age Days	Enzyme Activity (c.p.m./h/mg Protein)					
		PAPS-CST		S Gal T		C Gal T	
		Control	Mutant	Control	Mutant	Control	Mutant
Sciatic Nerve Quaking	30	3840	3700	—	—	—	—
Sciatic Nerve Quaking	46-50	—	—	1715	1130	1725	1280
Trigeminal Ganglia Quaking	30	4100	4280	—	—	—	—
Trigeminal Ganglia Jimpy	20-25	—	—	132 87	140 75	—	—

galactose (sphingosine or ceramide) galactosyl-transferase, PAPS-psychosine sulphotransferase and PAPS-cerebroside sulphotransferase develop during brain ontogenesis following a bell-shaped curve (Fig. 5).

FIGURE 5

CHANGES WITH AGE IN PAPS-CST ACTIVITY

OF NORMAL AND MUTANT MICE BRAINS

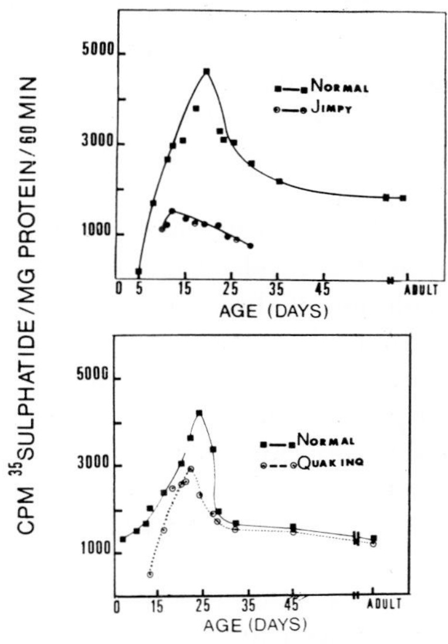

3′-phosphoadenosine 5′-phosphosulphate cerebroside-sulphotransferase (PAPS-CST) in brain of Jimpy (●———●) and normal (■———■) mice and in brain of Quaking (○- - - - -○) and normal mice (■———■).

In the mutants the values for these enzymes are close to normal during the premyelination period and much lower than normal during the period of intense myelination. The possibilities of an activating or inhibitory effect rather than changes of enzyme synthesis were again ruled out by performing crossed incubations of extracts from normal mice at different ages and from Jimpy and Quaking mice.

The sulphotransferase activities which during the premyelination period are similar in normal mice and mutants, but which differ markedly during myelination, can be interpreted in terms of two localizations of these enzymatic activities: one independent of the myelination process (perhaps neuronal or astrocytic), the other related to myelin formation (oligodendroglial). It is interesting to note that the PAPS-cerebroside sulphotransferase activity is normal in the kidneys of all the mutants. In the peripheral nerves of the mutants the sulphotransferase activities were also normal (Table 5).

DISCUSSION

The results show that mutant mice lack four enzymic activities involved in cerebroside and sulphatide synthesis (UDP-galactose sphingosine galactosyl-transferase, UDP-galactose cerebroside galactosyl-transferase, PAPS-psychosine sulphotransferase and PAPS-cerebroside sulphotransferase). The difference in behaviour of the galactosyl-transferases during purification and in the presence of different detergents are in favour of the existence of two different enzymes. Also the enzymic system producing long chain unsaturated and hydroxy fatty acids is disturbed as is the enzyme 2′, 3′-cyclic AMP 3′-phosphohydrolase. Crossed incubations suggest strongly a decrease in the rate of biosynthesis, although a definite answer will be obtained only when these enzymes are isolated and their identity immunologically confirmed. Furthermore Kandutsch and Saucier (1969) have shown that hydroxymethylglutarylacetyl coenzyme A reductase, the rate limiting enzyme for cholesterol biosynthesis in brain, has a much lower activity during the myelination period in Jimpy mice although premyelination activities are close to normal.

These mutants are also deficient in myelin specific-proteins, primarily proteolipids (Fig. 2) as well as in the myelin-localized enzyme 2′, 3′-cyclic AMP 3′-phosphohydrolase. The evolution of the different enzymic activities as well as of the proteolipid myelin proteins in normal mice and in the mutants, suggests strongly that the biogenesis of myelin is a coordinate phenomenon involving a whole series of enzymes or proteins that are transcribed and translated in parallel during a determined period of ontogenesis. This type of regulation suggests induction of a myelin enzyme operon or the possibility of a sequential transcription where the synthesis of one enzyme or its products regulates the transcription of enzymes involved in later stages of myelin biosynthesis. However there is no compelling evidence that the decrease of elongation of fatty acids is the primary defect in these mutants, as has been suggested by Baumann *et al.*, (1968).

It seems difficult to imagine that a mutation occurs at the same time in at least six or seven different structural genes, particularly since preliminary studies of the properties of the enzymes synthesized at a lower rate in the mutants show that they have physico-chemical properties (the Km and the pH optimum) similar to those of the normal mice. It seems more likely that the mutation is due either to an alteration of a regulatory gene, or to an impairment of oligodendroglial proliferation and differentiation.

We have observed under the light microscope the abnormalities of the central nervous system already described by Sidman, *et al.*, (1964). Moreover, we observed (Farkas, *et al.*, 1972) a lower density of glial cells, defective maturation of the oligodendrocytes, and defective positioning of these cells which failed to line up in rows. Differences in the glial populations of Jimpy and control mice become striking with age, and are especially marked in the regions which myelinate later. In the Jimpy mice, mature oligodendrocytes are almost completely absent from the corpus callosum, in 29 day-old animals. The scarcity of oligodendrocytes in the Jimpy was confirmed by histochemical methods.

Ultrastructural studies (Farkas *et al.*, 1970; Privat *et al.*, 1972) have shown that there is virtually no myelination in the Jimpy, and that those rare myelin sheaths which are observed are greatly reduced in size. In the Quaking mutants although some abnormalities in oligodendrocytes, such as vacuole-formation, have been observed, the number of oligodendroglia does not seem to be decreased. Thus it seems likely that in the Jimpy mutation the mutated gene controls the proliferation and maturation of oligodendroglia. The deficient myelination is a consequence of the deficient proliferation and maturation of oligodendroglia. This does not seem to be the case in the Quaking mutant where the defective gene is localized on another chromosome, although the resulting enzymic deficiencies are similar.

In summary, it may be stated that we are dealing with a situation in which enzyme activities absent due to a mutation in brain, are nevertheless present in the peripheral nervous system and in some other tissues. The genetic lesion in the Jimpy and MSD mutants appears to specifically affect the oligodendrocytes of the central nervous system. Both these mutants are sex-linked and the Jimpy mutation is known to be on chromosome XX. The Quaking mutant, an autosomal recessive mutation on chromosome IX, causes deficient myelination without marked effects upon the proliferation and migration of the oligodendrocytes. Moreover it would appear that there is segregation of control of myelination in the central and in the peripheral nervous systems and in addition myelination can be affected by mutations of several chromosomes and may be subject to very complex genetic control.

REFERENCES

Baumann, N. A., Jacque, C. M., Pollet, S. A. and Harpin, M. L. (1968) *J. Neurochem. 17,* 545

Drummond, G. I., Iyer, N. T. and Keith, J. (1962) *J. Biol. Chem. 237,* 3535

Farkas, E., Robain, O. and Mandel, P. (1972) *Acta Neuropathologica,* in press

Farkas, E., Zahnd, J. P., Nussbaum, J. L. and Mandel, P. (1970) in *Les Mutants Pathologiques chez l'Animal* (Sabourdy, M., éd.), n° 924, p. 21, Colloques du Centre National de la Recherche Scientifique, Paris

Folch, J. and Lees, M. (1951) *J. Biol. Chem. 191,* 807

Kandutsch, A. A. and Saucier, S. E. (1969) *Arch. Biochem. Biophys. 135,* 201

Kostic, D., Nussbaum, J. L. and Mandel, P. (1969) *Life Sci. 8,* Part II, 1135

Kostic, D., Nussbaum, J. L. and Neskovic, N. (1970) *Acta med. iug. 24,* 20

Kurihara, T., Nussbaum, J. L. and Mandel, P. (1969) *Brain Res. 13,* 401

Kurihara, T., Nussbaum, J. L. and Mandel, P. (1970) *J. Neurochem. 17,* 993

Kurihara, T. and Tsukada, Y. (1967) *J. Neurochem. 14,* 1167

Luse, S. A. (1956) *J. Biophys. Biochem. Cytol. 2,* 777

Mandel, P., Neskovic, N. M. and Nussbaum, J. L. (1971) *C. R. Soc. Biol. 165,* 474

Mandel, P., Nussbaum, J. L. Neskovic, N. M., Sarliève, L. L. and Kurihara, T. (1972) *Advances in Enzyme Regulation* (Weber, G., ed.), vol. 10, p. 101, Pergamon Press, New York

Meier, H. and McPike, A. D. (1970) *Exptl. Brain Res. 10,* 512

Mokrasch, L. C. (1969) in *Handbook of Neurochemistry* (Lajtha, A., ed.), vol. 1, p. 171, Plenum Press, New York

Neskovic, N. M., Nussbaum, J. L. and Mandel, P. (1969) *FEBS Lett. 3,* 199

Neskovic, N. M., Nussbaum, J. L. and Mandel, P. (1970) *FEBS Lett. 8,* 213

Nussbaum, J. L. and Mandel, P. (1972) *J. Neurochem. 19,* 1789

Nussbaum, J. L. and Neskovic, N. (1971) *3rd Intern. Meeting Neurochem. Budapest, Abstr.* p. 82

Nussbaum, J. L., Neskovic, N. M., Kostic, D. M. and Mandel, P. (1968) *Bull. Soc. Chim. Biol. 50,* 2194

Nussbaum, J. L., Neskovic, N. and Mandel, P. (1969) *J. Neurochem. 16,* 927

Nussbaum, J. L., Neskovic, N. and Mandel, P. (1971) *J. Neurochem. 18,* 1529

Privat, A., Robain, O. and Mandel, P. (1972) *Acta Neuropathologica,* in press

Sarliève, L. L., Neskovic, N. M. and Mandel, P. (1971) *FEBS Lett. 19,* 91

Sidman, R. L., Dickie, M. M. and Appel, S. H. (1964) *Science 144,* 309

Sudo, T., Kikuno, M. and Kurihara, T. (1972) *Biochim. Biophys. Acta 255,* 640

Suzuki, K., Poduslo, S. E. and Norton, W. T. (1967) *Biochim. Biophys. Acta 144,* 375

Waehneldt, T. (1971) *Anal. Biochem. 43,* 306

Waehneldt, T. and Mandel, P. (1970) *FEBS Lett. 9,* 209

Wolfgram, F. and Rose, A. S. (1961) *J. Neurochem. 8,* 161

Zanetta, J. P., Benda, P., Gombos, G. and Morgan, I. G. (1972) *J. Neurochem. 19,* 881

Gene Expression in Mitochondria and Chloroplasts

The Phenomenology of Cytoplasmic Genetics in Yeast: A Proposal for an Autonomy of Mitochondrial Membranes and the Determinism of Nucleo-cytoplasmic Genetic Interactions

ANTHONY W. LINNANE, C. L. BUNN, NEIL HOWELL, P. L. MOLLOY and H. B. LUKINS

Biochemistry Department, Monash University, Clayton, Victoria, Australia

Our laboratory has been interested for some years in the study of the biogenesis of mitochondria, particularly in the yeast *Saccharomyces cerevisiae*.

In recent years mitochondria have assumed increasing importance in studies of development and differentiation, following the recognition that the organelles possess some degree of genetic and synthetic autonomy. The formation of such a complex organelle as the mitochondrion involves the synthesis and coordinate assembly of a myriad of proteins, lipids, nucleic acids and small molecules, each with a specific spatial localization within the matrix and the two membrane systems of the organelle. The mitochondrion has very limited autonomy, only about 10% of its total protein is formed by the organelle. Its formation is under the control of both nuclear and cytoplasmic (mitochondrial) genetic systems which in addition to their direct effect on the mitochondrial localized systems also invoke the activities of non-mitochondrial enzyme systems and other organelles to contribute to the biosynthesis of the mitochondria. Biochemically the biogenesis of mitochondria is an immense problem; in essence, it is the study of the synthesis of an organelle whose detailed composition and molecular organization are at present largely unknown and a concerted attack employing a wide range of expertise and methodology is required.

In this paper aspects of cytoplasmic genetic phenomenology and nuclear-cytoplasmic interactions as they bear on the problem of mitochondriogenesis are discussed. Firstly some of the characteristics of the mitochondrial cytoplasmic genetic system in yeast and the role of mitochondrial DNA will be considered. The second part of the paper will deal with an analysis of mutants which are incompletely understood but which lead us to a very tentative proposal or concept of membrane determinism and gene expression. Cell differentiation is largely traceable to differences in genic expression; and during this symposium we have heard about differential genic activity as consisting of differential RNA and protein synthesis together with a variety of other factors such as hormones, chromosomal proteins and sigma factors, which influence these activities. Another ingredient of the process has been proposed by Sonneborn (1963, 1970) in which it may be envisaged that the assembly of genic products into organized structures is self-determined by the intrinsic macromolecular conformation or configuration of the particular structure. This concept should not be confused with that of self assembly which consists merely of random collision and union of concurrent products of genic action, such as, for example, aggregation of protein monomers into isoenzymes.

LIFE CYCLE AND GENETIC SYSTEMS OF *SACCHAROMYCES CEREVISIAE*

The characteristics of the yeast cell and its life cycle present an almost unique system for the genetic analysis of mitochondrial formation. The basic nuclear genetics of *Saccharomyces cerevisiae* is now well understood and the organism is readily manipulated by most of the techniques applied in the study of bacterial genetics. Perhaps the most outstanding attribute of yeast for genetic studies of mitochondrial development is the organism's capacity to dispense with functional mitochondria and to grow by the fermentation of carbohydrate substrates; any mutations resulting in the loss of important mitochondrial functions are hence not lethal to this organism. Further, the ability of yeast to grow on both fermentable and non-fermentable substrates provides an easy means for detecting mutations affecting such mitochondrial functions as respiration or energy conservation. Any cell-isolate able to grow on fermentable substrates but not on non-fermentable substrates, is taken as a presumptive mutant with defective mitochondrial function.

The life cycle of *Saccharomyces cerevisiae* consists of stable haploid and diploid phases (Fig. 1) a feature having a number of advantages both for genetic analysis and for the isolation of either nuclear or cytoplasmic mutations. Haploid cells, which can be either of "a" or "α" mating type, can multiply asexually by a process of budding. Under conditions where cells of both mating types are mixed, cells of opposite mating type can fuse to form a diploid zygote in which mixing of the cytoplasms of the two haploid parents takes place. Diploid cells can also multiply asexually by budding or can be induced to sporulate, a process involving a classis meiotic division and yielding four haploid ascospores contained within an ascus. The individual ascospores can be freed from the ascus sack, separated and germinated to regenerate the vegetative haploid phase. In this sexual cycle, any single pair of nuclear alleles segregate 2:2 in a tetrad ascus; for instance the mating type alleles "a" and "α" segregate to give two ascospores of "a" mating type and two of "α" mating type. The nuclear genetical system of yeast has recently been reviewed by Mortimer and Hawthorne (1969).

Segregation ratios of 2:2 for alleles describe the basic rules of inheritance by

FIGURE 1

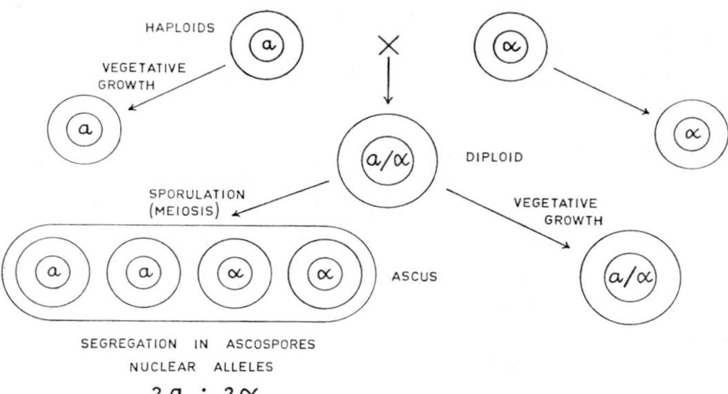

A cross between two haploid strains of opposite nulcear mating type (a/α) is shown. The subsequent segregation of these nuclear alleles during meiosis or vegetative growth is illustrated.

the nuclear or Mendelian genetic system of yeast. A complete description of the phenomenon of cytoplasmic inheritance in yeast has yet to be established, but several essential characteristics which distinguish this mode of inheritance from nuclear genetics are established.

THE CYTOPLASMIC GENETIC SYSTEM OF YEAST

1. General Properties

Some of the fundamental properties of the cytoplasmic genetic system were established over two decades ago by Ephrussi and colleagues in their studies of the cytoplasmic petite mutation and these are outlined in Fig. 2 (Ephrussi, 1953 for review). These characteristics are best explained by a description of the inheritance of these cytoplasmic mutants. Phenotypically the petite mutation was characterised by a loss of the mitochondrial cytochromes a, a_3, b and c_1, rendering the cells respiratory deficient, they are thus unable to utilize nonfermentable substrates, so that the colonies remain small on glucose media. Ephrussi and colleagues first reported that when a haploid respiratory competent cell (denoted grande and symbolized by ρ^+) is crossed with a haploid petite (symbolized by ρ^-), only respiratory competent diploids were observed and further that on sporulation and germination of the four haploid ascospores, all were found to be respiratory competent ($4\rho^+ : 0\rho^-$). Study of various back-crosses together with the work of Wright and Lederberg (1956) unequivocally established that the respiratory status of the progeny depended on extranuclear inheritance (for review, Linnane and Haslam, 1970).

2. The Phenomenon of Suppressiveness

The initial cytoplasmic petites described by Ephrussi were later shown by the same investigator to be a special class of petite (Ephrussi *et al.*, 1955, Ephrussi and Grandchamp, 1965).

FIGURE 2

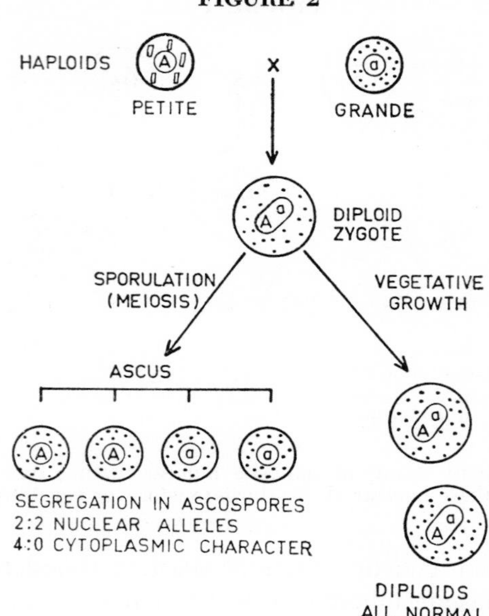

A cross between haploid petite and haploid grande strains is depicted. The dots represent the grande cytoplasmic determinants while the open rectangles represent the petite determinants. Note for this petite that none of its cytoplasmic determinants are transmitted to the diploids. A/a represent the genetic behavior of any pair of nuclear alleles.

Continued study of a series of cytoplasmic petites revealed that unlike the first mutants described, crosses between most petites and normal grande cells give rise to a percentage of both respiratory competent and deficient diploid cells. The imposition of respiratory deficiency on a fraction of the diploids in a cross between ρ^+ and ρ^- cells has been denoted suppressiveness and such petite cells denoted suppressive. The phenomenon is illustrated in Fig. 3. The essential characteristic of suppressiveness is that three types of zygotic clones are obtained following a cross. The zygote is initially unstable, most of the single zygotes give rise to both stable respiratory competent and stable respiratory deficient diploid cells as well as to further unstable diploids (Ephrussi *et al.*, 1966). The transitory state of the initial zygotes is further evidenced if sporulation is studied. Established respiratory deficient diploids cannot be sporulated. However if zygotes formed by crossing a suppressive petite and a normal cell are immediately induced to sporulate and tetrad analysis performed, both 0:4 and 4:0 segregations for respiratory competence: respiratory deficiency are obtained. On the other hand, if the induction of sporulation is delayed, the established respiratory deficient diploids do not sporulate and only 4:0 segregation for respiratory competence: respiratory deficiency are observed.

The degree of suppressiveness varies with the individual petite so that some petite cells may approach 100% suppressiveness, meaning that close to 100% of the diploid zygotes give rise to petite colonies. Conversely suppressiveness may

FIGURE 3

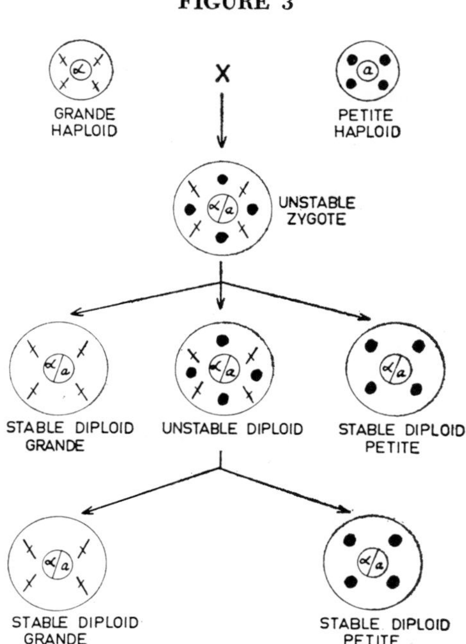

A cross between a suppressive petite haploid strain and a haploid grande strain is shown. Note that the petite cytoplasmic determinants (●) are transmitted into the diploids as are the grande cytoplasmic determinants (×). Unstable diploids contain both types of determinants; these diploids may distribute their cytoplasmic determinants to daughters in such a manner as to produce stable grande diploids, stable petite diploids or other unstable (mixed) diploids.

approximate 0% in which case, none of the diploids which arise are petites; such an initial haploid petite is then denoted a neutral or 0% suppressive and would correspond to the original petites described by Ephrussi and colleagues. The molecular basis for the phenomenon of neutrality in petites is discussed later.

3. Antibiotic Resistant Mutants

Progress in cytoplasmic genetics has been severely handicapped by the lack of genetic markers; indeed prior to 1967 the petite mutation was the only cytoplasmic mutation known that affected mitchondria. While the petite mutation has contributed greatly to our understanding of the mitochondrial genetic system it is far from ideal for genetic investigation, as established respiratory deficient diploid cells cannot be sporulated. An extensive class of mutants first discovered and characterised by Linnane and colleagues exhibit altered resistance to certain specific antibiotics. (Clark-Walker and Linnane, 1966, 1967; Linnane, *et al.*, 1968a; Linnane, *et al.*, 1968b). It was shown that the mitochondrial protein synthesizing system could be differentiated from that of the cytoplasmic ribosomes both *in vivo* and *in vitro,* by its sensitivity to a number of antibiotics (chloramphenicol, erythromycin, mikamycin, oleandomycin, carbomycin, spiramycin, lincomycin). *In vivo,* these antibiotics inhibit the synthesis of the cytochromes *a*, a_3, *b* and c_1, producing a near phenocopy of the cytoplasmic petite. Mutants resistant to the antibiotics are obtained by plating a lawn of sensitive cells on glycerol or ethanol media containing the appropriate antibiotic; only resistant cells will grow under these conditions, as a functional respiratory system is required for growth on these nonfermentable substrates. In this manner mutants resistant to a variety of antibiotics can be selected. The initial studies were carried out on mutants selected for erythromycin resistance and these simultaneously showed resistance to the macrolides spiramycin, carbomycin, and oleandomycin as well as lincomycin. In preliminary genetic studies these cross-resistances are found to accompany the erythromycin determinant.

Crosses between erythromycin resistant (ER^r) and sensitive (ER^s) cells revealed that the determinants were cytoplasmic. The essence of the data is illustrated in Fig. 4. The diploid zygote is unstable with regard to the two alleles

FIGURE 4

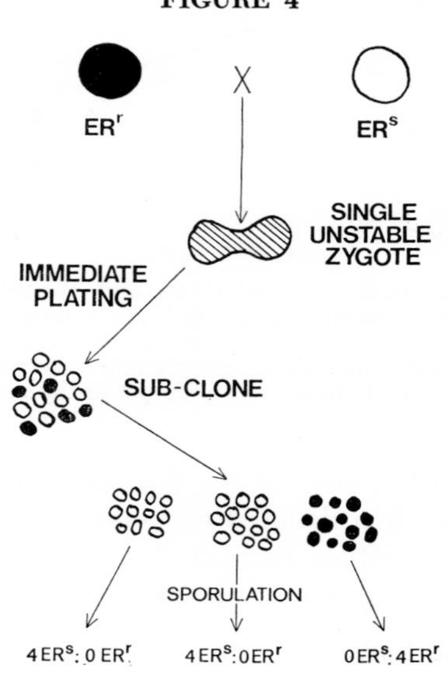

A cross between two haploid strains, one erythromycin sensitive (ER^s) the other resistant (ER^r) is depicted. The diploid zygotes are mixed for this pair of cytoplasmic alleles; immediate plating of the zygotes produced clones mixed for stable pure diploid resistant or sensitive cells which can be isolated by sub-cloning. The purified sensitive or resistant diploids on sporulation segregate 0 ER^r : 4 ER^s and 4 ER^r : 0 ER^s respectively.

ER^s and ER^r. If, immediately after crossing, the zygotes are plated, then the zygotic clones that arise are found to contain a mixture of ER^s and ER^r cells. On subcloning the mixed colonies, only pure ER^s and ER^r colonies are observed and they breed true for ER^s and ER^r. If the ER^r and ER^s cross is made and the diploid zygotes allowed to proliferate for some time (say 24 hours) before prototrophic selection of the diploid progeny is carried out, only pure colonies which are either ER^r or ER^s arise. On sporulation and tetrad analysis of such pure colonies they are found to show the characteristic cytoplasmic behaviour, viz. sensitive diploids show 4 ER^s : $0ER^r$ spore segregation and the converse is true for resistant diploids. The data are analogous to that obtained in a cross between a suppressive petite and a grande.

MITOCHONDRIAL DNA, CYTOPLASMIC GENES AND SUPPRESSIVENESS

Mounolou *et al.* (1966) first showed that the mutation from grande to cytoplasmic petite is frequently accompanied by a change in the buoyant density of mitochondrial DNA, suggesting that mitochondrial DNA is the locus of the cytoplasmic genetic information. A genetic study of the effect of the petite mutation on the retention of the erythromycin determinant emphasizes the non-unitary nature of the cytoplasmic petite mutation. The proof of this claim depends on the recognition that petites can be of three types with respect to the erythromycin determinant. The state of the erythromycin determinant cannot be directly estimated in petite cells but in crosses with appropriate ρ^+ cells and subsequent analysis of the resulting diploids its state can be determined.

The first type is the ρ^-ER^r cell, which after crossing with a ρ^+ER^s cell yields diploids that grow into mixed colonies of ρ^+ER^r and ρ^+ER^s cells. The second is the ρ^-ER^s cell, which after crossing with a ρ^+ER^r cell also yields diploids that grow into mixed colonies. The third type we have denoted ρ^-ER^o, when it is crossed with a ρ^+ER^r cell, only pure ρ^+ER^r diploid colonies arise. The ρ^-ER^o cell thus appears to have lost completely the erythromycin determinant while the petites ϵ^-ER^r retain it in its original form.

Petites induced chemically by treatment of either ER^r or ER^s cells with high concentrations of ethidium bromide are nearly all of the genotype ρ^-ER^o (Saunders *et al.*, 1971). However, petites which arise spontaneously from ρ^+ER^r and ρ^+ER^s frequently belong to the class of ρ^-ER^r or ρ^-ER^s, respectively, only a proportion being ρ^-ER^o. The fact that mutation to cytoplasmic petite does not necessarily lead to a loss of the erythromycin determinant demonstrates that cytoplasmic petites vary in their cytoplasmic information content, that is the cytoplasmic petite mutation is not unitary.

Saunders *et al.* (1971) have explored some of the characteristics of suppressiveness, making use of the erythromycin determinant as a probe and shown that the degree of suppressiveness of freshly isolated petites is influenced by the method of formation of the petite.

Spontaneously arising petites are found to fall in the range of 8-50% suppressive denoted as intermediate suppressiveness (see Linnane and Haslam, 1970, for discussion). The range of suppressiveness of petites induced with low concentrations of ethidium bromide (2 μg/ml of medium) approximates that of the spontaneously arising mutants but those induced by higher and more damaging levels of ethidium bromide (10 μg/ml of medium) are nearly all of low (2-8%) or neutral (0%) suppressiveness with a few being highly suppressive ($>$ 70%), while intermediate suppressiveness is observed only as a comparatively rare event. In correlating the degree of suppressiveness of a petite to ER^r retention it emerged that the petites of intermediate suppressiveness commonly retained the erythromycin determinant. However petites of very low and high suppressiveness rarely retained the ER^r gene. The detailed studies lead to the consideration that very high and very low suppressive petites contain very little cytoplasmic information content and further that the neutral petites might contain the most altered mitochondrial DNA (m-DNA) (Saunders *et al.*, 1971).

This suggestion required that the petites of 0% suppressiveness have m-DNA of such abnormal structure compared with grande cells that the petite mitochondrial DNA has no competitive survival for selection into progeny cells in the presence of normal m-DNA. An alternative and intellectually more satisfying possibility for the true neutral petite, which apparently transfers no cytoplasmic information, would be that it has no m-DNA. This indeed proved to be the case, an examination of a number of carefully prepared 0% suppressive petites failed to reveal the presence of detectable m-DNA in preparations either extracted from whole cells or from isolated mitochondria (Nagley and Linnane, 1970; Nagley and Linnane, 1972). Marmur and his colleagues simultaneously reported the discovery of DNA-less petites (Goldring *et al.*, 1970). The characterization of a 0% suppressive (neutral) petite as a cell devoid of m-DNA provides a ready rationalization of the phenomenon of neutrality. Thus in a genetic cross between a neutral petite and a wild-type cell, only wild-type diploids result as the neutral petites contribute no m-DNA. It may be further added that the absence of m-DNA from neutral petites has important biochemical implications; it indicates that those components of the mitochondrial organelle still present in the neutral petite are the gene products of the nucleus.

We have symbolized the m-DNA-less petite as ρ^o; three possible petite genotypes may be written in regard to any other single cytoplasmic determinant e.g. ρ^-ER^r (a petite containing m-DNA and the erythromycin gene), ρ^-ER^o (a petite containing m-DNA but has lost the erythromycin gene), ρ^oER^0 (a petite devoid of m-DNA and all mitochondrial cytoplasmic genetic information).

CYTOPLASMIC SEGREGATION AND RECOMBINATION

A complication of mitochondrial genetics is that there are many mitochondria per cell and thus there are a number of cytoplasmic chromosomes per haploid cell. The effective number of mitochondrial genomes per yeast cell is unknown although estimates varying from 1 to 20 have been made by target analysis of petite induction kinetics (for review see Linnane and Haslam, 1970).

The transfer of mitochondrial genetic information during cell division occurs by transmission to buds of mitochondria and associated mitochondrial DNA in a manner presumed (but not known) to be a random process. The occurrence of more than one mitochondrial genotype among the progeny of single zygotes indicates that more than one mitochondrial genome can be transmitted into such bud cells. Thus experiments on the segregation of genes for resistance and sensitivity in the progeny of the cross ER^r and ER^s have shown that at least as many as 6 successive buds taken from a single zygote are mixed for ER^r and ER^s and further that at least second and third generation buds of the primary bud can also contain a mixture of mitochondrial genotypes (Saunders *et al.*,

1971; Lukins, *et al.*, 1972). The persistence of the mixed cytoplasm in zygotes and zygote progeny for a number of cell divisions indicates that several cytoplasmic genomes can be involved in transmission from a mother cell to a newly formed bud. Similarly, the instability of suppressive petite cells (Ephrussi and Grandchamp, 1965; Saunders *et al.*, 1971) is indicative of multiple molecules of mitochondrial DNA being inherited from mother cell to bud (Nagley and Linnane, 1972).

MITOCHONDRIAL RECOMBINATION

The existence of two mitochondrial genotypes in zygotes and some zygotic buds creates a situation in which recombination of mitochondrial genomes would occur. Indications that mitochondrial recombination occurs were obtained from the generation of ρ^+ER^s cells in the cross $\rho^-ER^s \times \rho^+ER^r$ (Linnane, *et al.*, 1968b) and of ρ^+ER^r cells in the reciprocal cross $\rho^-ER^r \times \rho^+ER^s$ (Gingold *et al.*, 1969).

Subsequently, more evidence for recombination of mitochondrial genomes was obtained from bifactorial crosses involving two mitochondrial antibiotic resistance genes. Thomas and Wilkie (1968) reported the isolation of cytoplasmically inherited mutants resistant to paromomycin (PA^r) and from crosses of the type $ER^sPA^r \times ER^rPA^s$ they obtained recombinant cells of the type ER^sPA^s and ER^rPA^r in unequal proportions. In a more extensive study, Coen *et al.* (1970) have investigated recombination between erythromycin resistance genes and cytoplasmic genes for resistance to chloramphenicol, again, non-reciprocal proportions of recombinant types were observed.

A distinguishing feature of mitochondrial recombination in yeast is a polarity of both transmission and recombination of mitochondrial genomes in crosses between two haploid cells (Bolotin *et al.*, 1971). This polarity of transmission is observed both as the predominance among the diploid progeny of the mitochondrial genotype of the a parental strain (Coen *et al.*, 1970; Saunders *et al.*, 1971) and polarity of recombination as the production of non-equivalent proportions of the two recombinant genotypes (Coen *et al.*, 1970). For example, in a typical cross of the type $aC^rE^s \times {}^{\alpha}C^sE^r$ (where C^r, C^s, and E^r, E^s denote allelic forms of two mitochondrial genes), the percentage distribution of genotypes in the diploid progeny was 3.2% C^rE^s, 74.3% C^sE^r (parental genotypes) and 22.3% C^sE^s, 0.2% C^rE^r (recombinant genotypes) (Coen *et al.*, 1970). More recently Bolotin *et al.* (1971) have shown polarity to be separable from cellular mating type and have proposed that mitochondria may have an autonomous "sex" which they have designated ω^+ and ω^-. Crosses can be performed which, in terms of mitochondrial genomes, are described as heterosexual ($\omega^+ \times \omega^-$) or homosexual ($\omega^+ \times \omega^+$ or $\omega^- \times \omega^-$), and these crosses differ in respect to the production of recombinant genotypes. Thus, in heterosexual crosses, the polarity of transmission of parental genotypes and the polarity of recombination are high, and all recombinant mitochondrial genomes are ω^+. In homosexual crosses, the overall proportion of recombinants is lower than in heterosexual crosses, the recombinants retain the sex of the parental strains, and for the progeny of a large population of individual zygotes, no polarity is observed. The authors draw analogies with the characteristics observed for mitochondrial recombination and the properties of bacterial recombination during crosses of the type Hfr $\times$ F^-. However there are a number of objections which can be raised to the suggestions of Bolotin *et al.* (1971) and over-interpretation of data obtained at this time should be avoided, much is still to be learnt as to the nature of the system. Current results from our laboratory suggest a simple ω^+, ω^- involvement is insufficient to explain polarity (Howell, Lukins and Linnane, unpublished).

UNIQUE CHARACTERISTICS OF CYTOPLASMIC INHERITANCE IN YEAST

The foregoing brief outline of the characteristics of cytoplasmic inheritance can be briefly summarized:

(1) Cytoplasmic determinants distribute on meiosis in tetrads 4:0 or 0:4, only one allele is recovered in a given tetrad.

(2) Single diploid zygotes heteroplasmic for a pair of cytoplasmic alleles vegetatively segregate both alleles, this phenomen is known as cytoplasmic mixedness.

(3) There is an asymmetry of transmission and recombination of genetic information in crosses probably influenced by some intrinsic characteristic of the mitochondria ($\omega+$, $\omega-$) of the two cells involved in the cross.

(4) Cytoplasmic determinants may be readily eliminated by specific mutagens such as ethidium bromide and acridine.

(5) Cells devoid of mitochondrial DNA (ρ^o) and consequently of any mitochondrial genetic information are viable and available for study.

MIKAMYCIN RESISTANCE

Both the petite and a number of antibiotic resistance mutations unequivocally satisfy all the criteria for cytoplasmic genes. Recently we have been investigating the nature of resistance to the antibiotic, mikamycin, an inhibitor of both bacterial and mitochondrial protein synthesis (Bunn *et al.*, 1970, Dixon *et al.*, 1971). Biochemically the mikamycin resistance determinant appears to specify a mitochondrial membrane component. The various types of data leading to this conclusion, a point of importance to this paper, can be found in the papers of Bunn *et al.* (1970) and Mitchell *et al.* (1972). In this part of the paper the nature of the mikamycin resistance determinant is explored; it presents a paradox to us. On the one hand it shows cytoplasmic genetic characteristics and on the other hand apparently nuclear genetic characteristics. This point is illustrated in Fig. 5. Single diploid zygotes formed in the cross, MK^r $\times$ MK^s, give rise to mixed clones which on sub-cloning can be shown to be composed of both stable resistant and sensitive diploid cells. This is clearly the phenomenon of cytoplasmic mixedness. However on sporulation of *either* the resistant or sensitive diploids a 2:2 segregation of MK^r and MK^s is observed in both cases. A further point to be taken from this

FIGURE 5

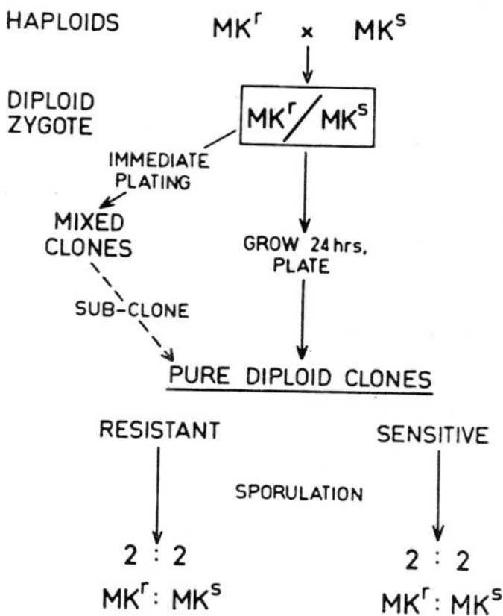

Apparent nuclear and cytoplasmic genetic behaviour by the mikamycin determinant. When a MK^r strain is crossed to a MK^s strain, the zygote diploids are unstable for the determinant and give rise to both pure resistant and pure sensitive diploids. This behaviour is analogous to that of the ER^r determinant. However, when pure diploids of *either* type are sporulated, the tetrads in both cases show nuclear segregation (2 : 2) for mikamycin resistance and sensitivity.

data is that there is apparently no consistent dominance as both resistant and sensitive diploids cells show 2:2 segregation of mikamycin genes.

Fig. 6 illustrates the effect of the petite mutation on retention of the mikamycin determinant. Cells of genotypes ρ^+MK^r on mutation to petite can be divided into three categories based on the occurrence of mitochondrial DNA in them and their reaction in subsequent crosses. These three petite types are ρ^oMK^o, ρ^-MK^o and ρ^-MK^r based on the results of crossing stable petite cells with a ρ^+MK^s cell. In the cross $\rho^oMK^o \times \rho^+MK^s$ there can be no contribution from mitochondrial DNA in the mutant; no mixedness is observed, all diploids are sensitive to mikamycin but on sporulation the haploids ρ^+MK^s and ρ^+MK^r emerge in the ratio 2:2, the same results are obtained from the cross $\rho^-MK^o \times \rho^+MK^s$ (Fig. 6). Based on these two crosses alone, the phenomenon would be explicable simply in terms of a single pair of nuclear alleles. This explanation does not hold, as in the cross $\rho^-MK^r \times \rho^+MK^s$ mixedness is observed. Subsequently on sporulation of stable diploids whether phenotypically sensitive or resistant, nuclear $2MK^r{:}2MK^s$ segregation is observed. If the original haploid ρ^-MK^r is treated extensively with ethidium bromide, elimination of the mikamycin resistant determinant is observed. It therefore follows that mitochondrial DNA has some contribution to make to the expression of the pair of nuclear alleles which are clearly determining mikamycin resistance or sensitivity. Apparently a contribution from the cytoplasm is required for the vegetative expression of mikamycin resistance or sensitivity, although no contribution appears to be required for meiotic segregational expression. We have been investigating considerable numbers of mikamycin resistant and sensitive strains for over two years and we have been confused by the data. A double mutation one nuclear, one mitochondrial does not

FIGURE 6

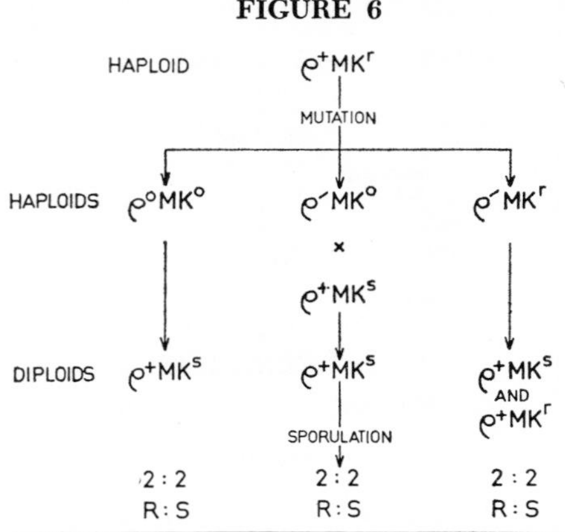

The effect of the petite mutation on retention of the mikamycin determinant. Analogously to the ER^r determinant three types of petites can be isolated from a ρ^+MK^r strain. Two types have lost the MK^r determinant as judged by assay of diploid cells; one type of petite lacks all cytoplasmic information (ρ^oMK^o) while the other possesses m-DNA but has lost the information necessary for expression of MK^r in vegetative diploids (ρ^-MK^o). The third type of petite (ρ^-MK^r) does retain the information necessary for vegetative expression as shown by the formation of mixed diploids. All respiratory competent diploids, formed from the crosses with the three petite types on sporulation yield tetrads showing nuclear segregation of the mikamycin determinant.

appear to provide a satisfactory answer, as all mutants initially isolated would be required to be double mutants; further in that case when the cytoplasmic gene is eliminated by ethidium bromide to become the new mutant $MK^r_{\text{nucleus}}\ \rho^o$ no further appearance of the phenotype mikamycin resistance would be predicted if both DNA systems directly contributed. However, following meiosis, mikamycin resistance does reappear. In considering possible alternate explanations we have turned to the work of Sonneborn.

Sonneborn (1963, 1970) has proposed that highly ordered macromolecular structures may have a degree of self determination. He has shown that isogenic strains of the single cell protozoan, *Paramecium aurelia,* can have distinctly different cortical structures, i.e. they are genetically identical but gross morphological differences in their outer complex membrane structure are observed. By microsurgical procedures he demonstrated that it is possible to remove part of the cortex structure of one paramecium and transfer it to another. The graft takes and can be maintained for many hundreds of cell divisions but during this whole process it will maintain its own distinct morphological characteristics and can clearly be recognized as different from the rest of the cortex of the organism. From studies of this kind and others, Sonneborn has proposed that various regions of the cortex may be considered to have macromolecules arranged in a highly ordered manner in a way which may be considered as a type of microcrystallinity, which is in part self-determining so that products formed by the organism are arranged within the cortex according to pre-existing microcrystallized structures. As a simple extension of this concept, we have been considering the possibility that membranes might also have a degree of autonomy independent of immediate direct genetic determination and from this a substantial model to explain cellular development and differentiation can be proposed.

We hypothesize, and we emphasize that our proposal must be considered a tentative one, that the apparent aspects of cytoplasmic inheritance of MK^r and MK^s is perhaps due to the transmission and segregation of pre-formed resistant or sensitive membranes among progeny cells, the membranes interacting with both cytoplasmic and nuclear gene products; all three needing to be independently and integratively considered (Mitchell *et al.*, 1972).

This proposition predicts that it should be possible to construct genetically identical diploid strains such that each strain has its own particular metastable membrane property (phenotypes). Thus, consider the haploids $MK^r_n\ \rho^+ MIKA^r_{mem}$ and $MK^s_n\ \rho^+ MIKA^s_{mem}$ where MK^r_n and MK^s_n are nuclear alleles whose products make up part of the mitochondrial membrane. These gene products constrain the membrane to a conformation (or microcrystallinity) characteristic of the particular allele, these phenotypes are represented as $MIKA^r_{mem}$ or $MIKA^s_{mem}$ The membrane conformation is such that $MIKA^s_{mem}$ allows mikamycin access to the mitochondrial protein synthesizing system which is inhibited; the converse is true for $MIKA^r_{mem}$.

In a diploid heterozygous for MK^r_n/MK^s_n, two gene products may be formed but the hypothesis predicts that only one product will be inserted into the mitochondrial membrane which one being determined by the pre-existing conformation of the mitochondrial membrane, that is $MIKA^s_{mem}$ preferentially accepts the gene product of MK^s_n and the converse for $MIKA^r_{mem}$ · i.e. the membrane microcrystallinity determines the phenotype of the diploid. Finally, it may be hypothesized as a consequence of this model that under appropriate physiological conditions which lead to particular perturbation of the membrane structure, it should be possible to so alter the membrane conformation (microcrystallinity) as to allow the incorporation of the alternate nuclear genetic allele product with the emergence of the corresponding alternative membrane phenotype. The mitochondrial DNA (ρ factor) in this scheme is not considered to be the repository of mikamycin resistance or sensitive genes. The cross $MK^s_n\ \rho^+ MIKA^s_{mem} \times MK^r_n\ \rho^+ MIKA^r_{mem}$ and its consequences are set out in Fig. 7 which serves to summarize the main points of the hypothesis.

The cross set out in Fig. 7 may then be described as follows: two haploid cells fuse to give rise to a single zygote heterozygous for the nuclear alleles, $MK^r n/MK^s n$ These zygotes are ρ^+ but in addition within the cytoplasm there are resistant and sensitive mitochondrial membranes which originated from the haploids. Cells of this type will give rise to mixed clones. On cytoplasmic purification the membranes which are resistant separate into some of the progeny while the membranes of the sensitive configuration separate into others. The end result is that stable resistant and sensitive diploid cells are formed which are isogenic with one another but are phenotypically either resistant or sensitive to the drug mikamycin. On undergoing sporulation and meiosis the haploid MK^r_n or MK^s_n state is re-

FIGURE 7

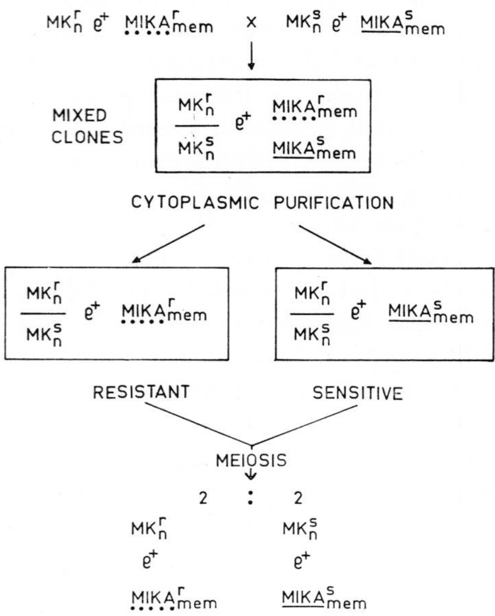

A model to explain the unusual genetic properties of MK^r/MK^s diploids is shown. Both alleles are considered to synthesize a mitochondrial membrane product. In the haploids, the product synthesized is incorporated into the membranes which have a certain conformation denoted as the phenotypes $MIKA^r_{mem}$ and $MIKA^s_{mem}$. The mixedness observed in the MK^r/MK^s diploids results from segregation not of cytoplasmic genes but of preformed mitochondria, one type being sensitive to mikamycin and the other type resistant to mikamycin. Meiosis involves the reestablishment of only one type of mitochondrial membrane, this being directed by the appropriate nuclear allele (MK^r or MK^s).

established and the ratio 2 resistant: 2 sensitive cells is again obtained independently of the original membrane conformation of the original diploid. Extensive membrane re-organization occurs in meiosis and the single nuclear gene product becomes available for possible expression.

Mitochondrial DNA has an important indirect role to play in the organization of the mitochondrial membrane conformation although it may be considered that there are no cytoplasmic mikamycin genes. The fact that ρ^0 cells contribute nothing in crosses to the subsequent vegetative progeny indicates that the mitochondrial DNA has some role to play in the genetic expression of the nuclear alleles, and hence ethidium bromide leads apparently to gene elimination. The ρ^- haploid cells which retain the potential for one of the nuclear alleles to be expressed (behaving analogously to a petite which retains a cytoplasmic gene e.g. $\rho^- ER^r$) are considered to possess mitochondria whose membranes in association with m-DNA can still maintain sufficient conformation to preserve the expression of the mikamycin nuclear allele, hence, mixedness is observed (Fig. 8). Considering the three variables: membrane, cytoplasmic genetic information (mitochondrial DNA) and nuclear determinants we can write the following three haploid genotype-phenotype combinations for haploid petite cells, MK^r_n

FIGURE 8

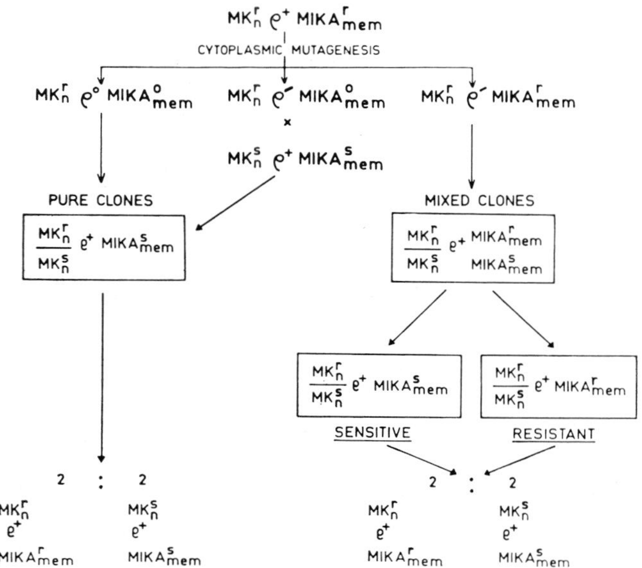

A proposal for membrane determinism and nuclear cytoplasmic genetic interactions is outlined. The data of Fig. 6 have been modified to incorporate the conformation of the mitochondrial membrane, the membrane conformations are represented as a form of cytoplasmic determinant. They may loose their "information" (i.e., resistance to mikamycin) or retain it after petite formation; if lost, only one type of membrane (sensitive) occurs in the diploids. If the information is retained, the two membrane types show the mixed behaviour described in Fig. 7.

ρ^-MIKA$^r_{mem}$, MK^r_n ρ^-MIKA$^o_{mem}$, MK^r_n ρ^o MIKA$^o_{mem}$; the same can be written for those petite cells carrying the nuclear allele, MK^s_n. Thus as set out in Fig. 6 it is envisaged that petite cells of the type MK^r_n ρ^-MIKA$^o_{mem}$ and MK^r_n ρ^oMIKA$^o_{mem}$ in crosses with haploids of the type MK^s_n ρ^+MIKA$^s_{mem}$ can only give rise to heterozygous diploids of the type (MK^r_n/MK^s_n) ρ^+MIKA$^s_{mem}$. However, in a cross of the type MK^r_n ρ^-MIKA$^r_{mem}$ $\times$ MK^s_n ρ^+MIKA$^s_{mem}$ mixed clones arise because the two membrane possibilities still existed at the time of the cross.

Having developed a rather elaborate hypothesis we considered one way to test the idea experimentally, and proceeded to do so as set out in Fig. 9. We reasoned that a diploid of the type (MK^r_n/MK^s_n) ρ^+MIKA$^r_{mem}$ can be manipulated by anaerobic growth, to produce gross mitochondrial membrane changes and thus possibly lead to the disruption of the microcrystallinity inherent in the MIKA$^r_{mem}$ conformation. Having disorganized the membrane microcrystallinity the anaerobic cells could then be aerated and examined for both resistant and sensitive diploid progeny. This experiment succeeded in that it was possible to isolate appreciable numbers of sensitive diploids from the strain without genetic change. As shown in Fig. 9, on sporulation both resistant diploids and sensitive diploids gave the characteristic nuclear 2:2 segregation for mikamycin resistance and sensi-

FIGURE 9

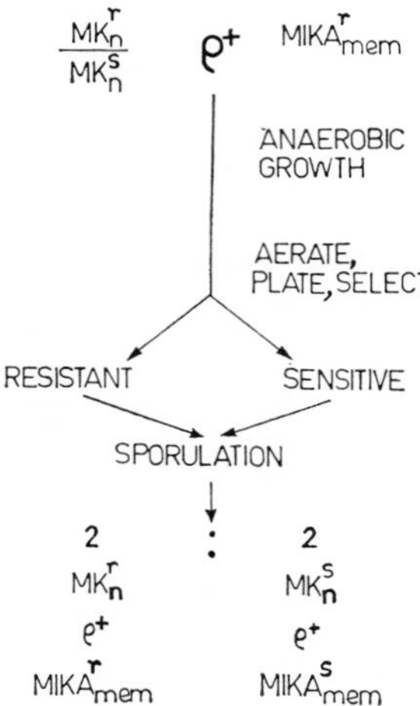

Mikamycin sensitive diploid clones derived from resistant diploid cells. Resistant diploid cells heterozygous for the nuclear mikamycin determinant were grown anaerobically, aerated for 16 hours and then stable sensitive diploid cells were isolated along with resistant ones. Sporulation of the diploids showed no nuclear genetic change had occurred.

tivity. It follows that one of the predictions of our model has been apparently satisfied.

A summary of this part of the paper and a differentiation model is proposed in Fig. 10. The haploid cell $\rho^+MK^r_n$ produces a gene product which we represent as a mitochondrial membrane protein of the R type, conferring resistance of the mitochondria to the effect of mikamycin. The mitochondrial membrane in order to accept the gene product must be in a particular conformation, which we denote R, and on accepting the MK^r_n gene product a conformation of $MIKA^r_{mem}$ is maintained in the presence of a single nuclear allele MK^r_n. Exactly the converse obtains in the haploid $\rho^+MK^s_n$. In a cross between the two aforementioned haploid types, a diploid is formed (MK^r_n/MK^s_n) ρ^+; two membrane proteins the gene products R and S can now be formed.

It is considered that competition between R and S for insertion in the mitocondrial membranes takes place and there is a preferential insertion of R into $MIKA^r$ membranes and S into $MIKA^s$ membranes, one excluding the other but this exclusion is not absolute; the conformation is considered to be metastable. Under certain conditions it is possible to insert one gene product in place of the other and the membrane conformations are altered. A differentiation model can then be constructed in which it is considered that as one gene product is inserted into a given structure, this conformation now acts in turn to discriminate between other possibilities. The established R and S configurations determine which of another pair of heterozygous gene products will be acceptable to the established conformation. The process then repeats itself, the second degree of microcrystallinity determining which of another pair of alleles is accepted into the macromole-

FIGURE 10

GENOTYPE	GENE PRODUCT	MITO MEMBRANE
$\rho^+ MK_n^r$	R MEMBRANE PROTEIN	R CONFORMATION ($MIKA_{mem}^r$)
$\rho^+ MK_n^s$	S MEMBRANE PROTEIN	S CONFORMATION ($MIKA_{mem}^s$)
$\rho^+ \frac{MK_n^r}{MK_n^s}$	TWO MEMBRANE PROTEINS (R, S)	R-S COMPETITION MIXEDNESS AND PURIFICATION

R EXCLUDES S ← META STABILITY → S EXCLUDES R

DIFFERENTIATION —— DE-DIFFERENTIATION MODEL

A summary of the basic features of mikamycin resistance and a simple differentiation model are outlined. Once mitochondrial membranes have incorporated a product of the MK^r (or MK^s) allele, they attain the R (or S) conformation. The membranes of any defined state (R or S) largely exclude the incorporation of the alternate gene product. Details of the metastability and differentiation-dedifferentiation model are elaborated in the text.

cular crystal. This process can be indefinitely repeated and it follows that what is initially a metastable situation in which one conformation may be readily converted into another is changed to one of increasing stability, as the process is not so simply reversed. For example, for step 1 to be reversed, possible steps 5, 4, 3 and 2 may all have to be reversed sequentially in that order, to enable 1 to be altered. In the forward full line direction we may have differentiation, in the reverse the process is dedifferentiation (Fig. 10).

In this paper we have taken the liberty to speculate a great deal, perhaps even to overtheorize. However we believe that the function of a symposium of this kind is to expose what might be new interesting ideas and concepts which cannot always be presented in more formal contexts. Perhaps the most exciting field in all present day biology is the structure-function relationships of membrane and the membrane-membrane recognition phenomena which when elaborated hold promise of revolutionizing the science of cell biology.

REFERENCES

Bolotin, M., Coen, D., Deutsch, J., Dujon, B., Netter, P., Petrochilo, E. and Slonimski, P. P. (1971) *Bull. Instit. Pasteur. 69,* 215

Bunn, C. L., Mitchell, C., Lukins, H. B. and Linnane, A. W. (1970) *Proc. Nat. Acad. Sci. U.S. 67,* 1233

Clark-Walker, G. D. and Linnane, A. W. (1966) *Biochem. Biophys. Res. Commun. 25,* 8

Clark-Walker, G. D. and Linnane, A. W. (1967) *J. Cell Biol. 34,* 1

Coen, D., Deutsch, J., Netter, P., Petrochilo, E. and Slonimski, P. P. (1970) *Soc. Exp. Biol. Symp. 24,* 449

Dixon, H., Kellerman, G. M., Mitchell, C. H., Towers, N. H. and Linnane, A. W. (1971) *Biochem. Biophys. Res. Commun. 43,* 780

Ephrussi, B. (1953) *Nucleo-Cytoplasmic Relations in Microorganisms* Oxford Univ. Press (Clarendon), London and New York

Ephrussi, B., Hottinguer, H. and Roman, H. (1955) *Proc. Nat. Acad. Sci. U.S. 41,* 1065

Ephrussi, B. and Grandchamp, S. (1965) *Heredity, 20,* 1

Ephrussi, B., Jakob, H. and Grandchamp, S. (1966) *Genetics 54,* 1

Gingold, E. B., Saunders, G. W., Lukins, H. B. and Linnane, A. W. (1969) *Genetics 62,* 735

Goldring, E. S., Grossman, L. I., Krupnick, D., Cryer, D. R. and Marmur, J. (1970) *J. Mol. Biol. 52,* 323

Linnane, A. W. (1968) in *Biochemical Aspects of the Biogenesis of Mitochondria,* (Slater, E. C., Tager, J. M., Papa, S. and Quagliariello, E. eds.) p. 437, Adriatica Editrice, Bari, Italy.

Linnane, A. W., Lamb, A. J., Christodoulou, C. and Lukins, H. B. (1968a) *Proc. Nat. Acad. Sci. U.S. 59,* 1288

Linnane, A. W., Saunders, G. W., Gingold, E. B. and Lukins, H. B. (1968b) *Proc. Nat. Acad. Sci. U.S. 59,* 903

Linnane, A. W. and Haslam, J. M. (1970) in *Current Topics in Cellular Regulation,* Vol. 2 p. 185, Academic Press Inc., New York

Lukins, H. B., Tate, J. R., Saunders, G. W. and Linnane A. W. (1972) *Molec. Gen. Genetics* in press

Mitchell, C. H., Bunn, C. L., Lukins, H. B. and Linnane, A. W. (1972) *J. Bioenergetics 4,* 161

Mortimer, R. K. and Hawthorne, D. C. (1969) in *The Yeasts,* (Rose, A. H. and Harrison, J. S. eds.) Vol. I, p. 386, Academic Press Inc., London

Mounolou, J. C., Jakob, H. and Slonimski, P. P. (1966) *Biochem. Biophys. Res. Commun. 24,* 218

Nagley, P. and Linnane, A. W. (1970) *Biochem. Biophys. Res. Commun. 39,* 989

Nagley, P. and Linnane, A. W. (1972) *J. Mol. Biol. 66,* 181

Saunders, G. W., Trembath, K., Gingold, E. B., Lukins, H. B. and Linnane, A. W. (1971) in *Autonomy of Chloroplasts and Mitochondria.* (Boardman, N. K., Linnane, A. W. and Smillie, R. M. eds.) North-Holland Publ., Amsterdam

Sonneborn, T. M. (1963) in *The Nature of Biological Diversity* (Allen, J. M. ed.) p. 165. McGraw-Hill, New York

Sonneborn, T. M. (1970) *Proc. Roy. Soc.* B. *176,* 347

Thomas, D. Y., and Wilkie, D. (1968) *Biochem. Biophys. Res. Commun. 30,* 368

Location of DNAs Coding for Various Kinds of Chloroplast Proteins

S. G. WILDMAN, N. KAWASHIMA[a], D. P. BOURQUE, FLOSSIE WONG, SHALINI SINGH, P. H. CHAN, S. Y. KWOK, K. SAKANO, S. D. KUNG, AND J. P. THORNBER

Department of Botanical Sciences, Molecular Biology Institute, University of California Los Angeles, California, USA

CYTOPLASMIC vs. MENDELIAN MODE OF INHERITANCE IN *NICOTIANA*

The genus *Nicotiana* is one of the few genera in the biological spectrum which allows for the production of a wide range of viable, interspecific F_1 hybrids. The genus consists of sixty or more species from which many different hybrid combinations have been made (Goodspeed, 1954). Several reciprocal hybrid combinations have provided a means for ascertaining whether the coding information for different kinds of proteins localized within chloroplasts is contained in chloroplast or nuclear DNA. The idea for using *Nicotiana* interspecific hybrids for this purpose is most easily explained with the aid of the photographs in Fig. 1. In the top row of plants, pollen from a variegated strain of *N. tabacum* fertilized the egg cells of the normal, unvariegated strain of *N. tabacum*. The result of this cross at the F_1 level shown in the photograph is the total absence of the variegation. Also any subsequent generations arising from self-pollination of the F_1 plant remain entirely free of the variegation. That pollen from the variegated plant did contain information necessary for the perpetuation of normal *tabacum* phenotypic characters during sexual reproduction is shown by the middle row of pictures. Pollen from the variegated strain readily fertilized the eggs of a different species, *N. glauca*, to produce the F_1 interspecific hybrid, *N. tabacum* x *N. glauca*. Many of the tobacco phenotypic traits are evident in the F_1 hybrid, but not a sign of the variegation character is evident. The interspecific hybrid is incapable of further sexual reproduction by self-pollination, a situation characteristic of

[a]Present address: Hatano Tobacco Experiment Station, Japan Monopoly Corporation, Hatano, Japan.

FIGURE 1

Illustration of Maternal and Mendelian Modes of Inheritance in *Nicotiana*

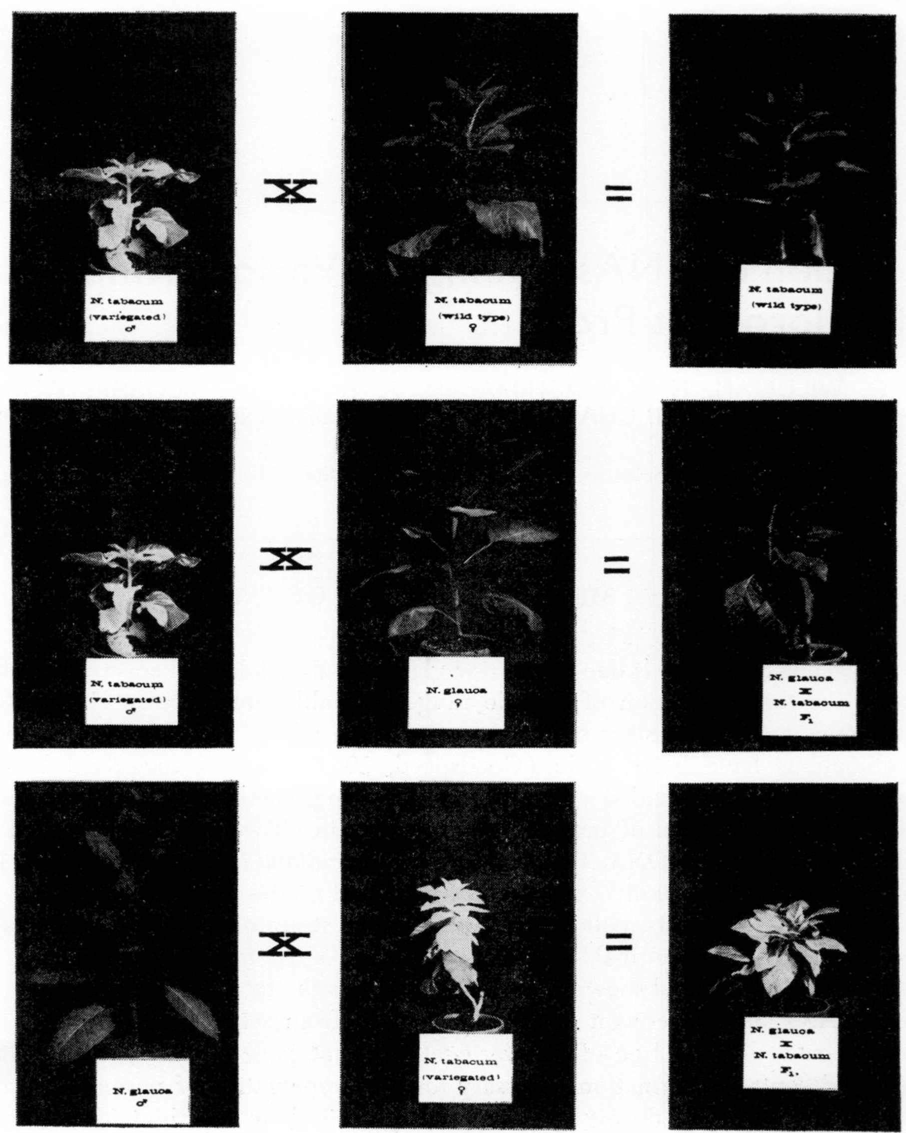

Variegated strain of *N. tabacum* var. Turkish Samsun appeared spontaneously and has been continuously propagated since 1952 by self-pollinated seeds harvested from conspicuously variegated flowers.

nearly all combinations of *Nicotiana* interspecific hybrids. That the variegation trait can, in fact, be carried into an F_1, *N. tabacum* x *N. glauca* hybrid is shown by the bottom row of pictures. What is essential is that the variegated

N. tabacum must be the female parent in the cross. Thus, it is clear that the information necessary to perpetuate this type of variegation can only be transmitted via the maternal line or, as commonly designated, the cytoplasmic mode of inheritance. It is independent of information contained in nuclear genes which are transmitted by pollen and form the basis for the Mendelian mode of inheritance.

DEFECTIVE CHLOROPLASTS IN VARIEGATED *NICOTIANA TABACUM*

(Liao, 1968)

The cause of variegation in *N. tabacum* is shown by the photomicrographs of a living palisade parenchyma cell in Fig. 2. The general plan of organization of the protoplasm in this cell is characteristic of all of the cells of the variegated leaf whether obtained from pure white, pure green, or green-white mosaic

FIGURE 2

Living Palisade Parenchyma Cell from a Variegated *N. tabacum* Leaf

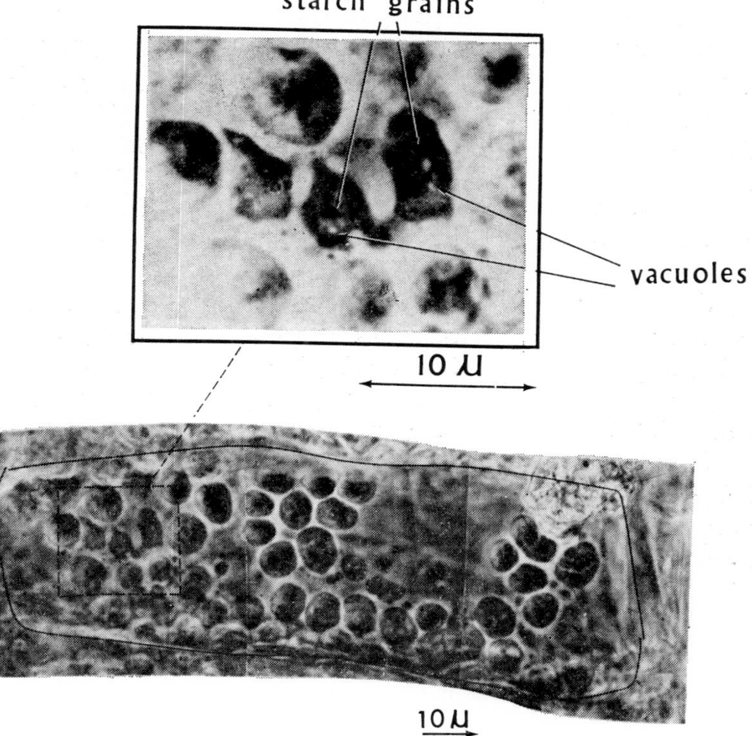

One face of the cell made from a composite of three photographs. Boundary of cell outlined in black for identification. Distribution of normal to defective chloroplasts representative of total chloroplast population in three dimensional cell. Inset: defective chloroplasts enlarged to show nature of their structure.

areas of the leaf. This particular cell contains chloroplasts that are identical to those found in wild-type *N. tabacum* in regard to numbers and appearance of grana, amount of chlorophyll pigments, abundance of starch grains, and appearance and behaviour of mobile phase. In contrast to wild-type *N. tabacum,* about one-third of the total chloroplasts in this cell are defective. The defective chloroplasts appear to be composed almost entirely of mobile phase; neither chlorophyll pigments nor grana are present. Defective chloroplasts are never found in wild-type *N. tabacum* nor in any of the other species of *Nicotiana* we have extensively observed. Thus, the cause of the variegation of these *N. tabacum* leaves is the presence of a mixture of defective chloroplasts and normal chloroplasts in the mesophyll cells of the leaves. The whiteness or greenness of a particular area of the mosaic of variegated leaf surface is simply the result of the ratio of the number of normal to defective chloroplasts in the cells composing the area of tissue being examined.

DEMONSTRATION OF MUTATION IN DNA OF DEFECTIVE CHLOROPLASTS
(Wong, 1972)

Defective chloroplasts contain DNA of the same density in CsCl (1.700 ± .001) and molecular complexity (6 x 10^7 daltons) as chloroplast DNA from wild-type *N. tabacum* leaves. However, the DNA from defective chloroplasts is slightly richer in GC content than DNA from wild type chloroplasts. This slight difference of less than one percent in base composition could only be revealed by simultaneous melting of wild type and defective chloroplast DNAs in a specially constructed apparatus. The device recorded only differences in the heterochromicity as melting occurred when one of the DNAs was placed in one beam of a Cary recording spectrophotometer and the other DNA in the other beam, the arrangement being that both DNAs could be melted at precisely the same rate of increase in temperature. Moreover, heteroduplexes could be visualised by electron microscopy in the presence of formamide after melting and reannealing mixtures of DNAs obtained from wild-type and defective chloroplasts. Small regions of nonhomology equivalent to about 500 base pairs were observed. These nonhomologous regions were of sufficiently rare occurrence as to suggest that there is only one grossly mutated region which differentiates the defective from wild-type chloroplast DNA. To our knowledge, this is the first rigorous evidence that a profound change in the structural organization of a chloroplast is directly correlated with a change in the physical-chemical properties of chloroplast DNA. Since the mutated chloroplast DNA is transmitted only by the maternal line, we can therefore feel confident that wild-type chloroplast DNA is also transmitted in a manner entirely independent of nuclear DNA. Consequently, the rationale used for interpretation of the experiments to be described below is:

1) When information regulating arrangement of amino acids in the primary structure of a chloroplast protein is transmitted by pollen to F_1 *Nicotiana* interspecific hybrids, the coding DNA is contained in the nucleus;

2) When transmission is strictly maternal, coding information for the chloroplast protein is contained in chloroplast DNA. The wide range of diversity exhibited by *Nicotiana* species made it possible to find those differences essential for successful application of this kind of genetic analysis. The differences served as phenotypic markers for ascertaining the mode of inheritance of the chloroplast proteins.

FRACTION I PROTEIN

Fraction I protein makes up approximately 40 percent of the total proteins of the mobile phase of mesophyll chloroplasts of higher plants, and is the enzyme which catalyzes the first step in photosynthetic carbon dioxide fixation in plants which operate by the Calvin cycle of carbon assimilation. When purified by repeated recrystallization to constant specific enzyme activity, the protein has an $s^{\circ}_{20,w}$ of 18.3 ± 0.05, a molecular weight of *ca.* 530,000 daltons, is completely colourless in solutions of more than 100 mg protein per ml., and is composed of two distinct kinds of subunits (Kawashima and Wildman, 1971; Kawashima *et al.*, 1971). The shape of the Fraction I protein molecule is probably like the shape of the dodecahedron protein crystals (Kawashima and Wildman, 1970). According to this view, the Fraction I protein molecule is composed of eight identical large subunits, each of about 50,000 daltons, organized as a cube composing 72% of the total mass of protein as depicted in Fig. 3. Attached to each face of the cube is a small subunit of 25,000 daltons, the six small subunits collectively making up 28% of the protein mass. X-ray analysis of Fraction I protein crystals has been less than successful so far because of the unusually low density of 1.058 g/cc for the crystals, eighty percent of the protein crystal consisting of water. However, high resolution electron microscopy of thin sections of Fraction I protein crystals has provided visualization of the molecules. Sections were cut perpendicular to the three different axes of the protein crystals. The various shapes of the individual molecules exposed in the different planes of sectioning conformed to predictions derived from the model (Kwok, 1972). It appears likely that the visualization of the arrangement of the two different subunits constituting the structure of the Fraction I protein molecule will occur with further refinements in the technique of sectioning the crystals. The large and small subunits are not linked together by S-S bonds, are dissociated by sodium dodecyl sulphate (SDS), urea, or alkali and can be resolved by Sephadex chromatography.

PURIFICATION OF FRACTION I PROTEIN BY DIRECT CRYSTALLIZATION FROM HOMOGENATES

(Chan *et al.*, 1972)

Discovery of an extremely simple method for purifying Fraction I protein by direct crystallization from leaf homogenates greatly facilitated work on the mode of inheritance of this chloroplast protein. After recovery of the protein crystals from the homogenate, two recrystallizations produce protein of constant specific

FIGURE 3

Model Proposed for the Organization of Fraction I Protein

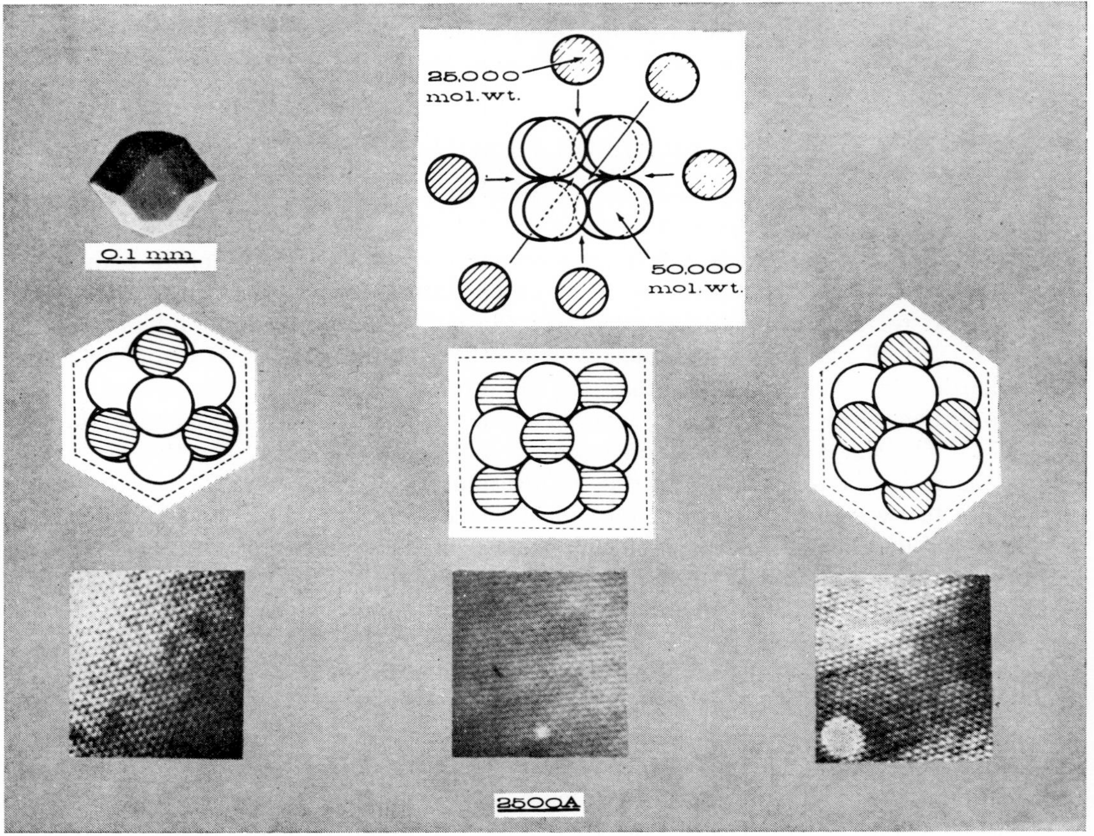

Upper left, protein crystal. Upper right, model derived from shape of crystal showing proposed arrangement of 8 large and 6 small subunits. Middle row, different aspects of the appearance of the model. Lower Row: Sections 100-200 Å thick cut perpendicular to the three different axes of a Fraction I protein crystal viewed by high resolution electron microscopy.

enzyme activity, chemical composition, and physical characteristics. Protein of this high degree of purity has been obtained by the simple method from the eight species and 16 reciprocal, interspecific hybrids of *Nicotiana* so far investigated.

CODE FOR SMALL SUBUNIT OF FRACTION I PROTEIN

(Kawashima and Wildman, 1972)

The composition of the tryptic peptides derived from the small subunit of Fraction I protein isolated from *N. tabacum* differs from the subunit obtained from *N. glutinosa* or *N. glauca.* Analysis of the reciprocal hybrids, *N. tabacum* x *N. glutinosa* showed that a *tabacum* extra peptide not found in *glutinosa* was nevertheless present in both hybrids. The information for synthesis of the extra peptide had been transferred by pollen and hence was contained in the genes of nuclear DNA. Two differences in tryptic peptides distinguished the 15 *N. tabacum* peptides from the 14 *N. glauca* peptides. The information for the *tabacum* type peptides was also transferred by pollen into the egg cells of *N. glauca.* However, the tryptic peptide composition of the small subunit from the *N. tabacum* ♂ x *N. glauca* ♀ hybrid was exclusively of the *tabacum* type. Consequently, not only was the information for the peptide conveyed via pollen, but the new *tabacum* information from nuclear DNA suppressed expression of the *glauca* type information already in the egg cells of the female parent of the cross. The type of dominance exhibited by this reciprocal hybrid combination is not always found, as will be indicated in reporting subsequent experiments. The peptides of the small subunit of Fraction I protein from *N. sylvestris* were shown to follow the pattern of inheritance in combination with *N. tabacum* as had been found for *N. glutinosa* x *N. tabacum* (Chan, 1972).

CODE FOR LARGE SUBUNIT OF FRACTION I PROTEIN

(Chan and Wildman, 1972)

In regard to the large subunit of Fraction I protein, 24 tryptic peptides were resolved from all four species of *Nicotiana* already mentioned. In addition, the same number was found for *N. rustica.* No significant differences in the tryptic peptides were found to serve as phenotypic markers. The notion that the large subunit of Fraction I protein is of very ancient origin and virtually unable to survive mutation has been presented previously (Wildman, 1971). According to Goodspeed (1954), the five species of *Nicotiana* used in this investigation had all been restricted to the Western Hemisphere until the advent of man. The ancestors of all present day *Nicotiana* species first appeared in Panagea—the primordial supercontinent whose breakup and drifting away of parts created the various continents of the world. Drifting apart began about 150 million years ago. At the time of separation, the progenitors of present-day *Nicotianas* had spread only so far as to have become isolated on those pieces of land that are now the continents of North and South America and Australia. In this manner, Australian species of *Nicotiana* have had a very long span of time to engage in a pathway of their own evolution entirely independent of the path of evolution of the Western Hemisphere species. On this account, three species of *Nicotiana* native to Australia were used as sources of Fraction I protein which might reveal desired differences. The tryptic peptide composition of the large subunit of the protein isolated from *N. excelsior, N. suaveolens* and *N. gossei* proved to be identical to each other. However, the Australian proteins contained 25 peptides compared to the 24 peptides found in the large subunit of the Western Hemisphere Fraction I proteins. Analysis of the peptides in the hybrids, *N. glauca* ♂ x *N. excelsior* or *N. gossei* ♀ revealed the extra Australian peptide to be present in the hybrids. The result was a strong suggestion that information for the primary structure of the large subunit was being transmitted exclusively by maternal inheritance. This suspicion was confirmed by analysis of the large subunit obtained from the reciprocal combination, *N. tabacum* x *N. gossei*. The results are summarized by the peptide maps shown in Fig. 4. The extra *N. gossei*

FIGURE 4

Peptide Maps showing the Cytoplasmic Mode of Inheritance of the Primary Structure of the Large Subunit of Fraction I Protein

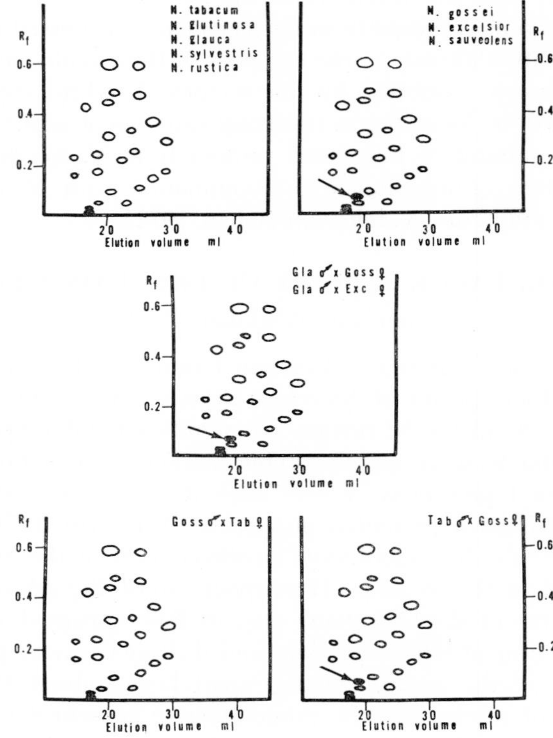

Information for an extra peptide (arrow) found in protein from Australian species of *Nicotiana* (upper right) is only found in hybrids where Australian species are female parents (middle and lower right).

peptide is only found when *N. gossei* is the female parent of the reciprocal hybrids. Consequently, the maternal mode of inheritance of information for the extra peptide requires chloroplast DNA to code for the sequence of amino acids forming the large subunit of Fraction I protein. This finding appears to be the first direct evidence which demonstrates that extranuclear DNA does, in fact, function in the biogenesis of organelles.

The new evidence that only one change in the peptide composition of the large subunit of Fraction I protein seems to have endured during the 150 million years of evolution separating the Australian and Western Hemisphere species of *Nicotiana* strongly reinforces the belief that nearly all mutations in chloroplast

DNA have been eliminated. Evidently, if mutation in the code for the large subunit occurs, the almost inevitable result is failure to synthesize an organelle competent for photosynthesis. In the case of a unicellular organism with a single chloroplast, the mutation would be lethal. In the higher plant, the mutated organelle could only survive in the presence of wild-type chloroplasts as found in the case of the defective chloroplasts in the variegated strain of *N. tabacum*. Even the defective chloroplasts in this strain would be gradually eliminated without the intervention of man to select and perpetuate the less efficient variegated plants. Whereas the sequence of DNA bases within the chloroplast coding for the large subunit of Fraction I protein seems virtually inviolate, it is an interesting commentary on nature that locating the coding information for the small subunit outside of the chloroplast has allowed for considerable mutation and consequent extensive diversity in composition of the small subunit.

CODING FOR THE RuDP CARBOXYLASE CATALYTIC SITE
(Singh, 1972)

After purification by recrystallization to constant specific enzyme activity, the Fraction I proteins from four Western Hemisphere and two Australian *Nicotiana* species exhibited nearly identical turnover numbers and K_ms for ribulose 1-5 diphosphate. Only the protein from *N. gossei* was different because it displayed a higher turnover number and different K_m. The data summarized in Table 1 illustrate the difference in turnover number between *N. gossei* compared to other species. The data also show that whenever *N. gossei* was the female parent of

TABLE 1

Cytoplasmic mode of inheritance of the specific Ribulose diphosphate carboxylase activity of crystalline Fraction I proteins isolated from different species and reciprocal interspecific hybrids of *Nicotiana* (after Singh, 1972)

Species	Protein per Assay μg	$H^{14}CO_3$ Input dpm × 10^5	$^{14}CO_2$ Fixed cpm/0.3 ml	Specific Activity μmoles CO_2 fixed/mg protein/min
N. glutinosa	20.3	7	7212	0.24
N. sylvestris	30.0	7	9645	0.23
N. tabacum	21.0	7	7200	0.24
N. suaveolens	21.8	7	7242	0.24
N. excelsior	28.5	7	9636	0.24
N. gossei	25.0	7	12600	0.36
Goss ♂ × Exc ♀	21.0	10	10500	0.25
Exc ♂ × Goss ♀	19.6	10	11400	0.29
Goss ♂ × Tab ♀	14.5	10	7200	0.24
Tab ♂ × Goss ♀	17.5	10	10500	0.29
Goss ♂ × Suav ♀	13.1	10	5787	0.22
Suav ♂ × Goss ♀	16.6	10	8802	0.28
Exc ♂ × Suav ♀	26.6	7	9312	0.25
Suav ♂ × Exc ♀	31.2	7	11200	0.25

several reciprocal hybrid combinations, the specific enzymatic activity of Fraction I protein was higher in this hybrid than in its reciprocal counterpart. We conclude therefore that chloroplast DNA contains information which regulates the conformation of that region of the catalytic site localized on the large subunit of the enzyme molecule. However, the turnover number was not as high for the enzyme isolated from the three hybrids where *N. gossei* was the female parent as for *N. gossei.* The reduction in activity is related to the formation of hybrid Fraction I protein molecules in the hybrid plants. As summarized by the data in Fig. 5, 15 tryptic peptides were resolved from the small subunit of both *N. gossei*

FIGURE 5

Tryptic Peptide Composition and Two of the Possible Arrangements of Subunits in Hybrid Fraction I Protein Molecules Isolated from Reciprocal Interspecific Hybrids of *Nicotiana.*

Species	Number of Tryptic Peptides	
	Large Subunit	Small Subunit
N. tabacum	24	15
N. gossei	24 + 1	15
Tab ♂ x Goss♀	24 + 1	17 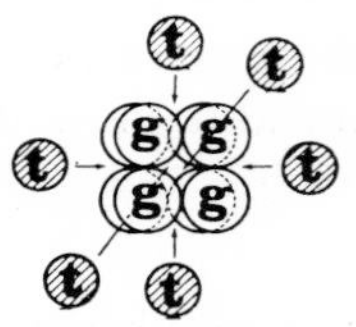
Goss ♂ x Tab ♀	24	17 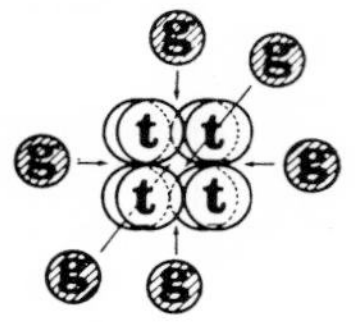

and *N. tabacum.* However, two out of the 15 peptides were found to have different compositions. The small subunit from both of the reciprocal hybrids, *N. gossei* and *N. tabacum,* contained 17 peptides. Therefore, it is evident that at least two different species of Fraction I protein are synthesized in the *N tabacum* x *N. gossei* reciprocal hybrid combinations. Since the primary structure of the large subunit is governed by information exclusive to the female parent, hybrid proteins of the composition indicated by the two models in Fig. 5 would seem likely prospects to explain the experimental findings. However, further analysis of protein from single plants will be required to determine whether the possibilities are limited to just these two kinds of hybrid protein. It is also possible that the macromolecules could be still more complicated if a mixture of *tabacum* and *gossei* small subunits could combine on the same Fraction I protein molecule. The latter situation could give rise to seven different kinds of hybrid Fraction I protein molecules appearing among the progeny resulting from the single act of fertilizing the flowers of one species of *Nicotiana* with the pollen of another species. We speculate that further research could provide a clue towards a molecular biological understanding of the nature of hybrid vigour, or the converse condition of growth inhibition, resulting from hybridization. Hybrid vigour and great uniformity in growth habit is exhibited by the individual plants composing the population of progeny derived from the reciprocal cross, *N. tabacum* x *N. glauca*. The uniformity and vigour is apparent irrespective of which parent provides pollen for fertilization. Correlated with vigour and uniformity, no evidence for the presence of mixtures of Fraction I protein molecules is found in these hybrids. The *N. tabacum* type dominates over *N. glauca*. In complete contrast, the *N. gossei* x *N. tabacum* reciprocal cross is characterized by the individual progeny displaying the most extreme variation. A very occasional plant in a large population will show enormous hybrid vigour, whereas the vast majority of individuals have the appearance of being dwarfed and malformed to various degrees. The extreme variability is exhibited regardless of which species provided pollen for fertilization. Thus, the great variation is also correlated with the presence of mixtures of Fraction I protein molecules. We imagine that these rules governing permission or suppression of mixtures of hybrid Fraction I protein molecules extend to other vital enzymes as well. Perhaps, certain combinations significantly increase the efficiency of an enzyme to perform its job and the greater efficiency is reflected in a more vigorous plant. Other combinations significantly reduce efficiency and stunted and grotesque progeny are the consequence.

CODING FOR PROTEINS OF CHLOROPLAST RIBOSOMES

(Bourque and Wildman, 1972)

Some of the proteins of the 70S chloroplast ribosomes isolated from *N. tabacum* can be distinguished from those from *N. glauca*. In contrast, the 32 discrete bands of protein which were resolved by gel electrophoresis at pH 4.5 in 8 M urea from the 80S cytoplasmic ribosomes isolated from the two *Nicotiana* species could not be told apart. The same situation pertained to the 23 protein bands resolved from the 30S subunit of the chloroplast ribosomes isolated from the two species of *Nicotiana*. However, in the case of the 50S chloroplast ribosome subunit, 25 bands were resolved for each species, but the subunit from *N. glauca* contained three polypeptides that were different from the *N. tabacum* proteins. Analysis of the proteins from 50S subunits obtained from the reciprocal hybrid, *N. tabacum* x *N. glauca* revealed that information for the differences in the ribosome polypeptides was transferred by pollen. One protein present only in *N. glauca* was found in both hybrids. The same condition was found for one protein present only in *N. tabacum*. Transmission of the information for these ribosomal proteins being independent of maternal inheritance, the necessary conclusion is that the information for some of the proteins of the chloroplast ribosomes is contained in nuclear

DNA. We suspect that further analyses may reveal that the information for the rest of the proteins of chloroplast ribosomes is closely linked and also coded by nuclear genes. The *Nicotiana* hybrids would appear to offer unusual opportunities to attempt to answer questions as to whether information for the approximately 50 different proteins comprising the structure of chloroplast ribosomes is contained in a single linkage group; whether fingerprinting ribosomal RNAs can reveal the source of their coding information; whether chloroplast and cytoplasmic ribosomes share polypeptides in common; whether ribosomes, as in the case of Fraction I proteins, become hybrids when dominance is lacking, etc.

CODING FOR THYLAKOID MEMBRANE PROTEINS

An improved purification procedure provided sufficiently large quantities of a homogenous, Photosystem II chlorophyll-protein complex from *N. tabacum* and *N. glauca* to allow comparison of the tryptic peptide fingerprints of the membrane protein (Kung and Thornber, 1971). The protein of the Photosystem II complex is characterized by having bound to it more than 50% of the total chlorophyll in *Nicotiana* chloroplasts. The chlorophyll a/b ratio of 1.0 ± 0.1 together with molecular weight estimates of 35,000 suggest that one molecule of each kind of chlorophyll is complexed to one molecule of the thylakoid membrane protein. Upon removal of chlorophyll, the protein was subjected to trypsin hydrolysis and the peptides resolved by two dimensional solvent chromatography and high voltage electrophoresis. One peptide was found only in *N. glauca.* It was not present in either of the reciprocal hybrids, *N. tabacum* x *N. glauca.* Another peptide out of about 35 for each species was found only in *N. tabacum* and both hybrids. Therefore, the conclusion was reached that the protein comprising a structural element in thylakoid membranes, where Photosystem II is located has its arrangement of amino acids regulated by information in nuclear genes (Kung *et al.,* 1972). The question of which DNA regulates the sequence of the other major thylakoid protein, Photosystem I chlorophyll-protein (Kung and Thornber, 1971), is currently being investigated by utilizing the same approach with interspecific hybrids of *Nicotiana.*

The diameters of individual grana in *N. excelsior* chloroplasts (*ca.* 0.5 μ on the average) as viewed in living palisade cells by high resolution phase microscopy are slightly larger than grana in *N. gossei* chloroplasts (*ca.* 0.4 μ on the average). Observation of grana in the reciprocal hybrids, *N. excelsior* x *N. gossei* indicates that information regulating the size of the grana in chloroplasts of the two hybrids is inherited as a maternal factor. Therefore this aspect of the formation of thylakoid membranes appears to be coded by chloroplast DNA. The photomicrographs in Fig. 6 illustrate these findings. The size of grana in all other species of *Nicotiana* which could also form reciprocal hybrids were too similar in size to serve as phenotypes for inheritance studies.

SUMMARY

The location of DNAs coding for six characters having to do with proteins uniquely localized within chloroplasts is summarized by the material compiled

FIGURE 6

Photomicrographs of Living Palisade Cells to Illustrate the Cytoplasmic Mode of Inheritance of the Size of Grana in Chloroplasts

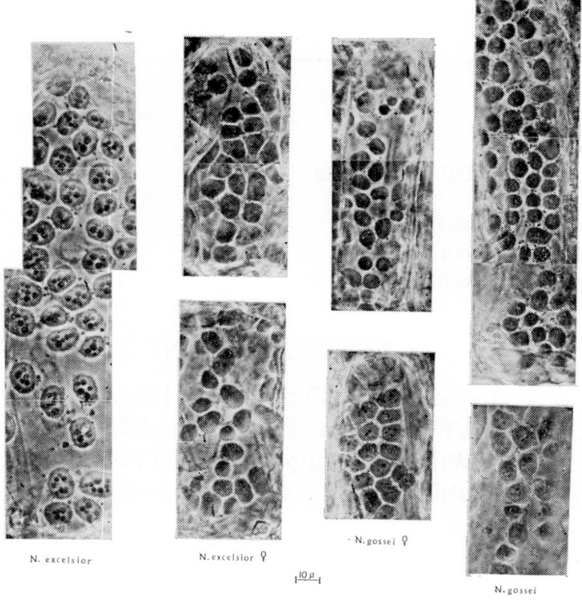

The objects much larger than grana in some chloroplasts are starch grains. Photographed through 10 X ocular and 100 X Neofluar phase objective by flash photography. Photographs of each species and hybrids encompass all chloroplasts on face of cell and thus represent *ca.* one-third of total chloroplasts in cell.

in Table 2. It is obvious that only the barest beginning has been made towards further understanding the role that chloroplast DNA performs in the biogenesis of chloroplasts and their perpetuation during sexual reproduction and alternation of generations. The impression is gained, however, that the function is probably highly circumscribed, as appears also to be true for mitochondrial DNA. Many of the proteins of the mitochondrion are known to be coded by nuclear DNA, but none so far as we are aware has been shown for certain to be coded by mitochondrial DNA. We are prone to believe that information in chloroplast DNA, while probably very limited in scope in terms of the total number of proteins it codes for, is nevertheless absolutely crucial. Even the slightest garble in the genetic message is a disaster for the new organelle about to be synthesized. Also, it becomes increasingly evident that chloroplasts are anything but autonomous entities as some dogmas insist. On the contrary, the chloroplast is highly dependent on interaction with other organelles in its protoplasmic environment for either informational macromolecules, or as a direct source for other kinds of polymers necessary to building the substance of a chloroplast.

TABLE 2

Location of DNAs having to do with some aspects of proteins uniquely localized within chloroplasts of *Nicotiana* species

Chloroplast Constituent	Coding DNA	
	Chloroplast	Nuclear
Fraction I Protein		
Large Subunit	+	
Small Subunit		+
Conformation of RuDP Carboxylase Catalytic site	+	17 and 2+
Proteins of 70S ribosomes		+
Protein of Photosystem II . . .		+
Area of grana thylakoid membrane .	+	

ACKNOWLEDGMENT

Supported by U.S. Atomic Energy Commission, Division of Biology and Medicine, Research Contract AT (11-1)-34, Project 8 and U.S. Public Health Service Grant, AI-00536 to S. G. Wildman; National Science Foundation Grant GB-31207 to J. P. Thornber; U.S. Public Health Service Biomedical Science Support funds grant.

REFERENCES

Bourque, D. P., and Wildman, S. G. (1972), in manuscript

Chan, Pak-Hoo (1972) The role of chloroplast and nuclear DNAs in coding for the large and small subunits of Fraction 1 proteins isolated from *Nicotiana* species and reciprocal, interspecific hybrids. Ph.D. thesis, University of California, Los Angeles

Chan, P. H., Sakano, K., Singh, S. and Wildman, S. G. (1972), *Science 176,* 1145

Chan, P. H. and Wildman, S. G. (1972), *Biochim. Biophys. Acta 277,* 677

Goodspeed, T. H. (1954), *The Genus Nicotiana,* Chronica Botanica Press, Waltham, Mass.

Kawashima, N., Kwok, S. Y. and Wildman, S. G. (1971) *Biochim. Biophys. Acta 236,* 578

Kawashima, N. and Wildman, S. G. (1970) *Biochem. Biophys. Res. Commun. 41,* 1463

Kawashima, N. and Wildman, S. G. (1971) *Biochim. Biophys. Acta 229,* 240

Kawashima, N. and Wildman, S. G. (1972) *Biochim. Biophys. Acta 262,* 42

Kwok, Shiu-Yen (1972) Studies on the crystal structure of ribulose diphosphate carboxylase isolated from tobacco leaves. Ph.D. thesis, University of California, Los Angeles

Kung, S. D. and Thornber, J. P. (1971) *Biochim. Biophys. Acta 253,* 285

Kung, S. D., Thornber, J. P. and Wildman, S. G. (1972) *FEBS Lett. 24,* 185

Liao, Catherine L. (1968) Macromolecular composition of defective chloroplasts from a cytoplasmic mutant of tobacco. Ph.D. thesis, University of California, Los Angeles

Singh, Shalini (1972) A study of the regulation of the catalytic activity of RuDP carboxylase isolated from various species and reciprocal, interspecific hybrids of *Nicotiana.* Ph.D. thesis, University of California, Los Angeles

Singh, S., Chan, P. H. and Wildman, S. G. (1972) *Biochim. Biophys. Acta* (in press)

Wildman, S. G. (1971) An approach towards ascertaining the function of chloraplast DNA in tobacco plants, In *Autonomy and Biogenesis of Mitochondria and Chloroplasts,* (Boardman, N. K., Linnane, A. W., and Smillie, R. M., eds.), p. 402, North-Holland Publishing Co.

Wong, Flossie Y. (1972) A comparative study of DNA satellites from a cytoplasmic, chloroplast mutant and wild-type tobacco. Ph.D. thesis, University of California, Los Angeles

Nuclear Genes Controlling Chloroplast Development in Barley

KNUD W. HENNINGSEN,[a] J. E. BOYNTON,[b]
D. VON WETTSTEIN AND N. K. BOARDMAN

Institute of Genetics, University of Copenhagen, Denmark and Division of Plant Industry, CSIRO Canberra, Australia

Studies over the past few years indicate that the synthesis and maintenance of a functional chloroplast is under the control both of nuclear genes, and extra-nuclear genes contained in chloroplast DNA. Earlier speakers in this session of the symposium have examined the possible role of chloroplast DNA in coding for some of the components of the chloroplast. Of particular significance are the findings of S. G. Wildman and his colleagues. These workers have studied the pattern of inheritance of some chloroplast components in a number of interspecific hybrids of tobacco. It is apparent from their results that there is a very close interaction between the informational contents of nuclear DNA and chloroplast DNA, even to the extent where the two different subunits of a single chloroplast protein (Fraction I) are coded for by different DNAs; one by nuclear DNA and the other by chloroplast DNA.

Photochemical studies on acetate-requiring mutants of *Chlamydomonas reinhardi* have shown that the assembly of the photosynthetic electron transport chain is under the control of nuclear genes (Levine, 1969). This is not to imply that the assembly is controlled solely by nuclear genes as it seems highly probable that future work will elucidate a role for chloroplast DNA.

The studies summarized in this paper demonstrate that nuclear genes hold a very tight control over pigment synthesis and chloroplast biogenesis. Nuclear gene mutants of barley have been found with lesions in the formation of the plastid of the dark-grown seedling (etioplast) from the proplastid or in the development of the mature, photosynthetically-competent chloroplast from the etioplast.

[a]*Present address: Division of Plant Industry, CSIRO, Canberra, A.C.T.*
[b]*Permanent address: Department of Botany, Duke University, Durham, North Carolina, U.S.A.*

MATERIALS AND METHODS

Plant material

The 198 recessive lethal chloroplast mutants induced in barley (*Hordeum vulgare* L.) have been classified (cf. Gustafsson, 1940) into six phenotypic categories. The distribution of the mutations among 86 gene loci were described earlier (von Wettstein *et al.*, 1971). The present investigation is concerned only with chloroplast mutants of the *xantha* (yellow to yellow-green leaves), *albina* (white leaves) and *viridis* (light green leaves) phenotypes. Wildtype barley of cultivar Svalöf's Bonus was used for comparison.

Seedlings were grown on vermiculite in darkness for either 6 days at 25°C or for 7 days at 23°C, before transfer to continuous white light (20, 50, 3200 or 4500 lx) from fluorescent tubes. Some of the *xantha* mutants were also illuminated for 10 days with a 16 hour photoperiod of 50 lx. Photoconversion of protochlorophyllide in dark grown leaves was accomplished by a 1 minute illumination with white light (50×10^3 J x m^{-2}) from a xenon lamp filtered through 6 cm of water. In some experiments, the shoots of 7-day dark-grown seedlings were cut 1-2 cm above the seed, and the cut ends immersed in 0.01M δ-aminolevulinic acid (ALA) in deionized water for 24 hours in darkness. The apical 10mm of each primary leaf was then discarded and the next 30mm used for pigment analyses.

Pigment analysis

In vivo absorption spectra of leaf segments were obtained with a Zeiss RPQ-20A recording spectrophotometer equipped with an RA-20 integrating sphere. Leaves treated with δ-aminolevulinic acid were extracted with 80% acetone and the pigments partitioned between petroleum ether and aqueous acetone. Protoporphyrin IX dimethyl ester (Sigma, crystalline grade I) and magnesium protoporphyrin di-K salt (kindly supplied by Dr. S. Granick) dissolved in methanol were used as reference compounds. Total chlorophyll ($a + b$) was measured in acetone extracts (Bruinsma, 1961).

Electron microscopy

Glutaraldehyde (4% in phosphate buffer, pH 7.2, for 2 to 4 hours) was used for fixation, followed by postfixation in osmium. Thin sections of epoxy resin-embedded material were poststained with uranyl acetate and lead citrate. Micrographs were taken with a Siemens Elmiskop I.

Oxygen evolution in leaves

For polarographic determination of oxygen evolution the upper surface of a single leaf segment was abraded (Kirk and Goodchild, 1972) and placed directly on the platinum electrode of an experimental set-up described by Kirk and Reade (1970). The leaf was exposed to saturating white light (3.5×10^2 J x m^{-2}) from a quartz iodine lamp, and the steady-state rate of oxygen evolution was measured.

RESULTS AND DISCUSSION

Nuclear genes controlling porphyrin biosynthesis in plastids

Chloroplast mutants of the *xantha* and *albina* phenotypes were analysed for their capacity to accumulate protochlorophyllide in darkness. Most mutants at the *xan-f,-g* and *-h* loci and several *albina* mutants failed to synthesize detectable amounts of protochlorophyllide (Boynton and Henningsen, 1967; Henningsen and Boynton, 1969a; von Wettstein *et al.*, 1971; Gough, 1972b). To investigate whether these mutants contained specific blocks in the biosynthetic pathway from ALA to protochlorophyllide, leaves from dark-grown seedlings were fed ALA. Granick (1961) had shown previously that ALA is readily taken up by detached leaves of dark-grown wildtype barley and converted to protochlorophyllide. Abnormal amounts of protochlorophyllide are accumulated by wildtype leaves in such a feeding experiment, but only a small fraction of the protochlorophyllide is converted to chlorophyllide *a* on brief illumination of the leaves (Fig. 1). Some of the protochlorophyllide remaining after the first exposure to light can be photo-converted in subsequent light exposures provided the leaves remain in the dark for suitable periods between the light treatments (Sundquist, 1969; Granick and Gassman, 1970). Most of the protochlorophyllide present in untreated leaves of wildtype barley is converted to chlorophyllide *a* on brief illumination (Fig. 1).

FIGURE 1

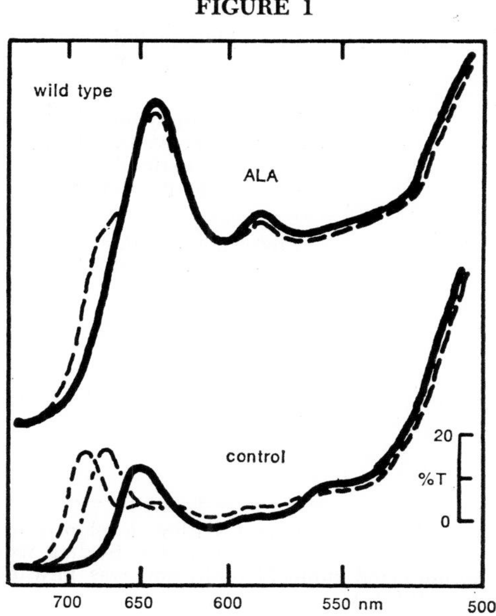

In vivo absorption spectra of wildtype barley leaves developed in darkness (control) and of leaf segments treated with 0.01 M ALA for 24 h. *Full lines*: Unilluminated leaves. *Dashed Lines*: Spectra recorded immediately after a 1 min illumination with white light (5×10^3 $J \times m^{-2}$). *Dashed-dotted Line*: Spectrum of the control leaves recorded after 15 min in darkness following the illumination. Most of the protochlorophyll(-ide) which accumulated in ALA-treated leaves was not photoconvertible.

Mutants at the *xan-f*,*-g* and *-h* loci which contained no detectable protochlorophyllide in untreated leaves accumulated large amounts of porphyrins on feeding with ALA (Table 1). Porphyrin accumulation in these mutants is illustrated by

TABLE 1

Genotype	Pigments accumulated	
	untreated	treated with ALA
wildtype; *xantha* mutants at several loci, e.g. *xan-b*^{18}, *-m*^{48} *albina* mutants e.g. *alb-f*^{17}	Protochlorophyllide (most photoconvertible)	Protochlorophyllide (most non-convertible)
xantha mutants at loci *xan-a*, *-i*, *-k*, *-n*	Protochlorophyllide (most photoconvertible)	Protochlorophyllide (most non-convertible + some protoporphyrin)
xantha mutants: *xan-f*^{10}, *-f*^{27}, *-f*^{40}, *-f*^{41}, *-f*^{58}; *xan-g*^{28}; *xan-h*^{30}, *-h*^{38}, *-h*^{56}	None	protoporphyrin IX
xan-f^{60}, *-g*^{37}, *-h*^{57}	None (or trace of protochlorophyllide)	protoporphyrin IX + some protochlorophyllide (photoconvertible)
xan-f^{26}, *-g*^{44}, *-g*^{45}	Protochlorophyllide (most photoconvertible)	protoporphyrin IX (+ other porphyrins)
xan-u^{21}	Protochlorophyllide (most photoconvertible) + uroporphyrinogen and/or coproporphyrinogen	protoporphyrin IX + uroporphyrin
xan-l^{35}	Protochlorophyllide (most photoconvertible)	Mg protoporphyrin (+ other porphyrins)
albina mutants at several loci, e.g. *alb-e*^{16}	Protochlorophyllide (small amounts)	protoporphyrin IX (+ other porphyrins)
vir-51, -59	(None?)	protochlorophyllide (photoconvertible) + protoporphyrin IX

Pigment accumulation in wildtype and mutant barley developed in darkness. The pigment content of the primary leaf was analyzed before and after feeding with 0.01 M ALA for 24 h. Photoconversion of the accumulated protochlorophyllide was measured by illumination with white light for 1 min (5×10^3 J $\times$ m^{-2}).

the *in vivo* absorption spectra of dark-grown leaves of *xan-f*^{27} taken before and after treatment with ALA (Fig. 2). All mutant alleles at the *xan-h* locus lack detectable amounts of protochlorophyllide in their untreated leaves. Some of the mutant alleles at the *xan-f* and *-g* loci form protochlorophyllide, either to the same extent as the wildtype or in smaller amounts. These are leaky mutants; they accumulate porphyrins on treatment with ALA in similar amounts to the completely blocked alleles at these loci (Fig. 2, spectra of *xan-g*^{45}).

FIGURE 2

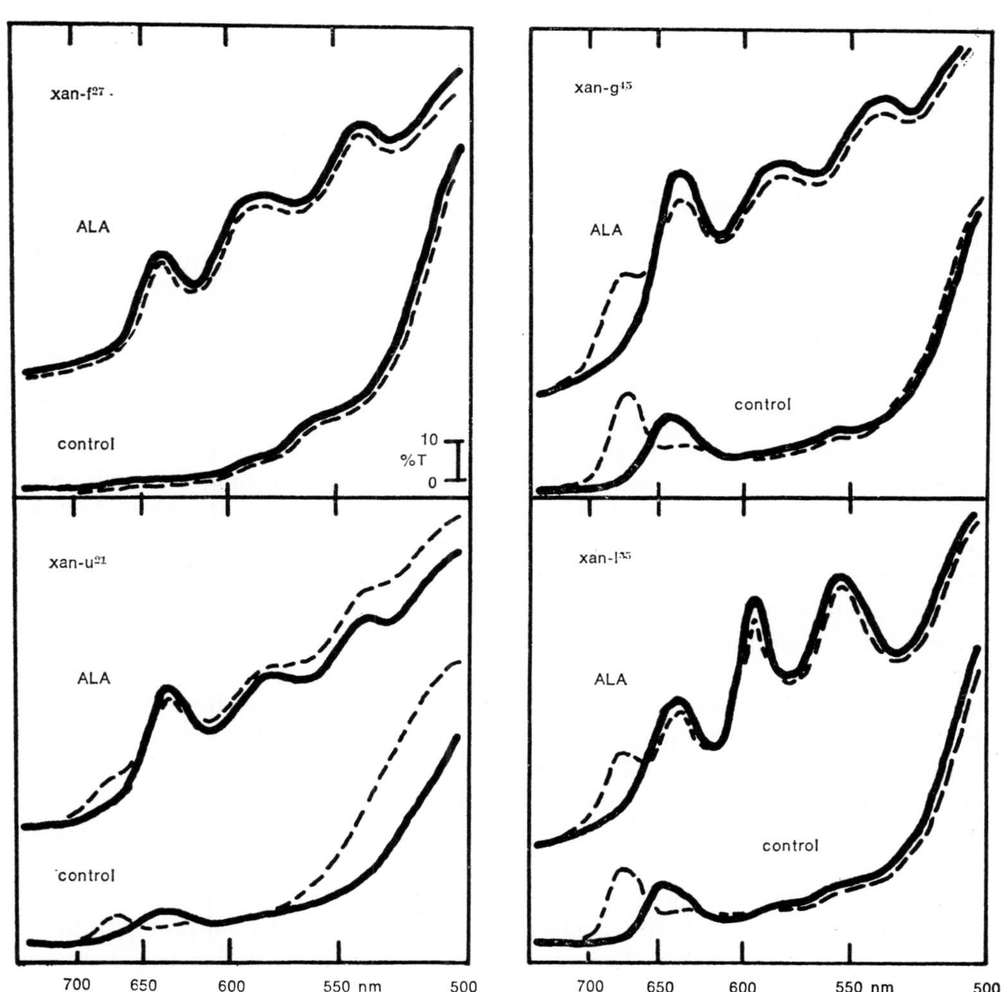

In vivo absorption spectra of leaves from mutant seedlings of *xan-f*^{27}, *-g*^{45}, *-u*^{21} and *-l*^{35}. *Control:* Leaves developed in darkness. *ALA*: Leaves developed in darkness and treated with 0.01 M ALA for 24 h. *Full Lines*: Unilluminated. *Dashed Lines*: After exposure to white light (1 min; 5×10^3 J $\times$ m^{-2}). Protochlorophyll(-ide) was not detectable in *xan-f*^{27}. The leaky mutants *xan-g*^{45}, *-u*^{21} and *-l*^{35} form protochlorophyllide, part of which is photoconvertible. Feeding with ALA results in accumulation of protoporphyrin IX in *xan-f*^{27}, *-g*^{45} and *u*^{21}, while *xan-l*^{35} accumulates magnesium protoporphyrin and other porphyrins.

Leaky mutants are also found at the *xan-u* and *-l* loci. A small amount of protochlorophyllide is formed in untreated leaves of *xan-u*^{21} (Fig. 2). Illumination of the leaves results in the photoconversion of protochlorophyllide and an abnormal increase in absorbance at wavelengths shorter than 540 nm. It is possible that part of the absorbance increase results from the photooxidation of a colourless precursor of protochlorophyllide (e.g. uroporphyrinogen or coprophyrinogen). The *in vivo* absorption spectrum of *xan-u*^{21} after treatment with ALA shows the accumulation of porphyrins (Fig. 2). From a comparison of the absorption spectrum of a leaf extract with that of known porphyrins it was concluded that the major pigment accumulated was protoporphyrin IX. This conclusion was confirmed by thin layer chromatography, which also showed the presence of some uroporphyrin (Gough, 1972b).

The *in vivo* absorption spectrum of leaves of *xan-l*^{35} after feeding with ALA (Fig. 2) resembles the *in vivo* spectrum of wildtype barley leaves which have been treated with α,α'-dipyridyl in combination with ALA. Under these conditions Mg-protoporphyrin monomethyl ester is accumulated in the wildtype (Granick, 1961). A comparison of the *in vivo* spectrum of ALA-treated *xan-l*^{35} with that of magnesium protoporphyrin is shown in Fig. 3. It seems evident that Mg-protoporphyrin is present in *xan-l*^{35}. Thin-layer chromatography of the pigments extracted from *xan-l*^{35} showed the presence of Mg-protoporphyrin monomethyl ester as well as Mg-protoporphyrin (Gough 1972a,b).

The absorption spectra of pigments extracted from ALA-treated leaves of a number of mutants are shown in Fig. 3. The spectrum of *xan-f*^{10} reveals only protoporphyrin IX, while the leaky mutant *xan-f*^{26} contains both chlorophyllide *a* (formed from protochlorophyllide during extraction of the leaves in the light) and protoporphyrin IX. Protoporphyrin IX is the major pigment in *xan-h*^{57} and -*g*^{37}. While these mutants do not form detectable amounts of protochlorophyllide in untreated conditions, trace amounts of chlorophyllide *a* (from protochlorophyllide) are formed during ALA feeding. Chromatographic analyses (Gough, 1972a) of extracts of mutants at the *xan-f,-g* and *-h* loci have shown that protoporphyrin IX is the major porphyrin accumulated in the ALA-treated leaves.

Small amounts of protoporphyrin IX accumulted in 10 of the 13 *albina* mutants investigated (cf. Table 1). Protochlorophyllide as well as protoporphyrin IX accumulated in mutants at the *xan-a,-i,-k,-n* loci and in the mutants *vir*-51 and -59.

It is concluded from these feeding experiments with ALA, that nuclear genes at the loci *xan-f,-g,-h* and *alb-e* control biosynthetic steps between protoporphyrin IX and Mg-protoporphyrin. The *xan-u* locus appears to control a step between uroporphyrinogen and protoporphyrin IX, and *xan-l* a step between Mg-protoporphyrin monomethyl ester and protochlorophyllide.

The relationship between pigment accumulation and the structure of the plastid membranes formed in darkness was investigated. Etioplasts in untreated wildtype barley contain crystalline prolamellar bodies with either wide or narrow spacing of the tubular membranes. The membranes in prolamellar bodies with wide tube-spacing are considered to have a higher pigment content than those with narrow spacing of the tubules (Henningsen and Boynton, 1969b). Treatment of leaves with ALA results in an increased number of prolamellar bodies with

FIGURE 3

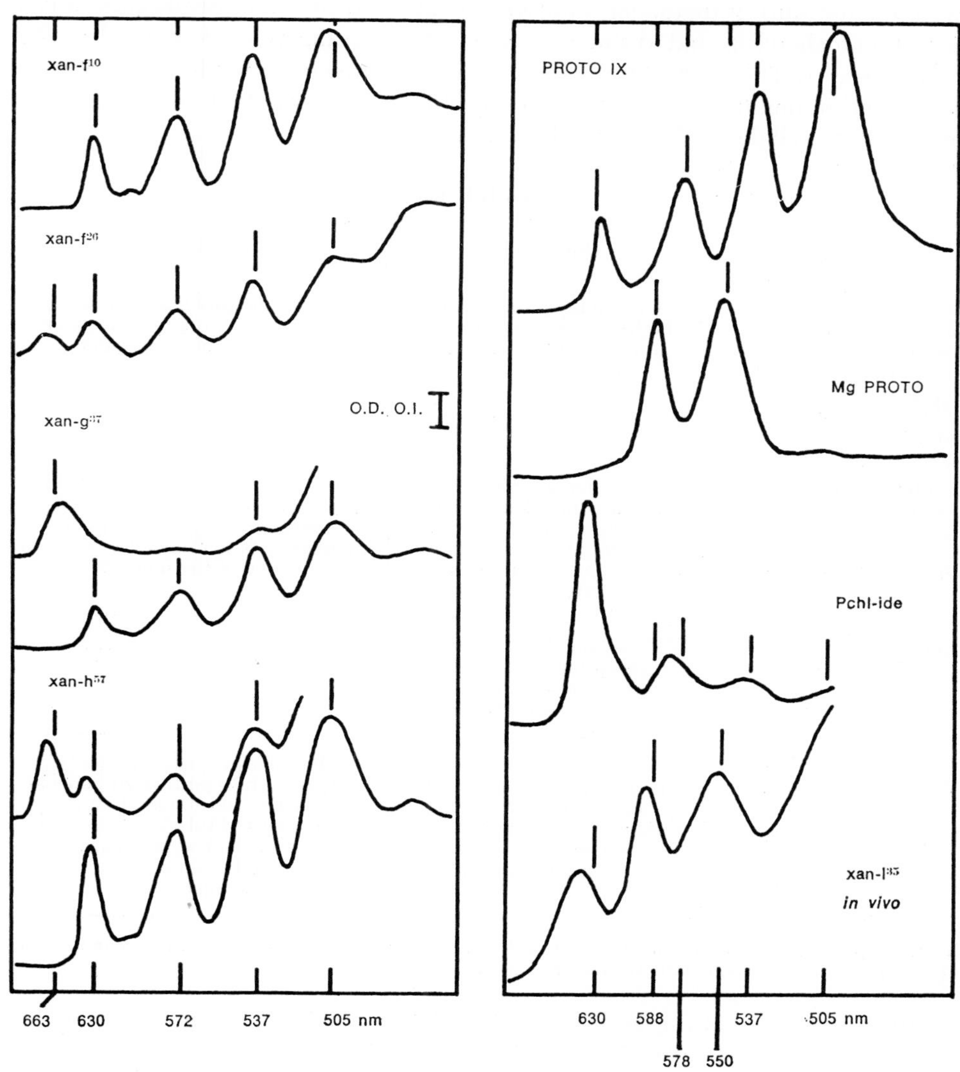

Spectra at left. Absorption spectra of pigments extracted from leaves of *xan-f*[10], *-f*[26], *-g*[37] and *-h*[57]. The leaves were treated with 0.01 M ALA in the dark for 24 h, and then exposed to white light during extraction with 80% acetone. Protoporphyrin IX (absorption maxima 626, 572, 537 and 502 nm) is accumulated in all these mutants. Chlorophyllide *a* (abs. max. 663 nm) is evident in the extract from the incompletely blocked mutant *xan-f*[26]. Small amounts of chlorophyllide *a* can also be detected in *xan-g*[37] and *-h*[57]. [Chlorophyllide and photoinactive protochlorophyll(-ide) were extracted into petroleum ether (upper set of spectra), while protoporphyrin IX remained in the aquous acetone (lower set of spectra)]. *Spectra at right.* Spectra of pure protoporphyrin IX dimethyl ester (PROTO IX) and magnesium protoporphyrin (Mg PROTO) in methanol, and protochlorophyllide in aqueous acetone. *In vivo* absorption spectrum of ALA-treated leaves from *xan-l*[35]. The presence of protochlorophyll-ide and magnesium protoporphyrin is evident from the absorption maxima at 638, 594 and 554 nm.

wide tube-spacing, suggesting that at least some of the protochlorophyllide accumulated during treatment with ALA is inserted into the membranes. Swelling of mitochondria and disintegration of their cristae is evident in the ALA treated leaves. These structural effects of ALA treatment of wild type barley are also found in mutants at the *xantha* and *albina* loci not involved in the control of protochlorophyllide synthesis (Tables 2 and 3).

TABLE 2

Genotype	Untreated	After feeding
wildtype; *xantha* mutants at loci *xan-a, -b, -c, -d, -e, -i, -j, -k, -m, -n, -o, -p, -q, -s, -t*; *albina* mutants, e.g. *alb-f*^{17}	Crystalline prolamellar bodies, wide and narrow tube spacing; primary lamellar layers	Crystalline prolamellar bodies; wide tube spacing more frequent primary lamellar layers
xantha mutants; *xan-f*^{10}, *-f*^{27}, *-f*^{40} *-f*^{41}, *-f*^{58}, *-f*^{60}; *-g*^{28}, *-g*^{37}; *-h*^{30}, *-h*^{38}, *-h*^{56}, *-h*^{57}	No prolamellar bodies; primary lamellar layers often in loose stacks; pairing between two lamellar layers rather frequent	No crystalline prolamellar bodies; increased number of stacked lamellar layers; increase in number of paired lamellar layers in some of the mutants
xan-u^{21}	Prolamellar bodies and primary lamellar layers are amorphous and abnormally structured	Similar to untreated leaves
xan-f^{26}; *-g*^{44}, *-g*^{45}; *-l*^{35}	Crystalline prolamellar bodies; primary lamellar layers often in loose stacks; pairing between two lamellar layers rather frequent	Prolamellar bodies partially dispersed into perforated lamellar layers; formation of long unconnected tubules; increased number of stacked lamellar layers; increased number of paired lamellar layers

Type of plastid membranes in wildtype and mutant barley developed in darkness before and after feeding leaves with 0.01 M ALA for 24 h.

Mutants at the *xan-f,-g* and *-h* loci completely blocked in protochlorophyllide synthesis lack prolamellar bodies (Fig. 4, Table 2). Several of the primary lamellar layers formed in these mutants are often arranged in a loose stack with frequent grana-like pairing between two adjacent lamellar layers. Treatment with ALA does not lead to the formation of prolamellar bodies. The leaky mutants at the *xan-f,-g* and *-l loci* form crystalline prolamellar bodies (Fig. 5, Table 2), the size and number of which appear to be related to the amount of protochlorophyllide. The loose stacks of lamellar layers and the grana-like pairing between lamellar layers are also found in the leaky mutants. Treatment with ALA results in partial dispersal of the prolamellar bodies (Fig. 6, Table 2) and reorganization of the membranes into perforated lamellar layers and many long unconnected tubules. The number of lamellar layers organized in loose stacks is increased and

TABLE 3

Genotype	Changes in structure of mitochondria	Formation of abnormal cytoplasmic membranes
wildtype; *xantha* mutants, e.g. *xan-b*^{18}, *-m*^{48}	Swelling	None
xantha mutants at locus *xan-f*, *-g*, *-h*	None	Some
xan-u^{21}	None	Many
xan-l^{35}	Swelling	None

Structural changes of mitochondria and cytoplasmic membranes in wild type and mutant barley resulting from feeding with 0.01 M ALA in darkness for 24 h.

FIGURE 4

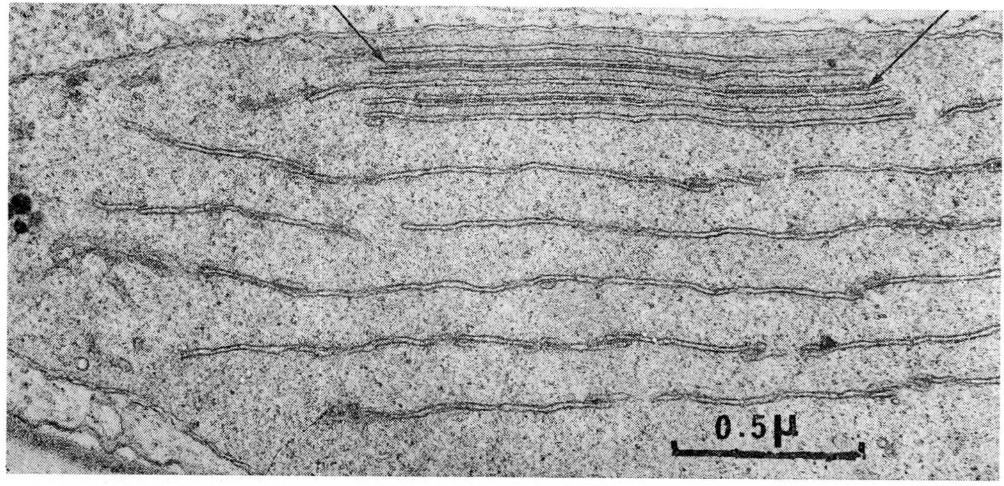

Part of a plastid in a leaf of *xan-h*^{56} grown in darkness. Primary lamellar layers are formed and long profiles of two layers in close apposition are frequent (arrows). Prolamellar body membranes are absent from the plastids and no protochlorophyllide can be detected in the leaves of this mutant.

grana-like arrangements consisting of several lamellar layers are frequent. The formation of the long unconnected tubules and the grana-like membrane arrangements might indicate that pigment is accumulated in these structures.

The effects of ALA treatment on the structure of mitochondria and cytoplasmic membranes in mutants at the *xantha* loci controlling steps in protochlorophyllide synthesis are summarized in Table 3. The mutant *xan-l*^{35}, where the block in protochlorophyllide synthesis occurs at a step later in the pathway than Mg-protoporphyrin, shows structural changes in the mitochondria. Changes in

FIGURE 5

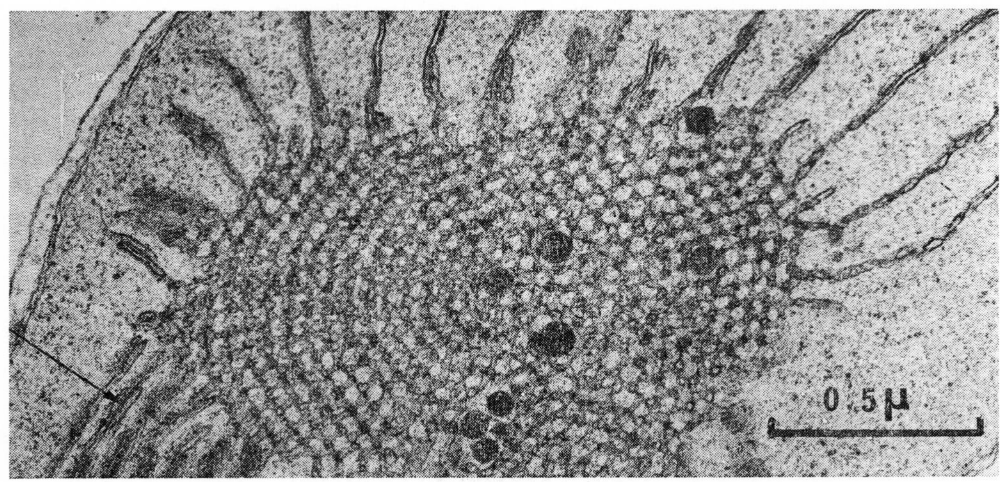

Plastid in a *xan-g*[45] seedling grown in darkness and treated with 0.01 M ALA for 24 h. *xan-g*[45] is a leaky porphyrin mutant which accumulates normal amounts of protochlorophyllide in darkness. The plastids contain large prolamellar bodies. Treatement with ALA results in the accumulation of protoporphyrin IX. Areas of the prolamellar body membranes are reorganized into primary lamellar layers, which often occur in pairs (arrow).

FIGURE 6

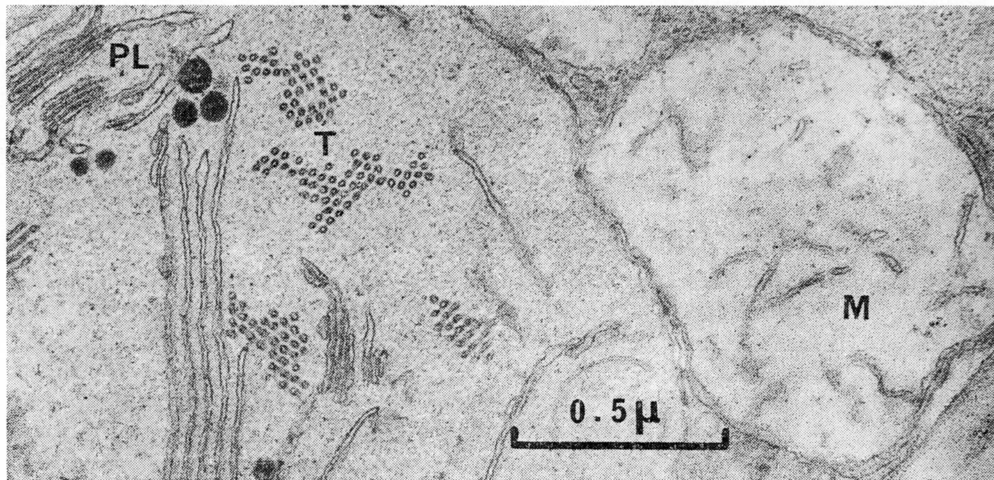

Plastid in a *xan-l*[35] seedling grown in darkness and treated with 0.01 M ALA for 24 h *xan-l*[35] is a leaky porphyrin mutant and large amounts of Mg-protoporphyrin are accumulated during the treatment with ALA. Pigment accumulation is accompanied by rearrangement of the prolamellar body membranes into long straight tubules (T) and stacks consisting of several paired lamellar layers (PL). Swelling of the mitochondria (M) and distortion of the christae result from ALA-treatment.

mitochondrial structure are not observed in mutants at the *xan-f,-g* and *-h* loci blocked at a step earlier in the pathway than Mg-protoporphyrin. These mutants on the other hand form abnormally structured cytoplasmic membranes. The formation of abnormal cytoplasmic membranes is pronounced in *xan-u*^{21}, which seems to be blocked at a step earlier in the pathway than protoporphyrin IX. If these structural changes are related to pigment accumulation, then it would appear that precursors earlier in the pathway than protoporphyrin IX are accumulated in the cytoplasm. Mutants at the *xan-f,-g* and *-h* loci blocked in a function required for the insertion of magnesium into the porphyrin molecule, accumulate protoporphyrin IX in the plastids. *xan-l*^{35}, with a block in the pathway later than magnesium protoporphyrin, also accumulates this porphyrin in the plastids. The wildtype resembles *xan-l*^{35}, except that protochlorophyllide is accumulated instead of magnesium protoporphyrin.

The absence of prolamellar body membranes in those mutants which are completely blocked in protochlorophyllide synthesis suggests that protochlorophyllide is essential for the assembly of prolamellar body membranes. As discussed in the following section, the formation of a protochlorophyllide-protein complex (termed protochlorophyllide holochrome) is essential for the photoconversion of protochlorophyllide to chlorophyllide *a*. Not only protochlorophyllide but also the protein moiety of the holochrome appears to be required for formation of prolamellar body membranes.

Mutations affecting membrane organization at early stages of plastid development

Photoconversion of protochlorophyllide in the wildtype is followed by structural alterations of the plastid membranes. The regular crystalline prolamellar body is changed into an irregular network of tubules (tube transformation), followed by a dispersal of the tubular membranes into perforated lamellar layers. In 7-day old wildtype barley seedlings, tube transformation occurs within 2 minutes after a one minute illumination of the dark-grown seedlings. Dispersal into lamellar layers is usually completed within 15 minutes.

A number of spectroscopic forms of chlorophyllide *a* have been observed in the early stages of plastid development and it will be convenient to consider these shifts in the absorption maximum of chlorophyllide *a* in relation to changes in membrane structure. Illumination of wildtype barley results in the photoconversion of more than 70% of the protochlorophyllide of dark-grown seedlings. Immediately following conversion, the absorption maximum of chlorophyllide *a* is at 678 nm and this rapidly shifts to 684 nm (Gassman *et al.*, 1968; Bonner, 1969). Then follows a spectral shift from 684 nm to 672 nm (Shibata shift after Shibata, 1957), the rate of which depends on the physiological age of the seedlings and temperature (Henningsen, 1970).

The *in vivo* absorption maxima of protochlorophyllide and newly formed chlorophyll(-ide) *a* in mutants of the *xantha* and *albina* phenotypes are given in Table 4. In leaves of 7-day old wildtype and in mutants at 11 *xantha* loci, the Shibata shift is completed in less than 15 min following a brief illumination. In mutants at the *xan-j* locus and in *alb-f*^{17} the chlorophyllide *a* absorption

TABLE 4

Genotype	Dark	1 min light + <2 min dark	1 min light + 15 min dark
wildtype and mutants at 11 *xantha* loci	645	680	671
xan-j; *alb-f*	648-650	683-684	680-684
xan-n, -t, -u; *alb-c*;	640-650	668-673	668-672
"Infrared"-5[a]	635	672	672
xan-a, -i, -k, -o, -q	644-650	675-678	670-674

In vivo absorption maximum of protochlorophyllide and newly formed chlorophyll(-ide)*a* in wildtype and mutant barley. Photoconversion of protochlorophyllide in 7 day old dark grown seedlings was accomplished by a 1 min illumination of white light (5×10^3 J $\times$ m^{-2}).
[a]From: Nielsen, (1970); von Wettstein *et al.* (1971).

FIGURE 7

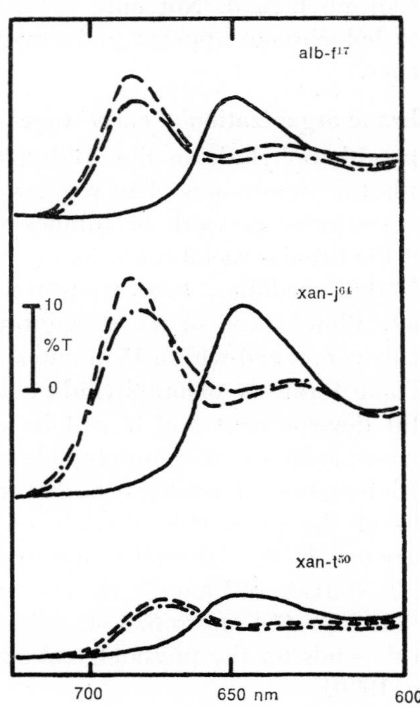

In vivo absorption spectra of leaves from mutant seedlings grown in darkness (full lines). The leaves were then illuminated for 1 min with white light (5×10^3 J $\times$ m^{-2}) and the spectrum recorded immediately (dashed lines) and following a subsequent dark period of 60 min (dashed-dotted lines). In *alb-f*[17] and *xan-j*[64] the chlorophyll(-ide) *a* absorption maximum does not show the shift to shorter wavelengths observed in the wildtype (cf. Fig. 1, lower spectra). The chlorophyllide *a* maximum in *xan-t*[50] is at the shorter wavelength (672 nm) immediately after illumination.

maximum fails to shift to shorter wavelengths (Fig. 7). Photoconversion of protochlorophyllide in mutants at the loci *xan-n,-t,-u* and *alb-c* gives chlorophyllide *a* with an absorption maximum around 670 nm (Table 4 and Fig. 7). In these latter mutants either the *in vivo* form of chlorophyllide *a* absorbing at 672nm is formed directly from protochlorophyllide, or the Shibata shift is completed rapidly. Chlorophyllide *a* with an absorption maximum at an intermediate wavelength (675 to 678 nm) is formed in mutants at the loci *xan-a,-i,-k,-o* and *q* (Table 4). The shift to shorter wavelengths occurs in these mutants.

In mutants investigated so far, photoconversion of protochlorophyllide is always followed by structural transformation of the prolamellar body i.e. tube transformation. In mutants at the *xan-j* locus and in $alb\text{-}f^{17}$ tube transformation is not followed by the usual dispersal into perforated lamellar layers. Although the crystalline prolamellar bodies formed in these mutants in the dark appear to be normal (Fig. 8), the prolamellar body membranes remain in the transformed

FIGURE 8

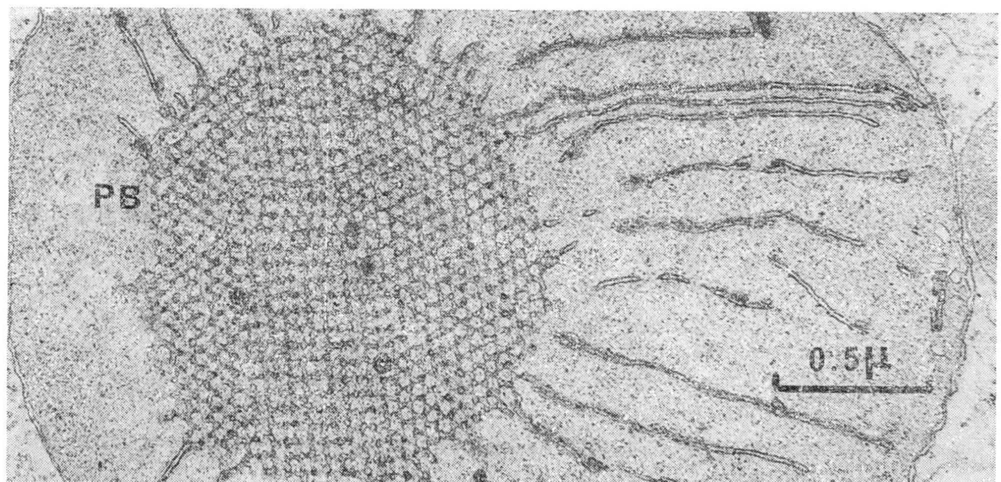

Plastid in a $xan\text{-}j^{59}$ seedling grown in darkness. Protochlorophyllide is accumulated by this mutant and the prolamellar body (PB) is organized in a normal crystalline configuration.

configuration for several hours in the light (Fig. 9). Prolonged illumination of $alb\text{-}f^{17}$ results in photo-destruction of the chlorophyll and derangement of the prolamellar body membranes. The chlorophyll formed in the *xan-j* mutants is less sensitive to light. An interesting feature of the *xan-j* mutants is the formation of many small grana-like membrane associations in the close vicinity of prolamellar bodies (Fig. 9). It is possible that chlorophyll may be translocated to these regions of the membrane system without a complete dispersal of the prolamellar body membranes.

The prolamellar bodies formed in mutants at the loci *xan-n,-t* and $alb\text{-}c^{7}$ are of crystalline configuration and not clearly distinguishable from the wildtype.

FIGURE 9

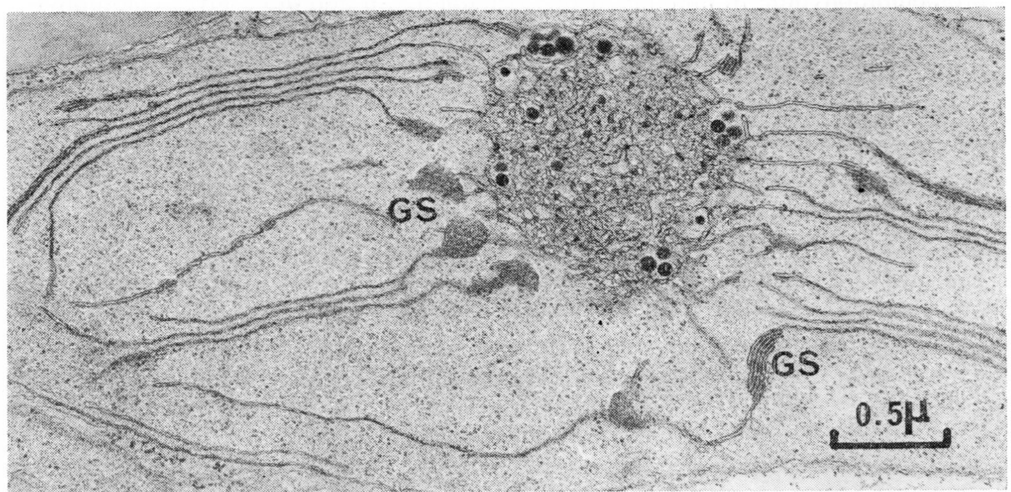

Plastid in a *xan-j*[64] seedling grown in darkness and then illuminated for 4 h with 3200 lx of white light. The prolamellar body (PB) remains un-dispersed, but several grana-like stacks of lamellae (GS) occur in close vicinity of the prolamellar body.

These mutants disperse their prolamellar bodies and some grana are eventually formed.

In darkness, *xan-u*[21] forms highly abnormal plastid structures. Both the primary lamellar layers and many of the prolamellar bodies consist of deranged membranes. Some long, usually unconnected tubules of amorphous texture are also formed. No structural changes are observed in *xan-u*[21] during illumination with high light intensity. When illuminated over a longer period in low intensity, *xan-u*[21] accumulates chlorophyll with *in vivo* absorption maxima at 674 and 745 nm. The *in vivo* form of chlorophyll absorbing at 745 nm is presumably identical to the chlorophyll formed in mutants of the "infrared" type (Nielsen, 1970; von Wettstein *et al.*, 1971). The plastid membranes formed in *xan-u*[21] under low light conditions resemble those found in "infrared"-5. Thus the following membrane configurations are found: giant grana with abnormal spacing of the lamellae, long profiles of merged and undulated membranes coated with osmiophilic material, membrane bound crystalloids consisting of regularly arranged particles, and also some deranged prolamellar bodies. Some characteristics of the plastid membranes formed in mutants at the *xan-a* locus should be mentioned. Although some crystalline prolamellar bodies are formed in darkness, the majority of the plastids in *xan-a* mutants contain only loosely connected tube elements and some honeycomb-like membrane arrangements, presumably derived from prolamellar body membranes.

Mutants at the *xan-a* locus have an abnormal lipid metabolism (Appelqvist *et al.*, 1968; von Wettstein *et al.*, 1971). The incomplete assembly of the tube elements into crystalline prolamellar bodies might indicate that these mutants fail to synthesize sufficient amounts of an essential lipid component of the prolamellar body membranes. During greening in low light intensity the *xan-a* mutants can develop elaborate membrane systems consisting of grana, crystalline prolamellar bodies and honeycomb-like structures. Some of the chlorophyll formed under these conditions absorbs at about 740 nm.

It has been suggested that the *in vivo* state of protochlorophyllide and newly formed chlorophyll (-ide) *a* are related to the configuration of the prolamellar body membranes (Butler and Briggs, 1966; Henningsen, 1970). Protochlorophyllide with an absorption maximum at 650 nm, present in crystalline prolamellar bodies, can be converted by physical treatments to a form absorbing around 635 nm. This shift of the absorption maximum is accompanied by a rupture of the prolamellar body membranes. The protochlorophyllide absorbing around 635 nm is photoconverted into chlorophyllide *a* absorbing at 672 nm. In untreated wildtype leaves the spectral shift from 684 to 672 nm precedes or coincides with the dispersal of prolamellar bodies into perforated lamellar layers (Henningsen and Boynton, 1969b; Henningsen, 1970).

The mutants, *xan-j* and *alb-f*[17], blocked in the spectral shift from 684 to 672 nm fail to disperse their prolamellar bodies. *Xan-u*[21] has abnormally structured prolamellar body membranes, and in this mutant protochlorophyllide absorbing around 640 nm is photoconverted to chlorophyllide *a* with an absorption maximum at 670 nm. The *in vivo* pigments and the structure of the plastid membranes found in the mutant "infrared"-5 (Nielsen, 1970) are similar to those observed in *xan-u*[21]. A relationship between the configuration of prolamellar body membranes and the *in vivo* state of chlorophyll is more difficult to establish in other mutants, e.g. *xan-t*[50] and *xan-a*. Mutants at the *xan-m* locus are able to complete the Shibata shift, but the prolamellar bodies remain undispersed. It seems that the change resulting in the spectral shift from 684 to 672 nm is necessary before membrane dispersal can take place, but other processes are also required.

Smith *et al.,* (1959) investigated chloroplast mutants in corn and concluded that no correlation could be established between the ability to perform the spectral shift and either phytolation of chlorophyllide *a* or content of carotenoids. Of the mutants investigated by Smith *et al.,* (1959), 3 *albina* mutants lacked the ability both for the spectral shift and for phytolation. A similar result was obtained with *alb-f*[17] (Henningsen, K.W., Boardman, N.K., and Thorne, S.W., unpublished results). In *xan-t*[50] and *-u*[21] where chlorophyll(-ide) absorbing around 672 nm is observed shortly after photoconversion, phytolation is completed more rapidly than in the wildtype. These observations support the hypothesis (Henningsen, 1970) that chlorophyllide *a* can be phytolated only after conversion to the *in vivo* form absorbing at 672 nm.

The positions of the *in vivo* absorption maxima of protochlorophyllide, chlorophyllide *a* and chlorophyll *a* are governed by the interaction of the pigments with other molecules; with neighbouring pigment molecules, and with protein and lipid. A mutation which markedly alters the binding of a pigment to these other components is likely to result in a shift in the absorption maximum of that pigment. It is apparent that many gene loci are involved in the synthesis of a complex macromolecular structure such as the chloroplast membrane. It is not surprising therefore, that the binding of the pigment molecules to the lipoproteins of the membranes, and therefore the spectral properties of the pigments, are under the control of many genes.

It was mentioned earlier that protochlorophyllide in a dark-grown seedling is complexed to a specific protein (holochrome). A photoactive protochlorophyllide holochrome (molecular weight 63,000) has been isolated from wildtype barley by the use of saponin (Henningsen and Kahn, 1971). On illumination, the protochlorophyllide holochrome with an absorption maximum at 644 nm, undergoes spectral changes which are similar to those observed with wildtype seedlings, i.e. the newly formed chlorophyllide *a* has an absorption maximum at 679 nm, which then shifts to 672 nm. The spectral shift from 679 nm to 672 nm is accompanied by a change in the circular dichroism of chlorophyllide *a* (Foster *et al.*, 1971, 1972).

During the early stages of chloroplast development the number of sites at which photoconversion of protochlorophyllide takes place remains constant (Nadler and Granick, 1970; Süzer and Sauer, 1971). This suggests that the newly formed chlorophyllide *a* molecules (or the phytylated derivative, chlorophyll *a*) are translocated from the holochrome to sites on the developing chloroplast membranes. A mutation in the holochrome protein may modify the binding of the protochlorophyllide and prevent the translocation of chlorophyllide *a* to the developing membrane, or a mutation in a membrane component may inhibit the translocation.

It is possible that *xan-j* and *alb-f*^{17}, which lack the spectral shift from 684 to 672 nm have altered structures which inhibit the translocation of chlorophyllide *a* from the holochrome to the site on the membrane. The lack of membrane dispersal could be a reflection of alterations at the molecular level. Protochlorophyllide holochromes isolated from *alb-f*^{17} (Henningsen, K.W. and Kahn, A., unpublished results) and from "infrared"-5 (Foster *et al.*, 1971; Nielsen, 1972) have properties similar to the holochrome from wildtype barley, and it seems, therefor, that the altered spectral properties of these mutants are dependent on their membrane organization.

Mutants with lesions in photosynthetic electron transport and in grana formation

In the search for chloroplast mutants with lesions in photosynthetic electron transport, it seemed desirable to screen mutants which were able to accumulate appreciable amounts of chlorophyll on illumination. Mutants of the *viridis* phenotype and some of the *xantha* phenotypes satisfied this criterion. A sensitive polarographic technique for measurement of oxygen evolution from leaves (Kirk and Goodchild, 1972) was found to be convenient for screening mutants for photosynthetic activity.

In agreement with earlier observations of Smith (1954), the first trace of oxygen evolution can be detected in some leaves of dark grown wildtype barley after 30 minutes illumination in 4500 lx (Henningsen, K.W. and Boardman, N.K.

unpublished results). After illumination for 1 hour oxygen evolution is detectable in all leaves of 6-day old seedlings. The average content of chlorophyll in wildtype leaves illuminated for 1 hour was 19μg per g fresh weight.

A summary of the results obtained with wildtype and mutant leaves illuminated for 24 hours is shown in Table 5. Based on their oxygen-evolving capacities

TABLE 5

Genotype	Rate of oxygen evolution at steady state; ΔmV/leaf area	Chl(a + b); μg/g fresh weight
Wildtype	1.3	890
Group I: *xan-m*, *-p*^{17}, *-q*^{51}; *vir-c*, *-e*, -27, -30, -47, -56, -63, -65, -68, -69	0.0	49-245
Group II: *xan-a*, *-b*, *-c*, *-d*, *-e*^{52}, *-f*^{26}, *-g*^{45}, *-l*^{35}, *-n*, *-t*^{50}; *alb-c*^{7}; *vir-a*, *-b*, *-d*, -13, -14, -15, -17, -19, -23, -29, -34, -35, -38, -42, -43, -44, -45, -46, -49, -60, -61	0.1-2.3	26-1263
Group III: *xan-h*, *-i*, *-j*, *-k*^{42}, *-o*, *-s*^{46}, *-u*^{21}; *alb-e*^{16}, *-f*^{17}; *vir*-51, -59.	0.0	0-25

Chlorophyll content and capacity for oxygen evolution in leaves from wildtype and mutant barley seedlings, developed in darkness and illuminated with either 50 or 4500 lx for 24 h (25 to 27°C). (Some of the *xantha* mutants were also studied after development for 10 days in 50 lx). In greening wildtype barley (1 h 4500 lx) detectable oxygen evolution (ΔmV/leaf area of 0.4) is obtained at a chlorophyll content of 19μg/g fresh weight.

and chlorophyll contents the mutants fall into three main groups. The first group comprise mutants which contained up to one-fourth of the chlorophyll present in the wildtype at 24 hours, but which did not evolve detectable amounts of oxygen. Such amounts of chlorophyll accumulated in the wildtype at shorter times of illumination, and were sufficient to give almost maximum rate of oxygen evolution. The chlorophyll in mutants at the *xan-m* and *-p* loci is partially bleached at high light intensity (4500 lx), but oxygen evolution was not detected even in low light intensity (20-50 lx) where substantial amounts of chlorophyll accumulated. Mutants at the *xan-m, vir-c* and *vir-e* loci, where several mutant alleles are available, were consistently defective in oxygen evolution. Eight of the *viridis* mutants in this group have not yet been assigned loci.

The lack of photosynthetic oxygen evolution in mutants of Group I may be related to their inability to form normal membrane structures, including grana. In 5 to 7-day old wildtype, grana are reasonably well developed after 4 hours of illumination at high light intensity (Henningsen and Boynton, unpublished results). The structure of plastid membranes in mutants at *xan-m* and *-p* loci after 24 hours of illumination are shown in Figs. 10 and 11. Mutants at the *xan-m*

FIGURE 10

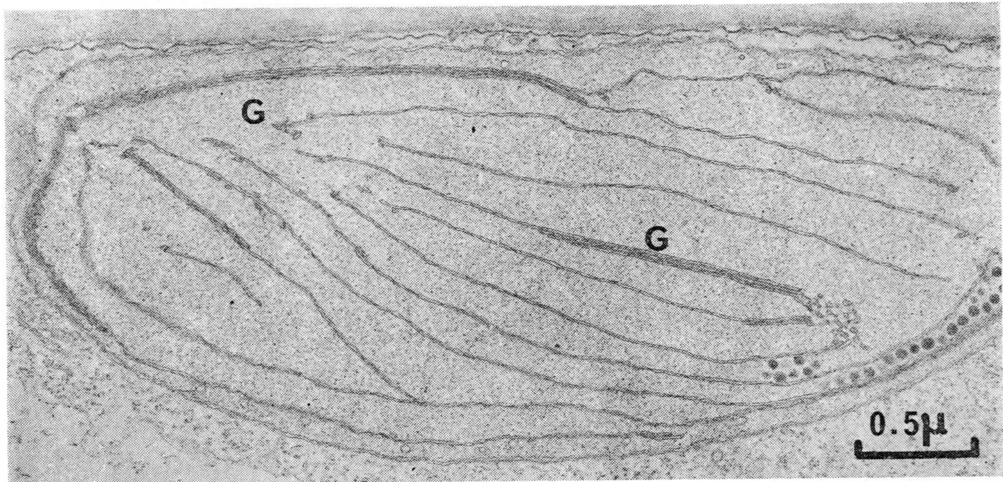

Plastid in a *xan-m*^{3} seedling grown in darkness and illuminated for 24 h with 20 lx of white light. Several giant grana (G) occur with abnormal pairing between the discs.

FIGURE 11

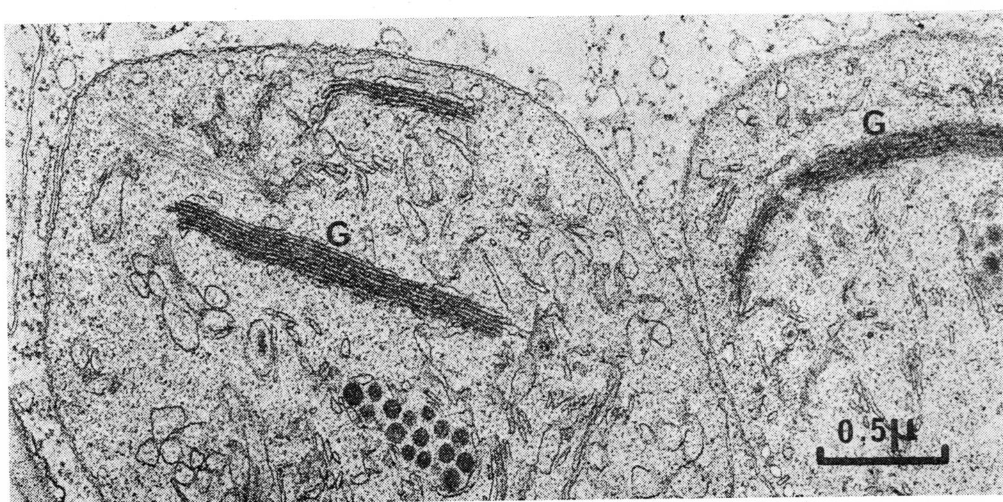

Plastid in a *xan-p*^{17} seedling grown for 10 days in a 16 h photoperiod with 20 lx of white light. Giant grana (G) and numerous vesicles occur. The spacing of the discs is abnormal.

locus do not disperse their prolamellar bodies in high light intensity, but form deranged membranes with numerous osmiophilic deposits. These mutants have an abnormal lipid metabolism (Appelqvist *et al.*, 1968). When greened at low light intensity, *xan-m* mutants form primary lamellar layers and some grana-like membrane associations, but the spacing between the membranes in adjacent lamellar layers is abnormal (Fig. 10). An abnormal spacing of the membranes also is observed in the grana-like structures in *xan-*p^{17} (Fig. 11). The chlorophyll formed in *xan-*q^{51} is not sensitive to high light. The prolamellar bodies in *xan-*q^{51} are slowly dispersed into primary lamellar layers. After 24 hours of illumination grana are present, but only a few short profiles of stroma lamellae are evident (Fig. 12).

FIGURE 12

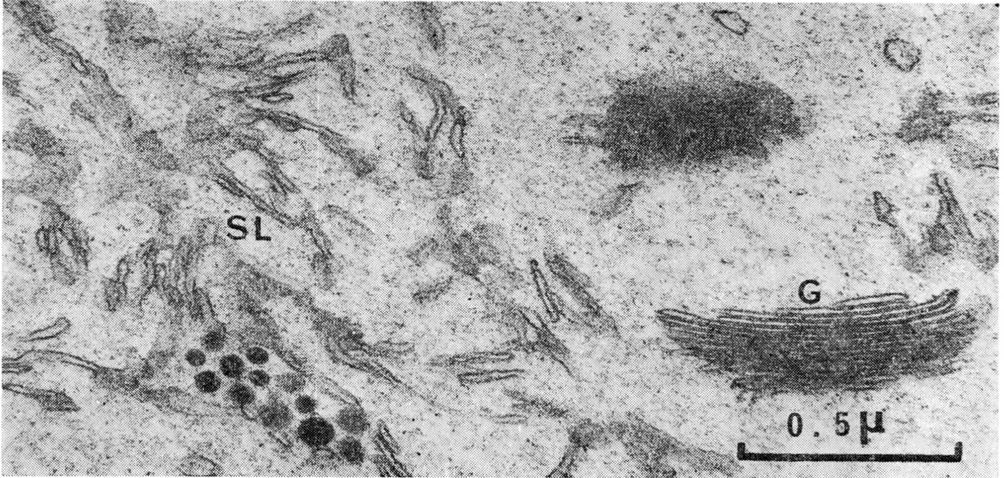

Plastid in a *xan-*q^{51} seedling grown in darkness and illuminated for 24 h with 3200 lx of white light. Several apparently normal grana (G) occur, while stroma lamellae (SL) are rare.

Group II is comprized of a large number of mutants which accumulate chlorophyll and are able to evolve oxygen. The amounts of chlorophyll formed depend on the conditions of illumination, and the values given in Table 5 show the range of chlorophyll content of the mutants, greened either at 50 lx or 4500 lx for 24 hours, or at 50 lx for 10 days for some of the *xantha* mutants. Mutants at the *xan-a,-e* and *-n* loci are sensitive to high light intensities. For example, *xan-a* mutants were almost devoid of chlorophyll if greened under high light and no oxygen evolution was detected. At high light intensities, the plastid membranes of *xan-a* mutants are disorganized and large amorphous bodies of osmiophilic material are formed (cf. Appelqvist *et al.*, 1968). When illuminated for longer periods at low light intensity, the *xan-a* mutants accumulate rather large amounts of chlorophyll and their rates of oxygen-evolution are reasonably high. The plastid membranes developed under low light consist of grana connected to other elaborate membrane configurations (Henningsen and Boynton, 1967; von Wettstein *et al.*, 1971). Mutants at the *xan-n* locus accumulate chlorophyll and develop some capacity for oxygen evolution only when greened for a long period in low light intensity. *Xan-e*^{52} is less sensitive to light and some oxygen evolution is observed even if the seedlings are greened under high light. *Alb-c*^{7} is temperature sensitive, but under the conditions used in this study it formed reasonably high amounts of chlorophyll and developed some capacity for oxygen-evolution.

A number of mutants in Group II accumulate at the temperature and light conditions used amounts of chlorophyll comparable to the wildtype (e.g. *vir-a* and *-d* and 10 *viridis* mutants at unidentified loci). In mutants at the *xan-b,-c,* and *-d* loci the plastids develop slowly, but after several days in the light both the chlorophyll content and capacity for oxygen evolution are high. The plastid lamellae in these mutants are organized into a few giant grana and stroma lamellae are rare. *Xan-b*^{18} will form chloroplasts with normal grana and stroma lamellae if illuminated at 30°C. *Xan-c*^{23} is temperature sensitive, but it can be normalized if the seedlings are supplied with leucine (Walles, 1963). The leaky mutants at the loci controlling protochlorophyllide synthesis (*xan-f,-g* and *-l*) are among the mutants where the rate of oxygen evolution is high. These mutants form numerous small sphaeroidal grana.

Mutants in Group III are unable to evolve oxygen, but their chlorophyll contents are at, or below, the limit for detectable oxygen in the wildtype. Partially and completely blocked mutants at loci controlling protochlorophyllide synthesis, and mutants with abnormal organization of their prolamellar body membranes, belong to this category.

It is possible that the mutants of Group I which accumulate sufficient amounts of chlorophyll, but lack the capacity to evolve oxygen may be blocked in the synthesis of certain electron transport components. But we must also consider the possibility that the primary lesion in some of these mutants is concerned with the synthesis of a structural component of the photosynthetic membranes. Some of the electron transport components, particularly those concerned with the electron flow from water to photosystem II may be an integral part of the membrane and a lesion in the synthesis of such a component may lead to an abnormal membrane structure.

A study of the photochemical activities of plastids isolated from mutants in Group I show that most of the mutants have good photosystem I activity. Mutants at the *xan-m,-p* and *-q* loci have been examined for photosystem II activity with diphenylcarbazide as electron donor and dichlorophenolindophenol as acceptor. Photosystem II activity was observed with mutants at the *xan-m* and *-q* loci, but not with *xan-p*^{17}. It appears that the lack of oxygen evolution in *xan-m* and *-q* mutants is due to a lesion in the electron transport pathway from water to photosystem II.

CONCLUSIONS

In these studies, a large number of nuclear genes that control chloroplast development in barley have been identified. It is apparent, therefore, that chloroplast biogenesis is tightly controlled by the nucleus, even though extranuclear genes are also involved. Nuclear genes apparently code for the enzymes required for the synthesis of protochlorophyllide. Possibly, they code also for the holochrome protein which enables protochlorophyllide to be photoconverted to chlorophyllide *a*. Nuclear genes control steps in the assembly of the membranes of the developing chloroplast, and as shown by the studies of R. P. Levine and his colleagues (Levine, 1969) the synthesis of particular components of the photosynthetic electron transport chain are under nuclear control.

SUMMARY

Recessive lethal mutants of the *xantha, albina* and *viridis* phenotypes of barley have been studied to identify chloroplast components synthesized under the control of nuclear genes. Many *xantha* and *albina* mutants have been tested for their ability to form protochlorophyllide and etioplast structures in the dark. Synthesis of protochlorophyllide is partially or completely blocked in mutants at the *xan-f,-g,-h,-l* and *-u* loci. Most *albina* mutants fail to form significant amounts of protochlorophyllide. Wildtype and mutants at many *xantha* loci accumulate large amounts of protochlorophyllide if supplied with δ-aminolevulinic acid. Leaky and slightly-leaky mutants at the *xan-f,-g,-h,-u* and *alb-e* loci accumulate protoporphyrin IX. In the leaky mutant *xan-l*^{35}, Mg-protoporphyrin monomethyl ester accumulates. These studies show that *xan-f,-g,-h,* and *alb-e* loci control steps in the biosynthesis of Mg-protoporphyrin monomethyl ester from protoporphyrin IX. The *xan-u* gene appears to control a biosynthetic step between uroporphyrinogen and protoporphyrin IX, while *xan-l* is involved with a step between Mg-protoporphyrin monomethyl ester and protochlorophyllide. Mutants almost completely blocked in the synthesis of protochlorophyllide do not contain crystalline prolamellar bodies in their etioplasts, while leaky mutants form prolamellar bodies capable of structural transformation in the light.

Other mutants were found to be defective in the early stages of chloroplast development. Mutants at *xan-j* and *alb-f* loci lack the spectral shift from 684 to 672 nm of the newly-formed chlorophyllide *a*, and fail to disperse their prolamellar bodies on illumination. In mutants at the *xan-t* and *-u* loci, protochlorophyllide is photoconverted to a form of chlorophyllide *a* absorbing at 672 nm. These observations suggest that in those mutants, there is an abnormal organization of the components of the prolamellar body membranes.

The photosynthetic capacity of mutant leaves was tested by polarographic measurements of oxygen evolution. Mutants at the *xan-m,-o,-p,-q, vir-c* and *-e* loci and 8 *viridis* mutants at unidentified loci accumulate substantial amounts of chlorophyll, but lack oxygen evolution. These nuclear genes in some way control the assembly of a functional photosynthetic electron transport system.

ACKNOWLEDGEMENTS

The financial support from the United States Public Health Service, National Institutes of Health (GM-10819) and the Danish Natural Science Council is gratefully acknowledged. The Danish Atomic Energy Commission has provided research facilities at Risö. The authors wish to thank Dr. J. T. O. Kirk for aid and use of polarographic equipment.

REFERENCES

Appelqvist, L. Å., Boynton, J. E., Henningsen, K. W., Stumpf, P. K. and von Wettstein, D. (1968) *J. Lipid Res.*, *9*, 513

Bonner, B. C. (1969) *Plant Physiol.*, *44*, 739

Boynton, J. E. and Henningsen, K. W. (1967) *Studia Biophys.*, *5*, 85

Bruinsma, J. (1961) *Biochim. Biophys. Acta*, *52*, 576

Butler, W. L. and Briggs, W. R. (1966) *Biochim. Biophys. Acta*, *112*, 45

Foster, R. J., Gibbons, G. C., Gough, S., Henningsen, K. W., Kahn, A., Nielsen, O. F. and von Wettstein, D. (1971) in *Proc. First Europ. Biophys. Congr.* (Broda, E., Locker, A. and Springer-Lederer, H., eds.) vol. IV p. 137. Verl. Wiener Med. Akad. Wien

Foster, R. J., Henningsen, K. W. and Kahn, A. (1972) in *Proc. II Intern. Cong. Photosynthesis Res.* (Forti, G., ed.) in press, Junk Publishers, Amsterdam

Gassman, M., Granick, S. and Mauzerall, D. (1968) *Biochem. Biophys. Res. Commun.*, *32*, 295

Gough, S. (1972a) in *Proc. II Intern. Cong. Photosynthesis Res.* (Forti, G., ed) in press, Junk Publishers, Amsterdam

Gough, S. (1792b) *Biochim. Biophys. Acta*, in press

Granick, S. (1961) *J. Biol. Chem.*, *236*, 1168

Granick, S. and Gassman, M. (1970) *Plant Physiol.*, *45*, 201

Gustafsson, A. (1940) *Kungl. Fysiografiska Sallskapets Handlinger, N.F. Biol.*, *51*, 1

Henningsen, K. W. (1970) *J. Cell Sci.*, *7*, 587

Henningsen, K. W. and Boynton, J. E. (1967) *Studia Biophys.*, *5*, 89

Henningsen, K. W. and Boynton, J. E. (1969a) in *Progress in Photosynthesis Research*, (Metzner, H., ed.) vol. III, p. 73. IUBS Tübingen

Henningsen, K. W. and Boynton, J. E. (1969b) *J. Cell Sci.*, *5*, 757

Henningsen, K. W. and Kahn, A. (1971) *Plant Physiol.*, *47*, 685

Kirk, J. T. O. and Goodchild, D. J. (1972) *Aust. J. Biol. Sci.*, *25*, 215

Kirk, J. T. O. and Reade, J. A. (1970) *Aust. J. Biol. Sci.*, *23*, 33

Levine, R. P. (1969) *Annu. Rev. Plant Physiol.*, *20*, 523

Nadler, K. and Granick, S. (1970) *Plant Physiol.*, *46*, 240

Nielsen, O. F. (1970) in *Microscopie Electronique 1970, Comm. Sept. congr. intern. Grenoble*, (Favard, P., ed.) vol. III. p. 179.: Soc. Franc. Microsc. Electr., Paris

Nielsen, O. F. (1972) *in Proc. II. Intern. Congr. Photosynthesis Res.* (Forti, G., ed.) in press Junk Publishers, Amsterdam

Shibata, K. (1957) *J. Biochem.*, (Tokyo) 44, 147

Smith, J. H. C. (1954) *Plant Physiol.*, *29*, 143

Smith, J. H. C., Durham, L. J. and Wurster, C. F. (1959) *Plant Physiol.*, *34*, 340

Sundqvist, C. (1969) *Physiol. Plant.*, *22*, 147

Süzer, S. and Sauer, K. (1971) *Plant Physiol.*, *48*, 60

Walles, B. (1963) *Hereditas*, *50*, 317

von Wettstein, D., Henningsen, K. W., Boynton, J. E., Kannangara, G. C. and Nielsen, O. F. (1971) in *Autonomy and Biogenesis of Mitochondria and Chloroplasts*, (Boardman, N. K., Linnane, A. W. and Smillie, R. M. eds.) p. 205, North-Holland Amsterdam

Gene Expression in Chloroplasts and Regulation of Chloroplast Differentiation

ROBERT M. SMILLIE, N. STEELE SCOTT AND D. G. BISHOP
Plant Physiology Unit, CSIRO Division of Food Research, and School of Biological Sciences, Macquarie University, North Ryde, Sydney, Australia

In most higher plants light is essential for the formation of fully-differentiated chloroplasts. However, the biochemical processes which are part of this differentiation, e.g. the synthesis of chloroplast RNA and protein, are also subject to complex interactions with other growth processes in the cell and are hence regulated, either in the light or in darkness, by factors which control cell development such as phytochrome and plant hormones. Because of the complexity of these interactions, and in an effort to find a meaningful experimental approach to elucidation of the processes underlying chloroplast differentiation, we have looked at two aspects which are at the opposite ends of the differentiation process. On the one hand we have asked the questions where are the genes for chloroplast development and what are some of their specific functions? In an effort to obtain a partial answer to this question the possible function of chloroplast DNA, as well as that of nuclear DNA, in the production of chloroplast ribosomal RNA is being investigated. On the other hand we are examining how gene expression in chloroplast differentiation is controlled with emphasis on those controls which might be directly related to the photosynthetic capacity of the plant. As an experimental approach to this problem the final differentiation step in the formation of the chloroplast, that of grana formation is being studied. Before investigating the regulation of grana formation, it was necessary to establish the biochemical changes associated with this structural change and their significance to the photosynthetic capacity of the cell.

NUCLEIC ACID CONTENT OF PLANT CELLS

The nucleic acid content of plant cells varies with factors such as species, age of tissue and ploidy. Table 1 summarizes data on the absolute amounts and estimates the relative amounts of nucleic acids in a range of plant cells. Comparative

TABLE 1

Nucleic Acid Content of Plant Cells

	DNA	RNA
per cell ($\times 10^{-12}$g)	2-30	2-100
per chloroplast ($\times 10^{-12}$g)	0.001-0.1	0.01-1
% in chloroplast	1-10	5-50

The range of values are obtained from data given by Kirk and Tilney-Bassett (1967), Smillie and Scott (1969), Scott, *et al.*, (1971a), Detchon and Possingham (1972) and Ingle and Sinclair (1972). The estimates of chloroplast RNA are derived from measurements of relative amounts of chloroplast ribosomal RNA and are made by assuming the ribosomal RNA is 80-90% of the total RNA of a cell.

studies suggest that individual chloroplasts contain about 0.001 to 0.01 pg DNA (Table 1) and that they contain more RNA than DNA. For example a chloroplast from *Euglena gracilis* has up to 1 pg of ribosomal RNA (unpublished data) and a spinach chloroplast 0.3 pg (Detchon and Possingham, 1972). The large numbers of chloroplasts in photosynthetic cells [e.g. up to 500 in tobacco (Tewari and Wildman, 1968)] means that up to 10% of the DNA and 50% of the RNA in a leaf may be located in the chloroplasts.

POTENTIAL OF THE CHLOROPLAST DNA

Many studies (see Smillie and Scott, 1969) have shown the importance of chloroplast ribosomal systems in protein synthesis, but how important is the DNA? Studies of the rate of renaturation of denatured chloroplast DNA have

TABLE 2

Kinetic Complexity of Chloroplast DNA

Species	Size of repeat unit (daltons[a])	DNA per chloroplast (daltons)	Number of copies per chloroplast	Reference
Lettuce	3×10^6 1.2×10^8	2×10^9	16	Wells and Birnstiel (1969)
Tobacco	1.14×10^8	3×10^9	25	Tewari and Wildman (1968)
Euglena gracilis	1.8×10^8	6×10^9	33	Stutz (1970)
Chlamydomonas reinhardi	1.9×10^8 $1\text{-}10 \times 10^6$	5×10^9	26	Bastia, Chiang, Swift and Siersma (1971)
Chlamydomonas reinhardi	2×10^8 $1\text{-}10 \times 10^6$	4.3×10^9	21	Wells and Sager (1971)

[a] 10^{-14}g $= 6 \times 10^9$ daltons.

suggested that the DNA within a chloroplast contains some 20 to 30 copies of the same molecule. The results of these studies are summarized in Table 2. Thus, while up to 10% of the DNA of a photosynthetic cell may be found distributed throughout the cytoplasm in chloroplasts, it is unlikely the 10% of the information in the cell is there also. For example, from the results summarized in Table 2 it can be calculated that a chloroplast DNA sequence has enough information to code for the chloroplast ribosomal RNA and about 300 to 400 proteins of average mol. wt., 30,000.

HYBRIDIZATION EXPERIMENTS WITH PLANT NUCLEIC ACIDS

When chloroplast DNA from *E. gracilis* is incubated with increasing concentrations of ^{32}P-chloroplast ribosomal RNA a plateau of hybridization, the "saturation" value is reached (Fig. 1), indicating that all the complementary base sequ-

FIGURE 1

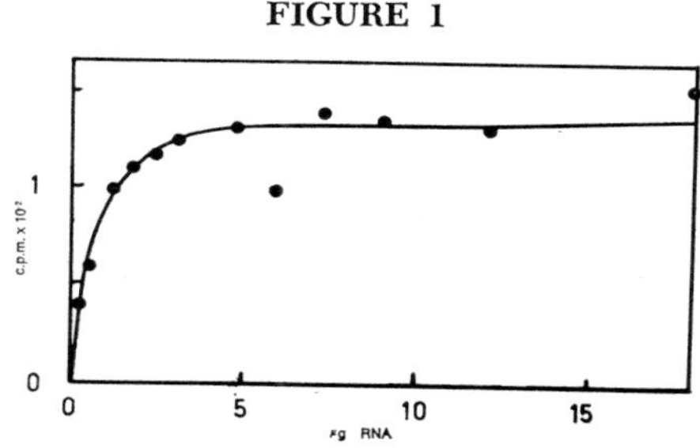

Hybridization of chloroplast r-RNA with chloroplast DNA. ^{32}P-chloroplast r-RNA (2,300 c.p.m./µg) prepared from 70S chloroplast ribosomes (Scott *et al.* 1971a) and chloroplast DNA (3.0 µg) from *E. gracilis* were hybridized in 5.0 ml of 2 × SSC (0.015 M NaCl, 0.15 M sodium citrate pH 7.0) at 70°C for 16 h as previously described (Scott *et al.* 1971a).

ences in the RNA-DNA complex are paired. If the two major components of the chloroplast ribosomal RNA are hybridized sequentially, then a plateau is obtained (Fig. 2) first with the larger ribosomal RNA molecule (1.1 x 10^6 mol. wt.), and then with the smaller molecule (0.56 x 10^6 mol. wt.), indicating that they anneal with different cistrons in the chloroplast DNA. Cytoplasmic ribosomal RNA does not anneal with chloroplast DNA but chloroplast ribosomal RNA also anneals with nuclear DNA. Similar types of analyses in algae and higher plants have given essentially the same results as shown in Fig. 3 (Scott and Smillie, 1967; Tewari and Wildman, 1968; Ingle *et al.*, 1970; Scott *et al.*, 1971a; Stutz, 1971). From these results two interpretations can arise—either that cistrons for chloroplast ribosomal RNA can be found in the nucleus as well as the chloroplast (Tewari and Wildman, 1968; Scott *et al.*, 1971a) or that the hybridization of chloroplast ribosomal RNA with nuclear DNA is an artefact (Ingle *et al.*, 1970; Stutz, 1971).

HYBRIDIZATION WITH NUCLEIC ACIDS FROM *EUGLENA GRACILIS*

The formation of the hybrids has been analysed in more detail by the use of CsCl gradient centrifugation to fractionate DNA.

FIGURE 2

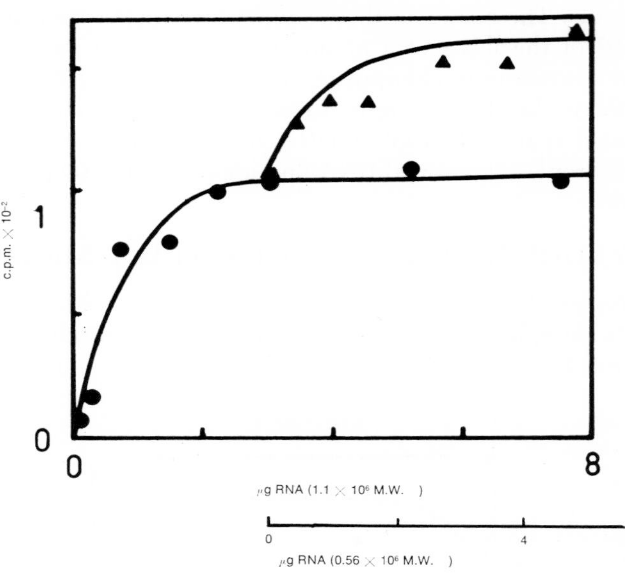

Hybridization of chloroplast r-RNA with chloroplast DNA. ^{32}P-chloroplast r-RNA was fractionated by polyacrylamide gel electrophoresis and the 1.1 × 10^6 mol. wt. species (2,700 cpm/μg) was hybridized with chloroplast DNA (3.0 μg) as described in Fig. 1. Samples which were saturated with the 1.1 × 10^6 mol. wt. r-RNA were then hybridized with the 0.56 × 10^6 mol. wt. r-RNA (2,700 c.p.m./μg).

FIGURE 3

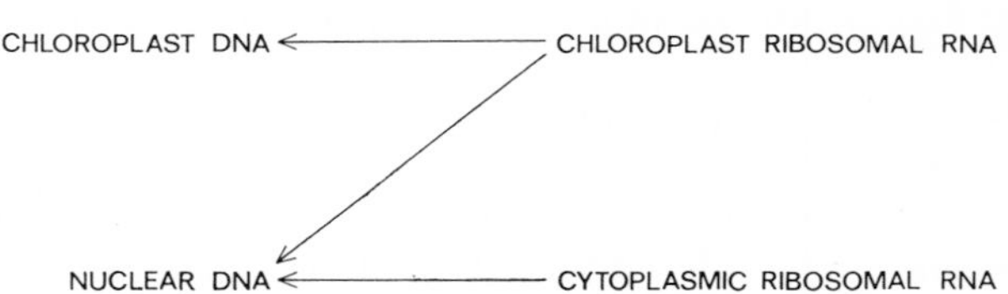

DNA-RNA hybrids formed in plants and algae. The hybridizations which have been observed between ribosomal RNA species and nuclear and chloroplast DNA, as described in the text, are indicated by arrows.

Chloroplast DNA from *E. gracilis* was fractionated on a CsC1 gradient, and each fraction halved and annealed with either purified chloroplast ribosomal RNA (1.1 x 10^6 mol. wt.) or cytoplasmic ribosomal RNA (0.56 x 10^6 mol. wt.) (Scott *et al.,* 1971a). Fig. 4 shows that while the chloroplast RNA anneals to a specific section of the DNA of average density 1.695 g cm^{-3} the cytoplasmic ribosomal RNA shows an insignificant level of hybridization along the whole gradient.

In a similar experiment (Fig. 5) cytoplasmic ribosomal RNA anneals with DNA from dark-grown cells of *E. gracilis* at a specific section of the gradient with an average density of 1.713 g cm^{-3}. The chloroplast ribosomal RNA species

FIGURE 4

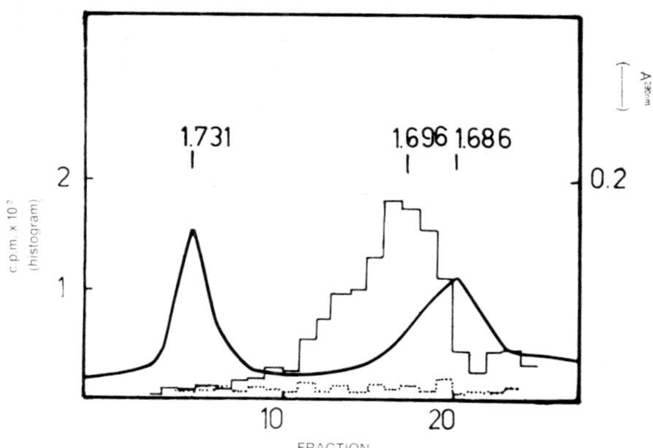

Hybridization of DNA from *E. gracilis* chloroplasts. Chloroplast DNA (ρ = 1.686 g cm^{-3}) was prepared from *E. gracilis* chloroplasts as previously described (Scott *et al.* 1971a). After centrifugation on a CsCl density gradient (Flamm *et al.*, 1966) together with a marker DNA (ρ = 1.731 g cm^{-3}, from *Micrococcus lysodeikticus*), the distribution of DNA along the gradient (————) was followed by measuring the absorbance at 260 nm in a flow cell and successive DNA fractions were collected. Each fraction was denatured and fixed on a nitrocellulose filter and the filter halved. One half was hybridized with 1.1×10^6 mol. wt. chloroplast r-RNA (21,000 c.p.m./μg) (————, histogram) and the other with 0.9×10^6 mol. wt. cytoplasmic r-RNA (40,000 c.p.m./μg) (· · · · · , histogram) in 6 × SSC (see legend to Fig. 1) at 70°C for 2 h at an RNA concentration of 3 μg/ml.

FIGURE 5

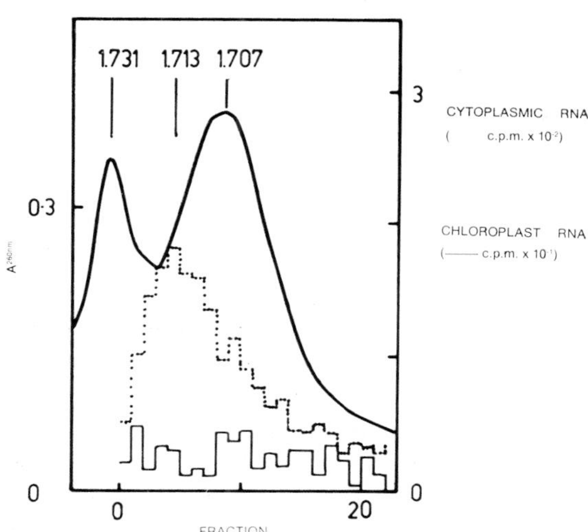

Hybridization of DNA from dark-grown *E. gracilis*. DNA (ρ = 1.707 g cm^{-3}) from dark-grown *E. gracilis* (————) was hybridized with 1.1×10^6 mol. wt. chloroplast r-RNA (21,000 c.p.m./μg) (————, histogram) or 0.9×10^6 mol. wt. cytoplasmic r-RNA (40,000 c.p.m./μg) (· · · · · , histogram) as described for Fig. 4.

does not anneal to the DNA at this point but shows a low level of hybridization along the length of the gradient.

These hybrids were further analysed by following their denaturation as the temperature was increased. Hybrids were formed between chloroplast ribosomal RNA and chloroplast DNA, and between cytoplasmic ribosomal RNA and DNA from dark-grown cells as well as the heterologous hybrid between chloroplast ribosomal RNA and the DNA from dark-grown cells. The melting curves of these hybrids are shown in Fig. 6. The two homologous hybrids show a sharp transition

FIGURE 6

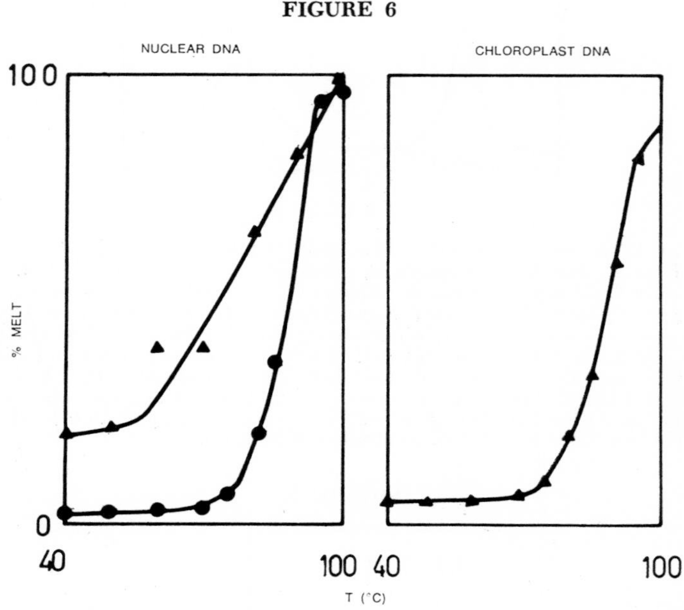

Melting curves of DNA-RNA hybrids. Hybrids formed between DNA and RNA species from *E. gracilis* as described in Figs. 4 and 5 were incubated at increasing temperatures in 1 × SSC (see legend to Fig. 1). After 10 min at each temperature the filters were transferred to fresh 1 × SSC and the radioactivity released at each temperature measured. The results are shown as a percentage of the total number of counts in each hybrid. ▲———▲, 1.1 × 10^6 mol. wt. chloroplast r-RNA hybrids; ●———●, 0.9 × 10^6 mol. wt. cytoplasmic r-RNA hybrids.

as the hybrid dissociates over a relatively narrow temperature range. By analogy with the melting of DNA duplexes (Waring and Britten, 1966) this sharp melting curve is regarded as characteristic of specific hybrids. The heterologous hybrid (Fig. 6) shows a spread of the melting curve, indicative of non-specific hybrid formation.

These results suggest that in dark-grown *E. gracilis* the cistrons for chloroplast ribosomal RNA are not found in the nucleus.

HYBRIDIZATION WITH NUCLEIC ACIDS FROM SWISS CHARD (*Beta vulgaris*)

The hybridization of chloroplast DNA and nuclear DNA with ribosomal RNA from Swiss chard after fractionation of the DNA on CsCl gradients has been studied by Ingle *et al.* (1970). It was found that the cytoplasmic ribosomal RNA annealed only with the nuclear DNA and that the cistrons had a density of 1.705 g cm^{-3}. The chloroplast ribosomal RNA also annealed at the same point on the CsCl gradient separation of nuclear DNA but annealed with the chloroplast DNA at a density of 1.699 g cm^{-3}. Cytoplasmic ribosomal RNA showed only a low level of hybridization to chloroplast DNA in Swiss chard.

The melting curves of these hybrids were followed in collaborative studies with Dr J. Ingle and are shown in Fig. 7. The homologous hybrids show sharp melting

FIGURE 7

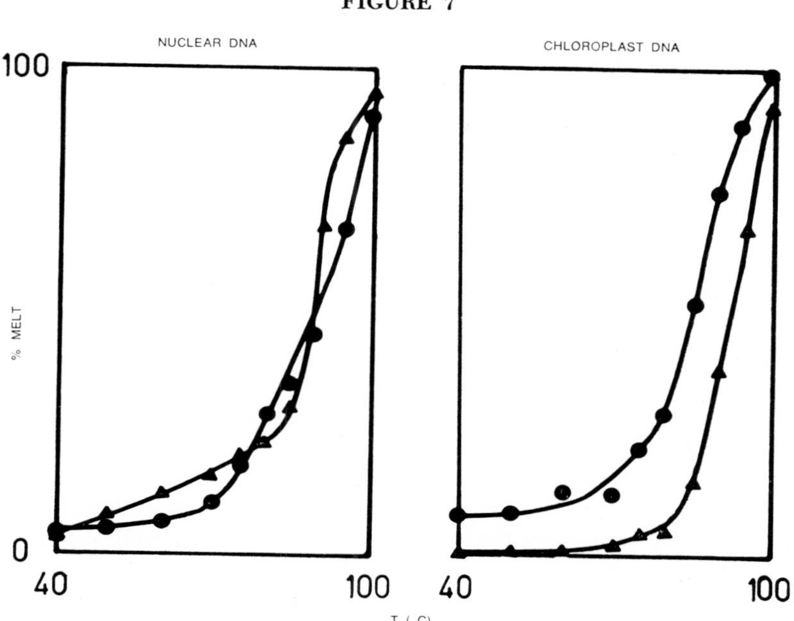

Melting curves of DNA-RNA hybrids. The melting curves of hybrids formed between DNA and RNA species from Swiss chard (*Beta vulgaris*) were determined as described in Fig. 6. ▲———▲, 1.1 × 10^6 mol. wt. chloroplast r-RNA; ●———●, 1.3 × 10^6 mol. wt. cytoplasmic r-RNA.

curves similar to those found for *E. gracilis,* and the curve for the heterologous hybrid between chloroplast ribosomal RNA and Swiss chard nuclear DNA is also similar. The other heterologous hybrid between chloroplast DNA and cytoplasmic RNA shows the less well-defined curve of a non-specific hybrid. These results suggest that the Swiss chard nuclear DNA contains cistrons for both cytoplasmic and chloroplast RNA.

The contrasting results of the experiments with *E. gracilis* and Swiss chard are summarized in Table 3 and indicate a difference in the information content of the nuclear genome of a photosynthetic alga and of a plant and suggest that the role of the nuclear genome in control of chloroplast development and metabolism is not the same throughout the plant kingdom.

TABLE 3

Specific DNA-r-RNA Hybrids Formed with Chloroplast and Nuclear DNA

	Chloroplast DNA	Nuclear DNA[a]
E. gracilis		
Chloroplast r-RNA	+	−
Cytoplasmic r-RNA	−	+
Swiss chard		
Chloroplast r-RNA	+	+
Cytoplasmic r-RNA	−	+

[a]The source of the DNA was dark-grown cells in the case of *E. gracilis* and root tissue in the case of Swiss chard.

CHLOROPLAST DIFFERENTIATION

As chloroplasts develop there is an increase in chloroplast RNA polymerase and RNA and a 70S ribosomal system for the synthesis of chloroplast proteins is formed (Smillie, *et al.*, 1972c). Soluble chloroplast enzymes including enzymes of the Calvin cycle are synthesized. In higher plants these synthetic processes do not require continuous light and are subject to regulatory substances which also influence cell growth such as photomorphogenetic pigments including phytochrome (Feierabend and Pirson, 1966; Smillie and Scott, 1969; Bradbeer, 1971; Graham, *et al.*, 1971; Scott, *et al.*, 1971b) and plant hormones (Feierabend, 1969). Chlorophyll synthesis in most higher plants requires continuous light and is associated with the structural differentiation of the chloroplast. When etiolated plants are illuminated single, primary lamellae containing the light absorbing pigments, lipids, electron transfer cofactors and membrane proteins develop first. In the final stages of chloroplast differentiation the single lamellae fuse to form stacks of appressed lamellae (grana). Appression does not occur along the entire length of many of the lamellae and sections of single lamellae remain joining the granal stacks. These are the stroma lamellae. As discussed in more detail below, the relative amounts of stroma to grana lamellae can vary considerably and in some organisms, chloroplasts containing few or no grana can be found.

Various control mechanisms operate to regulate gene expression in chloroplast development, such as repression of the synthesis of Calvin cycle and related enzymes, but these are largely metabolic in nature. Our recent studies on chloroplast differentiation have been mainly concerned with possible mechanisms of regulation of chloroplast development and function that could be more directly related to environmental factors such as light, temperature and nutrition. With this in view we have examined the significance of grana formation in terms of photochemical activities and have begun a study of the physiological and environmental factors which affect the extent of grana formation in the differentiating chloroplast.

CHLOROPLAST ULTRASTRUCTURE

Fig. 8a shows an electron micrograph of a chloroplast in a mesophyll cell of maize (*Zea mays* var. DS606A). This chloroplast is fairly typical of those found

FIGURE 8

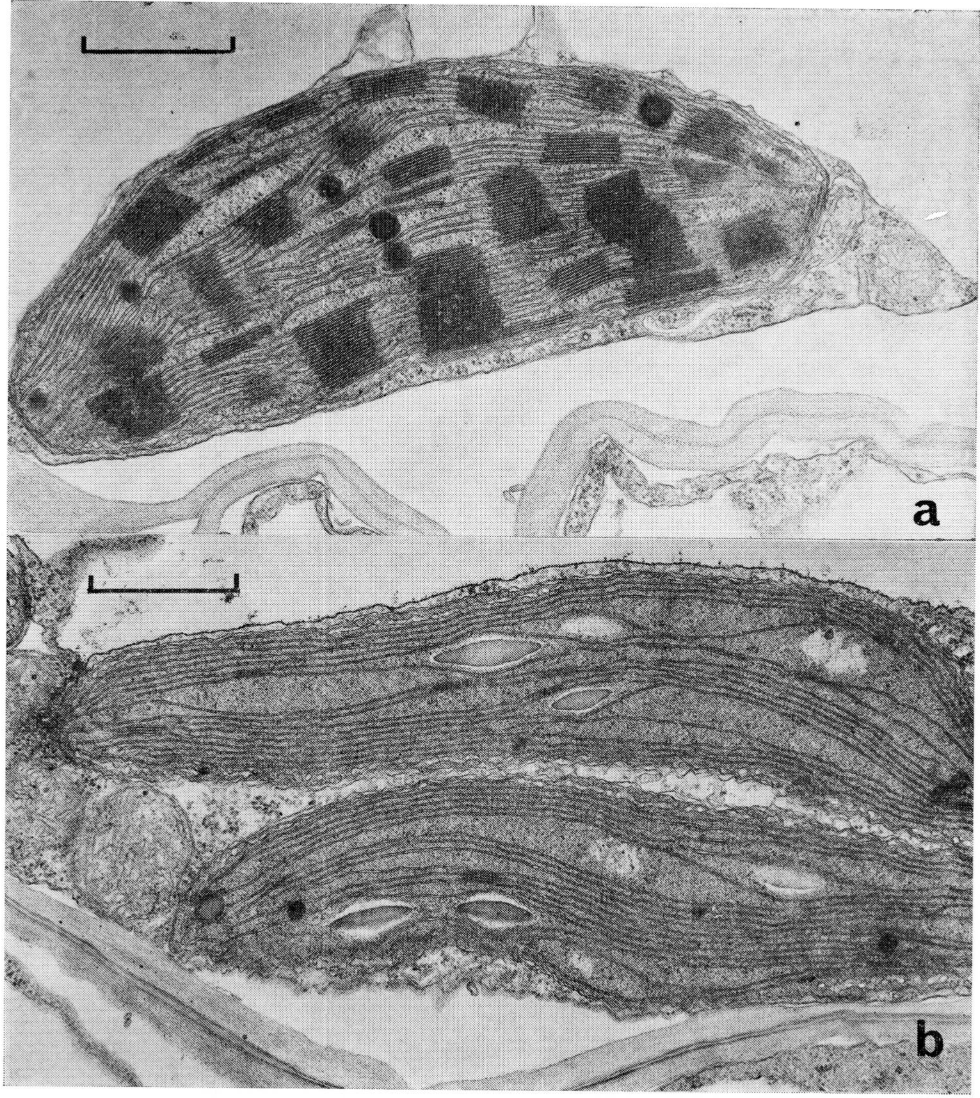

Electron micrograph of a chloroplast in (a) a mesophyll cell and (b) a bundle sheath cell of the secondary leaf of a 12-day-old maize plant. Samples for electron microscopy were fixed at 0°C in 6% (v/v) glutaraldehyde in 0.1 M phosphate buffer, pH 7.0, overnight and postfixed in 1% (w/v) osmium tetroxide for 1.5 h. After embedding in Epon, thin sections were stained with lead citrate and viewed in a Siemens Elmiskop I electron microscope. (Scale bar, 1.0 μm).

in the leaves of most higher plants, in that appressed grana lamellae and non-appressed stroma lamellae are clearly seen. The stroma lamellae are thought to contain chlorophyll and to be photochemically active. The relative amounts of grana and stroma lamellae are not constant and in mesophyll chloroplasts from younger leaves grana development is less evident (Fig. 9a). Maize, in common

FIGURE 9

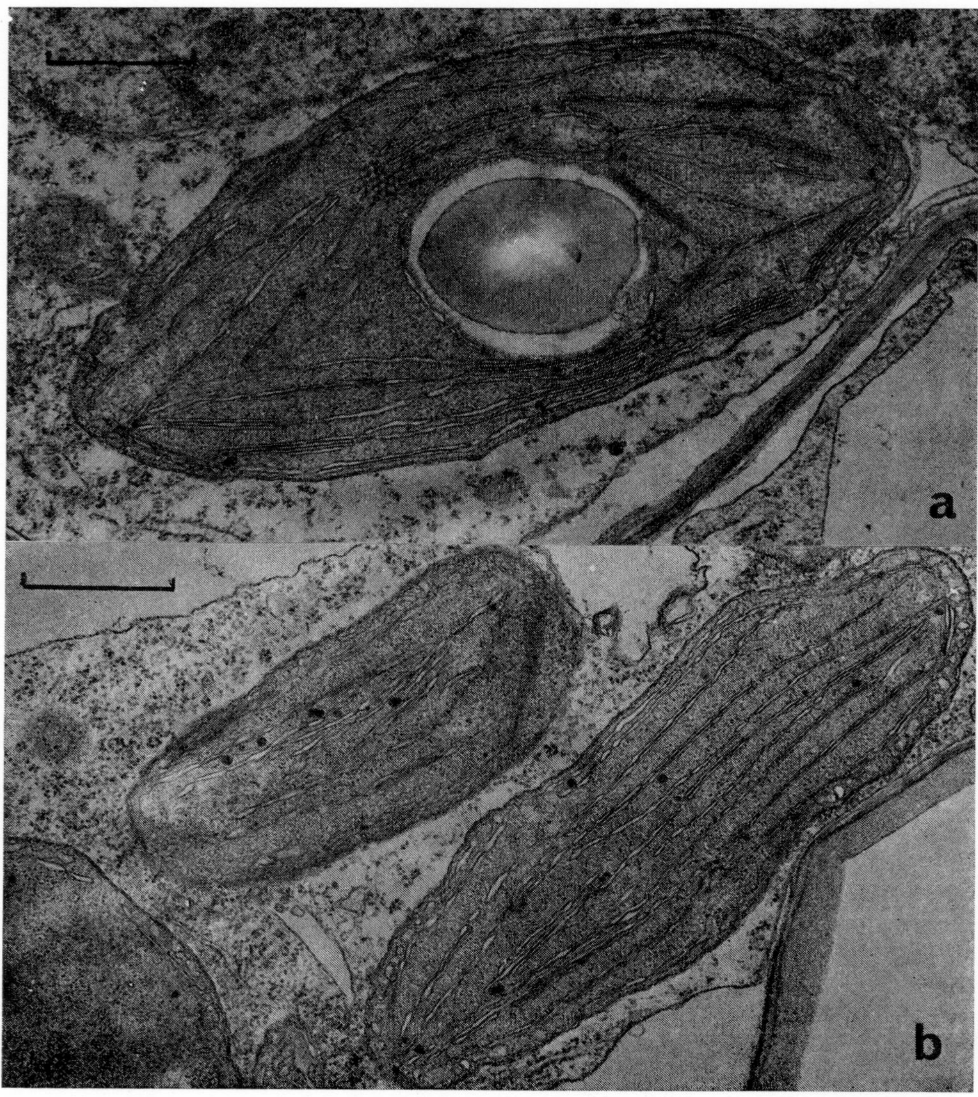

Electron micrograph of a chloroplast in (a) a mesophyll cell and (b) a bundle sheath cell of the secondary leaf of a 6-day-old maize plant. (Scale bar 1.0 μm).

FIGURE 10

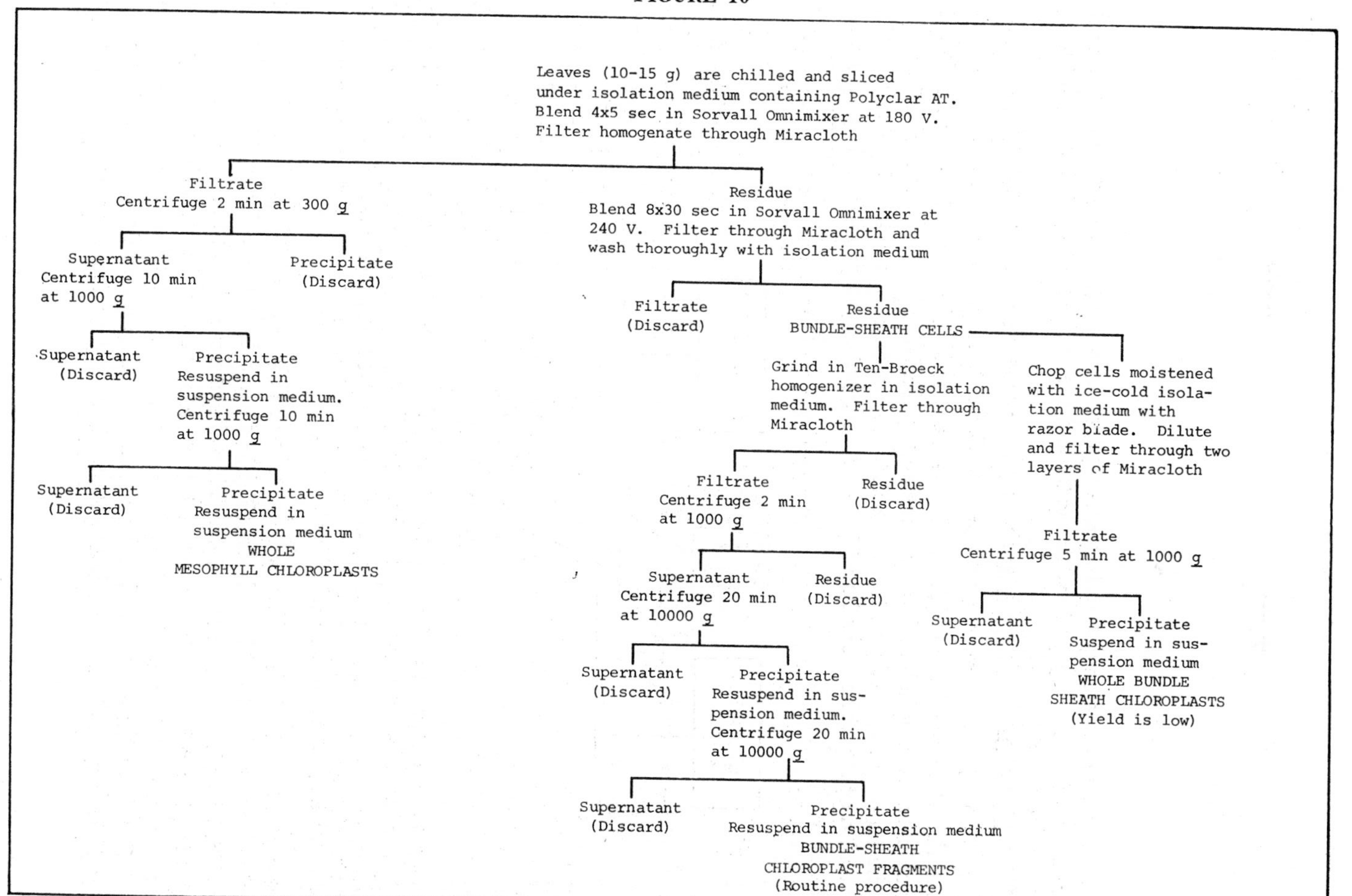

Preparation of maize mesophyll chloroplasts, maize bundle sheath cells and bundle sheath chloroplasts. The procedures follow those developed by Woo *et al.* (1970), Anderson *et al.* (1971a) and Anderson *et al.* (1971b). The composition of the isolation and suspension media are also given in these references.

with several other grasses which possess the C4-pathway of photosynthesis, has a second type of chloroplast in the cells of the bundle sheath. In the young leaf these chloroplasts are characterized by having many more stroma lamellae than grana lamellae (Fig. 9b) but as the leaf expands the chloroplasts become essentially agranal and only a very few small areas of appressed lamellae can be found in electron micrographs (Fig. 8b). A method to isolate the two types of maize chloroplasts represented in Fig. 8 is available (Fig. 10). By studying their individual photochemical activities a comparison can be made of the photosynthetic electron transfer system in a chloroplast containing single or non-appressed lamellae with a chloroplast in which extensive grana formation has taken place.

PHOTOCHEMICAL ACTIVITIES OF ISOLATED AGRANAL CHLOROPLASTS

At the time we began this work the essential differences in photochemical terms between single lamellae systems and grana lamellae were not clear in spite of the large number of publications relating structure to function in chloroplasts. Several recent publications had indicated that photosystem II (see Fig. 11) was

FIGURE 11

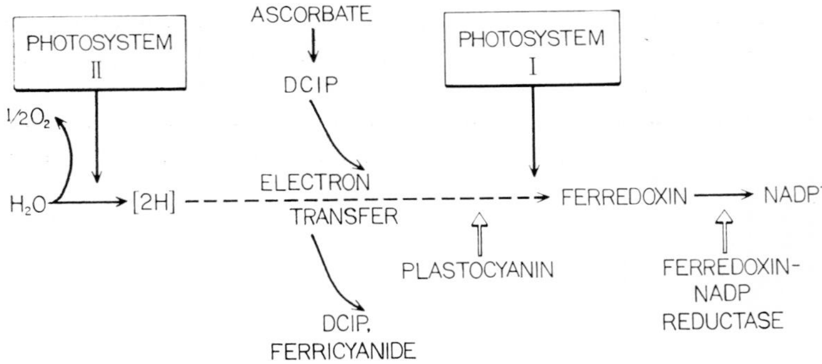

Photosynthetic electron transfer in chloroplasts. Photosystem II activity is measured by the light-dependent reduction of the electron acceptors ferricyanide or 2,6-dichlorophenolindophenol (DCIP). Photosystem I activity is measured by the light-dependent reduction of $NADP^+$ in the presence of 3-(3,4-dichlorophenyl)-1,1-dimethyl urea (DCMU) which inhibits photosystem II activity. Electron donation for photosystem I is provided by ascorbate and reduced 2,6-dichlorophenolindophenol.

not present in single lamellae systems of green algae and higher plants, including maize bundle sheath chloroplasts, (Table 4) and that grana formation might be essential for photosystem II activity. Thus the structural and biochemical changes resulting in the formation of grana might include the synthesis of the photosystem II pigment-protein complex. This, however, proved not to be the case in maize and *Sorghum* bundle sheath cells as chloroplast preparations containing very few if any grana showed photosystem II activity with a range of acceptors (Anderson *et al.*, 1971b; Bishop, *et al.*, 1971*a;* Andersen *et al.*, 1972). However, these chloroplast preparations did not photoreduce $NADP^+$ as had been reported previously by Woo *et al.* (1970) and their photosystem I activity was low. Since

TABLE 4
Absence of Photosystem II Activity in Single Lamellae Systems

Plant	Reference
Chloroplasts lacking grana	
Sorghum (bundle sheath cells)	Downton *et al.* (1970), Downton and Pyliotis (1970), Woo *et al.* (1970)
Maize (bundle sheath cells)	Woo *et al.* (1970)
Nicotiana tabacum mutant	Homann and Schmid (1967)
Bean plants grown in flashing light regime	Sironval *et al.* (1969), Argyroudi-Akoyunoglou *et al.* (1971)
Chlamydobotrys stellata grown on acetate	Wiessner and Amelunxen (1969a,b)
Stroma lamellae	
Spinach	Sane *et al.* (1970), Arntzen *et al.* (1972)

spectrophotometric measurements on intact bundle sheath cells indicated that both photosystems were active and that electron transfer between the two systems took place (Bishop, *et al.*, 1972), it appeared that some components of the electron transfer system might have been lost during isolation of the chloroplasts. When the isolated chloroplasts were supplemented with purified preparations of ferredoxin-$NADP^+$ reductase and the copper-protein plastocyanin which acts as an electron carrier between the two photosystems (Fig. 11), photoreduction of $NADP^+$ could be demonstrated (Smillie *et al.*, 1972a). As shown in Fig. 12, electron flow between the two photosystems in the isolated chloroplasts can be shown spectrophotometrically by oxidation and reduction of the added plastocyanin. When a mixture of chloroplasts and plastocyanin is illuminated the plastocyanin is reduced. The reduction occurs via photosystem II since the reaction is inhibited

FIGURE 12

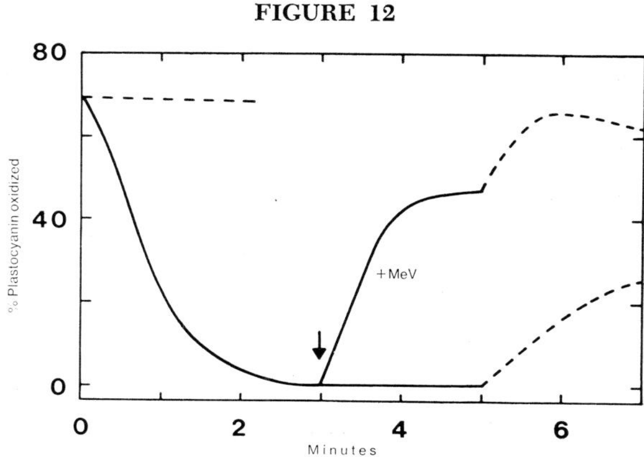

Photoreduction of plastocyanin by photosystem II and photooxidation by photosystem I in maize bundle sheath chloroplasts. The plastocyanin changes were recorded in samples illuminated with red light using an Aminco-Chance dual-wavelength recording spectrophotometer. The measuring wavelength was 590 nm and the reference wavelength 570 nm. The reaction mixture contained 4-μM plastocyanin (70% oxidized), 300-mM sorbitol, 10-mM potassium phosphate pH 7.4, 1-mM $MgCl_2$ and chloroplasts (containing 4.6 μg chlorophyll/ml.). Arrow: addition of methyl viologen (MeV) to give 1 mM; dashed lines: 2.5-μM DCMU was present. Data from Smillie *et al.* (1972a). (Reproduced with permission from *Plant Physiol.*)

by 3-(3,4-dichlorophenyl)-1,1-dimethyl urea (DCMU), a specific inhibitor of photosystem II. Upon the addition of an electron acceptor for photosystem I, methyl viologen, plastocyanin is photo-oxidized by photosystem I in a reaction which is insensitive to DCMU.

These experiments with intact bundle sheath cells and isolated chloroplasts show that the agranal chloroplasts, like the granal mesophyll chloroplasts, contain both photosystems and have the potential to photoreduce $NADP^+$ from water. Nevertheless, there are some significant differences between the photosystem II activities of the two types of chloroplasts when isolated, such as the pH optima (pH 7.0 for bundle sheath chloroplasts and pH 8.5 for mesophyll chloroplasts) and the extent of coupled photophosphorylation (Anderson, *et al.*, 1971*a*; Polya and Osmond, 1972). One striking difference is the light energy requirement for photosystem II in the two types of chloroplast.

LIGHT SATURATION CURVES FOR PHOTOSYSTEMS I AND II IN BUNDLE-SHEATH AND MESOPHYLL CHLOROPLASTS

Fig. 13 shows activity versus light intensity curves for photosystem I activity (DCMU-insensitive reduction of $NADP^+$ in the presence of ascorbate and reduced 2,6-dichlorophenolindophenol) and photosystem II activity

FIGURE 13

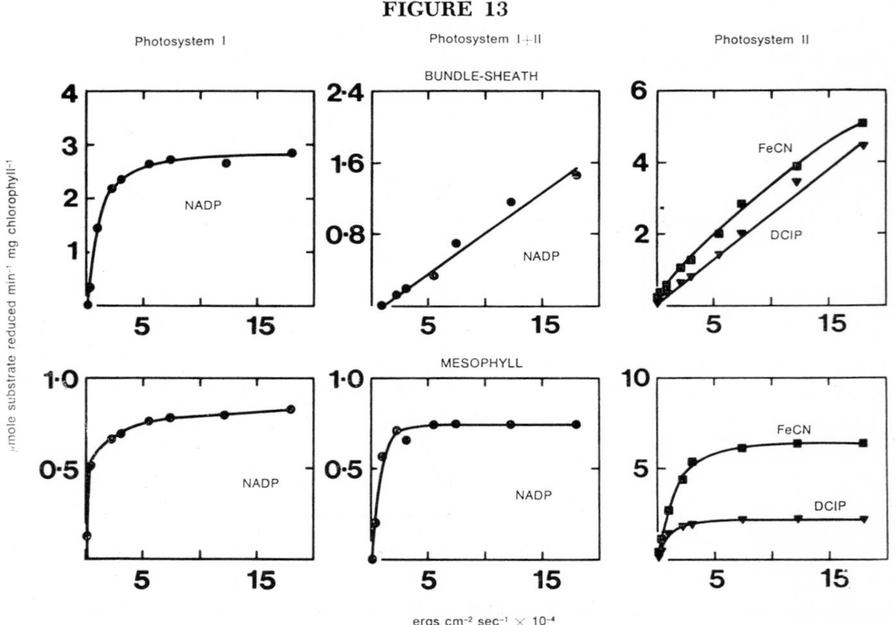

The effect of the intensity of red light upon the photochemical activities of maize bundle sheath and mesophyll chloroplasts. Assays were as described by Andersen *et al.* (1972) except that 6.4-μM plastocyanin was included in reaction mixtures where NADP reduction was measured. Actinic red light was obtained from the light of a tungsten lamp filtered through a Corning 2-60 red sharp cut filter and a Corning 1-69 heat filter. Potassium ferricyanide (FeCN). Data from Smillie *et al.* (1972b). (Reproduced with permission from *Proc. II Int. Cong. Photosynthesis Res.*, Dr W. Junk N.V. Publ., The Hague.)

(DCMU-sensitive reduction of potassium ferricyanide or 2,6-dichlorophenolindophenol) of maize bundle sheath and mesophyll chloroplasts. The activities for photoreduction of $NADP^+$ from water which involves both photosystems are also shown. Photosystem I activity in mesophyll chloroplasts becomes saturated at a slightly lower light intensity than photosystem II in the same chloroplasts, or photosystem I in bundle sheath chloroplasts, but the differences between the three curves are not great and resemble those found for both photosystems in chloroplasts containing grana in other plants. Photosystem II activity of the bundle sheath chloroplasts is quite different. The activity does not become light saturated at intensities at which photosystem II activity in mesophyll chloroplasts becomes saturated but continues to increase as the light intensity is further increased. Saturation was not obtained at the highest intensity of red light used in these experiments (4.2×10^5 erg cm^{-2} sec^{-1}).

The activity versus light intensity curve for the photoreduction of $NADP^+$ from water by bundle sheath chloroplasts resembles that obtained for photosystem II, suggesting that in these isolated chloroplasts photosystem II is the limiting light reaction for electron flow to $NADP^+$.

Light requirements for photosystems I and II similar to those shown in Fig. 13 were also found for mesophyll and bundle sheath chloroplast preparations from *Sorghum* leaves.

Several interesting considerations arising from the different light requirements for photosystem II shown by mesophyll and bundle sheath chloroplasts are discussed below.

(1) **Light requirement for photosystem II in agranal bundle sheath chloroplasts of maize.**

A disadvantage of the routine method employed to prepare mesophyll and bundle sheath chloroplasts (see Fig. 10) is that the bundle sheath chloroplasts are broken. The relationship between light intensity and photosystem II activity in the bundle sheath chloroplasts might be an artefact introduced during the isolation procedure but this is thought unlikely for the following reasons. The unusually high light requirement in the bundle sheath chloroplasts is not shown by both photosystems but only by photosystem II (Fig. 13). Also, breakage of maize mesophyll chloroplasts by sonication does not substantially alter the light saturation curve. Whole bundle sheath chloroplasts (Class I chloroplasts) can be obtained in low yield by chopping strands of bundle sheath cells with a sharp razor blade (Anderson *et al.*, 1971*a*, see Fig. 10) and these chloroplasts show essentially the same differences in the light intensity versus activity curves for photosystems I and II as do the broken bundle sheath chloroplasts. From the results obtained with bundle sheath chloroplasts it can be concluded that (i) grana formation is not a prerequisite for photosystem II activity, (ii) the bundle sheath chloroplasts may have a smaller photosynthetic unit for photosystem II than the maize mesophyll and grana-containing chloroplasts from other plants and (iii) the two light reaction centres in the agranal chloroplasts appear to be structurally separated in the lamellae.

(2) **Light intensity and photosynthesis in maize leaves.**

An important property of maize, *Sorghum* and a number of other plants which contain the C4-dicarboxylic acid pathway of photosynthesis is their high photosynthetic capacity at high light intensities, the rate of photosynthesis continuing to increase with increasing light intensity, up to full sunlight. In maize the maximum rate of photosynthesis is approached at a light intensity of about 10,000 ft-candles of sunlight (Stoy, 1965), equivalent to approximately 4.3×10^5 erg $cm^{-2} sec^{-1}$ (Gaastra, 1959), while in other plants light saturation is approached at about 2,000 ft-candles (Stoy, 1965; Gaastra, 1959; Lee *et al.*, 1970). It hence may be significant that photosystem II activity in isolated bundle sheath chloroplasts continues to increase with increasing light intensity up to very high intensities. Just how significant this is, would depend on how closely the relationship between light intensity and photosystem II activity of the chloroplast in the maize leaf resembles that of the isolated chloroplast. It is difficult to measure photosystem II activity of the bundle sheath cells in the intact maize leaf because photosystem II acceptors are not readily taken up by the leaves and because of difficulties in distinguishing between photosystem II activity in the bundle sheath cells with that in the other green cells of the leaf. It is possible, however, to measure photosystem II activity in isolated strands of intact bundle sheath cells using tetranitro blue tetrazolium (TNBT) as an acceptor. This dye, in contrast to the more common Hill oxidants such as 2,6-dichlorophenolindophenol and ferricyanide, is readily taken up by the cells. Fig. 14 shows that TNBT gives similar results to ferricyanide in measurements of photosystem II activity as a function of light

FIGURE 14

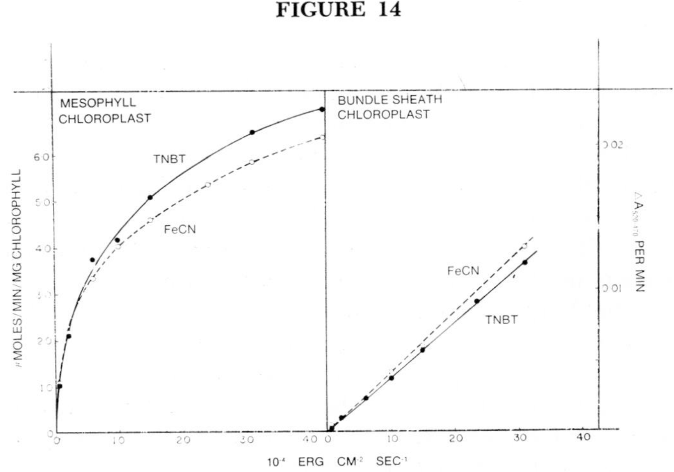

Comparison of the photoreduction of ferricyanide and TNBT at different intensities of red light by mesophyll and bundle sheath chloroplasts from maize. The reaction mixture for mesophyll chloroplasts contained in 0.75 ml, chloroplasts (3.5 μg chlorophyll), 100-mM NaCl, 2.5-mM KH_2PO_4, 5-mM $MgCl_2$, 0.05% (w/v) bovine serum albumin, 50-mM tricine buffer pH 8.5 and 66-μM potassium ferricyanide or 133 μg/ml TNBT. The same reaction mixtures were used for bundle sheath chloroplasts except that the buffer used was 50-mM phosphate pH 7.0.

intensity in isolated mesophyll and bundle sheath chloroplasts. In Fig. 15 it can be seen that the photosystem II activity, measured with TNBT, of isolated strands of bundle sheath cells increases linearly with light intensity up to 4.2 x 10^5 erg cm^{-2} sec^{-1} of red light.

FIGURE 15

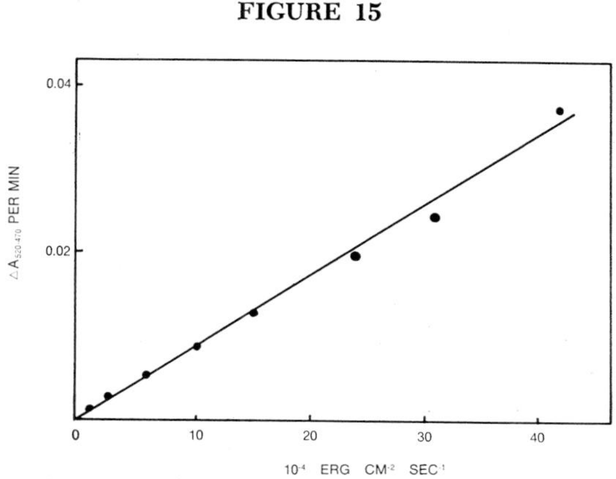

Photoreduction of TNBT by isolated strands of bundle sheath cells. The reaction was the same as used for bundle sheath chloroplasts in Fig. 14 except for the addition of Ficoll (0.35 g per 1 ml of reaction mixture). Cells were used at a concentration equivalent to 9.3 μg chlorophyll/ml.

(3) **Light intensity and photochemical activity in other non-appressed lamellar systems.**

Non-appressed chloroplast lamellae occur naturally in photosynthetic cells other than the bundle sheath cells of certain C4 plants, the most notable examples being the stroma lamellae of grana-containing chloroplasts and in the chloroplasts of red algae.

Sane *et al.* (1970) have prepared lamellae from spinach chloroplasts that are claimed to be largely derived from the stroma lamellae. These lamellae have photosystem I activity but lack photosystem II activity. Arntzen *et al.* (1972) have reported similar findings. Otherwise, there seems to be no reason to suppose that the non-appressed lamellae of maize bundle sheath cells are different from the stroma lamellae of the granal mesophyll chloroplasts or stroma lamellae in chloroplasts of other plants. Hall *et al.*, (1971) using a cytological assay concluded that stroma lamellae do in fact possess photosystem II activity. It is obviously important to establish this point with certainty since if the stroma lamellae contain a photosystem II with a light requirement similar to that shown in Fig. 15, the photosynthetic capacity of a chloroplast in relation to the light intensity will vary depending on the relative amounts of stroma and grana lamellae.

Red algae on the other hand contain chloroplasts which are completely devoid of grana lamellae. Our results with bundle sheath chloroplasts would predict that isolated chloroplasts from red algae would show the same relationship between light intensity and photosystem II activity once the red accessory pigment, phycoerythrin, had been removed. This has now been demonstrated for several different red algae and Fig. 16 shows results from an experiment with agranal chloroplasts isolated from the giant-celled red alga *Griffithsia monile* (Fig. 17).

FIGURE 16

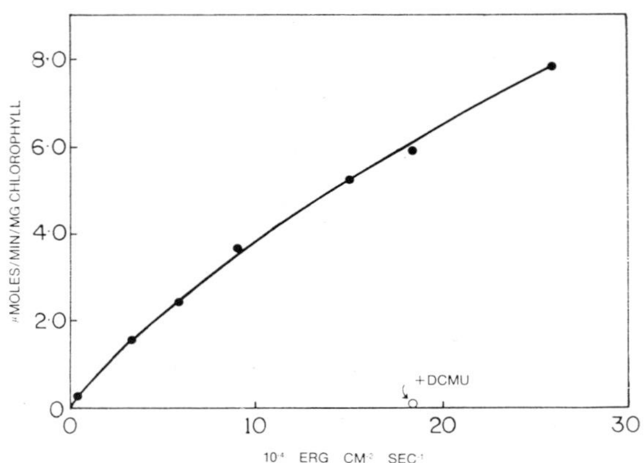

Photosystem II activity of chloroplasts from *Griffithsia monile* with increasing intensity of red light. Cells were broken in a Ten Broeck homogenizer in the medium used by Woo *et al.* (1970) and the chloroplasts were collected by centrifugation between 500 g and 3,000 g. The preparation was washed and centrifuged at 3,000 g until free of phycoerythrin. Photosystem II activity was measured by ferricyanide reduction using the reaction mixture given for bundle sheath chloroplasts in Fig. 14, except that the pH of the phosphate buffer was 7.5 and the ferricyanide concentration was 100 μM.

FIGURE 17

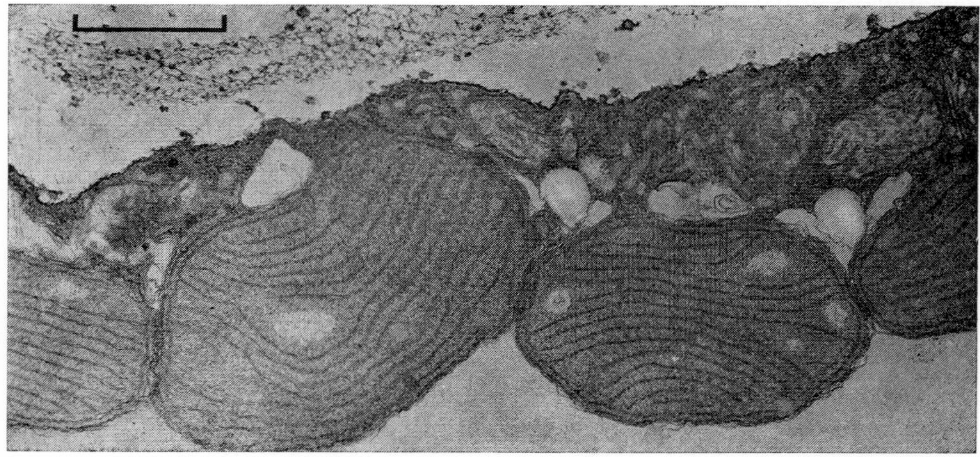

Electron micrograph of the agranal chloroplasts of the red alga *Griffithsia monile.* (Scale bar 1.0 μm).

GENETIC, DEVELOPMENTAL AND ENVIRONMENTAL FACTORS CONTROLLING THE FORMATION OF GRANA

(1) Genetic factors.

The development, structure and function of chloroplasts is under strict control of genes and many examples of mutations, both spontaneous and induced, are known in which the normal organization of the lamellar system has been arrested. Two types of mutants can be distinguished, those in which the mutation is in the nucleus and those in which mutations of extranuclear genetic factors are involved. A large number of nuclear mutations affect the chlorophyll content and ultrastructure of chloroplasts (e.g. Kirk and Tilney-Bassett, 1967; von Wettstein, 1967; Nasyrov, 1971; Walles, 1971) and mutants containing chloroplasts with no grana or rudimentary grana have been found, e.g. in mutants of *Helianthus* (Walles, 1965), *Lespedeza* (Clewell and Schmid, 1969), *Nicotiana tabacum* (Homann and Schmid, 1967) and *Chlamydomonas reinhardi* (Goodenough and Levine, 1969; Goodenough *et al.*, 1969). Chloroplasts of the ac-31 mutant of *C. reinhardi* studied by Goodenough *et al.* (1969) are of particular interest since they were completely devoid of grana yet showed higher photosystem II activity on a chlorophyll basis than chloroplasts from wild-type cells. Although photosystem I and II activities as a function of the light intensity were not measured, photosynthesis in the mutant cells saturates at a higher light intensity than in wild-type cells.

Mutations affecting plastid-localized genes also can affect grana formation (Dolzmann, 1968) as well as lamellae protein composition (Herrmann and Bauer-Stäb, 1969) and lamellae activity (Fork and Heber, 1968; Herrmann, 1971).

(2) Development and biogenesis.

In addition to external factors, lamellae differentiation is dependent upon the surrounding environment of the chloroplast such as the cell type, the kind of plant organ and its age. In many cases the level of chloroplast differentiation in the cells of a plant organ can be correlated with the photosynthetic activity of the organ (Tageeva, *et al.*, 1971).

When dark-grown leaves are illuminated the biochemical and structural changes culminating in grana formation occur in a stepwise fashion (Gyldenholm and Whatley, 1967; von Wettstein, 1967; Boardman *et al.*, 1971; Thorne and Boardman, 1971). Grana formation itself is not merely a structural rearrangement of chloroplast lamellae since it is accompanied by changes in the ratio of chlorophyll *a* to chlorophyll *b* and lipid (Bishop *et al.*, 1971*b*) and cytochrome contents (Woo *et al.* 1970). These observations suggest that it might be possible to arrest preferentially grana formation by exposing the developing chloroplast to specific inhibitors. Studies on pea leaves (Srivastava *et al.*, 1971) and *C. reinhardi* (Hoober *et al.*, 1969) indicate that chloramphenicol interferes with the fusion of lamellae. In *E. gracilis* chloramphenicol is a specific inhibitor of protein synthesis (Smillie *et al.*, 1971). It can be seen from Fig. 18 that chloramphenicol drastically inhibits lamellae fusion in *E. gracilis* although in this experiment, chlorophyll synthesis and formation of non-appressed lamellae are only partially inhibited. Cyclohexi-

FIGURE 18

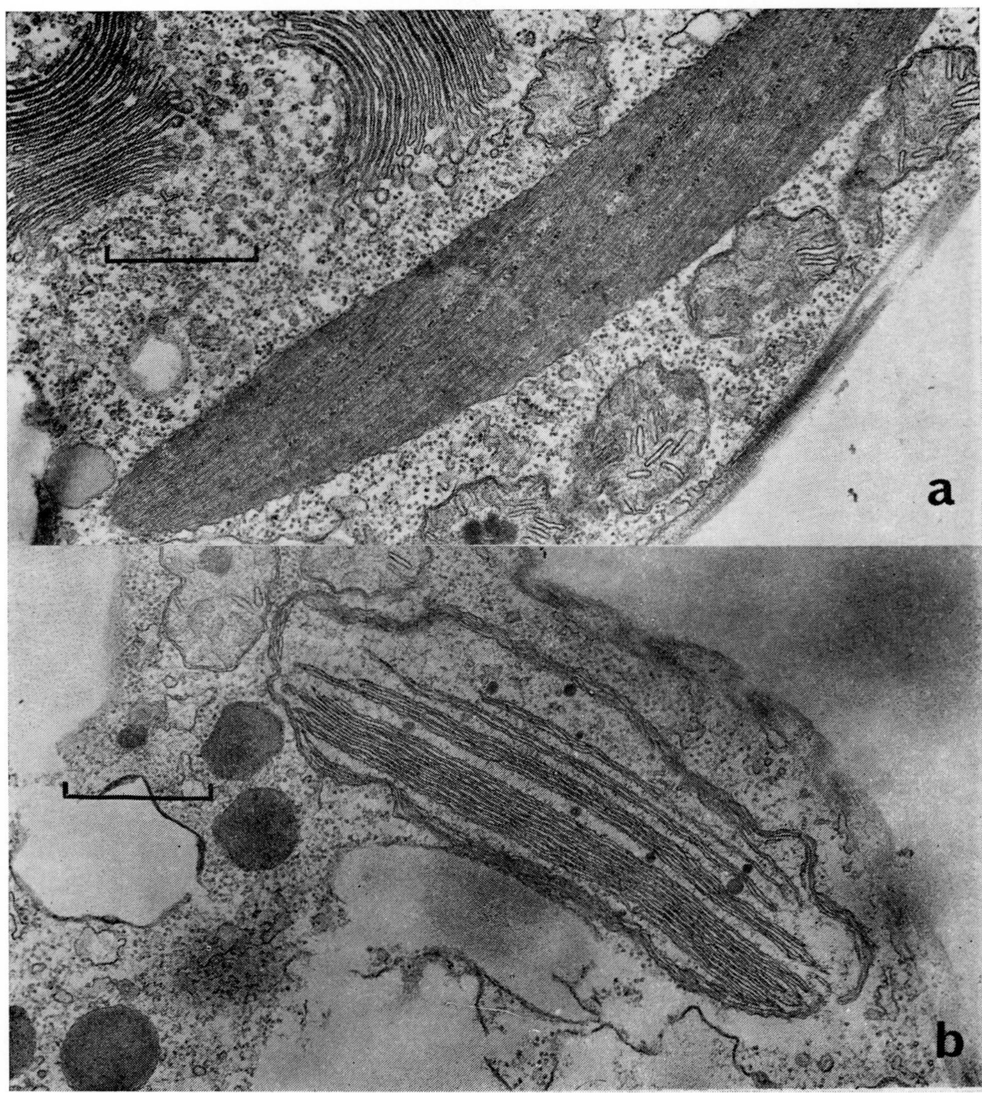

Electron micrograph of a chloroplast from a normal cell of *E. gracilis* (Fig. 18a) and one treated with chloramphenicol (Fig. 18b). In this experiment dark-grown cells were exposed to white light for 72 h in the absence or presence of chloramphenicol at 1 mg/ml. (Scale bar 1.0 μm).

mide, an inhibitor of protein synthesis on cytoplasmic ribosomes, does not preferentially inhibit fusion of the lamellae. Photosystem II activity can be demonstrated in chloroplasts isolated from chloramphenicol-treated cells, again showing that grana formation is not a prerequisite for function of photosystem II.

(3) **Environmental factors.**

Table 5 lists environmental factors which are known specifically to affect grana formation. The list is not an exhaustive one but is sufficient to indicate

TABLE 5

Environmental Factors Affecting Grana Formation

Environmental factor inhibiting grana formation	Plant	Reference
Light		
High intensity	Soybean	Ballantine and Forde (1970)
	Amaranthus	Lyttleton *et al.* (1971)
Red light	Water weed (*Elodea*)	Punnett (1971), Tageeva *et al.* (1969)
Far-red light	Bean	De Greef *et al.* (1971)
Flashing white light (1 msec every 15 min)	Bean	Sironval *et al.* (1969)
Temperature		
Low temperature (10°C)	*Sorghum* and other plants	Taylor and Rowley (1971) Taylor and Craig (1971)
Nutrition		
Growth on acetate	*Chlamydobotrys stellata*	Wiessner and Amelunxen (1969a,b)

that for a least a number of plants and algae, grana formation is subject to environmental control by factors including the intensity, spectral quality and duration of light, temperature and nutrition. High light intensity and low temperature have an additive effect in reducing the number and size of granal stacks in some plants (Ballantine and Forde, 1970; Taylor and Rowley, 1971; Taylor and Craig, 1971).

From these considerations it can be concluded that while chloroplast structure is under strict control by genes in the nucleus and chloroplast, gene expression can be modified by a variety of internal and external factors. In particular, alteration in the extent of lamellae appression brought about by these factors may change the photosynthetic activities and light requirements of the chloroplast. Whether these observations point to an important adaptive process in some plants whereby environmental factors such as light and temperature can combine to alter the characteristics of photosynthesis of a plant will be realized by future studies on the mechanism and control of the assembly of chloroplast lamellar membranes and their fusion to form grana.

SUMMARY

1. The function of chloroplast DNA in chloroplast development, factors controlling grana formation, the final step in the differentiation of chloroplasts, and

the possible significance of regulation of grana formation in optimization of photosynthetic capacity are examined in this paper.

2. Plant cells contain an average of 2-30 pg of DNA/cell and up to 100 pg RNA per cell. Although individual chloroplasts contain about 0.001 to 0.01 pg DNA and in the case of *Euglena gracilis* up to 1 pg RNA, the large numbers of chloroplasts in many photosynthetic cells means that up to 50% of the RNA and 10% of the DNA in a leaf may be located in the chloroplasts.

3. DNA-r-RNA hybridization studies have established that cytoplasmic r-RNA can anneal with nuclear DNA while chloroplast r-RNA can anneal to both nuclear and chloroplast DNA.

4. The specificity of hybrids was examined by hybridization of r-RNA with DNA fractionated on CsC1 gradients and by melting point curves of the hybrids. In experiments with nucleic acids from *Euglena gracilis,* cistrons for chloroplast r-RNA were not found in nuclear DNA fractionated on a CsC1 gradient. The homologous hybrids (chloroplast r-RNA/chloroplast DNA and cytoplasmic r-RNA/nuclear DNA) showed sharp melting curves, but the heterologous hybrid (chloroplast r-RNA/nuclear DNA) did not. In contrast, with nucleic acids of Swiss chard, hybrids of chloroplast r-RNA with chloroplast DNA and nuclear DNA both showed similar melting point curves which were characteristic of specific hybrids. These results show a difference in the information content of the nuclear genome of a plant and a photosynthetic alga and suggest that the role of the nuclear genome in the control of chloroplast metabolism is not uniform throughout the plant kingdom.

5. Photochemical changes accompanying differentiation of chloroplasts to form grana was investigated by comparing the photochemical properties of the dimorphic chloroplasts of maize. Both the mesophyll chloroplasts which contain grana and the bundle sheath chloroplasts which lack grana contain both photosystems of photosynthesis and have the capacity to photoreduce $NADP^+$ from water. The light saturation curves for photosystem I in both types of chloroplasts and for photosystem II (using ferricyanide or 2,6-dichlorophenolindophenol as the oxidant) in mesophyll chloroplasts were similar and light saturation was approached at intensities below 5×10^4 erg cm^{-2} sec^{-1} of red light. Photosystem II activity in bundle sheath chloroplasts did not saturate at intensities as high as 42×10^4 erg cm^{-2} sec^{-1} of red light.

6. Similar curves for photosystem II activity as a function of light intensity were obtained for both *intact* bundle sheath cells and isolated agranal bundle sheath chloroplasts using the dye tetranitro blue tetrazolium as the acceptor for photosystem II activity. It therefore appeared unlikely that the light intensity curve obtained for photosystem II activity in bundle sheath chloroplasts was an artefact arising during the isolation procedure. Agranal chloroplasts isolated from the red alga *Griffithsia monile* showed a very similar dependence upon light intensity for photosystem II activity to the bundle sheath chloroplasts. One conclusion arising from these results is that control factors which alter the extent of grana formation and the relative amounts of granal lamellae to single (non-

appressed) chloroplast lamellae in a plant may also change the light requirements for photosynthesis and photosynthetic capacity of the plant.

7. Genetic, developmental and environmental factors controlling grana formation are discussed. Grana formation also appeared to be dependent upon protein synthesis on the 70S chloroplast ribosomes since chloramphenicol, but not cycloheximide, prevented the appression of lamellae in *Euglena gracilis*.

ACKNOWLEDGMENTS

Many of the experiments reported here were carried out with Dr. Kirsten S. Andersen, Ann Bartsch, Jann Conroy, Vicki Home and Valerie Ryle. Dr. Joan Bain and David Gove provided the electron micrographs.

REFERENCES

Andersen, K. S., Bain, J. M., Bishop, D. G. and Smillie, R. M. (1972). *Plant Physiol. 49,* 461

Anderson, J. M., Boardman, N. K. and Spencer, D. (1971a). *Biochim. Biophys. Acta 245,* 253

Anderson, J. M., Woo, K. C. and Boardman, N. K. (1971b). *Biochim. Biophys. Acta 245,* 398

Argyroudi-Akoyunoglou, J. H., Feleki, Z. and Akoyunoglou, G. (1971) *Biochem. Biophys. Res. Common. 45,* 606

Arntzen, C. J., Dilley, R. A., Peters, G. A. and Shaw, E. R. (1972). *Biochim. Biophys. Acta 256,* 85

Ballantine, J. E. M. and Forde, B. J. (1970). *Amer. J. Bot. 57,* 1150

Bastia, D., Chiang, K., Swift, H. and Siersma, P. (1971) *Proc. Nat. Acad. Sci. U.S., 68,* 1157

Bishop, D. G., Andersen, K. S. and Smillie, R. M. (1971a). *Biochem. Biophys. Res. Commun. 42,* 74

Bishop, D. G., Andersen, K. S. and Smillie, R. M. (1971b). *Biochim. Biophys. Acta, 231,* 412

Bishop, D. G., Andersen, K. S. and Smillie, R. M. (1972). *Plant Physiol. 49,* 467

Boardman, N. K., Anderson, J. M., Kahn, A., Thorne, S. W. and Treffry, T. (1971). In *Autonomy and Biogenesis of Mitochondria and Chloroplasts,* (Boardman, N. K., Linnane, A. W. and Smillie, R. M., eds.), p. 70. North-Holland, Amsterdam.

Bradbeer, J. W. (1971). *J. Exp. Bot. 22,* 382

Clewell, A. F. and Schmid, G. H. (1969). *Planta 84,* 166

De Greef, J., Butler, W. L. and Roth, T. F. (1971). *Plant Physiol. 47,* 457

Detchon, P. and Possingham, J. V. (1972). *Phytochem. 11,* 943

Dolzmann, P. (1968). *Z. Pflanzenphysiol. 58,* 289

Downton, W. J. S., Berry, J. A. and Tregunna, E. B. (1970). *Z. Pflanzenphysiol. 62,* 194

Downton, W. J. S. and Pyliotis, N. A. (1970). *Can. J. Bot. 49,* 179

Feierabend, J. and Pirson, A. (1966). *Z. Pflanzenphysiol. 55,* 235

Feierabend, J. (1969). *Planta 84,* 11

Flamm, W. G., Bond, H. E. and Burr, H. E. (1966). *Biochim. Biophys. Acta 129,* 310

Fork, D. C. and Heber, U. W. (1968). *Plant Physiol. 4,* 606

Gaastra, P. (1959). *Mededelingen van de Landbouwhogeschool te Wageningen, Nederland 59,* 1

Goodenough, U. W., Armstrong, J. J. and Levine, R. P. (1969). *Plant Physiol. 44,* 1001

Goodenough, U. W. and Levine, R. P. (1969). *Plant Physiol. 44,* 990

Graham, D., Grieve, A. M. and Smillie, R. M. (1971). *Phytochem. 10,* 2905

Gyldenholm, A. O. and Whatley, F. R. (1967). *New Phytol. 67,* 461

Hall, D. O., Edge, H. and Kalina, M. (1971). *J. Cell Sci. 9,* 289

Herrmann, F. (1971). *Photosynthetica 5,* 358

Herrmann, F. and Bauer-Stäb, G. (1969). *Flora Jena 160,* 391

Homann, P. H. and Schmid, G. H. (1967). *Plant Physiol. 42,* 1619

Hoober, J. K., Siekevitz, P. and Palade, G. E. (1969). *J. Biol. Chem. 244,* 2621

Ingle, J., Possingham, J. V., Wells, R., Leaver, C. J. and Loening, U. E. (1970). *Symp. Soc. Exp. Biol. 14,* 303

Ingle, J. and Sinclair, J. (1972). *Nature* (*London*) *235,* 30

Kirk, J. T. O. and Tilney-Bassett, R. A. E. (1967). *The Plastids. Their Chemistry, Structure, Growth and Inheritance.* W. H. Freeman & Co., London and San Francisco.

Lee, S. S., Travis, J. and Black, C. C. (1970). *Arch. Biochem. Biophys. 141,* 676

Lyttleton, J. W., Ballantine, J. E. M. and Forde, B. J. (1971). In *Autonomy and Biogenesis of Mitochondria and Chloroplasts,* (Boardman, N. K., Linnane, A. W. and Smillie, R. M., eds.), p. 447. North-Holland, Amsterdam

Nasyrov, Yu. S. (1971). *Genetic Aspects of Photosynthesis,* (Nasyrov, Yu. S., ed.). Dushanbe, Donish (in Russian)

Polya, G. M. and Osmond, C. B. (1972). *Plant Physiol. 49,* 267

Punnett, T. (1971). *Science 171,* 284

Sane, P. V., Goodchild, D. J. and Park, R. B. (1970). *Biochim. Biophys. Acta 216,* 162

Scott, N. S., Munns, R., Graham, D. and Smillie, R. M. (1971a). In *Autonomy and Biogenesis of Mitochondria and Chloroplasts,* (Boardman, N. K., Linnane, A. W. and Smillie, R. M., eds.), p. 383. North Holland, Amsterdam

Scott, N. S., Nair, H. and Smillie, R. M. (1971b). *Plant Physiol. 47,* 385

Scott, N. S. and Smillie, R. M. (1967). *Biochem. Biophys. Res. Commun. 28,* 598

Sironval, C., Michel, J. M., Bronchart, R. and Englert-Dujardin, E. (1969). In *Progress in Photosynthesis Research,* (Metnzer, H., ed.), Vol I, p. 47. Int. Union Biol. Sci., Tübingen

Smillie, R. M., Andersen, K. S., Tobin, N. F., Entsch, B. and Bishop, D. G. (1972a). *Plant Physiol. 49,* 471

Smillie, R. M., Bishop, D. G. and Andersen, K. S. (1972b). In *Proc. II Int. Cong. Photosynthesis Res.,* (Forti, G., Avron, M. and Melandric, A., eds.), p. 779. Dr. W. Junk N.V., Publ. The Hague

Smillie, R. M., Bishop, D. G., Gibbons, G. C., Graham, D., Grieve, A. M., Raison, J. K. and Reger, B. J. (1971). In *Autonomy and Biogenesis of Mitochondria and Chloroplasts,* (Boardman, N. K., Linnane, A. W. and Smillie, R. M., eds.), p. 422. North-Holland, Amsterdam

Smillie, R. M., Munns, R., Graham, D., Scott, N. S. and Grieve, A. M. (1972). In *Proc. II Int. Cong. Photosynthesis Res.,* (Forti, G., Avron, M. and Melandric, A., eds.). Dr. W. Junk N.V., Publ., The Hague (in press)

Smillie, R. M. and Scott, N. S. (1969). In *Progress in Molecular and Subcellular Biology,* (Hahn, F. E., ed.), p. 136. Springer-Verlag, Berlin

Srivastava, L. M., Vesk, M. and Singh, A. P. (1971). *Can. J. Bot. 49,* 587

Stoy, V. (1965). *Physiol. Plant. Supp. IV,* 1

Stutz, E. (1970). *FEBS Letters* 8, 25

Stutz, E. (1971). In *Autonomy and Biogenesis of Mitochondria and Chloroplasts,* (Boardman, N. K., Linnane, A. W. and Smillie, R. M., eds.), p. 277. North-Holland, Amsterdam

Tageeva, S. V., Generosova, I. P., Derevyanko, V. G., Ladygin, V. G. and Semenova, G. A. (1971). In *Photosynthesis and Solar Energy Utilization,* (Zalensky, O. V., ed.), p. 144. USSR Academy of Sciences, Leningrad (in Russian)

Tageeva, S. V., Generosova, I. P. and Semenova, G. A. (1969). In *Progress in Photosynthesis Research,* (Metzner, H., ed.), vol. I, p. 21. Int. Union Biol. Sci., Tübingen

Taylor, A. O. and Craig, A. S. (1971). *Plant Physiol. 47,* 719

Taylor, A. O. and Rowley, J. A. (1971). *Plant Physiol. 47,* 713

Tewari, K. K. and Wildman, S. G. (1968). *Proc. Nat. Acad. Sci. U.S. 59,* 569

Thorne, S. W. and Boardman, N. K. (1971). *Plant Physiol. 47,* 252

Von Wettstein, D. (1967). In *Harvesting the Sun. Photosynthesis in Plant Life,* (San Pietro, A., Greer, F. A. and Army, T. I., eds.), p. 153. Academic Press, New York

Walles, B. (1965). *Hereditas, 53,* 247

Walles, B. (1971). In *Structure and Function of Chloroplasts,* (Martin Gibbs, ed.), p. 51. Springer-Verlag, Berlin

Waring, M. and Britten, R. J. (1966) *Science 154,* 791

Wells, R. and Birnstiel, M. (1969). *Biochem. J. 112,* 777

Wells, R. and Sager, R. (1971). *J. Mol. Biol. 58,* 611

Wiessner, W. and Amelunxen, F. (1969a). *Arch. Mikrobiol. 66,* 14

Wiessner, W. and Amelunxen, F. (1969b). *Arch. Mikrobiol. 67,* 357

Woo, K. C., Anderson, J. M., Boardman, N. K., Downton, W. J. S., Osmond, C. B. and Thorne, S. W. (1970). *Proc. Nat. Acad. Sci., U.S. 67,* 18

Products of Chloroplast DNA-Directed Transcription and Translation

P. R. WHITFELD, D. SPENCER AND W. BOTTOMLEY
Division of Plant Industry, CSIRO, Canberra, Australia

INTRODUCTION

The problem which concerns us in this section of the symposium is to define the role which chloroplast DNA plays in the development and biochemical functioning of chloroplasts. To put the problem in perspective it is appropriate to consider briefly, a few pertinent facts concerning the information content of chloroplast DNA (Table 1).

TABLE 1.

Information content of chloroplast DNA

Total amount of DNA per chloroplast	5×10^{-15}g [a]
Maximum molecular weight	3×10^{9} daltons
Molecular size from kinetic complexity analysis	1-2 x 10^{8} daltons[b,c]
Approximate number of base-pairs	2×10^{5}
Potential information content	200 proteins of molecular weight 40,000 daltons

[a]Spencer and Whitfeld (1969)
[b]Wells and Birnstiel (1969)
[c]Tewari and Wildman (1970)

The total amount of DNA present in a higher plant chloroplast is around 5×10^{-15}g (Spencer and Whitfeld, 1969) and if all this DNA were in a single molecule it would have a molecular weight of 3×10^{9} daltons. Rate of renaturation studies, however, indicate that chloroplast DNA has a kinetic complexity of only 1 to 2×10^{8} daltons (Wells and Birnstiel, 1969; Tewari and Wildman, 1970). Thus there may be as many as 15 to 30 molecules of DNA of this size per chloroplast or there may be fewer molecules, but of a larger size, containing tandem repeats of a basic unit of 10^8 daltons. DNA of molecular weight 10^8 daltons has a length of 50 μm and it is interesting to note that circular DNA molecules of approximately this size have been isolated from *Euglena gracilis* chloroplasts (Manning *et al.*, 1971).

In terms of potential genetic information, chloroplast DNA is roughly equivalent to phage T4 DNA; that is, there are at least 2 x 10^5 base-pairs and these presumably could code for approximately 200 proteins of 40,000 molecular weight. Despite considerable effort in many laboratories we can at present attribute specific functions to less than 10% of this genetic information. DNA-RNA hybridization studies have shown that 2%-3% of the chloroplast DNA can be accounted for as cistrons coding for chloroplast ribosomal RNAs (Scott and Smillie, 1967; Tewari and Wildman, 1968), and that approximately 1% of the DNA codes for chloroplast-specific transfer RNAs (Tewari and Wildman, 1970). The only *proteins* whose synthesis has been convincingly linked with chloroplast DNA function by the analysis of plastid mutants are two protein components of chloroplast membranes (Herrmann, 1971), and the large subunit of Fraction I protein (Wildman, 1972). Several other proteins involved in photosystem II activity of chloroplasts may also be coded for by chloroplast DNA but the evidence is as yet inconclusive (Fork and Heber, 1968). Allocation of a further 5% of the genetic information of the chloroplast DNA would more than take care of these products, still leaving, however, 90% unaccounted for.

In contrast to this paucity of information concerning the involvement of chloroplast DNA in the specification of chloroplast components, there is abundant evidence from the genetic analysis of biochemical lesions affecting chloroplasts that *nuclear* DNA is very much involved in directing the synthesis of chloroplast specific proteins. For example, many of the steps leading to the synthesis of chlorophyll, of carotenoids, and of the photosynthetic electron transport system are affected by nuclear mutations (see Kirk and Tilney-Bassett, 1967).

From considerations such as these, then, we became interested in trying to ascertain what the role of the major portion of the chloroplast DNA might be. Is the DNA mainly concerned in the coding of, as yet, unidentified chloroplast structural and enzymic proteins or is its main function a control one? In this paper we shall describe some of the experiments we have been doing both with respect to the transcription of spinach chloroplast DNA and also with respect to the characterization of products of chloroplast protein synthesis.

TRANSCRIPTION OF CHLOROPLAST DNA

One possible explanation for the fact that no more than 10% of the potential genetic information content of chloroplast DNA has yet been accounted for is simply that most of the DNA in chloroplasts is repressed or "switched-off". For instance, if the main function of chloroplast DNA is to code for chloroplast structural proteins, then maximum demand on the information content of the DNA might be expected to occur during the early stages of chloroplast development and transcription would be largely shut off in the later stages. Most of our studies have involved chloroplasts which are derived from spinach leaves 3cm to 6cm long and which can therefore be regarded as being fully developed. So before investing time trying to identify other chloroplast DNA-coded products we thought to ask whether the DNA of mature spinach chloroplasts is available for transcription and, if it is, then what proportion of the chloroplast DNA is in fact being transcribed *in vivo*.

We have used two approaches to assess the proportion of chloroplast DNA which is available for transcription in mature spinach chloroplasts. The first involves the use of *E. coli* RNA polymerase to measure template availability, and the second is to determine the accessibility of the DNA in chloroplasts to deoxyribonuclease digestion. The successful application of these procedures relies on the fact that, if spinach chloroplasts are gently shocked by exposure to hypotonic buffer containing Mg^{++}, the outer membrane is removed, but all the chloroplast DNA and the RNA-synthesizing activity remain closely associated with the chloroplast structure. This is illustrated in Table 2 where it can be seen that the DNA content per mg chlorophyll remains constant through two washes of the chloroplasts with hypotonic buffer.

TABLE 2.

Membrane-bound nature of DNA and the RNA-synthesizing activity of chloroplasts

Number of washes	DNA content[a] (w/w of chlorophyll)	RNA synthesis[b] (c.p.m./mg chlorophyll)
None	2.33×10^{-2}	47,500
1	2.38×10^{-2}	38,200
2	2.38×10^{-2}	45,150
3	—	46,600

[a]Spencer and Whitfeld (1969)
[b]Semal *et al* (1964)

Template availability

The basis of this method is to titrate the amount of DNA in chloroplasts which is available for transcription with *E. coli* RNA polymerase (Marushige and Bonner, 1966). The level of RNA synthesis which chloroplasts will support is measured and compared with that which purified chloroplast DNA will support. This value is then related to the total amount of DNA present in the chloroplasts as measured by direct chemical assay. Conditions of time and temperature of incubation with RNA polymerase are selected such that the amount of RNA synthesized in the presence of excess template is a linear function of the amount of enzyme added.

In Fig. 1 it can be seen that, as the amount of template added, whether in the form of purified DNA or of chloroplasts, is increased, the amount of RNA synthesized increases until a point is reached at which the enzyme is saturated. With chloroplasts as template, saturation of RNA polymerase is not reached until

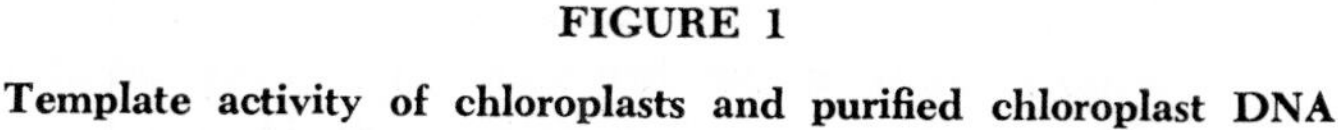

FIGURE 1

Template activity of chloroplasts and purified chloroplast DNA

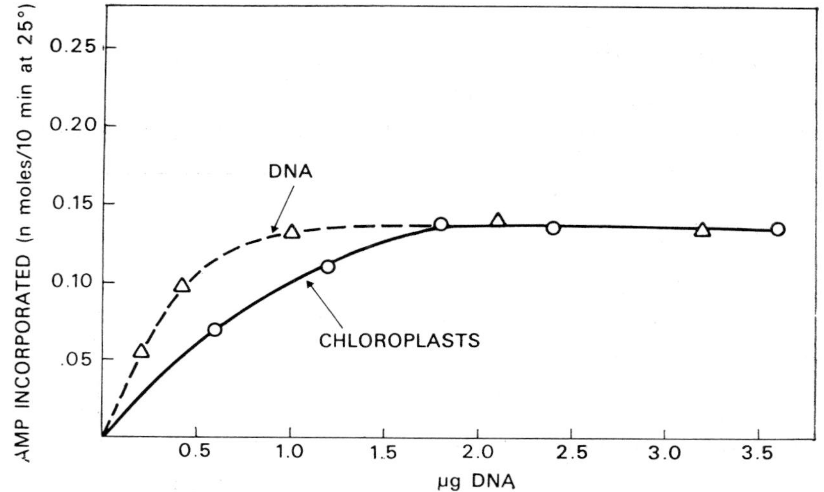

Increasing amounts of spinach chloroplasts (o——o) or purified chloroplast DNA (Δ - - Δ) were incubated with CTP, UTP, GTP, $[^{14}C]$ ATP and 0.5 units of *E. coli* RNA polymerase for 10 min at 25°C and the amount of labelled AMP incorporated into a cold trichloroacetic acid precipitable fraction was determined. Total DNA content of the chloroplasts was measured by the diphenylamine reaction.

the equivalent of approximately 1.8 μg DNA has been added, whereas with free chloroplast DNA as template, saturation is achieved with about 1.3 μg DNA. Thus it would appear that more than two thirds of the DNA in mature spinach chloroplasts is available for transcription by RNA polymerase.

Accessibility of the DNA in chloroplasts to deoxyribonuclease

Although we considered that the above result was probably a true reflection of the template availability of DNA in chloroplasts we though that confirmation of the finding by an alternative approach would be advisable. We therefore turned to the method used by Clark and Felsenfeld (1971) to measure the template availability of liver chromatin. The assumption is made that DNA which is masked in some way so that it is not able to serve as a template for RNA polymerase will be relatively resistant to attack by deoxyribonuclease.

In order to measure the deoxyribonuclease sensitivity of DNA in chloroplasts it is necessary to make use of chloroplasts whose DNA has been specifically labelled with radioactive thymidine, for the reason that the absolute amount of DNA in chloroplasts is very low. When labelled chloroplasts containing the equivalent of approximately 1 μg DNA were incubated with deoxyribonuclease and the change in acid-precipitable counts measured, it was found (Fig. 2) that 75% of the DNA was rendered acid-soluble in the first 10 minutes. After 60 minutes' incubation this value had increased to almost 90%. Control incubations in which purified labelled DNA was added to unlabelled chloroplasts and digested with deoxyribonuclease under the same conditions showed a drop of 87% in acid-precipitable counts in the first 10 minutes, increasing to more than 95% after 40 minutes (Fig. 2).

FIGURE 2

Hydrolysis of DNA in chloroplasts and of purified chloroplast DNA by deoxyribonuclease

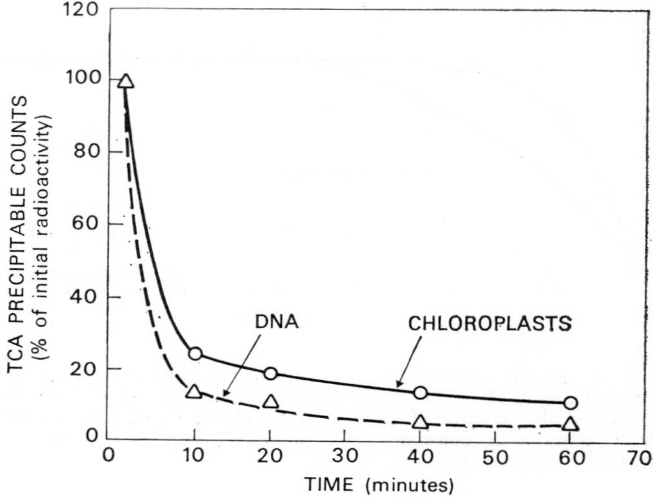

Chloroplasts (○——○) whose DNA was labelled with [^{3}H] thymidine were digested with deoxyribonuclease at 37°C and the change in cold trichloroacetic acid precipitable counts followed with time. Purified labelled chloroplast DNA (△ - - △) was treated with deoxyribonuclease under the same conditions, in the presence of unlabelled chloroplasts.

The pattern of degradation is very similar both for the free DNA and for the DNA in chloroplasts and we conclude that as much as 90% of the DNA in chloroplasts is accessible to deoxyribonuclease. If we equate deoxyribonuclease susceptibility of the DNA in chloroplasts directly to template availability for RNA polymerase then a discrepancy exists between the estimates of 90% and 70% which were obtained by the two procedures. There are many possible explanations to account for this, such as the fact that deoxyribonuclease is a much smaller molecule than RNA polymerase and so can penetrate more readily into the chloroplast matrix. For the purposes of this exercise, however, the point is irrelevant because the question we are really interested in is whether a *major* part of the DNA in mature chloroplasts is in an unmasked form. The answer to this is clearly yes, it is.

Although we conclude that most of the DNA in chloroplasts is available for transcription it does not necessarily follow that it is in fact being transcribed. Control of transcription may well be exercised through specific RNA polymerases, or factors associated therewith, and relate only indirectly to gross template availability. So how then can we get some measure of the extent to which the chloroplast genome is being transcribed *in vivo*?

"Model" system for transcription of chloroplast DNA

We have tried to answer this question with the aid of what we refer to as a "model" system for chloroplast RNA synthesis and which, in fact, is simply a purified chloroplast DNA—*E. coli* RNA polymerase transcription system. Such a system has the potential of being able to provide us with relatively large amounts of RNA transcripts complementary to all regions of the chloroplast genome.

We use *E. coli* polymerase because spinach chloroplast RNA polymerase has proved an extremely intractable enzyme to solubilize and purify. Chloroplast DNA is a very effective template for *E. coli* RNA polymerase and reaction conditions can readily be adjusted so that the ratio of the amount of RNA synthesized to the amount of template provided varies from 0.1 to 10. Examples of the type of RNA product which can result from this system are shown in Fig. 3. Fig. 3a shows the polyacrylamide gel electrophoresis pattern of RNA synthesized

FIGURE 3

Polyacrylamide gel electrophoresis patterns of RNAs synthesized in "model" systems

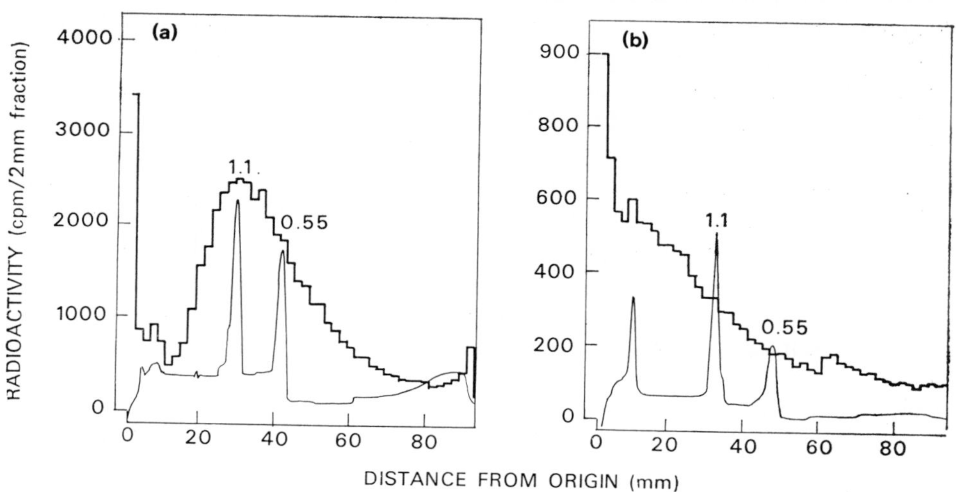

Electrophoresis was carried out in 2.4% polyacrylamide gels according to Loening (1967). Distribution of the radioactive RNA product is shown by the histograms and the positions of *E. coli* ribosomal RNA markers, as detected by u.v. scanning of the gels, are indicated by the line-graphs.

(a) RNA synthesized on 2μg chloroplast DNA template by 1 unit of *E. coli* RNA polymerase in 20 min at 25°C in the absence of salt;

(b) RNA synthesized on 8μg chloroplast DNA by 0.5 units of RNA polymerase in 60 min at 33°C in 0.07 M NH_4Cl. The u.v. absorbing peak close to the origin on the gel is due to the template DNA.

in 20 minutes at 25°C in the absence of salt, conditions which we have normally used when working with an isolated chloroplast RNA synthesizing system. The product is heterogeneous ranging in size from 2 x 10^6 to 2 x 10^4 daltons with a marked peak in the region of 1.1 x 10^6, i.e., coinciding roughly with the uv absorption peak of marker *E. coli* 23 S ribosomal RNA. This pattern is remarkably similar to the patterns we obtain for the RNA synthesized *in vitro* by isolated chloroplasts (Spencer *et al.,* 1971). Fig. 3b is a gel electrophoresis pattern of RNA synthesized in 60 minutes at 33°C in the presence of 0.07MNH_4Cl, conditions which we use in the coupled transcription-translation system which will be discussed later. The product here is clearly much larger than that synthesized at the lower temperature and raises the possibility that some chain initiation or termination signal is being by-passed.

If this model system, where chloroplast DNA is acting as template for a heterologous RNA polymerase, is to be of any assistance in the analysis of RNA synthesis in chloroplasts it must meet several requirements and we shall deal with these in turn.

Does the *E. coli* RNA polymerase preferentially transcribe from only one strand of chloroplast DNA? The first requirement is that the RNA polymerase should preferentially transcribe from only one strand of chloroplast DNA. If *E. coli* polymerase is randomly transcribing from both strands of chloroplast DNA then a population of self-complementary RNA molecules will result. Complementary RNA strands form a ribonuclease-resistant double helix if held under annealing conditions, so the percentage of ribonuclease-resistant counts present in the annealed product is a reflection of the extent to which the RNA polymerase has transcribed both strands of the DNA.

The results of the analysis of an RNA product of the "model" system, synthesized under conditions such that the ratio of RNA produced to DNA template added was 0.75, are given in Table 3. Untreated RNA, and heated, quickly cooled RNA are both 23% resistant to pancreatic ribonuclease digestion. This value seems

TABLE 3.

Self-complementarity of the RNA product of "model" system

Sample	Ribonuclease-resistant counts (% of total)
Untreated RNA	23%
RNA heated, quickly cooled	23%
RNA heated, annealed	26%
% increase in ribonuclease-resistant counts after annealing	3%
Proportion of product arising from asymmetrical transcription	97%

Model product RNA was synthesized in a reaction containing (final volume 1 ml): 0.05M TES, pH 7.8, 0.01 M $MgCl_2$ 0.07 M NH_4Cl, 0.01 M mercaptoethanol, 0.5 mM of each of ATP, CTP, GTP and UTP, 20μCi [^{14}C] ATP, 80μg spinach chloroplast DNA and 5 units of *E. coli* RNA polymerase (Burgess, 1969). Incubation was at 33°C for 60 min. The RNA was extracted with phenol-SDS, precipitated with ethanol, treated with deoxyribonuclease and purified by passage through a Sephadex G50 column. The ribonuclease-resistant assay was carried out as described by Bishop and Robertson (1969).

high but may be a result of a high content of adenylic acid residues which would reflect the 62% A-T composition of chloroplast DNA (Whitfeld and Spencer, 1968). The number of ribonuclease-resistant counts increased to 26% when the RNA was annealed. Assuming that only the difference in the ribonuclease-resistant counts of the annealed and non-annealed samples is a measure of self-complementarity, we can see that 3% of the product results from transcription of both strands of chloroplast DNA.

This result, namely that 97% of the RNA produced represents transcription from one strand only, or from non-complementary regions of both strands, is encouraging in that it implies that the *E. coli* RNA polymerase can exercise a considerable degree of strand selectivity.

Saturation hybridization of product RNA to chloroplast DNA. Another requirement that we have of the RNA product of the "model" system is that it should represent a substantial fraction of all the nucleotide sequences present in the coding strand of chloroplast DNA. Information on this question can be obtained by determining the saturation level in an RNA-DNA hybridization experiment.

FIGURE 4

Hybridization of "model" system RNA product to chloroplast DNA

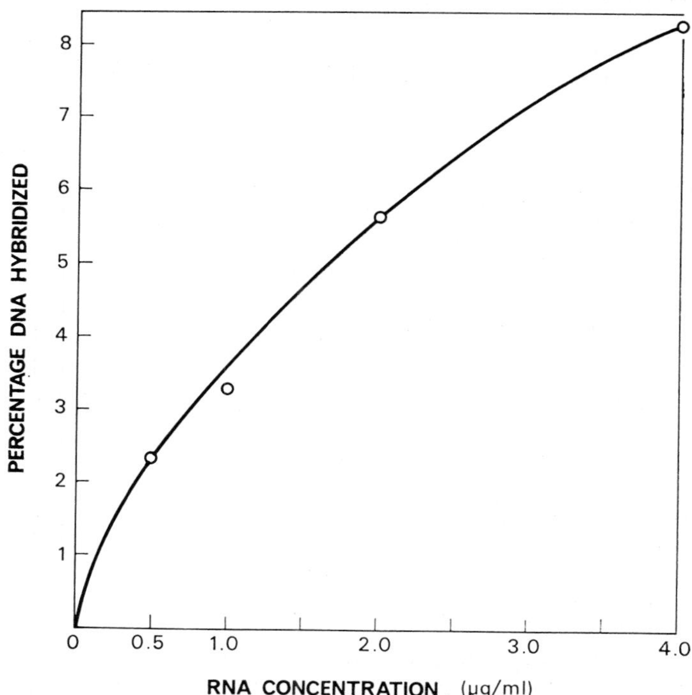

Hybridization was carried out according to the procedure of Gillespie and Spiegelman (1965). Product RNA (60,000 c.p.m./μg) was prepared as described in Table 3. Each membrane filter contained 2μg chloroplast DNA.

When increasing amounts of product RNA are hybridized to chloroplast DNA the curve shown in Fig. 4 is the result. Although saturation has not been achieved under the experimental conditions used, if the data are subjected to the analysis developed by Bishop *et al.* (1969) we can calculate that the total amount of hybridizable DNA is 30%. In other words, the RNA product contains nucleotide sequences representative of 30% of those present in the chloroplast DNA.

As we know from the self-complementarity data presented above that 97% of the RNA product is transcribed from only one of the DNA strands, we can conclude from the hybridization data that the bacterial polymerase is copying the equivalent of 60% of the nucleotide sequences of only one of the DNA strands.

We have done little as yet in the way of manipulating reaction conditions to try and achieve 100% transcription of one DNA strand. All we know is that if the ratio of RNA polymerase to template is increased too much then asymmetry of transcription tends to be lost.

Hybridization competition using *in vivo* chloroplast RNA. The last requirement we have for the RNA product of the "model" system is that it should contain nucleotide sequences complementary to the natural coding strand of chloroplast DNA. To answer this question we turned to competition hybridization. If RNA which has been isolated from spinach chloroplasts effectively competes with the "model" RNA product for hybridization to chloroplast DNA then we can assume that the species of RNA synthesized by bacterial polymerase on purified chloroplast DNA have a counterpart in the intact leaf cell. When samples of labelled "model" product RNA, containing increasing amounts of unlabelled chloroplast RNA, were hybridized to chloroplast DNA the amount of label which bound to the DNA decreased until a plateau was reached at which approximately 70% of the product was competed for. Very high levels of chloroplast RNA (ratio of competing RNA to labelled product RNA in excess of 500) were necessary to achieve maximum competition. This suggests that the ribosomal and transfer RNAs, which are the major components of chloroplast RNA preparations, were not a significant factor in the hybridization competition, and this would be in keeping with the fact that the ribosomal and transfer RNA cistrons account for only a few percent of the chloroplast genome. The RNA species which are principally effective in the competition must be present in only trace amounts in chloroplasts, as would be expected if they were of the messenger RNA class.

From this result we can conclude that *E. coli* RNA polymerase preferentially transcribes *in vitro* the same strand of chloroplast DNA as the chloroplast RNA polymerase transcribes in the intact leaf cell.

Proportion of chloroplast genome transcribed *in vivo*

This information has some bearing on the question posed above, namely to what extent is the chloroplast genome transcribed *in vivo*. Under conditions of limiting enzyme concentration the bacterial polymerase preferentially transcribes one strand, or possibly non-complementary regions of both strands of chloroplast DNA. The equivalent of 60% of the nucleotide sequences of one of the DNA strands is represented in the RNA product and 70% of this RNA product is competed for by chloroplast RNA in hybridization experiments. Therefore, approximately 40% (70% of 60%) of the chloroplast DNA information is present as RNA transcripts in mature chloroplasts. This of course is a minimal estimate because the "model" system RNA product does not contain sequences complementary to all the coding strand of chloroplast DNA.

As mentioned in the Introduction less than 10% of the potential genetic information of chloroplast DNA has been accounted for in terms of identifiable products. We now know that at least 40% of the information is likely to be expressed in mature chloroplasts and we can go ahead with some confidence and try to characterize other chloroplast DNA coded products.

PROTEIN SYNTHESIS

Protein Synthesis in a chloroplast DNA-directed "model" system

In parallel with our experiments on the transcription of chloroplast DNA we have been involved with the problem of trying to identify those proteins which are coded for by chloroplast DNA. A logical extension of the work just described on transcription of chloroplast DNA by *E. coli* RNA polymerase is to couple this process to a protein synthesizing system. The resultant protein products would of necessity be coded for by chloroplast DNA and, in this sense, the system has an advantage over an endogenous protein-synthesizing system of isolated chloroplasts. In the isolated chloroplast system (as described in the next section) protein synthesis is not dependent on concomitant transcription but reflects the continued translation of preexisting messenger RNAs, the origin of which could be either chloroplast or nuclear DNA.

The coupled transcription-translation system we have adopted is basically that which was developed and applied so successfully by Hayashi and his colleagues (Bryan *et al.*, 1969) to the phage DNA-directed synthesis of specific phage proteins. Purified components prepared from *E. coli* (ribosomes, high speed supernatant and RNA polymerase) are added to chloroplast DNA along with the four nucleoside triphosphates, an ATP-generating system and radioactive amino acids. Under these conditions at least 90% of the resultant protein synthesis is dependent on the presence of chloroplast DNA.

If the reaction mixture is electrophoresed on a polyacrylamide gel at pH 8.4 the distribution of radioactivity is found to be quite similar to that which is obtained for the product of an endogenous, isolated chloroplast system. However, if the product is electrophoresed on a polyacrylamide gel in the presence of sodium dodecyl sulphate (SDS), that is under conditions where electrophoretic mobility reflects molecular weight of the individual sub-units, a very different picture is obtained. In Fig. 5a we see that a significant proportion of the released polypeptides (i.e., those in the 144,000 g supernatant of the reaction mixture) are of low molecular weight, migrating almost as rapidly as the bromphenol blue marker. The remainder of the soluble product is of larger molecular weight, ranging up to 30,000 daltons, but discrete radioactive bands are not apparent

FIGURE 5

SDS-polyacrylamide gel fractionation of protein products synthesized in a chloroplast DNA-directed transcription-translation system

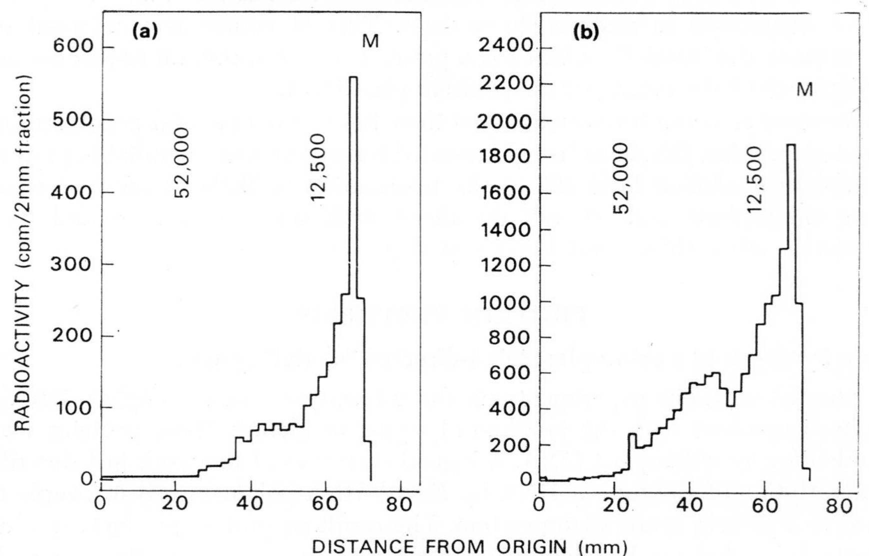

Chloroplast DNA was incubated with purified *E. coli* components (RNA polymerase, ribosomes, high speed supernatant), appropriate substrates and radioactive amino acids for 60 min at 33°C. The reaction mixture was centrifuged at 144,000g for 60 min and the supernatant and pellet electrophoresed on 9% polyacrylamide gels in SDS. (a) 144,000g supernatant; (b) 144,000g pellet. M indicates position of bromphenol blue marker. The migration of marker proteins of molecular weight 52,000 and 12,500 daltons, is also shown.

and the whole pattern is suggestive of a rather heterogeneous population of fairly small polypeptides. A very similar pattern (Fig. 5b) is obtained for the 144,000 g pellet fraction (i.e., the nascent, ribosome-associated peptides)—no well-defined sharp bands, and a large proportion of the product having a molecular weight of 8,000 to 10,000 daltons. A check on the RNA synthesized in the coupled system showed it to be much the same as reported in the preceding section for the transcription step alone, that is, a heterogeneous product, but of a size generally greater than 0.5 x 10^6 daltons. Thus it could readily code for a protein of molecular weight 50,000 or more. These results seem to indicate that either much of the transcribed RNA is not destined to code for protein or that the RNA is not being read accurately by the translation machinery.

Efforts to find conditions where a larger protein product is formed have so far been without success. Addition of folinic acid as a source of formyl groups and of partially purified protein factors prepared by salt extraction of chloroplast ribosomes was without effect on the system. It seemed possible that transcriptional controls which exist in the chloroplast could be lost or destroyed during isolation of free chloroplast DNA. We therefore tried using isolated chloroplasts themselves as a source of DNA template in the coupled "model" system, but again the proteins synthesized were heterogeneous and, if anything, of lower molecular weight than those synthesized with purified chloroplast DNA as template.

At present then we are unable to draw any conclusions from these results. The evidence presented earlier would suggest that a major fraction of the RNA transcripts produced by *E. coli* RNA polymerase from chloroplast DNA is meaningful in that it is competitive with a naturally occurring component of chloroplast RNA. Thus it is likely that the translation step in the coupled system is at fault and we are examining ways of overcoming this. However, one should not lose sight of the remote possibility that the result we have obtained is the correct one and that the shortcoming lies in our interpretation of it.

Endogenous protein synthesis in isolated chloroplasts

(a) **Lysed chloroplasts.** In this final section we shall describe some very recent experiments on the synthesis of proteins by isolated chloroplasts, because here there is evidence for the identification of at least one specific protein product.

Our standard procedure for studying endogenous protein synthesis in isolated chloroplasts has been to isolate chloroplasts from young spinach leaves in a

FIGURE 6

SDS-polyacrylamide gel fractionation of proteins synthesized in isolated, disrupted spinach chloroplasts

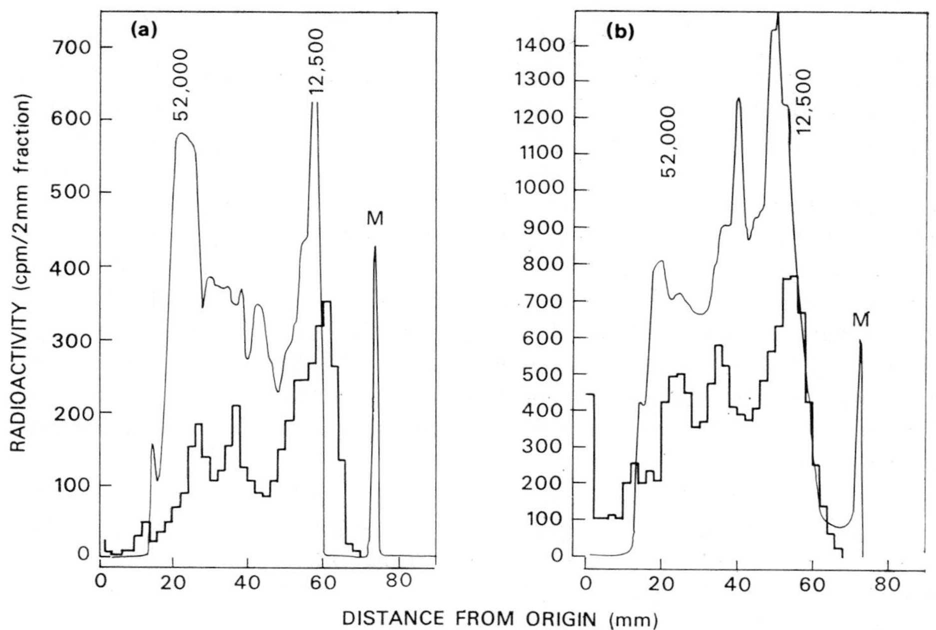

Spinach chloroplasts were incubated in TES, Mg^{++}, mercaptoethanol buffer with GTP, ATP, an ATP-generating system and radioactive amino acids for 45 min at 25°C. The reaction mixture was centrifuged at 144,000g for 30 min and fractionated on SDS-polyacrylamide gels. (a) 144,000g supernatant; (b) 144,000g pellet. Histograms show distribution of radioactivity; line-graphs are the scans of the stained chloroplast proteins.

buffered ficoll-dextran-sucrose medium and then to assay them in a hypotonic medium containing TES, $MgCl_2$ and mercaptoethanol (Spencer *et al.*, 1971). When supplemented with labelled amino acids, GTP, ATP and an ATP-generating system these lysed chloroplasts incorporate amino acids into protein. When the reaction mixture is fractionated into 144,000 g supernatant and pellet and the products are electrophoresed on SDS polyacrylamide gels the distribution of chloroplast proteins (as detected by staining) and of radioactive products is as shown in Fig. 6. Three points can be made about these scans. The first is that these patterns are clearly very different from those obtained with the chloroplast DNA-directed transcription-translation system (Fig. 5). There is no dominant peak of radioactivity in the low molecular weight region of the gel, comparable to that seen in the "model" system. Secondly, relatively discrete protein species appear to become labelled; at least three peaks of radioactivity occur both in the soluble, 144,000 g supernatant (Fig. 6a) and in the 144,000 g pellet (Fig. 6b). The two larger products, of approximate molecular weight 44,000 and 30,000 daltons, may be common to both pellet and supernatant fractions, but the small molecular weight products have slightly different electrophoretic mobilities.

The third point is that in no fraction is there any obvious coincidence between radioactive peaks and stained bands of chloroplast proteins.

(b) **Intact chloroplasts.** This then is what we thought was the normal pattern of protein synthesis in isolated chloroplasts. We were therefore very interested to read the recent report by Blair and Ellis (1972) in the Proceedings of the Biochemical Society that isolated pea chloroplasts incorporate radioactive leucine into only one soluble protein, tentatively identified by them as the large subunit of Fraction I protein (ribulose diphosphate carboxylase) in a reaction which is completely light-dependent. While the synthesis of Fraction I protein by chloroplasts might not seem too surprising because it is the major protein component of chloroplasts accounting for 33% of the soluble protein we, and a number of other groups, have tried repeatedly but without success to demonstrate its synthesis in isolated chloroplasts. Two other aspects of the system reported by Blair and Ellis (1972) were contrary to our experience. The first was that the synthesis was unaffected by ribonuclease and the second was that only one soluble protein became labelled. We have invariably found that a number of soluble protein species become labelled.

Because these results were qualitatively so different from our own we decided to examine the Blair and Ellis, light-driven, protein synthesizing system. Following their prodecure, spinach chloroplasts were isolated in a sucrose buffer, resuspended in a 0.2 M KCl buffer and incubated at 20°C in the dark or light with radioactive amino acids. No other additions, such as ATP, were made; this system is entirely dependent on endogenous components and endogenous photophosphorylation. Direct assay of trichloracetic acid precipitable counts showed that incubation of the chloroplasts in the light stimulated incorporation 10 to 15 fold and that the reaction is insensitive to actinomycin D, deoxyribonuclease, ribonuclease, rifamycin and cycloheximide but sensitive to chloramphenicol and DCMU (Table 4).

TABLE 4.

Effect of inhibitors on light-dependent protein synthesis in isolated spinach chloroplasts

Inhibitor	Concn. (μg/ml)	% inhibition
Actinomycin D	20	0
Deoxyribonuclease	2	0
Ribonuclease	4	0
Rifamycin SV	80	0
Chloramphenicol	50	78
Cycloheximide	60	0
DCMU[a]	5	100

[a]1-(3,4-dichlorophenyl)-3,3 dimethyl urea

0.5 ml chloroplasts (containing approximately 200μg chlorophyll) in 0.06M TRICINE, 0.007M $MgCl_2$, 0.2M KCl, pH 8.3, were incubated with 0.5μCi ^{14}C-amino acid mix (54 mCi/m atom C) at 20° C for 30 min in the dark or light (7,500 foot-candles white light). Hot trichloroacetic acid insoluble counts were assayed as described elsewhere (Spencer, 1965). Control incubations incorporated approximately 5,000 c.p.m./mg chlorophyll in the dark and 75,000 c.p.m./mg chlorophyll in the light.

If the chloroplasts are lysed after incubation, then fractionated into a soluble 144,000 g supernatant and a 144,000 g pellet and the products analysed by electrophoresis on SDS polyacrylamide gels, the patterns shown in Fig. 7 are obtained. It can be seen (Fig. 7a) that in the 144,000 g supernatant fraction only one protein has become labelled, and that, by reference to the tracing of the stained gel, the protein which is labelled coincides exactly with the large subunit of Fraction I protein. We have not as yet attempted to verify the identity of the labelled protein as Fraction I subunit by finger-print analysis but we believe Blair and Ellis have now done this. Fig. 7b shows the distribution of radioactivity and of chloroplast proteins from the 144,000 g pellet. This fraction would include membrane proteins, ribosomal proteins and incomplete polypeptide chains associated with ribosomes. Here four peaks of radioactivity are apparent. The peak having the largest molecular weight is probably the large subunit of Fraction I protein which can appear in the 144,000 g pellet because of incomplete lysis of the chloroplasts. There is no obvious coincidence between stained protein bands and the other three radioactive peaks, and at present we can offer no clue as to their identity.

These results, which are so different from our earlier observations with lysed chloroplasts, completely confirm the findings of Blair and Ellis (1972) and lead us to wonder why it is that the endogenous protein product synthesized in a light-dependent reaction is not the same as that synthesized in a supplemented, ATP-dependent reaction. The most likely explanation is that the specific synthesis of the large subunit of Fraction I protein depends on the physiological state of the

FIGURE 7

SDS-polyacrylamide gel fractionation of proteins synthesized in isolated, intact spinach chloroplasts

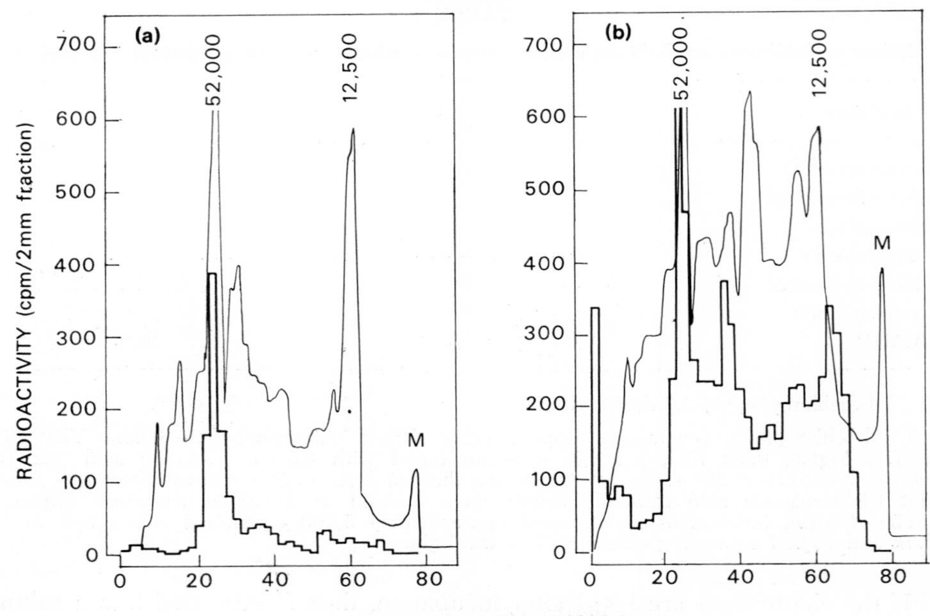

Spinach chloroplasts were incubated in TRICINE, Mg^{++}, 0.2M KCl buffer with radioactive amino acids for 45 min at 20°C in the light. The chloroplasts were then lysed by dialysis against 2MM KCl buffer and centrifuged and fractionated as described in Fig 6. (a) 144,000g supernatant; (b) 144,000g pellet. Histograms show distribution of radioactivity; line-graphs are the scans of the stained chloroplast proteins.

TABLE 5.

Effect of isolation and assay media on light-dependent protein synthesis in spinach chloroplasts

Isolation medium	Assay medium	
	Sucrose, TRIS, Mg^{++}, 0.01 M KCl, mercaptoethanol[a]	TRICINE, Mg^{++}, 0.2 M KCl[b]
	Light-dependent amino acid incorporation (c.p.m./mg chlorophyll)	
Ficoll, dextran, sucrose, TRIS, Mg^{++}, mercaptoethanol[c]	680	22,900
Sucrose, HEPES, EDTA, isoascorbate[b]	3,570	80,000

[a]Spencer and Whitfeld (1967)
[b]Ellis and Hartley (1971)
[c]Spencer (1965)

isolated chloroplasts. The two points of difference between our original system and that of Blair and Ellis lie in the composition of the medium used for chloroplast isolation and in the composition of the medium used for assay. Chloroplasts isolated in our usual ficoll-dextran-sucrose buffer and then assayed in buffered sucrose show very little light-dependent protein synthesizing activity (Table 5), but if such chloroplasts are assayed in the Blair-Ellis, 0.2 M KCl buffer then there is a substantial increase in the amount of light-dependent protein synthesis. Chloroplasts isolated in the Blair-Ellis, sucrose-EDTA buffer and assayed in

FIGURE 8
Electron micrograph of spinach chloroplasts resuspended in buffered sucrose

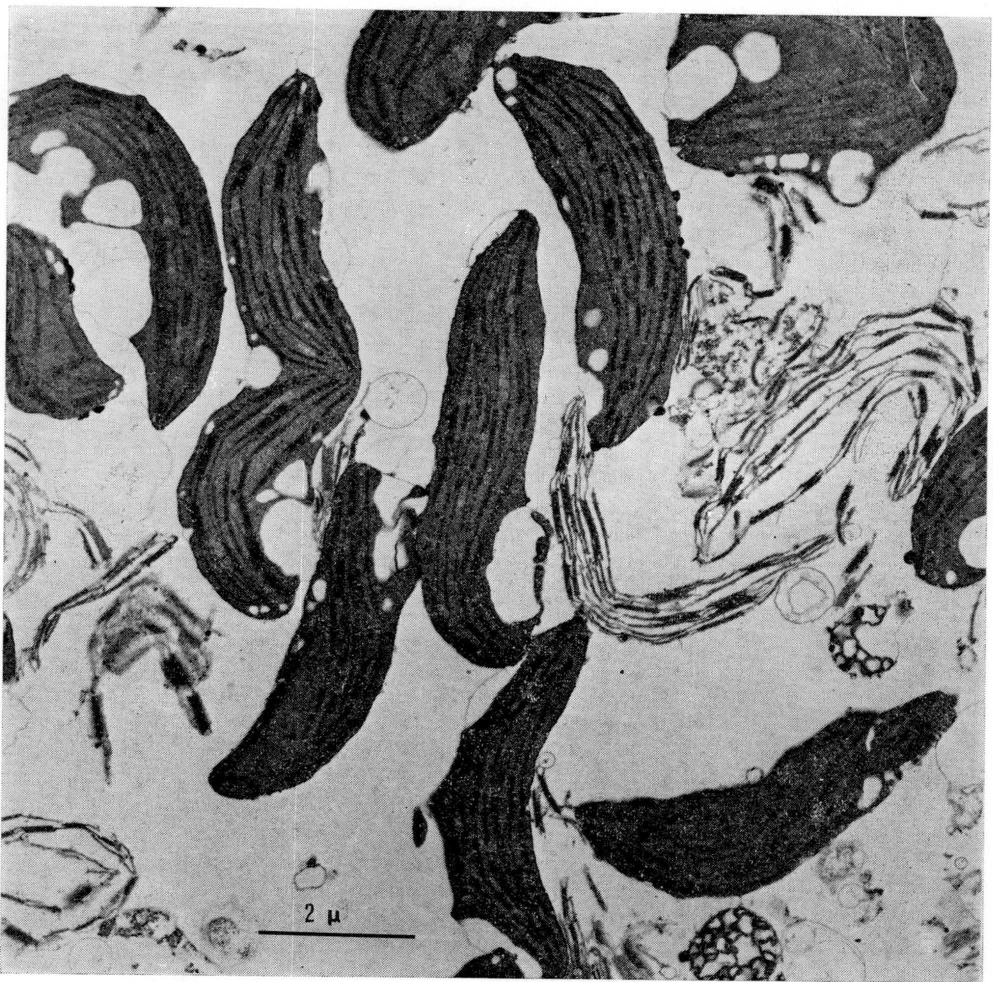

Chloroplasts were isolated in buffer A (0.35M sucrose, 25mM HEPES—NaOH, 2mM EDTA, 2mM isoascorbate, pH 7.6) and resuspended in 0.4M sucrose, 50mM TRIS, pH 7.8, 10mM $MgCl_2$, 10mM KCl and 4mM mercaptoethanol. The chloroplasts were then pelletted, fixed in glutaraldehyde—osmium, sectioned and stained with uranyl acetate and lead.

buffered sucrose show more activity than do chloroplasts isolated in ficoll-dextran-sucrose but, if the chloroplasts isolated in sucrose-EDTA are assayed in the Blair-Ellis, 0.2 M KCl buffer then a startling increase in light-dependent protein synthesis is seen (Table 5). It thus appears that the inclusion of 0.2 M KCl in the assay buffer might be the major factor in the successful functioning of the system.

If these various chloroplast preparations are examined by light and electron microscopy some interesting differences are revealed. Fig. 8 is an electron micrograph of a section taken through a population of chloroplasts isolated in the

FIGURE 9

Electron micrograph of spinach chloroplasts resuspended in buffered KCl

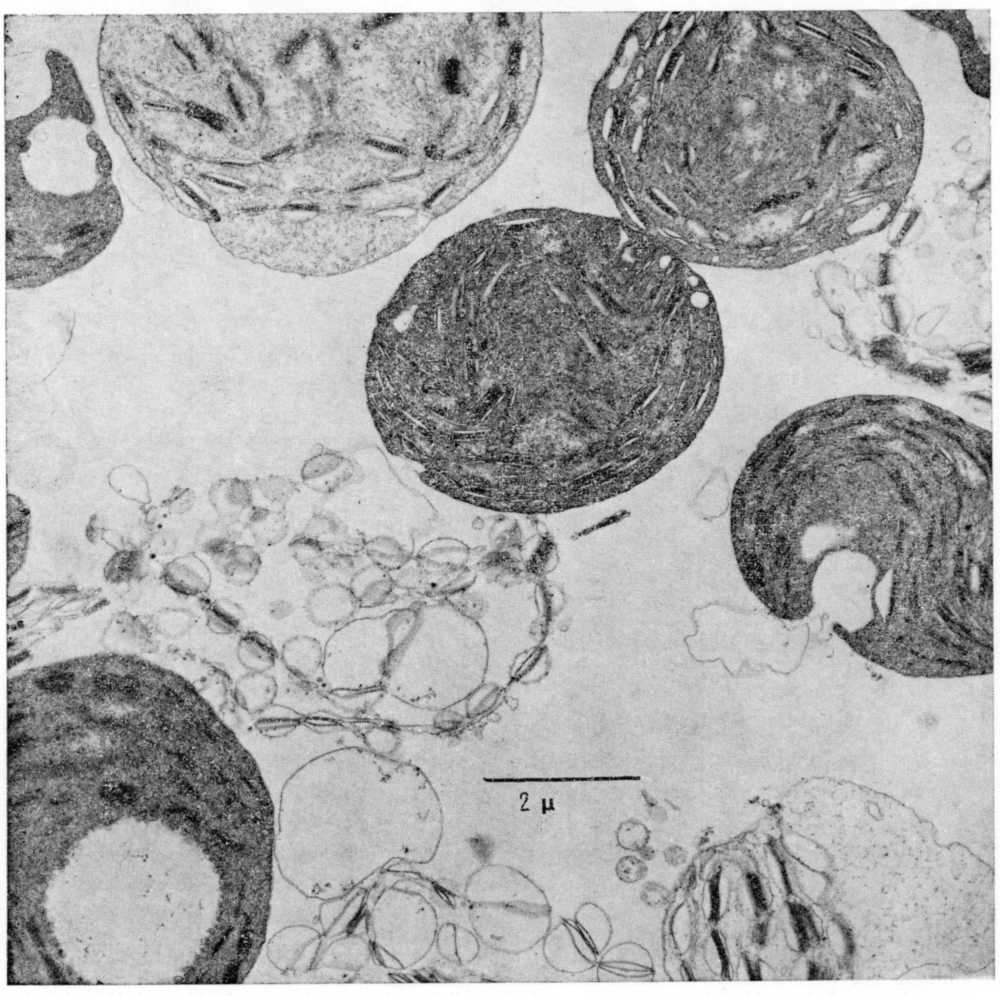

Chloroplasts were isolated in buffer A (see Fig. 8) and resuspended in 0.2M KCl, 20mM TRICINE, and 7mM $MgCl_2$, pH 8.3. Other procedures as in Fig. 8.

sucrose-EDTA buffer of Blair and Ellis and resuspended in our buffered sucrose medium. These chloroplasts appear intact (i.e. complete with outer membrane), and their internal grana and stroma lamellae are in a state quite comparable to that seen in chloroplasts in whole leaf sections. Yet, in spite of their well-preserved appearance, these chloroplasts have relatively little light-dependent protein synthesizing activity. On the other hand, if the chloroplasts are resuspended in buffered 0.2 M KCl medium, as in the Blair and Ellis procedure, they are found to be somewhat swollen (Fig. 9). Although still retaining their outer membrane they have assumed a spherical shape (in contrast to the thickened saucershape seen in Fig. 8) and the internal lamellar compartments, especially those of the grana stacks show distinct swelling. And these are the chloroplasts which exhibit the highest light-dependent amino acid incorporation.

Thus it would appear that the ability of isolated chloroplasts to incorporate amino acids into the large subunit of Fraction I protein in a light-dependent reaction is not a reflection of the intactness of the chloroplasts and, in fact, the most efficient system is that in which the chloroplasts are quite swollen.

It clearly is important to establish whether the soluble protein species which are synthesized in lysed chloroplasts are incomplete chains of the large subunit of Fraction I protein or whether they are quite unrelated to this protein. If it should turn out that synthesis of different proteins can be demonstrated to occur in isolated chloroplasts of differing degrees of intactness, then the possibility of compartmentalization of protein synthesis sites within chloroplasts would have to be considered.

CONCLUSIONS

Evidence has been presented in this paper that at least 40% of the potential genetic information in chloroplast DNA is being expressed in mature chloroplasts, at least in so far as the production of RNA transcripts is concerned. This is in contrast to the 10% of genetic information which can at present be accounted for in the way of identifiable products (See Introduction).

Isolated chloroplasts synthesize one soluble protein, the large subunit of Fraction I protein, and probably three other proteins which are associated with the ribosome—membrane (144,000 g pellet) fraction. Although it has not yet been possible to demonstrate that synthesis of these proteins is coupled to transcription of chloroplast DNA, there is evidence from other studies (Wildman, 1972) that chloroplast DNA codes at least for the large subunit of Fraction I protein. We might therefore assume that the other three proteins whose synthesis also occurs in isolated chloroplasts are likewise coded for by chloroplast DNA.

If this is the case then there is a striking parallel between the number of proteins whose synthesis can be demonstrated in isolated chloroplasts and the number of proteins which are known to be coded for by chloroplast DNA on the basis of other lines of evidence (see Introduction). This tends to reinforce the view that chloroplast DNA is involved in coding for only a few structural and enzymic proteins and that the major function of the DNA is likely to be one concerned with control. Finally, it is apparent that, if the role of the greater part of chloroplast DNA is to be elucidated, then recourse to an approach other than that of the isolated chloroplast system will be necessary.

ACKNOWLEDGEMENTS

We are very grateful to Dr. D. J. Goodchild for the electron micrographs and to Denise Brownbill, Inara Licis and Lois Moore for skilled technical assistance.

REFERENCES

Bishop, J. O. and Robertson, F. W. (1969) *Biochem. J. 115,* 353

Bishop, J. O., Robertson, F. W., Burns, J. A. and Melli, M. (1969) *Biochem. J. 115,* 361

Blair, G. E. and Ellis, R. J. (1972) *Biochem. J. 127,* 42p

Bryan, R. N., Sugiura, M. and Hayashi, M. (1969) *Proc. Nat. Acad. Sci. U.S. 62,* 483

Burgess, R. R. (1969) *J. Biol. Chem. 244,* 6160

Clark, R. J. and Felsenfeld, G. (1971) *Nature (London) 229,* 101

Ellis, R. J. and Hartley, M. R. (1971) *Nature, New Biol. 233,* 193

Fork, D. C. and Heber, U. W. (1968) *Plant Physiol. 43,* 606

Gillespie, D. and Spiegelman, S. (1965) *J. Mol. Biol. 12,* 829

Herrmann, F. (1971) *FEBS Lett. 19,* 267

Kirk, J. T. O. and Tilney-Bassett, R. A. E. (1967) *The Plastids,* Freeman, London and San Francisco

Loening, U. (1967) *Biochem. J. 102,* 251

Manning, J. E., Wolstenholme, D. R., Ryan, R. S., Hunter, J. A. and Richards, O. C. (1971) *Proc. Nat. Acad. Sci. U.S. 68,* 1169

Marushige, K. and Bonner, J. (1966) *J. Mol. Biol. 15,* 160

Scott, N. S. and Smillie, R. M. (1967) *Biochem. Biophys. Res. Comm. 28,* 598

Semal, J., Spencer, D., Kim, Y. T. and Wildman, S. G. (1964) *Biochim. Biophys. Acta 91,* 205

Spencer, D. (1965) *Arch. Biochem. Biophys. 111,* 381

Spencer, D. and Whitfeld, P. R. (1967) *Arch. Biochem. Biophys. 121,* 336

Spencer, D. and Whitfeld, P. R. (1969) *Arch. Biochem. Biophys. 132,* 477

Spencer, D., Whitfeld, P. R., Bottomley, W. and Wheeler, A. (1971) in *Autonomy and Biogenesis of Mitochondria and Chloroplasts* (Boardman, N. K., Linnane, A. W. and Smillie, R. M. eds) p. 372, North-Holland, Amsterdam.

Tewari, K. K. and Wildman, S. G. (1968) *Proc. Nat. Acad. Sci. U.S. 59,* 569

Tewari, K. K. and Wildman, S. G. (1970) *Symp Soc. Exp. Biol. 24,* 147

Wells, R. and Birnstiel, M. L. (1969) *Biochem. J. 112,* 777

Whitfeld, P. R. and Spencer, D. (1968) *Biochim. Biophys. Acta 157,* 333

Wildman, S. G. (1972) This Symposium

Gene Expression and the Immune Response

The Relevance of Immunology to the Biochemistry of Gene Expression

G. J. V. NOSSAL

The Walter and Eliza Hall Institute of Medical Research, Melbourne, Victoria, Australia[a,b]

The central problem of immunology is to reconcile the apparently endless diversity of the vertebrate immune response with the clearly finite amount of structural genetic information in the fertilized ovum. *Inter alia,* this rapidly becomes a number game, and one of our great embarrassments is that we do not know the total number of antibodies which any individual can make. All we know is that the number is large. Much of the discussion that will follow my introduction will relate to the question of whether it is plausible to conceive of structural genes in the germ line for all the heavy and light immunoglobulin (Ig) chains which an animal can make; or whether we must call on some *somatic* phenomenon such as somatic mutation or somatic recombination to give us the diversity that we need. My contribution is designed to be two-fold. First, I wish to put the more modern theories for the origin of antibody diversity that you will hear from Drs. Edelman and Cohn into historical perspective; secondly I wish to sketch the outlines of the cell biological realities of the immune system which must provide the background for a realistic genetic analysis.

The early writings of Ehrlich (1900) conceived of antibodies as normal constituents of a cell, but produced in excess after infection or immunization. Antibody formation was thought of as a "Mehrleistung"—a greater or accelerated metabolic achievement. Landsteiner (1946), however, introduced the art of making antibodies to simple chemical determinants or haptens, not normally encountered in biology. He showed that animals could not only make antibodies to these haptens, but that the antibody system could display exquisite specificity

[a]Supported by grants from the National Health and Medical Research Council and the Australian Research Grants Committee, Canberra, Australia; and from the United States Public Health Service (AI-O-3958).

[b]This is publication No. 1698 from the Walter and Eliza Hall Institute.

in discriminating between haptens that were chemically closely related. Landsteiner believed that he had refuted Ehrlich's claim of "Mehrleistung" and was convinced that antibody formation was an "Andersleitstung"—a formation of a new, different molecule. Haurowitz (1936) articulated the direct template hypothesis, in which it was believed that antibody molecules were made to fit specifically to intracellular antigens, and Pauling gave this greater precision when he introduced the idea that antigen guided the correct folding of the antibody polypeptide chain.

These theories antedated the molecular biology revolution. They were not seriously challenged till 1955 when Jerne (1955) introduced his natural selection hypothesis. This theory contained the revolutionary concept that a full library of antibody patterns, adequate for the recognition of any conceivable antigenic determinant (be this natural or unnatural) was already pre-existent in the serum of a normal, antigenically unstimulated animal. Antigen was conceived as accelerating the synthesis of that particular randomly-generated antibody which happened to fit with the relevant antigenic determinant. Some aspects of Jerne's theory were clumsy, and the mechanism by which accelerated antibody synthesis was caused has a rather implausible ring today. The theory was modified by Burnet (1957), who added the concept that the unit of selection was the individual lymphoid cell. Burnet also postulated that antibody patterns were randomly generated, and argued for a process of somatic mutation amongst lymphocytes as the mechanism of generation of diversity. In his clonal selection theory, Burnet postulated that individual lymphocytes had on their surface antibody-like receptors of one unique pattern. The library of patterns dictated the response capacity of the whole animal. When an antigen was introduced, all it had to do was to find the "right" cell, which would then be stimulated to clonal proliferation and to differentiation towards full antibody secretory capacity. In other words, antigen was thought of as a genetic regulator of antibody synthetic rate. The only choice which a cell had to make was a simple binary one—whether to be stimulated by a given antigen or not.

With this historical background, let us now add a series of key cellular facts about the immune response. It is convenient to consider these under three headings: Cell proliferation, Cell performance and Cell co-operation.

CELL PROLIFERATION

The other contributors will deal with the evidence for diversity amongst lymphocytes. Clearly, the process of diversification, whatever it may be, depends on cell division, and we must therefore consider in depth the cell proliferation history of immunologically competent cells. It is convenient to think of this at three levels. The first deals with the period *before* the cell lineage concerned becomes lymphocyte in nature. The second deals with so-called primary lymphoid organs. The third deals with cells exported from these organs and capable of reacting with antigens.

The first level concerns itself with the genesis and maintenance of a stem cell pool. It is now generally accepted that a common haematologic stem cell acts as the precursor of all the formed elements of the blood. These multipotent, self-maintaining stem cells first arise in the foetal yolk sac, and later colonize the foetal liver and the bone marrow (Metcalf and Moore, 1971). In adult animals they live chiefly in the bone marrow and the spleen. We do not know whether the "generator of diversity", be this mutation, recombination or some other mechanism, is operative already in this phase of the cells' history. It is believed that stem cells become committed to a particular pathway of differentiation through the influence of "inducers" manufactured by other cells. In the case of the lymphoid system, these inducers appear to be manufactured in the so-called primary lymphoid organs.

This brings us to the second level of cell proliferation. There are two organs known to be responsible for the genesis of immunologically competent cells, but known not to be sites of antibody production. These are the thymus and the avian bursa of Fabricius. As we shall see later, the types of lymphocytes generated by these two organs are different, but the basic principles by which the organs function are similar. These organs receive an inflow of stem cells from the bone marrow via the blood stream. When the stem cell arrives, it undergoes a series of rapid, sequential divisions. During this period, definitive commitment to lymphocyte morphology and function occurs. The cells lose their multipotency, develop identifiable surface immunoglobulin receptors, and manifest other specific surface properties recognizable as antigens which are useful in tracing the fate of the cells in various experimental systems. After several cycles of division and concomitant maturation, a proportion of the cells leave the primary lymphoid organs and seed into the so-called secondary lymphoid organs such as the spleen and the lymph nodes.

It is tempting to assume that the primary lymphoid organs are centrally involved in the generation of diversity. Within them, a high mitotic rate prevails and a large total number of cells is generated. Neither the rate of cell division nor the pathways of differentiation are influenced by antigenic stimulation. However, it must be admitted that we do *not* know whether the thymus and the bursa are specific mutant-breeding organs or not. We *do* know that the cells which emerge and are seeded out to the periphery are diversified, and are equipped with receptors for antigen (Nossal and Ada, 1971). We currently have no way of finding out whether mutations and/or recombinations in structural genes for immunoglobulins occur preferentially in primary lymphoid organs, or whether perhaps such events occur already in the multipotent haemopoietic stem cells that have the job of colonizing primary lymphoid organs. This is an important point, because it bears on the length of time and therefore the number of generations over which the postulated generator of diversity may work.

The third level of multiplication of cells to concern us relates to the secondary lymphoid organs. These lymphoid collections, such as the spleen, the lymph nodes, the Peyer's patches, the tonsils or the appendix, are the sites at which antigens meet lymphoid cells. Here, mitosis and generation of immunological effector cells is intimately dependent on antigen as a driving force. In fact, such organs manifest a low mitotic rate and a low output of lymphocytes in germ-free animals, and it is probable that, in these organs, there would be virtually no cell turnover in totally antigen-free animals. The cells seeded into these secondary lymphoid organs by the primary lymphoid organs perform no active secretory function in their resting state. Rather, they circulate widely throughout the body, exhibiting marked dynamism both *in vivo* and *in vitro,* presumably in the expectation of meeting some antigen with which they can react. When an appropriate antigen comes along, they are stimulated to divide and simultaneously to differentiate. A clone of progeny cells results, which are the real immunological effector cells. These either form antibodies or mediate cellular immune phenomena such as delayed hypersensitivity or graft rejection.

CELL PERFORMANCE

It is clear that immunological effector cells are highly specialized in their performance (Nossal and Lederberg, 1958). As regards antibody forming cells, there is every indication that the antibody produced is a monomolecular species of protein molecules (Marchalonis and Nossal, 1968). In fact, in some antibody forming cells, crystals of antibody actually form within the cisternae of the endoplasmic reticulum. A variety of specialized single cell techniques has shown that one cell produces only one type of antibody specificity and usually only one immunoglobulin class and light chain type. Moreover, in animals heterozygous for immunoglobulin allotype, each antibody forming cell confines itself to the expression of only one of the two available parental genes. Information on the performance of effector cells of cell mediated immunity is less precise, but the probability is that such cells are equally specialized.

Another aspect of cell performance relates to the great heterogeneity in the response of cells to antigen. In a given animal, the antibody formed in response to even a simple haptenic determinant will show great heterogeneity in immunoglobulin class, electrophoretic mobility and affinity of binding. This is consistent with a particular antigen stimulating not just one, but a great number of different individual cell clones in a given animal. Moreover, when the same antigen is given to a variety of different animals of the same species, the detailed characteristics of the resulting antibody will differ markedly from animal to animal. In other words, each animal has many different ways of making a given antibody; and different animals exhibit different spectra of reactivity. This argues for an element of randomness in the generation of immune patterns.

CELL CO-OPERATION (Miller and Mitchell, 1969)

We have come to recognise that lymphocytes fall into two radically different groups, called T and B lymphocytes as a convenient shorthand. T lymphocytes are derived from the thymus, and their primary concern is with cell mediated immunity. These cells, on meeting antigen, undergo clonal expansion and produce progeny cells which can kill foreign graft or tumor cells by direct complement-independent cytotoxic killing. They also release a variety of pharmacologically active agents on meeting antigen, which play important roles in inflammation. T lymphocytes do not secrete antibody in the classical sense. B lymphocytes are derived from the bursa of Fabricius in birds, or from the bone marrow in mammals. On antigenic stimulation, they produce a progeny family of antibody secreting cells, either plasma cells or modified lymphocytes.

The new finding is that, for many antibody responses, a collaboration between T and B lymphocytes is required. For example, in mice totally lacking in T cells because of a congenital defect, antibody formation to most test antigens is severely depressed. If T lymphocytes do not themselves produce antibody, how can they collaborate? This is the topic of much work and theorizing in the current immunological literature. Most available data are consistent with the role of the T cell in antibody formation being a more effective presentation of antigen to the B cell. It has become clear that most antigenic molecules consist of a mosaic of different antigenic determinants. Each B cell will form antibody to only one component of the antigenic mosaic. It has now become evident that T cells with a recognition capacity for *other* determinants on the same molecule can help the immune response to a given antigenic determinant on the molecule. Convenient model systems for study of this phenomenon involve linking a small hapten such as dinitrophenol (DNP) on to a protein. One can show that T cells activated against this carrier protein can "help" B cells with specificity for the DNP to form anti-DNP antibody, whereas in the absence of carrier specific T cells, the DNP-protein conjugate may be quite lacking in ability to induce an immune response (Feldmann and Nossal, 1972; Mitchison, 1971).

The exact mechanism of T-B collaboration is the subject of intense debate at the present time. In our Unit, we have developed a working hypothesis to explain the basis of collaboration, based on our own experiments, most of which are still unpublished. The work concerned was performed mainly by Feldmann (1972) and his collaborators, and by Marchalonis and his colleagues (see p. 629). Feldmann's approach involved a system of antibody formation *in vitro*, combined with fractionation and cell biological procedures capable of yielding T cells, or B cells, or macrophages in relatively pure form. First, it was shown that appropriately stimulated T cells manufactured two separable factors of importance to a B cell immune response. The first factor, deemed to be of lesser physiological significance, is antigen non-specific. It is of low molecular weight, as it can pass through a dialysis membrane. It is incapable of acting on B cells in the absence of added antigen, but in the presence of an antigen appropriate for the triggering of B cells, it allows the development of a great final antibody response. Though this factor has not been characterized at all in our laboratory, it can be thought of as a non-specific B cell mitogen, conceivably acting through lowering the triggering threshold for individual B cells.

The second factor appears to be even more interesting. To understand it, some of Feldmann's experimental data must be briefly described. Let us consider the case where T cells, primed to carrier protein C1, are placed into tissue culture with B cells, but are separated from them by a cell-impermeable membrane. One can then antigenically stimulate the culture with a hapten coupled to C1, or to an unrelated carrier protein C2. The finding is that antibody production will be much greater with the hapten linked to the homologous carrier. However, it soon became apparent that collaboration across cell impermeable barriers works only when each of two conditions is fulfilled. First, the barrier must allow the passage of macromolecules; second, macrophages must be present in the B cell population.

This allowed Feldmann to develop a double-transfer system of *in vitro* antibody production. First, T cells were stimulated with antigen, and were placed in tissue culture together with macrophages but separated by a nuclepore membrane of 0.1 microns diameter. Then, after 24 to 48 hours, macrophages were washed and were placed into tissue culture with B cells. It was found that, without any further addition of antigen, the B cells could form antibody against the hapten. The factor which passed from T cell to macrophage was antigen specific; in other words, when T cells primed against carrier 1 were stimulated with hapten-linked carrier 2, no collaborative factor was generated, at least in the assay system under consideration. The T cells could be stimulated with antigen in the cold, could be washed, and could then be placed into culture with macrophages but without the further addition of antigen. In this case, collaborative factor still attached to the macrophages. In other words, this suggested that the factor which was passing from T cells to macrophages was the cell surface receptor complexed to antigen.

The working hypothesis thus states that B cell triggering requires an appropriately spaced matrix of antigenic determinants to be presented to the B cell surface. In artificial *in vitro* systems, this can be achieved by the use of polymerised protein antigens. However, *in vivo*, it appears that the matrix-generating entity is a T cell acting in collaboration with a macrophage. We believe that the T cell, after union with the carrier antigen, is stimulated to produce anti-carrier IgM which attaches to receptors on the surface of macrophages, and if a hapten is linked to the carrier, there is created a matrix of haptenic determinants in an environment appropriate for B cell stimulation. The evidence that the T cell product is indeed IgM will be presented by Dr. Marchalonis and colleagues later in this Symposium.

Whether the details of this formulation are correct or not, the important thing to realise from the viewpoint of this Symposium is that an appropriate immune response to an antigen may involve simultaneous and independent recognition by the reactive cell system of more than one antigenic determinant on the one molecule. In many cases, triggering of antibody formation does not occur unless two antigenic determinants on the same molecule are recognised as foreign by the system. This has implications for the deep problem of the tolerance of self components. It is unlikely that two antigenic determinants in a "self" antigen would mutate simultaneously, and the need for cell collaboration in antibody production may therefore be some kind of safeguard against phenomena of autoimmunity.

It should be added that we do not yet know whether T-T collaboration of essentially similar nature is involved in cell mediated immunity. Some evidence consistent with this view has been presented, and the possibility remains open. Our concepts of the regulation of antibody responses have been revolutionised by the realisation of the importance of cell collaboration, and the field is now ripe both for more precise molecular probing and for exploration of possible clinical applications.

REFERENCES

Burnet, F. M. (1957), *Austral. J. Sci. 20,* 67

Ehrlich, P. (1900), *Proc. Roy. Soc.* (*B*) *66,* 324

Feldmann, M. (1972), *Cellul. Immunol.* (in the press)

Feldmann, M. and Nossal, G. J. V. (1972), *Quarterly Review of Biology 46* (in the press)

Haurowitz, F. (1936), *Z. Physiol. Chem. 245,* 23

Jerne, N. K. (1955), *Proc. Nat. Acad. Sci., U.S., 41,* 849

Landsteiner, K. (1946), *The Specificity of Serological Reactions.* Harvard University Press.

Marchalonis, J. J. and Nossal, G. J. V. (1968). *Proc. Nat. Acad. Sci., U.S., 61,* 860

Metcalf, D. and Moore, M. A. S. (1971), *Haemopoietic Cells.* North-Holland Publishing Co., Amsterdam.

Miller, J. F. A. P. and Mitchell, G. F. (1969), *Transplant. Rev. 1,* 3

Mitchison, N. A. (1971), *Europ. J. Immunol. 1,* 10 and 68

Nossal, G. J. V. and Ada, G. L. (1971) in *Antigens, Lymphoid Cells and the Immune Response.* (Dixon, F. J. and Kunkel, H. G., eds.) Academic Press, New York and London

Nossal, G. J. V. and Lederberg, J. (1958), *Nature,* (*London*), *181,* 1419

The Reaction of Antigen With Lymphocytes

G. L. ADA, M. G. COOPER AND R. LANGMAN

Department of Microbiology, The John Curtin School of Medical Research, Australian National University, Canberra, Australia

From an academic viewpoint, two important questions facing immunologists are:

1. Are immunocompetent lymphocytes so restricted that individual cells have on their surface immunoglobulin (Ig) receptors of a single specificity (amino acid sequence) as far as the binding site for antigen is concerned?
2. If so, how is this situation achieved; i.e. how is the diversity generated?

These two questions are recurrent themes in this Meeting. This contribution is concerned mainly with the first question. We shall be dealing with two aspects —some of the evidence suggesting that there is restricted competence at the level of the individual lymphocyte and what this tells us about the nature of the immunoglobulins at the cell surface.

THE SPECIFICITY OF INDIVIDUAL LYMPHOCYTES

The concept of individual specificity of lymphocytes is primarily due to Burnet who proposed a number of postulates which became known as the Clonal Selection Theory, as long ago as 1957 (Burnet, 1957). It took 10 years to gain general acceptance and the last five years to obtain virtually complete acceptance, at least as far as major alternative theories were concerned. We are just now beginning to realize that in some ways, the theory is an oversimplification. For example, it was originally thought that there existed only one class of lymphocyte. Now we know that there are two classes of lymphocytes and that the participation of both is necessary for many immunological reactions to occur. In the first part of the presentation, we wish to discuss the contributions one approach has made to our knowledge of the individuality of lymphocytes.

THE BINDING OF ANTIGEN TO LYMPHOCYTES

The technique of antigen-binding to lymphocytes was first used in its present form by Naor and Sulitzeanu (1967). They exposed suspensions of cells from mouse spleens to a low concentration of bovine serum albumin, which had been labelled with radioactive iodide (iodide-125). Two important findings emerged from these first experiments. 1. Only a small proportion of the cells present took up sufficient radioactive protein so that they could be identified as labelled cells; 2. within this labelled population some cells bound more antigen than others. This result was entirely in accord with predictions one would make from Burnet's Theory. However, there was nothing to suggest whether the results were simply a laboratory curiosity or whether the labelled cells had functional significance.

My colleague, Pauline Byrt and I looked at this reaction more closely. The technique adopted (Byrt and Ada, 1969) has since been used widely and briefly is as follows. A suspension of normal mouse lymphoid cells (10^8 cells/ml) held at 0°-4°C in a protein medium containing 15mM-sodium azide is allowed to react for 30 minutes with small amounts (1-500 ng) of a protein antigen which is labelled to high specific activity (10-200 μCi/μg) with carrier-free iodide-125. The cells are thoroughly washed and examined by radioautography. The following are the main points which were found:

1. A minor proportion (usually less than 1%) of the total cell population took up the antigen. Only cells which appeared by light microscopy to be lymphocytes were counted.
2. Examination of the labelled cells in the electron microscope (Mandel, Byrt and Ada, 1970) showed the following two main points—
   (a) Most of the cells were small lymphocytes.
   (b) The distribution of the grains in the radioautograph was consistent with the antigen being present at the cell surface.
3. The binding of the antigen to the cell surface could be inhibited in two ways.
   (a) Pretreatment of the cells with excess unlabelled antigen of the same specificity as the labelled antigen inhibited the subsequent uptake of labelled antigen.
   (b) The second method of inhibition involved pretreatment of the cell with anti-Ig sera. This approach was based on the notion that the receptor for antigen on the cell surface was an immunoglobulin molecule and that reaction of this receptor with anti-Ig sera would block access by the antigen. This proved to be the case. The use of anti-sera which were specific for either light or the different heavy chains indicated that for mouse spleen cells the major Ig present on the surface of the cells was IgM (Warner *et al.*, 1970). This finding has been amply confirmed since (Raff, 1970; Rabellino *et al.*, 1971; Bankhurst and Warner, 1971; Vitetta *et al.*, 1971). IgM is also the predominant Ig present on sheep lymphocytes (P. Ey, unpublished results) and rabbit lymphocytes (Pernis *et al.*, 1970) whereas the most common Ig present on guinea pig lymphocytes is IgG (Davie and Paul, 1971).

4. At any given antigen concentration, there was a hierachy of cells within the population of labelled cells. Some bound more than others. At least for two antigens which were studied, it was found that the *number* of cells which labelled with antigen varied according to the concentration of the antigen within the reaction mixture. The upper limit appeared to be 1-2% of cells becoming labelled. As far as could be determined, cells which bound a given antigen could exist at a frequency of between 10^{-2} and at least 10^{-5} (Ada, 1970).

INACTIVATION OF LYMPHOCYTES BY RADIOACTIVE ANTIGEN

Abrogation of antibody production

Though these were interesting findings, a test was needed which would show that this binding of antigen was functionally important. This was provided by an experiment which has subsequently become known as the "suicide" experiment. The rationale behind this was as follows. 1. Some of the cells which bound labelled antigen were necessary for an antibody response to occur. 2. It was known that during an antibody response, the precursor lymphocyte underwent a number of cell divisions. 3. Therefore it was argued that the emissions from radioactive antigen localized at the cell surface might irreversibly damage the cell concerned (but not other cells) so that it could not take part in an immune response. For this, iodide-125 was an ideal isotope as the path length of the β emission is about 10μ, which is close to the diameter of the small lymphocytes.

This approach was first used on spleen cells from normal mice (Ada and Byrt, 1969). Cells were exposed to antigen labelled either with iodide-125 (radioactive) or iodide-127 (non radioactive). The cells were then injected into X-irradiated mice which were challenged with non radioactive antigen of the same serological specificity together with antigen of a different serological specificity. The results were quite clear cut. Pre-exposure of cells to labelled antigen of a given serological specificity greatly decreased or abolished the ability of that cell population to give an antibody response to that antigen but not to other antigens. Three important conclusions could be drawn from this result.

1. At least some of the cells which bound antigen were necessary for an antibody response to occur.
2. If only a single population of lymphocytes were concerned in the production of antibodies, then all of them must be able to recognise antigen.
3. The specificity of the suicide was such as to be consistent with the Theory postulated by Burnet—that individual cells made antibody molecules of only one antigen-binding (i.e. serological) specificity.

With regard to this last point, the antigens used in these experiments were chosen rather carefully. They were polymerized forms of a soluble protein, flagellin of mol. wt. 40,000 isolated from the flagella of different strains of *Salmonella* organisms (Nossal and Ada, 1971). According to the Kaufmann-White scheme, these antigens are serologically distinct. They were used in these experiments because of the general similarity in their physical and chemical properties. We will return to this point later.

This approach thus provided a means of testing the function of lymphocytes in immune responses. The first extension was obvious. Antibody responses to many antigens depend upon the co-operation of two classes of lymphocytes. These are the bone marrow derived (Bursa-equivalent) or B cell and the thymus derived, or T cell (see review, Miller, *et al.,* 1971). There is abundant evidence which shows that the B cell is the precursor of the antibody secreting cell so that the T cell must do the co-operating. The precise mechanism by which this co-operation is achieved is not clear and only certain aspects of this are of interest to this Meeting. But it is important to realize that one of the two main types of evidence leading to the concept of cellular co-operation was provided by what is now known as the hapten-carrier effect. Haptens, low molecular weight compounds, are non-immunogenic by themselves. If attached to a suitable carrier, such as an immunogenic protein or polysaccharide, the complex is immunogenic and antibody is formed against both the hapten and the carrier. As a general statement, we can say in fact that both the B and T cells recognise both hapten and carrier: there is some preference for the B cell to recognise the hapten rather than the carrier but T cells almost invariably fail to recognise the hapten. That is, there seems to be a difference in the quality of recognition by these two classes of lymphocytes. This is an important point which we will again raise later.

The question of immediate interest was: is it possible to suicide both T and B cell functions in a co-operative antibody response, or could only one cell type be suicided? Two groups investigated this problem using slightly different approaches. Basten, *et al.,* (1971) chose to use as their antigen the protein, fowl γ-globulin. Antibody to this protein is made only if both T and B cells are present and can co-operate. X-irradiated mice were reconstituted with both T and B cells, the source of T cells being thymocytes from normal, syngeneic mice, and the source of B cells being the spleens of mice which had been thymectomised, X-irradiated and then reconstituted with bone marrow. Such mice, known as ATxBM mice, have only B and no T cells in their spleens. Thus, the procedure was to inject X-irradiated mice with both these cell types, one of which (either B or T cells) had been exposed to fowl γ-globulin prelabelled with iodide-125. Recipient mice were then challenged with the same antigen (unlabelled) and a *completely* unrelated antigen, horse erythrocytes, and the antibody secreting cells to each antigen enumerated some days later. The result was quite clear. Both T and B cells could be specifically suicided.

Roelants and Askonas (1971) used a slightly different approach, a classical hapten-carrier system in which the complex was either DNP-haemocyanin or DNP-ovalbumin. X-irradiated mice were reconstituted with spleen cells from syngeneic mice immunized either with DNP-ovalbumin or with haemocyanin. The reconstituted mice were then challenged with DNP-haemocyanin. Only those mice receiving cells from the haemocyanin-primed mice produced high titres of anti-DNP antibody, showing that the spleens of these donor mice must have contained T cells *"activated"* against the carrier, haemocyanin. If, prior to such a transfer, the donor "activated" T cells were incubated with [^{125}I]haemocyanin under suiciding conditions, this helper effect was abrogated. This was entirely consistent with the findings of Basten *et al.,* (1971). It should be noted that the difference between a thymocyte or a T cell (neither having met an antigen) and an "activated" T cell (having been "activated" by antigen) is not clear. Cell for cell, activated T cells may be more efficient than T cells at helping B cells; however, it seems more likely at present that the process of activation simply results in the production of a *greater* number of T cells of the appropriate specificity without a change in their individual properties.

Abrogation of a cell mediated immune response

The evidence was thus quite clear that B cells and helper T cells could be suicided by labelled antigen and that this was specific. One further approach was desirable. As well as helping B cells, recent evidence shows that T cells can themselves (in the absence of B cells) initiate an immune response—the so-called cell mediated immune response. Under appropriate circumstances, T cells may "kill" other (target) cells; or together with phagocytic cells such as monocytes, give rise to hypersensitivity reactions of the delayed type. Whether the same T cell which helps B cell also mediates either one of these other functions is not known. The question thus was—can T cells taking part in a cell-mediated immune response be suicided?

A model system was set up in the mouse for measuring delayed type hypersensitivity (DTH) to the flagellar antigens (Cooper, 1972a, b). X-irradiated mice were injected with thymic lymphocytes and the antigen, polymerized flagellin. Six days later, lymphoid cells, harvested from the spleens and lymph nodes of these mice, were injected into normal, syngeneic mice. The same (or other) antigens were injected into the hind foot pads of these recipient mice and the delayed type sensitivity estimated (footpad swellings) at 20-30 hrs post injection. Highly significant sensitivity reactions were found and classified as such by their histological appearance, and their ability to be transferred only by activated T cells and not by B cells or specific serum antibody. The finding of particular interest was that the specificity of this reaction differed from that previously seen with antibody responses to these antigens. Whereas the antibody response clearly differentiated between polymerized flagellin from two different strains of *Salmonella* organisms, the DTH reaction could not distinguish between them. T cells sensitized to one polymerized flagellin could be activated equally well by the *same* or by a *different* polymerized flagellin from *Salmonella* organisms but not by a completely unrelated antigen, haemocyanin. Similarly it was found (Cooper and Ada, 1972) that T cells, which had been exposed to ^{125}I-labelled polymerized flagellin of one strain (i.e. suicided) could no longer initiate a DTH reaction when tested either with the same antigen or with a polymerized flagellin from other strains of *Salmonella.*

However, a DTH reaction to an unrelated antigen haemocyanin could not be abrogated if, prior to their transfer, the cells activated by haemocyanin were exposed to ^{125}I-labelled polymerized flagellin of either specificity.

It was further found that either thymocytes or activated T cells could be suicided upon exposure to labelled antigen.

The results of all experiments on lymphocyte inactivation by radioactive antigen are summarized in Table I and discussed in the next section.

COMMENTS ON THE SUICIDE OF T AND B CELLS

The implications of these results can now be discussed. The first point to be made is that the general finding in a number of laboratories is that lymphocytes can be divided into two classes—those that have many Ig molecules on their surface (e.g. Vitetta *et al.*, 1971; Cone *et al.*, 1971; Nossal *et al.*, 1972) and those which appear to possess very few. The main methods of detection have been (1) labelling cells directly with iodide-125, solution of the membrane, isolation of labelled Ig chains (light and heavy) by co-precipitation techniques followed by identification by polyacrylamide gel electrophoresis; and (2) direct uptake of labelled anti-immunoglobulin antibody onto the surface of lymphocytes, followed (sometimes) by the isolation of the Ig-anti-Ig complex as above. It is the consensus of most workers in the field that B cells have many Ig molecules on their surface (estimates vary from 30,000-200,000; e.g. Nossal *et al.*, 1972) whereas thymocytes or T cells have very few, if any. It is also widely found that antigens bind readily to B cells and far less readily (some claim not at all) to thymocytes or T cells (e.g. Ada, 1970).

The properties of B cells can be discussed first. Because of the high density of Ig molecules on the surface of these cells and because it can be shown (Byrt and Ada, 1969) that B cells reacting with a given antigen can bind many molecules of that antigen there is reason to expect that suicide of these cells would be readily accomplished. The D_{37} value for B cells is about 80 rds (Kennedy *et al.*, 1965). Using antigen substituted with iodide-125 to the extent achieved in the suicide experiments, it has been calculated (Pye, personal communication) that a lethal dose of radiation would be delivered by the isotope to those cells which bind most antigen; that is, to those cells which occur at frequencies of c.10^{-5}. Complete dose response curves have not been established but from the available information, the amount of radioactive isotope which has been found to cause suicide is probably not excessive. Whether cell inactivation is due to replicative failure because of chromosomal damage or to interphase death is not clear. A distinction at this point is not essential.

Although up to about 1% of B cells may bind a particular labelled antigen, the above and other evidence suggests that only those cells occurring at low frequency (possibly 10^{-5}) are important in an antibody response to the antigen (Ada, 1970; Cooper *et al.*, 1972). At this low frequency, it is rather difficult to plan experiments which will enable a clear interpretation of the events leading to the activation of the cells. Although most immunologists in this field are agreed that the receptors on lymphocytes for antigens are immunoglobulins, there is as yet no formal proof that reaction of antigens with Ig at the surface of the B cell is a necessary, though perhaps not sufficient occurrence, for that cell to be stimulated.

The suicide of B cells by antigen is plausible in view of the density of the receptors and the amount of antigen known to bind to these cells. The suicide of thymocytes and of T cells is not so easy to understand. Investigators who have exposed thymocytes or activated T cells to labelled antigens report either an absence or very low frequency (10^{-4}—10^{-5}) of binding cells, (Byrt and Ada, 1969; Humphrey and Keller, 1970; Basten *et al.*, 1971; Cooper and Ada, 1972). Yet the activity of thymocytes or activated T cells has been abrogated using these different labelled antigens (Table 1). Thymocytes are reported to contain either no immunoglobulins (e.g. Raff, 1970; Rabellino *et al.*, 1971) or very few (perhaps 500) molecules per cell (Nossal *et al.*, 1972). Yet both Basten *et al.* (1971), and Cooper and Ada (Table 1) (1972) found that the suicide of thymus cells or of activated T cells could be prevented if the cells were exposed to rabbit anti-mouse Light chain IgG prior to treatment with the labelled antigen. This prevention of suicide can only be interpreted to indicate the presence of Ig molecules on the surface of these cells and it implies that they act as receptors for antigen. If the few Ig molecules claimed to be present on thymocytes (Nossal *et al.*, 1972) do act as receptors, only a very small amount of labelled antigen would become attached to thymus cells: their suicide could then only be explained if these cells were extraordinarily sensitive to irradiation damage. There are in fact many reports to the contrary (e.g. Smith and Vos, 1963; Cunningham and Sercarz, 1971). There thus seems to be a conundrum. This could possibly be resolved if the thymus was found to contain a proportion of cells which contained a high density of Ig molecules at their surface. Two very recent reports indicate this is so (Hammerling and Rajewsky, 1971; Perkins *et al.*, 1971) and Miss Kirov and I have found that a small proportion of thymus cells have a high density of Ig Light chains on their surface. The significance of these cells is at present being assessed.

THE SPECIFICITY OF THE LYMPHOCYTE RECEPTORS FOR ANTIGEN

It is clear from Table 1 that individual lymphocytes have receptors of limited specificity. Two of these reports throw more light on the degree of this restriction. At first sight, it seems that the receptors on thymocytes or T cells for polymerized flagellins show a broader specificity than those on B cells. That is, thymocytes and T cells possessed Ig molecules of more than one specificity (amino acid sequence). An alternative explanation would be if the T and B cells recognised *different* antigenic determinants. It has now been found (Langman, 1972a,b, unpublished results) that flagellins from different *Salmonella* organisms possess two classes of determinants—one called, Hc, which is common to the different polymerized flagellins and the other, called Hv, which is unique for the polymerized flagellins of each strain. To explain the *specificity* of the suicide experiments it is merely necessary to postulate that thymocytes or T cells recognise only the Hc determinants, whereas B cells preferentially recognise the Hv determinants. This concept is consistent with all the data now available. If correct, it has interesting implications. 1. Polymerized flagellin behaves essentially as do

TABLE 1

Summary of Experiments Demonstrating Inactivation ("*Suicide*") of Mouse Lymphocyte Function by Radio-labelled Antigen

Source and/or Type of Lymphocyte	Response Measured	Function Involved	Labelled Antigen	Control Antigen	Reference
Spleen (B)	Antibody production	Differentiation, replication	POL 1338 POL 870	POL 870 POL 1338	Ada and Byrt, 1969
Activated T lymphocytes	Antibody production	Helper activity	Haemocyanin	Ovalbumin	Roelants and Askonas, 1970
Thymocytes	Antibody production	Helper activity	Fowl IgG	RBC	Basten *et al.*, 1971
B lymphocytes	Antibody production	Differentiation, replication	Fowl IgG	RBC	Basten *et al.*, 1971
Thymocytes	Delayed type hypersensitivity	Replication	POL 870	POL 1338 Haemocyanin	Cooper and Ada, 1972
Activated T lymphocytes	Delayed type hypersensitivity	Secretion of active factors	POL 1338	POL 870 Haemocyanin	Cooper and Ada, 1972
			Haemocyanin	POL 870 POL 1338	

artificial hapten-carrier complexes; determinants can be preferentially recognised by either T or B cells. 2. This would seem to be a rational situation; if the antigen has to react with both B and T cells for cellular co-operation to occur, it would be unwise for both cell types to compete for the same determinant. 3. The observation should be considered with two other observations (a) T cells do not recognise pneumococcal polysaccharide (Howard *et al.*, 1971), and (b) T and B cells appear to recognise different parts of the glucagon molecule (Senyk *et al.*, 1971). These three observations suggest strongly that the expression of diversity in immunoglobulin specificities differs for T and B cells. 4. Finally, as thymocytes and T cells are not only stimulated by polymerized flagellin but preferentially recognise certain determinants, it seems likely that the antibody response to polymerized flagellin *in vivo* is T-dependent. Reports to the contrary, based on *in vitro* experiments (e.g. Feldmann and Basten, 1971) may not be relevant *in vivo*.

CONCLUSION AND SUMMARY

The evidence from antigen binding studies, and particularly from inactivation experiments, leave little doubt that individual B lymphocytes respond to a very restricted number of antigens. The absolute extent of restriction is unknown. By analogy with the myeloma situation, the simplest and most appealing concept is that individual lymphocytes have on their surface Ig receptors which possess only one amino acid sequence in the V regions of the light and heavy chains. Most evidence points towards this interpretation but it has not been established.

It has also yet to be shown that reaction of antigen with the Ig at the cell surface is a necessary initial step in cell activation. It may not of course be a sufficient step; other factors such as the number, degree of closeness and inter-linkage of the Ig molecules after the reaction with antigen are probably important (Ada, 1970).

An outstanding feature of the immune response is its specificity. We believe that this specificity is controlled by the events happening at the cell surface. A subsequent penetration by the bound antigen into the cell (if it occurs) is probably irrelevant in this context. The activation process within at least all B cells would involve common pathways.

It is anticipated that a similar series of events will occur in T cells, but it seems at present that the elucidation of these will be more difficult. Obviously T cells play a dominant role—as controlling cells in regulating B cell activation and as initiators of cell mediated immune responses. They provide a challenge for the immunologist.

REFERENCES

Ada, G. L. (1970). *Transplant Rev.* *5,* 105-129

Ada, G. L. and Byrt, P. (1969). *Nature,* (*London*), *222,* 1291

Bankhurst, A. and Warner, N. L. (1971). *J. Immunol.* *107,* 368

Basten, A., Miller, J. F. A. P., Warner, N. L. and Pye, J. (1971). *Nature New Biol.* *231,* 104

Burnet, F. M. (1957), *Austral. J. Sci. 20,* 67

Byrt, P. and Ada, G. L. (1969). *Immunology 17,* 503

Cone, R. J., Marchalonis, J. J. and Rolley, R. T. (1971). *J. Exp. Med. 134,* 1373

Cooper, M. G. (1972a). *Scand. J. Immunol. 1,* 167

Cooper, M. G. (1972b). *Scand. J. Immunol. 1,* 237

Cooper, M. G. and Ada, G. L. (1972). *Scand. J. Immunol. 1,* 247

Cooper, M. G., Ada, G. L. and Langman, R. (1972). *Cell. Immunol.* (in the press)

Cunningham, A. J. and Sercarz, E. E. (1971). *Eur. J. Immunol. 1,* 413

Davie, J. M. and Paul, W. E. (1971). *J. Exp. Med. 134,* 495-516

Feldmann, M. and Basten, A. (1971). *J. Exp. Med. 134,* 103

Hammerling, U. and Rajewsky, K. (1971). *Eur. J. Immunol. 1,* 447

Howard, J. G., Christie, G. H., Courtenay, B. M., Leuchars, E. and Davies, A. J. S. (1971). *Cell. Immunol. 2,* 614

Humphrey, J. H. and Keller, H. U. (1970), in *Developmental Aspects of Antibody Formation and Structure* (Sterzl, J. and Riha, I., eds.) vol. 2, p. 485, Academia, Prague

Kennedy, J. C., Till, J. E., Siminovitch, L. and McCulloch, E. A. (1965). *J. Immunol. 94,* 715

Mandel, T., Byrt, P. and Ada, G. L. (1970). *Exp Cell. Res., 58,* 179

Miller, J. F. A. P., Basten, A., Sprent, J. and Cheers (1971). *Cell. Immunol., 2,* 469

Naor, D. and Sulitzeanu, D. (1967). *Nature (London), 214,* 687

Nossal, G. J. V. and Ada, G. L. (1971) in *Antigens, Lymphoid Cells and the Immune Response.* (Dixon, F. J. and Kunkel, H. G., eds.) Academic Press, New York and London

Nossal, G. J. V., Warner, N. L., Lewis, H. and Sprent, J. (1972). *J. Exp. Med. 135,* 405

Perkins, W. D., Karnovsky, M. J. and Unanue, E. R. (1971). *J. Exp. Med. 135,* 267

Pernis, B., Forne, L. and Amante, L. (1970). *J. Exp. Med. 132,* 1001

Rabellino, E., Colon, S., Grey, H. M. and Unanue, E. R. (1971). *J. Exp. Med. 133,* 156

Raff, M. (1970). *Immunol. 19,* 637

Roelants, G. E. and Askonas, B. A. (1970). *Europ. J. Immunol. 1,* 151

Senyk, G., Williams, E. B., Nitecki, D. E. and Goodman, J. W. (1971). *J. Exp. Med. 133,* 1294.

Smith, L. H. and Vos, O. (1963). *Radiation Research 19,* 485

Vitetta, E. S., Baur, S. and Uhr, J. W. (1971). *J. Exp. Med. 134,* 242

Warner, N. L., Byrt, P. and Ada, G. L. (1970). *Nature (London) 226,* 942

Synthesis, Transport and Secretion of Immunoglobulins in Lymphoid Cells

FRITZ MELCHERS
Basel Institute for Immunology, Basel, Switzerland

Thymus-derived (T-) and bone marrow-derived (B-) cells cooperate in the response to an antigen. Both cells recognize antigen as carrier and hapten. Both cells contain receptor molecules in their outer membrane which recognize antigen. These receptor molecules are immunoglobulins (Ig) (reviewed in Progress in Immunology, 1971). Antigen attaches to these receptors and thus stimulates T- and B-lymphocytes to proliferate into clones of cells. Clones of B-cells produce immunoglobulin identical in polypeptide structure to that of the original receptor immunoglobulin. One B-cell makes only one structure of immunoglobulin in many copies. During proliferation and differentiation a clone may switch from the production of IgM to the production of IgG with the same specificity for antigen (Pernis *et al.*, 1971). B-cells, but not T-cells, differentiate into plasma cells, the prime antibody-producing cells. During this differentiation B-lymphocytes progressively lose surface immunoglobulin and turn to synthesize and secrete increased amounts of immunoglobulin. The regulatory status of immunoglobulin synthesis, transport and secretion in B-cells changes from that in small lymphocytes where immunoglobulin is synthesized, transported to and then held in the outer membrane, to that in plasma cells, where immunoglobulin is synthesized, transported *and* secreted.

This differentiation of B-cells can be monitored by the morphological changes occurring in lymphocytes after antigenic stimulation. Small lymphocytes have little cytoplasm with apparently little membranous structures visible in the electron microscope. Plasma cells, during their maturation develop more and more cytoplasm with membranous structures such as the rough and the smooth endoplasmic reticulum, including Golgi apparatus (for a selected list of references of work on the morphology of lymphocytes, see Melchers, 1971a; also: Whaley *et al.*, 1972).

The paper will review some of the biochemical reactions which occur in plasma cells, when immunoglobulins are synthesized and secreted. The contributions of the author to these studies were initiated at the Salk Institute for Biological Studies, La Jolla, California, U.S.A., in the laboratory of Dr. E. S. Lennox (supported by a U.S. Public Health Service Grant No. Al-06544 to Dr. E. S. Lennox) and represent the joint efforts of Drs. E. S. Lennox, P. M. Knopf, Y. S. Choi, B. Parkhouse and the author and were continued by the author first at the Max Planck Institute for Molecular Genetics, Abteilung Trautner, Berlin, Germany, and later at the Basel Institute for Immunology. These biochemical reactions have recently been followed up for immunoglobulin synthesis in mitogen-stimulated small lymphocytes (Parkhouse *et al.*, 1972; Andersson and Melchers, unpublished). This work is being done in collaboration with Dr. J. Andersson at the Basel Institute for Immunology. A detailed description of the experimental results will be given elsewhere. The regulatory status of synthesis and secretion in small B-lymphocytes stimulated by locally concentrated concanavalin A (c-Con A) will be compared to that in plasma cells with the hope to define biochemical and molecular parameters for the differentiation of an antibody-forming cell clone from an antigen-sensitive small lymphocyte to an antibody-producing plasma cell.

STRUCTURE OF IMMUNOGLOBULINS

Immunoglobulins are glycoproteins formed of heavy (μ in IgM, mol. wt. 75,000; γ in IgG, mol. wt. 50,000) and light chains (mol. wt. 22,500). The chains are linked by disulphide bonds. Carbohydrate (mol. wt. 2000-3000) is attached at several points in the constant portions of μ-chains and usually at one point

FIGURE 1

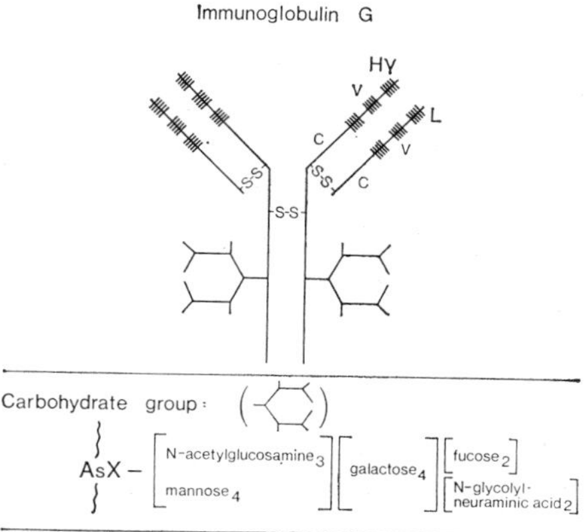

Schematic representation of the structure of immunoglobulin G. For details see *Progress in Immunology, 1971* and other pertinent publications. It should be emphasized that the structure of the carbohydrate moieties is an artist's view pointing to the branched nature of the groups. For exact analysis of one such structure the reader is referred to Kornfeld *et al.* (1971).

within the F_c-portion of γ-chains. Carbohydrate has been found attached to aspartic acid or asparagine probably via N-acetyl-glucosamine in N-glycosidic linkages around sequences of the general type AsX-X-Ser/Thr (Melchers, 1969a; Sox and Hood, 1970). This general sequence may be recognized by an enzyme attaching the first sugars to Ig. The carbohydrate groups in Ig closely resemble those found in other glycoproteins and are probably all of branched structure. N-acetyl-glucosamine and mannose residues form a core, while galactoses occupy semi-terminal and N-glycolylneuraminic acids and fucoses occupy terminal positions within these carbohydrate groups. This is schematically summarized for the structure of IgG in Fig.1, for the structure of IgM in Fig. 2 (for references

FIGURE 2

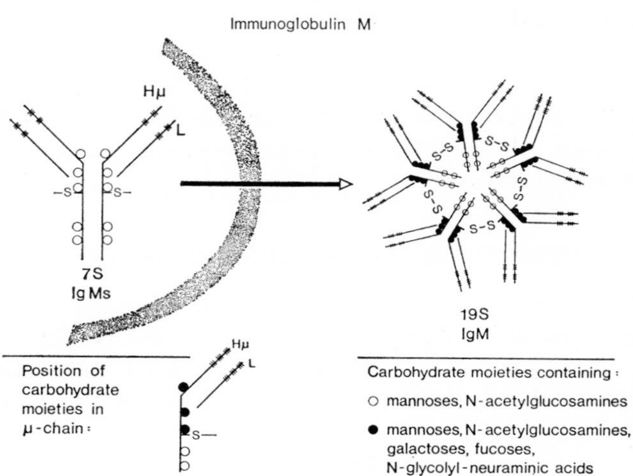

Schematic representation of the structure of immunoglobulin M. Indicated is the polymerization of 7S intracellular IgMs subunits to 19S IgM during the secretion from the plasma cell (Askonas and Parkhouse, 1971) the localization of carbohydrate moieties in μ-chains as determined for one human myeloma IgM and their carbohydrate composition in 19S molecules isolated from the serum (Shimizu *et al.*, 1971), and the acquisition of galactose, fucose and neuraminic acid residues during secretion and polymerization to 19S IgM (Parkhouse and Melchers, 1971; Melchers, 1972). It should be pointed out that detailed structural analysis on the carbohydrate composition and localisation of individual glycopeptides within μ-chains has only been determined for human, but not for mouse myeloma IgM. For an analysis of possible conformational differences between intracellular and extracellular 7S-subunits of IgM the reader is referred to Melchers (1972).

see Progress in Immunology, 1971). A detailed structural analysis of a glycopeptide from human IgG myeloma protein has been undertaken by Kornfeld *et al.* (1971).

SYNTHESIS, TRANSPORT AND SECRETION OF IMMUNOGLOBULINS IN PLASMA CELLS

Biochemical analyses of the modes of synthesis, transport and secretion of immunoglobulin in normal lymphoid cell populations are difficult since these cells represent mixtures both of morphologically and functionally different cells. However the availability of a series of myeloma plasma cell tumours transplantable in inbred strains of mice has made possible such biochemical studies. Each tumour can be regarded as a clone of plasma cells producing and secreting a homogenous immunoglobulin (Potter, 1967).

In plasma cells the polypeptide chains of immunoglobulins are synthesized on membrane-bound polyribosomes of the rough endoplasmic reticulum. Assembly into SS-bonded molecules appears immunoglobulin-class-and-subclass-specific, begins on polyribosomes and continues, while immunoglobulin chains traverse the cell. This subject has been reviewed recently by Scharff *et al.* (1970) and by Bevan *et al.* (1972).

IgM inside plasma cells assembles only to its 7S subunit (IgMs) form and polymerizes to 19S IgM shortly before or during secretion (Askonas and Parkhouse, 1971).

KINETICS OF SYNTHESIS AND SECRETION IN PLASMA CELLS

The characteristics of synthesis and secretion of immunoglobulin by plasma cells, first measured by Helmreich *et al.* (1961), are illustrated in Fig. 3. In

FIGURE 3

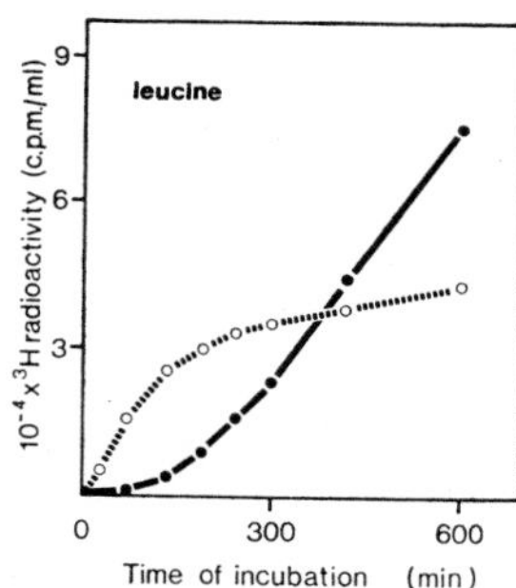

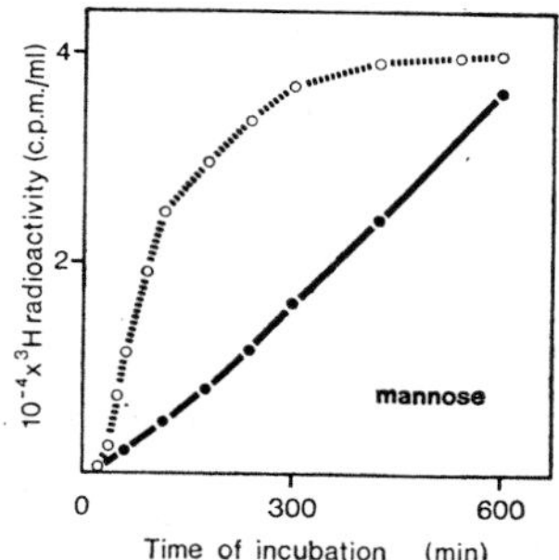

Time-course of incorporation of L—[4, 5—$^3H_2$]leucine (left side) and D—[1—^{3}H]mannose into intracellular (— — — —) and secreted (———) IgG_1 myeloma protein produced by the myeloma plasma cell tumour MOPC 21. For details see Melchers (1970).

cells labelled with radioactive leucine, protein synthesis is followed by the incorporation of radioactivity into acid precipitable material, while immunoglobulin synthesis and secretion is monitored by radioactivity serologically precipitated with immunoglobulin-specific antisera. In different plasma cell tumours producing different classes and subclasses of immunoglobulin between 5% and 40% of the total protein synthesis is devoted to immunoglobulin synthesis. Newly synthesized Ig accumulates inside the cell, reaching a constant specific activity between 2 and 5 hours after addition of radioactive leucine to the cell suspension. Newly synthesized Ig appears outside the cell after a lag of 20 to 30 minutes and reaches a steady state in the rate of secretion after 2 to 5 hours. Kinetics of synthesis and secretion in plasma cells are very similar for IgM (Askonas and Parkhouse, 1971), IgG_1, IgG_{2a}, K-light chain (Choi *et al.*, 1971a; Melchers, 1970) and IgA (Melchers, unpublished observations).

From data such as those illustrated in Fig. 3, the synthetic and secretory activity of a plasma cell under such experimental conditions can be calculated as the number of molecules of radioactive myeloma protein secreted per unit time with the assumptions made earlier (see Melchers 1970 and 1971b). One plasma cell secretes between 50 and 200 molecules of Ig per second and contains between $1x10^6$ and $1x10^7$ molecules Ig.

The kinetics of synthesis and secretion follow a simple precursor-product relationship, where the cells are the source of the immunoglobulin found secreted into the medium. The half-time to reach a steady state of secretion, between 1 and 3 hours in different plasma cells, represents the average time for an immunoglobulin molecule to be synthesized and secreted. Only 1-2 minutes are required to synthesize the polypeptide portion of immunoglobulin (Knopf *et al.*, 1967), hence almost all the time is needed to transport immunoglobulin from its site of synthesis through and out of the cell. We have focused our interest on what happens to immunoglobulin inside the cells during this lag period and whether the time span of the lag period is dependent on the state of differentiation of a lymphoid cell.

Since only between 5% and 40% of all the synthesized protein inside a plasma cell is immunoglobulin, while over 90% of the secreted material can be identified as immunoglobulin, the question arises of how plasma cells achieve the selection of immunoglobulin as the only protein to be secreted from the cells.

SUBCELLULAR DISTRIBUTION OF IMMUNOGLOBULIN

The cytoplasm of plasma cells contains membranous structures visible in the electron microscope. The role of such structures, the rough and smooth endoplasmic reticulum and the Golgi apparatus, in synthesis, transport and secretion of glycoproteins has been studied in plasma cells as well as in other cells (for references see Melchers, 1971a; and Whaley *et al.*, 1972). Immunoglobulin has been seen in the rough and the smooth endoplasmic reticulum, the Golgi apparatus and on the outer (plasma) membrane.

Disruption of plasma cells by tissue homogenizer or by ultrasonication yields subcellular fractions, which can be separated in sucrose gradients according to their density. Fractions have been obtained which are enriched in membranous structures either of the rough or of the smooth endoplasmic reticulum and Golgi apparatus (Choi *et al.*, 1971b; Melchers, 1971c). Immunoglobulin has been found in these membranous fractions (designated RM- and SM- fractions) as well as in a nonsedimentable cytoplasmic supernatant fraction (CS-fraction). For IgG_1, IgG_{2a} and K-light chain myeloma protein the distribution between RM-, SM- and CS-fractions varies between 55% to 75%, 10% to 25% and 3% to 30%. The cytoplasmic supernatant fraction may be an artefact arising in the preparation of the cell homogenates (Melchers, 1971c; Knopf, Sasso, Destree and Melchers, unpublished). Therefore all immunoglobulin molecules inside plasma cells may be associated with membranous structures inside the cell.

A number of myeloma plasma cells have immunoglobulin on their surface, which by immunofluorescence has been shown to be of the same class as the class secreted by the cells (Pernis, Loor and Melchers, unpublished observations). This is contrasted by the finding that normal mature plasma cells do not contain any detectable surface immunoglobulin (de Petris *et al.,* 1963). Until more detailed analyses of the occurrence of surface immunoglobulin on outer membranes of maturing lymphocytes have been carried out, one can only speculate, that plasma cells during their differentiation may progressively lose their surface immunoglobulin.

MIGRATION OF IMMUNOGLOBULIN INSIDE PLASMA CELLS

Plasma cells are pulse-labelled with radioactive leucine for a short time (5 min.) and then transferred into medium containing unlabelled leucine. Thereby the labelled immunoglobulin is chased through and out of the cells. The changing distribution of serologically identifiable radioactive immunoglobulin in the different subcellular fractions is measured. Such experiments have been carried out with light chain-producing myeloma plasma cells (Choi *et al.,* 1971a) and

FIGURE 4

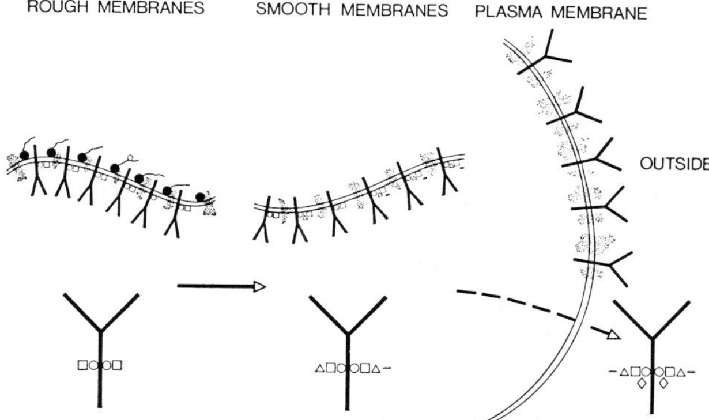

Schematic drawing of the subcellular distribution, carbohydrate composition and intracellular transport of immunoglobulin. The immunoglobulin is synthesized on membrane-bound polyribosomes within the rough membranes. A small part of all growing polypeptide chains of immunoglobulin may acquire a glucosamine residue (Melchers and Knopf, 1967), before it is released from polyribosomes and drawn into the membranous channels of the rough endoplasmic reticulum. There the population of immunoglobulin molecules obtain most of their glucosamine (O) and mannose (□) residues.
Immunoglobulin then migrates into smooth membranous structures like Golgi apparatus, where it acquires galactose (<>) and neuraminic acid (—) residues. The population of immunoglobulin molecules has on the average half of all possible carbohydrate residues of galactose and neuraminic acid. Secreted immunoglobulin is complete in its carbohydrate moieties, with all mannose, glucosamine, galactose, neuraminic acid and fucose (Δ) residues attached to it.
Receptor immunoglobulin in the outer (plasma) membrane has an unknown carbohydrate composition. In this representation results obtained for synthesis, transport and secretion of IgG, but not of IgM, are taken into account.

with IgG_1—producing myeloma plasma cells (Uhr and Schenkein, 1970; Melchers, unpublished observations). From these experiments it is concluded that immunoglobulin passes from rough membranes to smooth membranes and from there out of the cell. A schematic representation of the flow of immunoglobulin through plasma cells is given in Fig. 4. The rate-limiting step in the overall process of secretion in these plasma cells has been found to be the transfer from the rough to the smooth membranes.

At present it is not known whether immunoglobulin destined to be secreted from the cells can be found in the outer (plasma) membrane in transition and before being released from the cell, or whether two routes of migration exist for immunoglobulin to be incorporated into the outer membrane and for immunoglobulin to be secreted.

SPECIFICITY AND KINETICS OF INCORPORATION OF RADIOACTIVE SUGARS

Radioactive mannose, glucosamine, galactose and fucose are incorporated into immunoglobulins (Melchers, 1970; Melchers, 1971d). The kinetics of incorporation of radioactive mannose into IgG_1 shown in Fig. 3 are similar to those of the incorporation of leucine. However, sugar-labelled IgG, in contrast to leucine-labelled IgG, is secreted from the first minutes on. This indicates that sugar residues are added long after synthesis of the polypeptide chain and shortly before secretion of immunoglobulin.

D-mannose labels all residues of the carbohydrate moieties, e.g. mannose, glucosamines, galactose, fucoses and neuraminic acids. In some plasma cells a large pool of glucosamine exists, which prevents the labelling of glucosamine and neuraminic acid residues in immunoglobulins (Parkhouse and Melchers, 1971). D-glucosamine more specifically labels only glucosamine and neuraminic acid residues (Knopf, Sasso, Destree and Melchers, unpublished). D-galactose labels only galactose residues, L-fucose only fucose residues. Naturally other glycoproteins within plasma cells are labelled by these sugars. Electron microscopic radioautographic pictures of cells labelled with these sugars (Zagury *et al.*, 1970) should therefore be interpreted cautiously.

Remarkable differences are observed in the kinetics of incorporation of these different sugars into immunoglobulin inside plasma cells and the secretion of them (for experimental details see Melchers, 1970; Melchers, 1971d). In IgG-producing tumours, mannose, glucosamine and galactose label intracellular and secreted immunoglobulin, while fucose labels secreted, but not intracellular, immunoglobulin. It suggests that fucose residues are added to IgG while the protein is at its final stage of secretion. The kinetics of incorporation indicate at least three sequential precursor-product relationships between different intracellular forms and the secreted form of IgG myeloma protein (Melchers, 1970).

In an IgM-producing plasma cell tumour mannose and glucosamine are added at an early stage in the process of secretion, while galactose and fucose residues are added just before, or at the time, IgM leaves the cell. Neuraminic acid has not been assayed for in these analyses (Parkhouse and Melchers, 1971). The main difference between the incorporation of sugars into IgG and IgM is, therefore, the attachment of galactose residues. It is remarkable, that IgM acquires 35%-40% of its total carbohydrate residues shortly before, or at the time of secretion, concomitant with its polymerization to the 19S form. This final carbohydrate attachment may be concomitant with a conformational change from intracellular 7S IgMs to 7S subunits of extracellular 19S IgM (Melchers, 1972).

The incorporation of fucose into IgG_1 and the secretion of fucose-labelled IgG_1 from plasma cells is abolished after 2 to 3 hours in glucose-free medium, while leucine-labelled IgG_1 molecules are synthesized and secreted until 6 to 8 hours after starvation in glucose-free medium (Melchers, 1971d). Thus myeloma plasma cells appear to secrete, for a limited period of time, IgG_1 molecules complete in its polypeptide portion, but incomplete in its carbohydrate portion. Heterogeneity within the carbohydrate moieties of glycoproteins (discussed by Gottschalk (1969)) may in some cases be the result of incomplete biosynthesis, rather than of degradation subsequent to biosynthesis and secretion.

Synthesis and secretion of fucose-labelled IgG_1 can be restored in glucose-starved cells by the addition of 250 μM D-mannose, D-glucosamine and D-galactose. With the addition of these sugars, plasma cells will synthesize and secrete even more IgG_1 than in medium containing only glucose. It indicates that the addition of sugars, which are known to be incorporated into the carbohydrate portion of immunoglobulin, to the medium facilitates synthesis and secretion.

CARBOHYDRATE COMPOSITION OF IMMUNOGLOBULIN ISOLATED FROM DIFFERENT SUBCELLULAR FRACTIONS

Incorporation of labelled sugars into intracellular immunoglobulin shows that carbohydrate residues are added stepwise while the protein migrates through the cell (Choi *et al.*, 1971a; Melchers, 1970; Melchers, 1971c; Uhr and Schenkein, 1970). Mannose and glucosamine residues are added in the rough membranes, some of the galactose and neuraminic acid residues (Knopf, Sasso, Destree and Melchers, unpublished) in the smooth membranes and the rest of the galactose and neuraminic acid residues, as well as the predominant part of the fucose residues, while IgG is leaving the cell. The distribution of enzymic activities transferring the sugars onto immunoglobulin within the plasma cells are in accord with this labelling pattern (Schenkein and Uhr, 1970; Melchers, 1969b). It seems interesting to note that all such enzyme activities appear membrane-bound.

Chemical analysis of the carbohydrate content of IgG_1 and of a carbohydrate-containing K-light chain purified from different subcellular fractions of plasma cells (Melchers, 1971c) show that:

1. myeloma protein associated with rough membranes contains glucosamine and mannose residues, but only traces of galactose and no fucose residues. In the average population of molecules within this subcellular fraction only 10%-20% of the mannose and glucosamine are still missing, when compared to the secreted

portion. It argues that most of the molecules have obtained all the mannose and glucosamine residues and are piling up at the end of the rough membranes waiting to migrate to the smooth membranes.

2. myeloma protein associated with smooth membranes contains glucosamine, mannose, galactose and neuraminic acid residues, but no fucose residues.

3. myeloma protein associated with the cytoplasmic supernatant contain glucosamine, mannose, galactose, and neuraminic acid residues, and traces of fucose residues. The cytoplasmic supernatant fraction is probably an artefact in the preparation of subcellular fractions.

The results support the conclusions reached from studies on the kinetics of incorporation of radioactive sugars (see above).

Recently a method has been developed for the fractionation of glycopeptides from immunoglobulin (Knopf, Sasso, Destree and Melchers, unpublished). Three extracellular glycopeptides differing in their neuraminic acid content (Melchers, 1969a) and four intracellular glycopeptides differing in their galactose and neuraminic acid content have been distinguished in a pronase-digest of a carbohydrate-containing light chain. More refined separation methods promise to resolve even further the expected heterogeneity of glycopeptides from different forms of intracellular immunoglobulin and may be used to investigate precursor-product relationships between these different forms. With the aid of such analyses it should be possible to define more clearly the steps of assembly of the carbohydrate moieties of immunoglobulins, and to see whether one or more routes of migration exist within plasma cells.

Similar chemical analyses have been done of the carbohydrate composition of intracellular 7S and extracellular 19S IgM (Melchers, 1972). These analyses support the conclusions reached from kinetics of incorporation of radioactive sugars into IgM and suggest that the intracellular 7S IgMs contains only mannose and glucosamine residues—again 80%-90% complete, when compared to extracellular IgM—and acquires galactoses, fucoses and possibly neuraminic acids, while leaving the cell and polymerizing to 19S IgM.

Migration of immunoglobulin through subcellular membranous structures and out of the cell and the parallel stepwise acquisition of carbohydrate residues at these different subcellular sites is schematically shown in Fig. 4.

SECRETION OF IMMUNOGLOBULIN CHAINS WITHOUT ATTACHMENT OF CARBOHYDRATE

It has been suggested that attachment of carbohydrate to immunoglobulin may be requisite for the selection of immunoglobulin to be secreted and for the regulation of its migration through the cell (Melchers and Knopf, 1967; Swenson and Kern, 1968). This hypothesis has been supported by the findings described above showing that parallel to the migration through plasma cells immunoglobulins acquire residues of their carbohydrate moieties stepwise at different subcellular sites.

Two cases of synthesis of immunoglobulins in plasma cells argue against this hypothesis. In the one case, described above, IgG_1 has been found to be secreted from plasma cells in glucose-free medium with an incomplete carbohydrate moiety lacking terminal fucose residues. It could still be argued that selection of immunoglobulin for secretion is accomplished early on after synthesis, when the molecule is released from polyribosomes and pulled into the membranous channels of the rough endoplasmic reticulum. If this transport into the cisternae of the rough membranes would require the addition of core sugars only, like glucosamine and/or mannose, later sugars may not be requisite for the final stages of secretion in differentiated plasma cells.

However, two other myeloma plasma cells both secrete a light chain of immunoglobulin without detectable amounts of carbohydrate attached to the light chains (Parkhouse and Melchers, 1971; Melchers, 1971b). In the population of these light chains either from inside the cells or as secreted proteins, less than 0.2 residues of either mannose, galactose or fucose were found. Since in one of the two cells the pool of glucosamine has been found to be large, since in the other cell such a pool of glucosamine cannot be ruled out at present, and since even amino acid sequence determination of one of the light chains has not been able to rule out some glucosamine attached to it (Hood, 1967), it may be still possible that one or more residues of glucosamine are attached to the light chain. The attachment of this or these residues of glucosamine may be sufficient to select it from all other intracellular proteins for secretion.

Other, yet unknown, processes must occur in connection with transport and secretion of immunoglobulins in plasma cells. If two modes of synthesis, transport and secretion exist in plasma cells, packaging into membranous structures (Jamieson and Palade, 1966) could be one way by which large amounts of protein are secreted even in the absence of the antigenic stimulus. Attachment of carbohydrate may be the other mode of transport which could be tightly regulated by membrane-bound sugar-transferring enzymes. Such an assembly line, in which the last enzyme is turned on after antigenic stimulation, would allow for a controlled synthesis, transport and secretion of immunoglobulins. As lymphoid cells differentiate into plasma cells they may progressively evade this control mechanism. Sugar attachment would still occur, since immunoglobulin still has the protein structure to accept carbohydrate, and since the enzymes transferring sugars may be still existing in plasma cells as remnants of the differentiation from small lymphocytes.

It is evident that biochemical studies similar to those done with plasma cells should be undertaken with small lymphocytes to clarify these questions.

SYNTHESIS AND SECRETION OF IMMUNOGLOBULIN IN SMALL LYMPHOCYTES

A major problem for studies on the modes of synthesis, transport and secretion in small lymphocytes stimulated by antigen arises from the fact that only very few lymphocytes recognize a given antigen. This has been shown by the binding of radioactive antigen (0.02% of all lymphocytes, see Ada, 1970), of a mutant β-galactosidase, which is activated by specific antibodies to enzymic activity (1 in 10^6 lymphoid cells; Melchers and Messer, unpublished) and by stimulation of lymphoid cells into clones of antibody-forming cells in micro-cultures (5.5 in 10^6 cells for sheep red blood cells as antigen, Lefkovits, 1972). For biochemical studies, however, a cell population is needed, where the predominant part reacts to an antigenic stimulus. Cell separation methods (Hulett *et al.*, 1969; Henry *et al.*, 1972) and cloning of antigen-reactive cells in irradiated hosts in the presence of antigen (Askonas *et al.*, 1970) have so far not yielded cell preparations suitable for biochemical studies. These methods also have the disadvantage that the antigen is used to purify a given cell. The cell is thus exposed to the antigen for the time of purification and prior to the studies of biochemical processes.

Unspecific stimulation of small lymphocytes, not triggered by antigen, promises to yield at least information on the reaction of the cells to an external stimulus. Mitogens unspecifically stimulate small lymphocytes. It has been found that phytohaemagglutinin (Janossy and Greaves, 1972) and concanavalin A (Andersson *et al.*, 1972b) stimulate T-cells, while pokeweed mitogen (Janossy and Greaves, 1971) lipopolysaccharide (Andersson *et al.*, 1972b) and locally concentrated phytohaemagglutinin (Greaves and Bauminger, 1972) and concanavalin A (Andersson *et al.*, 1972a) stimulate B-cells to increased DNA-synthesis and proliferation. Mitogen-stimulated B-cells produce and secrete only IgM but do not switch to the production of other classes of immunoglobulins (Parkhouse *et al.*, 1972; Andersson and Melchers, unpublished).

The kinetics of synthesis and secretion of IgM in B-cells from nude (athymic) mice stimulated for 72 hours with c-Con A are strikingly similar to those observed with a myeloma plasma cell tumour producing IgM. Secreted IgM is in the polymeric 19S form, intracellular IgM is mostly IgMs (7S), with some precursor material of lower S-values (probably L and μL; Parkhouse, 1971; Parkhouse *et al.*, 1972) and a small amount of 19S IgM.

Carbohydrate attachment to IgM is also very similar to that observed with the IgM-secreting tumour plasma cells. Intracellular IgMs is only labelled by radioactive mannose, probably in mannose and glucosamine residues, while galactose and fucose label is associated predominantly with the secreted 19S form of IgM. We conclude for the structurally and functionally heterogeneous population of IgM molecules synthesized and secreted after c-Con A-stimulation the same as we did for a homogeneous myeloma IgM: the semiterminal and terminal carbohydrate residues of IgM, galactose, fucose and neuraminic acid, e.g. 35%-40% of the total weight of carbohydrate in IgM, are acquired shortly before or while IgM leaves the cell and is polymerized into the 19S form.

INDUCTION OF IgM-SYNTHESIS AND SECRETION IN B-CELLS BY LOCALLY CONCENTRATED CONCANAVALIN A

Rates of protein synthesis and of IgM synthesis and secretion were measured after various time periods of exposure of B-cells to c-Con A by the incorporation of radioactive leucine for 4 hours. In the first 8 to 10 hours after stimulation no significant increase in the rate of total protein synthesis or of intra cellular IgM synthesis was observed. In these first 8 to 10 hours IgM was found only inside the cells, but not secreted. At 8 to 10 hours after stimulation the rates of total protein synthesis and of intracellular IgM synthesis began to increase, when compared to the unstimulated cell cultures. Secretion of IgM however was detectable only 24-32 hours after c-Con A-stimulation.

The ratio of IgM synthesis to total protein synthesis changed from 0.015 at 10 hours to 0.17 at 75 hours after stimulation. The ratio of IgM secretion to secretion of total protein changed from 0.1 at 20 hours to 0.5 at 75 hours after stimulation. It shows that other proteins than IgM are secreted by B-cells early after stimulation. However, with increasing length of stimulation IgM synthesis and secretion increases over total protein synthesis. This ratio represents a parameter of the differentiation into IgM-secreting plasma cells. In mature plasma cells this ratio should approach 1, e.g. IgM then remains the only protein to be secreted.

If c-Con A stimulates B-cell through binding to glycoproteins other than immunoglobulins on the outer membrane (Allan *et al.*, 1972) the observed preferential synthesis and secretion of IgM could mean that antigen senstitive B-cells are determined prior to stimulation to run through proliferation and differentiation into antibody-producing cells no matter whether the stimulus is specific, e.g. antigen binding at the receptor immunoglobulin, or unspecific, e.g. mitogen binding at some other membrane glycoprotein (Ginsburg and Kobata, 1971). However, since immunoglobulins are glycoproteins it remains to be clarified whether or not concanavalin A attaches to immunoglobulin receptors on the outer membranes of small B-lymphocytes.

In summary the induction of B-lymphocytes by locally concentrated concanavalin A involves first an increased intracellular synthesis of IgM at 8 to 10 hours, followed by the initiation of secretion of IgM at 24-32 hours after stimulation. Differentiation into antibody-producing and secreting cells after stimulation is manifested by 1) a rate of IgM-synthesis, which increases continuously over total protein synthesis and 2) an increased rate of IgM secretion with concomitant decrease of secretion of other proteins.

REFERENCES

Ada, G. L. (1970). *Transplant. Rev. 5,* 105

Allan, D., Auger, J. and Crumpton, M. J. (1972). *Nature (London) 236,* 23

Andersson, J., Edelman, G. M., Möller, G. and Sjöberg, O. (1972a). *Eur. J. Immunol. 2,* in press

Andersson, J., Möller, G. and Sjöberg, O. (1972b). *Cellular Immunology,* in press

Askonas, B. A. and Parkhouse, R. M. E. (1971). *Biochem. J. 123,* 629

Askonas, B. A., Williamson, A. R. and Wright, B. E. G. (1970). *Proc. Nat. Acad. Sci., U.S., 67,* 1398

Bevan, M. J., Parkhouse, R. M. E., Williamson, A. R. and Askonas, B. A. (1972). *Progr. Mol. Biophys.,* in press

Choi, Y. S., Knopf, P. M. and Lennox, E. S. (1971a). *Biochemistry 10,* 668

Choi, Y. S., Knopf, P. M. and Lennox, E. S. (1971b). *Biochemistry 10,* 659 de Petris, S., Karlsbad, G. and Pernis, B. (1963). *J. Exper. Med. 117*, 849

Ginsburg, V. and Kobata, A. (1971) in: *Structure and Function of Biological Membranes*, (ed. Rothfield, L. I.), p. 439, Academic Press, New York

Gottschalk, A. (1969). *Nature (London) 222*, 452

Greaves, M. F. and Bauminger, S. (1972). *Nature (London) 235*, 67

Helmreich, E., Kern, M. and Eisen, H. N. (1961). *J. Biol. Chem. 236*, 464

Henry, C., Kimura, J. and Wofsy, L. (1972). *Proc. Nat. Acad. Sci., U.S., 69*, 34

Hood, L. (1967). *Cold Spring Harbor Sympos. Quant. Biol. 32*, 262

Hulett, H. R., Bonner, W. A., Barrett, J. and Herzenberg, L. A. (1969). *Science 166*, 747

Jamieson, J. D. and Palade, G. E. (1966). *Proc. Nat. Acad. Sci., U.S., 55*, 424

Janossy, G. and Greaves, M. F. (1972). *Clin. Exper. Immunol. 9*, 483

Knopf, P. M., Parkhouse, R. M. E. and Lennox, E. S. (1967). *Proc. Nat. Acad. Sci., U.S., 58*, 2288

Kornfeld, R., Keller, J., Baenziger, J. and Kornfeld, S. (1971). *J. Biol. Chem. 246*, 3259

Lefkovits, I. (1972). *Eur. J. Immunol. 2*, in press

Melchers, F. (1969a). *Biochemistry 8*, 938

Melchers, F. (1969b). *Behringwerk-Mitteilungen 49*, 169

Melchers, F. (1970). *Biochem. J. 119*, 765

Melchers, F. (1971a). *Histochemical J. 3*, 389

Melchers, F. (1971b). *Eur. J. Immunol. 1*, 330

Melchers, F. (1971c). *Biochemistry 10*, 653

Melchers, F. (1971d). *Biochem. J. 125*, 241

Melchers, F. (1972). *Biochemistry*, in press

Melchers, F. and Knopf, P. M. (1967). *Cold Spring Harbor Sympos. Quant. Biol. 32*, 255

Parkhouse, R. M. E. (1971). *Biochem. J. 123*, 635

Parkhouse, R. M. E., Janossy, G. and Greaves, M. F. (1972). *Nature (London) 235*, 21

Parkhouse, R. M. E. and Melchers, F. (1971). *Biochem J. 125*, 235

Pernis, B., Forni, L. and Amante, L. (1970). *J. Exper. Med. 132*, 1001

Pernis, B., Forni, L. and Amante, L. (1971). *Ann. N.Y. Acad. Sci.*, in press

Potter, M. (1967) in: *Methods in Cancer Research*, Vol. 2, (ed. Busch, H.) p. 105, Academic Press, New York

Progress in Immunology, 1971 First International Congress of Immunology, (ed. Amos, B.) Academic Press, New York and London

Scharff, M. D., Bargellesi, A., Baumal, R., Buxbaum, J., Coffino, P. and Laskov, R. (1970). *J. Cell. Physiol. 76*, 331

Schenkein, I. and Uhr, J. W. (1970). *J. Immunol. 105*, 271

Shimizu, A., Putnam, F. W., Paul, C., Clamp, J. R. and Johnson, I. (1971). *Nature (London) 231*, 73

Sox, H. C., Jr. and Hood, L. (1970). *Proc. Nat. Acad. Sci., U.S., 66*, 975

Swenson, R. M. and Kern, M. (1968). *Proc. Nat. Acad. Sci., U.S., 58*, 546

Uhr, J. W. and Schenkein, I. (1970). *Proc. Nat. Acad. Sci., U.S., 66*, 952

Whalcy, W. G., Dauwalder, M. and Kephart, J. E. (1972). *Science, 175*, 596

Zagury, D., Uhr, J. W., Jamieson, J. D. and Palade, G. E. (1970). *J. Cell. Biol. 46*, 52

Molecular and Cellular Mechanisms of Clonal Selection

GERALD M. EDELMAN
The Rockefeller University, New York, U.S.A.

Immunological thinking in the last ten or fifteen years has been profoundly altered by two major developments: elaboration of the theory of clonal selection (Jerne, 1955; Burnet, 1959) and analysis of the structure of antibodies (Cairns, 1967; Killander, 1967; Edelman and Gall, 1969). The clonal selection theory has provided a general framework for understanding immunological phenomena. The analysis of antibody structure has allied immunology with molecular biology (Edelman, 1971) and has provided the major means for understanding the evolution, genetics and function of antibodies.

One of the main results of structural analysis has been to clarify the basis for the enormous diversity of antibodies required by the clonal selection theory. Differences in the amino acid sequences of variable domains of light and heavy chains and the assortment of light chain-heavy chain interactions permit the formation of an enormous number of different antibodies. There is therefore no longer any reason to question this aspect of clonal selection.

There are a number of mechanisms that remain to be understood, however. These include the genetic mechanisms of commitment which result in production by a given cell of antibodies of only one specificity, the means by which antigen is presented to lymphoid cells and the mechanisms by which a cell is triggered to mature and divide to produce more of its specific type of antibody. The clonal selection theory is not quantitative in its present form and little is known about the relative numbers of cells of a given specificity, binding capacity and triggering threshold present in unimmunized and immunized animals. A description of these parameters in quantitative terms would yield a more predictive theory particularly in respect to the population dynamics of the immune cell population. A complete theory of clonal selection should be able to trace the pathways from antibody gene expression to cell triggering, yielding a description of the composition of the lymphoid cell population at any time, as well as a description of the molecular mechanisms at each stage of development.

In the present paper, I will consider some studies and hypotheses bearing upon several of these mechanisms. The subjects to be considered include:
1) the evolution of antibody genes to yield a special arrangement in the genome,
2) the separation by fibre fractionation of lymphoid cell populations according to the specificity and affinity of their immunoglobulin (Ig) receptors,
3) the behaviour of the lymphocyte membrane and its immunoglobulin receptors after interaction with lectins and antigens.

ARRANGEMENT AND EVOLUTION OF ANTIBODY GENES: A UNIFYING HYPOTHESIS

This subject is important for several reasons. The genetic apparatus for the immune system is a late evolutionary development, having appeared first in the vertebrates, and its relation to other vertebrate genes such as those of the histocompatibility system is of general interest. Moreover, the number and arrangement of immunoglobulin genes is critical in assessing theories of antibody diversity. Finally, both genetic and structural data suggest that a special event is required to construct a complete structural gene specifying an immunoglobulin polypeptide chain, and this event may be essential in committing a cell to a single type of antibody molecule.

The fundamental genetic unit underlying immunoglobulin production appears to be a cluster of linked nucleotide sequences specifying the amino terminal or variable regions (Fig. 1) of the immunoglobulin chains (V genes) followed by a cluster of nucleotide sequences specifying the carboxyl terminal or constant

FIGURE 1

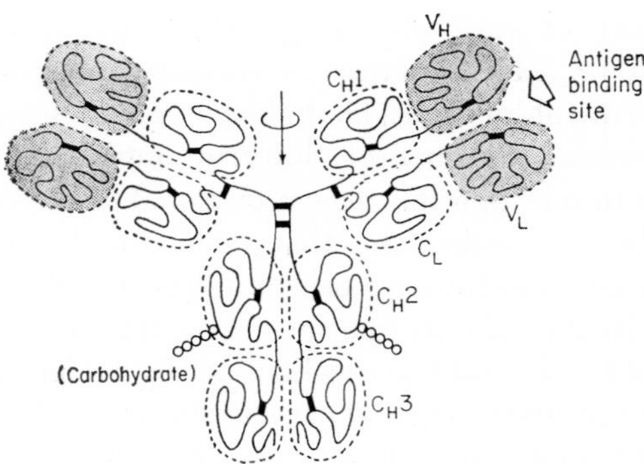

A model of the structure of a human IgG molecule. The variable regions of heavy and light chains (V_H and V_L), the constant region of the light chain (C_L), and the homology regions in the constant region of heavy chain (C_H1, C_H2 and C_H3) are thought to fold into compact domains (delineated by dotted lines), but the exact conformation of the polypeptide chains has not been determined. The vertical arrow represents the two-fold rotation axis through the two disulphide bonds linking the heavy chains. A single interchain disulphide bond is present in each domain. Carbohydrate prosthetic groups are attached to the C_H2 regions.

portions (C genes). The entire cluster (Fig. 2) is termed a translocon (Cunningham *et al.*, 1971; Gally and Edelman, 1970; Gally and Edelman, 1972).

FIGURE 2

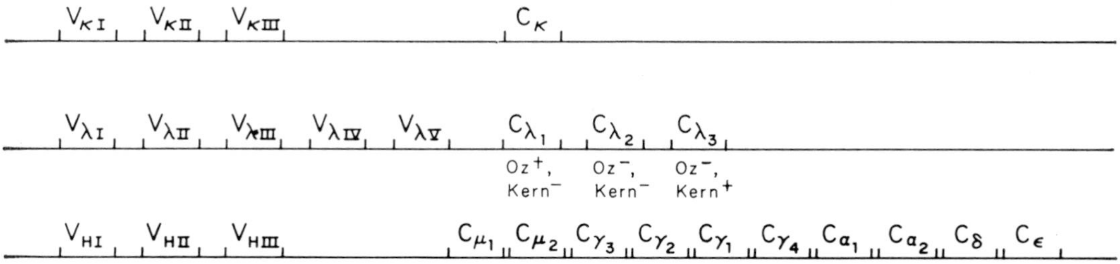

$V_{\kappa I}$ $V_{\kappa II}$ $V_{\kappa III}$ C_{κ}

$V_{\lambda I}$ $V_{\lambda II}$ $V_{\lambda III}$ $V_{\lambda IV}$ $V_{\lambda V}$ C_{λ_1} C_{λ_2} C_{λ_3}

Oz^+, $Kern^-$ Oz^-, $Kern^-$ Oz^-, $Kern^+$

V_{HI} V_{HII} V_{HIII} C_{μ_1} C_{μ_2} C_{γ_3} C_{γ_2} C_{γ_1} C_{γ_4} C_{α_1} C_{α_2} C_{δ} C_{ϵ}

Arrangement of antibody genes in clusters. V and C genes in the translocons proposed for a normal human germ cell. The exact number and arrangement of genes within the translocons is not known. It is probable that additional V region subgroups and C region subclasses will be discovered.

A mammalian cell appears to contain at least three translocons, each within a separate non-homologous autosome, one for kappa chains, one for lambda chains and one for heavy chains.

The evidence for this hypothetical gene arrangement comes from a variety of sources and has been recently reviewed (Gally and Edelman, 1970; Gally and Edelman, 1972). Here I will mention without documentation only the main lines of evidence, which are derived from analysis of the amino acid sequences of antibodies as well as from immunogenetic studies of immunoglobulins. The latter studies indicate that genetic markers on C regions are specified by single genes which behave in a classical Mendelian fashion. They also indicate that each haploid portion of the genome contains multiple non-allelic V genes, specifying different V region subgroups each with distinctive amino acid sequences. Immunoglobulin chains with different V region subgroups (each specified by different genes) can have identical C regions specified by a single C gene. The evidence therefore suggests that immunoglobulin chains, unlike other proteins, are specified by two genes, V and C.

Additional evidence comes from the finding that heavy chains from different immunoglobulin classes which are specified by distinct but genetically linked C genes all share the same V gene subgroups (Köhler *et al.*, 1970). Moreover, in the rabbit, there are similar genetic markers in the V regions of heavy chains but separate markers for C regions. The genes coding for these markers are all closely linked to each other as indicated by a low frequency of recombination between V and C loci (Mage, 1971).

It is not surprising that the organization of a molecule for both variability and constancy would require a special evolutionary development such as a translocon. Amino acid sequence analysis provides support for linear regional differentiation of function within the immunoglobulin molecule (Fig. 1): V region domains carry out antigen binding and C region domains carry out effector functions such as complement fixation. Recent studies have strongly supported the domain hypothesis (Edelman, 1970) which states that each homology region forms a separate compact three-dimensional domain carrying out a particular function. A major requirement in immunoglobulin evolution was to develop a means of coupling and controllably expressing antigen-binding functions in a coordinated way. The arrangement of genes in a translocon is a particularly economical way of doing this and at the same time assuring that one cell will be restricted to making antibodies of only one specificity as required by the clonal selection theory.

Analysis of the structure of various immunoglobulins suggests that the various translocons arose both singly and as a group by processes of gene duplication (Hill, *et al.*, 1966; Singer and Doolittle, 1966) and chromosomal translocation (Gally and Edelman, 1970; Gally and Edelman, 1972). There is, for example, strong evidence for sequence homologies among V regions as well as homologies among all C regions (Fig. 3). The simplest hypothesis consistent with both

FIGURE 3

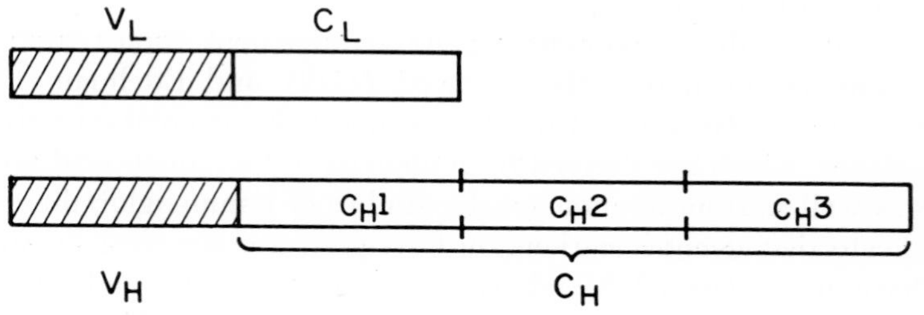

Internal homologies in the structure of γG1-immunoglobulin. Variable regions V_H and V_L are homologous. The constant region of the heavy chain (C_H) is divided into three regions, C_H1, C_H2, and C_H3, which are homologous to each other to the C region of the light chain.

the structural and the genetic evidence would therefore suggest that it is the translocon (rather than the immunoglobulin structural gene) which is the fundamental evolutionary unit for immunoglobulin structure (Fig. 4).

After introduction into the genome, each translocon could serve as a separate unit of Ig evolution whose components were modified in a coordinate fashion. The translocon is thus analogous to "supergenes" or closely linked genetic loci that have been reported to function and evolve together in eukaryotic genomes. Less certain aspects of the evolutionary scheme shown in Fig. 4 are: 1) whether or not the original V and C genes were themselves evolutionarily homologous; 2) the function of the original nucleotide sequences from which the translocons evolved; and 3) the exact order and timing of the various gene duplication events.

FIGURE 4

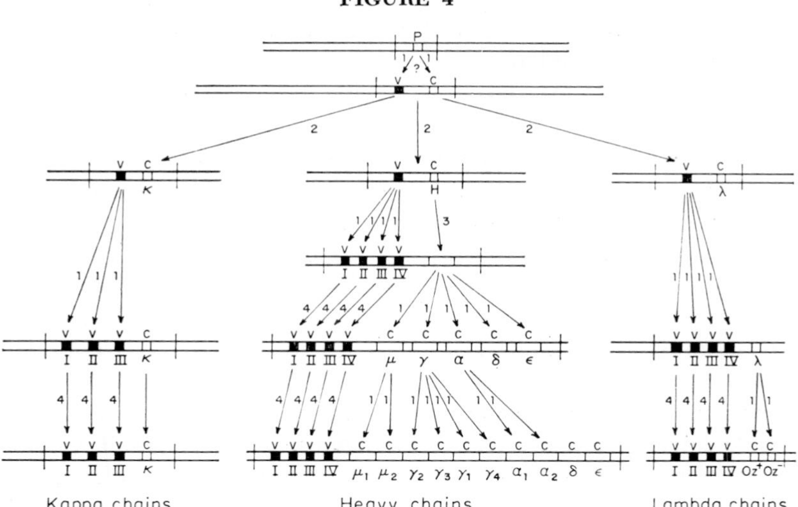

The hypothetical chain of events during translocon evolution. Little or no evidence is available to indicate the exact order of the gene duplication events, or of the detailed arrangement of genes within translocons, so that the arrangement shown here is arbitrary. It is possible, for example, that extensive V and C gene duplication occurred within the original single translocon. Each of the shaded V genes might represent a single, somatically hypermutable locus, or a set of tandemly-duplicated, frequently-recombining nucleotide sequences. The genetic mechanism employed at each step of this evolutionary scheme is indicated by the number next to each arrow. These signify:

1—gene duplication by inter-cistronic, unequal crossing-over (either *inter-* or *intra-*chromosomal).

2—gene duplication by polyploidization or chromosomal translocation.

3—gene enlargement by intra-cistronic, unequal crossing-over (either *inter-* or *intra-*chromosomal).

4—co-evolution of multiple, tandem genes, possibly by democratic gene conversion (Gally and Edelman, 1970).

TRANSLOCATION AND PHENOTYPIC RESTRICTION

A gene arrangement of the kind shown in Fig. 2 is obviously unusual and requires the existence of means for combining the information from a given V gene with that from a given C gene to make a complete VC structural gene. This could take place at the polynucleotide level or by joining separately synthesized V and C peptides. Against the latter possibility are the findings that immunoglobulin chains appear to be synthesized from a single starting point (Fleischman, 1967; Lennox, 1967) and that they can be synthesized from messenger RNA in frog oöcytes (Gurdon, *et al.,* 1971). There is some indirect evidence that the translocation of information may be at the DNA level. This is provided by analysis of several unusual myeloma polypeptide chains seen in "Heavy Chain Disease" (Ein, 1968) as well as in light chains (Smithies, *et al.,* 1971). These polypeptide chains contain sequence deletions and in many cases the deletion occurs in the region at which V and C regions ordinarily join. This can most easily be accounted for by assuming errors in joining or recombination of DNA strands.

It is now clear that one lymphoid cell makes one antibody of a given specificity at one time (Makëla and Cross, 1970) and aside from determining the origin of antibody variability, perhaps the most important problem to be analyzed is the relationship between translocon expression in lymphoid cells and the cellular commitment to only one kind of antibody (phenotypic restriction). There are several levels at which such restriction can operate: 1) class restriction, (e.g. IgG or IgM), 2) V region subgroup restriction, 3) allelic exclusion (the expression of only one of two possible allelic variants of V or C regions by a single cell). A useful working hypothesis is that a major stage in restriction is translocation of a V to a C gene with the creation of a single VC gene for a given immunoglobulin polypeptide chain. Moreover, translocation provides an additional mechanism for switching an antigen-combining site of a particular specificity from one class of immunoglobulin to another. In this way, a cell may be phenotypically restricted to producing one set of V regions in association with one set of C regions and then change later to another set of C regions. The net result would be the maintenance of specificity of antigen binding with a change in effector functions. The recent finding that a considerable number of cells convert from IgM to IgG production with retention of V region genetic markers (Pernis, *et al.*, 1971) strongly supports this hypothesis. Additional evidence has come from the fact that identical idiotypes (unique antigenic markers in V regions) can occur in IgM antibodies and later in IgG antibodies in immunized rabbits (Oudin and Michel, 1969).

Direct molecular evidence that the same V region can be attached to two separate C regions has been obtained in a study of a rare case of multiple myeloma in which two clones of malignant Ig-producing cells were present, one making IgG and the other IgM (Wang, *et al.*, 1970). The light chains isolated from the two myeloma globulins were found to be chemically very similar or identical. More significantly, the first 35 residues of the amino terminal sequence of the chain were identical to those of gamma chain, and the two globulins shared idiotypic determinants. This strongly suggests that one malignant clone was derived from the other by a V gene translocation event.

Although translocation appears to be a necessary prerequisite for the synthesis of normal Ig polypeptide chains by lymphoid cells, it certainly cannot be considered as the only, or even as the most decisive, factor in determining whether a cell will eventually differentiate to produce antibody molecules. Obviously, other control mechanisms must exist to express VC genes for both light and heavy chains independently. Some factors that may play important roles in this process include genes other than those in the translocon, antigen processing and presentation, cells other than those which secrete antibodies (T cells, macrophages) and other physical and chemical properties of the local cellular environment. Once these factors operate, certain lymphoid cells may become antigen binding cells, expressing a particular immunoglobulin at their surface and some of these cells may be triggered to clonal proliferation by binding antigen.

The key questions that remain to be answered in the context of clonal selection are: 1) How many cells are there in each population and what is the specificity, affinity and relation to triggering of their receptor types? 2) How many receptors are there per cell and to what Ig class do they belong? 3) What is the state of the receptor in the membrane and how does that state relate to the capacity of the cell to be stimulated or triggered?

In the remainder of this paper, I shall consider some experiments concerned with certain aspects of these questions.

QUANTITATION OF CLONAL SELECTION AND SPECIFIC FRACTIONATION OF LYMPHOID CELLS

Lymphocytes are heterogeneous at a variety of levels and an analysis of their population dynamics in quantitative terms is essential to a complete understanding of clonal selection. For these purposes we have been using the recently developed general method of fibre fractionation (Edelman *et al.*, 1971; Rutishauser *et al.*, 1972). Before reviewing some of the results it may be useful to consider certain aspects of cellular expression and heterogeneity of lymphocyte receptors as they pertain to clonal selection.

Almost all immunoglobulin-producing cells whose products have been characterized are derived from precursor cells that have been stimulated to proliferate (by antigen or otherwise) to form a clone of cells, each committed to the production of the same immunoglobulin (Fig. 5). The phenotypic expression of

FIGURE 5

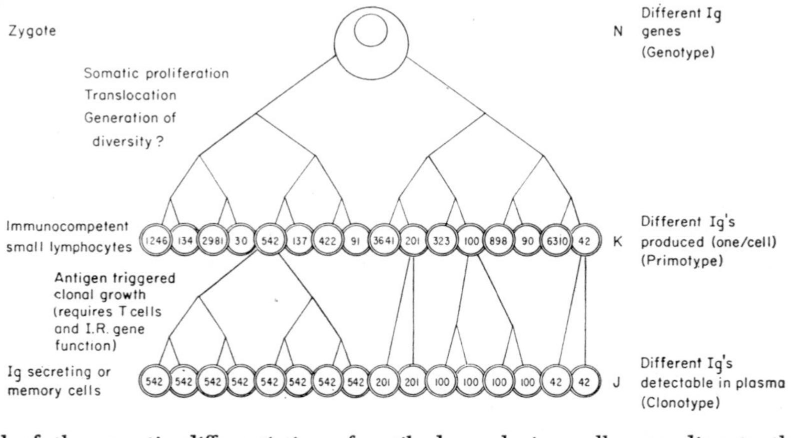

A model of the somatic differentiation of antibody-producing cells according to the clonal selection theory. The number of Ig genes, which equals N in the zygote, may increase during somatic growth so that in the immunologically mature animal K different cells are formed, each committed to the synthesis of a structurally distinct Ig (indicated by the arabic number). A small proportion of these cells proliferate upon antigenic stimulation to form J different clones of cells, each clone producing a different antibody. $N<J<K$

the immunoglobulin genes is therefore mediated in the animal by somatic division of precommitted cells. Since an animal is capable of responding specifically to an enormous number of antigens to which it is usually never exposed, it follows that the animal must contain genetic information for synthesizing a much larger number of different immunoglobulin molecules than actually appear in detectable amounts in the blood stream. The clonal selection theory suggests, therefore, that large numbers of cells may commit themselves to the production of immunoglobulin molecules with binding sites that are not complementary to any known antigen. In other words, the immunoglobulin molecules whose properties we can examine may represent only a minor fraction of those for which genetic information is available.

To avoid confusion concerning the various levels of expression of immunoglobulin genes, I shall employ three specific terms (see Fig. 5): 1) the genotype: the genetic information stored within the zygote which can be passed on to future generations, specifically the three translocons (Fig. 2) and their regulatory genes. 2) the primotype: the sum total of different, structurally distinct immunoglobulin molecules generated within an organism during its lifetime. While the number of different Ig molecules in the primotype may be much less than that which is *potentially* present in the genotype, it is probably orders of magnitude greater than the number of different *effective* antigenic determinants to which the animal is ever exposed. 3) the clonotype: the different immunoglobulin molecules that can be detected and classified according to antigen binding specificity, class, antigenic determinants, primary structure, allotype, or any other experimentally measurable molecular property. This definition takes explicit account of the fact that, in general, for these properties to be detected, a large number of identical molecules must be present, usually far more than can be obtained from a single cell.

As a class, the clonotype is smaller than the primotype and wholly contained within it. So far, those immunoglobulin molecules whose production has not been clonally amplified cannot be examined and one can only assume without evidence that they are similar to those that can be examined. Since certain ways of stimulating clonal proliferation may considerably bias the clonotype observed, however, the immunoglobulin molecules that can be analyzed may be quite unrepresentative of the actual class of receptors in the primotype. The proportion of the various V gene subgroups observed in a collection of myeloma globulins, for example, may differ considerably from those in normal pooled IgG and these may both differ from purified antibody molecules. None of these necessarily reflect the frequencies of the subgroups occurring in the primotype, and their relationship with the number of V genes within the genotype is purely conjectural.

Failure to take these various levels of gene expression into account when interpreting structural data can lead to dubious conclusions. The fact that different animals might synthesize chemically distinct antibodies, for example, does not necessarily indicate that these animals differ in either genotype or primotype. The absence of a particular class or subgroup of immunoglobulin in the clonotype does not entail an absence in the primotype. Moreover, a change in the clonotype of an animal which occurs during prolonged immune response does not imply any corresponding change in genotype or primotype.

Although some view of the clonotype is afforded by analysis of humoral antibodies, we know very little about the primotype. It is therefore important to attempt to fractionate the cells of the immune system according to their receptors. A method of separating cells on nylon fibres which appears to be effective for this purpose has recently been developed (Edelman, *et al.*, 1971; Rutishauser, *et al.*, 1972). A scheme of the details of the method is shown in Fig. 6.

FIGURE 6

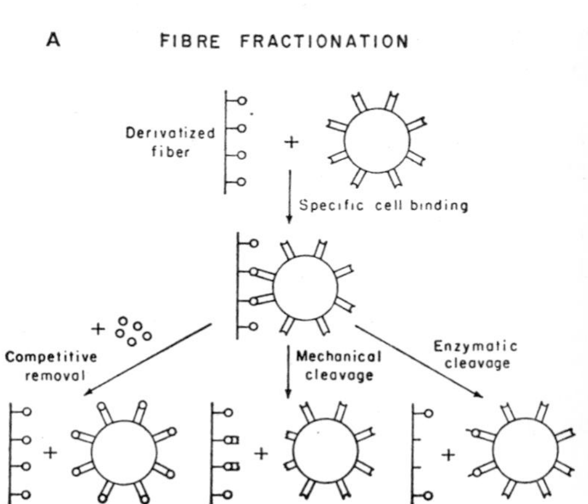

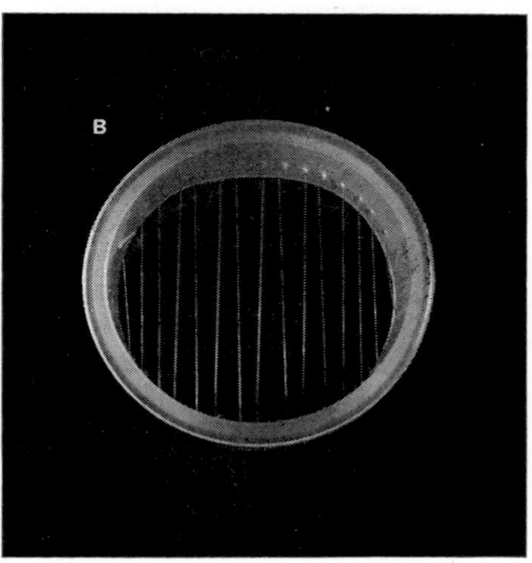

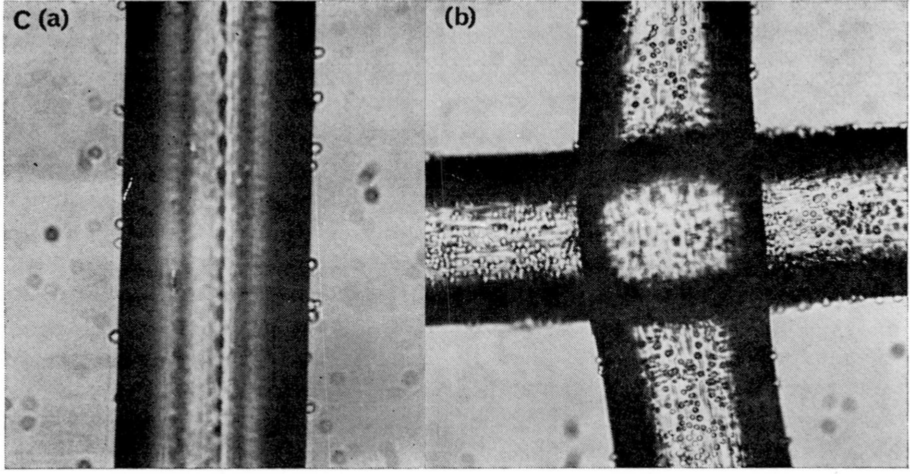

Fibre fractionation of cells. A. General scheme. B. Tissue culture dish with nylon fibres in frame. C. Mouse spleen cells bound to DNP-BSA nylon in monofilament (a) and mesh form (b).

TABLE 1

Binding of Mouse Spleen Cells to Antigen Derivatized Fibres

Fibre Antigen[a]	Number of Cells Bound[b]	
	Immunized[c]	Unimmunized
DNP-BSA		
Exp't 1	802	285
2	1004	301
3	654	283
Tosyl-BSA		
Exp't 1	353	143
2	297	130
BSA		
Exp't 1	173	65
2	112	58
Stroma		
Exp't 1	160	75
2	145	70

[a]See materials and methods for details of derivatization. The DNP-BSA and tosyl-BSA fibres were coated with 10^{11} and 2 x 10^{11} antigen molecules per cm respectively.

[b]Expressed as number of cells bound to both edges of a 2.5 cm fibre segment.

[c]Secondary responses to DNP-BGG, tosyl-BSA, BSA and sheep erythrocytes respectively. Cells from three mice were pooled.

A variety of antigens, including haptens, proteins and cell membranes can be used to derivatize the fibres (Table 1). The total number of bound cells in a dish was determined by counting the cells bound to a short segment of the total fibre length. Individual dishes containing 25 cm of fibre specifically bound from 10^5 to 10^6 cells. Greater binding capacities (up to 5 x 10^6 cells) were achieved by the use of fibre meshes. As shown in Table 1, the number of isolated cells varied with the particular antigen used, the extent to which the fibres were substituted, and the immunological history of the cell donors.

The binding of cells from immunized and unimmunized mice to an antigen-derivatized fibre was specifically inhibited by the addition of soluble antigen to the cell suspension during the incubation period but immunologically distinct molecules did not inhibit binding (Table 2). Preincubation of the spleen cells with the immunoglobulin G fraction of rabbit antiserum directed against mouse immunoglobulin G prevented the binding of cells to any antigen substituted fibre; no inhibition was obtained when normal rabbit IgG was used. These data indicate that the attachment is specific and is mediated by immunoglobulin receptors on the cell surface.

In order to examine the topographical distribution and specificity of the immunoglobulin receptors, spleen cells bound to fibres derivatized with dinitrophenylated bovine serum albumin (DNP-BSA) were tested for their ability to

TABLE 2

Specific Inhibition of Spleen Cell Binding to Derivatized Fibres

Fibre Antigen	Immunogen	Inhibitor[a]			
		DNP	Tosyl	Stroma	Anti-Ig
DNP-BSA	DNP-BGG	91%[b]	1%	—	93%
DNP-BSA	NONE	73%	2%	—	72%
Tosyl-BSA	Tosyl-BSA	3%	87%	—	90%
Tosyl-BSA	NONE	6%	59%	—	63%
Stroma	Stroma	<5%	<5%	70%	80%
Stroma	NONE	<5%	<5%	50%	45%

[a]DNP_8-BSA and $Tosyl_{20}$-BSA at 200μg/ml; sonicated stroma and rabbit anti-mouse immunoglobulin at 1 mg/ml.

[b]Expressed as % inhibition.

form rosettes *in situ* with trinitrophenyl-coated sheep erythrocytes. About 50% of the fibre binding cells from immunized mice and 17% from unimmunized mice were capable of forming rosettes (Table 3). An unfractionated spleen cell population from immune and non-immune mice contained 1-5% and 0.2-0.4% rosette-forming cells respectively as determined by the centrifugation-resuspension assay. Formation of fibre-rosettes could be inhibited by both DNP-BSA and anti-immunoglobulin, but not by tosyl-BSA. Microscopic examination of the rosettes

TABLE 3

Formation of DNP-specific Rosettes by Spleen Cells Bound to DNP-derivatized Fibres

	FBC[a]	% Fibre-RFC[b] Inhibitor of Rosette Formation[c]			
		None	DNP	Tosyl	Anti-Ig
Immunized[d]	603	54	6	49	2
Unimmunized	296	17	5	18	1

[a]2.5 cm edge count.

[b]Expressed as (fibre-rosette-forming cells/fibre-binding cells) x 100

[c]DNP_8-BSA and $Tosyl_{20}$-BSA at 200 μgr/ml; rabbit anti-mouse immunoglobulin at 0.5 mg/ml added during rosette formation.

[d]Secondary response to DNP-BGG.

showed that the red cells were evenly distributed over the lymphocyte surface. Moreover, estimation of the number of antibody molecules bound to immunoglobulin receptors on each using ^{125}I-labelled rabbit anti-mouse immunoglobulin showed that both normal and immune cells contained similar numbers of receptors.

Fibre binding cells do not include plaque forming or antibody secreting cells. A cell population depleted of specific fibre-binding cells was obtained by incubating spleen cells with a large amount of antigen-derivatized fibre and collecting the unbound cells. This process removed less than 5% of the original cells. When the number of DNP-specific antibody secreting cells in the unabsorbed and depleted cell populations was compared by the plaque assay, no difference was observed. Similar results were obtained for both IgG and IgM secreting cells from immunized and unimmunized animals. Thus, although the biological role of fibre binding cells remains undefined at present, there is clear evidence that they are specific antigen binding cells and that they do not include plaque forming cells.

Lymphoid cells could be fractionated according to their affinity for a fibre hapten or antigen by addition of different amounts of free hapten to the cells as they were incubated and shaken with the fibres. Once cells are bound to the fibres, they cannot be removed by addition of soluble antigen (Rutishauser, *et al.*, 1972). Under these conditions, binding of a cell with high affinity for antigen to the fibre can be blocked by lower concentrations of free antigen than binding of a cell with low affinity receptors.

In Fig. 7 is shown the effect of various antigen concentrations on the number of spleen cells bound to DNP-BSA derivatized fibres. Although the total number of fibre binding cells detected in single dishes with cell suspensions from immunized mice was only 2 to 4 times that obtained with non-immune animals, this study revealed that the two cell populations differ greatly in their susceptibility to inhibition by DNP-BSA conjugates. Similar results were obtained using the monovalent antigen, ε-DNP-lysine, as an inhibitor. In the immune animal, 500 fibre binding cells were inhibited by antigen concentrations of less than 4μg/ml whereas the number of fibre binding-cells from the unimmunized animal decreased

FIGURE 7

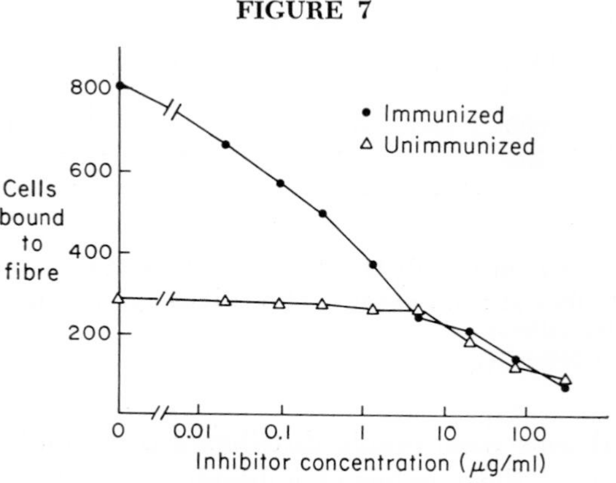

Inhibition by free dinitrophenyl bovine serum albumin of spleen cell binding to dinitrophenyl bovine serum albumin derivatized fibres. Cell numbers represent fibre edge counts for a 2.5 cm fibre segment. Spleens from immunized mice were removed at the height of a secondary response to dinitrophenyl bovine gamma globulin and cells from several mice were pooled.

by less than 20 at these antigen concentrations. At higher concentrations, the curves are identical (Fig. 7).

These differences were not due to variation in the number of receptors per cell as shown by two findings: 1) the number of anti-immunoglobulin molecules bound to fibre fractionated cells from non-immunized animals was about the same as that found with cells from immunized mice. Although these results do not account for possible differences in the classes of immunoglobulin receptors, it is evident that immunization does not result in a *large* increase of receptors. 2) The inhibition curves obtained with the monovalent antigen ε-DNP-lysine were similar to those using the multivalent DNP_8-BSA inhibitor. If immunization had resulted in an increase in the number of receptors, the binding of cells from non-immune animals should have been inhibited by lower concentrations of monovalent inhibitor than those of the immunized donor. This was not the case and the varying concentrations of inhibitor required are more likely to reflect differences of binding constants of a given set of receptors for the antigen. At present the biological role of fibre binding cells has not yet been defined. If it is assumed, however, that fibre binding cells include precursors of plaque forming cells, then it is clear that only precursor cells of higher affinity are stimulated.

The most striking aspect of the results obtained in these studies on the fractionation of immune cell populations is the high percentage of specific fibre binding cells found in non-immune animals (Rutihauser, *et al.*, 1972). If it turns out that fibre binding cells include precursor cells, the finding that 1-2% of the spleen cells from an unimmunized animal can bind to a single antigenic determinant would pose a serious paradox. This paradox would be resolved if it is postulated that specificity in a clonal system requires the operation of at least two factors. The first factor, is the *a priori* existence of a vast repertoire of cells in the primotype having different specific receptors for antigen (Fig. 5). The second factor, suggested in the present studies, as well as those of others, (Ada, 1970; Siskind and Benacerraf, 1969; Davie and Paul, 1972) is the necessity of a binding threshold for stimulation or triggering of each set of antigen-binding cells. Only those cells from each of these sets which have a sufficiently high binding constants for the antigen would be stimulated to undergo division and differentiation. Specificity in the clonotype is therefore assumed to arise from binding of an antigen to a set of cells in the primotype followed by triggering of only a small subset of binding cells to mature and divide. Corroboration of this or alternative models of specificity by detailed analysis of populations of cells of different functions is one of the outstanding tasks remaining in the development of a truly quantitative selective theory.

THE STATE OF Ig RECEPTORS IN THE LYMPHOCYTE MEMBRANE AND THE MECHANISM OF TRIGGERING

One of the fundamental tenets of the clonal selection theory is that the antigen serves to stimulate the lymphoid cell after specific binding to the Ig receptors on its surface. That this is not a simple event is suggested by the analysis of the requirement of T cells and antigen for stimulation of B cells in a model of cell cooperation (Gowans *et al.,* 1971). Both thymus-derived (T) and bone marrow-derived (B) lymphocytes are required for the induction of a primary immune response to many but not all antigens. There is evidence to suggest that T cells concentrate antigen and present it in a form capable of activating B cells. There is also evidence that antigen activated T cells secrete humoral factors that are necessary for B cell proliferation after contact with antigen (Dutton, *et al.,* 1971).

Whatever the absolute requirement for additional factors, it is clear that specific binding to Ig receptors is required. It therefore becomes pertinent to ask about the state of these receptors and their relationship to the mechanism of cellular triggering.

FIGURE 8

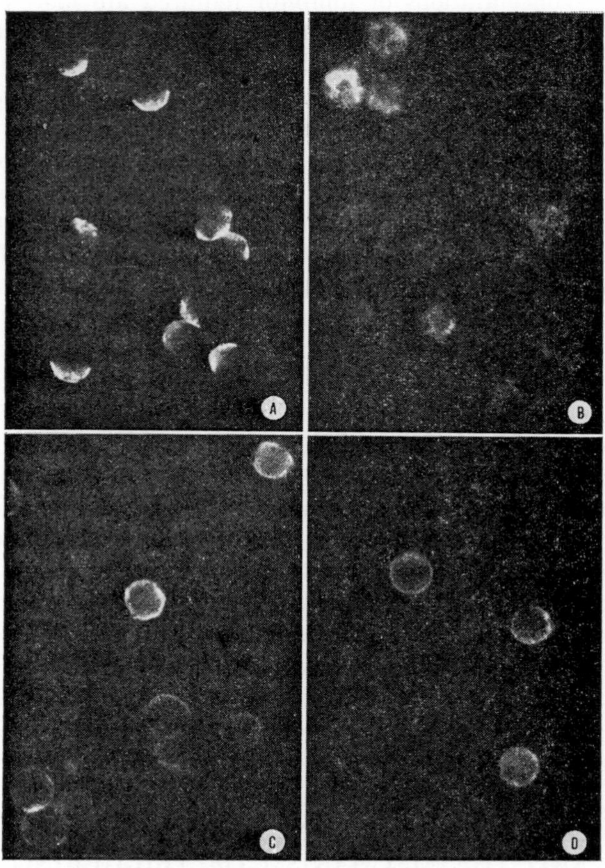

Labelling patterns of cells with fluorescein-labelled anti-immunoglobulin and with fluorescein-labelled Con A. (A) Cells incubated with fluorescein-labelled anti-immunoglobulin ($80\mu g/ml$) at 21°C for 30 minutes showing caps. (B) Prior addition of sodium azide (1×10^{-2}M) showing patches. (C) Prior addition of Con A ($100\mu g/ml$) showing diffuse patterns. (D) Patterns after incubation with fluorescein-labelled Con A alone ($100\mu g/ml$) at 21°C for 30 minutes.

Recently it has been shown (Taylor, *et al.*, 1971) that Ig receptors (presumably on B cells) are mobile in or on the lymphocyte membrane and that addition of anti-immunoglobulin results in patch formation by these receptors followed by cap formation. Cap formation presumably occurs by coalescence of patches via a metabolically controlled process for it is inhibited by sodium azide (NaN_3). Obviously, it is important to know the significance of these phenomena for immune induction and tolerance. Recent experiments in our laboratory (Yahara and Edelman, 1972) provide further information on the mechanism of patch and cap formation and the mobility of the immunoglobulin receptors of mouse lymphocytes.

The lectin from the jack bean, concanavalin A (Con A) which is itself mitogenic for lymphocytes was found to inhibit the formation of patches and caps that would otherwise form after addition of antibodies to immunoglobulin receptors (Fig. 8). The inhibition by Con A of cap formation and patch formation by immunoglobulin receptors was concentration dependent and reversible (Fig. 9a). Moreover, Con A binding did not appreciably diminish the number of antibodies bound to immunoglobulin receptors (Fig. 9b). These data suggest the possibility that the lymphocyte receptors fall into at least two classes: mobile Ig receptors and less mobile lectin receptors. Apparently, binding of Con A restricts the mobility of the Ig receptors which remain in a diffuse state.

FIGURE 9

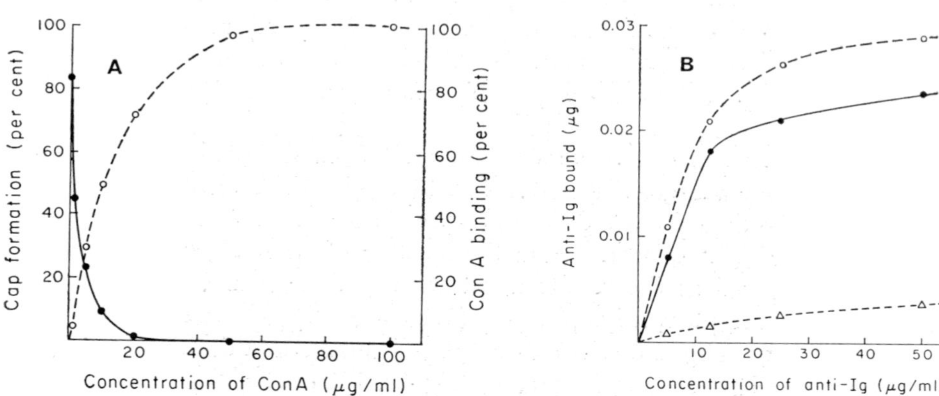

Effect of Con A on cap and patch formation and on binding of anti-immunoglobulin to cells. A. Effect of Con A on cap formation (—●—●—): cells (2 x 10^7/ml) were incubated with various concentrations of Con A in PBS-BSA at 21°C for 10 minutes. Fluorescein-labelled-anti-Ig was added to the mixture to give a concentration of 80μg/ml, and the mixture was incubated for 30 minutes at 21°C. Cap formation was determined after washing cells. Binding of [^{125}I] Con A to cells: (—○—○—), cells (2 x 10^7/ml) were incubated with various concentrations of [^{125}I] Con A (2.4 x 10^4 c.p.m./μg) at 21°C for 30 minutes in PBS-BSA. Cells were washed with PBS-BSA and the radioactivity was determined. B. Effect of Con A binding of anti-Ig to cells: Cells (1 x 10^7/ml) were incubated at 21°C for 30 minutes with various concentrations of [^{125}I] anti-Ig (7.6 x 10^4 c.p.m./μg) in 0.5 ml of PBS-BSA with (—●—●—) or without (—○—○—) Con A (100μg/ml). They were then washed on filters with PBS-BSA, and the radioactivity was determined. Binding of [^{125}I] Ig from an unimmunized rabbit (—Δ—Δ—).

A proposed mechanism (Yahara and Edelman, 1972) for patch and cap formation and the Con A effect is shown in Fig. 10. According to this model, addition of anti-Ig leads to three successive processes: a) binding of divalent antibodies

FIGURE 10

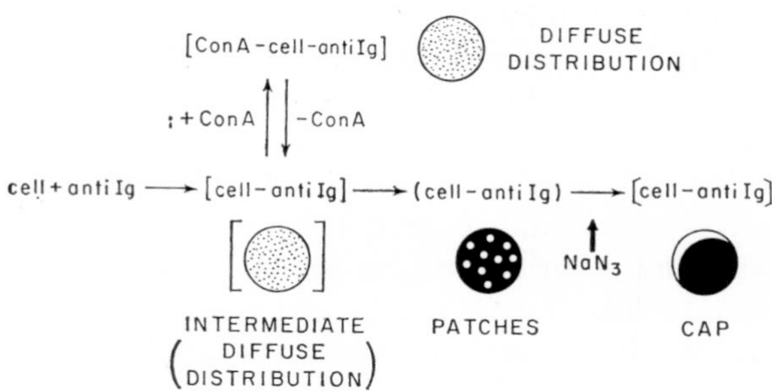

A model for patch and cap formation and their inhibition by Con A.

to Ig molecules on the cell surface b) formation of patches of antibody-Ig receptor complexes, and c) cap formation over one pole of the cell. Processes (a) and (b) are not inhibited by NaN_3, since binding was found to be unaffected by this reagent and patches can be observed in its presence; NaN_3 therefore inhibits process (c). The time course of cap formation after the removal of NaN_3 from the medium indicated that cap formation usually occurs within at most 10 minutes after forming patches at 21°C. Cap formation therefore appears to be dependent on cell metabolism and possibly requires a minimal concentration of ATP in the cell.

In contrast, patch formation appears to result from diffusional motion of Ig receptors which form larger aggregates after binding with divalent antibody. The resultant patches then do not diffuse at appreciable rates. A calculation using the known dimensions of Con A and the location of its binding sites (Becker, *et al.*, 1971) as well as the results of titration (Edelman and Millette, 1971) of the number of receptors of lymphocytes (1.4×10^6/cell) indicates that, at saturation, Con A occupies no more than 1% of the cell surface in a diffuse distribution. Inasmuch as the Ig receptors (approximately 4×10^4/cell) also occupy no more than 1% of the surface, binding of Con A cannot directly block the diffuse movement of Ig receptors. It is therefore suggested that Con A binding leads to a change in the cell surface or the cell membrane which results in alteration of either the anchorage or the free path of Ig receptors. One possibility, currently under test, is that the Con A interacts with the carbohydrate of the Ig receptor thus fixing it to the immobile Con A receptor site. Alternatively, the change could be due to aggregation of intramembranous protein particles linked to Con A receptors, either with formation of a tortuous path for motion of Ig receptors or, more likely, the binding of Ig receptors to particle-associated structures. Finally, binding of Con A to cell surface glycoproteins may result in secondary interactions of the Con A with the membrane and a change in membrane fluidity.

What is the significance of these findings for the mechanism of lymphoid cell triggering? More recent studies (Yahara and Edelman, unpublished results) have indicated that at 37°C and long times, Con A at low concentrations (5 μg/ml) itself leads to cap formation in T and B cells. The total number of cells does not exceed 15% of the population, however, and the inhibition of Ig receptor mobility still occurs in 40% of the cells under these conditions. Inasmuch as concanavalin A in the free state stimulates T cells and in an aggregated state stimulates mitogenesis in B cells (Andersson *et al.*, 1972), cap formation may not be required for cell stimulation. As suggested by Taylor *et al.*, (1971), patch and cap formation may play a central role in antigenic modulation. Moreover, cap formation could be a major factor in the induction of immune tolerance. This suggestion has the appeal that circulating antibodies could act as an amplifier in immune tolerance (Edelman and Gall, 1969): Antibodies interacting with the antigen already bound to immunoglobulin receptors on lymphocytes could form caps and sweep away receptor molecules so that the cell is unresponsive to further antigenic stimulation.

Although microscopically visible patch formation may not be necessary for cell triggering in the immune response, local submicroscopic aggregation of mobile receptors may still turn out to be of signal importance. Such a local aggregation may permit cooperative interactions among intramembranous particles linked to enzymes or institute changes in membrane permeability resulting in the appropriate stimulus for cell maturation and division. Elucidation of these molecular aspects of mitogenesis would clearly be a major step towards the development of a complete theory of clonal selection.

SUMMARY

A complete understanding of clonal selection requires a knowledge of the origin of antibody diversity, a quantitative analysis of the population dynamics and specificity of lymphoid cells, and an understanding of the mechanism of triggering of cell division after encounter with the antigen.

Analysis of antibody structures indicates that their diversity results from differences in the amino acid sequences of their variable (V) regions, from assortment of different V regions of light and heavy polypeptide chains which interact to make a complete molecule, and from the association of the same V region with different constant (C) regions. Both structural and genetic studies suggest that immunoglobulins have evolved by gene duplication with the emergence of three gene clusters or translocons, each on a different autosome and each consisting of separate V and C genes. It has been proposed that the formation of a complete VC gene for an immunoglobulin chain restricts the number of types of immunoglobulin that can be made by the cell and expressed at its surface as a receptor for antigen. Combination of antigen with this receptor immunoglobulin of a single specificity provides the trigger for clonal proliferation of the cell.

In addition to the analysis of humoral antibodies, there is a need to study antibody receptors on cells. This requires the separation of antigen binding cells into populations of different specificities and affinities for antigens. Using a new method, fibre fractionation, lymphoid cell populations have been analyzed in terms of their cross reactivity, affinity, receptor behaviour and number both before and after immunization. The results suggest that although there is a considerable number of cells of varying affinities for a given antigen in an unimmunized animal, only cells of relatively high affinities are triggered to multiply in the immune response.

Understanding of the mechanism of specific triggering by antigen is facilitated by studies of lectins which are themselves mitogenic for lymphocytes. Concanavalin A, a lectin from the jack bean, has been found to inhibit free diffusion of immunoglobulin receptors on lymphoid cells, and to prevent patch and cap formation after addition of antibodies to these receptors. These findings suggest that a major step in triggering lymphocytes may be alteration of the lymphoid cell membrane with aggregation of proteins and possibly decreased fluidity within regions of the membrane.

ACKNOWLEDGMENTS

The work of the author was supported by USPHS Grants AI 09273, AI 09921 and AM 04256.

REFERENCES

Ada, G. L. (1970), *Transplant. Rev., 5,* 105

Andersson, J., Edelman, G. M., Möller, G. and Sjöberg, O. (1972), *Eur. J. Immunol., 2,* 233

Branton, D. (1969), *Annu. Rev. Plant Physiol., 20,* 209

Becker, J. W., Reeke, G. N., Jr., and Edelman, G. M. (1971), *J. Biol. Chem., 246,* 6123

Burnet, F. M. (1959), *The Clonal Selection Theory of Acquired Immunity,* Vanderbilt University Press, Nashville, Tennessee

Cairns, J. (1967), *Cold Spring Harbor Symp. Quant. Biol., 32*

Cunningham, B. A., Gottlieb, P. D., Pflumm, M. N. and Edelman, G. M. (1971) in *Progress in Immunology* (B. Amos, ed.), p. 3, Academic Press, N.Y.

Davie, J. M. and Paul, W. E. (1972), *J. Exp. Med., 135,* 660

Dutton, R. W., Ralkoff, R., Hirst, J. A., Hoffman, M., Kappler, J. W., Kettman, J. R., Lesley, J. F. and Vann, D. (1971) in *Progress in Immunology,* (B. Amos, ed.), p. 355, Academic Press, N.Y.

Edelman, G. M. (1970), *Biochemistry, 9,* 3197

Edelman, G. M. (1971), *Ann. N.Y. Acad. Sci.* (S. Kochwa and H. G. Kunkel, eds.), Vol. 190, p. 5

Edelman, G. M. and Gall, W. E. (1969), *Annu. Rev. Biochem., 38,* 415

Edelman, G. M. and Millette, C. F. (1971), *Proc. Nat. Acad. Sci., U.S., 68,* 2436

Edelman, G. M., Rutishauser, U. and Millette, C. F. (1971), *Proc. Nat. Acad. Sci., U.S., 68,* 2153

Ein, D. (1968), *Proc. Nat. Acad. Sci., U.S., 60,* 982

Fleischman, J. (1967), *Biochemistry, 6,* 1311

Gally, J. A. and Edelman, G. M. (1970), *Nature (London), 227,* 341

Gally, J. A. and Edelman, G. M. (1972), *Annu. Rev. Genet.,* in press

Gowans, J. L., Humphrey, J. H. and Mitchison, N. A. (1971), *Proc. Roy. Soc., London, B, 176,* Number 1045, pp. 369-481

Gurdon, J. B., Lane, C. D., Woodland, H. R. and Marbaix, G. (1971), *Nature, (London), 232,* 177

Hill, R. L., Delaney, R., Fellows, R. E., Jr., and Lebovitz, H. E. (1966), *Proc. Nat. Acad. Sci., U.S., 56,* 1762

Jerne, N. K. (1955), *Proc. Nat. Acad. Sci., U.S., 41,* 849

Killander, J. (1967), *Gamma Globulins,* Nobel Symp., 3rd, Almqvist and Wiksell, Stockholm

Köhler, H., Shimizu, A., Paul, C., Moore, V. and Putnam, F. W. (1970), *Nature, (London), 227,* 1318

Lennox, E. and Cohn, M. (1967), *Annu. Rev. Biochem., 36,* 365

Mage, R. G. (1971), in *Progress in Immunology* (B. Amos, ed.), p. 47, Academic Press, N.Y.

Mäkela, O. and Cross, A. W. (1970), *Prog. Allergy, 14,* 145

Oudin, J. and Michel, M. (1969), *J. Exp. Med., 130,* 619

Pernis, B., Forni, L. and Amante, L. (1971), *Ann. N.Y. Acad. Sci.,* Vol. 190, p. 420

Rutishauser, U., Millette, C. F. and Edelman, G. M. (1972), *Prov. Nat. Acad. Sci., U.S. 69,* 1596

Singer, S. J. and Doolittle, R. E. (1966), *Science, 153,* 13

Siskind, G. P. and Benacerraf, B. (1969), *Adv. Immunol., 10,* 1

Smithies, O., Gibson, D. M., Fanning, E. M., Percy, M. E., Parr, D. M. and Connell, G. E. (1971), *Science, 172,* 574

Taylor, R. B., Duffus, P. H., Raff, M. C. and de Petris, S. (1971), *Nature New Biology, 233,* 225

Wang, A. C., Wilson, S. K., Hopper, J. E., Fudenberg, H. H. and Nisonoff, A. (1970), *Proc. Nat. Acad. Sci., U.S., 66,* 337

Yahara, I. and Edelman, G. M. (1972), *Proc. Nat. Acad. Sci., U.S., 69,* 608

Antibody Diversification: the Somatic Mutation Model Revisited

MELVIN COHN
The Salk Institute for Biological Studies, San Diego, California, U.S.A.

I would like to re-examine a previously proposed model for V gene diversification (Cohn, 1970, 1971; Weigert *et al.*, 1970).

Each animal carries in its germ-line of the order of 50 V_L and 50 V_H genes and therefore expresses prior to somatic antigenic selection roughly 2500 immunoglobulins (50 $V_L \times$ 50 V_H). A small number of these, about 50, are specifically directed against surface determinants on common pathogens. These immunoglobulins coded for by germ-line V genes are diversified further by somatic mutation and selection by antigen. The selection proceeds sequentially following each amino acid replacement in the immunoglobulin molecule.

In order to focus this problem I am going to assume an extreme form of the model of the V_LV_H domain proposed by Kabat and Wu (1972). The polypeptide coded for by a single V_L or V_H gene has amino acid residues which either 1) make contact with the antigenic determinant or 2) provide the structural framework in that they correctly position in space the contact-residues, permit the V_LV_H interaction producing a combining site and transmit the conformational signal for paralysis (Bretscher and Cohn, 1970; Cohn, 1971) that results from the specific binding of a ligand, etc. Somatic mutations which result in replacements of the contact-residues will alter the combining specificity and will be selected upon by antigen. Somatic mutations which introduce replacements in framework-residues ordinarily will not be detected because in large measure they will be either neutral or deleterious. If we were to look at the family of variants selected from the somatic descendants of *one* germ-line V gene, the contact-residues would appear "hypervariable" whereas the framework-residues would appear "invariant". According to this view, two V sequences which differ in framework-residues are derived from two germ-line V genes whereas if they differ in contact-residues only, they are probably derived somatically from one germ-line V gene. In an inbred population like the mouse the two germ-line V genes coding for different framework-residues would be non-allelic. In a random bred population, for example man, these two germ-line V-genes could be either allelic or non-allelic.

How did this model for the gene-product relationship arise?

The ratio of kappa (κ) to lambda (λ) subunits expressed by different species varies 1000-fold (Hood *et al.*, 1967). In horse the κ/λ ratio is < 0.05, in man, it is 2 and in mouse, it is around 20. The implication is that these two subunits are completely interchangeable for function. Presumably, the potential range of different specificities which could be generated in either class is the same. The selective pressure is largely on the total number of V_L germ-line genes whether they be κ or λ, not the ratio of them. This suggests that the κ/λ ratio in normal immunoglobulin might reflect in a rough way the ratio of V_κ to V_λ germ-line genes. Since mouse expresses λ in 5% of its serum immunoglobulin, the amino acid sequences of V_λ in myeloma proteins might reveal that they were derived from one germ-line V_λ gene. This appears to be the case. (The reason that I have used the phrase "reflect in a rough way" is that when a given V_L class, V_λ in this extreme case is represented as a single germ-line gene, there could be strong selection pressure to keep it because of the specificities it codes. This means that this gene would be disproportionately expressed in the animal due to antigenic selection and the proportionality with V_κ/V_λ germ-line genes would be distorted somewhat).

Seventeen V_λ sequences have been completed (Weigert *et al.*, 1970; Appella, 1971; Cesari and Weigert, 1972; Weigert, unpublished work). Twelve are identical, three differ by one amino acid, one by two amino acids and one by three amino acids. Of the eight amino acid replacements seven can be accounted for by single base changes at the level of DNA; one is due to two base changes, a total of nine replacements. The twelve identical sequences are assumed to be coded by one germ-line V_λ gene.

The number of replacements in this family of subunits are too few to delineate the pattern of variability. Another study (Wu and Kabat, 1970) established the pattern.

When the frequency of replacement in the total number of known V_L sequences, κ and λ of mouse and man, is analyzed by a statistical normalizing procedure, the three regions of hypervariability which appear are between positions 24-34, 50-56 and 89-97. The invariant disulphide bridge is between cysteines 23 and 88. All nine mutations in V_λ leading to amino acid replacements which are selected for by immunogenic encounters are in the hypervariable regions (Fig. 1). If it is estimated that 1) the V cistron is 350 nucleotides long, and 2) of the 60 nucleotides in codons for contact-residues, only mutations in 40 lead to amino acid replacements, then there is a 10% chance that a random mutation in the V gene will lead to an amino acid replacement which can be selected for by antigen. Since all nine replacements are in the hypervariable region, there is a less than 1% chance that a replacement outside of this region will affect the specificity and be selected upon by antigen. In other words, the remainder of the V sequence is effectively invariant.

The analysis of the hypervariable regions in V_H has led to similar conclusions

FIGURE 1

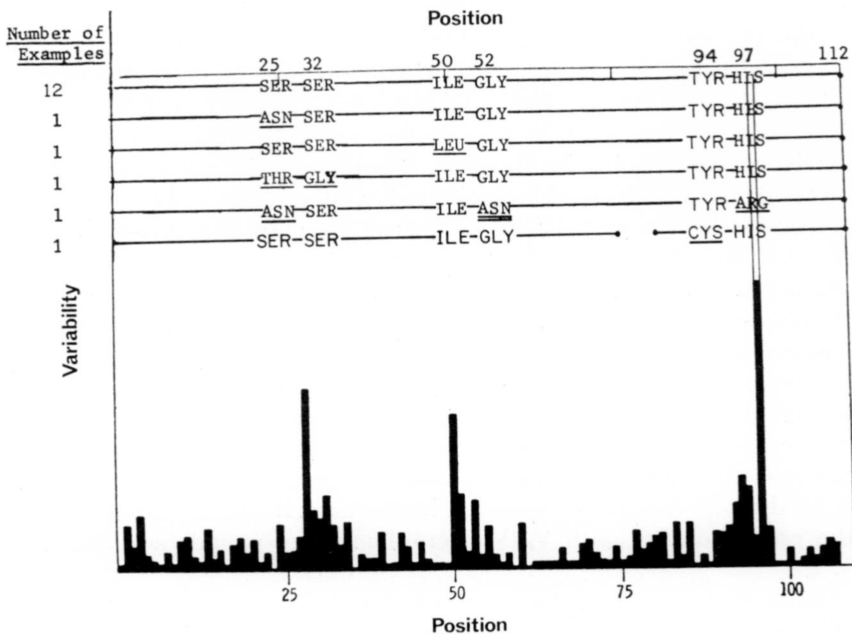

Diagram of mouse λ sequences superimposed on a plot of the extent of variability versus position in the variable region sequences of human light chains, κ and λ. The portions of the mouse λ proteins in which no differences could be detected are indicated by a line. The positions at which differences were found are indicated by the residues involved. The underlining indicates the number of base changes needed to get the replacement. The gap indicates a region for which the sequence has not been completed. The plot is taken from Wu and Kabat (1970). The sequences are taken from Weigert *et al.*, 1970; Appella, 1971; Cesari and Weigert, 1972; Weigert (unpublished).

(Kabat and Wu, 1971; Capra, 1971). There are three regions between positions 31-37, 86-91 and 101-110. The invariant cysteines forming the disulphide bridge are at positions 22 and 98.

This interpretation of the invariant and variable residues of the V sequence as framework and contact-residues respectively, has been greatly strengthened by Kabat and Wu's (1972) elegantly simple approach to the problem of the relationship between primary and tertiary structure. They developed an empirical method based on the following considerations.

From proteins of known tertiary structure as determined by X-ray crystallography, they tabulated the configuration around the alpha carbon (C_α) of any given middle amino acid when it is flanked by various combinations of other amino acids. Assuming the peptide bond to be planar then the polypeptide chain has only two degrees of freedom per residue: the angle of twist around the C_α-N bond axis, ϕ, and that about the C_α-C axis, ψ. A list of all of the (ϕ, ψ) values for all residues completely defines the path in space of the chain. A set of criteria to choose between various possible (ϕ, ψ) angles when a given tripeptide present in V_L was not found in a known protein, were decided upon for the model building. The underlying assumption is that framework-residues play similar roles in all proteins and therefore assume the same (ϕ, ψ) configurations. This assumption which will be formulated in a more precise way when more data are available, changes the way in which we can discuss immunoglobulin evolution, germ-line and somatic.

The model of the molecule of V_L constructed following these ground rules surprisingly enough fulfilled two exigent criteria:

(1) The sulphydryl residues at positions 23 and 88 were juxtaposed close enough to be able to form the disulphide bridge.

(2) The contact-residues were on the same side of the molecule and near enough to each other so that they would be in the combining site.

As a first approximation, it might be argued that the framework-regions of all V_L or V_H sequences are functionally equivalent. This approximation would imply an extreme and most provocative extrapolation. If one were to clip out the contact-regions of one immunoglobulin and sew them into another framework, e.g. from $V\kappa$ into $V\lambda$ or V_{HI} into V_{HII}, the resultant immunoglobulin would function and have a similar specificity. Unfortunately, it is likely that this extrapolation will need some correcting when the details are known. Nevertheless, it is worth speculating (in order to pose the problem unambiguously) whether an animal could function perfectly well if all of its germ-line V_L or V_H genes coded for identical framework-residues but different contact residues. If gene duplication is a major factor in V gene evolution then such situations will arise.

It can be seen in Fig. 1 that the statistical analysis of Wu and Kabat (1970) suggests a spread of some 5 to 10 contiguous residues in each of the three contact-regions. Yet when one looks (Fig. 1) at the replacements in the somatic products of one germ-line gene, in this case $V\lambda$, position 25 has 3 replacements and position 52 has 2 replacements (one observed and one surmised) out of the 9 which are known. This suggests that the spread in the positions of contact-residues determined by the statistical analysis of the degree of variability may be the result of each unique framework exposing different contact-residues within the contact-region. This means that different V_L or V_H genes might code for subunits with different spacings and arrangements of contact-residues within the statistically defined regions. In this way the framework-residues might permit an animal to express a broader spectrum of specificities than the extreme picture would permit. A possible example of this chosing by framework of contact-residues might be seen in the key $V\kappa$ sequences of McKean, Potter and Hood (reported in Table 12 of Potter (1972) which might be looked upon as the product of a single $V\kappa$ gene. Residues different from those found in $V\lambda$ (still within the contact regions) are hypervariable. All this having been said, I still wonder how close future experiments will come to confirming the extreme idealized picture before we have to seek second order corrections.

The framework-region of any given V gene then undergoes a germ-line evolution no different from that of a C gene or for that matter any protein such as lysozyme, cytochrome *c* or haemoglobin. V_L genes, e.g. $V\kappa$ and $V\lambda$, of the same

TABLE 1

Mouse V_K Sequences[a]

V_K-gene	Number of Sequences per gene	ASP	ILE	VAL	MET	THR	GLN	SER	PRO	SER	SER	LEU	SER	ALA	SER	LEU	GLY	GLU	ARG	VAL	THR	ILE	SER	CYS
I	1	PCA	—	—	LEU.																			
II	1	—	—	GLN	—	ILE	—	—	—	—	—	MET	PHE	—	—	ILE	—	ASP	GLX	—	SER	—	—	—
III	1	—	—	GLN	—	—	—	—	—	ASX	TYR	—	—	—	—	VAL	—	—	THR	—	—	—	THR	—
IV	1	—	—	GLN	—	—	—	—	—	—	—	—	—	—	—	—	—	—	—	—	SER	LEU	THR	—
V	1	—	—	GLN	—	—	—	THR	THR	—	—	—	—	—	—	—	—	ASP	(?)	—	—	—	(?)	—
VI	1	—	—	—	—	—	—	—	GLN	—	PHE	MET	—	THR	—	VAL	—	ASP	—	—	SER	VAL	THR	—
VII	1	—	—	—	—	—	—	—	—	—	—	MET	GLN	—	—	ILE	—	—	LYS	—	—	—	—	—
VIII	1	GLU	VAL	—	—	—	—	THR	—	LEU	—	—	ALA	VAL	—	—	—	(?)	GLX	ALA	SER	(?)	(?)	(?)
IX	1	GLU	VAL	—	LEU	—	—	—	—	ALA	ILE	MET	—	—	—	—	—	LEU	—	—	SER	MET	—	—
X	1	GLU	ASN	—	LEU	—	—	—	—	ALA	ILE	MET	—	—	—	PRO	—	—	—	—	—	MET	THR	—
XI	1	ASN	—	—	—	—	—	—	—	LYS	—	MET	—	MET	—	VAL	—	—	—	—	—	LEU	THR	—
XII	2	GLU	THR	THR	VAL	—	—	—	—	ALA	—	—	—	MET	ALA	ILE	—	—	LYS	—	—	—	—	—
XIII	2	—	VAL	—	—	—	—	THR	—	LEU	THR	—	—	VAL	THR	ILE	—	—	PRO	ALA	SER	LEU	—	—
XIV	2	—	VAL	LEU	—	—	—	THR	—	LEU	—	—	PRO	VAL	(SER)	—	—	ASP	GLU	ALA	(?)	—	—	—
XV	2	—	—	GLN	—	—	—	—	—	ALA	—	—	—	—	—	VAL	—	—	—	—	—	—	THR	—
XVI	2	—	—	—	LEU	—	—	—	—	ALA	THR	—	—	VAL	THR	PRO	—	ASP	SER	—	SER	LEU	—	—
XVII	2	—	—	—	ILE	—	—	ASX	GLU	LEU	—	ASP	PRO	VAL	THR	SER	—	—	(SER)	—	SER	—	THR	—
XVIII	2	—	—	GLX	—	—	—	—	—	—	—	—	—	—	—	—	—	ASP	THR	—	—	MET	THR	—
XIX	2	—	VAL	VAL	VAL	—	—	THR	—	LEU	—	—	—	VAL	—	MET	—	ASP	—	—	SER	—	THR	—
XX	3	—	—	—	—	—	—	—	—	—	—	—	—	VAL	—	ALA	—	—	LYS	—	—	MET	—	(?)
XXI	4	—	—	—	LEU	—	—	—	—	ALA	—	—	ALA	VAL	—	—	—	GLN	—	ALA	—	—	—	—
XXII	4	—	—	—	—	—	—	—	—	THR	PHE	—	ALA	VAL	THR	ALA	SER	LYS	LYS	—	—	—	—	—

[a]Data taken from Potter (1972), Hood (unpublished), Appella (unpublished).

or even different species may differ by as many as 50% of their residues in framework-regions, yet the S-S bonds and the complementarity determining residues are similarly positioned in all of them.

This conclusion justifies my previous estimate of the total number of V_L genes in the germ-line of the mouse (Cohn, 1970). The assumption was that replacements in framework-regions define germ-line V genes just as they would if we considered any protein, e.g. α and β chains of haemoglobin.

I have tabulated thirty-eight sequences on mouse V_κ which cover the framework-region positions 1-23 (Table 1) (Potter, 1972; Hood, unpublished work; Appella, unpublished work). Since I have assumed that any replacement in these regions defines a germ-line V gene, there are 22 directly countable germ-line V_κ genes coding for the 38 sequences listed in Table 1. If one calculates what minimum number of total V_κ genes these 38 sequences represent, the answer is 26. The number is minimal because only the first 23 residues are being considered. If the V region is 110 residues long, 90 are framework and therefore the maximum number (assuming extrapolation by length were valid) of V_κ genes is 100. Since, however, there are linked replacements in the framework-regions, the value of 100 is likely to be high. I will assume therefore that there are about 50 germ-line V_κ genes.

A similar analysis of the small number of known mouse V_H sequences places the total number of germ-line V_H genes at $\geqslant$ 25 so I have illustrated my argument using the figure of 50 only because it is reasonable that the numbers of V_L and V_H genes should be roughly the same (see later discussion of V_H^B and V_H^T genes).

In the human population similar calculations show that there are $\sim$ 300 germ-line V_κ genes, $\sim$ 400 V_λ genes and $\sim$ 100 V_H genes. The individual like the mouse only has presumably 50 V_L (33 V_κ and 17V_λ) and 50 V_H germ-line genes.

Are V_L and V_H genes equivalent? It is possible that a functioning immunoglobulin would result if the V_H genes were expressed linked to C_L and the V_L genes linked to C_H. The V_HC_H and V_LC_L combinations are the result of their evolution as an L or H locus. What is less clear to me is whether any two V_L or V_H polypeptides could function as a domain together. The long evolutionary selection for effective complementation between V regions to form a site might have led to a divergence of V_L and V_H genes for this property so that today V regions are not equivalent and there is a lock and key relationship between V_L and V_H due to framework-residues. There are two regions of complementation holding the L and H chains together, one a V_LV_H interaction and the other a C_LC_H interaction. Since the latter is constant in sequence whereas the former varies, it is probable that some effect of framework-replacements might be seen on the strength of the V_LV_H interaction. In general, I would expect this variation in the strength of the V_LV_H interaction to be minor because a functioning immune system requires that most V_LV_H combinations be compatible and in fact this seems to be the case (Smith and Dorrington, 1972). The observed preferential fit (to whatever extent it exists) must be the result of an immunogenic selection for molecules in which the V_LV_H interactions are at the high end of the distribution. In other words, even if certain V_LV_H combinations can be observed by sensitive techniques to vary in how well they interact, this contribution to the LH interaction, in most but not all cases, would not be large enough to affect the functioning of the immunoglobulin. The energy of the V_LV_H interaction compatible with function probably varies over a narrow range and in large measure resides in the configuration of framework-residues. The fit between V_L and V_H due to this configuration must vary slightly from one V_L or V_H gene product to the other. Since presumably the $V\kappa$ and $V\lambda$ genes function equally well with V_H genes the framework-residues can diverge quite extensively and yet maintain a functioning structure. This allelic and non-allelic polymorphism is seen as the background variation outside the regions of hypervariability in the Wu-Kabat plot (Fig. 1). Of course, I am giving you a telescopic view. When we look through the microscope it is possible that not all germ-line V_L and V_H products complement properly and some waste results. We do not see this population of non-functioning immunoglobulins. How effective selection would be in minimizing this waste determines the upper size of the V locus.

Each locus $V\kappa$, $V\lambda$ or V_H then evolves from one species to the next as would be expected, by conserving a few of the V genes, duplicating a few and deleting the excess. The most informative aspect to discuss is the V gene duplication. The classical (now outmoded) division of human $V\kappa$ sequences for example, into three (or four) subgroups $V_{\kappa I}$, $V_{\kappa II}$ and $V_{\kappa III}$, reflects the germ-line evolutionary origin of each subgroup from the gene duplication of one. Whereas the $V_{\kappa I}$, $V_{\kappa II}$, $V_{\kappa III}$ subgroup classification tells us about the pathway of evolution, it says nothing about how many V genes are in the germ-line.

As an example, Gally and Edelman (1972) point out that "it is possible to determine which subgroup a chain belongs to by examining its primary structure and looking for 'family resemblances'". This implies that there is a relationship of parenthood for the origin of each of the three V subgroups thus defined. It does not imply that "somatic hypermutation theories of diversity postulate the existence of only one germ-line gene for each V region subgroup in a translocon" (Gally and Edelman, 1972) because their criterion for classification into subgroups does not relate automatically subgroup to germ-line V genes. What must be added is a theory of diversification which ties the two together. A somatic mutation model which uses framework-replacements to count germ-line V genes divides the $V\kappa$ sequences into 50 $V\kappa$ subgroups. The gene-conversion model for the evolution of germ-line V-genes (Gally and Edelman, 1972) equates the classification of $V\kappa$ sequences into three $V\kappa$ subgroups with three groups of recombining V genes (recombinons). This is valid classification provided one reinterprets the data illustrated in Fig. 1 that have led Wu and Kabat (1970), Kabat and Wu (1971, 1972) and ourselves (Cohn, 1970, 1971, 1972a; Weigert *et al.*, 1970) to consider the V region, for all practical purposes, to be made up of contact-residues and framework-residues.

The origin of a V locus from the duplication of very few genes can be illustrated by two additional cases.

(1) It is known that the three alleles at the rabbit V_H locus are due to amino acid replacements in framework-residues. The peptides carrying the amino acid

markers can be isolated from serum immunoglobulin in yields as high as 70% (Mole *et al.*, 1971). This implies that a major part of the immunoglobulin of one homozygous individual carries identical replacements at given positions that are different in the allele. Since it is reasonable that the rabbit H-locus has > 20 V_H genes and that the proportion of V_H genes genetically marked by the given replacement is high i.e. $\geqslant 15$, I would conclude that these genes were derived from one which duplicated after the allelic replacements had been introduced. (For further discussion, see Cohn, 1971).

(2) V_H sequences of man, mouse, cat and dog each contain unique framework-residues which are present in a large proportion of the independently arising myelomas and of the serum immunoglobulin. These linked residues differ in the four species (Kehoe and Capra, 1972). Using the same arguments as above, I would conclude as do Kehoe and Capra (1972) that many of the germ-line V_H genes were derived in each species by duplication of one which already had the linked residues.

What is the selection pressure on largely identical duplicated germ-line V genes? The selection pressure regulating the total number around a mean might be visualized as follows: Suppose the total number of V genes were too high. Mutational drift would inactivate some of the germ-line V genes. This would be felt as an increase in the number of functionless antigen-sensitive cells. When this proportion became too high the selection pressure would be for animals which deleted or corrected part of the genome. If a deletion reduced the total to well below the mean, then the initiation of the somatic selection would be slowed and the animal would be selected against. Such a view of the gene duplication and deletion process implies that all animals will carry some functionless V genes which are expressed in antigen-sensitive cells that can never be induced and are therefore never revealed in serum immunoglobulin or in myeloma proteins. It also implies that sometimes certain V genes will be in at least "functionally" multiple copies so that deletion can rid functionless V genes without entirely wiping out a recognition unit, essential to the survival of the animal. The sensitivity of the selective pressure mechanism to overshooting and undershooting the mean determines in part what proportion of the mean will be functional. When I described a way of estimating V_L genes from sequences, I was referring of course to functional V_L genes, i.e. those expressed in plasma cells which result from induction, not to the total which would include the functionless ones also. This point deals largely with framework-residues. Of course, the process of mutational drift which inactivates germ-line V genes also introduces neutral replacements into others presumably one source of the replacements we see in framework-residues. The other source of replacements is the consequence of the correction of one deleterious replacement by another mutation leading to a replacement elsewhere. This event would be selected for constantly as would deletions of functionless V genes. Seesawing correcting mutations made possible by a functioning multigenic system would be the source of the linked residues which characterize a subgroup.

The second selection pressure, given a functioning framework, is for germ-line immunoglobulins with activities of immediate survival value to the animal. This selection pressure unlike the first affects the contact-residues. The limitation imposed here is that all of the possible V_LV_H combinations cannot be selected upon for specificity. The reason is that two given subunits, V_L and V_H selected upon for anti-Salmonella activity for example, also participate in other combinations. A replacement in either subunit conferring the anti-Salmonella activity would destroy any selected for activity in another combination in which that given V_L or V_H subunit participates. Thus, of the 2500 immunoglobulins coded for in the germ-line (50 $V_L \times$ 50 V_H) only about 50 could be selected upon for specific activities of survival value. However, the selection for 50 functional V_LV_H combinations guarantees that most of the other possible V_LV_H combinations will also be functional.

Can we guess what these activities might be? The most likely candidates would be anti-capsular, anti-cell-wall and anti-membrane constituents of the pathogens. The choice of antibody to the carbohydrate in cell or viral surfaces permits the animal to collect an armamentarium of germ-line antibodies directed against a battery of target determinants which are widely shared by microorganisms and from which the pathogen would have great difficulty in escaping by mutation.

There are several lines of evidence pointing to this view particularly as a consequence of the work of Krause (1970) and Haber (1971) on restricted responsiveness to carbohydrate or cell wall constituents. Our own studies derive from the analyses of the combining activities of mouse myeloma proteins carried out by Potter (1972) and ourselves (Hirst *et al.*, 1971).

I will illustrate this with two systems in mice, the response to the α-1,3 glucosyl determinants in dextrans and phosphorylcholine determinants in the pneumococcal C-carbohydrate. Roughly five out of a thousand BALB/c mouse myelomas produce immunoglobulins which are 1) directed against α-1,3 glucosyl or phosphorylcholine determinants and 2) indistinguishable by sequence, combining activity and idiotype (Potter, 1972). These immunoglobulins are in different heavy chain classes. The anti-α-1,3 glucosyl activity is always associated with a λ chain of unique sequence (Weigert, *et al.*, 1970; Cesari and Weigert, 1972; Weigert, unpublished work) and the anti-phosphorylcholine activity with a κ very probably of unique sequence (Potter, 1972; Sher *et al.*, 1971). [I am simplifying this story in a way which does not affect the point I am making. In the case of myeloma proteins with anti-phosphorylcholine activity I am considering those which are indistinguishable by idiotype and partial sequences. These comprise about 75% of the total number. In the case of myeloma proteins with anti-α-1,3 glucosyl activity the specificities and idiotypes are similar (not identical). Since some of them share identical λ chains, the V_H sequences must be different, thus opening the possibility that two germ-line V_H genes code for anti-α-1,3 glucosyl activity when their products are complemented with a λ chain.]

Immunization of BALB/c mice with dextran (containing α-1,3 glucosyl determinants) or C-carbohydrate (containing phosphorylcholine determinants) induces a restricted response solely in the IgM class and of a unique idiotype similar to that of a reference IgA myeloma. However, immunization of other strains of mice results in an anti-α-1,3 glucosyl or anti-phosphorylcholine response of another idiotype (Blomberg *et al.*, 1972; Sher and Cohn, 1972). Although the rate of the response is different in these strains, I will comment on the rates of the response (high and low responders) later.

The anti-α-1,3 glucosyl response of BALB/c is of a restricted idiotype and in the λ class whereas that of C57BL/6 is of another idiotype and in the κ class. The anti-phosphorylcholine responses of the BALB/c and A/J mice are indistinguishable as far as combining activity is concerned, but differ completely in the idiotype of the resultant antibody (Table 2). One of the genes controlling

TABLE 2

Characteristics of Response[a]

Mouse Strain	Response	Reference Idiotype	Light Chain Class
BALB/c(C)	anti-α-1, 3 glucoside	+	λ
C57BL/6(B)	anti-α-1, 3 glucoside	—	κ
(B x C)F_1	anti-α-1, 3 glucoside	+	λ
BALB/c(C)	anti-phosphoryl-choline	+	κ
A/J(A)	anti-phosphoryl-choline	—	κ
(A x C)F_1	anti-phosphoryl-choline	+	κ

[a]The data are taken from Blomberg *et al.*, 1972; Sher and Cohn, 1972.

the idiotype of the induced antibody is in both cases linked to the heavy chain allotype (Tables 3 and 4). The assumption then is that germ-line V_H genes are involved, one or two coding for α-1,3 glucosyl specificity when its product is comple-

TABLE 3

Linkage on Anti-α-1, 3 Dextran Response to Heavy Chain Allotype[a]

Mouse Strain[b]	Known Locus		Phenotype of Response
	H-2	C_H Allotype	
CXBD	C	B	B
CXBE	B	B	B
CXBG	B	C	C
CXBH	C	B	B
CXBI	B	B	B
CXBJ	B	C	C
CXBK	B	B	B

[a]Data taken from Blomberg *et al.*, 1972.

[b]These are recombinant inbred lines obtained by inbreeding of various pairs of (C x B)F_2 mice. The resultant inbred lines are then characterized for various allelic markers of B and C, only two of which, H-2 and the C_H allotype are involved here.

C = BALB/c; B = C57BL/6.

TABLE 4

Linkage of Anti-phosphorylcholine Response to C_H Allotype[a]

Mouse Strain[b]	H-2 Type	C_H Allotype	Phenotype of Response[c]
BALB/c	d	a	+
C57BL/10	b	b	+
A/WY, A/J	a	e	—
AL/N	a	d	—
C.AL	d	d	i
A.BY	b	e	i
B10.A	a	b	+

[a]Data taken from Sher and Cohn, 1972.

[b]C.AL is congenic with BALB/c but carries the C_H allotype locus of AL/N.
A.BY is congenic with A/WY but carries the H-2 locus of C57BL/10.
B10.A is congenic with C57BL/10 but carries the H-2 locus of A/WY.
These are congenic, lines references for which are found in Sher and Cohn (1972).

[c]The idiotype of the response is determined as percent inhibition of expression of plaque forming centers by a reference anti-idiotype serum prepared against an anti-phosphorylcholine myeloma protein.

+ = idiotype indistinguishable from that of BALB/c, e.g. C57BL/10
— = idiotype distinct from that of BALB/c, e.g. A/WY
i = idiotype intermediate between BALB/c and A/WY.

mented with the proper germ-line V_λ chain and the other for phosphorylcholine specificity when complemented with the proper germ-line V_κ chain. In the case of the anti-α-1,3 glucosyl response it can be seen in Table 3 that the phenotype of the response, BALB/c (C) or C57BL/6 (B), depends on whether it carries the C or B heavy chain allele not upon the C or B H-2 allele. The situation seems to be slightly more complex in the case of the anti-phosphorylcholine response because two factors may be involved. One of them is linked to the C_H locus because the congenic line C.AL identical to the positive phenotype (BALB/c) except for the chromosome carrying the C_H allele of a negative phenotype (AL/N) expresses an intermediate phenotype (see below). Backcrosses (Sher and Cohn, 1972; Weigert, unpublished work) have confirmed the linkage to the C_H locus of the idiotype of both the anti-α-1, 3 glucosyl and anti-phosphorylcholine responses.

An intermediate, not a negative, phenotype results when the heavy chain locus from a negative phenotype is put into a positive phenotype background e.g. C.AL or A.BY (Table 4). One interpretation is that a second genetic locus contributing to the idiotype is involved. This could be the kappa locus which it is tempting to speculate is the H-2 linked Ir-1 locus. The reason that I am underlining weak data is to point out that there is no formal evidence that Ir-1 does not code for the kappa locus. However, there is suggestive evidence of two kinds against Ir-1 coding for kappa.

1. Edelman and Gottlieb (1970) have isolated a V_κ genetic marker called the I_B peptide (positions 19-24). This is probably a result of a lysine to arginine replacement at 24 in several V_κ subgroups. I_B^+ strains have lysine at position 24

whereas I_B^- strains have arginine. Eighteen strains of mice of all H-2 alleles are I_B^- whereas three strains are I_B^+. All I_B^+ strains are H-2^κ but not all H-2^κ strains are I_B^+. Since H-2^κ strains can be either I_B^+ or I_B^-, an argument that the kappa locus is unlinked to H-2 can be made.

2. In man, InV, a kappa constant region marker is unlinked to the HL-A locus. Since HL-A and H-2 are homologous, by extension the kappa locus and H-2 are unlinked. Further in the rat the kappa and major histocompatibility loci are unlinked.

I have analyzed elsewhere the theory that Ir-1 codes for a new heavy chain locus expressed in thymus-derived cells (Cohn, 1971, 1972a, 1972b) but I would be more at ease in discussing this question if it were unambiguous that Ir-1 does not code for kappa.

Five comments might be made at this point:

1. I picture the idiotypic determinant on an immunoglobulin to be made up of the interacting framework-residues of V_L and V_H modulated in some cases by the replacements in contact-residues. Assuming that the germ-line V genes code for $> 1{,}000$ immunoglobulin molecules then there would be at least 1000 different idiotypic determinants on germ-line encoded immunoglobulins. A first approximation operationally defines an anti-idiotypic serum to a given antibody or immunoglobulin as one which does not react at a given level of sensitivity with "normal" serum immunoglobulin or with a large number of randomly chosen antibodies or myeloma proteins. This operational definition does not imply that a given anti-idiotypic serum only recognizes immunoglobulins of identical V region sequences. It is however very restrictive. If the anti-idiotypic serum recognized determinants formed by the interaction of germ-line encoded V_LV_H framework-residues only, then there would be a less than 0.001 chance that any randomly chosen antibody molecule or myeloma protein would react with it. The observed degree of specificity of most anti-idiotypic sera could be accounted for by assuming that all of the somatic derivatives from one combination of germ-line encoded V_LV_H polypeptides possessed similar idiotypes due to the interacting framework-residues from V_L and V_H. Brient and Nisonoff (1970) discovered that at least part of the antibody in an anti-idiotypic serum can no longer interact with the corresponding immunoglobulin when the latter is specifically bound to its ligand. Within the present formulation, the interpretation would be that the interaction of a hapten with antibody results in a conformational change revealed as a "melting" of one of the idiotypic determinants. Carson and Weigert (1973) have shown that, if framework-residues determine the idiotypic determinant, the configuration of that determinant can be modulated by certain replacements in contact-residues. Using an anti-idiotypic serum directed uniquely against "meltable" determinants present on an anti-α-1,3 glucosyl myeloma protein, they assayed its interaction with reconstructed molecules containing the same heavy chain as the myeloma protein but different λ chains each possessing replacements in contact-residues only (Fig. 1). The substitution of the original λ chain by a λ chain with one or two replacements did not result in immunoglobulins distinguishable in their reaction with the anti-idiotype. How-

ever, the molecule reconstructed with a λ chain possessing three replacements, had only 15% of the affinity for the anti-idiotype as the original protein. The substitution of a κ chain completely ($< 0.3\%$) destroyed the ability to react. If the "meltable" idiotypic determinant is composed of framework-residues this experiment shows that certain substitutions in contact-residues can modulate it, whereas other substitutions in contact-residues can pass undetected. This result requires a slight (but expected) modification of the extreme form of the Kabat-Wu model of the V_LV_H domain as I have formulated it here. Some modulation of the configuration of the framework occurs when certain contact-residues are replaced. The framework must be somewhat flexible. Further as I discussed earlier, it is also *a priori* obvious that some somatic replacements in framework-residues might affect combining site specificity by altering the relative configuration of contact-residues or by determining which of the residues in the statistically defined hypervariable region are actually contact-residues for a given V-gene product. However, as I pointed out at the beginning, most framework-replacements in the *somatic* products of a single V-gene are neutral or deleterious to function, whereas most if not all, contact-replacements simply alter the specificity. Therefore, these somatic framework-replacements are rarely, if ever, seen.

2. Dominant responsiveness can be the consequence of polymorphism in structural germ-line V genes. The evidence comes from the response to α-1,3 dextran. BALB/c is a high responder whereas C57BL/6 is a low responder. High responsiveness is always associated with antibody of the reference idiotype and in the λ class. Low responsiveness is always associated with antibody *not* carrying the reference idiotype and in the κ class. Both characteristics, level of response and idiotype are controlled by a gene locus closely linked to the heavy chain allotype. It is therefore reasonable to assume that level of response as well as the idiotype is determined by a structural V_H gene. The initial rate of the response depends upon the number of antigen-sensitive cells with specificities for the test antigen. The fewer the number of somatic mutations required to generate a given antibody specificity from the germ-line V_LV_H combinations, the more rapid will be the rate of response.

3. The genetics of responsiveness to microbial surface constituents, dextran and phosphorylcholine shows that the antibody specificity is determined by germ-line V genes and supports the hypothesis that one of the selection pressures on germ-line V genes is for antibody activity of protective value. This selection pressure tends to maximize the proportion of functioning V genes in the germ-line.

4. If we assume that Ir-1 does *not* code for the kappa locus, then the extension of the above argument is that Ir-1 codes for a new heavy chain locus expressed in T-cells ($V_H^TC_H^T$) as contrasted to the allotype marked heavy chain locus expressed in B-cells ($V_H^BC_H^B$). If the selection pressure on the germ-line is for given V_LV_H combinations, then $V_LV_H^T$ and $V_LV_H^B$ combinations could code different specificities. Since V_L genes are expressed in both T and B cells, there might be more of them than either V_H^B or V_H^T genes, i.e. $V_L \geqslant V_H^B + V_H^T$ or $V_L \sim 50, V_H^B \sim 25, V_H^T \sim 25$.

5. Since the heavy chain locus $V_H^TC_H^T$ (Ir-1) is expected to show allelic exclusion

then this phenomenon cannot be due to inactivation or rearrangement of the entire chromosome since both alleles of H-2, to which Ir-1 is linked, are expressed in each T-cell (Cohn, 1971, 1972a).

Given this picture of germ-line V genes, most of the variability must be derived from somatic antigenic selection. The replacements in contact-residues of V_λ sequences suggests mutation as the simplest mechanism. Two factors must be analyzed:
1) the rate at which variants arise; 2) the priming of antigenic selection.

A rough approximation of the rate of generating new variants can be made, if one envisages the generation of antigen-sensitive cells as occurring in a "chemostat" of total population size, N. The rate at which mutants (M) arise is—

$$\frac{dM}{dt} = 0.69\ a\mu N$$

where a = mutational frequency and μ = divisions/day.

One estimate of the parameters which seems reasonable is—

$N = 10^8$ cells
$\mu = 3\text{-}4$ divisions/day
$a = 10^{-3} - 10^{-5}$ functional variants/division

Since the choice of a is the major factor I should discuss this number. In bacteria in which repair mechanisms are inactive so that the true replication error rate is measured, the mutational frequency per base pair per division is 10^{-7}. Since we are interested in replacements in contact residues only, the frequency of mutations leading to replacements affecting the combining specificity of the immunoglobulin (V_LV_H) expressed in each antigen sensitive cell is roughly 10^{-5}, i.e. roughly 100 base pairs are involved. It is well known that at certain specific regions of the genome the mutational frequency can be increased 10-100 times, a point which has been discussed by Mäkelä and Cross (1970). I would expect strong evolutionary selection for this differential rate of mutation affecting the contact-residues. In that case, I estimate the value of a to fall between 10^{-3} —10^{-5} functional variants/division. It should be stressed however that values of $a < 10^{-6}$ would be incompatible with a somatic mutation model.

The total number of mutants increases linearly with time, so that the "chemostat" would contain between 2×10^4 and 2×10^6 functional variants in 10 days in the absence of selection. Ten days is the minimum time that it could take an immune system to become mature.

The newly generated somatic variants as well as the cells expressing germ-line genes are acted upon by tolerogenic encounters which rid anti-self cells and allow anti-nonself cells to accumulate. However, this affects a small number of cells for the major population expresses germ-line V genes and it is this population which must be eliminated. One simple assumption is that virgin antigen-sensitive cells are short-lived but upon induction become memory cells and are long-lived. Thus memory cells resulting from immunogenic encounters are siphoned off and accumulate whereas the remainder of the population is generated at a steady state and simply turns over.

How many variants are required to make immunogenic encounters possible? This problem arises because the immunogenic encounter has an obligatory requirement for the associative recognition of at least two determinants on an antigen (Bretscher and Cohn, 1970) one by the antigen-sensitive cell and the other by associative antibody (the origin and mode of action of which need not worry us at the moment.) If an animal can recognize 10^4-10^6 different determinants, what is the probability that a given foreign antigen would be immunogenic? This is the priming problem.

In my previous analysis of this somatic mutation model (Cohn, 1970) I estimated the probability that two different antigen-sensitive cells would arise with receptor-specificity for any given antigen. I looked upon this criterion as a rough way of evaluating whether a randomly chosen antigen would be immunogenic. These estimates showed that immunogenic encounters in the emerging embryonic immune system were reasonably probable. However, the calculations are in reality too conservative and I would like to restate the selection problem here in a more optimistic and luckily realistic way. If the selection for 50 germ-line V_LV_H combinations is evolutionarily judicious then the immune system faces common foreign determinants generally repeating on polymeric structures. This assures induction of the 50 germ-line immunoglobulins. Consequently a memory antigen-sensitive cell population is accumulated among which mutants appear and upon which further selection can occur. The remaining unselected > 1000 V_LV_H germ-line combinations have a peculiar relationship to the "chosen fifty". They have hybrid combining sites with one subunit complementary to a portion of one key determinant and the other complementary to a portion of another key determinant. The combining site is pictured here as the jaws of pliers, each jaw (V_LV_H) with a different arrangement of the teeth. If there is some order in the configurations of naturally occurring determinants, then the unselected germ-line V_LV_H combinations will see certain kinds of cross-reacting structures with a high probability. In other words, by optimizing the selection of 50 V_LV_H pairs, not only is the animal initially protected against the most common pathogens but the other unselected > 1000 V_LV_H combinations have a higher than average chance of being induced. These considerations added to those previously analyzed (Cohn, 1970) make it reasonable that in this model the induction process can be primed.

One last point needs mentioning. We have postulated that normal induction requires the recognition of two determinants on an antigen, one by the receptor on the antigen-sensitive cell (T or B) and the other by associative antibody. Further we have argued that the associative antibody is thymus-derived and cytophilic for a third party cell (Bretscher and Cohn, 1970; Cohn, 1971). This poses a "chicken and egg" paradox, since the associative antibody must be induced and induction requires associative antibody. We have suggested two ways of getting started. Either initially associative antibody is maternally derived or *virgin* antigen-sensitive cells leak it at a very limited rate. Today, I would favour the latter proposal because the former puts the animal in too high risk of auto-immune disease. The latter, if controlled, allows nonself associative antibody to reach a plateau value at which induction is primed. This is so because the half life of associative antibody must be short in the absence of the cell making it. Since anti-self cells are eliminated and the anti-nonself cells accumulate, autoimmune risk is minimized.

My closing comments are directed towards two competing theories, the Hood-Talmage germ-line model and the Gally-Edelman hyper-recombination model. Both models generate, in the absence of somatic antigenic selection, the total potential of the animal to respond, i.e. the entire range of specificities is expressed independent of antigenic selection. The antigen-independent expression of the total range of specificities is called the "primotype" by Gally and Edelman (1972). Unlike the somatic mutation model described here the range does not increase with time of antigenic selection. The burden of diversification is placed on germ-line evolution. The essential point is, that under a somatic mutation model, there is an increase with time in the range of specificities present in the individual and this increase is antigen-dependent. In other words, the number of different specificities expressed in the "primotype" increases with time and is antigen-dependent. It is this difference between the theories which is liable to enable us to make the distinction between them experimentally. Hood—Gally—Edelman look upon somatic genesis as an act of creation by GOD, whereas I think GOD is learning all the time.

The discussions by Hood and collaborators (Hood and Talmage, 1970; Smith *et al.*, 1971) and Gally and Edelman (1972) need to be put in mesh. No one doubts that recombination, mutation, gene-repair (conversion), duplication (expansion), deletion (contraction), etc. *can* occur in both germ-line and somatic cells. These are well-known mechanisms of gene variation upon which we are all leaning for our arguments. The discussions concern the relative rates and extents of the contribution of each process to the evolution of V genes, both germ-line and somatic. Further the selection pressures must be and have not been considered.

The germ-line and hyper-recombination models are explicit. In the first, the somatic contribution to the total variability is negligible (not zero). In the second, the germ-line contribution to the total variability is negligible (not zero). The fundamental distinction between the Hood-Talmage model and the one I propose lies in the consequences following the decision as to where to draw the line between the contributions of germ-line and somatic genes to the total variability. There is nothing in what I have said concerning the germ-line evolution of tens of V genes with which Hood and his collaborators would be required to disagree. We part company when we begin to interpret the genetics of responsiveness and the evolution of framework and contract-residues (Fig. 1). Gally and Edelman have analyzed the arguments against the Hood-Talmage germ-line model and in general I agree with them. However, their arguments do not apply to the somatic mutation model which I have proposed here (Cohn, 1970, 1971). I have analyzed (Cohn, 1971) the strength and weaknesses of the explicit Gally-Edelman hyper-recombination model. There is no need to repeat this analysis here.

If one starts with the assumption that the three V_κ subgroups (defined as sequences with "family resemblances") are represented in the κ-"translocon" as three V_κ "recombinons" (corresponding recombining V genes with "family resemblances") then, of course, a model for variability can be generated. The rules of degree of resemblance which reduce recombination between "recombinons" in a "translocon" to a negligible level are set empirically because "inter-recombinon" ($V_{\kappa I} \times V_{\kappa II} \times V_{\kappa III}$) recombinants are not found. It is probable that this apparent lack of recombination is due both to a low recombination rate as well as the failure of antigenic selection that requires a functioning immunoglobulin. These same rules are adjustable to apply to the V_κ genes of Table 1. Under a somatic mutation model, recombination between V_κ genes is of either too low frequency or in general yields a non-functioning subunit. How often a functioning subunit would emerge if a V_L sequence were made up of the N-terminal part of one chain and the C-terminal part of another chain (the "centaur") is a very important structural question which is posed by the Kabat-Wu (1972) analysis. Since their model of the V-region is built by putting together short segments, the "centaurs" might turn out to be "fertile" and this strengthens the Gally-Edelman "V_λ-recombinon" classification as being meaningful for somatic diversification. However, the demonstration that the "centaur" is "sterile" would explain why we do not see them on earth today. The corrections on the extreme picture which I discussed earlier might be important here, namely that different V_κ genes could code for frameworks which require different positions to become contact-residues within of course the regions defined by the Wu-Kabat plot (Fig. 1). "Centaur" molecules might not function for this reason because incompatible positions would be activated as contact-residues.

It might be argued that high frequency intracistronic recombination is operating to generate diversity, but is being overlooked because of the lack of complete sequences as well as of my operationally extreme definition of a V_κ gene which is incompatible with hyper-recombination mechanisms. In that case I expect that Gally and Edelman will reclassify the sequences in Table 1 into what they feel are the correct subgroups or "recombinons" and with that model of the genome account for the findings on the genetics of dominant specific responsiveness (Tables 2, 3, 4) and the pattern of variability (Fig. 1). We will then be able to weigh again the updated models.

At this point in the game, both the germ-line and hyper-recombination models must be reevaluated in the light of our present concepts concerning the different selective pressures operating on framework and contact-residues for without such considerations these models no longer come to grips with the problem of the origin of diversity.

SUMMARY

The arguments are presented for a somatic mutation model with the following characteristics:

The germ-line of each animal codes for approximately $40 V_L$ and $40\ V_H$ genes. This conclusion is derived from a consideration of the sequence variation in the variable regions of mouse light chains (V_L).

One V_LV_H pair of germ-line genes is expressed in each antigen-sensitive cell whether it be bone-marrow or thymus-derived. These virgin antigen-sensitive cells are turned over at a constant rate in the absence of antigenic stimulation.

Fifty out of the possible 2500 V_LV_H combinations ($50V_L \times 50V_H$) are selected during germ-line evolution to code for combining specificities of immediate and most likely survival value to the animal, i.e. directed against surface components of common pathogens. The genetics of responsiveness to dextran and C-carbohydrate are described to support this postulate. This selection for 50 V_LV_H combinations optimizes the number of functional combinations the germ-line could code for.

Somatic mutations in V_L and V_H genes which result in replacements in specific regions (contact-regions) are selected for by antigenic encounter in a stepwise and sequential fashion. Somatic mutations leading to replacements in framework-residues are rarely seen ($< 1\%$) because these are usually either neutral or deleterious and do not affect a change in combining specificity.

If the short-lived virgin antigen-sensitive cell is induced it becomes a long-lived cell. This permits the siphoning off from the virgin population which is accumulating mutants linearly, a long-lived population which is the source of further evolution. The distinction between other theories and this one is that the former postulate the expression of the total range of specificities prior to antigenic encounter whereas the latter requires antigenic selection for the expression of the total range.

ACKNOWLEDGEMENTS

This work was supported by a National Institutes of Health Grant (A-105875) and a Training Grant (A1-00430).

As usual, I am indebted to the elves Peter Bretscher, William Geckeler and Roy Riblet whose unkind comments about the first draft made me reorient my point of view. Having done that, they continued to wield the shillelagh until I was forced to retreat from "must" to "might", "is" to "could be" and "identical" to "similar". While it is a lucky man who finds that he is not a hero in his own laboratory, I cannot help wonder whether my "PV = nRT" version might not have been a more welcome cup of tea.

REFERENCES

Appella, E. (1971). *Proc. Nat. Acad. Sci., U.S., 68,* 590

Benacerraf, B. and McDevitt, H. (1972). *Science 175,* 273

Blomberg, B., Geckeler, W. R. and Weigert, M. (1972). *Science, 177,* 178

Bretscher, P. and Cohn, M. (1970). *Science 169,* 1042

Brient, B. W. and Nisonoff, A. (1970). *J. Exp. Med. 132,* 951

Capra, J. D. (1971). *Nature New Biology 230,* 61

Carson, D. and Weigert, M. (1973). *Proc. Nat. Acad. Sci. U.S.,* in press

Cesari, I. M. and Weigert, M. (1972). *Proc. Nat. Acad. Sci., U.S.,* in press

Cohn, M. (1970). *Cell. Immunol. 1,* 461

Cohn, M. (1971). *Ann. N.Y. Acad. Sci. 190,* 529

Cohn, M. (1972). *Cell Immunol., 5,* 1

Cohn, M. (1972a). In *"Genetic Control of Immune Responsiveness"* (McDevitt, H. and Landy, M., eds.), in press. Academic Press, New York

Edelman, G. M. and Gottlieb, P. D. (1970). *Proc. Nat. Acad. Sci., U.S., 67,* 1192

Gally, J. A. and Edelman, G. M. (1972). *Annu. Rev. Genetics,* in press

Haber, E. (1971). *Ann. N.Y. Acad. Sci. 190,* 285

Hirst, J. W., Jones, G. G. and Cohn, M. (1971). *J. Immunol. 107,* 926

Hood, L., Gray, W. T., Sanders, B. G. and Dreyer, W. J. (1967). *Cold Spring Harbor Symp. 32,* 133

Hood, L. and Talmage, D. W. (1970). *Science 168,* 325

Kabat, E. A. and Wu, T. T. (1971). *Ann. N.Y. Acad. Sci. 190,* 382

Kabat, E. A. and Wu, T. T. (1972). *Proc. Nat. Acad. Sci., U.S., 69,* 960

Kehoe, J. M. and Capra, J. D. (1972). *Proc. Nat. Acad. Sci., U.S., 69,* 2052

Krause, R. M. (1970). *Adv. in Immunol. 12,* 1

Mäkelä, O. and Cross, A. M. (1970). *Prog. Allergy, 14,* 145

Mole, L. E., Jackson, S. A., Porter, R. R. and Wilkinson, J. M. (1971). *Biochem. J. 124,* 301

Potter, M. (1972). Physiol. Reviews, *52,* 631

Sher, A., Lord, E. and Cohn, M. (1971). *J. Immunol. 107,* 1226

Sher, A. and Cohn, M. (1972). *Europ. J. Immunol., 2,* 319

Smith, G., Hood, L. and Fitch, W. (1971). *Annu. Rev. Biochem. 40,* 969

Smith, B. R. and Dorrington, K. J. (1972). *Biochem. Biophys. Res. Comm. 46,* 1061

Weigert, M. G., Cesari, I. M., Yonkovich, S. J. and Cohn, M. (1970). *Nature (London) 228,* 1045

Wu, T. T. and Kabat, E. A. (1970). *J. Exp. Med. 132,* 211

An Alternate Mechanism for Immune Recognition

K. J. LAFFERTY

Department of Immunology, John Curtin School of Medical Research, Australian National University, Canberra, A.C.T., Australia

The immune system functions to maintain the integrity of the living organism. Amongst the vertebrates, it is elements of the lymphoid tissue that are concerned with the fine degree of discrimination between normal 'self' and other foreign components, that is essential for the effective operation of the immune system. Thus we may define immune reactions as those events involved in the recognition of foreign material by the lymphoid system and the subsequent reactions that result from this initial recognition event.

Much information concerning immune reactivity has been gathered by studying the response of the immune system to contact with foreign antigens such as proteins or polysaccharides. As a result, we have come to consider that all forms of immunological reactivity involve a response of immunocompetent cells to contact with foreign antigen; a process that may be mediated by immunoglobulin or immunoglobulin-like receptors on the surface of lymphocytes. While there is no doubt that many immune reactions follow this pattern of recognition and response, I would like to present evidence in this paper that there is another system of immune recognition that comes into play when the immune system interacts with foreign tissues. This alternate system of immune reactivity may utilize recognition mechanisms that are quite distinct to those operating when immunocompetent cells respond to foreign antigen.

RESPONSE OF THE IMMUNE SYSTEM TO ANTIGEN

Fig. 1 summarizes what we know of the way in which the immune system responds to antigen. Immune responses are dependent on the reactivity of lymphocytes (Gowans, 1968-69). Cells of this type that can respond to antigenic stimulation will be referred to as antigen reactive cells. Antigen reactive cells can be divided into two sub-populations, those cells involved in the production of humoral antibody, the B, or bone marrow derived lymphocytes; and cells involved in the generation of cell mediated immunity, the thymus derived, or T, cells (Miller *et al.*, 1971; Parrott, 1970). Cells of both the T and B lines carry immunoglobulin or immunoglobulin-like receptors in their surface. These receptors are more dense on the surface of B cells and there are also qualitative differences between the receptors on the surface of B and T cells (Greaves and Hogg, 1971).

FIGURE 1

RESPONSE OF THE IMMUNE SYSTEM TO ANTIGEN

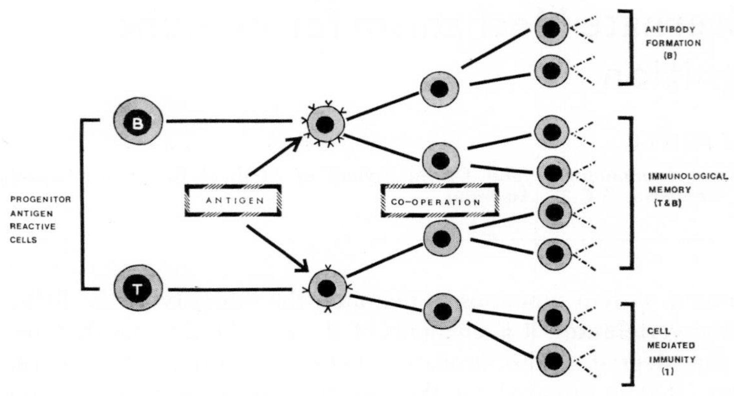

Summary of cellular events that result from the antigenic stimulation of progenitor antigen reactive cells of either T or B origin.

It is likely, although not proven, that such receptors form the recognition units of the antigen reactive cells.

An animal is unresponsive to 'self' antigens either because clones of cells reactive towards self antigens have been eliminated (Burnet, 1959) or in some way repressed (McCullagh, 1972). However, the population of antigen reactive cells possessed by an animal, give it the capacity to respond to a large number of foreign antigens. In general, the animal responds most vigorously to those antigenic patterns that show least similarity to normal 'self' components.

In the case of antibody formation, antigen reactive cells of any particular specificity occur at very low frequencies (approximately 1 in 10^5 cells) amongst the total lymphocyte population (Playfair *et al.*, 1965; Kennedy *et al.*, 1966; Armstrong and Diener, 1969). Individual antigen reactive cells appear to respond to only one or to a limited number of antigenic patterns, since it is possible to specifically inactivate cells reactive towards one antigen without affecting the ability of the population as a whole to respond to another unrelated antigen (Ada and Byrt, 1969; Basten *et al.*, 1971; Roelents and Askonas, 1971).

The progenitor antigen reactive cells, cells which will give rise to antibody forming cells following antigenic stimulation, are not mitotically active (Syeklocha *et al.*, 1966). However, following antigenic stimulation, antigen reactive cells rapidly move into a phase of proliferation (clonal expansion) which leads to the eventual production of antibody forming cells (Szenberg and Cunningham, 1968).

Similar proliferative events occur amongst the T cell population following their stimulation by foreign antigen and as a result these cells give rise to the effectors of cell mediated immunity (Davies *et al.,* 1966, 1969; Parrott, 1970). In the case of certain antigens, stimulated T cells can cooperate with B cells to help in the generation of antibody forming cells (Miller *et al.,* 1971).

Following antigenic challenge, the immunized animal shows a state of immunological memory. Memory may be expressed in cells of both the B and T populations (Cunningham and Sercarz, 1971) and probably results from the expansion of the population of antigen reactive cells specific for the challenge antigen.

Since the response of the immune system to foreign antigen involves intense cell proliferation, both antibody formation (Kennedy *et al.,* 1965) and cell mediated immune responses (Smith and Vos, 1963) are radiosensitive processes. The D_0 for each of these processes was found to be of the order of 70-80 rds which is similar to the radiosensitivity of the proliferative capacity of mammalian cells (Puck and Marcus, 1956).

In summary, the antigen responsive capacity of the immune system has the following features:

1. Antigen reactive cells (lymphocytes) allow the discrimination between 'self' and foreign antigens, and the animal responds most vigorously to those antigenic patterns that show least similarity to 'self' components.
2. Antigen reactive cells of any one specificity occur at very low frequencies (approximately 10^{-5}) amongst an animal's lymphocyte population.
3. The response to antigen is adaptive. Immunization leads to the expansion of cell clones reactive towards the challenge antigen.
4. The response to foreign antigen involves the proliferation of antigen reactive cells, and hence is a radiosensitive process.

The remaining section of this discussion will examine the way immunocompetent lymphocytes interact with foreign tissues.

THE GRAFT-VERSUS-HOST (GvH) REACTION

The chicken embryo provides a convenient system to study the way immunocompetent cells interact with foreign tissues. When lymphoid cells from adult birds are introduced into the embryo, the recipient, because of its immunological immaturity, is unable to reject the donor cells. The donor cells, however, can mount an immunological attack against the recipient's tissues. Reactions of this type are usually referred to as graft-versus-host (GvH) reactions. Two approaches can be used to quantitate the activity of grafted cells. Cells placed directly on the chorioallantoic membrane (CAM) of the embryo produce focal proliferative lesions within 3-4 days. There is evidence to suggest that one active donor cell is able to initiate one pock and that the number of pocks is thus directly proportional to the number of active cells inoculated (Coppleson and

Michie, 1966). The intensity of GvH reactions can also be quantitated by estimating the degree of splenomegaly that results from the introduction of immunocompetent cells into the recipient embryo (Simonsen, 1962). Using one or both of these assays it has been possible to show that this reaction is an immune phemonenon:

1. GvH reactions are only produced when adult lymphoid cells are introduced into the recipient embryo (Simonsen, 1962).
2. The reaction shows the 'self' 'not self' discrimination that is characteristic of immune reactions since lymphoid cells produce no response in syngeneic embryos (animals of the same inbred strain) but induce strong reactions in allogeneic embryos (animals of the same species but of a different genotype).

ANOMALOUS FEATURES OF GvH REACTION

There are, however, some features of the GvH reaction that are quite distinct from those associated with antigen responsiveness of the immune system. By using pock forming activity or splenomegaly as an index of GvH activity, it has been possible to estimate the proportion of cells in the lymphocyte population of a given donor that can initiate a GvH reaction. In the case of pock formation a figure of 1 cell in 30 has been obtained using purified lymphocyte populations (Simons and Fowler, 1966). Nisbet *et al.* (1969), using splenomegaly as an index of reactivity, estimated that 2% of a given donor's lymphocyte population were able to react against the tissues of a particular allogeneic recipient. It should be emphasized that these figures represent a minimum estimate of the proportion of reactive lymphocytes possessed by a particular donor animal. Thus the effective units for the initiation of a GvH reaction can occur at a frequency of 10^{-2} to 10^{-1} amongst the lymphocyte population of the donor animal. This figure is some 3 orders of magnitude higher than the frequency of antigen reactive cells (approximately 10^{-5}).

Another anomalous feature of these reactions is the finding that they are not always adaptive. When the donor and recipient differ at a major histocompatibility locus, pre-immunization of the donor against the recipient does not increase the reactivity of the donor's lymphoid cells (Lind and Szenberg, 1961). In fact, it may in some cases, result in a diminution of their GvH activity (Warner, 1964a; Seto and Albright, 1965).

When discussing antigen responsiveness we emphasized that the response of lymphocytes, in either the T or B line, to contact with foreign antigen always involved the proliferation of the antigen reactive cells. This is not always so in the case of GvH reactions. Very strong GvH reactions may be initiated in the absence of any donor cell proliferation (Nisbet and Simonsen, 1967). However, since the pathogenesis of the GvH reaction may vary depending on the developmental status of the recipient embryo (Walker *et al.*, 1972), in certain GvH situations donor cell proliferation is also observed (Nisbet and Simonsen, 1967).

TISSUE SPECIFICITY OF GvH REACTION

By using the experimental system illustrated in Fig. 2 we were able to study the interaction of adult lymphoid cells with various tissues obtained from allogeneic embryos (Lafferty and Jones, 1969).

FIGURE 2

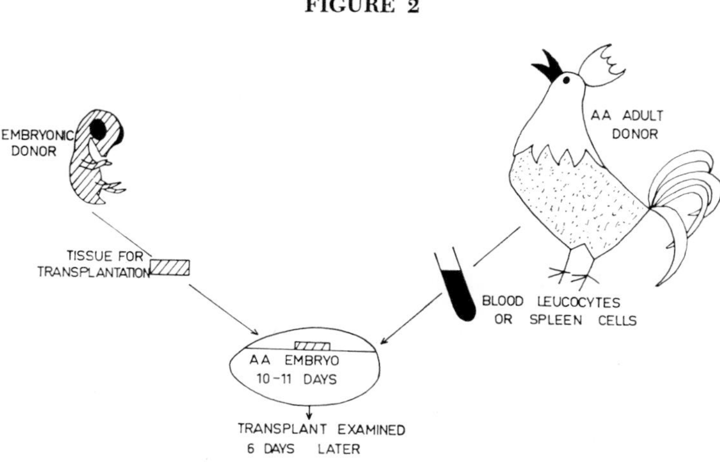

Experimental system used to study the tissue and species specificity of reactions of the GvH type.

Whenever the target tissue contained haemopoietic elements (bone, spleen, liver) very violent reactions were observed. However, if we transplanted a tissue that was free of blood or blood forming elements (heart muscle) the tissue survived well and was not attacked. On the basis of these studies, we suggested that the primary interaction occurring in this system involved the donor lymphocytes and recipient haemopoietic cells. More recent studies concerning the pathogenesis of GvH reactions (Walker, *et al.*, 1972) have confirmed this suggestion and shown that different elements of the embryo's haemopoietic tissue respond in a different manner to contact with adult immunocompetent cells. Primitive stem cells of the yolk sac blood islands are inactivated as a result of this interaction, while more differentiated derivatives of these cells enter a phase of rapid and uncontrolled proliferation as a result of the same stimulus.

SPECIES SPECIFICITY OF GvH REACTION

Another striking characteristic of these reactions is their species specificity. Table 1 shows the spleen weight of chicken embryos that were injected intravenously with lymphoid cells obtained from adult allogeneic or syngeneic donors, or donors of another species (duck, goose, pigeon or sheep). Splenic enlargement was only observed when the embryos were inoculated with allogeneic cells. Xenogeneic (cells from another species) or syngeneic cells produce no detectable splenomegaly. Similar findings were obtained using the CAM assay (Lafferty and

TABLE 1

Splenomegaly Produced by the Intravenous Inoculation of Lymphoid Cells Into the Chicken Embryo

Cells Injected	No. Cells Injected	Mean spleen weight (mg) 6 days after inoculation. The bracketed figure is the number of embryos inoculated.
Chicken leucocytes	10^6	133.8(5)
Duck leucocytes	10^7	7.8(6)
Goose leucocytes	10^7	9.4(6)
Pigeon spleen cells	10^7	8.4(5)
Sheep lymphocytes	10^6	10.1(5)
Syngeneic leucocytes	10^7	9.9(8)
Medium alone	—	10.6(49)[a]

[a]The standard deviation of the population was ± 3.8 mg.

Jones, 1969). Thus we see that reactions of the GvH type are essentially the the result of allogeneic interactions. It is not strictly true to state that xenogeneic cells never produce GvH reaction in the chicken embryo. The more extensive studies of Payne and Jaffe (1962) show that while the most violent reactions are obtained with allogeneic combinations, some reactivity is observed when cells from closely related donor species are introduced into recipient embryos. However, the degree of reactivity decreases rapidly as the phylogenetic separation of donor and recipient is increased.

The species specificity of the GvH reaction was first observed by Murphy in 1916. The observation prompted him to suggest that such reactions were not immunological since it was well known at that time that the lymphoid system usually responded more vigorously to foreign antigens than it did to antigens obtained from animals of the same or a closely related species. However, when Simonsen (1957) demonstrated that a GvH reaction only occurred when there was a genetic and hence antigenic difference between the donor and recipient animals, the immunological nature of this reaction seemed to be established. The failure of xenogeneic cells to produce a response was thought to be due to their failure to survive, or possibly function, in the environment of the chicken embryo. This explanation, however, is inadequate since it has been demonstrated that xenogeneic cells can survive and function in the chicken embryo (Lafferty and Jones, 1969).

The transplantation system shown in Fig. 2 was used to transplant bones from either chicken or pigeon embryos onto the CAM of recipient AA embryos. When adult pigeon or adult AA chicken lymphoid cells were introduced into the embryos carrying transplanted bones, it was found that chicken cells could only destroy the chicken bone. Adult pigeon cells, on the other hand, did not attack either the recipient embryo or the transplanted chicken bone, but reacted very violently against bones of pigeon origin. Clearly the xenogeneic cells can both survive and function in the environment of the chicken embryo.

CHARACTERISTICS OF REACTIONS OF THE GvH TYPE (ALLOGENEIC INTERACTIONS)

On the basis of the above information we can make the following statement about reactions which involve the interaction of lymphocytes with foreign embryonic tissues:

1. They are immune reactions because they involve the differential recognition of 'self' and foreign tissues, and this process is mediated by mature lymphocytes. However their recognition spectrum is severely limited. The most violent reactions are observed with allogeneic combinations and the intensity of these reactions diminishes rapidly as the phylogenetic separation of the lymphocyte donor and the donor of the target tissue is increased. For this reason we shall refer to reactions of this type as *allogeneic interactions*.
2. When the interacting cells are derived from animals that differ at a major histocompatibility locus, allogeneic interactions are not adaptive.
3. Allogeneic interactions involve the interaction of mature lymphocytes and elements of the embryonic haemopoietic system. Primitive stem cells are inhibited by such interaction whilst more differentiated derivatives of these cells respond by pathological proliferation.
4. A relatively large proportion of an animal's lymphocytes may be reactive towards the tissues of an allogeneic strain of animal. These cells occur at a frequency of approximately 10^{-2} amongst the lymphocyte population.
5. An initiation of an allogeneic interaction does not require the proliferation of the adult immunocompetent lymphocyte.

If we compare these features of allogeneic interactions with those of the antigen responsive capacity of the immune system, we find striking differences. Clearly allogeneic interactions must represent either a very atypical form of antigen responsiveness or alternatively may involve a completely different form of immunological reactivity.

Jerne (1971) has postulated that allogeneic reactivity represents the primary antigen recognition system from which other forms of antigen responsiveness have developed by a process of somatic mutation. We have proposed a more radical hypothesis to explain these events (Killby, *et al.*, 1972). This hypothesis postulates that allogeneic interactions represent a quite distinct class of immunological reactivity. We have suggested that the interaction between normal immunocompetent cells and the allogeneic leucocytes or embryonic haemopoietic cells is stimulated in an epigenetic manner by the transfer of RNA between the interacting cells. According to this hypothesis, it is the cells from amongst the immunologically competent population that provides the stimulus in allogeneic interactions. We shall call this immune function, *allogeneic stimulation*. Our theory of immunological reactivity postulates that antigen responsiveness and allogeneic stimulation represent two distinct classes of immunological reactivity (Fig. 3). In the case of GvH reactions initiated in young chicken embryos we are observing immune reactions that result from allogeneic stimulation (Killby *et al.*, 1972; Lafferty *et al.*, 1972).

FIGURE 3

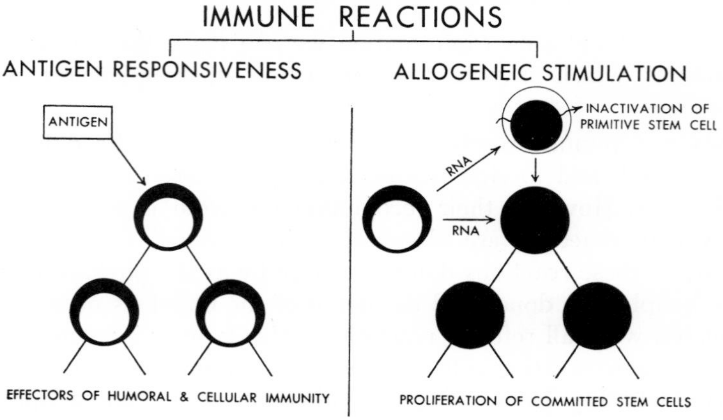

IMMUNE REACTIONS

ANTIGEN RESPONSIVENESS

ALLOGENEIC STIMULATION

ANTIGEN

INACTIVATION OF PRIMITIVE STEM CELL

RNA

RNA

EFFECTORS OF HUMORAL & CELLULAR IMMUNITY

PROLIFERATION OF COMMITTED STEM CELLS

Diagrammatic representation of the two classes of immune reactivity. *Antigen responsiveness* is the capacity of immuno-competent cells to proliferate and differentiate, following antigenic stimulation, and generate effector cells of humoral and cellular immunity. *Allogeneic stimulation* is the capacity of immuno-competent cells to provide a stimulus that results in the inactivation of primitive stem cells but causes the proliferation of their more differentiated derivatives.

DISCRIMINATION BETWEEN THE THEORIES

It is possible to discriminate between the two alternative explanations of the way immunocompetent cells interact with embryonic haemopoietic cells. That is, whether these reactions are the result of antigen responsiveness or allogeneic stimulation.

Warner (1964b) developed a system that can be used to analyse the interaction between adult blood lymphocytes and spleen cells of allogeneic embryos. He showed that when blood from inbred AA chickens or embryonic spleen cells from non-inbred (NI) embryos were inoculated onto the CAM of AA embryos, few or no pocks developed. However, if the two cell populations were mixed and placed on the CAM of AA embryos, a large number of proliferative lesions was produced. These lesions resulted from the interaction between lymphocytes of the blood donor and allogeneic embryonic spleen cells (indirect pock formation). According to the theory of antigen responsiveness, such lesions would be due either to the simple proliferation of antigen stimulated lymphocytes or to both proliferation of donor lymphocytes and host embryonic cells following the release of non-specific factors by antigen stimulated cells (Warner, 1964b). In contrast, the theory of allogeneic stimulation sees indirect pock formation to be dependent on the activation of the embryonic cells following their contact with adult immunocompetent lymphocytes.

If the formation of indirect pocks is due to the stimulation of the adult AA lymphocytes by antigen on the surface of NI spleen cells, treatment of the embryonic cells with mitomycin-C, a mitotic inhibitor, should not affect this process. The following experiment was designed to examine this point.

NI embryonic spleen cells were treated with mitomycin-C at a concentration of 50μg per ml and then mixed with 1 ml of AA blood diluted 1:1 with Alsever's solution. 0.1 ml of this mixture was then inoculated onto the CAM of recipient AA embryos. Control preparations, consisting of normal NI embryonic spleen cells mixed with AA blood, and AA blood or NI embryonic spleen cells alone, were inoculated onto the CAM of AA eggs at the same time. A further control was included to determine whether leakage of mitomycin-C from the treated embryonic cells could interfere with the activity of the blood leucocytes in the mixture. Mixtures of AA blood and NI embryonic spleen cells, treated with mitomycin-C at 50μgm per ml, were inoculated onto the CAM of eggs that were responsive to AA leucocytes. The activity of this mixture was compared with that observed when AA blood alone was inoculated onto the CAM of eggs of this type.

TABLE 2

Pock Formation by Mixtures of AA Blood and NI Embryonic Spleen Cells (Normal or Mitomycin-C Treated)

Test Reaction Mixture	Recipient CAM	Mean Pock Count ± S.E.
0.1 ml AA blood[a] containing 5×10^6 NI embryonic spleen cells	AA	75.0 ± 4
0.1 ml AA blood[a]	AA	0.0
0.1 NI embryonic spleen cells (5×10^7/ml in Eagle's medium)	AA	0.0
0.1 ml AA blood[a] containing 5×10^6 NI embryonic spleen cells (Mitomycin-C treated)	AA	0.8 ± 0.4
0.1 ml AA blood containing 5×10^6 NI embryonic spleen cells (Mitomycin-C treated)	NI	35 ± 6
0.1 ml AA blood[a]	NI	38 ± 5

[a]AA blood diluted 1:1 with Alsever's solution.

The results of this experiment were summarised in Table 2. As previously demonstrated by Warner (1964b), mixtures of AA blood and NI embryonic spleen cells initiated pock formation on AA CAMs, whilst either cell population alone produced very few, or no pocks, on the membranes of AA embryos. Prior treatment of the embryonic cells in the mixture with mitomycin-C reduced this pock forming activity to background levels. Moreover when mixtures containing mitomycin-C treated NI embroyonic spleen cells and AA blood were titrated on embryos that were not syngeneic with the adult blood donor, their pock forming activity was not impaired by the presence of the mitomycin-C treated embryonic cells.

Since mitomycin-C does not affect the expression of transplantation antigen on the cell surface (Schellekens and Eijsvoogel, 1970), this study shows that indirect pock formation does not result from the simple activation of antigen reactive cells by transplantation antigen on the surface of the embryonic spleen cell. These findings do not support the proposition that indirect pock formation is a form of antigen responsiveness.

RADIO-SENSITIVITY OF ANTIGEN RESPONSIVENESS AND ALLOGENEIC STIMULATION

Further support for the theory of allogeneic stimulation comes from the study of the radio-sensitivity of the capacity of lymphoid cells to cause the inactivation of allogeneic stem cells or to stimulate the proliferation of yolk-sac-derived stem cells. If the ability of immunocompetent cells to provide an allogeneic stimulus does not depend on their proliferative capacity, one might expect allogeneic reac-

FIGURE 4

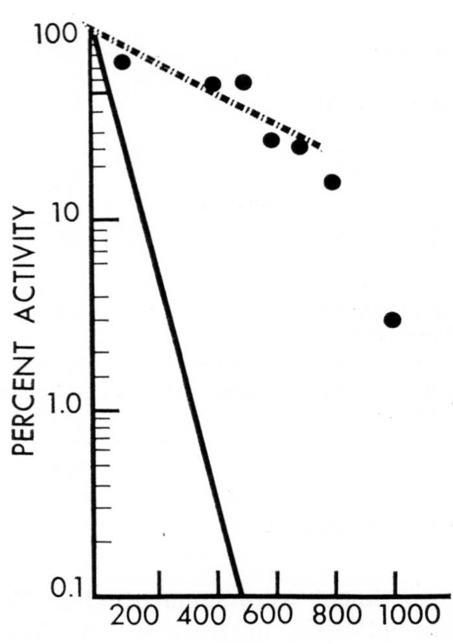

Radio-sensitivity of antigen responsive capacity and the capacity for allogeneic stimulation.

(————) Gamma-ray survival for the capacity of mice to produce antibody forming cells (from Kennedy *et al.*, 1965).

(—·—·—) Gamma-ray survival for the capacity of mouse lymphocytes to inactivate allogeneic mouse stem cells from (Manyko, 1971).

(●) Gamma-ray survival of the capacity of chicken leucocytes to stimulate the proliferation of spleen cells from allogeneic embryos.

tions to be more radio-resistant than the antigen responsive capacity of the immune system which involves cell proliferation and is thus a relatively radio-sensitive function.

Fig. 4 is a composite of data taken from several sources. The unbroken line shows the radio-sensitivity of the capacity of antigen reactive cells to respond to antigenic stimulation and go on to produce antibody (Kennedy *et al.,* 1965). The broken line shows the radio-sensitivity of the capacity of mouse lymphocytes to inactivate allogeneic mouse haemopoietic stem cells (Manyko, 1971) and the experimental points show the results obtained (Killby *et al.,* 1972) when we measured the radio-sensitivity of the capacity of AA leucocytes to stimulate indirect pock formation.

The process of stem cell inactivation and indirect pock formation have remarkably similar radiation sensitivites, over the dose range of zero to about 800 rds. The studies of Manyko (1972) were only continued over a dose range of 0 to 600 rds. However our own studies, which extended to doses of 1,000 rds, indicated that the dose response curve may show a downturn at high radiation levels. Over the range of 0 to 800 rds both these capacities were considerably more radio-resistant than the capacity of antigen reactive units to respond to antigeneic stimulation. The greater radio-resistance of the allogeneic interactions may indicate that immunocompetent cells can express either of these effects without undergoing cell divisions. Such a concept is consistent with the findings of Nisbet and Simonsen (1967) that intense GvH reactions can be elicited with no evidence of significant donor cell proliferation.

POSSIBLE INVOLVEMENT OF RNA IN ALLOGENEIC INTERACTIONS

The theory of allogeneic stimulation postulates that these reactions result from the transfer of an RNA molecule or an RNA containing component from the immunocompetent to the responsive haemopoietic cell. Some evidence in favour of the involvement of RNA in these interactions comes from the study of dermal reactions mediated by lymphocytes or RNA extracted from these cells (Jones and Lafferty, 1968). The normal lymphocyte transfer reaction, a dermal inflammatory reaction that results from the intradermal injection of lymphocytes into the skin of allogeneic recipients, is another example of an allogeneic interaction (Jones and Lafferty, 1969; Jones *et al.,* 1969). We argued that if these reactions are mediated by RNA transfer between the stimulating and responsive cells, RNA isolated from normal lymphocytes should produce a similar reaction following its injection into the skin of normal allogeneic recipients. The theory of allogeneic stimulation would maintain that RNA must be active in the skin of allogeneic animals but not in the skin of the RNA donor. Moreover RNA isolated from the lymphoid cells of the xenogeneic animal should not be active when tested intradermally.

Fig. 5 shows the results obtained when sheep lymphocyte RNA was injected into the skin of the lymphocyte donor or of an allogeneic animal. This figure also shows the dermal response following the injection of chicken spleen RNA into the

FIGURE 5

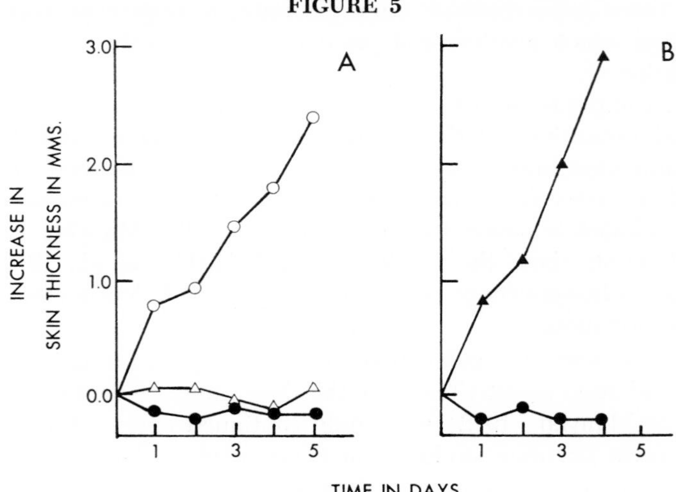

A. Dermal response to the injection of 30μg allogeneic (○), autothonous (●) or xenogeneic (△) RNA.

B. Dermal response following the injection of 100μg allogeneic RNA before (▲) and after (●) RNAse treatment, (from Jones and Lafferty, 1969).

skin of a recipient sheep. It can be seen from this figure that only the allogeneic RNA produces a significant dermal response. Fig. 5b shows that the dermal reactivity of sheep spleen RNA can be destroyed by RNAse treatment prior to its injection into the skin of the recipient. Thus it would appear that RNA could be the stimulating agent involved in allogeneic interactions. These experiments do not, however, prove that such a mechanism is operating in the *in vivo* situation.

CONCLUSIONS

On the basis of the studies outlined above it seems likely that there are two independent forms of immunological reactivity. Both involve the activity of immunologically competent lymphocytes. Antigen responsiveness depends on the proliferation and differentiation of these cells to produce the effectors of both humoral and cellular immunity. In contrast, allogeneic stimulation does not result from the response of the immunocompetent cell to some external stimulation, but rather, is dependent on the ability of these cells to provide a stimulus that results in the inactivation of primitive stem cells, or the pathological proliferation of their more differentiated derivatives. Although allogeneic stimulation is mechanistically distinct from antigen responsiveness, the two immune functions can influence one another. Thus, if the immune system is confronted with an allogeneic stimulus at the same time as it is challenged with foreign antigen, the allogeneic interaction can have a marked adjuvant effect on the response to antigenic challenge (see Lafferty *et al.*, 1972). It is this interaction of these immune functions that must be taken into account in any attempt to understand transplantation reactions.

REFERENCES

Ada, G. L. and Byrt, P. (1969). *Nature (London)* *222,* 1291

Armstrong, W. D. and Diener, E. (1969). *J. Exp. Med.* *129,* 371

Basten, A., Miller, J. F. A. P., Warner, N.L. and Pye, J. (1971). *Nature New Biol.* *231,* 104

Burnet, F. M. (1959). *The Clonal Selection Theory of Acquired Immunity.* Cambridge University Press.

Davies, A. J. S., Leuchars, E., Wallis, V. and Koller, P. C. (1966). *Transplantation* *4,* 438

Davies, A. J. S., Carter, R. L., Leuchars, E. and Wallis, V. (1969). *Immunology* *17,* 111

Coppleson, L. W. and Michie, D. (1966). *Proc. Roy. Soc. Series B* *163,* 555

Cunningham, A. J. and Sercarz, E. E. (1971). *Eur. J. Immunol.* *1,* 413

Gowans, J. L. (1968-69). *The Harvey Lectures, Series* *64,* 87

Greaves, M. F. and Hogg, N. N. (1971). In *Progress in Immunology* (B. Amos ed.), p. 111. Academic Press, New York and London.

Jerne, N. K. (1971). *Eur. J. Immunol.* *1,* 1

Jones, M. A. S. and Lafferty, K. J. (1968). *Proc. 11th Congr. Int. Soc. Blood Transf.*, Sydney, 1966; Bibl. Haemat., No. 29, part 2, p. 635, Karger, Basel/New York

Jones, M. A. S. and Lafferty, K. J. (1969). *Aust. J. Exp. Biol. Med. Sci.* *47,* 159

Jones, M. A. S., Yamashita, A. and Lafferty, K. J. (1969). *Aust. J. Exp. Biol. Med. Sci.* *47,* 325

Kennedy, J. C., Till, J. E., Siminovitch, L. and McCulloch, E. A. (1965). *J. Immunol.* *94,* 715

Kennedy, J. C., Till, J. E., Siminovitch, L. and McCulloch, E. A. (1966). *J. Immunol.* *96,* 973

Killby, V. A. A., Lafferty, K. J. and Ryan, M. A. (1972). *Aust. J. Exp. Biol. Med. Sci.* *50,* 309

Lafferty, K. J. and Jones, M. A. S. (1969). *Aust. J. Exp. Biol. Med. Sci.* *47,* 17

Lafferty, K. J. Scollay, R., Walker, K. Z. and Killby, V. A. A. (1972). *Transp. Rev.* In the press.

Lind, P. E. and Szenberg, A. (1961). *Aust. J. Exp. Biol. Med. Sci.* *39,* 507

McCullagh, P. (1972). *Aust. J. Exp. Biol. Med. Sci.* *50,* 49

Manyko, V. M. (1971). *Folia Biol., Praha* *17,* 365

Miller, J. F. A. P., Basten, A., Sprent, J. and Cheers, C. (1971). *Cell. Immunol.* *2,* 469

Murphy, J. B. (1916). *J. Exp. Med.* *24,* 1

Nisbet, N. W. and Simonsen, M. (1967). *J. Exp. Med.* *125,* 967

Nisbet, N. W., Simonsen, M. and Zaleski, M. (1969). *J. Exp. Med.* *129,* 459

Parrott, D. M. (1970). In *New Concepts in Allergy and Clinical Immunology* (U. Serafini, A. W. Frankland, C. Masala and J. M. Jamar, eds.). Excerpta Medica, Amsterdam, London and Princeton, p. 61

Payne, N. L. and Jaffe, W. P. (1962). *Transpl. Bull.* *30,* 20

Playfair, J. H. L., Papermaster, B. W. and Cole, L. J. (1965). *Science* *149,* 998

Puck, T. T. and Marcus, P. I. (1956). *J. Exp. Med.* *103,* 653

Roelants, G. E. and Askonas, B. A. (1971). *Eur. J. Immunol.* *1,* 151

Schellekens, P. Th. A. and Eijsvoogel, V. P. (1970). *Clin. Exp. Immunol.* *7,* 229

Seto, F. and Albright, J. F. (1965). *Develop. Biol.* *11,* 1

Simons, M. J. and Fowler, R. (1966). *Nature (London)* *209,* 588

Simonsen, M. (1957). *Acta. Path. Microbiol. Scand.* *40,* 480

Simonsen, M. (1962). *Prog. Allergy* *6,* 349

Smith, L. H. and Vos, O. (1963). *Radiat. Res.* *19,* 897

Syeklocha, D., Siminovitch, L., Till, J. E. and McCulloch, E. A. (1966). *J. Immunol.* *96,* 472

Szenberg, A. and Cunningham, A. J. (1968). *Nature (London)* *217,* 747

Walker, K., Schoefl, G. I. and Lafferty, K. J. (1972). *Aust. J. Exp. Biol. Med. Sci.* In the press.

Warner, N. L. (1964a). *Aust. J. Exp. Biol. Med. Sci.* *42,* 417

Warner, N. L. (1964b). *Brit. J. Exp. Path.* *45,* 459

The Clonal Development of Antibody-Forming Cells

A. J. CUNNINGHAM

Department of Microbiology, John Curtin School of Medical Research, Australian National University, Canberra, A.C.T., Australia

The stimulation of immunologically competent cells by antigen begins a process of cell division and differentiation which generates specialised antibody-forming cells. This process is of central interest in immunology, but it also has a number of advantages as a model system for studying differentiation and gene expression. The inducing stimulus, antigen, may be defined and its application controlled. The end product, antibody, is a well-characterised protein which is easily detected in minute amounts. Very small differences in structure of the V-region of different antibodies may be detectable as large differences in their antigen binding properties, so that antibody provides a useful "window" through which to detect small changes in the genetic material of cells. In addition, it seems that there may be a switch from production of μ to γ chain attached to the same V-region of the antibody produced by some clones (Nossal *et al.*, 1971), a particularly interesting example of a change in gene expression, perhaps analogous to the switch in type of haemoglobin produced by single erythrocyte precursors (Matioli and Thorell, 1963).

While only a small proportion of lymphoid cells respond to any one antigen, such a response involves proliferation of a fairly large number of distinct clones. Obviously it would be an advantage to be able to follow single clones. This would give a clearer picture of the developmental history of antibody-forming cells, and allow comparisons between clones of such properties as the number of divisions they undergo, and whether or not a μ to γ chain switch takes place. It would also allow a comparison of the effect of different stimuli on identical daughter cells within a clone: where large populations of cells are tested it is usually not possible to say whether various stimuli affect identical cells in different ways, or whether they select out slightly different clones preferentially.

How can such single clones of antibody-forming cells be grown? The ideal way would be to isolate a single precursor cell in a microdrop, stimulate it with antigen, and watch it divide repeatedly. This has not yet been done, partly because of the difficulty of growing small numbers of cells in isolation (a problem which might be overcome by using "feeder" layers), and partly because specific precursor cells cannot be identified with certainty. It is also clear now that stimulating such precursor cells with antigen is a complex process often requiring the co-operation of thymus-derived cells and macrophages (Miller *et al.*, 1971).

An alternative, if less convenient way of growing single clones of antibody-forming cells is to inject small numbers of lymphoid cells, together with antigen, into irradiated syngeneic mice, and examine their spleens some days later for foci of antibody production. This approach was pioneered by Kennedy *et al.*, (1966) and by Playfair *et al.*, (1965), and has been used, with modifications, by a number of other workers (Celada and Wigzell, 1966a and b; Cunningham, 1969a and b; Bosma and Weiler, 1970; Klinman and Aschinazi, 1971). When approximately 10^6 normal spleen cells are injected, a proportion (about 10%) settle in the spleen, and roughly 1 in 10^5 of these can be stimulated by antigen (foreign erythrocytes) to proliferate into a localised colony of antibody-plaque-forming cells (PFC). These colonies are best recognised, several days later, by cutting the spleen transversely into 8 pieces, and assaying each piece separately with the haemolytic plaque technique (Jerne and Nordin, 1963). Fig. 1 illustrates the experimental protocol, and Table 1 contains some examples of the uneven distribution of PFC

TABLE 1

Spleen Number	Number of PFC in Segment							
	1	2	3	4	5	6	7	8
1	<u>4800</u>	<u>80</u>	0	4	8	8	36	24
2	28	28	4	0	20	12	<u>144</u>	0
3	40	<u>2720</u>	<u>18560</u>	<u>7040</u>	112	28	40	24
4	<u>48</u>	8	4	8	<u>20</u>	<u>40</u>	12	8

Four examples of the distribution of direct plaque-forming cells in the spleens of irradiated mice injected intravenously with 1-2 x 10^6 normal spleen cells, and sheep erythrocytes as antigen. Spleens of recipients were collected 5-8 days later and cut transversely into 8 segments, each being assayed separately; colonies are underlined.

in such spleens reconstituted with limiting numbers of spleen cells (Cunningham, 1969a and b). Because all spleens contain small numbers of "background" PFC, colonies must be rather arbitrarily defined as local concentrations of PFC where at least 1 of the segments involved has a count greater than 3 times background.

Studies of this kind have given the following results:

(1) Spleen colonies are almost certainly clones. This is suggested by their localisation, and strongly supported by the observation that within any one colony, PFC are strikingly uniform in their properties, as would be expected in a number of immediate descendants of a single cell. When a mixture of sheep

FIGURE 1

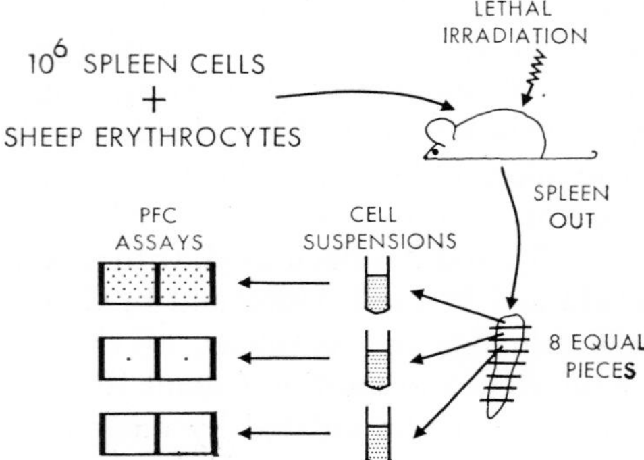

Growing single clones of antibody-forming cells *in vivo.* Approximately 10^6 normal spleen cells are injected intravenously into lethally irradiated syngeneic mice. About 7 days later, these mice are killed, and their spleens removed and cut transversely into 8 segments, each of which is assayed for plaque-forming cells.

and horse red cells is used as antigen, colonies containing anti-sheep or anti-horse PFC appear in separate segments. When PFC from an anti-sheep colony are tested on a mixed monolayer of sheep + goat or of sheep + cow erythrocytes, all antibody-forming cells lyse both red cell types, or all lyse only the sheep erythrocytes, giving partial plaques. In other words, the specificity of the antibody produced by each cell within a colony is probably the same (Table 2,

TABLE 2

	Number of Plaques	
	Clear	Partial
Control	65	196
Colony 1	0	42
2	0	40
3	57	5

The proportions of clear and partial plaques found when samples from 3 colonies were tested on mixed red cell monolayers (sheep + goat red cells). The control was a mixture of cells from a large number of colonies.

Fig. 2). This statement has to be qualified by admitting that there is always a varying number of background PFC in the spleen which contaminate clones, but the general pattern is clear. The uniformity of plaques extends to their size, and to various more subjective properties such as the sharpness of their borders.

FIGURE 2

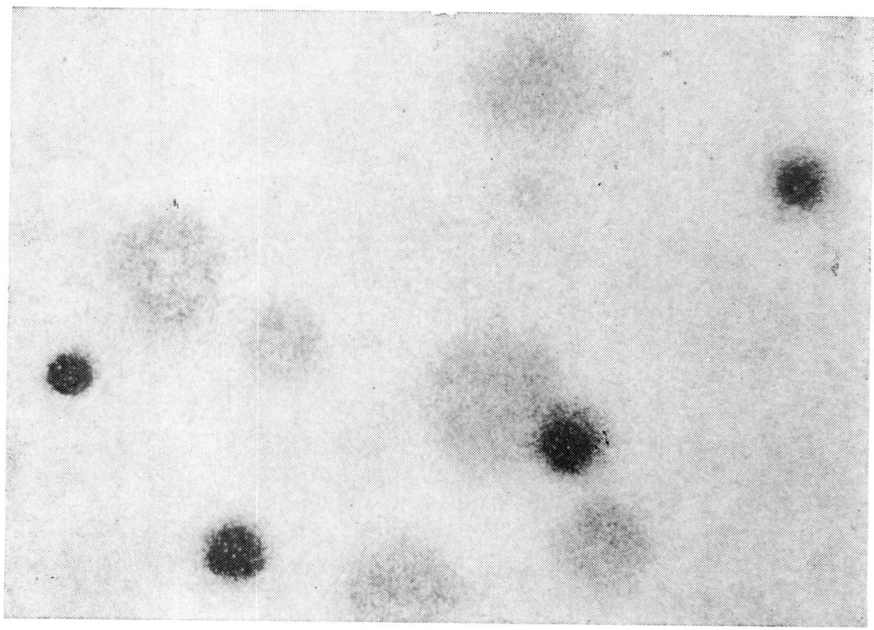

Clear and partial plaques in a monolayer containing a mixture of sheep and goat red cells. The illumination is dark-ground, so clear plaques appear black. The largest plaques are about 1 mm diameter.

(2) Numbers of PFC per clone vary from 28,000 down to background levels of less than 40. Maximum sizes are reached in 7 days (Fig. 3).

(3) The average area of spleen through which a clone extends is about 0.5 mm in diameter, and the maximum about 1.5 mm, (estimated from the frequency with which clones overlap 2 segments). This suggests that PFC remain near the site of generation, and that the size of clones is probably not underestimated.

(4) Clones of indirect (IgG) PFC are much less common under the conditions used than direct (IgM) plaques (about 15% of the total, Fig. 3). However, half of these indirect clones mapped precisely on clones of direct PFC, a highly significant association supporting the idea of a switch from IgM to IgG production within at least some clones (Nossal *et al.*, 1971).

(5) When the morphology of individual PFC from a clone is examined (using micromanipulative techniques, Cunningham *et al.*, 1966), cells are often very similar, a finding which hints at some microenvironmental control of differentiation which may synchronise development.

The next step in studying the clonal expansion of antibody-forming cells would seem to be to define factors which control the process, in particular the size of clones, why they stop growing, what determines switches in the class of antibody produced, and why some members of the clone become B-line memory cells instead of antibody-formers (Cunningham and Sercarz, 1971). *In vitro* models are likely to be the best for this sort of analysis. Osoba (1969) cultured

FIGURE 3

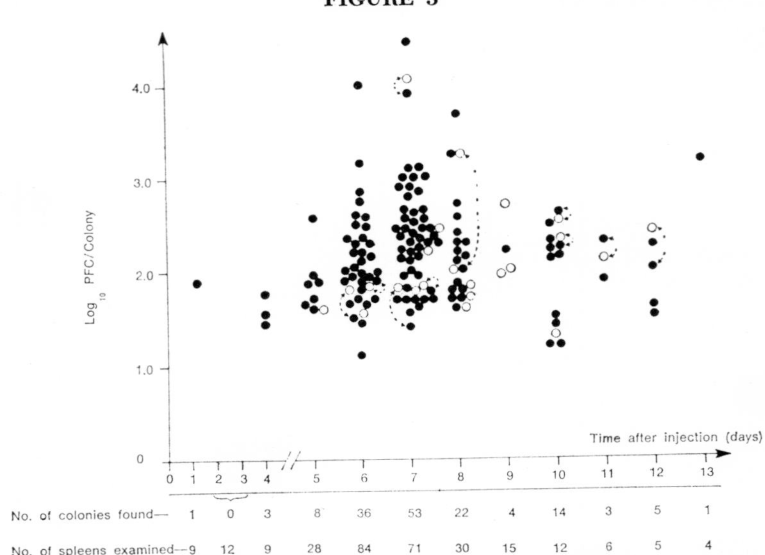

Development of colonies of antibody-forming cells in irradiated spleens. Each circle represents one colony. Black circles show numbers of cells producing direct plaques (IgM antibody); open circles record indirect (IgM) plaques. Where direct and indirect colonies were found in the same segment these have been connected by a dashed line. (Reproduced from *Aust. J. Exp. Biol. Med. Sci.*, with permission).

small numbers of normal spleen cells in the presence of large numbers of irradiated cells, allowing initiation of single clones of one specificity. Marbrook and Haskill (1971) describe a technique which is much more convenient. Rafts of polyacrylamide gel float on a dish of medium. Every raft has a number of separate "dimples" each containing about 10^5 spleen cells. By adjusting the number of cells per dimple, the frequency of responding dimples can be varied.

We have some preliminary results with this technique (P. J. Russell and A. J. Cunningham, unpublished). Analysis of the specificity of antibody produced within individual dimples using mixed red cell monolayers confirms that these probably belong to single clones. The mean size of clones after 3 days in culture varies from about 4 to 9 plaques in different experiments (calculated after correcting for overlap), although many are larger than this. Table 3 gives an example of how clones were distributed among -the dimples on a raft in one experiment when 10^5 normal spleen cells were cultured in each dimple. When spleen cells from previously primed animals are used, the frequency of responding dimples increases considerably.

This technique is a more convenient way of measuring absolute numbers of clonal precursors than the spleen colony method. It also has the advantage of being a "closed" system: all cells put in stay in the raft, whereas only a fraction of the cells injected into an irradiated animal settle in the spleen. (The efficiency with which clones are stimulated is unknown in both techniques). However, clones do not grow to nearly the same size *in vitro* as *in vivo*: presumably

TABLE 3

1	27	0	0	0	4	3	3
7	0	0	1	1	6	3	0
0	4	0	0	0	0	0	0
0	0	0	0	0	6	2	0
5	18	0	0	6	4	0	0
0	0	2	6	18	0	0	0
0	0	0	0	2	0	0	0

Distribution of PFC from a raft with 56 "dimples," each of which was seeded with 10^5 normal spleen cells and 5 x 10^4 sheep red cells 3 days previously.

the irradiated mouse provides a more natural environment for cell proliferation. The polyacrylamide rafts should allow a study of the influence of varying numbers of specific helper T cells (Mitchell and Miller, 1968) on the size and chance of initiation of clones. They should also be particularly suitable for analysing the conditions which will induce a clone to switch from IgM to IgG antibody production.

In summary, the development of single clones of antibody-forming cells may be followed both *in vivo* and *in vitro*. These clones vary enormously in the number of divisions which they undergo. In some, a switch from production of IgM to IgG antibody of the same specificity seems to occur. These clones are now sufficiently well defined to act as a useful model system for studying factors which influence differential gene expression in cells.

REFERENCES

Bosma, M. and Weiler, E. (1970). *J. Immunol. 104,* 203
Celada, F. and Wigzell, H. (1966a). *Immunology 10,* 231
Celada, F. and Wigzell, H. (1966b). *Immunology 11,* 453
Cunningham, A. J. (1969a). *Aust. J. Exp. Biol. Med. Sci. 47,* 485
Cunningham, A. J. (1969b). *Aust. J. Exp. Biol. Med. Sci. 47,* 493
Cunningham, A. J., Smith, J. B. and Mercer, E. H. (1966). *J. Exp. Med. 124,* 701
Cunningham, A. J. and Sercarz, E. E. (1971). *Eur. J. Immunol. 1,* 413
Jerne, N. K. and Nordin, A. A. (1963). *Science 140,* 405
Kennedy, J. C., Till, J. E., Siminovitch, L. and McCulloch, E. A. (1966). *J. Immunol. 96,* 973
Klinman, N. R. and Aschinazi, G. (1971). *J. Immunol. 106,* 1338
Marbrook, J. and Haskill, J. S. (1971). In "Cellular Interactions in the Immune Response," p. 66, Karger, Basel
Matioli, G. and Thorell, B. (1963). *Blood 21,* 1
Miller, J. F. A. P., Basten, A., Sprent, J. and Cheers, C. (1971). *Cellular Immunol. 2,* 469
Mitchell, G. F. and Miller, J. F. A. P. (1968). *Proc. Nat. Acad. Sci., U.S., 59,* 296
Nossal, G. J. V., Warner, N. L. and Lewis, H. (1971). *Cellular Immunol. 2,* 41
Osoba, D. (1969). *J. Exp. Med. 129,* 141
Playfair, J. H. L., Papermaster, B. W. and Cole, L. J. (1965). *Science 149,* 998

Immunoglobulin Gene Expression in Murine Lymphoid Cells[a]

NOEL L. WARNER AND ALAN W. HARRIS[b]

Laboratory of Immunogenetics, The Walter and Eliza Hall Institute of Medical Research, Melbourne, Australia

Immune responses of all types depend fundamentally on the expression of immunoglobulin structural genes in lymphoid cells. The complete *in vivo* manifestation of the various types of immunity, such as delayed hypersensitivity, transplantation immunity, allergy, resistance to infectious disease, is dependent on the type of lymphoid cell activated and on secondary processes which are initiated by a primary interaction between the antigen concerned and one of the types of structural immunoglobulin gene products. In many situations it is the particular type of immunoglobulin produced that determines these subsequent events; for example, production of IgE antibody leads to mast cell fixation of an antigen-IgE complex which in turn initiates release of vasoactive amines and causes an anaphylactic or allergic reaction (Bloch, 1967; Austen and Becker, 1971). Our concern here is to discuss the factors that are responsible for the primary event, that is, the activation of the particular immunoglobulin gene.

In recent years, it has become recognised that not only are there multiple types of immunoglobulin molecules controlled by different structural genes (Lennox and Cohn, 1967; Hood and Prahl, 1971; Edelman and Gall, 1969; Fudenberg and Warner, 1970), but there are also several distinct categories of lymphoid cells that can be involved in the immune response (Raff, 1971; Craddock *et al.*, 1972; Warner 1972). A selective expression of different immunoglobulin genes in the various types of lymphoid cells may constitute a valid model for broader problems of control of gene expression in higher organisms. The following discussion considers the expression of immunoglobulin structural genes in relation to the differentiation of various members of the lymphoid cell series.

[a]The experimental work described in this paper was supported by research grants from the U.S. Public Health Service, AM-11234 and the Australian Research Grants Committee. This is publication Number 1753 from the Walter and Eliza Hall Institute.

[b]Queen Elizabeth II Research Fellow.

Genetic control of immunoglobulins in mice

The immunoglobulin molecule is composed of two types of polypeptide chains, termed light and heavy. Amino acid sequences of a large number of light chains and a few heavy chains have recently been determined and appear to show that each of these polypeptide chains is structurally determined by at least two different genes, a variable region and a constant region gene (Hood and Prahl, 1971; Hilschmann, 1967; Wang *et al.*, 1971; Köhler *et al.*, 1970). Variable region genes code for the N-terminal half of the light chains, and for approximately the N-terminal quarter of heavy chains. At least three groups of V region genes are presently recognised, $V\lambda$, V_K, and V_H, each respectively coding for the V regions associated with λ or K type light chains or for all heavy chains. Constant region genes code for the remaining part of the polypeptide chains, and for light chains, three constant region genes have been recognised, λ, K, and a third type found so far, only in MOPC-315 myeloma protein (Schulenberg *et al.*, 1971).

In mice, up to the present, seven heavy chain constant region genes have been identified which code for the heavy chains of immunoglobulin classes IgM, IgA, IgG1, IgG2a, IgG2b, IgG3 and reaginic immunoglobulin (Fudenberg and Warner, 1970; Grey *et al.*, 1971; Vaz and Provost-Danon, 1969). Genetic polymorphism of four of these genes has been recognised, and multiple alleles of IgG1, IgG2a, IgG2b and IgA can be recognised by the use of specific antiallotype antisera (Herzenberg and Warner, 1968; Potter and Lieberman, 1967).

Lymphoid cell types

Although cell types are usually classified on the basis of morphological appearance, the use of functional properties and surface allo-antigenic markers has

FIGURE 1

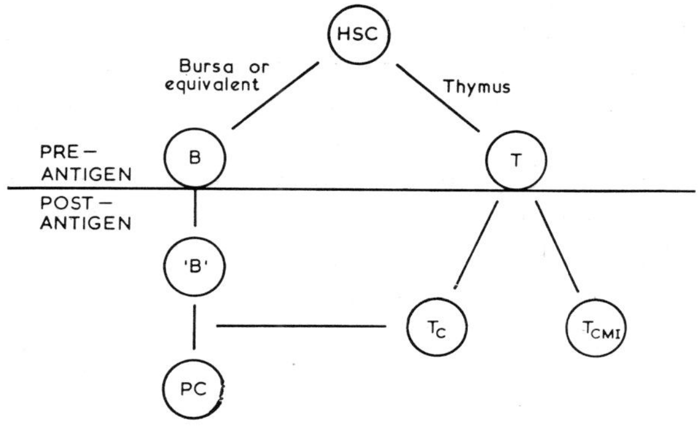

The haemopoietic stem cell (HSC) differentiates under the influence of a bursal factor (or its equivalent) into a B lymphocyte (B). Antigen then stimulates this cell into activation ('B') and to plasma cell formation (PC).

Alternatively, HSC can under thymic influence differentiate into T lymphocytes (T), which on exposure to antigen may become T cells capable of stimulating B cells into antibody secretion (Tc) or into cells involved in cell mediated immunity (Tcm,). It is possible that Tc and Tcm, are one and the same cell.

recently led to a more practical classification of lymphoid cell types. This is presented in simplified form in Fig. 1. Haemopoietic stem cells are the precursors of all members of the lymphoid and plasma cell series, as well as of all other haemopoietic cells such as erythrocytes and granulocytes (Metcalf and Moore, 1971). Under the influence of the bursa of Fabricius in birds (Warner, 1967; Cooper *et al.*, 1966) or some equivalent tissue in mice (Cooper and Lawton, 1972), differentiation of the stem cell to the B lymphocyte is induced (B for bursal or bone marrow derived). This stage of differentiation is antigen independent and occurs initially during embryonic life. Further differentiation to an activated B cell stage (recognised as immunoblasts or plasmablasts) is initiated by antigen, and specific antibody secretion at this stage is initiated. Continued differentiation with cell division ultimately leads to the formation of plasma cells which are generally considered to be end cells incapable of further division or differentiation.

An alternate pathway of differentiation for the stem cell involves induction by the thymic environment and leads to the formation of T lymphocytes (thymus-derived) (Miller and Osoba, 1967). Following antigenic stimulation, this pathway leads to two functionally recognised types of T cells, one involved in cell mediated immune reactions, such as delayed hypersensitivity or transplant rejection, and the other to a cell type which is involved in directing B cell differentiation into the plasma cell series. It should be noted that these two functional properties of T cells have not been proven to represent two different differentiation pathways for T cells, and the possibility exists that they are different activities of only one type of activated T cell.

In any attempt to determine the immunoglobulin strucutural genes that may be activated in a lymphoid cell of a particular type, either purified lymphoid cells of that type must be used, or else the study must also involve identification of the cell by means of a functional or antigenic marker. Both approaches have been used. There are several heterologous and alloantigenic markers available for distinguishing B cells (Raff, 1971; Nussenzweig *et al.*, 1971; Basten *et al.*, 1972), T cells (Reif and Allen, 1964; Boyse *et al.*, 1968), and plasma cells (Takahashi *et al.*, 1970, 1972), and in the last few years several sources of pure cell populations have become available (Basten *et al.*, 1972, Sprent and Miller, 1972; Shortman, 1972). Other sources of homogeneous lymphoid cell populations are lymphomas and plasma cell tumors, and in the following sections, studies with both normal and malignant populations of cells will be considered.

Haemopoietic stem cells

Although the stem cell is capable of differentiation into immunoglobulin synthesizing and secreting cells, there is no evidence that immunoglobulin gene expression occurs in the stem cell itself. As murine haemopoietic stem cells have not been completely purified, and no specific stem cell antigen has been recognised as yet, studies of this problem have employed a functional identification of stem cells. The injection of cell populations containing stem cells, such as spleen or bone marrow, into lethally irradiated mice, causes the formation of macroscopically visible colonies in the recipient spleen, each of which is composed of differentiated haemopoietic cells derived from a single stem cell (Becker *et al.*, 1963). Treatment of spleen cell suspensions with anti-immunoglobulin sera, under conditions that lead to inactivation or suppression of lymphoid cell activity, do not affect the formation of colonies from stem cells. Thus, treatment of spleen cells with anti-light chain sera can inhibit T cell function (see below) without affecting the colony forming activity of that population (Warner, 1971) (Table 1). Similarly, inhibition of B cell-induced immune responses by pretreatment of the spleen cells with a high specific activity ^{125}I-labelled preparation of

TABLE 1

Lack of Inhibition of Stem Cell Function by Anti-immunoglobulin Pretreatment

Expt.	Test Cells and Assay	Serum Pretreatment	% Suppression[c]
1[a]	Stem cells—CFU	anti-mu	10-30
	B cells—anti-SRBC-PFC	anti-mu	50-70
2[b]	Stem cells—CFU	anti-κ	5-10
	T cells—GVHR	anti-κ	80-100

[a]High specific activity ^{125}I-labelled purified anti-mu chain antibody was used to treat normal mouse spleen cells. Aliquots of the cell suspension were then assayed for their ability to transfer a primary response (by plaque test, PFC) to sheep red cells (SRBC) or for colony forming unit (CFU) content.

[b]Normal spleen cells were treated with anti-κ chain serum. Aliquots of the cells were then tested for CFU activity or for ability to induce a graft versus host reaction (GVHR).

[c]Results show percent suppression of either anti-SRBC PFC, CFU, or GVHR, in the test groups as compared with untreated control groups.

anti mu chain purified antibody was not associated with any marked suppression of spleen colony forming activity (Herrod and Warner, 1972) (Table 1).

B lymphocytes

Studies on man, mouse, rabbit and chicken (e.g. Coombs *et al.*, 1969; Pernis *et al.*, 1971; Raff, 1970; Rabellino *et al.*, 1971; Cooper *et al.*, 1972) have all clearly shown that B cells synthesize immunoglobulin that is held transiently on the cell surface. Several results of our own recent work on this problem are summarized in Table 2.

TABLE 2

Surface Immunoglobulin on B Lymphocytes

(i) Embryonic chicken bursal cells synthesize IgM at a time before any other lymphoid organ shows immunoglobulin synthesis (Thorbecke *et al.*, 1968).

(ii) Spleens of normal adult chickens, but not of adult hormonally bursectomised chickens, contain approximately 24% of cells with surface IgM (Bankhurst *et al.*, 1972).

(iii) Antigen binding cells (ABC) (binding radio-iodinated soluble antigens) are present in the spleens of athymic mice (Dwyer *et al.*, 1971). ABC detection can be inhibited by prior treatment of spleen cells with anti-κ and anti-mu chain antisera (Warner *et al.*, 1970).

(iv) Anti-κ serum is cytotoxic for B lymphocytes (Miller *et al.*, 1972).

(v) Thoracic duct lymphocytes of athymic mice (97% B cells) have a high surface density of immunoglobulin; 76% of cells with IgM (Bankhurst and Warner, 1972).

(1) Short term culture of bursal fragments from embryonic chickens with ^{14}C-amino acids followed by analysis of the cells for labelled immunoglobulin has demonstrated that IgM synthesis can be detected before the time of hatching (Thorbecke *et al.*, 1968). This is the first site in chickens to show immunoglobulin production. More detailed studies by Cooper *et al.* (1972) have shown the presence of surface immunoglobulin on bursal cells, using fluorescent labelled class-specific antiglobulins. Bursal cells are developmentally the first lymphoid cells in the chicken to carry surface immunoglobulin. In that IgM (both light chain and mu chain) on the bursal cell surface was detected within 24 hours after the time of arrival (Moore and Owen, 1966) of yolk sac stem cells in the embryonic bursa. Similarly, in assays for antigen-binding cells in embryonic chicken lymphoid tissues, it was found that bursal cells from 14 day embryos could specifically bind antigen (Dwyer *et al.*, 1972). These results therefore strongly suggest that both V region and C region (light and mu) genes are first expressed under the influence of bursal induction after only a very short period of residence by the stem cell in the inductive environment.

(2) Spleens from normal adult chickens contain approximately 24% of cells bearing readily detectable surface IgM immunoglobulin. No cell surface immunoglobulin was detectable on any spleen cells from embryonically bursectomized agammaglobulinaemic chickens, indicating that the positive cells in normal chickens were bursa derived (Bankhurst *et al.*, 1972).

(3) Antigen binding lymphocytes can be detected in the spleens of congenitally athymic nude (nu/mu) mice in proportions similar to those found in normal mice (Dwyer *et al.*, 1971). This has been observed with four different antigens, and indicates that this technique detects antigen binding B cells. Pretreatment of spleen cells with anti-light chain or anti-mu chain antibodies markedly suppressed the detection of antigen binding cells from unimmunized animals (Warner *et al.*, 1970). Treatment with other anti-heavy chain sera had little effect. These results are consistent with the view that the antigen binding receptor on most if not all unimmunized B cells is an IgM molecule.

(4) Treatment of cell suspensions from various sources (antigen-activated T cells, normal T cells, thoracic duct lymphocytes from normal or athymic nude mice) with anti-kappa chain sera and followed by incubation with complement, indicates that B cells can be killed by this procedure while T cells cannot (Miller *et al.*, 1972). These results suggest either that there is quantitatively greater expression of light chain on B cells than on T cells, or that the membrane localization of immunoglobulin on the surface of the two cell types is different. Comparable tests with anti-heavy chain sera have not yet been performed.

(5) Pure populations of B cells can be obtained from the thoracic duct of athymic (nude) mice. Examination of these cells by autoradiography using radiolabelled globulins (Bankhurst and Warner, 1972) shows that the majority of cells (92% react with anti-kappa chain) have surface immunoglobulin and that 76% have IgM.

These results are all consistent with the interpretation that the majority of B cells have surface IgM globulin that acts as the antigen receptor. The observation that a proportion of cells may also have other classes of immunoglobulin is discussed below under activated B cells.

Malignant B cells

Monoclonal lines of B cell lymphomas would be of considerable advantage for studies of the chemical nature and function of B cell surface immunoglobulin. At present it appears that in mice, such tumours may be rare.

A functional surface marker for B cells is their ability to bind to antigen-antibody complexes, and is detected by the use of either radiolabelled soluble antigen-antibody complexes, (Basten *et al.*, 1972a) or by rosette formation with anti-erythrocyte antibody-coated red cells either with (Nussenzweig *et al.*, 1971) or without (Cline *et al.*, 1972) addition of complement. Assays on human chronic lymphocytic leukaemia (CLL) cells for their ability to rosette with globulin coated red cells and complement, have shown that most CLL cells have this property (Pincus *et al.*, 1972); most of these also have readily detectable surface immunoglobulin. This type of observation would appear to be entirely in accord with the previously mentioned studies on normal B cells.

However, our studies with murine lymphoid cells have suggested that unusual patterns of B and T cell surface markers may occur in malignant cell populations. A series of tumour lines adapted to continuous cell culture have been examined for their ability to rosette with immunoglobulin-coated sheep erythrocytes. Two lymphoidal lines were found to form rosettes and if this is indeed a strict marker of B cells, (it has been reported (Basten *et al.*, 1972a,b) to be absent from T cells and plasma cells) then we would conclude that these lines were B cell in type, and their immunoglobulin synthesizing ability might therefore be that of B cells. Some characteristics of these two lines and two other lines are given in Table 3. One of the rosette-forming lines is a radiation induced thymoma which carries the theta antigen (Harris *et al.*, 1972), and the other is a variant line derived from a plasma cell tumour that originally secreted IgA globulin (Cline *et al.*, 1972). Both of these lines synthesize some immunoglobulin and, were it not for the fact that one of them is theta positive and the other is derived from a plasmacytoma, it might be concluded that they were typical B cells with surface immunoglobulin and antigen-antibody complex receptor sites. The other two lines described are in contrast to these in that HPC-108 cells show similar immunoglobulin synthesizing properties to HPC-6, but differ quantitatively in the amount of immunoglobulin synthesized WEHI-105 differs from WEHI-22 in the class of immunoglobulin synthesized. Two general possibilities concerning these observations might be considered.

(1) Does HPC-6 represent a dedifferentiation from a plasma cell to a B lymphocyte, thereby reacquiring the receptor for antigen-antibody complexes, but still retaining its plasma cell type of immunoglobulin synthesis? and

(2) Does WEHI-22 represent a B cell lymphoma with surface IgM globulin but showing an unusual expression of theta possibly due to its residence within the thymus (at least at the time of malignant transformation)?

TABLE 3

Rosette Formation and Immunoglobulin Gene Expression by Lymphoid Tumour Cells

Tumour Line	Origin	Immunoglobulin Synthesis and Secretion		Rosette[c] Formation
		Primary *In Vivo*[a]	Cultured Line[b]	
HPC-6	Plasmacytoma	IgA serum myeloma	Trace IgA ($<1\mu g/ml$)[d]	77%
HPC-108	Plasmacytoma	Trace IgA synthesis. Marked κ chain BJP	$30\mu g/ml$ IgA	1%
WEHI-22	Radiation induced thymoma	Weak IgM synthesis	Weak IgM[e] synthesis	58%
WEHI-105	Radiation induced thymoma	Weak κ chain synthesis	Not tested	1%

[a]Qualitative assessment of immunoglobulin synthesis and secretion in primary tumour and early *in vivo* transplanted generations.

[b]Secreted immunoglobulin concentration in tissue culture fluid from cultured cells. IgA concentration determined with α chain specific radioimmunoassay.

[c]Percent of cells forming rosettes with mouse IgG-coated sheep erythrocytes. Methods and experimental details as described elsewhere (Cline *et al.*, 1972).

[d]Although no IgA was detected in the culture fluids by radioimmunoassay, a rabbit immunised with HPC-6 cells from culture produced a specific anti-α chain serum, indicating the presence of at least some IgA production.

[e]Detected by lactoperoxidase labelling of cell surface followed by polyacrylamide gel electrophoresis (Marchalonis *et al.*, 1972b).

The first possibility is consistent with electron microscopic observations on these cell lines, (Fig. 2) in that HPC-6 is lymphoblastoid in appearance with an occasional strand of rough endoplasmic reticulum, while HPC-108 possesses more cytoplasm and much more endoplasmic reticulum. In view of the quantitative difference observed in immunoglobulin secretion between these two lines, an alternative explanation to the proposal above (1) might be made. Rather than there being an actual loss of the surface receptor for complexes during differentiation of plasma cells from B cells, it may be that the increased rate of immunoglobulin secretion results in a steric masking of the receptor. The low rate of secretion by HPC-6, would therefore permit detection of the receptor.

In regard to the problem concerning WEHI-22 posed above (2) other explanations include the possibility that the IgM globulin on the surface of the WEHI-22 cells has an antiglobulin specificity which binds to the globulin-coated red cells (Harris *et al.*, 1972).

There is of course, a general question as to whether malignant populations of lymphocytes can be used as models for immunoglobulin gene expression by normal lymphocytes. It would be naive to simply dismiss studies of malignant cells in this regard because they are malignant. Unusual observations such as the two mentioned above do not necessarily indicate that an aberrant expression of certain genes is occurring in these malignant cells. Instead, these cell lines may be of considerable value in highlighting cell types that are present in normal lymphoid organs, but are in a small minority as compared to other lymphoid cell

FIGURE 2

Electron micrographs of cultured myeloma cells

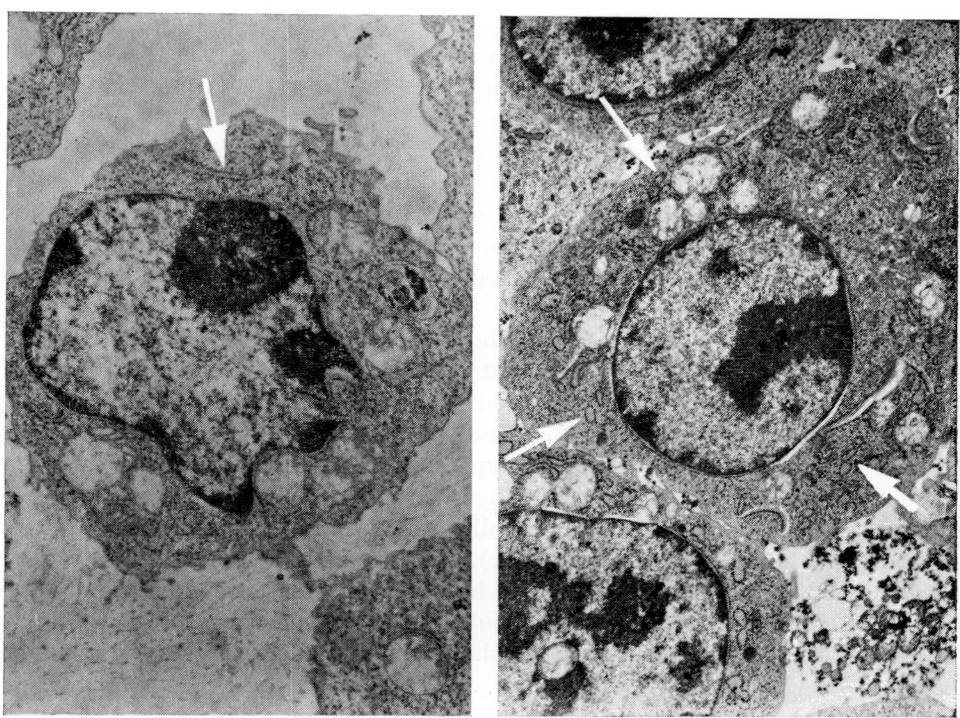

HPC-6 (on the left) shows an occasional strand of rough endoplasmic reticulum (arrow); HPC-108 (right) possesses much more rough endoplasmic reticulum (arrows). (Magnifications = HPC-6, 8000 x; HPC-108, 6000 x).

types. Indeed, WEHI-22 is unique amongst 19 radiation induced thymomas examined (see below) and may represent a minor population of B cells that are resident in the thymus. Similarly, HPC-6, although being a variant cell line derived during serial transplantation of a typical plasma cell tumour, may reveal more about the differentiation history of the receptor for complexes than can normal lymphoid cell populations at present.

Activated B cells and the constant region gene "switch" problem

The observations cited above indicate that IgM is synthesized in unprimed resting B lymphocytes. However, when pure populations of B cells (such as thoracic duct cells from athymic nude mice) are examined for the presence of other immunoglobulin classes, approximately 40% are found to carry IgG and 40% IgA (Bankhurst and Warner, 1972). When the value for IgM (76%) is also considered, it is evident that at least some individual cells must have more than one class of immunoglobulin on their surface. On the simplest hypothesis that the excess number represents cells bearing all major classes (rather than several types having assortments of 2 of the 3 classes examined), our results would indicate that about 40% of cells express all major classes of immunoglobulin, and most of the remainder express only IgM.

Two explanations for these observations might be considered.

(1) Some B cells may bind immunoglobulin derived from the serum to their surface, and the results may not therefore directly relate to immunoglobulin gene expression by those individual cells, or (2) that the cells bearing multiple classes of surface immunoglobulins are not typical unprimed B cells, but rather represent a subpopulation of B cells that have been antigenically activated and have now commenced expression of other constant region genes, in prelude to further differentiation into the different classes of immunoglobulin-secreting plasma cells.

In considering the first type of hypothesis, Greaves (Greaves and Hogg, 1972) has shown using heterozygous cells, that not only are several classes of immunoglobulins expressed in B cells, but that both allelic genes at the IgG2a locus are expressed. From studies involving passive serum transfer and *in vitro* mixtures of cells with sera from different allotype sources, Greaves could not find any significant evidence of cytophilic binding of IgG2a globulin to B cells. Studies in rabbits, however, have indicated that allelic exclusion of immunoglobulin genes operates in B cells (Pernis *et al.*, 1971). Further analysis of this problem with chimeric cell populations (such as can be obtained from tetraparental mice) as compared with heterozygous cell populations should show more conclusively whether there is multiple immunoglobulin gene (and allele) expression in any B lymphocyte.

Although B cells from athymic nude mice contain proportions of cells with surface expression of the various immunoglobulin genes similar to those from normal mice, they show a marked inability to develop a normal IgG antibody response, and to produce normal levels of serum IgA globulin (Crewther and Warner, 1972). The immune response of athymic nude mice to formalin-killed *Brucella abortus* is normal for IgM antibody production, but markedly depressed in IgG antibody production (Crewther and Warner, 1972). Furthermore, in mice which have a genetic defect, expressed in T cells, in their ability to respond to the synthetic copolymer antigen (Tyr-Glu)-Ala-Lys, a normal IgM antibody response is observed, but the IgG response is grossly deficient (Mitchell *et al.*, 1972). In considering both of these results, together with the observations on multiple gene expression in some B cells, it might be proposed that antigenic activation of B cells, which have already expressed IgM, results in an intracellular expression of IgA and IgG constant region genes, which also causes multiple class localization on the membrane. For full differentiation of these cells into IgA and IgG *secreting* plasma cells, a further activation step is then necessary. The results with athymic mice and nonresponder mice suggest that this activation requires some action of antigenically stimulated T cells.

This interpretation implies that all plasma cells are derived by division and differentiation of cells that initially had expressed only IgM. In other words, it is essential in this concept that an intracellular selective and ordered activation, or "switch" of immunoglobulin structural constant region genes occurs.

It has been observed that a small proportion of antibody-forming cells can secrete antibodies of more than one class (Nossal *et al.*, 1971). It has also been reported that some rabbit lymphoid cells have surface bound IgM and intracellular IgG (Pernis *et al.*, 1971). Furthermore, amino acid sequence studies on an IgM and an IgG myeloma protein derived from a single individual has shown that the same variable heavy region gene product is associated with both a mu and a gamma constant region (Wang *et al.*, 1970). Although these results suggest that more than one constant region gene can be activated within a single cell, they do not necessarily imply that a sequential activation has occurred. However, some evidence suggesting a *sequential* activation of constant region genes in B cell populations has been obtained from experiments in which anti-mu chain treatment has interfered with the subsequent expression of IgG and IgA production. Injection of purified anti-mu chain globulin of appropriate type into embryonic chickens (Kincade *et al.*, 1970) or newborn germfree mice (Lawton *et al.*, 1972) led to a significant reduction in the subsequent formation of serum IgG (and IgA in mice). Similar results have been obtained in our laboratory from experiments in which spleen cells from mice of a certain allotype were injected into sublethally irradiated recipients that were congenic to the donor for immunoglobulin allotype (Herrod and Warner, 1972). The subsequent production of donor type immunoglobulin in the recipient was measured by assessing serum levels of donor allotype at various time intervals. The results given in Table 4 show that if the spleen cells were first pretreated with an anti-mu chain immunoglobulin, then the subsequent elaboration of donor type IgG1, IgG2a and IgG2b was severely inhibited. This is a specific effect of the anti-mu chain antibodies in the antiserum used, since removal of the anti-mu chain antibody from the antiserum also prevented the effect. These results may indicate that

TABLE 4

Effect of Anti-mu chain Pretreatment on the Cellular Transfer of IgG Production Between Allotype Congenic Mice

Treatment of Donor Spleen Cells	Percent Donor Immunoglobulin in Recipients[a]		
	IgG2a	IgG2b	IgG1
Normal rabbit serum	56%	21%	10%
Rabbit anti-mu chain serum	9%	3%	1%

[a]The amount of donor type immunoglobulin is given as a percentage of that found in a serum pool from adult donor strain mice. The results given are the mean values from three separate experiments involving three different pairs of congenic strains: serum taken 3 weeks after cell transfer. Detailed results are given in Herrod and Warner (1972).

B cells bearing IgM differentiate into IgG producers although as the experiments only measure the products of cell populations, they do not provide conclusive information about gene activation in any one cell line within the population. For example, it is possible that T cells bearing IgM (and therefore susceptible to the antiserum) may be required for differentiation of B cells into IgG producing plasma cells.

Malignant plasma cells

It was first demonstrated many years ago that a large proportion of humoral antibody production is carried out by plasma cells. Although this has been confirmed by many biosynthetic studies, it has also been found that a considerable proportion of serum antibody can be derived from cells at an earlier stage of the plasma cell development pathway, lymphoblasts or plasmablasts, particularly in situations of local lymph node antigenic stimulation (Harris *et al.*, 1966; Morris, 1971).

Immunofluorescent staining of normal plasma cells has clearly revealed that the great majority of these cells individually secrete immunoglobulin of only one class and that only one allele of the particular locus concerned is expressed by each of these cells (Cebra *et al.*, 1966; Pernis *et al.*, 1965).

In this regard, malignant plasma cells behave exactly like their normal counterparts. A series of plasma cell tumours have been induced in our laboratory in F1 hybrid mice whose parental strains were of different allotypes. Antigenic analysis of the myeloma proteins with anti-allotype sera has shown that only one allele of the particular heavy chain constant region gene is expressed in the monoclonal plasma cell. This has been found for IgA, IgG2a and IgG2b myeloma proteins (Table 5).

TABLE 5

Monoclonal Myeloma Proteins Produced by Plasma Cell Tumours from Heterozygous Mice

Myeloma protein	Strain of origin of tumour	Immunoglobulin class	Allotype detected[a] in the purified myeloma protein
HPC-38	(BALB/c x NZB)F_1	IgA	Ig-2^e
-24	(BALB/c x NZB)F_1	IgA	Ig-2^e
- 4	(BALB/c x NZB)F_1	IgA	Ig-2^a
-11	(BALB/c x NZB)F_1	IgA	Ig-2^a
-82	(NZB x NZC)F_1	IgG_{2a}	Ig-1^a
-98	(BALB/c x NZB)F_1	IgG_{2a}	Ig-1^a
-85	(BALB/c x NZB)F_1	IgG_{2a}	Ig-1^e
GPC- 7	(BALB/c x NZB)F_1	IgG_{2a}	Ig-1^e
HPC- 8	(BALB/c x NZB)F_1	IgG_{2b}	Ig-3^a
-89	(BALB/c x NZB)F_1	IgG_{2b}	Ig-3^a
-97	(BALB/c x NZB)F_1	IgG_{2b}	Ig-3^a
-96	(BALB/c x NZB)F_1	IgG_{2b}	Ig-3^e
-117	(BALB/c x NZB)F_1	IgG_{2b}	Ig-3^e
-100	(NZB x NZC)F_1	IgG_{2b}	Ig-3^e

[a]Inbred strains BALB/c and NZC carry the Ig-a allotypes, and NZB carries Ig-e.

Although all observations of this type on either normal or malignant plasma cells are quite consistent and are without exception, it might be noted that an alternative to the more usual interpretation of allelic exclusion does exist. It is possible, that in plasma cell development, cell fusion *in vivo* followed by pairing of homologous chromosomes and then reduction division, could produce homozygosity for immunoglobulin structural genes. This therefore would lead to a plasma cell that was expressing only one allotype, but both chromosomal genes would be operative in the cell.

T Lymphocytes

Although the presence of immunoglobulin on the surface of B lymphocytes has been readily demonstrable by many workers, there is considerable controversy over the possible presence of immunoglobulin on T cells. Many studies have demonstrated that the recognition of antigen by T cells is specific in that immunological tolerance (Chiller *et al.*, 1970), memory (Miller and Sprent, 1971) and radioactive antigen suicide (Basten *et al.*, 1971) can be induced in T cells. The simplest interpretation of these results would be that the same mechanism for the recognition of antigen is operating in both T and B cells, i.e. immunoglobulin molecules are acting as antigen receptors. This interpretation would require that immunoglobulin be synthesized by T cells and be present on the T cell membrane. Experimental evidence for this has been sought by three approaches in our laboratory and as these have mostly been reported elsewhere, will only be summarized here. They are tests for binding of antiglobulins to T cells, inhibition of T cell functions by antiglobulins and synthesis of immunoglobulins by T cells. A fourth technique, namely radio-iodination of the T cell surface followed by physical and serological analysis of labelled material, has recently demonstrated an IgM-like molecule on T cells, and this is discussed in detail elsewhere in this volume (Marchalonis *et al.*, 1973).

Direct labelling of T cells with ^{125}I-labelled anti-immunoglobulins

As mentioned previously in this paper, surface immunoglobulin containing both light and heavy chains can be readily demonstrated on B cells. In our own work (Bankhurst and Warner, 1971) this involved autoradiographic examination of cells treated with radiolabelled antiglobulins. Conditions of autoradiography were established in which B cells labelled, but T cells did not. However, when the sensitivity of the assay was increased by prolonging the exposure of autoradiography, up to 17% of thymus cells and 49% of an H-2 antigen-activated population of T cells showed labelling with an anti-κ globulin but not with a polyvalent anti-heavy chain globulin (Bankhurst *et al.*, 1971). These results suggest that at least light chain antigenic determinants are expressed on at least some T cells.

Inhibition of functional activity of T cells by antiglobulin

Radiolabelled antigen-induced suicide of T cells can be inhibited by pretreatment of the thymus cells with anti-κ chain globulin (Basten *et al.*, 1971), although anti-heavy chain sera have not yet been tested in this regard. In studies of the inhibition of graft versus host reactions (GVHR) in mice (Mason and Warner, 1970) and chickens (Rouse and Warner, 1972), we have found that *some* anti-light chain sera will inhibit this immune response. As there are, however, several documented examples of *negative* results from this experimental approach (Crone *et al.*, 1972), three points should be considered. In both murine and avian studies, when an antiserum containing anti-light chain activity was found to depress T cell-induced GVHR, this activity of the serum could be removed by prior absorption with purified light chains or whole immunoglobulin. Many sera which contained similar amounts of anti-light chain activity (as shown by *in vitro* precipitation tests) did not inhibit, and thirdly, no serum with only anti-heavy chain activity was found to inhibit the reaction.

From the results of these first two sections four general points might be made.

(1) These results do not rule out the possibility that the inhibitory antisera contain an antibody activity directed against a non-immunoglobulin T cell membrane component which happens to contaminate the purified light chain preparations used for antiserum production and absorption of active sera.

(2) The failure to demonstrate inhibitory activity with anti-heavy chain sera does not prove that a complete immunoglobulin molecule is not present on the cell surface. Rather, as has been discussed elsewhere (Marchalonis *et al.*, 1972), the Fab region may be the only part of the immunoglobulin molecule that is accessible *in situ* to experimentally admixed antiglobulins whose anti-heavy chain activity is mostly directed against the Fc region.

(3) The selective action of only certain antisera having anti-light chain activity may indicate that only anti-variable region antibodies will inhibit T cell function.

(4) None of these results distinguish between synthesis of immunoglobulin by the T cells, and the extrinsic binding of small amounts of serum-derived immunoglobulin, which, when mixed with antiglobulins produces steric inhibition of the true receptor, which itself may not be an immunoglobulin.

Immunoglobulin synthesis by T cells

In an attempt to resolve problems (2) and (4) above, we have commenced a series of tests for possible immunoglobulin synthesis by T cell preparations using incorporation of ^{14}C-amino acids and subsequent radioimmunoelectrophoresis. Our preliminary results are as follows:

Six of 19 radiation-induced thymomas synthesize very small, but detectable, amounts of light chain (Harris *et al.*, 1972). In one of these instances, WEHI-105, the tumour cell line involved has been cloned in tissue culture and possesses the θ antigen, but not the B cell receptor for antigen-antibody complexes. Another of the 19 tumour lines, WEHI-22, also cloned in culture and demonstrated to be θ positive, synthesizes IgM in small amounts, but displays it in relatively high concentration on its surface (Harris *et al.*, 1972). As this line also appears to have the B cell characteristic receptor for immune complexes, it cannot at present be considered as unequivocal evidence of T cell IgM synthesis, and this tumour is discussed in more detail elsewhere (Harris *et al.*, 1972).

In 8 of 25 cultures of spleens from agammaglobulinaemic hormally bursectomised chickens (lacking B cells) (Bankhurst *et al.*, 1972), synthesis of an immunoglobulin component that behaves electrophoretically as light chains, and not as intact IgG or IgM, was detected.

CONCLUSIONS AND SUMMARY

B cell immunoglobulin gene expression

Three specific levels of induction of immunoglobulin gene expression in B cells might be proposed, and these are schematically indicated in Fig. 3.

(1) The haemopoietic stem cell, on entering the bursa of Fabricius in chickens and perhaps on encountering a similar "bursal factor" in mammals, differentiates into a B cell with expression of variable region genes of light and heavy types, and of constant region genes of light chain and mu chain types. This cell synthesizes IgM at a relatively low rate, and much of the IgM is held on the surface to act as receptor for antigen. This proposal therefore contemplates the existence of a specific inducer of differentiation, probably of epithelial origin.

(2) Antigen is then the inducer of further B cell differentiation, and following contact of the B cell immunoglobulin with specific antigen (in the absence of any T cell effect at this stage) two events occur. Firstly the rate of synthesis and secretion of the B cell IgM is increased, causing detectable IgM antibody to appear in the serum, and secondly, a low level of expression of the other constant region genes occurs, which leads to the appearance of several immunoglobulin classes on the cell surface.

(3) The final induction of B cell differentiation into plasma cells then requires a third inducer, which is of T cell origin and in some at present unknown

FIGURE 3

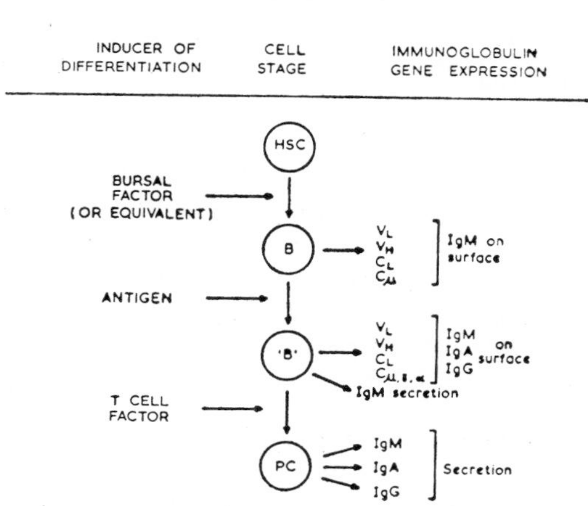

Schematic representation of the proposed pattern of immunoglobulin gene expression during differentation of the B lymphocyte series.

manner, causes B cells to become antibody secreting plasma cells, each of which expresses only one constant region gene type.

This proposal therefore requires an immunoglobulin class switch mechanism, in that the μ chain constant gene is expressed first (in the absence of antigen), and this is followed, after induction by antigen by expression of other C region genes, which in turn become *selectively* expressed with further antigenic stimulation and T cell help. The sequence of selective gene expression could represent either changes occurring within individual cells, or changes in the relative numbers of different cells within the population. For example, instead of an intracellular switch mechanism, the change in immunoglobulin class expression within the cell population may reflect the differential activation of separate cell lines which require T cell help to a greater degree.

T cell antigen receptor

Two central questions are still quite unresolved at present. First, is the T cell receptor for antigen an immunoglobulin? All the experimental evidence derived from our work indicates that T cells synthesize and bear surface immunoglobulin, albeit in very small amounts. These results in a sense fall between those studies (Uhr, Unanue, reported in (McDevitt and Landy, 1972)) which fail to find any T cell immunoglobulin and those that find as much immunoglobulin on T cells as on B cells (Marchalonis *et al.*, 1972, 1973; Hämmerling and Rajewsky, 1972).

Second, if the receptor is an immunoglobulin, is there only one type of immunoglobulin gene expression permissible in T cells, or are there divergent pathways related to the two recognised functional activities of T cells—namely cell-mediated immune activity, and ability to help B cells to differentiate into antibody secreting plasma cells. If there is divergence within T cell differentiation, it may be based on differential expression of different immunoglobulin genes.

REFERENCES

Austen, K. F. and Becker, E. L. (eds.) (1971). *Biochemistry of the Acute Allergic Reactions.* Blackwell Scientific Publications.

Bankhurst, A. D., Rouse, B. T. and Warner, N. L. (1972). *Int. Arch. Allergy Appl. Immunol. 42,* 187

Bankhurst, A. D. and Warner, N. L. (1971). *J. Immunol. 107,* 368

Bankhurst, A. D. and Warner, N. L. (1972). *Austral. J. Exp. Biol. Med. Sci.,* 50, in press

Bankhurst, A. D., Warner, N. L. and Sprent, J. (1971). *J. Exp. Med. 134,* 1005

Basten, A., Miller, J. F. A. P., Sprent, J. and Pye, J. (1972a). *J. Exp. Med. 135,* 610

Basten, A., Miller, J. F. A. P., Warner, N. L. and Pye, J. (1971). *Nature (London) 231,* 104

Basten, A., Warner, N. L. and Mandel, T. (1972b). *J. Exp. Med. 135,* 627

Becker, A. J., McCulloch, E. A. and Till, J. E. (1963). *Nature (London) 197,* 452

Bloch, K. J. (1967). *Prog. Allergy. 10,* 84

Boyse, E. A., Mujazawa, M., Aoki, T. and Old. L. J. (1968). *Proc. Roy. Soc. B. 170,* 175

Cebra, J. J., Colberg, J. E. and Dray, S. (1966). *J. Exp. Med. 123,* 547

Chiller, J. H., Habicht, G. S. and Weigle, W. O. (1970). *Proc. Nat. Acad. Sci., U.S., 65,* 551

Cline, M. J., Sprent, J., Warner, N. L. and Harris, A. W. (1972). *J. Immunol. 108,* 1126

Coombs, R. R. A., Feinstein, A. and Wilson, A. B. (1969). *Lancet 2,* 1157

Cooper, M. D. and Lawton, A. R. (1972) in *Contemporary Topics in Immunobiology* (Hanna, M. G., ed.) vol 1, p. 49, Plenum Press, New York

Cooper, M. D., Lawton, A. R. and Kincade, P. W. (1972) in *Contemporary Topics in Immunobiology* (Hanna, M. G., ed.) vol 1, p. 33, Plenum Press, New York

Cooper, M. D., Peterson, R. D. A., South, M. A. and Good, R. A. (1966). *J. Exp. Med. 123,* 75

Craddock, C. G., Longmire, R. and McMillan, R. (1972). *New Engl. J. Med. 285,* 324, 378

Crewther, P. and Warner, N. L. (1972). *Austral. J. Exp. Biol. Med. Sci. 50,* in press

Crone, M., Koch, C. and Simonsen, M. (1972). *Transpl. Rev. 10,* 36

Dwyer, J., Mason, S., Warner, N. L. and Mackay, I. R. (1971). *Nature (London) 234,* 252

Dwyer, J., Warner, N. L. and Mackay, I. R. (1972). *J. Immunol. 108,* 1439

Edelman, G. M. and Gall, W. E. (1969). *Annu. Rev. Biochem. 38,* 415

Fudenberg, H. H. and Warner, N. L. (1970). *Adv. Human Genetics 1,* 131

Greaves, M. F. and Hogg, N. M. (1972). *Progr. Immunol.* (B. Amos, ed.) vol. 1, p. 111 Acad. Press

Grey, H. M., Hirst, J., Wegman, J. and Cohn, M. (1971). *J. Exp. Med. 133,* 289

Hammerling, U. and Rajewsky, K. (1972). *Europ. J. Immunol.* (1971) *1,* 447

Harris, A. W., Bankhurst, A. D., Mason, S. and Warner, N. L. (1972). In press

Harris, T. N., Hummeler, K. and Harris, S. (1966). *J. Exp. Med. 123,* 161

Herrod, H. and Warner, N. L. (1972). *J. Immunol.,* in press

Herzenberg, L. A. and Warner, N. L. (1968). In *Regulation of the Antibody Response* (B. Cinader, ed.). C. C. Thomas, Springfield

Hilschmann, N. (1967). In *Nobel Symposium 3, Gammaglobulins* (J. Killander, ed.). p. 33, Almquist and Wiksell, Stockholm

Hood, L. and Prahl, J. (1971). *Adv. Immunol. 14,* 291

Kincade, P. W., Lawton, A. R., Bockman, D. E. and Cooper, M. D. (1970). *Proc. Nat. Acad. Sci., U.S., 67,* 1918

Kohler, H., Shimizu, A., Paul, C., Moore, V. and Putnam, F. W. (1970). *Nature (London) 227,* 1318

Lawton, A. R., Asofsky, R., Hylton, M. B. and Cooper, M. D. (1972). *J. Exp. Med. 135,* 277

Lennox, E. S. and Cohn, M. (1967). *Annu. Rev. Biochem. 36,* 365

McDevitt, H. O. and Landy, M. (eds.) (1972). *Genetic Control of Immune Responsiveness.* In press.

Marchalonis, J. J., Cone, R. E. and Atwell, J. L. (1972). *J. Exp. Med. 135,* 956

Marchalonis, J. J., Cone, R. E., Atwell, J. L. and Rolley, R. T. (1973). These proceedings p. 629

Mason, S. and Warner, N. L. (1970). *J. Immunol. 104,* 762

Metcalf, D. and Moore, M. A. S. (1971). *Haemopoietic Cells,* North Holland, Amsterdam

Miller, J. F. A. P. and Osoba, D. (1967). *Physiol. Rev. 47,* 437

Miller, J. F. A. P. and Sprent, J. (1971). *J. Exp. Med. 134,* 66

Miller, J. F. A. P., Sprent, J., Basten, A. and Warner, N. L. (1972). *Nature (London), 237,* 18

Mitchell, G. F., Grumet, C. and McDevitt, H. O. (1972). *J. Exp. Med. 135,* 126

Moore, M. A. S. and Owen, J. J. T. (1966). *Devlop. Biol. 14,* 40

Morris, B. (1971). In *Proc. 2nd Meeting of Asian-Pacific Div. of Intern. Soc. of Hematol.,* p. 169

Nossal, G. J. V., Warner, N. L. and Lewis, H. (1971). *Cellular Immunol. 2,* 41

Nussenzweig, V., Bianco, C., Dukor, P. and Eden, A. (1971). *Progr. Immuno.* (B. Amos, ed.) vol. 1, p. 73 Acad. Press

Pernis, B., Chiappino, G., Kelus, A. S. and Gell, P. G. H. (1965). *J. Exp. Med. 122,* 853

Pernis, B., Forni, L. and Amante, L. (1971). *Ann. N.Y. Acad. Sci. 190,* 420

Pincus, S., Bianco, C. and Nussenzweig, V. (1972). *Fed. Proc. 31,* 775

Potter, M. and Lieberman, R. (1967). *Adv. Immunol. 7,* 91

Rabellino, E., Colon, S., Grey, H. M. and Unanue, E. R. (1971). *J. Exp. Med. 133,* 156

Raff, M. C. (1970). *Immunology 19,* 637

Raff, M. C. (1971). *Transpl. Rev. 6,* 52

Reif, A. E. and Allen, J. M. V. (1964). *J. Exp. Med. 120,* 413

Rouse, B. T. and Warner, N. L. (1972). *Cellular Immunol. 3,* 470

Schulenberg, E. P., Simms, E. S., Lynch, R., Bradshaw, R. A. and Eisen, H. N. (1971). *Proc. Nat. Acad. Sci., U.S., 68,* 2623

Shortman, K. (1972). *Ann. Rev. Biophys. and Bioeng. 1.* In press.

Sprent, J. and Miller, J. F. A. P. (1972). *Cellular Immunol. 3,* 385

Takahashi, T., Old, L. J. and Boyse, E. A. (1970). *J. Exp. Med. 131,* 1325

Takahashi, T., Old, L. J., Hsu, C-J. and Boyse, E. A. (1972). *Europ. J. Immunol.,* in press

Thorbecke, G. J., Warner, N. L., Hochwald, G. M. and Ohanian, S. H. (1968). *Immunology 15,* 123

Vaz, N. M. and Prouvost-Danon, A. (1969). *Progr. Allergy. 13,* 111

Wang, A. C., Fudenberg, H. H. and Pink, J. R. L. (1971). *Proc. Nat. Acad. Sci., U.S., 68,* 1143

Wang, A. C., Wilson, S. K., Hopper, J. E., Fudenberg, H. H. and Nisonoff, A. (1970). *Proc. Nat. Acad. Sci., U.S., 66,* 337

Warner, N L. (1967). *Folia Biol. 13,* 1

Warner, N. L. (1971). *Transpl. Proc. 3,* 848

Warner, N. L. (1972). In *Immunogenicity* (F. Borek, ed). *Frontiers in Biology* (North Holland) *25,* 467

Warner, N. L., Byrt, P. and Ada, G. (1970). *Nature* (*London*) *226,* 942

Structure and Function of Lymphocyte Surface Immunoglobulin

JOHN J. MARCHALONIS, ROBERT E. CONE, JOHN L. ATWELL AND RONALD T. ROLLEY

Laboratory of Molecular Immunology, The Walter & Eliza Hall Institute of Medical Research, Royal Melbourne Hospital, Parkville, Victoria, Australia

The nature and chemical properties of the lymphocyte receptor for antigen have puzzled immunologists for over 70 years. On the basis of the exquisite binding specificity which antibodies showed for antigen, Ehrlich proposed in 1900 that the cell surface receptor for antigen must be antibody. The bulk of present evidence supporting this hypothesis has been obtained in an indirect manner by using antisera directed against immunoglobulins to inhibit binding of antigen by lymphocytes (Warner *et al.*, 1970; Davie *et al.*, 1971; Greaves and Hogg, 1971) and a variety of immune phenomena (Lesley and Dutton, 1970; Feldmann and Diener, 1971; Mason and Warner, 1970; Basten *et al.*, 1971; Riethmuller *et al.*, 1971). Fig. 1 shows a lymphocyte from an unimmunized guinea pig which has

FIGURE 1

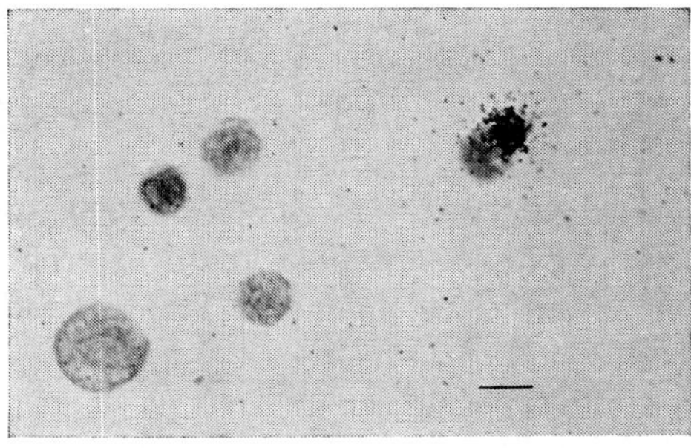

Autoradiograph of guinea pig lymph node lymphocytes incubated with ^{125}I-labelled-dinitrophenylated guinea pig haemoglobin. The cell in the upper right corner is an antigen binding cell which has antigen limited to one relatively small area of the periphery. The bar in the lower right hand corner represents 10 microns.

bound radioactively labelled dinitrophenylated haemoglobin. Such binding is specific for the dinitrophenol hapten (DNP) inasmuch as it can be inhibited by excess amounts of unlabelled DNP-proteins. Moreover, the presence of such antigen binding cells in a population of lymphocytes is abrogated by rabbit antiserum directed against the light chain of guinea pig immunoglobulins (Rolley and Marchalonis, 1972a). This evidence is consistent with Ehrlich's proposal but is not free from objections in the interpretation. If immunoglobulin is not the receptor for antigen but is located close to it on the lymphocyte surface, antibodies to immunoglobulin might sterically hinder the approach of antigen to the receptor. Similar objections can be marshalled against other studies using antiglobulin reagents to inhibit antigen binding by lymphocytes.

We approached the lymphocyte receptor problem in a direct biochemical fashion by isolating cell surface immunoglobulin from normal lymphocyte populations and from populations of specifically activated lymphocytes. We then determined the capacity of these isolated molecules to combine with antigens. Operationally, we proposed three criteria to establish that cell surface immunoglobulin constituted the lymphocyte receptor for antigen as follows: (1) If immunoglobulin is indeed the receptor, it should be possible to isolate immunoglobulin from the surfaces of all lymphocytes which can bind antigens specifically. (2) Immunoglobulin molecules isolated from lymphocyte populations characterized by a known binding specificity should possess the same specificity. (3) Combination of antigen with this immunoglobulin at the cell surface should initiate biochemical events reflecting activation and differentiation of the cells. Our studies were designed to provide information on these issues and on the structural properties of lymphocyte surface immunoglobulin.

ANALYSIS OF CELL SURFACE PROTEINS

Since the cell membrane constitutes only 2-5% of the cell mass (Dowben, 1969) and a given receptor molecule would probably comprise less than 5% of the membrane protein or approximately 10^{-13}g/cell, extremely sensitive techniques were required to enable the identification and isolation of the lymphocyte surface antigen receptor. We employed the enzyme lactoperoxidase to catalyze the covalent binding of radioactive iodide to accessible tyrosines on the surfaces of living lymphocytes (Marchalonis, *et al.*, 1971). Lactoperoxidase-catalyzed iodination was originally developed in this laboratory as a gentle method for the trace iodination of serum proteins (Marchalonis, 1969). Various modifications have subsequently been applied to radioiodinate external proteins of erythrocytes (Phillips and Morrison, 1971), viruses (Stanley and Haslam, 1971) and lymphocytes (Baur, *et al.*, 1971; Marchalonis *et al.*, 1971). As shown in the electron microscopic autoradiograph presented in Fig. 2, it was possible to label the surface proteins of living lymphocytes with [^{125}I] iodide (Marchalonis *et al.*, 1971; Baur *et al.*, 1971). The silver grains are located only along the surface of the cell. Sufficient radioactivity was incorporated into surface proteins to enable the isolation and partial characterization of these molecules. Cell surface protein

FIGURE 2

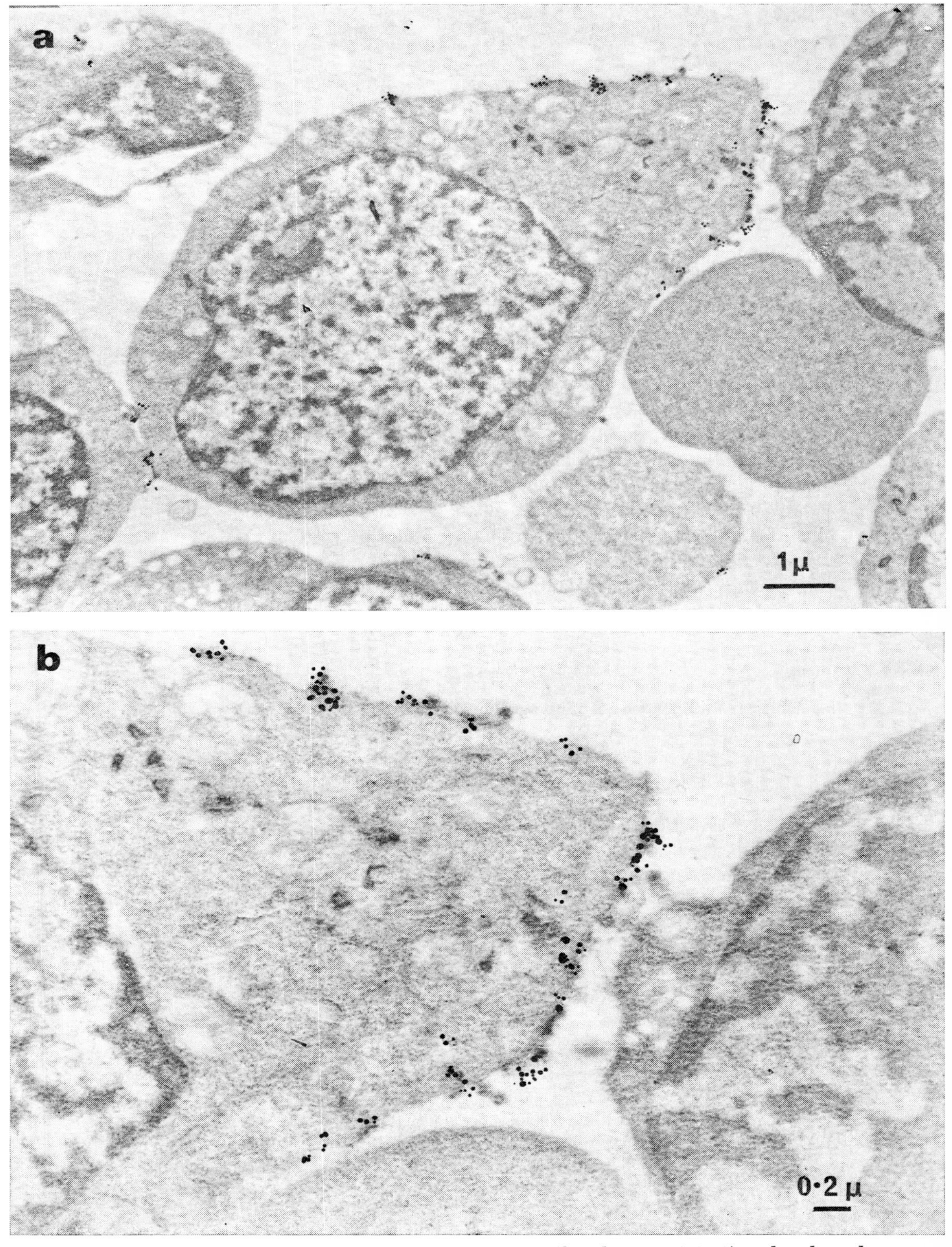

Electron-microscopic autoradiography of lymphocytes (lymphoma SIAT. 4) radioiodinated with [^{125}I] iodide by the lactoperoxidase method. The sections are unstained to maximize contrast between the grains and the cells. The silver grains are located along the periphery of the cell (from Marchalonis *et al,* 1971.)

was obtained in solution by two techniques; (1) the cells were dissolved in dissociating solutions such as 9M urea, 10% acetic acid and the soluble material dialyzed against physiological buffers (Marchalonis *et al.*, 1971), or, alternatively, (2) the cells were incubated under tissue-culture conditions and released surface proteins into the medium (Cone *et al.*, 1971; Marchalonis *et al.*, 1972a). Lymphocytes actively release and replace cell surface proteins including immunoglobulins (Cone *et al.*, 1971; Marchalonis *et al.*, 1972a; Marchalonis *et al.*, 1972b; Lerner *et al.*, 1972; Vitetta and Uhr, 1972). All proteins were not released at the same rate but a general inverse correlation between molecular weight and rate of release was observed for ^{125}I-labelled surface proteins (Cone, *et al.*, 1971). This observation was consistent with studies of turnover of membrane components using other techniques (Kiehn and Holland, 1970). We will consider immunoglobulin turnover in greater detail below.

It was necessary to obtain the cell surface protein in soluble form because the major purification procedure entailed specific immunological co-precipitation (Sherr and Uhr, 1970; Marchalonis *et al.*, 1972a, b). Conditions were employed which brought about precipitation of 80-90% of unlabelled mouse immunoglobulin by rabbit antiserum (Marchalonis *et al.*, 1972a, b). Aliquots of radio-iodinated cell surface protein were added to this precipitating system. Control experiments to verify the specificity of the reaction were (1) use of normal rabbit serum or antiserum to heterologous antigens, (2) a control to test the tendency of radio-iodinated material to adhere nonspecifically to antigen antibody complexes which

TABLE 1

Immunologic Precipitation of Surface Immunoglobulin Extracted from Lymphoid Cells

Lymphocyte source	Specific Precipitate (% of high molecular weight ^{125}I-labelled protein)	Ratio Specific Precipitate to Control
Thymus (CBA × C57Bl)F_1	4.1	4.2
Activated thymus cells (CBA)	4.5	5.3
Spleen (CBA × C57Bl)F_1	4.8	3.0
Spleen (nu/nu)	4.3	3.1
Thoracic duct (nu/nu)	5.9	5.3
WEHI 22 (Thymoma)	14.9	15.1
WEHI 105 (Thymoma)	0.3	1.1
SIAT 4 (Lymphoma)	0.5	1.3
Test globulin	*% Precipitated*	
Mouse γG (^{125}I-labelled)	82.3	90

Under conditions used here, the amount of radioactivity passively adsorbed to indifferent antigen-antibody complexes ranged from 0.9 to 1.6%. Data are expressed as specific precipitates, i.e. the amounts precipitated by indifferent precipitating systems were subtracted from the experimental values. The rabbit anti-serum used in these experiments possessed binding activity for light chains and μ chains of normal mouse immunoglobulin (based upon Marchalonis *et al*, 1972b).

consisted of an indifferent precipitate such as *Limulus* haemocyanin plus rabbit antiserum to this antigen, and (3) "blocking" experiments which showed that precipitation of ^{125}I-labelled surface immunoglobulin was inhibited by addition of excess carrier to bring the reaction into the zone of antigen excess. The relative amounts of immunoglobulin present in dialyzed or high-molecular weight fractions of surface proteins from a variety of lymphocytes are shown in Table 1. The results given are the means of triplicate analyses. In all thymus and spleen lymphocyte preparations, approximately 5% of the high molecular weight protein was specifically precipitated. Although thymoma WEHI 22 possessed substantial quantities of surface immunoglobulin, the thymoma WEHI 105 and the lymphoma SIAT.4 lacked detectable immunoglobulin. The tumors included here possess the θ antigenic determinant (Harris, Bankhurst, Mason and Warner, unpublished) which has been taken as a marker for thymus-derived lymphocytes (Raff, 1971).

It was necessary to study thymus and other lymphoid populations because recent evidence has established that lymphocytes can be divided into two categories; namely, thymus-derived (T-cells) and bone-marrow-derived lymphocytes (B-cells). The former cells function in cellular immunity such as graft rejection, whereas the latter type are active in production of circulating antibodies (Miller, 1972; W.H.O. Technical Bulletin, 1969). Moreover, antibody formation to certain antigens requires collaboration between T-cells and B-lymphocytes (Miller and Mitchell, 1969). Both types of cells possess specific receptors for antigen (Basten *et al.*, 1971; Greaves and Hogg, 1971; Mitchison *et al.*, 1970).

These data establish that detectable amounts of immunoglobulin could be isolated from lymphocyte surfaces but did not give information on the immunoglobulin class because the antisera used reacted against light chains and, consequently, precipitated mouse immunoglobulins of all classes. Different immunoglobulin classes are defined by the presence of class-distinctive heavy chains; light chains are common to all classes (Edelman and Benacerraf, 1962; Edelman and Gall, 1969). In order to ascertain if murine lymphocyte surface immunoglobulin contained both light chains and heavy chains, the specifically co-precipitated protein was dissolved in 9M urea and reduced and alkylated to cleave interchain disulphide bonds and prevent reformation of these covalent links (Edelman and Marchalonis, 1967). The reduced alkylated protein was analyzed by polyacrylamide gel electrophoresis (PAGE) in acid urea under conditions which resolved light chains and heavy chains (Parish and Marchalonis, 1970). This approach, furthermore, enabled a tentative identification of the heavy chain types because γ, α and μ chains are resolved by this technique (Marchalonis and Schonfeld, 1970). Fig. 3 presents the fractionation patterns obtained by PAGE of the polypeptide chains of surface immunoglobulins obtained from thymus lymphocytes of CBA mice. Surface immunoglobulin from the thymus contains components which resemble light chain and μ chains in gel penetration; no significant components with a mobility comparable to γ chain were present. Components similar to light chains, γ chains and μ chains were resolved in the surface immunoglobulin isolated from splenic lymphocytes of (CBA x C57Bl)F_1 mice

FIGURE 3

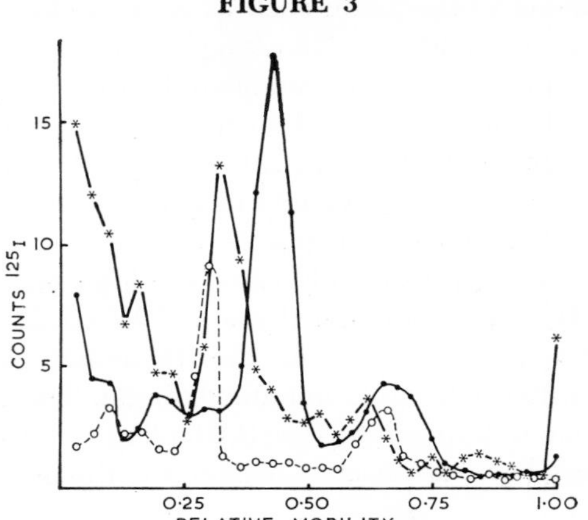

Comparison by disc electrophoresis in acid urea of ^{125}I-labelled polypeptide chains of surface immunoglobulin from mouse thymus lymphocytes with polypeptide chains of purified γM and γG immunoglobulins. Thymus cells from CBA/H/WEHI mice were activated *in vitro* to BALB/c histocompatibility antigens. ⋆——⋆, surface immunoglobulin of activated thymus cells (c.p.s.); •——•, polypeptide chains of murine γG immunoglobulin (c.p.m. $\times$ 10^{-2}); o — o, polypeptide chains of human γM myeloma immunoglobulin (c.p.s. $\times$ 10^{-1}). Relative mobilities of standard chains: light, 0.55-0.75; γ, 0.40-0.48; μ, 0.27-0.33. All samples were specific co-precipitates which were reduced and alkylated in the presence of 9M-urea. (From, Marchalonis *et al*, 1972b).

(Fig. 4). The number of counts in the μ chain peak was at least twice that found in the γ region. The thymus cells used in this experiment were activated speci-

FIGURE 4

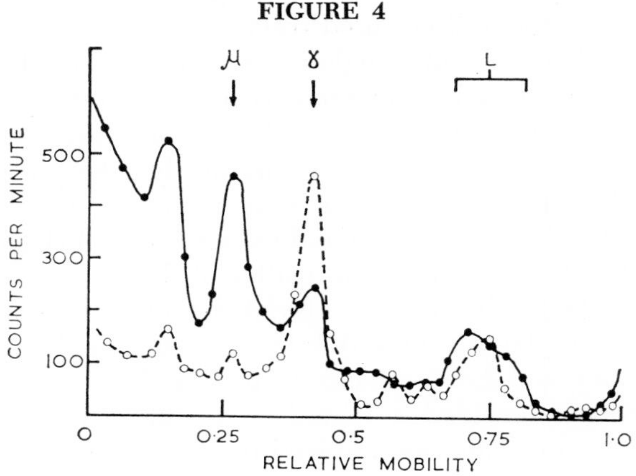

Analysis by disc electrophoresis in acid urea of polypeptide chains of immunoglobulin isolated from spleen lymphocytes of (CBA $\times$ C57Bl)F_1 mice. •——•, surface iodinated lymphocytes were incubated in tissue culture medium at 37°C for 3h. The tissue culture fluid containing released surface immunoglobulin was specifically co-precipitated and the precipitate reduced and alkylated in the presence of 9M urea prior to electrophoresis. o — o, polypeptide chains of immunoglobulin secreted by the spleen lymphocytes. Aliquots of tissue culture fluid were radioiodinated and co-precipitated specifically. The precipitates were reduced and alkylated in the presence of 9M urea. The positions of co-precipitated standard immunoglobulin polypeptide chains are indicated as follows: μ, μ chain; γ, γ chain and L, light chain.

fically to histocompatibility antigens (see Table 2). Similar studies were carried out using normal thymus cells of (CBA x C57Bl)F_1 mice and thymus-derived cells activated to erythrocyte (Cone and Marchalonis, unpublished) and protein antigens (Feldmann, Cone and Marchalonis, unpublished). These cells, likewise, possessed only light chains and μ-type heavy chains. We also studied splenic and thoracic duct lymphocytes from congenitally athymic nu/nu mice observing that these bone-marrow derived, or B-lymphocytes, possess chiefly γM-type immunoglobulin on their surfaces (Marchalonis *et al.*, 1972b). Baur *et al.* (1971) and Vitetta and her colleagues (1971) observed only γM immunoglobulin on the surfaces of splenic lymphocytes of BALB/c mice. Despite the fact that electron micrographic autoradiography showed that only the surface proteins of lymphocytes were radioactively labelled, the possibility existed that the detected immunoglobulin might be a secretion product released by a small number in the population of the cells actively secreting antibodies. To ascertain the nature of the secreted antibody, we iodinated aliquots of the serum-free medium in which the cells were incubated. As shown in Fig. 4, the immunoglobulin present in the supernatant fluid was predominantly γG in contrast to that isolated from the surface which was primarily γM.

The preceding polyacrylamide gel electrophoretic analysis showed that lymphocytes of both bone-marrow and thymus origin possessed immunoglobulins consisting of light chains and heavy chains on their surfaces but provided no information regarding the size of the intact protein. This problem was approached by dissolving specific co-precipitates in acid-urea or detergents and fractionating the unreduced protein by PAGE or by gel filtration on Sepharose 6B. Fig. 5 illustrates the polyacrylamide gel electrophorogram of surface immunoglobulin from activated CBA thymus cells. The surface immunoglobulin migrated as a

FIGURE 5

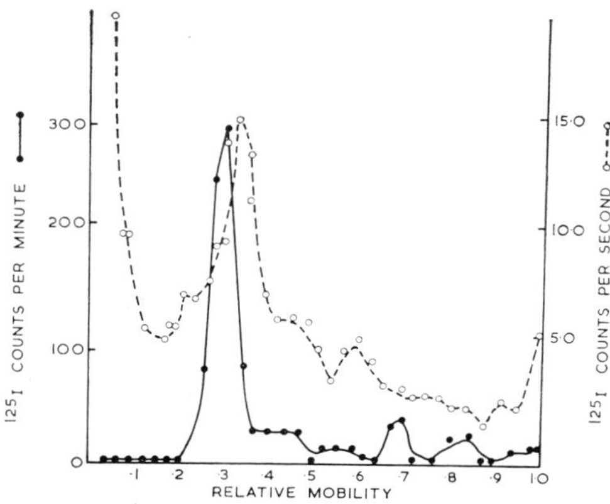

Analysis of intact lymphocyte surface immunoglobulin by disc electrophoresis in acid urea. Surface immunoglobulin was isolated from thymus-derived lymphocytes activated to keyhole limpet haemocyanin. ●——●, thymus lymphocyte immunoglobulin; O — O, mouse γG immunoglobulin standard. Both samples were specific co-precipitates dissolved in urea-acetic acid.

single component which is slightly retarded relative to the co-precipitated mouse γG standard (mol. wt. 150,000). This result indicated that the cell surface γM existed as a molecule of molecular weight approximately 180,000 rather than as the macroglobulin from (mol. wt. 900,000) usually found in serum. The high molecular weight form is a pentamer of the 180,000 unit (Miller and Metzger, 1965); it is too large to penetrate the acrylamide gel used in this experiment. The low molecular weight nature of the cell surface IgM was confirmed for both thymus-derived and bone-marrow derived lymphocytes by gel filtration on Sepharose 6B which provides optimum resolution between the IgM pentamer and γG immunoglobulin (Marchalonis and Schonfeld, 1970). Our results were consistent with the findings of Baur *et al.* (1971) and Vitetta *et al.* (1971) that splenic lymphocytes of BALB/c mice possessed 7SγM immunoglobulin on their surfaces. Moreover, Klein and Eskeland (1971) isolated such low molecular weight γM immunoglobulin from lymphoid cells obtained from patients suffering from chronic lymphocytic leukaemia and 'Burkitt' lymphoma cells grown in tissue culture.

The above results were obtained with immunoglobulin prepared by solubilizing the radioiodinated lymphocytes in urea-acetic acid. We found that such treatment often denatured surface immunoglobulin and rendered it unsuitable for use in studies of its binding specificity. The fact that living cells release their surface components by a process of turnover proved quite useful as a gentle means of isolating surface immunoglobulin. Fig. 6 illustrates the kinetics of release of surface immunoglobulin from a variety of lymphocyte populations. Spleen lymphocytes of nu/nu mice exhibit the fastest rate of release and activated

FIGURE 6

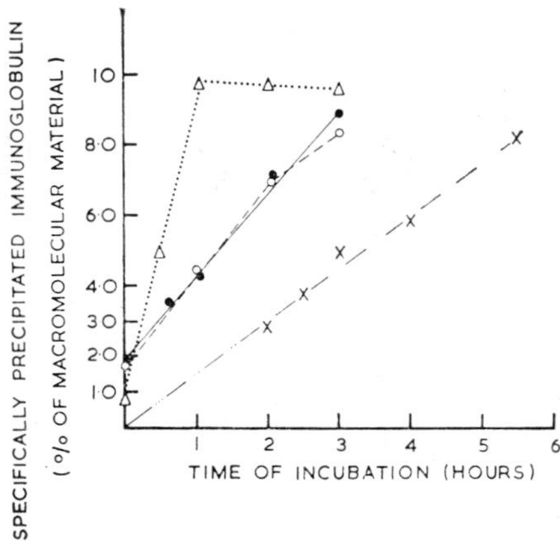

Kinetics of release of lymphocyte surface immunoglobulin. ^{125}I-labelled lymphocytes were incubated at 37°C in tissue culture medium and aliquots were removed at various times for determination of immunoglobulin content by specific co-precipitation. Δ-----Δ, surface immunoglobulin of spleen lymphocytes of nu/nu mice (B-cells); ●——●, spleen lymphocytes of (CBA × C57Bl) F_1 mice; ○—○, thymus lymphocytes of (CBA × C57Bl) F_1 mice, ×——-——×, surface immunoglobulin of thymus-derived cells of CBA mice which were activated against BALB/c histocompatibility antigens.

thymus cells are the slowest. Thymus and spleen lymphocytes of (CBA x C57Bl)F_1 mice show similar turnover kinetics which are intermediate in behaviour. The curve obtained for nu/nu spleens represents that of a pure B-lymphocyte population (DeSousa *et al.*, 1969; Sprent, personal communication). The relatively low immunoglobulin release rate of activated thymus cells probably reflects the fact that these cells are actively dividing because data obtained for L cells showed that actively dividing cells have a decreased membrane turnover rate relative to resting cells (Warren and Glick, 1968). Our rates obtained by the lactoperoxidase catalysed iodination technique compared favourably with turnover rates calculated by other techniques (Lerner *et al.*, 1972; Wilson *et al.*, 1972). This process of turnover of surface immunoglobulin occurs with a rate at least 100 fold lower than that of the release of immunoglobulins by plasma cells and may be considered a "resting state" level. In fact, Lerner *et al.* (1972) believe that immunoglobulin synthesis and turnover of surface immunoglobulin are two completely distinct processes.

Despite the number of observations that a variety of T-cell mediated immune functions can be abrogated by antisera to immunoglobulins (Lesley *et al.*, 1971; Mason and Warner, 1970, Greaves and Hogg, 1971; Basten *et al.*, 1971), it has been extremely difficult to demonstrate immunoglobulin on the surfaces of these cells by use of fluorescent (Rabellino *et al.*, 1971) or radioiodinated antiglobulins (Bankhurst and Warner, 1971; Nossal *et al.*, 1972). Prolonged autoradiographic exposures employing antisera specific for light chains do, however, indicate the presence of surface immunoglobulins on T-lymphocytes (Bankhurst and Warner, 1971; Nossal *et al.*, 1972). These approaches readily detect immunoglobulins on the surfaces of bone-marrow derived lymphocytes. Our data provide evidence that approximately equal amounts of immunoglobulin can be isolated from the surfaces of T-cells and B-cells, indicating that it is unlikely that the presence

FIGURE 7

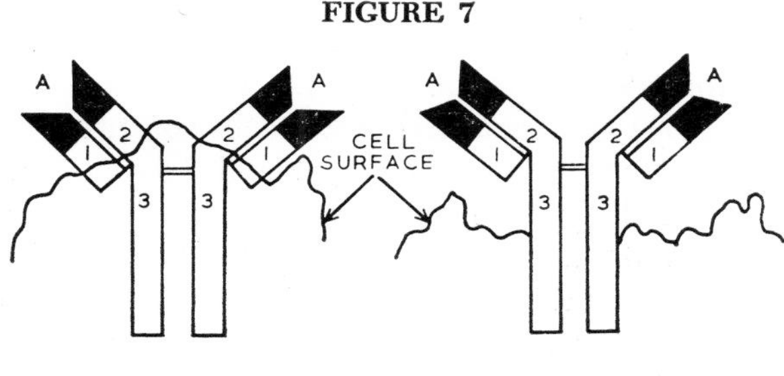

Hypothetical model for the presentation of immunoglobulin upon the lymphocyte surface. The diagram on the left illustrates a situation where the F_c portion of the μ-chain is surrounded by surface components. This condition may obtain for unactivated thymus derived lymphocytes. The diagram on the right presents a situation in which all of the Fab piece and most of the F_c region of the heavy chain are exposed. This situation probably occurs with activated thymus cells and bone-marrow derived lymphocytes. The dark regions represent the variable regions of the light and heavy polypeptide chains which share the combining site for antigen. 1, constant region of light chain; 2, F_d piece; 3, F_c region of heavy chain; A, combining site for antigen (modified from Marchalonis *et al.*, 1972b.)

of T-cell immunoglobulin results from contamination with small numbers of B-lymphocytes. As we will present below, functional data also support this conclusion. Fig. 7 illustrates a tentative model for the presentation of immunoglobulin on the surfaces of T-cells and B-cells (Marchalonis *et al.*, 1972b). Since most of the μ chain is obstructed by cell surface components other than immunoglobulin (left diagram), antiglobulins directed against μ chain antigenic determinants (F_c) could not bind, but antisera to light chains might bind with some difficulty. This situation may occur on unstimulated thymus cells. In contrast, a large portion of the F_c determinant may be exposed on B-lymphocytes, making this determinant as well as the light chain readily accessible to antiglobulin molecules. The problem of steric hindrance in binding of antiglobulins may be acute because these molecules are large (mol. wt. 150,000) and must find a small portion of the target molecule in order to combine specifically. In lactoperoxidase catalyzed iodination at neutral pH, radioiodination of accessible tyrosines can be performed by short-lived iodine free radicals (mol. wt. 125) or molecular iodine (mol. wt. 250) formed by the action of the enzyme on iodide (Bayse and Morrison, 1971). Therefore, more of the surface immunoglobulin molecule might be accessible to this reagent than to antiglobulins. The relative proportions of heavy chains and light chains labelled with [^{125}I] iodide give general support to the proposed model because the ratio of μ chain to light chain is 1/1 in normal thymus cell populations and approximately 2/1 for B-lymphocytes and activated T cells (Marchalonis *et al.*, 1972b).

BINDING SPECIFICITY OF ISOLATED LYMPHOCYTE SURFACE IMMUNOGLOBULIN

Although immunoglobulin could be isolated from the surfaces of both thymus-derived and bone marrow derived lymphocytes, proof that such molecules serve as receptors for antigen requires that immunoglobulin isolated from cells of known binding specificity, should exhibit the same specificity. Three problems arise which make this demonstration difficult. Firstly, the immune response is selective (Burnet, 1959) and specific receptor cells constitute only a small fraction of the total lymphocyte population (Byrt and Ada, 1970). This condition requires a means of enriching the relative fraction of specifically reactive cells. In the second place, antibody released by a small number of contaminating plasma cells, might be detected instead of surface immunoglobulin. Moreover, the detection by affinity labelling of tyrosines in the active sites of many antibodies (Singer and Doolittle, 1966) raises the possibility that radioiodination of cell surface immunoglobulin may result in destruction of the combining site for antigen. We attempted to circumvent the first two difficulties by working chiefly with populations of activated T-lymphocytes since such preparations do not secrete antibodies in the classical sense and can be obtained virtually free of B-lymphocytes (Sprent and Miller, 1971), which are precursors of antibody-forming cells. Furthermore, the percentage of cells within an activated population which are specifically reactive to histocompatibility antigens may range from 20 to 80 percent of total lymphocytes (Cheers and Sprent, 1972). The question of blocking antigen combining sites by insertion of radioactive iodide could be assessed only in retrospect. However, use of binding assays which required only the presence of one combining site facilitated our experimental approach to the elucidation of the binding specificity of isolated cell surface immunoglobulin since molecules having only one functional antigen binding site could be detected.

Three types of T-cell activation systems were used. Since the results obtained were substantially similar, only the conceptually most straight-forward one will be described in detail here. Thymus cells from CBA/H/WEHI mice were specifically activated to react to histocompatibility antigens of the BALB/c strain as follows: (CBA x BALB/c)F_1 hybrids were heavily irradiated (750 rds) to suppress their immune capacity. They were then injected with thymus lymphocytes of the CBA parental strain which reacted to the BALB/c antigens by giving a graft versus host reaction (Sprent and Miller, 1971). After four days, thoracic duct lymphocytes were collected by cannulation. This cell population consisted pri-

TABLE 2

Specificity of T. TDL ^{125}I-Cell Surface Proteins for Activating Antigens

Cell surface proteins	Treatment	Radioactivity (cpm) binding to Thymus cells		
		C57Bl	BALB/c	CBA
BALB/c-activated CBA T. TDL	none	7900 ± 2500 p < .05	17180 ± 4700 p < .01	3730 ± 550
BALB/c-activated CBA T. TDL	co-precipitated with anti-immunoglobulin antiserum	1300 ± 300 N.S.	1800 ± 400 N.S.	1400 ± 200 N.S.
BALB/c-activated CBA T. TDL	co-precipitated with normal rabbit serum	n. d.	12200 ± 2400 p < .01	2400 ± 800
C57Bl-activated CBA T. TDL	none	1544 ± 252 p < .001	630 ± 30 p < .001	416 ± 35
C57Bl-activated CBA T. TDL	co-precipitated with anti-immunoglobulin serum	46 ± 30 N.S.	107 ± 97 N.S.	33 ± 20
C57Bl-activated CBA T. TDL	co-precipitated with NRS	900 ± 100 p <.001	240 ± 10 N.S.	180 ± 40

p-values given here represent comparisons of counts bound to the cells indicated with those bound to the CBA control. The Student's t-test modified for small sample size was used in the calculations. For BALB/c activated T. TDL, the p-value for the comparison BALB/c vs. C57Bl was < .05. For C57Bl activated T. TDL, the p-value for the comparison C57Bl vs. BALB/c was < .001.

marily (99%) of T-lymphocytes (Sprent, personal communication). Cheers and Sprent (1972) estimate that at least 20% of these lymphocytes responded specifically to BALB/c antigens. CBA T-cells were also activated to C57Bl strain histocompatibility antigens in a similar fashion. These procedures were developed by Sprent (Sprent and Miller, 1971) who collaborated with us in this venture (Cone *et al.*, 1972).

CBA cells activated against BALB/c antigens contained low molecular weight γM-type immunoglobulin on their surfaces. This was the only class detectable. Because we desired this protein to retain binding capacity for antigen, cell surface protein of these activated cells was collected by metabolic release into tissue culture medium. As shown in Table 2, a significant amount of ^{125}I-labelled cell surface protein bound to BALB/c cells and a lesser quantity bound to C57Bl cells. This cross-reactivity proved to be important because BALB/c and C57Bl share certain histocompatibility antigens (Snell and Stimpfling, 1966) and this result correlates well with specificity shown by the cells themselves (Sprent and Miller, 1971) and known fraction of shared antigens (Snell and Stimpfling, 1966). Furthermore, this specific binding was completely abolished by removing immunoglobulin in our standard co-precipitation system. Similar treatment using normal rabbit serum had no effect on binding of ^{125}I-labelled CBA T. TDL surface protein to BALB/c cells. Exactly parallel results were obtained when CBA T-cells were "educated" to C57Bl cells.

The two other T-cell systems we used were cells specific for sheep erythrocyte antigens (Cone and Marchalonis, unpublished) and for the protein keyhole limpet haemocyanin (Feldmann, Cone and Marchalonis, unpublished). In both cases, the activated T-cells served as helper cells in collaboration with B cells. Antibody formation to these antigens in mice requires interaction between T-cells, termed helper cells, and B-cells which are the actual synthesizers of antibody (Miller and Mitchell, 1969). Since both classes of cells possess specific receptors for antigen (Miller and Mitchell, 1969; Mitchison *et al.*, 1970; Greaves and Hogg, 1971; Basten *et al.*, 1971), it was possible for us to study binding activity of surface immunoglobulin. Moreover, we were able to investigate the mechanism of collaboration between T-cells and B-cells *in vitro* (Feldmann, Cone and Marchalonis, unpublished). All the specifically-activated populations of T-lymphocytes which we studied possessed γM-type surface immunoglobulin characterized by a molecular weight of approximately 180,000 (Marchalonis *et al.*, 1972b; Cone and Marchalonis, unpublished; Cone *et al.*, 1972; Feldmann, Cone and Marchalonis, unpublished). Binding specificity of this immunoglobulin for the respective antigens was demonstrated as above. Fig. 8 presents the scheme for collaboration between T-cells and B-cells hypothesized by Feldmann (1972a) on the basis of tissue culture studies. Macrophages are an obligatory part of this system because these provide the substrate upon which the collaborative T-cell factor is collected and displayed. We have shown in collaboration with Feldmann that this factor which is elaborated by activated T-cells is low molecular weight γM immunoglobulin (Feldmann, Cone and Marchalonis, unpublished). Direct isolation and analysis of polypeptide chains by acrylamide gel electrophoresis was consistent with results indicating that collaboration between T-cells and B-cells was inhibited by rabbit antiserum to mouse K-type light chains and μ heavy chains (Feldmann, 1972b). Further work (Cone, R.E., Marchalonis, J.J. and Nossal, G.J.V., unpublished observations) suggest that γM surface immunoglobulin of B-cells does not bind to macrophages.

FIGURE 8

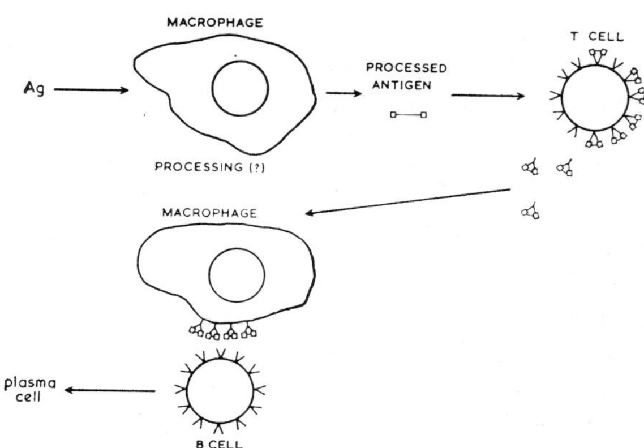

Theoretical scheme for the collaboration of thymus-derived lymphocytes (T-cells) and bone marrow-derived lymphocytes (B-cells). Macrophages play two roles in this scheme. One is to process antigen. The second function is to concentrate the collaborative factor elaborated by T-cells and present it, bound to antigen, to the unstimulated B-cells. (After Feldmann, 1972a).

We investigated the capacity of B-cell surface immunoglobulin to function as the receptor for antigen by studying lymphocytes of unimmunized animals which specifically bound dinitrophenylated haemoglobin (Rolley and Marchalonis, 1972b). Since approximately 1% of lymphocytes from mouse lymph nodes reacted specifically to DNP-haemoglobin (murine) sufficient [^{125}I] iodide counts could be affixed to lymphocytes to ascertain binding activity of cell surface components released by metabolic turnover and limited proteolysis with trypsin. In double label experiments utilizing [^{125}I] labelled cell surface proteins, we found that (1) use of antiglobulin reagents brought down antigen, thereby suggesting that antigen released from the cell surface was combined with antibody, and (2) precipitation of antigen by specific antiserum brought down cell surface immunoglobulin, confirming that this protein was in combination with the antigen. Furthermore, the ability of released cell surface protein to bind labelled antigen was abrogated by co-precipitation with antiglobulin reagents, but not by normal rabbit serum or by indifferent precipitating systems. Another interesting point emerged from this study is relevant to recent reports, that antiglobulin reagents bound to lymphocyte surfaces are collected to one portion of the cell surface by a metabolic process and endocytosed (Taylor *et al.*, 1971). This may not be the major means of lymphocyte handling of antigen because we were able to recover 85-90% of ^{125}I-labelled antigen associated with the lymphocyte surface by a procedure consisting of incubation in tissue culture medium for 1 h followed by gentle treatment of the cells with trypsin.

BIOCHEMICAL MECHANISMS OF LYMPHOCYTE ACTIVATION BY ANTIGEN

We have presented data showing that lymphocytes possess surface immunoglobulin which exhibits binding specificity for antigens used in challenging the cells. The manner in which immune differentiation is triggered by contact of such cell surface molecules with antigen remains to be determined. Reasoning from systems studied by endocrinologists, it is possible to suggest two general models. Antigen may act in a fashion analogous to either a steroid hormone[a] such as progesterone (Spelsberg *et al.*, 1971) or to a peptide hormone such as glucagon (Langan, 1971). In this example of steroid hormones, the progesterone itself enters the cell, combines with a binding protein in the cytoplasm, and finally enters the nucleus (O'Malley *et al.*, 1972), where it may function as a derepressor in activating certain genes. In the latter case, combination of glucagon with the liver cell surface receptor activates the enzyme adenyl cyclase which forms cyclic adenosine monophosphate which then functions as an intermediary messenger activating histone kinases (Langan, 1970). The phosphorylated histones presumably are altered in their function as repressors of the genome and certain genes are activated thereby allowing differentiation to occur.

Although direct evidence in support of these alternative mechanisms is lacking, a variety of indirect arguments can be brought to bear on this issue. The steroid type model requires that antigens enter the cell and act at the level of the nucleus. Studies by Nossal *et al.*, (1965) using sensitive autoradiographic methods have shown that the vast majority of antibody forming cells contain fewer than 4 molecules of antigen. In contrast, Taylor *et al.* (1971) have shown that antiglobulin antibodies bound to the lymphocyte surface tend to gather over one pole of the cell and are subsequently internalized by a process of pinocytosis. These workers have suggested that such internalization may play a major role in the triggering of lymphocyte differentiation. However, Greaves and Bauminger (1972) have reported that mitogens coupled to Sepharose beads retain their capacity to induce lymphocyte division although they cannot be taken into the cells. The equivalence of the processes of lymphocyte activation by mitogens and by antigens remains to be established. A further point which militates against the function of antigen itself as an internal derepressor is the theoretical argument that no known mechanism of protein synthesis can account for the translation of three dimensional information from a specific carbohydrate or substituted benzene ring (e.g. DNP) into the linear language of DNA. Thus, the simpler hypothesis maintains that the combination of antigen with immunoglobulin receptor at

---

[a]See Sekeris C. E. and Schmid and Baxter, J. D., *et al.*, (This symposium) for detailed discussions of steroid functions in cell regulation.

the cell surface brings about the release of an intermediary messenger with the eventual results that genes for immunoglobulin synthesis are activated within the clonally restricted lymphocyte. The event which mediates the release of an intermediary messenger may be a conformational change in the receptor upon combination with antigen (Marchalonis and Gledhill, 1968). Cessation of turnover of membrane immunoglobulin may also constitute a powerful stimulus for differentiation of lymphocytes (Cone *et al.,* 1971). The role of multivalency of antigen and the requirement for carriers must also be considered on functional (Feldmann, 1972b) and thermodynamic grounds.

DISCUSSION

Our results establish that immunoglobulin can be isolated from the surfaces of both thymus derived and bone marrow derived lymphocytes. Furthermore, T-cell immunoglobulin shows binding specificity towards antigens used to activate the lymphocytes. The T-cell surface immunoglobulin consists of γM-type molecules characterized by a molecular weight of about 180,000. Not only did these molecules function in binding antigen, but they served as a means of collaboration in an *in vitro* immune response requiring interaction between T-cells, macrophages and B-cells. This observation that collaboration between T-cells and B-cells is mediated by a released amplifier, obviates a major difficulty of the original model for cell interaction which required the improbable event of direct contact between two extremely rare cell types (Mitchison *et al.,* 1970). Our data provide experimental support for the prediction of Bretscher and Cohn (1968) that cell interaction required immunoglobulin as a mediator.

The results discussed here are consistent with a number of indirect observations that functions of T-cells (Lesley and Dutton, 1970; Basten *et al.,* 1971; Mason and Warner, 1970; Greaves and Hogg, 1971) and B-cells (Warner *et al.,* 1970; Feldmann and Diener, 1971; Greaves and Hogg, 1971) can be abrogated by treatment with antisera to immunoglobulins. Although it is relatively simple to demonstrate immunoglobulins on B-lymphocyte surfaces by the use of radioactively labelled antisera, more rigorous conditions are required to detect T-cell immunoglobulin by this means (Bankhurst and Warner, 1971; Nossal *et al.,* 1972). Hammerling and Rajewsky (1971), however, using virus labelled antiglobulins have found γM immunoglobulin on both T and B cells from murine lymph nodes. In our hands, approximately equal numbers of immunoglobulin molecules (1-3x10^5 molecules/cell) could be isolated from both T-cells and B-cells (Marchalonis *et al.,* 1972b) by application of the lactoperoxidase-catalyzed iodination technique and specific immunological coprecipitation. We proposed that the difference in labelling of T-cells and B-cells by antiglobulin reagents reflected differences in accessibility of cell surface immunoglobulin to the labelling reagents rather than an absolute difference in number of molecules. This concept is illustrated graphically in Fig. 7.

T-lymphocytes and B-lymphocytes differed in three major properties. Firstly, we detected only γM-type molecules on T-cells; whereas B-cells possessed both

γM and γG immunoglobulin. In the latter case, the γM-type immunoglobulin always constituted the major class. The predominance of low molecular weight γM immunoglobulin was consistent with other studies of spleen lymphocytes (Baur *et al.*, 1971; Vitetta *et al.*, 1971) and lymphoid tumour cells (Klein and Eskeland, 1971).. In the second place, B-lymphocytes from congenitally athymic mice showed an immunoglobulin turnover rate which was three-fold faster than that of activated T-cells. This finding concurred with a study by Wilson *et al.* (1972) who investigated the release of antiglobulin reagents from B-cells and T-cells. Our turnover rates for B-cells and spleen lymphocytes (mixtures of T-cells and B-cells) correlated well with those calculated by Lerner *et al.* (1972) for myeloma tumour cells. The half-life for immunoglobulin turnover was slightly less than one hour. A third difference between T-cells and B-cells was reflected in binding properties of their isolated immunoglobulins. Preliminary evidence suggests that 7S γM from B-cells differs from that of T-cells in lacking the capacity to bind to macrophages. This observation is consistent with the inability of B-cells to collaborate in immune responses which require B-cell and T-cell interaction.

Specifically activated populations of T-cells provided excellent material for studies of antigen binding capacity of isolated surface immunoglobulins. These systems such as the T. TDL's "educated" to histocompatiblity antigens, a possible model for graft rejection and tumour immunity, contained extremely few, if any, contaminating B-cells[b] and were highly enriched in specific cells. Although the evidence is clear that γM-type immunoglobulin functions as the receptor for antigen on activated T-cells, the question if this is the case for "virgin" T-cells must be considered. Since thymus lymphocytes from normal mice and neonatal humans (Marchalonis *et al.*, 1972a) also possess 7S γM immunoglobulin, this function appears likely. Less μ chain appeared to be expressed on normal thymus lymphocytes than on activated T-cells.

In the 72 years since Ehrlich proposed his "side-chain" hypothesis (1900), the immunological picture has become complex because of the division of lymphocytes into two functional categories; namely T-cells and B-cells. Nevertheless, both types of lymphocytes possess specific receptors for antigen (Basten *et al.*, 1971) and a variety of indirect approaches suggested that these receptors were immunoglobulins. We have provided direct evidence by isolation and determination of binding specificity that lymphocyte surface immunoglobulin can function as the cell receptor for antigen in cellular and humoral immunity. The exact nature of the process of specific activation of lymphocytes remains problematic.

SUMMARY

Immunoglobulins were isolated from the surfaces of a variety of murine lymphocyte populations including thymus, spleen, thoracic duct lymphocytes, and

[b]A detailed discussion regarding the possibility of B-cell contamination is given in Cone *et al.* (1972)

thymus-derived lymphocytes activated to cellular and protein antigens. Cell surface proteins were labelled with [^{125}I] iodide by lactoperoxidase-catalyzed iodination and recovered in solution either by solubilization in dissociating solvents or active metabolic release. Immunoglobulins were identified by specific immunological co-precipitation and characterized by polyacrylamide gel electrophoresis. The polypeptide chain structure of immunoglobulins isolated from lymphocyte surfaces was analyzed by polyacrylamide gel electrophoresis of reduced alkylated samples in acid-urea. Thymus lymphocytes of normal mice and specifically activated thymus-derived cells possessed only γM immunoglobulin on their surfaces. This protein contained light chains and μ-type heavy chains and was characterized by a molecular weight of approximately 180,000. Murine splenic lymphocytes from (CBA x C57Bl)F_1 animals and congenitally thymic aplastic (nu/nu) mice possessed both γM and γG immunoglobulin on their surfaces. The ratio of μ chain to γ chain was about 3/1.

We provide evidence that the γM immunoglobulin isolated from activated thymus cells possesses binding specificity for the antigens used in the sensitization process. We present data indicating that low molecular weight γM immunoglobulin from T cells functions as the collaborative factor in the activation of bone marrow derived lymphocytes in the initiation of antibody-formation. Cell surface immunoglobulin isolated from antigen binding bone-marrow-derived cells of unimmunized mice contained sub-populations of molecules showing binding specificity for certain antigens.

Our results provide direct evidence that immunoglobulin is present on the surfaces of lymphocytes of all classes and that the isolated immunoglobulin exhibits binding specificity for antigen. These conclusions are consistent with Ehrlich's proposal stated in 1900, that antibodies serve as cell surface receptors for antigen. The precise nature of the mechanisms of immune activation remains problematic.

ACKNOWLEDGEMENTS

This paper is publication No. 1703 from The Walter & Eliza Hall Institute of Medical Research. We thank Miss P. Smith, Miss J. Gamble and Miss L. Ptschelinzew for technical assistance. This research was supported by the Australian Research Grants Committee, the National Health and Medical Research Council of Australia, and the American Heart Association (72-1050). R.E.C. is a post-doctoral Fellow of the Damon Runyan Memorial Fund, J.L.A. is the recipient of an Australian Commonwealth Postgraduate Award, and R.T.R. is a special Fellow of the National Institute of Arthritis and Metabolic Diseases (U.S.P.H.S.).

REFERENCES

Bankhurst, A. D. and Warner, N. L. (1971). *J. exp. Med. 134,* 1005

Basten, A., Miller, J. F. A. P., Warner, N. L. and Pye, J. (1971). *Nature New Biol. 231,* 104

Baur, S., Vitetta, E. S., Sherr, C. J., Schenkein, I. and Uhr, J. W. (1971). *J. Immunol. 106,* 1133

Bayse, G. S. and Morrison, M. (1971). *Arch. Biochem. Biophys. 145,* 143

Bretscher, P. A. and Cohn, M. (1968). *Nature, (London) 220,* 444

Burnet, F. M. (1959). *The Clonal Selection Theory of Immunity,* Vanderbuilt University Press, New York

Byrt, P. and Ada, G. L. (1969). *Immunology, 17,* 503

Cheers, C. and Sprent, J. (1972) Proc. Fourth, Int. Conf. on Lymphatic Tissue and Germinal Centres in Immune Responses, in the press

Cone, R. E., Marchalonis, J. J. and Rolley, R. T. (1971). *J. exp. Med. 134,* 1373

Cone, R. E. and Marchalonis, J. J. (1972). In press

Cone, R. E., Sprent, J. and Marchalonis, J. J. (1972). In press

Davie, J. M., Rosenthal, A. S. and Paul, W. E. (1971). *J. exp. Med. 134,* 517

DeSousa, M. A. B., Parrot, D. M. V., and Pantelouris, E. M. (1969). *Clin. exp. Immunol. 4,* 637

Dowben, R. M. (1969). *General Physiology—A Molecular Approach,* p. 369. Harper and Row, New York

Edelman, G. M. and Benacerraf, B. (1962). *Proc. Nat. Acad. Sci. U.S. 48,* 1035

Edelman, G. M. and Marchalonis, J. J. (1967). In *Methods in Immunology and Immuno-Chemistry,* (Ed. by Williams, C. A. and Chase, M. W.), Vol. 1, p. 405. Academic Press Inc., New York

Edelman, G. M. and Gall, W. E. (1969). *Annu. Rev. Biochem. 38,* 415

Ehrlich, P. (1900). Croonian Lecture, *Proc. Roy. Soc. B, 66,* 424

Feldmann, M. and Diener, E. (1971). *Nature, (London) 231,* 183

Feldmann, M. (1972a) *J. exp. Med.,* In the press

Feldmann, M. (1972b). *J. exp. Med.,* In the press

Feldmann, M., Cone, R. E. and Marchalonis, J. J. (1972). In press

Greaves, M. F. and Hogg, N. W. (1971). In *Cell Interactions and Receptor Antibodies in Immune Responses;* (Ed. by Makela, O., Cross, A. and Kosunen, T. U.) p. 145. Academic Press Inc. New York

Greaves, M. F. and Bauminger, S. (1972). *Nature New Biol. 235,* 67

Hammerling, U. and Rajewsky, K. (1971), *Eur. J. Immunol. 1,* 447

Harris, A. W., Bankhurst, A. D., Mason, S. and Warner, N. L. (1972). In press

Klein, E. and Eskeland, T. (1971). In *Cell Interactions and Receptor Antibodies in Immune Responses,* (Ed. by Makela, O., Cross, A. and Kosunen, T. U.) p. 91. Academic Press, New York

Kiehn, E. D. and Holland, J. J. (1970). *Biochemistry, 9,* 1716

Langan, T. A. (1970). *Adv. Biochem. Psychopharmacology, 3,* 307

Lerner, R. A., McConahey, P. J., Jansen, I., and Dixon, F. J. (1972). *J. exp. Med., 135,* 136

Lesley, J. F. and Dutton, R. W. (1970). *Science, 169,* 487

Lesley, J. F., Kettman, J. R. and Dutton, R. W. (1971). *J. exp. Med., 134,* 618

Marchalonis, J. J. (1969). *Biochem. J. 113,* 299

Marchalonis, J. J., Atwell, J. L. and Cone, R. E. (1972a). *Nature New Biol. 235,* 240

Marchalonis, J. J., Cone, R. E. and Atwell, J. L. (1972b). *J. exp. Med. 135,* 956

Marchalonis, J. J., Cone, R. E. and Santer, V. (1971). *Biochem. J. 124,* 921

Marchalonis, J. J., and Gledhill, V. X. (1968). *Nature, (London), 220,* 608

Marchalonis, J. J. and Schonfeld, S. A. (1970). *Biochim. Biophys. Acta, 221,* 604

Mason, S. and Warner, N. (1970). *J. Immunol. 104,* 762

Miller, J. F. A. P. (1972). *Int. Rev. Cytol.,* in press

Miller, J. F. A. P. and Michell, G. F. (1969). *Transplant. Rev. 1,* 3.

Miller, F. and Metzger, H. (1965). *J. Biol. Chem., 240,* 3325.

Mitchison, N. A., Rajewsky, K. and Taylor, R. B. (1970). In *Developmental Aspects of Antibody Formation and Structure* (Ed. by Sterzl, J. and Riha, I.) p. 547. Academic Press, New York

Nossal, G. J. V., Ada, G. L. and Austin, C. M. (1965). *J. exp. Med. 121,* 949

Nossal, G. J. V., Warner, N. L., Lewis, H. and Sprent, J. (1972). *J. exp. Med.* 135, 405.

O'Malley, B. W., Spelsberg, T. C., Schrader, W. T., Chytil, F., and Steggles, A. W. (1972). *Nature, (London), 235,* 141

Parish, C. R. and Marchalonis, J. J. (1970). *Anal. Biochem. 34,* 436

Phillips, D. R. and Morrison, M. (1971). *Biochemistry, 10,* 1766

Rabellino, E., Colon, S., Grey, H. M. and Unanue, E. R. (1971). *J. exp. Med. 133,* 156

Raff, M. C. (1971). *Transplant. Rev. 6,* 52

Riethmuller, G., Rieber, E.-P. and Seeger, I. (1971). *Nature, New Biol., 230,* 248

Rolley, R. T. and Marchalonis, J. J. (1972a). *Transplantation, 14,* 118

Rolley, R. T. and Marchalonis, J. J. (1972b). *Transplantation,* in press

Sherr, C. J. and Uhr, J. W. (1970). *J. exp. Med., 133,* 901

Singer, S. J. and Doolittle, R. F. (1966). *Science. 153,* 13

Snell, S. D. and Stimpfling, J. H. (1966). In *Biology of the Laboratory Mouse,* (Ed. by Green, E. L.) p. 457. McGraw Hill, New York

Spelsberg, T. C., Steggles, A. W. and O'Malley, B. W. (1971). *J. Biol., Chem., 246,* 4186

Sprent, J. and Miller, J. F. A. P. (1971). *Nature New Biol. 234,* 195

Stanley, P. and Haslam, E. A. (1971). *Virology, 46,* 764

Taylor, R. B., Duffus, W. P. H., Raff, M. C. and de Petris, S. (1971). *Nature, New Biol. 233,* 225

Vitetta, E. S., Baur, S. and Uhr, J. W. (1971). *J. exp. Med. 134,* 242

Vitetta, E. S. and Uhr, J. W. (1972). *J. Immunol. 108,* 577

Warner, N. L., Byrt, P. and Ada, G. L. (1970). *Nature, (London) 226,* 942

Warren, L. and Glick, M. C. (1968). *J. Cell. Biol. 37,* 729

Wilson, J. D., Nossal, G. J. V. and Lewis, H. (1972). *Eur. J. Immunol.,* In press

Wld. Health Org. tech. Rep. Ser., 1969, No. 423. *Cell-Mediated Immune Responses*

Index

Absorption spectra of leaves from mutant barley seedlings 461
Actinomycin D,
  administration and superinduction 398
  effect on induction of TAT 217
Adeno-specific RNA transcription 117
Adenovirus DNA replication 117
Adenovirus-specific RNA labelling 117, 118
ALA synthetase,
  antibody to 371
  effect of AIA 373
Aleurone, response to gibberillic acid 340
Allogeneic,
  interactions 599
  interaction, involvement of RNA 603
  responsiveness 600, 602
  stimulation 599, 600, 602
α-amanittin 91
  and adenovirus transcription 118
    protein synthesis 229
  effect on RNA synthesis 228, 229
Amber mutants 66
Amino acid incorporation into mitochondrial proteins 316
  replacements and corresponding single base-pair changes 66
  replacements in *iso*-l-cytochromes *c* 64, 66
  sequence of Vκgenes 579
δ-aminolevulinic acid synthetase induction 369
α-amylase induction in barley 339
Anaemia induction with phenylhydrazine 382
Antibody diversification 574
  forming cells 595
  clonal development of 606
Antibody genes,
  evolution 556
  arrangement in clusters 557
Antibody-plaque-forming cells 607
Antibody-producing cell and somatic differentiation 561
Antiglobulin binding and steric hindrance 638
Antigen,
  binding to lymphocytes 532
  reaction with lymphocytes 532
Antigenic activation of B cells 620
'A' protein localization 263
Athymic nude mice, immune response 620
Bacteriophage R17 RNA 40
  nucleotide sequences 43
B-cell differentiation 625
  suicide of 537
Biosynthesis of α-lactalbumin 260
B lymphocyte 594, 615
  malignant 617
  surface immunoglobulin 615
Bonemarrow (Bursa)-derived lymphocytes 633
Brain phospholipid components 414
Bundle-sheath, light saturation curves 492
Bursal fragments, culture of 616
Calliphora salivary glands 321
Calluses, transgenosised 30
cAMP see Cyclic AMP
Cap formation, 569
  and concanavalin A 569
Cell,
  adhesiveness, induction of 209
  co-operation 528
  mediated immunity 595
Cell surface protein, 630
  immunoglobulin precipitation 632
  purification 632
Cerebrosides, biosynthesis of 417
Chain initiation mutants 67, 80
Chloramphenicol, effects on mitochondrial RNA polymerase 113
Chlorophyll content of leaves from barley seedlings 473
Chloroplast development and nuclear genes 457
  in barley 457
  differentiation 479, 486
  DNA,
    accessibility to deoxyribonuclease 507
    availability for transcription 506
    hybridization with RNA 511
    membrane-bound nature of 506
    potential of 480
    proteins coded by 513
    transcription and translation of 504
  genome, proportion transcribed *in vivo* 512
  isolated, endogenous protein synthesis in 515
  ribosomes 453
  RNA synthesis 509
  spinach, structure of 519
  ultrastructure 487
Chondroblast 297, 298
Chondrogenesis 296, 301
Chondrogenesis and DNA synthesis 287
Chondroitin sulphate 296, 298, 299

Chromatid sub-units, polarity of 5
Chromatin,
  characterization of RNA 193
  chemical composition of 143
  effect of cortisol on 226, 230
  evolution 149
  fractionation 145
  isolated 142
  isolation of non-histone proteins from 238
  preparation of 199
  removal of histone from 147
  template activity of 193
Chromosomal components in avian red blood cells 19
  DNA, semiconservative mode of duplication 4
  frequencies of ring types 12
  proteins properties of 145
  RNA 153
    preparation 192
  sub-unit polarity, continuity of 6
Chromosome,
  a continuous DNA molecule 3
  sister, sub-unit reunion 11
  structure 3
Circularisation of phage genome 35
Clonal development of antibody-forming cells 606
Clonal selection,
  molecular and cellular mechanisms 555
  specific fractionation of lymphoid cells 561
  theory 526
Clones, detection *in vitro* 608
Concanavalin A,
  and induction of IgM-synthesis 552
  inhibition of patch or cap formation 569
  low concentration and cap formation 571
Control of α-lactalbumin biosynthesis 260
Cordycepin, and disappearance of m-RNA from cytoplasm 132
Cortisol, effect on,
  chromatin 230
  extranucleolar and nucleolar RNA 227
  RNA polymerase 231
  RNA synthesis 225, 228
  template activity of chromatin 226
  transcription 233
  negative control of transcription 238
Cotyledon growth 358
Cyclic AMP 242, 275, 283
  affecting chromosome structure 99
    initiation factors for replication or transcription 99
    transcriptive or replicative enzymes 99
  induction of δ-aminolevolinic acid synthetase by 369
  occurrence and distribution in plants 336
  3′-phosphohydrolase in developing brain 412
    in myelin fractions 411
  possible modes of RNA polymerase regulation 99
Cyclic mononucleotide, effect on flowering 340, 341, 342
Cyclic mononucleotide-gibberillin interactions 336
cy-31 mutant 63, 74, 77
  association with amino-terminal region of *iso*-1-cytochrome *c* 58
  isolation of 56, pp.
  mutational events leading to 70
  nucleotide alterations 77,
*cy*l strain revertants 60
Cytocholasin-B 291
Cytochrome P-450 induction, need for steroid hormones 400
Cytology of gene action 330
Cytoplasmic genes and suppressiveness 431
  membranes, structural changes of 465
  segregation and recombination 432
Defective chloroplasts 446
Deoxyribonucleases 50
Development and biogenesis of grana 494
Dexamethasone,
  enzyme induction by 209
  kinetics of binding of 212
Dicentric chromosome 9
  in Chinese hamster cells 9
Dicentrics, unpaired 10
Differentiating liver, protein synthesis and degradation 274
Differentiation
  of chloroplasts 479
  of chromosomal components in red blood cells 191
  model for mitochondrial membrane conformation 440
Dinitrophenylated haemoglobin 630
Diversification, of antibody 574
DNA,
  changes in redundancy 350
  content in higher plant chloroplast 504
  content of nuclear binding sites 214
  continuity of chromosome 3
  from cotyledons, density of 363
  density in roots of *Vicia faba* 363
  in genome of *Drosophila* 14
  fragment, cyclization of 17
  highly repetitious 144
  interaction with histones 201
  location coding for chloroplast proteins 443
  middle repetitious 144
  mutation of defective chloroplasts 446
  sequences 45
  synthesis in cotyledons 357
DNA from cucumber,
  effect of cAMP and gibberillic acid 349
  nucleotide sequence diversity 348
  renaturation kinetics 346

DNA-less petite yeasts 432
DNA-RNA hybrids, melting curves 484, 485
Drosophila melanogaster, genome 13
Drugs, induction of δ-aminolevolinic acid synthetase by 369, 374
Electron transfer, photosynthetic in chloroplasts 490
Endoplasmic reticulum, 393
 associated enzymes 394
 formation during perinatal period 306
 formation of smooth ER 395
 and Ig production 619
 proliferation of membranes 402
 reorganization in the newborn rat 395
Environmental factors affecting grana formation 499
Erythrogenesis 287, 292, 301
Eukaryotic ribosome cycle 267
Fat body of *Calliphora* 321, 322
Fibre fractionation of cells 563
Fraction 1 protein,
 model 448
 ribulose diphosphate carboxylase activity 451
Fractionation of ribonuclease T_1digest 39
Frameshift mutants 71
Framework-residues of V_H sequences of man, mouse cat and dog 581
Galactolipid/phospholipid ratio in subcellular fractions of mouse brain 415
β-galactosidase protein, definitive test for presence of 32
Galactosyl transferase in mouse brain 418
Ganglioside distribution in normal, quaking and jimpy mouse brains 416
Gene activity,
 in blowfly larval tissue 320
 control by hormones 236
Genetic systems of *Saccharomyces cerevisiae* 426
Genetic unit of immunoglobulin production 556
Germ-line V genes 581
 amino acid sequence coding 579
Giant grana 474
Glucagon 275
Glucocorticoid,
 action, early events 216
 hormone action on mammalian cells 206
 influence on level of active m-RNA for TAT 216
Glucocorticosteroids, regulation of transcription by 225
Gluconeogenesis in neonatal rat liver 275
Glucose effect on ALA synthetase 375
Glucose 6-phosphatase 394
 changes during development 310
 changes of kinetics with alloxan 397
 development in neonatal rat 306
 increase in ER of newborn rat 395
 modification by enzyme-membrane interaction 310
 effect of Triton 309
Glucose 6-phosphate phosphohydrolase 305
α-1, 3 glucosyl determinants in dextrans, reponse to 582
Graft-versus-host reaction 595
 anomalous features 596
 species specificity 597
 tissue specificity 597
Grana 471
 genetic, developmental and environmental factors controlling formation 497
Guanosine residues in virus associated RNA 126
Haem, effect on ALA synthetase 371, 375
Haem biosynthesis, role in monooxygenase induction 400
Haemoglobin,
 dinitrophenylated 630
 incorporation studies of A-C switch 381
 tetramers, extracellular exchange of subunits 383
Haemoglobin A,
 intracellular exchange of tetramers with subunits 380, 384
 switch to haemoglobin C 379
Haemopoietic stem cell, 614
 differentiation into B-lymphocyte 613, 625
Halflives of foetal rat liver mitochondria, 312
Haploid callus, 24
 cultures, production of 21
Haploid plant cells, as recipients for bacterial genes 21
Hapten-carrier effect 535
Hertones, 142, 145, 151
 chemical characteristics 154
 heterogeneity 152
 tissue-specific patterns 153
Histones, 144
 acetyl groups turnover 167
 acetylation 165, 166
 amino acid sequences 149
 of avian erythroid cells 199
 -DNA association 146
 fractions, preparation 165
 incorporation of lysine 186
 incorporation of phosphate 186
 interaction with DNA 201
 metabolism 202
 methylation 165, 168
 modifications 150
 number of gene copies coding for 150
 phosphorylation 165, 170, 189
 protease from rat liver 158
 radiophosphorus turnover 171
 role in avian erythropoiesis 177
 structural modifications 164

Hn-RNA synthesis by RNA polymerase II 118
Homochromatography 40
Hormonal control of gene activity 236
Hormonal modulation and plant development 333
HTC cells, receptors of for glucocorticoids 208, 221
Hybrid ribosomes, activity 267
Hybridization of chloroplast r-RNA,
  with chloroplast DNA 481
  with nucleic acids from *Euglena gracilis* 481
  with nucleic acids from *Beta vulgaris* 484
Immune recognition, an alternative mechanism 593
Immune response to antigen 593
Immune response of athymic nude mice 620
Immunochemical studies of the haemoglobin A-C switch 385
Immunofluorescent staining during the haemoglobin A-C switch 388
Immunoglobulin,
  amino acid sequence 613
  association with membranes 547
  binding specificity 639
  carbohydrate composition 549
  chemical analysis of IgG_1 549
  electrophoresis 633, 634
  evolution 558
  fucose incorporation into IgG_1 549
  genetic control in mice 612
  genetic unit for production 556
  half-life 644
  heavy chain, amino acid sequence 613
  light chain, amino acid sequence 613
  of the lymphocyte surface 629
  in lymphoid cells 542
  migration inside plasma cells 547
  model 636
  molecular weights 636
  in plasma cells 544
  precipitation from cell surface 632
  receptors in the lymphocyte membrane 567
  reduction and alkylation 633
  secretion 550
  structure 543
  sugar incorporation 548
  subcellular distribution 546
  synthesis 559
  synthesis and secretion in small lymphocytes 551
  synthesis by T cells 624
  time required for synthesis 546
Immunoglobulin G,
  carbohydrate prosthetic group 556
  heavy chain structure 556
  light chain structure 556
  production and cellular transfer 621
Immunoglobulin gene,
  clonotype 562
  genotype 562
  primotype 562
Immunoglobulin M,
  assembly 545
  carbohydrate composition 550
  induction in B-cells 552
  secretion in B-cells 552
  synthesis and secretion in plasma cells 545
Immunological effector cells 528
Immunological memory 595
Immunology, relevance to biochemistry of gene expression 525
Inactivation of lymphocytes by radioactive antigen 534
Inducers sub-optimal 211
Induction of TAT, effect of actinomycin D on 217
Inorganic pyrophosphate-glucose phosphotransferase 305
Iodination, lactoperoxidase-catalyzed 630
Irradiation, effect on chromosomes 9
Iso-l-cytochrome *c*,
  altered forms 81
  association of 21 *cyl* mutants with amino-terminal region 59
  from *cyl*-183 revertants, frequency of 75
  gene of yeast, mutations of 56
  genetic control of levels 83
  peptide maps of 62
Isoenzymes of tyrosine aminotransferase 244
Jimpy mice, galactosyl transferase content 418
Jimpy mice, proteolipid protein content 412
Jimpy mouse brains, phospholipid components 414
Jimpy mutation 411
lac operon 27
lac^+ transgenosis 24
α-lactalbumin,
  and lactose synthesis 261
  m-RNA 270
  purification of 264
  synthesis, role of ribosomes in 263
Lactating mammary gland, activity of polyribosomes 265
Lactoperoxidase-catalyzed iodination 630
Lactose synthetase 261
Large RNA, retention in nucleus 123
Lead acetate, effect on ALA synthetase 376
Lead, induction of δ-aminolevulinic acid synthetase by 369
Legumin 357, 359
Legumin content of cotyledons 359
Life cycle of *Saccharomyces cerevisiae* 426

Light saturation curves 492
Liver endoplasmic reticulum, 393
  associated enzymes 394
Liver-microsomal monooxygenase system, drug-induced synthesis 397
Liver regeneration, microsomal enzyme levels 396
Lymphocyte,
  activation by antigen 642
  autoradiography at E.M. level 631
  binding of antigen to 532
  collaboration between T and B cells 529
  co-operation of two classes 535
  generation of 527
  inactivation by radioactive antigen 534
  membrane, immunoglobulin receptors 567, 569
  receptors, specificity for antigen 538
  specificity 532
  surface immunoglobulin 629
Lymphoid cells,
  fractionation 566
  immunoglobulins 542
  in mice 612
Lymphoid tumour cell, rosette formation 618
Macromolecular structures, selfdetermination of 436
Mammalian cells, effect of glucocorticoid 206
Membrane assembly, dependence of gene expression on 305
Membrane organization,
  affected by mutation 467
  in *Saccharomyces cerevisiae* 105
Membrane, origin of non-histone protein 198
Membrane structure, effect on expression of G6 Pase 305
Mesophyll chloroplasts, light saturation curves 492
Messenger-RNA,
  decay due to puromycin 134
  half-life in exponentially growing HeLa cells 133
  kinetics of labelling 134
  for α-lactalbumin 270
  rate of initiation 138
  stability 130
  stability in presence of actinomycin D 133
Methionine aminopeptidase mutants 83
Mikamycin, resistance 434, 441
Microcrystallinity 436, 437
Microsomal monooxygenase system drug-induced synthesis 397
Microsomal TAT release factor 256
Mitochondria,
  halflives of 312
  inner membrane maturation 311
  structural changes of 465
  transcriptional system 92
Mitochondrial
  biosynthesis, cyclic pathway 314
  DNA, role in membrane conformation 438
  DNA and suppressiveness 431
  genetic information, transfer 432
  inner membranes, leakiness 313
  membrane, and protein synthesis 314
  membranes, autonomy 425
  protein, decay 313
  proteins amino acid incorporation 316
Mixedness, cytoplasmic 434
Monooxygenase system, drug-induced synthesis 397
Mucopolysaccharide from notochord-somite cultures 300
Multiple base changes 67
  from *cyv*-9 and *cyv*-179 revertants 65
Murine lymphoid cells 612
Mutagenesis 79
Mutagens and resulting *cyl* mutants 58
Mutants,
  methionine aminopeptidase 83
  suppressed nonsense 81
Myelin,
  composition of 410
  lipids 413
  proteins 412
  synthesis deficient mutants 411
Myelin-specific molecules, biosynthesis of 417
Myeloma protein,
  galactose content 550
  glucosamine content 550
  mannose content 550
  neuraminic acid content 550
Myoblast,
  daughter 289, 290
  presumptive 289, 290
Myogenesis 288, 301
Neurological mutants, as experimental models 410
Nicotiana
  cytoplasmic inheritance 443
  defective chloroplasts 445
  mendelian mode of inheritance 443
  variegated 445
Non-histone,
  chromosomal protein 142
  origin from membrane proteins 198
  protein of avian erythroid cells 196
    analysis 197
  proteins, isolation from chromatin 238
Notochord 296
Nuclear binding activity, destruction by DNase 214
Nuclear binding sites, DNA content of 214

Nuclear RNA, kinetics of hybridization 119
Nucleic acid,
content of plant cells 479, 480
sequences 38
synthesis in cotyledon 360, 362
Nucleo-cytoplasmic interactions 425
Nucleolar function 329
Nucleoli
development 325
in salivary gland 326
of larval fat body 327
Nucleolus, role of 320
Ochre mutants 63
Oxygen evolution in barley leaves 473
Peptide maps of normal *iso*-cytochrome *c* 62
Perfusion of liver 371
Petite yeast, 432
DNA-less 432
extranuclear inheritance 427
mutation 428
suppressiveness 428
3′-phosphoadenosine 5′-phosphosulphate cerebroside-sulphotransferase in brains of mutant mice 419
Phosphoenolpyruvate carboxykinase,
activity in foetal rat liver 275
antigen and activity 277
changes in degradation rate 278
elimination of increase at birth by actinomycin D 278
half times of 279
inactivation in liver from newborn rats 281
rates of synthesis 278
Phospholipid components in Jimpy mouse brains 414
Phospholipid distribution in subcellular fractions of mouse brain 415
Phosphorylcholine determinants in pneumococcal C-carbohydrate, response to 582
Photosynthesis in maize leaves, and light intensity 494
Pigment accumulation in wildtype and mutant barley 460
Plant development, effects of hormones and environment 333
Plant nucleic acids, hybridization 481
Plaque-forming cells 607
Plasma cells,
and immunoglobulin 544
malignant 622
maturation 553
pulse-labelling 547
Plastids, porphyrin biosynthesis in 459
Polyacrylamide gel raft, for measurement of clonal expansion 610
Polyadenylic acid, terminal of m-RNA 131
Polarity of chromosomal sub-units 6
Polyribosome,
from adenovirus-infected cells, 135
cycloheximide, effect on
comparison between pregnant and lactating mammary gland 265
injection into frog oocytes 270
from lactating mammary gland 263
Polysomes, decay of in presence of actinomycin D 131
Polytene chromosomes, wide distribution of 15,
Polytene nuclei, development of 323
Porphyrin biosynthesis in plastids 459
Postnatal development of TAT 243
Post-transcriptional regulation of TAT synthesis 219
Pregnant mammary gland, activity of polyribosomes 265
Progesterone, kinetics of binding 212
Prolamellar bodies,
crystalline configuration 469
crystalline, formation in darkness 470
formation 464
un-dispersed 470
Protamines, core common sequence of 149
Protein degradation,
in cytosol fraction of newborn rat liver 282
in newborn rat liver *in vitro* 280
Protein, mitochondrial decay of 313
Protein synthesis, termination step 242
Proteolipid proteins, in Jimpy and Quaking mice 412
Protochlorophyllide,
accumulation of 459, 462
lack of in mutant 465
photoconversion 469
Puromycin, effect on TAT 245
Pyrimidine oligonucleotides, homology 48,
Pyrimidine sequences 46
Quaking mice, galactosyl transferase content 418
proteolipid protein content 412
Quaking mutation 411
Quantal cell cycles 288, 290
Receptor,
for glucocorticoids in HTC cells 208
proteins, binding to chromatin 236
Receptor-steroid complex,
binding to DNA 215
binding to nucleus 212
Regulation of chloroplast differentiation 479
Replication of adenovirus DNA 117
Resistance to transgenosis 33
Respiratory control, development 316
Response of cells to antigen, heterogeneity 528

Response to phosphorylcholine determinants in the pneumococcal C-carbohydrate 582
Reticulocyte RNA, analysis 195
Ribonuclease T_1, for degrading RNA 39,
Ribosome,
couples 268
cycle, of eukaryotes 267
mitochondrial
role of mitochondrial membrane in controlling assembly and attachment of 107
role of in synthesis of α-lactalbumin 263
subunits 268
Ribulose diphosphate carboxylase 451
Ring chromosome, induction of 7, 8,
RNA
adenovirus-specific labelling of 118
chicken reticulocyte analysis 195
extranucleolar, effect of cortisol 227
fractionation, by gel electrophoresis 125
gel electrophoresis 509
hybridization to chloroplast DNA 511
RNA polymerase,
cortisol effect on 231
homologies between eukaryotic and prokaryotic 92
initiation specificity of DNA templates 94
levels during sea urchin embryogenesis 102
mitochondrial, and chloramphenicol 113
mitochondrial DNA-dependent
effect of UFA and sterol depletion 110
mitochondrial membrane in controlling function of 107
multiplicity 90
nuclear 91
properties 109
regulation by c-AMP 99
regulation of rate by specific "factors" 97
selective gene transcription 89
site specificity 97
transcriptive specificity 97
turnover 101
RNA polymerase activity,
effects of aeration 111
effects of changes in lipid composition 111
effect of oestrogen treatment 101
stimulation 98
RNA polymerase I,
involvement in r-RNA synthesis 94
localization 93
response to physiological state of cell 100
RNA-polymerases I and II, specificity for various templates 95
RNA polymerase I, II and III, nuclear localization and transcriptive function 93
RNA polymerase II,
inhibition of by α-amanitin 93
molecular weight 92
phosphorylation 99
sensitivity to ionic strength 120
RNA synthesis,
chloroplasts, in 509
during growth in cotyledons 357
effect of Mn^{2+} 121
rat liver, and cortisol 225, 226
sites in salivary gland nuclei 329
RNA, viral associated 122
Saccharomyces cerevisiae, life cycle and genetic systems 426
Salivary gland in *Calliphora,*
protein synthesis 322
Selection pressure 581
Selfdetermination of macromolecular structures 436
Sister chromatid exchange 11
Small RNA, release from nuclei 123
Steroid hormones and cytochrome P-450 induction 400
Steroids as allosteric effectors, 210
effect on ALA synthetase 374
induction of δ-levulinic acid synthetase by 369
Suicide of T and B cells 537, 539
Sulphatides, biosynthesis 418
Supergene 558
supF$^+$ transgenosis 25
Suppressed nonsense mutants 81
Tandem duplications 74
T-B collaboration 529
T cell,
antigenic receptor 626
and ^{125}I-labelled anti-immunoglobulin 623
immunoglobulin synthesis 624
suicide of 537
Termination step in protein synthesis 242
Thylakoid membrane proteins 454
Thymus-derived lymphocytes 633
Thymus-derived lymphocyte, 633
formation 614
interaction with B-cells 641
T lymphocyte 594, 623
Transcription
adeno-specific RNA of 117
asymmetrical in chloroplasts 510
chloroplast DNA of 505
mammalian cells in 117
negative control by cortisol 238
regulation by glucocorticosteroids 225
Transducing phage, vector for the transfer of genes 23
Transferase, effect of Triton on pH curve 309

Transgenosised callus,
  low level of *E. coli* β-galactosidase in 34
Transgenosis, 23
  bacterial genes to plant cells 21
  bacterial nitrogen fixation 36
  E. coli galactose operon to haploid tomato callus 28
  gal genes 29
  for survival 27
  practical applications 36
  resistance to 33
  transducing phages, use of in 35
  whole plants 35
  z gene of *lac* operon from *E. coli* 30
Translation in mammalian cells, 117, 130
  rate of adenovirus m-RNA 136
Translocon, 557
  evolution 559
Turnover of RNA-polymerases 101
Tyrosine aminotransferase, 216, 242
  actinomycin D effect 246
  adrenalin effect 245
  cascade release effect 251
  formation 275
  multiple forms 244
  postnatal development 243
  polysome 245
  puromycin effect on post-natal development 245
  pyridoxal and cascade effect 252
  release by cyclic AMP *in vitro* 250
  release system 248
  synthesis, post-transcriptional regulation of 219
V gene, 556
  diversification 574
Vicia faba, isolabelling in 4,
Vicilin 357, 359
Vinblastine 290
Viral-associated RNA 122
Virus associated RNA,
  guanosine residues in 126
  hybridization to adenovirus DNA 126
  hybridization to HeLa DNA 126
  manganese, effect on 129
Yeast mutants, resistant to antibiotics 429